THE ADVANCED COMPOSITION EXPLORER MISSION

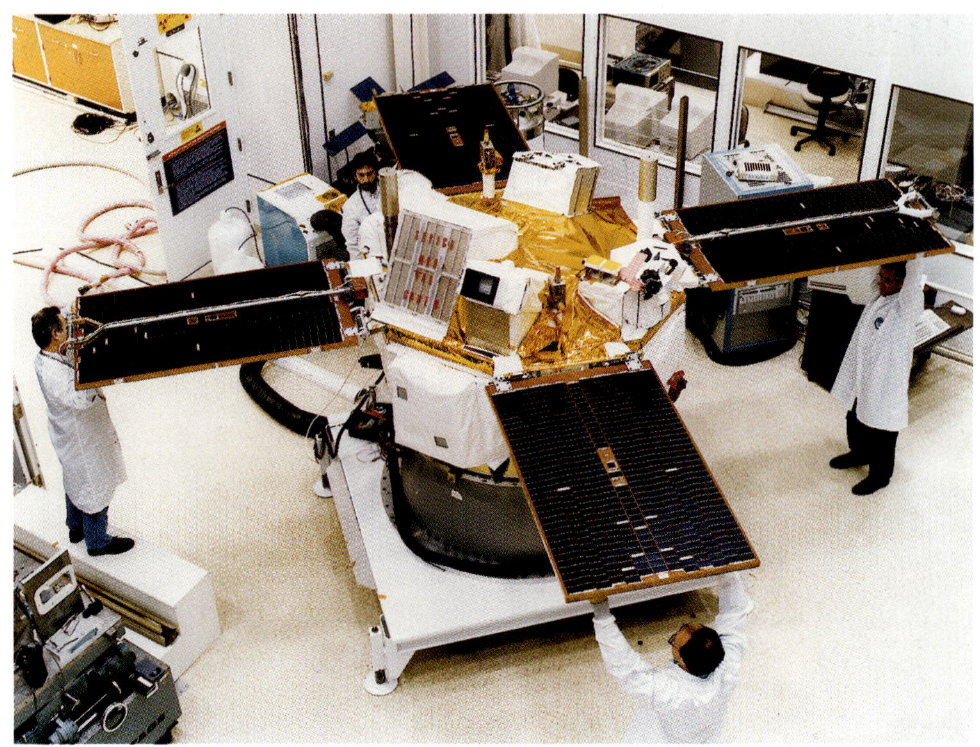

The ACE spacecraft was built by the Applied Physics Laboratory of the John Hopkins University (JHU/APL).

(Photo courtesy of JHU/APL)

THE ADVANCED COMPOSITION EXPLORER MISSION

Edited by

C. T. RUSSELL

University of California, Los Angeles, CA, U.S.A.

R. A. MEWALDT

California Institute of Technology, Pasadena, CA, U.S.A.

and

T. T. VON ROSENVINGE

NASA/Goddard Space Flight Center,
Greenbelt, MD, U.S.A.

Reprinted from *Space Science Reviews*, Volume 86, Nos. 1–4, 1998

KLUWER ACADEMIC PUBLISHERS
DORDRECHT / BOSTON / LONDON

A C.I.P. Catalogue record for this book is available from the Library of Congress.

ISBN 0-7923-5530-X

Published by Kluwer Academic Publishers,
P.O. Box 17, 3300 AA Dordrecht, The Netherlands

Sold and distributed in North, Central and South America
by Kluwer Academic Publishers,
101 Philip Drive, Norwell, MA 02061, U.S.A.

In all other countries, sold and distributed
by Kluwer Academic Publishers,
P.O. Box 322, 3300 AH Dordrecht, The Netherlands

Printed in acid-free paper

Printed in the Netherlands

TABLE OF CONTENTS

FOREWORD

The composition of the sun is the standard of reference for studies of the origin and evolution of matter in the solar system and beyond. Yet, much of what is known about 'Solar Abundances' is derived from terrestrial, meteoritic, and lunar material rather than from samples of the sun. Fortunately, the solar wind and energetic solar particles are accessible samples of the solar atmosphere comprised of accelerated individual ions. Other accelerated ions coursing through interplanetary space are samples of the local interstellar medium and of matter from nearby regions in the Galaxy. The primary objective of the Advanced Composition Explorer (ACE) is the determination and comparison of these three distinct samples of matter. The observational challenge has been to determine the masses of individual nuclei with sufficient resolution and in adequate number to allow abundance measurements of even relatively rare isotopes.

To carry out these observations ACE includes six state-of-the-art high-resolution spectrometers that measure the elemental, isotopic, and ionic charge-state composition of nuclei from H to Ni ($1 \leq Z \leq 28$). These observations cover more than five decades in energy, from solar wind energies (~ 1 keV nucl^{-1}) to galactic cosmic-ray energies (~ 500 MeV nucl^{-1}). ACE also carries three instruments that provide the context for ion composition studies by monitoring the state of the interplanetary medium. The wide range of scientific objectives that can be addressed by studies of these three samples of matter are summarized in the first paper of this book, along with an overview of the scientific payload.

In January of 1997 a three-day workshop was held at Caltech to review the scientific goals of ACE and to identify and assess progress that has been made by other missions such as Voyager, *Ulysses*, SAMPEX, and Wind. The invited review papers from this workshop that are included in this volume provide an excellent introduction to the scientific context in which ACE began its observations. The next group of papers includes a description of the ACE spacecraft and its capabilities, followed by eight papers that provide details of the design, development, and new capabilities of the instruments on ACE.

ACE was launched from Cape Canaveral on August 25, 1997 and is now in a halo orbit about the L_1 Lagrangian point, ~ 1.5 million km sunward of Earth. In response to long-standing interests by the Air Force Geophysical Laboratory and NOAA, Dr Stan Shawhan, then Director of NASA's Space Physics Division, requested in 1989 that the ACE team consider providing solar wind measurements in real time to aid in forecasting geomagnetic storms. In 1994 a joint NOAA/NASA agreement was reached to implement this system and on January 21, 1998, the ACE Real-Time Solar Wind System was inaugurated. These data are also available in real time to the public at http://sec.noaa.gov/ace/ACErtsw_home.html, as described in the penultimate paper in this volume.

Space Science Reviews **86:** vii–viii, 1998.
© 1998 *Kluwer Academic Publishers. Printed in the Netherlands.*

The final paper describes the ACE Science Center (ASC) at Caltech, which is responsible for preliminary processing, and for distribution and archiving of ACE data. The ASC was the dream of Dr. Thomas Garrard, a Staff Scientist at Caltech and ACE co-investigator who recognized the importance of making ACE data accessible to the broader community and accepted the challenge of making this happen. As a result, preliminary data from all nine instruments are available on the World Wide Web within days of receipt (see http://www.srl.caltech.edu/ACE/), to be followed by higher level data products produced by the instrument teams. Unfortunately, Tom Garrard suddenly passed away just two months before the ACE launch, and he was unable to share in the success of his team's efforts. This volume is dedicated to his memory.

All contributions in this volume have undergone peer review, generally by two experts, one from within and one from outside the ACE community. We would like to thank all the referees for their insightful comments that have improved both the readability and content of the articles herein.

The authors too deserve our thanks for their prodigious efforts undertaken at a very busy time for the mission. We gratefully acknowledge the help of Anne McGlynn in the editorial process.

Many individuals have contributed to the success of the ACE programs. We particularly wish to acknowledge the Explorer Program Office at NASA Headquarters, the ACE Project team at Goddard Space Flight Center, the ACE spacecraft team at the Johns Hopkins University Applied Physics Laboratory, the ACE Payload Management Office at Caltech, and the ACE Science Team, led by Principal Investigator E. C. Stone.

C. T. RUSSELL, R. A. MEWALDT, T. T. VON ROSENVINGE

THE ADVANCED COMPOSITION EXPLORER

E. C. STONE, A. M. FRANDSEN, R. A. MEWALDT
California Institute of Technology, Pasadena, CA 91125, U.S.A.

E. R. CHRISTIAN, D. MARGOLIES, J. F. ORMES and F. SNOW
Goddard Space Flight Center, Greenbelt, MD 20771, U.S.A.

Abstract. The Advanced Composition Explorer was launched August 25, 1997 carrying six high-resolution spectrometers that measure the elemental, isotopic, and ionic charge-state composition of nuclei from H to Ni ($1 \leq Z \leq 28$) from solar wind energies (~ 1 keV nucl^{-1}) to galactic cosmic-ray energies (~ 500 MeV nucl^{-1}). Data from these instruments is being used to measure and compare the elemental and isotopic composition of the solar corona, the nearby interstellar medium, and the Galaxy, and to study particle acceleration processes that occur in a wide range of environments. ACE also carries three instruments that provide the heliospheric context for ion composition studies by monitoring the state of the interplanetary medium. From its orbit about the Sun–Earth libration point ~ 1.5 million km sunward of Earth, ACE also provides real-time solar wind measurements to NOAA for use in forecasting space weather. This paper provides an introduction to the ACE mission, including overviews of the scientific goals and objectives, the instrument payload, and the spacecraft and ground systems.

1. Introduction

Particle acceleration is ubiquitous in solar system and astrophysical plasmas. As a result, interplanetary space is filled with diverse populations of energetic nuclei that have been accelerated in the solar wind, by solar flares and coronal shocks at the Sun, in interplanetary space, and in the Galaxy. These accelerated particle populations, illustrated schematically in Figure 1, provide samples of matter originating in other regions of the solar system and Galaxy which cannot be reached for *in situ* measurements. Advances in space instrumentation have made it possible to determine not only the elemental, but also the isotopic and ionic charge-state composition of these samples of matter with much greater precision than was previously achievable. As a result, it is now possible to undertake comparative studies of the origin and evolution of these distinctly different samples of matter, as well as studies of acceleration processes occurring in distinctly different plasma environments.

Because these various components of solar, interplanetary, local interstellar, and galactic cosmic-ray particles are observed over a wide range of energies and intensities, as illustrated in Figure 2, comparative studies of these components are beyond the dynamic range of a single instrument. Rather, these studies require

Space Science Reviews **86**: 1–22, 1998.

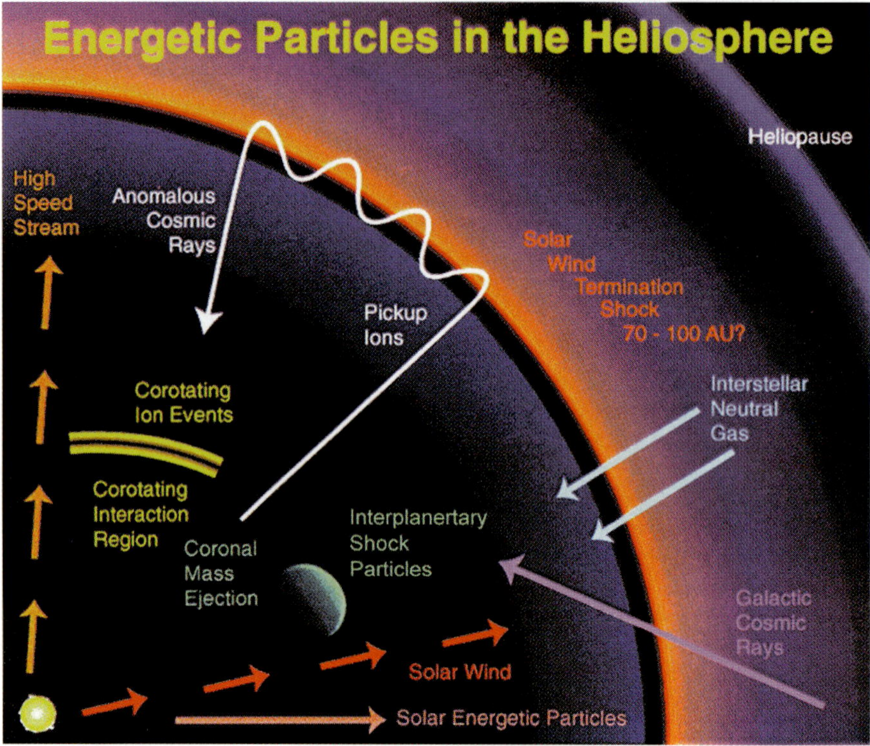

Figure 1. Schematic illustration of some of the various energetic particle populations that can be observed in the heliosphere.

a coordinated approach with a suite of instruments designed to obtain definitive measurements over as broad an energy range as possible.

In response to these opportunities, and recognizing these requirements, the Advanced Composition Explorer (ACE) was proposed in 1986 as part of the Explorer Concept Study Program. ACE is designed to make coordinated measurements of the elemental and isotopic composition of accelerated nuclei from H to Zn ($1 \leq Z \leq 30$) spanning six decades in energy per nucleon, from solar wind to galactic cosmic-ray energies, with sensitivity and with charge and mass resolution much better than heretofore possible. Following a Phase-A definition study, ACE was selected for development in 1989, and began construction in 1994. On August 25, 1997, ACE was successfully launched from Cape Canaveral Air Station by a Delta II rocket. The August 1997 launch was originally scheduled back in 1993.

The ACE observatory, a spinning spacecraft (5 rpm), will orbit around the Sun-Earth L1 libration point at 240 R_{E} sunward of the Earth. The spin axis of the cylindrically-shaped spacecraft points generally towards the Sun, with 156 kg of science instrumentation mounted both on the flat, Sun-facing side, and on the side of the cylinder. A high-gain antenna for data communications is mounted on the

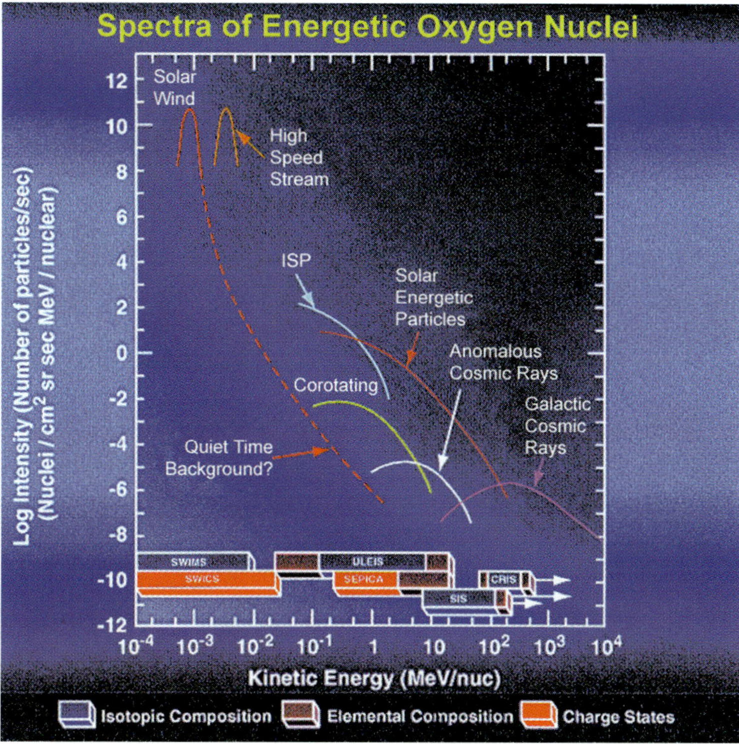

Figure 2. Typical energy spectra of energetic oxygen nuclei resulting from the various particle populations illustrated in Figure 1. The solid curves represent 'steady state' components that vary slowly
over the solar cycle, while the dashed curves are for 'transient' phenomena. The energy ranges of
the various ACE instruments are indicated for resolution of isotopes, elements only, and ionic charge
states. Other species generally have spectra that are similarly shaped when plotted as a function of
energy/nucleon.

back side pointing towards the Earth. The spacecraft will operate in the vicinity of
the L1 point with >90% duty cycle for 2 to 5 years studying solar wind and solar
energetic particles, particles energized by interplanetary shocks, partially ionized
particles from the local interstellar medium (pickup ions and anomalous cosmic
rays), and galactic cosmic rays. Sensors on ACE are designed to measure the
properties of ions from solar wind energies of 100 eV up to several hundred MeV
galactic cosmic rays (GCR), determining the mass and charge of incident particles
during both solar quiet and solar active periods.

   ACE has nine science instruments that are identified below. The various articles
in this volume discuss the contributions each will make to meeting ACE's scientific
objectives. ACE will make possible simultaneous measurements over a very broad
energy range, with large-area, high-resolution instruments, each optimized for a
specific energy range. In addition, ACE will provide real time *in situ* monitoring of

the solar wind that will allow NOAA to make timely forecasts of impending space weather.

The ACE mission development was managed by the Goddard Space Flight Center (GSFC) Explorer Projects Office of the Flight Projects Directorate. The spacecraft was developed by the John's Hopkins University Applied Physics Laboratory (JHU/APL). Instrument development was the responsibility of the California Institute of Technology (Caltech) under contract to NASA. The development of the ACE payload, spacecraft, and ground system was managed under a strict cost cap and schedule constraint, both of which were satisfied. Following integration of the instruments onto the spacecraft at APL, the observatory was taken to the GSFC for final testing and checkout and then shipped to the Kennedy Space Center for launch.

In addition to describing the scientific goals for ACE, this paper outlines the capabilities of the scientific payload, describes the mission operations plan, and discusses the unique aspects of the spacecraft and its supporting ground test and operating equipment. It is meant as a brief overview of the more detailed papers in this special edition of *Space Science Reviews*.

2. Scientific Goals

The prime objective of ACE is to determine and compare the elemental and isotopic composition of several distinct samples of matter, including the solar corona, the interplanetary medium, the local interstellar medium, and galactic matter. Some of the processes undergone by this material as it flows to 1 AU are illustrated in Figure 3, along with the populations of energetic nuclei that result.

Matter from the Sun will be studied directly by measuring the composition of the solar wind, of coronal mass ejections (CMEs; large plasma clouds ejected by the Sun), and of solar energetic particles (SEPs) accelerated in impulsive solar flares and by coronal and interplanetary shocks initiated by CMEs. Matter from the local interstellar medium enters the solar system as interstellar neutral particles that are subsequently ionized, picked up by the solar wind to become solar-wind 'pickup ions' and then accelerated to cosmic-ray energies at the solar wind termination shock. Both the pickup ions and the accelerated nuclei (known as 'anomalous cosmic rays' or ACRs) provide a sample of matter from the local interstellar medium (LISM). Galactic cosmic rays (GCRs) provide a sample of matter from more distant regions of the Galaxy that is believed to be accelerated by supernova shock waves. Each of these samples of matter has undergone a distinctly different history: the pickup ions and anomalous cosmic rays sample the present-day interstellar medium; galactic cosmic rays provide a sample of matter from the Galaxy that is thought to have been accelerated $\approx 10^7$ years ago; and matter in the Sun represents a still older sample of interstellar matter that has been stored in the Sun for the last 4.6 billion years.

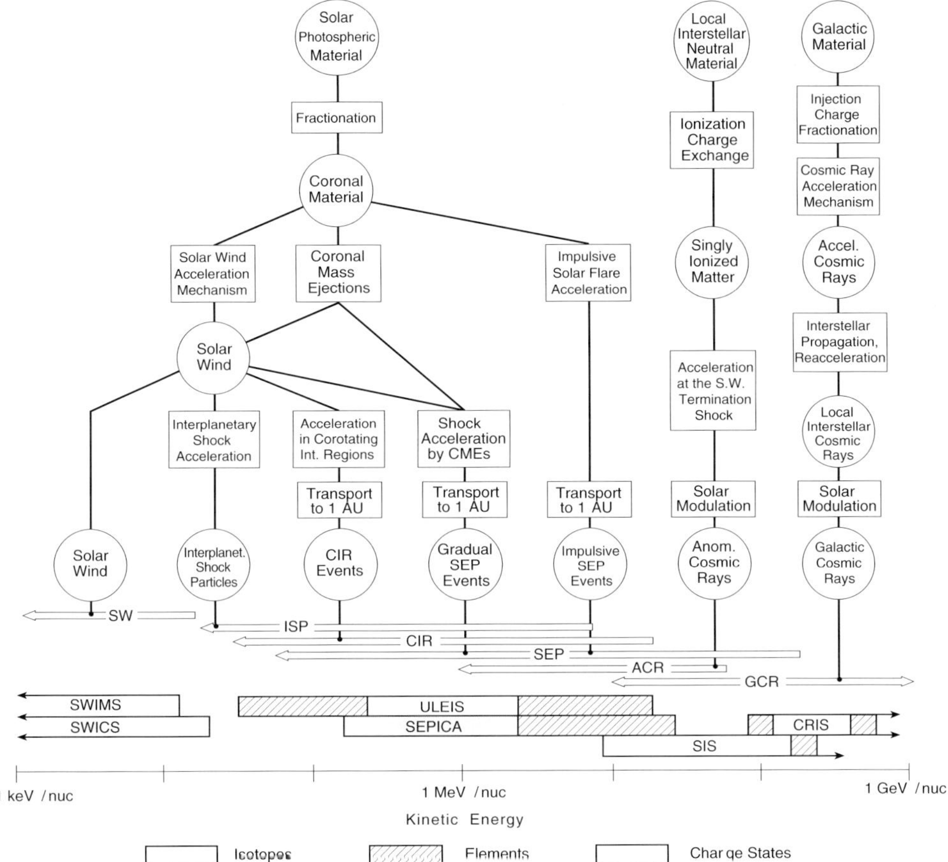

Figure 3. Diagram showing how energetic particles provide a flow of material to 1 AU from the solar photosphere, the neutral interstellar medium, and galactic cosmic-ray sources. Circles represent particle populations, rectangles represent fractionation, acceleration, and/or transport processes. Also indicated are the energy ranges over which the ACE instruments can measure the elemental, isotopic, and ionic charge-state composition of these particle populations.

The comparison of these samples of matter will be used to study the origin and subsequent evolution of both solar system and galactic material by isolating the effects of fundamental processes that include nucleosynthesis, charged and neutral-particle separation, bulk plasma acceleration, and the acceleration of suprathermal and high-energy particles. Specifically, these observations would allow the investigation of a wide range of fundamental problems in the following four major areas:

(1) The elemental and isotopic composition of matter. A major objective is the accurate and comprehensive determination of the elemental and isotopic composition of the various samples of 'source material' from which nuclei are accelerated. These observations will be used to:

– Generate a set of solar isotopic abundances based on direct sampling of solar material.

– Determine the coronal elemental and isotopic composition with greatly improved accuracy.

– Establish the pattern of isotopic differences between galactic cosmic-ray and solar system matter.

– Measure the elemental and isotopic abundances of interstellar and interplanetary 'pickup ions'.

– Determine the isotopic composition of the 'anomalous cosmic-ray component', which represents a sample of the local interstellar medium.

(2) Origin of the elements and subsequent evolutionary processing. Isotopic 'anomalies' in meterorites indicate that the solar system was not homogeneous when formed. Similarly, the Galaxy is neither uniform in space nor constant in time due to continuous stellar nucleosynthesis. ACE measurements will be used to:

– Search for differences between the isotopic composition of solar and meteoritic material.

– Determine the contributions of solar wind and solar energetic particles to lunar and meteoritic material, and to planetary atmospheres and magnetospheres.

– Determine the dominant nucleosynthetic processes that contribute to cosmic-ray source material.

– Determine whether cosmic rays are a sample of freshly synthesized material (e.g., from supernovae) or of the contemporary interstellar medium.

– Search for isotopic patterns in solar and Galactic material as a test of galactic evolution models.

(3) Formation of the solar corona and acceleration of the solar wind. Solar energetic particle, solar wind, and spectroscopic observations show that the elemental composition of the corona is differentiated from that of the photosphere, although the processes by which this occurs, and by which the solar wind is subsequently accelerated, are poorly understood. The detailed composition and charge-state data provided by ACE will be used to:

– Isolate the dominant coronal formation processes by comparing a broad range of coronal and photospheric abundances.

– Study plasma conditions at the source of solar wind and solar energetic particles by measuring and comparing the charge states of these two populations.

– Study solar wind acceleration processes and any charge or mass-dependent fractionation in various types of solar wind flows.

(4) Particle acceleration and transport in nature. Particle acceleration is ubiquitous in nature and understanding its nature is one of the fundamental problems

of space plasma astrophysics. The unique data set obtained by ACE measurements will be used to:

– Make direct measurements of charge and/or mass-dependent fractionation during solar energetic particle and interplanetary acceleration events.

– Constrain solar flare, coronal shock, and interplanetary shock acceleration models with charge, mass, and spectral data spanning up to five decades in energy.

– Test theoretical models for ^{3}He-rich flares and solar γ-ray events.

In January of 1997, a three-day workshop was held at Caltech to review the ACE scientific goals, and to identify and assess progress that has been made by other on-going missions such as Voyager, *Ulysses*, SAMPEX, and Wind. A number of the invited papers from this workshop are included here, along with papers that describe the capabilities of the individual ACE experiments and the ACE mission. These invited review papers provide an excellent introduction to the scientific context in which ACE will begin its investigations.

During the period since ACE was first selected for flight, Voyager-1 and Pioneer-10 both moved to beyond 65 AU in the outer heliosphere, and *Ulysses* completed its first two passes over the solar poles. This armada of spacecraft has provided us with our first three-dimensional view of energetic particles and plasma in the large-scale heliosphere and they have revealed a variety of new heliospheric phenomena. For example, the SWOOPS and SWICS instruments on *Ulysses* (the forerunners of those on ACE) have demonstrated that there are distinct differences between fast- and slow-speed solar wind that provide important clues to their origin in the corona, how they may have been fractionated, and how these components were accelerated (e.g., Fisk et al., 1998). ACE will build on these studies by adding simultaneous measurements of isotopic mass to the elemental and ionic charge state measurements made on *Ulysses*.

Pioneer-10 first discovered that extensive particle acceleration occurs when high-speed solar wind streams overtake slow-speed streams to form co-rotating interaction regions (CIRs). *Ulysses* has found that there are a number of mysteries that arise in our understanding of these events when the third dimension of the heliosphere is explored. ACE will be able to track interplanetary acceleration events continuously over more than four decades in energy per nucleon.

Although the Sun has been remarkably quiet at energies > 10 MeV over the past few years of solar minimum (1993–1996), new instrumentation on the Wind spacecraft (Lin, 1998) has demonstrated that the Sun is never really quiet at lower energies. While there are very few large solar particle events near solar minimum, there continue to be small 'impulsive' solar events that are rich in ^{3}He, electrons, and heavy elements. An important goal of ACE will be to try to understand the energy-release and particle-acceleration mechanisms in these impulsive events (Miller, 1998), as well as those in larger 'gradual' events that are apparently caused by shocks associated with coronal mass ejections. ACE will be ideally suited for these studies because it can measure elemental, isotopic, and ionic charge states from solar wind energies to several MeV nucl^{-1}, and because the greatly improved collecting power of its instruments can provide the statistical accuracy required to study a wide range of species in these events.

The elemental composition of particles in gradual SEP events is observed to be similar to that of the corona. A primary goal of the SEP studies (as well as solar wind studies on ACE) will be to determine the elemental and isotopic composition of the solar corona, and, ultimately, the solar photosphere, by direct sampling of solar material. These studies will complement spectroscopic studies and composition measurements of other solar system bodies such as meteorites and comets, which form the backbone of our understanding of the origin and evolution of solar system material.

Over the past decade there has been significant progress in understanding the sequence of events by which interstellar neutral atoms become solar wind pickup ions, some fraction of which are then accelerated to become anomalous cosmic rays. *Ulysses* has measured directly neutral interstellar He and the composition of solar wind pickup ions that result when the neutral interstellar gas is ionized (Gloeckler and Geiss, 1998). In addition, SAMPEX has shown that while most ACRs are singly-charged as expected, there are also multiply-charged ACRs that dominate at higher energies (> 20 MeV nucl^{-1}) and there is now considerable indirect evidence that the bulk of the acceleration of ACRs occurs at the solar wind termination shock (see Jokipii, 1998). ACE will make improved measurements of the elemental and isotopic composition of both pickup ions and ACRs in an effort to relate these to the composition of the local interstellar medium (see Frisch, 1998). It now appears that there are also other sources of solar wind pickup ions, including interplanetary and perhaps interstellar dust (Gloeckler and Geiss, 1998).

Galactic cosmic-rays are believed to result from shock acceleration processes on yet a larger scale, in this case powered by supernova shocks. While there appears to be a consensus that supernova shocks provide the necessary energy, there remains considerable debate on the origin of the accelerated material. In a recent model (Meyer et al., 1998; Ellison et al., 1998) all but the volatile elements in cosmic rays result from the sputtering products of dust grains that have themselves been accelerated to energies approaching 1 MeV nucl^{-1} by supernova shocks. This model helps explain the observed cosmic-ray elemental composition but requires an additional source to explain the overabundance of ^{22}Ne in cosmic rays. ACE will test this and other models with detailed measurements of the elemental and isotopic composition of cosmic-ray source material (Webber, 1998), building on recent progress by *Ulysses* (Simpson, 1998). A second objective of these studies is accurate measurements of various radioactive clocks that can measure cosmic-ray time scales for nucleosynthesis, acceleration, and transport (Ptuskin and Soutoul, 1998).

3. Mission Requirements and Description

In keeping with the 'Better, Faster, Cheaper' philosophy of modern day NASA, the ACE mission was developed in a non-traditional, low-cost manner in which the Principal Investigator and his team at Caltech had responsibility for the development of the instrument suite with minimal NASA oversight. With the Project Management, Spacecraft, and Instrument teams all keeping in close contact, it was Caltech's responsibility to ensure that the science payload was delivered within cost and on time. If any instrument was to have gotten into serious development trouble, a series of agreed-to descope options was available, including the possibility of leaving an instrument off the spacecraft, since the scientific goals of ACE are broad enough that no single instrument was deemed critical to its mission success. This philosophy permitted the instruments to be developed under a somewhat lower level of quality assurance requirements, and helped reduce instrument costs. In the end, no instruments required significant descoping, and this approach led to the successful development of all nine instruments.

The focus of ACE studies will depend on the phases of the solar cycle which ACE will be able to observe. The mission was launched in August, 1997, near the minimum of solar activity. The high fluxes of galactic cosmic rays and the presence of the ACR component make this the ideal time to study particles from more distant sources as well as solar wind from the quiet Sun. In the years following launch, solar activity will increase and solar energetic particles can be studied. While the instrument and spacecraft design requirements were for a 2-year mission, ACE has enough hydrazine and proportional counter gas to last more than 5 years. During this time, ACE will be able to study phenomena important at all phases of the solar cycle.

In order to provide continuous measurements of the solar wind, low-energy solar and interplanetary particles, and cosmic rays, ACE must be stationed well outside of the Earth's magnetosphere. The modified halo orbit about the Earth–Sun interior libration point, L1 (see Figure 4) meets this requirement. This orbit is similar to that originally attained by the ISEE-3 mission in 1978.

4. Instrumentation

To address the objectives described in Section 2 requires coordinated, high-precision measurements of the elemental, isotopic, and ionic charge-state composition of energetic nuclei over a broad energy range, with time resolution adequate to investigate the dynamical processes affecting the composition. Table I summarizes properties of the nine instruments that will provide these measurements on ACE. They are as follows:

CRIS is a Cosmic-Ray Isotope Spectrometer designed to measure the elemental and isotopic composition of galactic cosmic rays over the energy range from

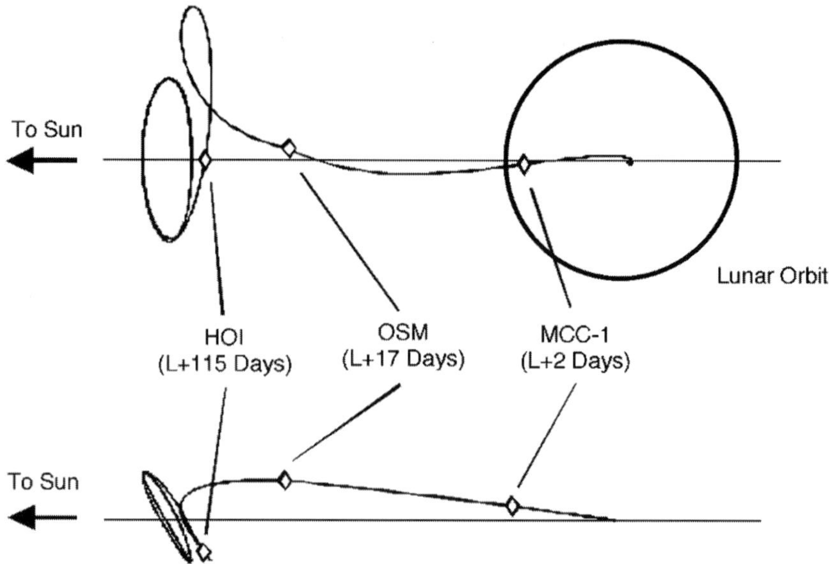

Figure 4. Schematic of the ACE transfer trajectory out to the Earth–Sun interior libration point (L1), and subsequent halo orbit about L1. The actual modified halo orbit is a complicated lissajous-like path about L1 with a major axis of about 150,000 km and a minor axis of about 75,000 km. The diamonds signify a mid-course correction (MCC-1) orbit-shaping maneuver (OSM), and halo-orbit intersection (HOI) at the indicated intervals following launch (L).

approximately 100 to 600 MeV nucl^{-1} (see Stone et al., 1998a). CRIS uses a scintillating optical-fiber trajectory system viewed by two CCD cameras to track the trajectories of energetic nuclei that stop in four co-aligned stacks of large-area silicon detectors.

SIS is a Solar Isotope Spectrometer that will measure the elemental and isotopic composition of energetic nuclei from $\approx$ 10 to $\approx$ 100 MeV nucl^{-1} using two identical stacks of silicon detectors (see Stone et al., 1998b). Particle trajectories are tracked with position-sensitive silicon-strip detectors instrumented with custom low-power VLSI circuitry. The energy range covered by SIS includes transient fluxes of energetic nuclei accelerated in large solar particle events, as well as anomalous cosmic rays and low-energy galactic cosmic rays.

ULEIS is an Ultra Low Energy Isotope Spectrometer that measures the mass and kinetic energy of nuclei from He to Ni by combining precise measurements of their time-of-flight over a 50 cm flight path and a measurement of their total kinetic energy (see Mason et al., 1998). The energy range covered by ULEIS includes solar energetic particles, particles accelerated by interplanetary shocks, and low-energy anomalous cosmic rays.

SEPICA is a Solar Energetic Particle Ionic Charge Analyzer designed to measure the charge state, kinetic energy, and the nuclear charge of energetic ions from $\approx$ 0.2 to $\approx$ 3 MeV nucl^{-1} (see Möbius et al., 1998). Particles entering SEPICA's

TABLE I

ACE instrumentation

Sensor	Full name	Measured species	Measured quantities	Typical energy (MeV nucl^{-1})	Technique
CRIS	Cosmic-ray isotope spectrometer	$2 \leq Z \leq 30$	Z, M, E	≈ 200	$dE/dx - E$
SIS	Solar isotope spectrometer	$2 \leq Z \leq 30$	Z, M, E	≈ 20	$dE/dx - E$
ULEIS	Ultra low energy isotope spectrometer	$2 \leq Z \leq 28$	M, E	≈ 1	$TOF - E$
SEPICA	Solar energetic particle ionic charge analyzer	$2 \leq Z \leq 28$	Q, Z, E	≈ 1	E/Q $dE/dx - E$
EPAM	Electron, proton and alpha monitor	H, He, e^-	Z, M, E	≈ 0.3	$dE/dx - E$
SWIMS	Solar wind ion mass spectrometer	$2 \leq Z \leq 30$	$M, E/Q$	≈ 0.001	E/Q $TOF - E$
SWICS	Solar wind ion composition spectrometer	$2 \leq Z \leq 30$	Z, E	≈ 0.001	E/Q $TOF - E$
SWEPAM	Solar wind electron, proton and alpha monitor	H, He, e^-	E/Q dist.	≈ 0.001	E/Q
MAG	Magnetometer	B	B_x, B_y, B_z		Triaxial fluxgate

E = energy, M = mass, Z = nuclear charge, Q = ionic charge, B = magnetic field.

multi-slit collimator are electrostatically deflected between six sets of electrode plates carrrying high voltages up to 30 kV. Measurements of SEP charge states will provide information on the temperature of the source plasma, as well as possible charge-to-mass dependent acceleration processes. SEPICA will also make direct measurements of ACR charge states.

EPAM is an Electron, Proton, and Alpha Monitor designed to characterize the dynamic behavior of electrons and ions with ≈ 0.03 to ≈ 5 MeV that are accelerated by impulsive solar flares and by interplanetary shocks associated with CMEs and CIRs (see Gold et al., 1998). EPAM was developed from the flight spare of the Hi-SCALE instrument flown on *Ulysses*, and it includes two telescope assemblies with five separate aperatures.

SWIMS is a Solar Wind Ion Mass Spectrometer with excellent mass resolution ($M/dM > 100$) that will measure solar wind isotopic composition in all solar wind conditions (see Gloeckler et al., 1998). The design is based on that of similar instruments on Wind and SOHO. Mass resolution is achieved by measuring the time-of-flight of nuclei through a specially designed electrostatic potential in which their flight-time is proportional to the square root of the mass-to-charge ratio.

SWICS is a Solar Wind Ion Composition Spectrometer that determines the elemental and ionic charge-state composition of all major solar wind ions from H to Fe using a combination of electrostatic deflection, post-acceleration, time-of-flight, and energy measurements (see Gloeckler et al., 1998). SWICS is the flight spare of an essentially identical instrument flown on *Ulysses*, where it has been particularly useful for measuring the characteristics of solar wind pickup ions.

SWEPAM is a Solar Wind Electron, Proton, and Alpha Monitor that is designed to measure the three-dimensional characteristics of solar wind and suprathermal electrons from $\approx$ 1 to 900 eV and ions from 0.26 to 35 keV (see McComas et al., 1998). It consists of modified versions of the spare solar wind electron and ion sensors from the *Ulysses* mission.

MAG is a twin triaxial flux-gate Magnetometer that will measure the dynamic behavior of the vector magnetic field, including measurements of interplanetary shocks, waves, and other features that govern the acceleration and transport of energetic particles (see Smith et al., 1998). MAG is a flight spare of the magnetometer instrument flown on Wind.

The instrumentation on ACE was designed to provide several key capabilities. First, it was necessary that the ensemble of the instruments provide coordinated measurements of charge and mass of the elements from H to Ni ($1 \leq Z \leq 28$) over a broad range in energy in order to isolate contributions from the various particle populations, and to relate the observed composition patterns back to the appropriate acceleration process and sample of source material. The energy/nucleon range over which the ACE sensors can measure elemental and isotopic composition is shown in Figure 5.

To resolve adjacent isotopes of heavy nuclei typically requires mass resolution of < 0.3 amu. CRIS, SIS, ULEIS, and SWIMS will each use approaches that have been developed and proven in earlier missions such as ISEE-3, *Ulysses*, SAMPEX, and Wind. In ACE these approaches have been optimized to achieve excellent mass resolution over a broad dynamic range with greatly improved collecting power.

Measurements of the ionization states of energetic nuclei are important for determining mass/charge fractionation during acceleration and transport and for characterizing the temperature of the source plasma. SWICS will measure the charge states of solar wind and pickup ions from ~ 0.1 to ~ 60 keV charge^{-1}, while SEPICA can measure ionic charge states of particles accelerated on the Sun and in interplanetary space between ~ 0.2 and ~ 3 MeV nucl^{-1}.

To ensure statistically significant measurements of rare isotopes, it is necessary to have large geometry factors and to process and transmit measurements of a sufficient number of individual particles under conditions ranging from solar quiet time to the most intense solar particle events. The collecting power of CRIS, SIS, ULEIS, and SEPICA typically improves on that of earlier isotope and charge state instruments by factors of ≈ 10 to 100. This improvement leads to a corresponding

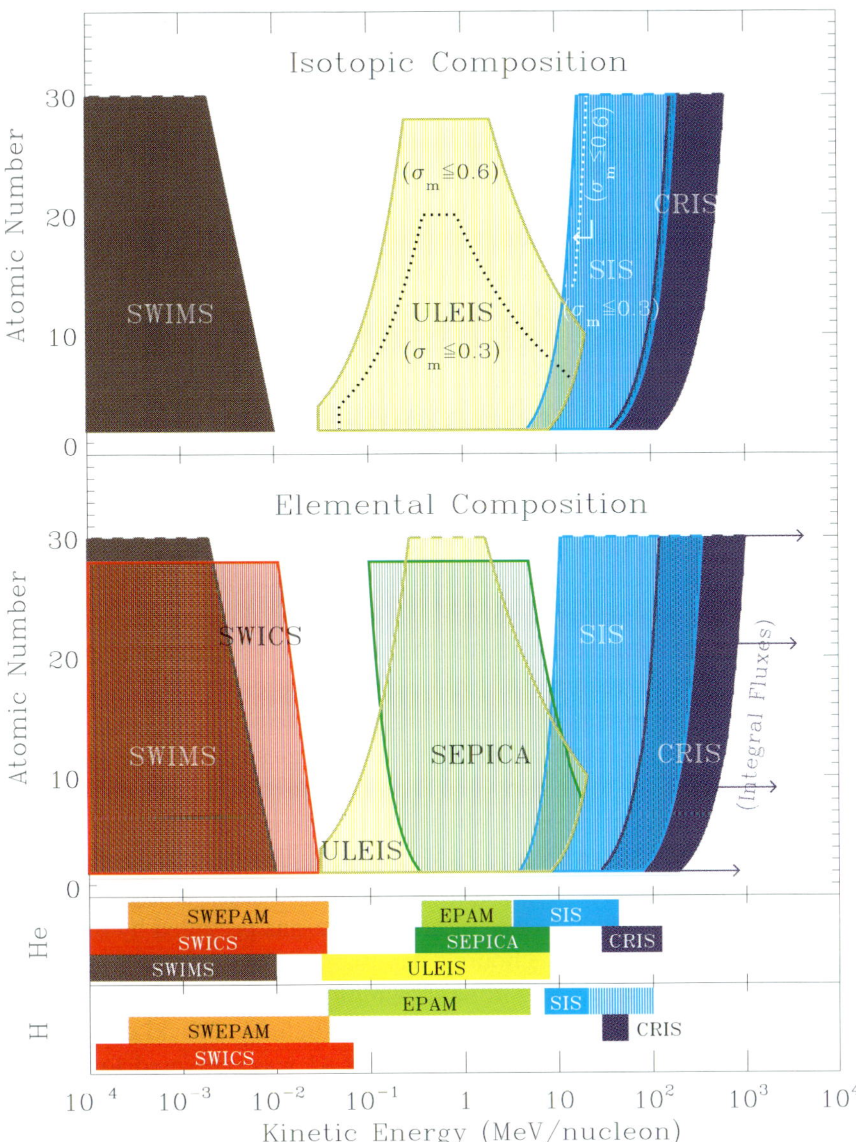

Figure 5. A plot of the energy-per-nucleon range over which the ACE sensors will measure the elemental and isotopic composition of nuclei from He to Zn ($Z = 2$ to 30), with exploratory measurements extending to $Z = 40$ in some cases. In the top panel, boundaries for an rms mass resolution of ≤ 0.3 amu and ≤ 0.6 amu are indicated (note that many studies, e.g., ^{22}Ne/^{20}Ne, do not require resolution of adjacent isotopes). A 400 km s^{-1} solar wind velocity corresponds to $\sim 0.8 \times 10^{-3}$ MeV nucl^{-1}. Note that the SWICS and SEPICA sensors will also measure ionic charge-state composition over the energy and element ranges indicated above.

gain in the time resolution with which temporal variations in composition can be measured.

High quality measurements of solar wind protons, alpha particles, and electrons, and of the interplanetary magnetic field are essential for determining solar wind characteristics that relate solar plasma at 1 AU to its origin in the corona, and energetic particle electron, proton, and alpha particle fluxes are important to characterize transient particle populations in the interplanetary medium. Measurements from SWEPAM, EPAM, and MAG will characterize the interplanetary environment for studies of heavier nuclei.

From its position 1.5×10^6 km in front of Earth, ACE is in excellent position to measure the interplanetary magnetic field and the properties of the solar wind before it impacts the Earth's magnetosphere. The Real Time Solar Wind (RTSW) monitoring system on ACE will sample and transmit interplanetary data in real time that can provide as much as one hour's warning of geomagnetic storms that result from the interaction of the interplanetary and terrestrial magnetic fields (see Zwickl et al., 1998). Such storms are known to affect ground-based power and communications systems, as well as satellite operations, and can lead to sudden increases in the trapped particle populations in the Earth's magnetosphere. The National Oceanic and Atmospheric Administration (NOAA) has responsibility for forecasting space weather for a wide range of government and commercial customers.

Realizing the advantages of the ACE location at L1, NASA and NOAA completed an interagency agreement in 1994 which specifies that (1) ACE will broadcast real-time measurements of solar wind parameters continuously, so that (2) NOAA can track and interpret these data on the ground for use in forecasting space weather. The data to be provided include solar wind velocity, density, and temperature measurements from SWEPAM, magnetic field vectors from MAG, and energetic particle fluxes from EPAM and SIS. The typical time resolution is 1 min. These data will be broadcast continuously at 434 bps during 21 hours of the day and tracked by a worldwide grid of ground stations provided by the Air Force, Japan and England. During the 3 hours a day that ACE is in contact with the Deep Space Network (DSN), the data will be transmitted to NOAA by ground link. The RTSW system is described in more detail in the article by Zwickl et al. (1998).

The final flight element of the science payload was delivered to Goddard 35 months after the start of Phase C/D implementation. Following integration onto the spacecraft and testing of the complete observatory, the science payload was declared flight ready.

5. The ACE Spacecraft

The ACE spacecraft was designed and built by JHU/APL so as to meet mission requirements in as simple and cost-efficient a manner as possible (see Chiu et al., 1998). The team began with a design based upon the Active Magnetospheric Parti-

TABLE II

ACE resources summary

Name	Mass (kg)	Peak power (W)	Nominal power (W)	Data rate (bps)	Field of view
CRIS	31.6	16.6	12.2	464	$120° \times 120°$
SIS	22.4	22.4	17.5	1992	$104°$ conical
ULEIS	21.9	15.1	14.6	1000	$24° \times 20°$
SEPICA	38.3	17.5	16.5	608	$61.5° \times 17.5°$
EPAM	12.8	4.2	3.8	168	$60°$ conical ($\times 5$)
SWICS	6.0	6.1	5.0	504	$82° \times 10°$
SWIMS	8.6	7.8	6.8	512	$62° \times 62°$
SWEPAM	6.8	6.1	5.8	1000	$160° \times 30°$ (E)
					$80° \times 10°$ (I)
MAG	4.1	2.4	1.8	304	
S^3DPU	3.9	3.5	2.5		
Spacecraft	586	248	177	392	
Other/heaters	8.6	34.3	21.5		
Total	751	384	285	6944	

Note: The spacecraft mass includes 195 kg of hydrazine.

cle Tracer Explorer (AMPTE) and other programs. Subsystem heritage was derived from as many as six different flight projects.

The spacecraft is designed to operate in a spin-stabilized mode at 5 rpm. The main body of the spacecraft is a cylinder 2 m in diameter and 1.9 m in length. The ends are inscribed octagonal aluminum honeycomb decks joined by eight flat side panels. With the solar panels and magnetometer booms extended, ACE has a wingspan of approximately 8.3 m (see Figure 6).

To meet observing requirements and to simplify access to the instruments and spacecraft subsystems, all components except the propulsion system are mounted on the external surfaces of the body. Six of the instruments are mounted on the top (sunward facing) deck, and two are mounted on the sides. The instruments are positioned so as to keep their many fields-of-view clear of one another, while retaining a reasonable balance of weight around the spacecraft. Figure 6 shows an exploded view of the spacecraft in which the location of each instrument is indicated. The fields-of-view and the resources required by the instruments are given in Table II. The magnetometer has its two sensors mounted on booms that extend from two of the four solar panels to reduce the magnetic effects from the spacecraft.

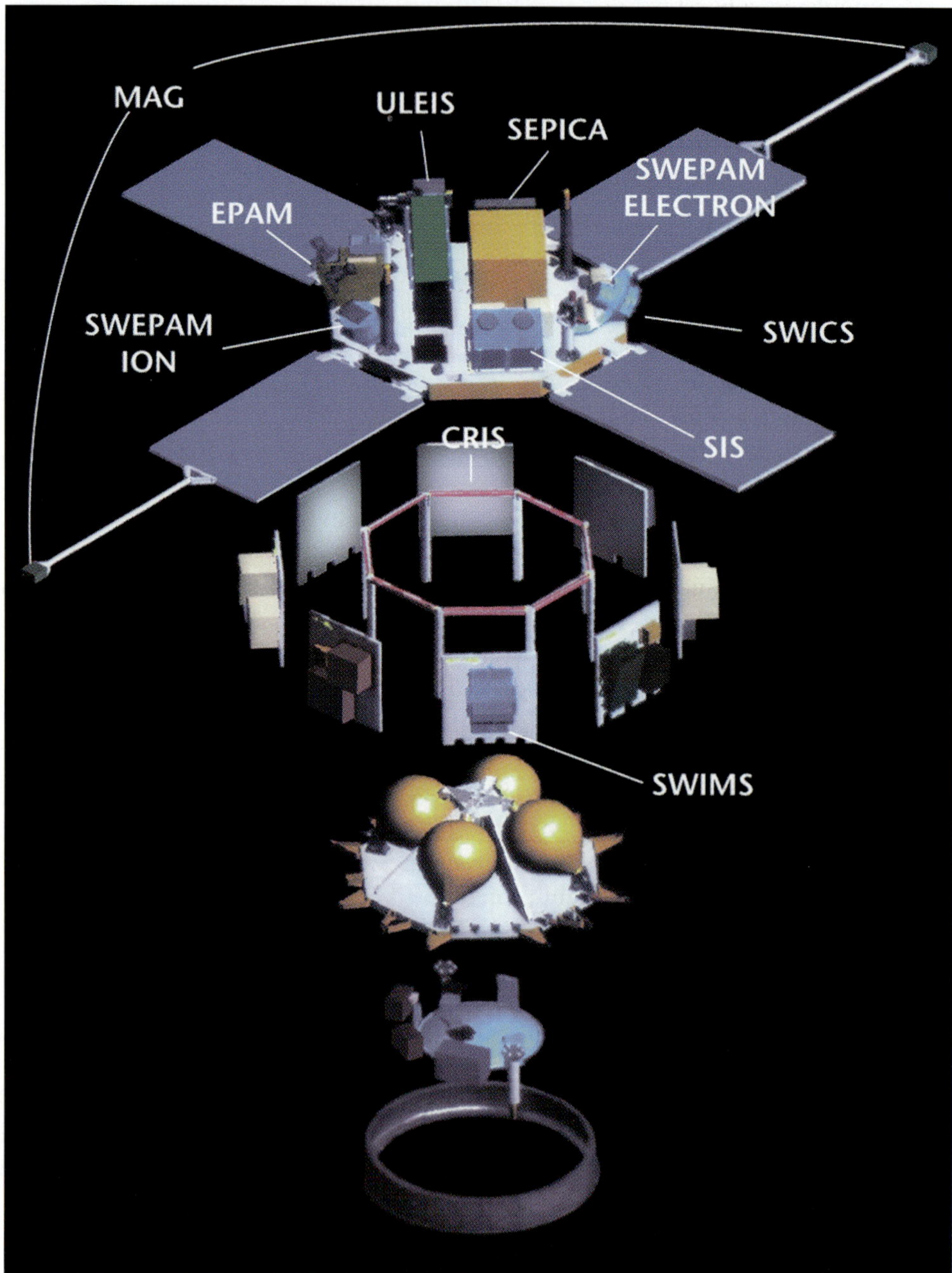

Figure 6. Exploded view of the ACE spacecraft structure. The Z axis is the spin axis of the observatory and will be aligned within 20 deg of the Earth–Sun line. Six of the scientific instruments are on the $+Z$ (sunward) deck, two are on side panels, and the ninth is mounted on the Y-axis booms. Only the four propellent tanks of the propulsion subsystem are mounted in the spacecraft interior. The $-Z$ (earthward) deck has the fixed high-gain antenna for communication.

A propulsion subsystem was needed to correct launch vehicle dispersion errors, make trajectory corrections, inject the spacecraft into its orbit about L1, adjust the orbit, adjust spin axis pointing, and maintain a 5 rpm spin rate. The subsystem is a monopropellant hydrazine blowdown design that uses nitrogen gas as a pressurant. There are four axial thrusters for velocity control along the spin axis and six radial thrusters for spin plane velocity and spin rate control. ACE launched with 195 kg of hydrazine onboard, more than sufficient for a 5-year mission.

Since ACE is spin stabilized with the spin axis always within 20° of the Sun, the power subsystem is quite simple. An array of four silicon solar panels unfold after launch. With quartz covers to provide radiation shielding, the power system should provide in excess of 440 W after 5 years in space, with an observatory peak power load of approximately 425 W.

Maintaining the proper thermal environment for the ACE instruments was complicated by the fact that several of the instruments were inherited from *Ulysses*, which had a different orientation and thermal environment. The thermal design accommodates the planned 4° to 20° angle between the spin-axis and the direction of the Sun with adequate margin.

The main constraint for the attitude control system is to keep the Earth within the radiation pattern of the high-gain antenna. Attitude is determined with a star scanner and redundant Sun sensors. The gyroscopic stability of the spinning ACE spacecraft, two fluid-filled ring nutation dampers, and the ten thrusters of the propulsion subsystem provide for the attitude control.

Because orbital manuevers are required approximately every five days to reorient the spin axis, and several of the instruments use the spinning of the spacecraft to scan the sky, nutation is a concern. The induced nutation of the spacecraft is limited by segmenting the manuevers into many short thruster pulses. This restricts the maximum nutation to less than 1 deg, and the remaining nutation is corrected using on-the-ground analysis and post-processing of the data.

Communications are through the Deep Space Network (DSN) operated by the Jet Propulsion Laboratory (JPL). The DSN receives ACE data and transmits spacecraft commands with the appropriate link margins. The RF system operates at S-band. On board ACE there are two identical and independent communications subsystems, a single high-gain, dual-polarized, parabolic reflector antenna and two sets of broad-beam antennas. In addition to receiving commands and transmitting data, the RF subsystem has the capability of supporting two-way ranging from DSN sites.

The ACE mission is designed to communicate with the DSN for one pass per day of approximately 3 hours duration. It is also designed to be able to miss one pass without the loss of any data. Data storage is accomplished via two solid state recorders, developed for both the ACE and NEAR missions, which have a capacity > 1 Gb and are one of the 'state of the art' components on ACE.

The spacecraft was built and ready for instrument integration 30 months after permission to proceed was received from NASA. JHU/APL met all of its required milestones and delivered the spacecraft at a cost much less than that allocated at the start of the contract.

6. The ACE Common Ground-Support Systems

The ACE mission, driven in part by requirements to keep total mission costs within a given cap, took some common sense (but uncommon in practice) steps in the development of mission ground support and testing equipment. The three major subsystems, the ACE Science Center (ASC), the Mission Operations Center (MOC), and the spacecraft Integration and Testing (I&T) ground-support system, have many common functions and were able to share software, databases and testing procedures. Smaller programs, especially balloon and sounding rocket experiments, have often taken advantage of this synergy to design, test, and operate one system, but for larger missions, including Explorers, this has not done. For ACE, the ground-support and testing teams used a common system architecture that supports all three ground elements.

Historically, larger satellites have been developed and operated by separate teams. In 1993, when the ACE Project sought innovative approaches to reduce its budget, it was realized that an important step would be to provide a basic backbone for a common ground system. It was also recognized that this approach required conquering political and financial barriers (until recently, budgets within NASA for operations were kept separate from project development funds).

In this common system architecture approach the three ground-support system elements each utilize the same computational platforms (i.e., HP 715's and HP 748's) and, to the extent possible, use a common set of software modules. To make this possible, the performance requirements for instrument checkout and data verification had to be understood far earlier than is usually the case and procedures had to be developed on a pace that matched the spacecraft development.

The ground-support equipment for the integration and test of the spacecraft and its instruments is known as the Integration and Test Operations Control Center (ITOCC). Spacecraft operations and level zero processing of payload data are performed in the mission operations center (MOC) at GSFC. The data is then transmitted to the ACE Science Center (ASC) at Caltech (see Garrard et al., 1998). The ASC is responsible for initial processing of the data and for making the data accessible to the co-investigators and guest investigators located at their home institutions. Instrument teams are located at ACE Science Analysis Remote Sites (ASARS) where they can review and analyse data and evaluate instrument performance and, as necessary, forward instrument commands to the ACE MOC for transmission to the satellite. In practice, three teams (including contractors) at JPL, GSFC and Caltech had to work together to make this common ground-support system plan work.

A trade study early in the mission development cycle led to the conclusion that Transportable Payload Operations Control Center (TPOCC), a Unix-based system which has been a foundation system for the past several years for most spacecraft control centers at the GSFC, could provide much of the needed functionality in the ITOCC, the MOC, and the ASC. A generic ACE TPOCC, including all functionality common to the three subsystems, was adapted from the X-ray Timing Explorer (XTE) version of TPOCC. Separate copies of the generic system were then augmented individually with unique capabilities needed in the three subsystems. Most of the MOC unique capabilities will eventually be transferred to the ASC, making it possible to extend the duration of the ACE mission by operating it at low cost from the ASC.

Functions of the MOC system being included in the ASC include mission planning and scheduling, and attitude determination. Special requirements for the ITOCC included in the generic system are for Ground Support Equipment (GSE) commanding and data collection, the counting of relay state changes, and spacecraft system and subsystem run time processing. The ASC will use essentially the same system as the MOC with the addition of the science data analysis software.

A single database was used for spacecraft development and testing and is now being used for spacecraft and instrument operations. Using the TPOCC System Test and Operations Language (TSTOL) for all procedures and displays allows the instrument teams to use common elements throughout all phases of instrument development, test, checkout and monitoring during spacecraft integration and testing, and flight operations. Developmental phase checkout procedures are the same as activation procedures during initial phases of the mission, and the instrument development teams have had the opportunity to 'practice' and debug the approaches to monitoring the health and status of their instruments.

Members of the Flight Operations Team (FOT) were included in the design and development from the early phases of the mission. This enabled a smooth transition between spacecraft design and ground operations and avoided the frequently encountered situation where ground operations can be made unnecessarily complicated by spacecraft design decisions. The concept of the Integrated Team approach working across spacecraft design, integration and test, and operations has worked successfully on smaller spacecraft, but this is the first time NASA has attempted this approach on a program of this magnitude. The approach worked very well and is recommended for future missions.

The combined approach has worked in spite of the different institutional responsibilities for mission operations and I&T. The development of TPOCC was contracted to Computer Sciences Corporation (CSC). This task also included development of the MOC-specific enhancements. JHU/APL let a separate contract to CSC for development of the ITOCC-specific enhancements. Many personnel at CSC worked on both the ITOCC and MOC contracts, providing common ground and approaches for both developments. This effort required close cooperation between all users and the developers.

Significant benefits have accrued from the multiple use of the generic system, saving NASA millions of dollars. Although the generic TPOCC task required more lines of code than needed for the MOC alone, it was substantially less than the lines of code required for the three separate systems. A similar savings in maintenance will be realized; enhancement or correction in one system will enhance or correct all three systems. The approach led to the deletion of the software training simulator for the FOT. Instead, the FOT received high fidelity training with the spacecraft at JHU/APL, thus reducing risk to mission operations. The one database utilized for all three systems eliminated time consuming, costly translations from one database to another while increasing the reliability of the operations database. Also, the content of the TSTOL procedures, developed for spacecraft I&T, is easily adapted for the ACE MOC and the ASC.

7. Conclusion

The ACE mission promises to return exciting new results on a broad range of topics related to two of the NASA science themes: the structure and evolution of the Universe, and the Sun–Earth Connection. The fixed cost imposed by NASA for the development of ACE has led to new management approaches for the development of the instrumentation and the integration and testing of the spacecraft. The final cost of developing ACE was significantly less than had been allocated by NASA, and considerable funding was returned to the Explorer Program for other future missions. The discipline and dedication of the large team from many institutions and companies in the US and Europe also enabled the spacecraft and its instruments to meet the challenging schedule. The common ground-support system architecture supported observatory integration and testing, and will support the flight operations and science analysis.

A major factor in the success of the ACE development was the close cooperation of all members of the various teams: the instrument teams, the contractors, the spacecraft development team at JHU/APL, and GSFC and Caltech. Working as an integrated team, the challenges of each subsystem were recognized, and the talented scientists, engineers, and system developers provided cost-effective system solutions. ACE has successfully reengineered the mission development processes, resulting in contained development costs for the ACE Project, and lower-cost and more reliable flight operations.

We look forward to an exciting mission that will pursue the scientific questions posed here and discussed in more detail in the other articles of this special edition of *Space Science Reviews*.

Acknowledgements

The ACE Mission was designed and developed by many organizations and individuals under the support of the Explorer Program in the NASA Office of Space Science. Special thanks go to the many members of the spacecraft team at JHU/APL, the ACE Payload Management Office at Caltech, the NASA/GSFC ACE Project team at GSFC, and the nine instrument teams that are acknowledged in the accompanying papers. Guidance by W. Vernon Jones, the ACE Program Scientist, throughout all phases of the ACE Project is greatly appreciated.

References

Chiu, M. C., Von-Mehlem, U. I., Willey, C. E., Betenbaugh, T. M., Maynard, J. J., Krein, J. A., Conde, R. F., Gray, W. T., Hunt, Jr., J. W., Mosher, L. E., McCullough, M. G., Panneton, P. E., Staiger, J. P., and Rodberg, E. H.: 1998, 'ACE Spacecraft', *Space Sci. Rev.* **86**, 257.

Ellison, D. C., Drury, L. O'C., and Meyer, J-P.: 1998, 'Cosmic Rays from Supernova Remnants: A Brief Description of the Shock Acceleration of Gas and Dust', *Space Sci. Rev.* **86**, 203.

Fisk, L. A., Schwadron, N. A., and Zurbuchen, T. H.: 1998, 'On the Slow Solar Wind', *Space Sci. Rev.* **86**, 51.

Frisch, P. C.: 1998, 'Interstellar Matter and the Boundary Conditions of the Solar System', *Space Sci. Rev.* **86**, 107.

Garrard, T. L., Hammond, J. S., Davis, A. D., and Sears, S. R.: 1998, 'The ACE Science Center', *Space Sci. Rev.* **86**, 651.

Gloeckler, G. and Geiss, J.: 1998, 'Interstellar and Inner Source Pickup Ions Observed with SWICS on *Ulysses*', *Space Sci. Rev.* **86**, 127.

Gloeckler, G., Bedini, P., Bochsler, P., Fisk, L. A., Geiss, J., Ipavich, F. M., Cani, J., Fischer, J., Kallenbach, R., Miller, J., Tums, O., and Wimmer, R.: 1998, 'Investigation of the Composition of Solar and Interstellar Matter Using Solar Wind and Pickup Ion Measurements with SWICS and SWIMS on the ACE Spacecraft', *Space Sci. Rev.* **86**, 497.

Gold, R. E., Krimigis, S. M., Hawkins, S. E., III, Haggerty, D. K., Lohr, D. A., Fiore, E., Armstrong, T. P., and Holland, G.: 1998, 'Electron, Proton, and Alpha Monitor on the Advanced Composition Explorer Spacecraft', *Space Sci. Rev.* **86**, 541.

Jokipii, J. R.: 1998, 'Insight Into Cosmic-Ray Acceleration from the Study of Anomalous Cosmic Rays', *Space Sci. Rev.* **86**, 161.

Lin, R.: 1998, 'WIND Observations of Superthermal Electrons in the Interplanetary Medium', *Space Sci. Rev.* **86**, 61.

Mason, G. M., Gold, R. E., Krimigis, S. M., Mazur, J. E., Andrews, G. B., Daley, K. A., Dwyer, J. R., Heuerman, K. F., James, T. L., Kennedy, M. J., LeFevere, T., Malcolm, H., Tossman, B., and Walpole, P. H.: 1998, 'The Ultra Low Energy Isotope Spectrometer (ULEIS) for the ACE Spacecraft', *Space Sci. Rev.* **86**, 409.

McComas, D. J., Bame, S. J., Barker, P., Feldman, W. C., Phillips, J. L., Riley, P., and Griffee, J. W.: 1998, 'Solar Wind Electron Proton Alpha Monitor (SWEPAM) for the Advanced Composition Explorer', *Space Sci. Rev.* **86**, 563.

Meyer, J-P., Drury, L. O'C., and Ellison, D. C.: 1998, 'A Cosmic Ray Composition Controlled by Volatility and A/Q Ratio. SNR Acceleration of Gas and Dust', *Space Sci. Rev.* **86**, 179.

Miller, J. A.: 1998, 'Particle Acceleration in Impulsive Solar Flares', *Space Sci. Rev.* **86**, 79.

Möbius, E., Hovestadt, D., Klecker, B., Kistler, L. M., Popecki, M. A., Crocker, K. N., Gliem, F., Granoff, M., Turco, S., Anderson, A., Arbinger, H., Battell, S., Cravens, J., Demain, P.,

Distelbrink, J., Dors, I., Dunphy, P., Ellis, S., Gaidos, J., Googins, J., Harasim, A., Hayes, R., Humphrey, G., Kastle, H., Kunneth, E., Lavasseur, J., Lund, E. J., Miller, R., Murphy, G., Pfeffermann, K., Reiche, K.-U., Sartori, E., Schimpfle, J., Seidenschwang, E., Shappirio, M., Stockner, K., Taylor, S. C., Vachon, P., Vosbury, M., Wiewesiek, and Ye, V.: 1998, 'The Solar Energetic Particle Ionic Charge Analyzer (SEPICA) and the Data Processing Unit (S3DPU) for SWICS, SWIMS and SEPICA', *Space Sci. Rev.* **86**, 449.

Ptuskin, V. S. and Soutoul, A.: 1998, 'Cosmic Ray Clocks', *Space Sci. Rev.* **86**, 225.

Simpson, J. A.: 1998, 'Challenges for the Experiments on the Advanced Composition Explorer', *Space Sci. Rev.* **86**, 23.

Smith, C. W., Acuña, M. H., Burlaga, L. F., L'Heureux, J., Ness, N. F., and Scheifele, J.: 1998, 'The ACE Magnetic Fields Experiment', *Space Sci. Rev.* **86**, 613.

Stone, E. C., Cohen, C. M. S., Cook, W. R., Cummings, A. C., Gauld, B., Kecman, B., Leske, R. A., Mewaldt, R. A., Thayer, M. R., Dougherty, B. L., Grumm, R. L., Milliken, B. D., Radocinski, R. G., Wiedenbeck, M. E., Christian, E. R., Shuman, S., Trexel, H., von Rosenvinge, T. T., Binns, W. R., Dowkontt, P., Epstein, J., Hink, P. L., Klarmann, J., Lijowksi, M., and Olevitch, M. A.: 1998a, 'The Cosmic Ray Isotope Spectrometer for the Advanced Composition Explorer', *Space Sci. Rev.* **86**, 285.

Stone, E. C., Cohen, C. M. S., Cook, W. R., Cummings, A. C., Gauld, B., Kecman, B., Leske, R. A., Mewaldt, R. A., Thayer, M. R., Dougherty, B. L., Grumm, R. L., Milliken, B. D., Radocinski, R. G., Wiedenbeck, M. E., Christian, E. R., Shuman, S., and von Rosenvinge, T. T.: 1998b, 'The Solar Isotope Spectrometer for the Advanced Composition Explorer', *Space Sci. Rev.* **86**, 357.

Webber, W. R.: 1998, 'What are the Limits for ACE Galactic Cosmic Ray Isotope Studies?', *Space Sci. Rev.* **86**, 239.

Zwickl, R., Doggett, K., Sahm, S., Barrett, B., Grubb, R., Detman, T., Raben, V., Smith, C. W., Riley, P., Gold, R., Mewaldt, R. A., and Maruyama, T.: 1998, 'The NOAA Real-Time Solar Wind (RTSW) System Using ACE Data', *Space Sci. Rev.* **86**, 633.

CHALLENGES FOR THE ADVANCED COMPOSITION EXPLORER *

J. A. SIMPSON

Enrico Fermi Institute and Department of Physics, The University of Chicago, IL 60637, U.S.A.

Abstract. This report is a brief introduction to some of the vital contributions that the Advanced Composition Explorer Mission will make towards our understanding of the origins of matter and acceleration of particles on a wide range of solar and astrophysical scales. Examples of these contributions are drawn from two broad areas of the space sciences. They are: (1) Dynamical phenomena at the Sun and in the inner heliosphere; and (2) The elemental and isotopic composition of matter in the solar wind, solar accelerated ejecta, galactic cosmic radiation and the anomalous nuclear component in the heliosphere. Some current problems with theories intended to account for these phenomena are discussed, including interpretations of the stable and radioactive isotopes in the galactic cosmic rays.

1. Introduction

The Advanced Composition Explorer (ACE) opens a new era for investigations that will expand our knowledge of solar activity, the solar-terrestrial connection, heliospheric dynamics and the origins of both cosmic ray and interstellar matter in our galaxy. Over the past 40–50 years we have made dramatic progress in all these disciplines, especially after research in space became a reality. However, recent discoveries have demonstrated that many current ideas and theories are inadequate for an understanding of the fundamental physics of solar, heliospheric and galactic phenomena.

In view of the recent developments and discoveries in these fields, what are some of the current problems that ACE investigators will undoubtedly be able to solve?

In this brief introduction to the ACE Science Workshop, I shall attempt to provide a few examples of how ACE researchers could make vital contributions to our current state of knowledge in these areas of space science. Two broad research disciplines enable us to illustrate the importance of the ACE investigations; they are:

(1) Dynamical phenomena in the inner heliosphere, such as Corotating Interaction Regions (CIRs), charged particle acceleration and modulation, coronal mass ejections (CMEs), etc., and

* Introductory lecture for the ACE Science Workshop at the California Institute of Technology (7–9 January 1997).

Space Science Reviews **86**: 23–50, 1998.

(2) The elemental and isotopic composition of matter in the cosmic radiation, the anomalous nuclear component, the solar wind and solar ejecta (e.g., flares, CMEs, etc.).

This report is divided into these two disciplinary areas with examples of basic problems in these disciplines. I have drawn my examples mainly from the energetic charged particle measurements by the CRIS and SIS instruments on ACE, but from these examples it will be clear to the reader that these examples, in many cases, will also require magnetic field and plasma measurements from ACE. Some additional examples are mentioned in Section 4 (Concluding Remarks) concerned with solar flare and interplanetary plasma questions that may be answered by the SIS and SWICS experiments.

2. Examples of Dynamical Phenomena in the Inner Heliosphere

The ACE spacecraft will join a flotilla of spacecraft both at the Lagrangian libration point between the Sun and Earth, such as the SOHO Mission (Fleck et al., 1995), Wind (Acuña et al., 1995), and in Earth orbit, such as IMP-8 (King, 1982) and SAMPEX (Baker et al., 1993). Their observations, combined with the high time resolution and collecting power of ACE, will yield major advances for understanding physical processes at the Sun and their impact on Earth. However, it is also important to recognize unique opportunities in the years 1998 through 2001 for three-dimensional and radial investigations of transient high energy solar phenomena and many other dynamical phenomena in the inner heliosphere – opportunities that are not likely to be repeated in the foreseeable future of space flight.

These unique windows of opportunity for ACE occur at times when the *Ulysses* spacecraft is either at a high solar latitude or near the equatorial plane at $\sim$5 AU. The *Ulysses* spacecraft trajectory is shown in Figure 1, (Smith et al., 1995) for its initial prime mission period through completion of the first south-north solar polar pass. This trajectory will be extended and repeated as the approved *Ulysses* Solar Maximum Mission, ending the last quarter of year 2001. The radius, latitude and longitude of *Ulysses* (with respect to Earth and ACE) are shown in Figure 2 and compared with the time interval for ACE measurements, where I have assumed a six year mission time for ACE. Unique opportunities for correlated investigations at ACE and *Ulysses* include the late 1997 through 1998 period with *Ulysses* at 5 ± 0.4 AU within $\sim$20° of the equator for radial investigations such as cosmic-ray modulation, solar wind and heliospheric current sheet phenomena. For three-dimensional investigations *Ulysses* will be at latitudes above $\sim$40° from year 2000 to end of mission, except for the rapid south to north transit through the equator. The following examples may illustrate the value of these unique periods:

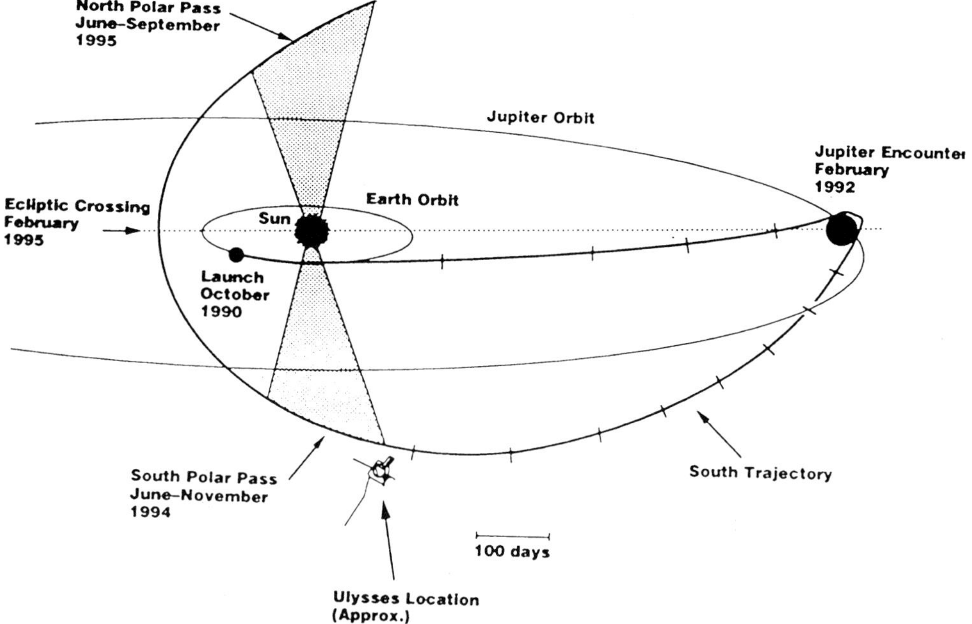

Figure 1. The *Ulysses* prime mission trajectory viewed from 15° above the ecliptic plane. The *Ulysses* Solar Maximum Mission will return again to the south solar pole (2000) and north solar pole (2001) (from Smith et al., 1995).

2.1. A CORONAL MASS EJECTION IN THREE DIMENSIONS

Ulysses investigators discovered that a coronal mass ejection (CME) and a solar flare that occurred at 9° N on 20 February 1994 extended over a vast latitude range in both the north and south hemispheres (Gosling et al., 1994; Simpson et al., 1995). Protons were measured at IMP-8, as shown in Figure 3(C). Several days later this radiation, with the accompanying CME, arrived at *Ulysses*, located at 54° south latitude and 3.5 AU from the Sun. Between the time of the forward and the arrival of the reverse shock at *Ulysses* (Figure 3(A)), as reported by Gosling et al. (1994), fluxes of protons and electrons over a wide range of magnetic rigidities arrived without significant time dispersion, as shown in Figure 3(B). These observations suggest continual charged particle acceleration within the bounds of the CME. Obviously, this new class of CME presents several unsolved problems in particle acceleration, propagation and magnetic field structures. It is the kind of solar event that can only be fully studied in the future with the instrumentation capabilities of ACE and *Ulysses* for a 3-D analysis and interpretation during those periods when *Ulysses* is far out of the ecliptic (Figure 1), mainly during the years 1999–2003.

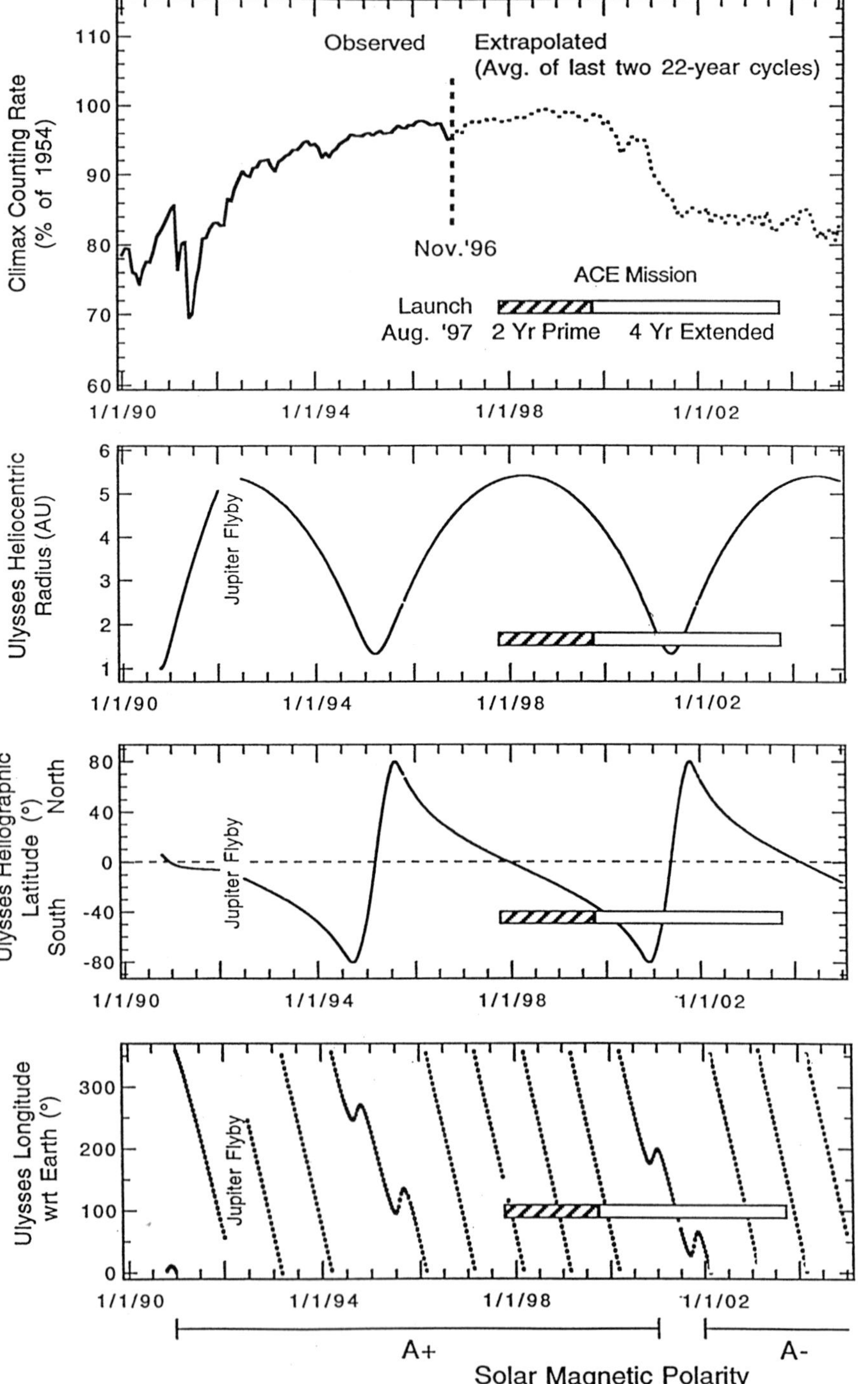

Figure 2. Top panel: the change in modulation level of the cosmic-ray nuclear component: ⎯⎯ observed; ·········· extrapolated. Bottom panels: *Ulysses* position.

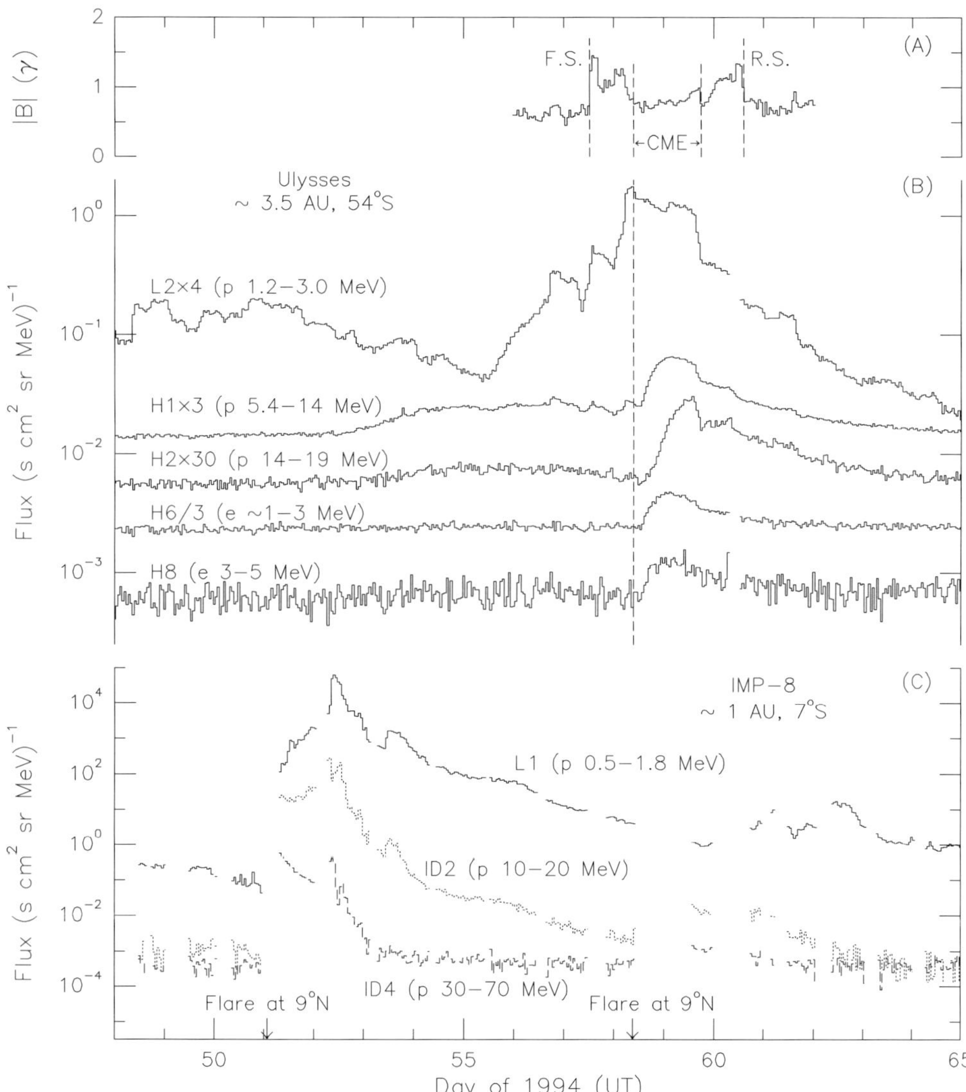

Figure 3. The solar flare-CME event of February 1994 as observed in the ecliptic plane by the *Ulysses'* LET (L) and HET (H) instruments (B) and IMP-8 (C); p, protons; e, electrons. The locations of the forward (F) and reverse (R) shock pair associated with the CME and the magnetic field strength (A) are from Gosling et al. (1994) and Simpson et al. (1995).

2.2. RECURRENT MODULATION EXTENDING TO SOLAR POLAR LATITUDES

From the equator to approximately 40° south latitude the well-known recurrent interplanetary magnetic field compression and the solar wind velocity enhancements associated with Corotating Interaction Regions (CIRs) were observed in the *Ulysses* magnetometer and solar wind measurements, as shown in Figures 4(A)

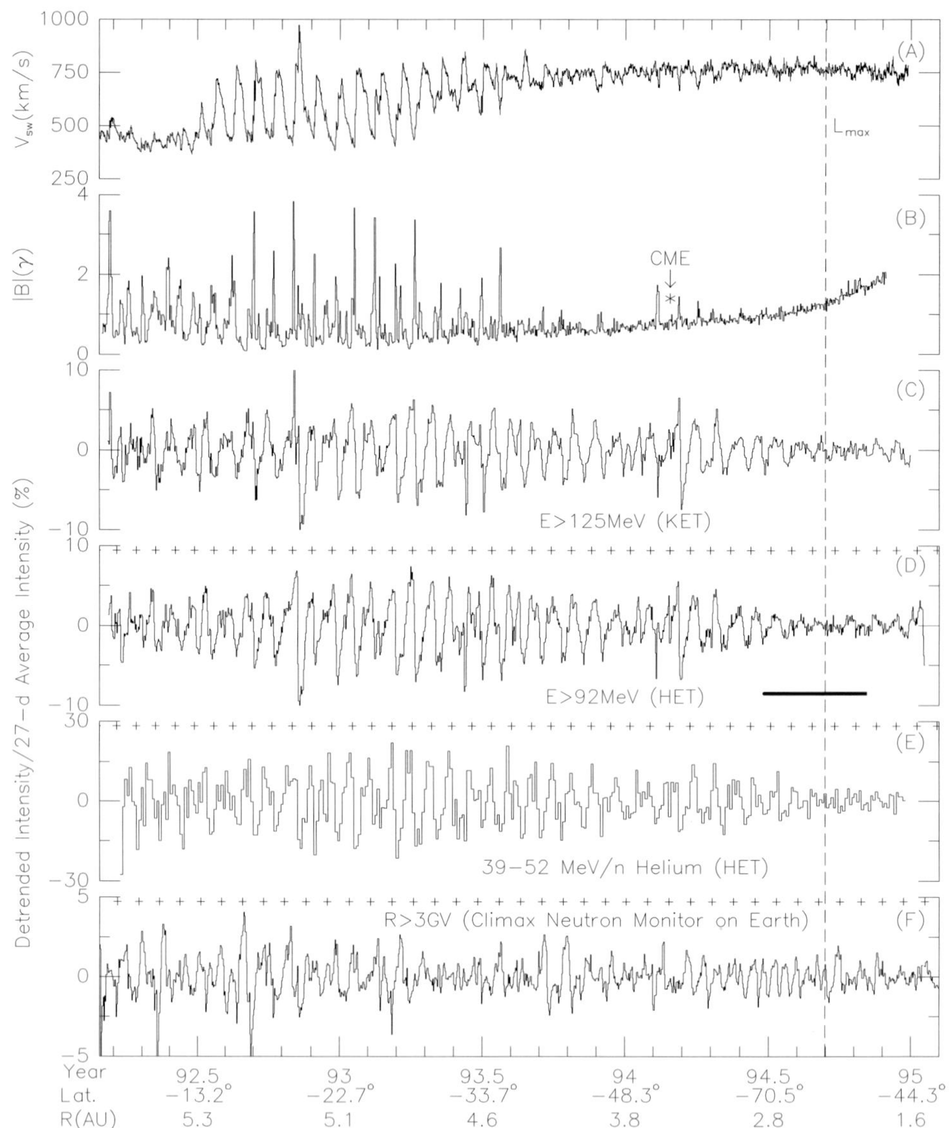

Figure 4. (A) 6-hour averages of solar wind speed from Phillips et al. (1995); (B) Daily averages of magnetic field B from Balogh et al. (1995); (C and D) Daily average intensity after detrending to remove the long-term increase in cosmic-ray intensity for the integral intensity of cosmic rays as measured by the KET and the HET. (E) Three-day average detrended helium flux measured by the HET, consisting of a mixture of cosmic-ray and anomalous helium. (F) Daily average detrended Climax, Colorado, neutron monitor intensity variations from primary protons with energy $> 3 \times 10^9$ eV (from Simpson et al., 1995).

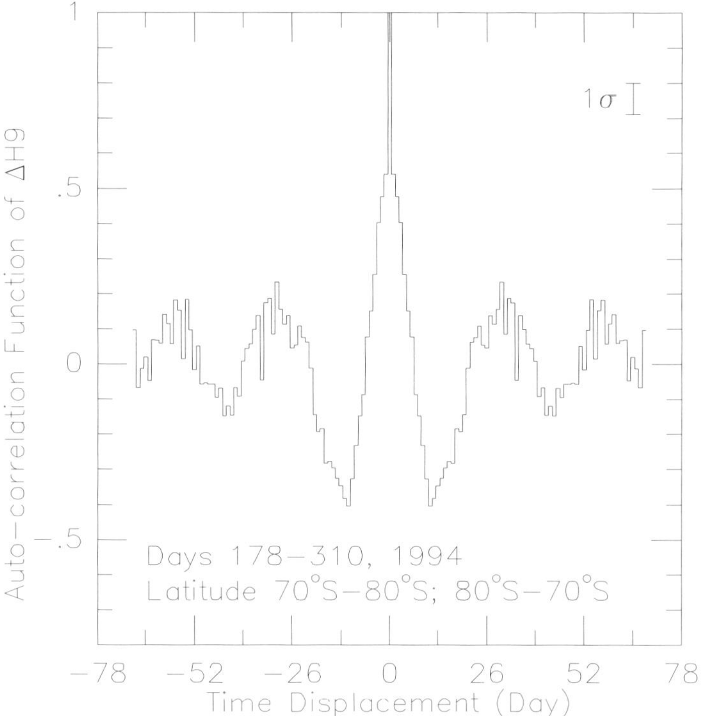

Figure 5. Determination of the average recurrence period of cosmic ray > 92 MeV proton intensity modulation near −80° latitude (bar interval in Figure 4(D)) representing days 178 to 310, 1994 and latitudes from −70° to −80° and back to −70° (from Simpson et al., 1995).

and 4(B). These low latitude results conformed to current theory for CIR's. It was, therefore, surprising to discover that – as the magnetic field recurrent compressions and solar wind enhancements disappeared above ∼40° latitudes – the 26 day recurrent modulation of galactic cosmic rays and the anomalous nuclear component associated with CIR's at low latitudes continued to the *Ulysses* maximum solar polar latitude of 80° south. This phenomenon is shown in Figures 4(C–E) for galactic cosmic-ray modulation (McKibben et al., 1995; Kunow et al., 1995). This modulation phenomenon is observed in both the north and south solar polar regions with a 26 ± 2-day recurrence that is not significantly latitude dependent (Figure 5). Since this recurrent modulation of cosmic rays by CIRs extends beyond 13 GV in magnetic rigidity, it is clearly a large scale global phenomenon of the inner heliosphere. It was recognized immediately that the source of this recurrent modulation occurred beyond the radial range of *Ulysses*, probably at low latitudes.

This experimental discovery dramatically changes models and theories that had accounted for the physical phenomena underlying the 26-day modulation of the galactic cosmic rays and the anomalous nuclear component. These new findings also relate to the propagation to high latitude of low energy nucleons and electrons accelerated by CIR shocks (Pizzo, 1994). At the present time two different models

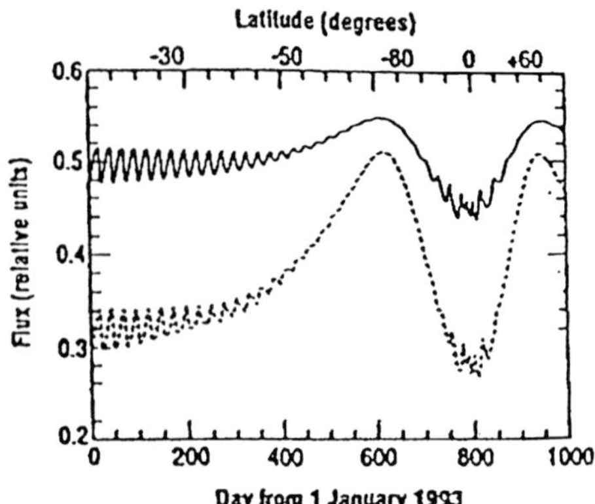

Figure 6. Computed variation of the flux of 1 GeV cosmic-ray protons along the trajectory of *Ulysses* for the heliospheric model in Kóta and Jokipii (1995), their Figure 1. Calculations were carried out for larger ($\kappa_\perp/\kappa_\parallel = 0.5$, solid line) and lower ($\kappa_\perp/\kappa_\parallel = 0.02$, dashed line) values of the perpendicular diffusion coefficient.

represent the principal approaches for attempts to explain the experimental find-ings. On the one hand Kóta and Jokipii (1995) by means of a three-dimensional, time-dependent computer code demonstrated that recurrent modulation by the he-liospheric current sheet could result in particle modulation to high latitude provided there was strong latitudinal particle diffusion across magnetic field lines (Figure 6). On the other hand, Fisk (1996) pointed out that the interplay between the dif-ferential rotation of the footprints of magnetic field lines in the photosphere and the subsequent nonradial expansion of these same field lines with the solar wind from rigidly rotating coronal holes may result in extensive latitudinal excursions of heliospheric magnetic field lines, i.e., field lines at high latitudes near the Sun may be connected directly to CIR's at low latitudes at large radial distances permitting fast charged particle transport to high latitudes (Figure 7). These models and some tests for their validity have been discussed by Simpson (1998).

The unique opportunity to exploit the configuration of the dual spacecraft posi-tions in the equatorial region of the inner heliosphere – namely, ACE at the libration point and *Ulysses* at 5 ± 0.4 AU from the Sun from late 1997 into 1999 (Figure 2) – will depend on the extent to which approximately solar minimum conditions pre-vail in this time interval. In the top panel of Figure 2 we have extrapolated the level of solar modulation measured by the Climax Neutron Monitor by averaging the modulation levels measured over the last two 22-year solar cycles. It is hazardous to project the future behavior of the Sun and its generation of the solar wind and magnetic fields modulating the galactic cosmic rays and the anomalous nuclear component. Nevertheless, it appears from Figure 2 that – during the phase when the solar magnetic field vector potential, A, is positive – the modulation level is likely to be close to minimum modulation through 1998 and probably into 1999. Under these conditions it will be possible to study the evolution of corotating inter-

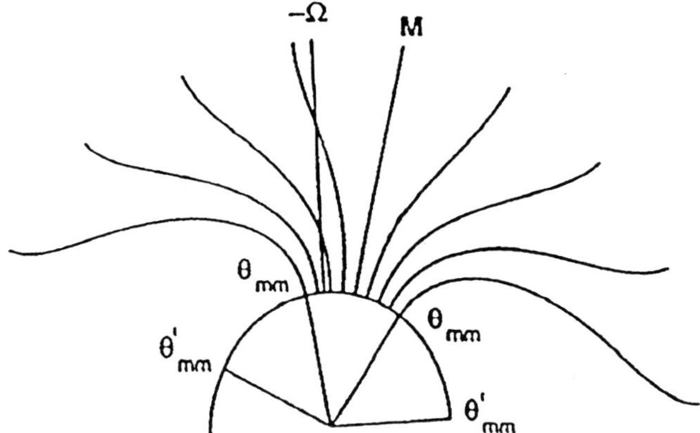

Figure 7. A schematic illustration of the expansion of magnetic field lines from a polar coronal hole (from Fisk, 1996).

action regions, the heliospheric current sheet and the access of energetic charged particles to and from high latitude magnetic fields of solar origin that may extend to the equator at 5 AU. These studies would test the relative importance of charged particle cross field diffusion *versus* charged particle transport along field lines as models to explain the discovery of the extension of 26-day recurrent modulation to solar polar latitudes, as discussed above.

The examples illustrated above for dual investigations in the equatorial zone or in three dimensions are but a subset of possible inner heliospheric studies directed towards developing a 3-D theory for cosmic ray and anomalous nuclear component modulation and for transport of energetic particles through the inner heliosphere. For further information on the present state of charged particle modulation see, for example, McKibben, et al., 1998; Simpson, 1998; Heber et al., 1997.

3. Isotopic Composition of the Cosmic Radiation

3.1. INTRODUCTION

The principal goal of the experiments on the ACE spacecraft is the determination of the elemental and isotopic composition of matter over a wide range of particle energies and astrophysical sites, including the Sun, the interplanetary medium and galaxy. To illustrate how ACE may advance our knowledge of the cosmic abundances of matter and the importance of this goal for high energy astrophysics, I shall – in this brief report – focus on a few of the possible advances in understanding the galactic source composition of the cosmic radiation and the role for ACE in testing some models of nucleosynthesis in the Galaxy.

Recently, we have observational proof that the cosmic radiation permeates the entire galaxy. This evidence is based on the EGRET instrument (Hunter et al.,

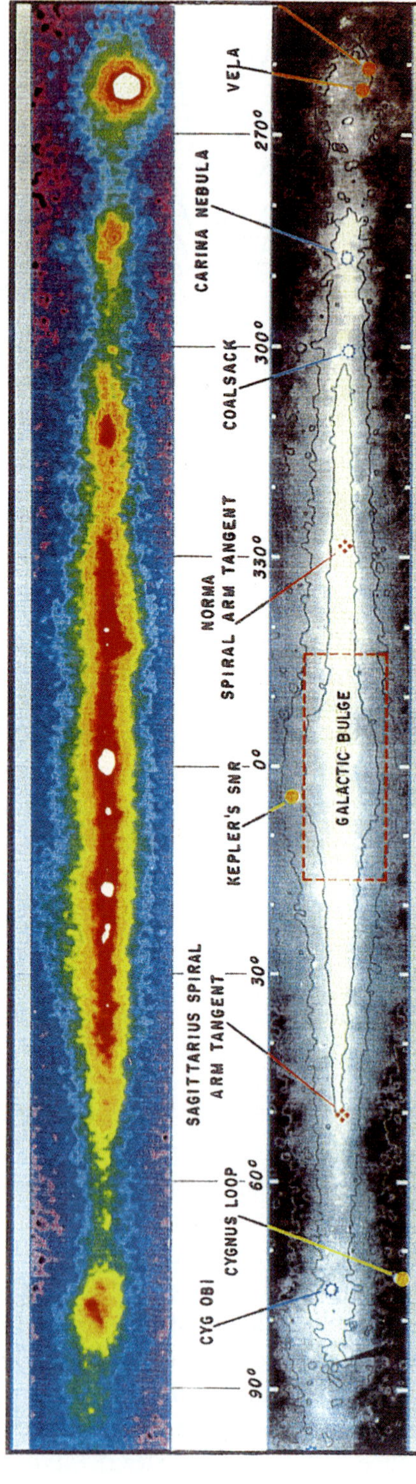

Figure 8. The Milky Way. *The lower panel* displays the light observed at night mainly as photons produced by excitation of the electronic structure of hydrogen atoms. *The upper panel* displays the EGRET instrument observations on the Compton Gamma-Ray Observatory in the diffuse light of ~100 MeV gamma rays. The gamma rays are the product of interstellar nuclear breakup by cosmic-ray nuclei. Thus the lower panel displays the atomic galaxy, whereas the upper panel displays the nuclear or cosmic-ray galaxy (from Hunter et al., 1997).

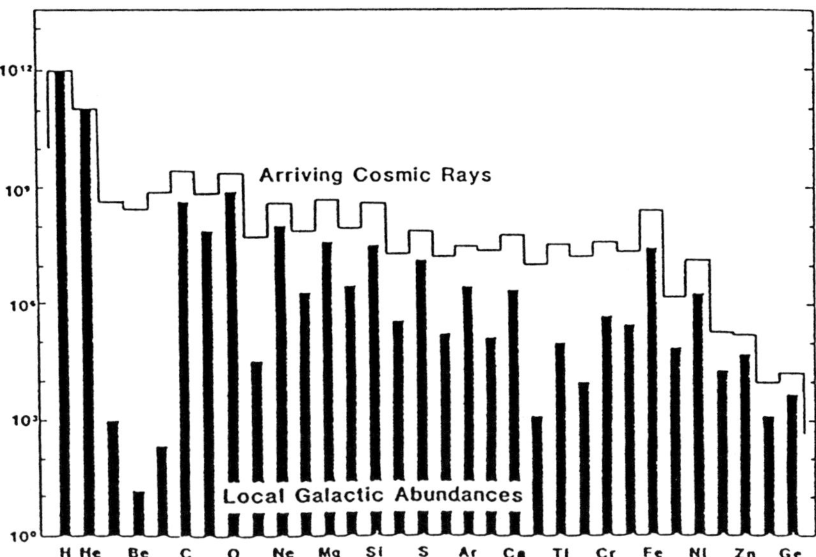

Figure 9. Elemental abundances of arriving cosmic rays (line) compared with local galactic abundances (bars). The normalization is at hydrogen (from Lund, 1989).

1997) on the Compton Gamma-Ray Observatory observing the distribution of the diffuse ~ 100 MeV γ-ray emission that is produced mainly by cosmic-ray proton collisions with interstellar nuclear matter (Figure 8). These collisions produce π^0 mesons that rapidly decay into pairs of >70 MeV γ-rays observed throughout the Galaxy. Local regions of enhanced γ-ray emissions in Figure 8 are also evident as sources of high-energy nuclei. Thus, our Galaxy is a cosmic-ray galaxy. This cosmic-ray matter from hydrogen to uranium is a direct sample of matter from the Galaxy.

The arriving cosmic radiation is a mixture of accelerated source matter and the nuclear products of secondary nuclei from cosmic-ray interactions with the atoms in the interstellar medium, as shown in Figure 9. Furthermore, the relative abundances of the arriving elemental composition display a bias – probably introduced during acceleration. This bias is currently described as a dependence on the first ionization potential of the accelerated atom. (See, however, an alternate proposal to explain this bias by Meyer et al. (1997) at this Workshop. Fortunately, the isotopic composition of these source and secondary cosmic-ray nuclei do not display any known bias. The isotopic abundances of cosmic-ray nuclei convey not only the 'imprint' of their nucleosynthetic origins at astrophysical sites in the Galaxy, but also their radioactive nuclides reveal the time required for their acceleration and the history of their propagation in interstellar magnetic fields. For example, Table I lists some of the radionuclides that measure the time between nucleosynthesis and acceleration, acceleration and detection or the model dependent confinement time in the magnetic fields of the galactic disk (e.g., 'leaky box' or homogeneous models) and/or the galactic magnetic halo (e.g., diffusion models).

How do we relate the source composition of the isotopes of elements, especially the less abundant isotopes to specific kinds of nucleosynthesis? To illustrate the possibilities, I have compiled in Table II the isotopic 'signatures' from two different nucleosynthetic scenarios believed to exist in galactic sources. They are:

(1) The evolution of massive stars (masses greater than 10 solar masses; Type II supernovae) – they are the main source of the isotopes of Ne to Ca;

(2) Explosive silicon-burning isotopes.

This compilation in Table II is a collection derived from many sources in Arnett (1996) and Truran (1997). (Any errors in Table II are by this writer.)

3.2. EVOLUTION OF INSTRUMENTS FOR ISOTOPIC MEASUREMENTS

With more than 95% of the cosmic-ray flux arriving in the inner solar system with energies below a few GeV per nucleon, the principal measurement technique is a charged particle telescope in which energy loss and residual energy of the stopping particles are measured. With access to space, isotope measurements were first undertaken with an all solid-state telescope on the IMP-I-II-III spacecraft series beginning in 1963 (King, 1982). For the IMP series this instrument development led to the discovery of deuterium in the cosmic rays (Fan et al., 1966). Since ^{2}H is readily destroyed in nucleosynthetic processes, the observed ^{2}H is produced entirely in nuclear reactions during cosmic ray propagation and – with ^{1}H, ^{3}He, ^{4}He – forms a quartet to 'monitor' secondary nuclear production during interstellar propagation (Simpson, 1971). A formalism was developed by Stone and Wiedenbeck (1979) for deriving cosmic ray source isotopic abundances from observed local abundances using secondary nuclides as tracers of spallation production.

A major factor limiting the isotopic mass resolution of a charged particle telescope is the inherent spread in pathlength in the energy-loss detector due to the isotropic incidence of the cosmic rays. By reducing the width of the average pathlength using curved semiconductor detectors (Perkins et al., 1969) a four-fold increase in isotopic resolution was achieved in space flight missions by the early 1970's (e.g., *Pioneer 10/11*; IMP 5-6-7-8). For the IMP series this extended the measurements of the stable and radioactive isotopes from Be through to Si. The *ISEE-3* satellite included instruments that measured the trajectory of each particle using either semiconductor strip detectors (Althouse et al., 1978; Mewaldt et al., 1981) or electron drift chambers (Greíner et al., 1978) with a larger geometrical factor than was achieved with silicon detector technology. For a review of thin ($\sim$50 μm) two-dimensional position sensing Si detectors, see Tuzzolino (1988).

Reviews by Simpson (1983), Simpson and Garcia-Muñoz (1988), and Mewaldt (1989) have summarized the stable and radioactive isotope research that was based mainly on the above evolution of instrumentation extending into the 1980's (see also Cosmic Abundances of Matter, in Waddington, 1989).

TABLE I

Radioactive nuclides

Time between nucleosynthesis and detection	Time between nucleosynthesis and acceleration[†]	Confinement time in galactic disc plus halo
^{40}K ($\tau_{1/2} = 1.3 \times 10^9$ yr)	^{56}Ni ($\tau_{1/2} = 6.10d$)	^{10}Be ($\tau_{1/2} = 1.5 \times 10^6$ yr)
^{60}Fe ($\tau_{1/2} = 3 \times 10^5$ yr)	^{57}Co ($\tau_{1/2} = 270d$)	^{26}Al ($\tau_{1/2} = 7.4 \times 10^5$ yr) $\rightarrow 8.7 \times 10^5$ yr*
	^{59}Ni ($\tau_{1/2} = 8 \times 10^4$ yr [β^- capture])	^{36}Cl ($\tau_{1/2} = 3 \times 10^5$ yr) $\rightarrow 3.08 \times 10^5$ yr*
	$\tau_{1/2} > 8 \times 10^6$ yr [β^+ decay])	
		^{54}Mn ($\tau_{1/2} \sim 1$–2×10^6 yr) [†]

[†]Only observed if the time for electron stripping during acceleration is less than the electron capture half-life.

*Electron capture mode suppressed for fully stripped isotope in the interstellar medium.

[†] See also Chan et al. (1997).

TABLE II

Isotopic 'signatures' for two types of nucleosynthesis models

---

Evolution of massive stars ($M > M_0$; type II SN) are the main source of isotopes Ne–Ca

^{16}O + [[perhaps some ^{18}O]]
^{20}Ne = [[perhaps some ^{21}Ne and ^{22}Ne]]
^{23}Na
^{24}Mg, ^{25}Mg, ^{26}Mg
^{27}Al
^{28}Si, ^{29}Si, ^{30}Si
^{31}P
^{32}S, ^{33}S, ^{34}S - [[not ^{36}S]]
^{35}Cl, ^{37}Cl
^{36}Ar, ^{38}Ar - [[not ^{40}Ar]]
^{39}K, ^{41}K - [[perhaps some ^{40}K]]
^{40}Ca, ^{42}Ca, ^{43}Ca, ^{44}Ca - [[not ^{46}Ca or ^{48}Ca]]

Note here that the neutron-rich isotopes ^{36}S, ^{40}Ar, ^{46}Ca, and ^{48}Ca are not made in the standard carbon and oxygen burning phase reactions but are believed to result from neutron capture reactions.

Explosive silicon burning nucleosynthesis isotopes include:

^{45}Sc
^{46}Ti, ^{47}Ti, ^{48}Ti, ^{49}Ti - [[not ^{50}Ti]]
^{51}V + [[perhaps some ^{50}V]]
^{50}Cr, ^{52}Cr, ^{53}Cr - [[not ^{54}Cr]]
^{55}Mn
^{54}Fe, ^{56}Fe, ^{57}Fe - [[not ^{58}Fe]]
^{59}Co
^{58}Ni, ^{60}Ni, ^{61}Ni, ^{62}Ni - [[not ^{64}Ni]]
^{63}Cu, ^{65}Cu

Truran (1997); Arnett (1996)

The development of large area, Li-drifted silicon detectors in the 1970s (Lamport et al., 1976) with conducting strips, arranged as shown in Figure 10, made possible another step in achieving higher isotopic resolution. Figure 11 is a cross section of the charged particle telescope that was developed in the late 1970s with these position sensing detectors. For a detailed description of its performance, calibration and scientific goals see Simpson et al. (1992). This instrument, the High Energy Telescope (HET), is included in the *Ulysses* mission launched in October 1990, and is part of the COsmic ray and Solar Particle INvestigation (COSPIN) consortium (Simpson et al., 1992).

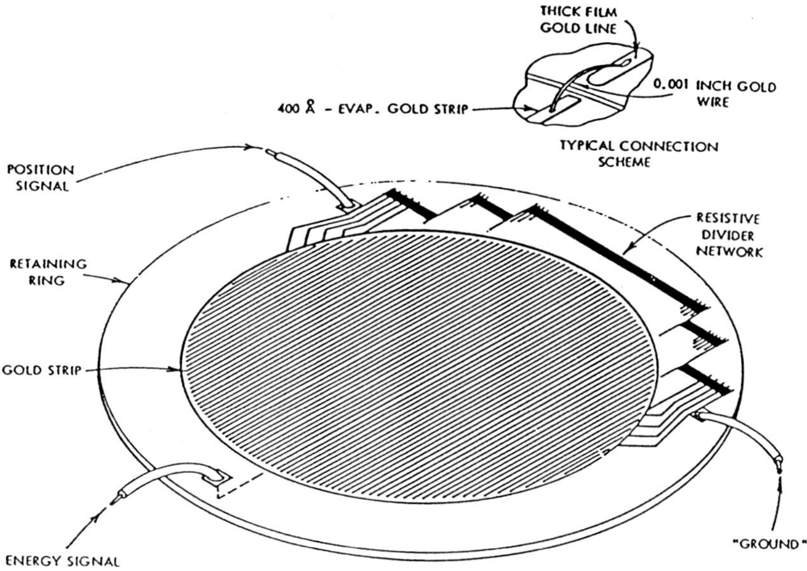

Figure 10. Perspective view of silicon position sensitive detector developed for the High Energy Telescope (HET) D1-D3 detectors carried by the *Ulysses* spacecraft (Simpson et al., 1992). The strip pattern contains 223 gold strips. The strip width is 0.190 mm and gap width is 0.127 mm.

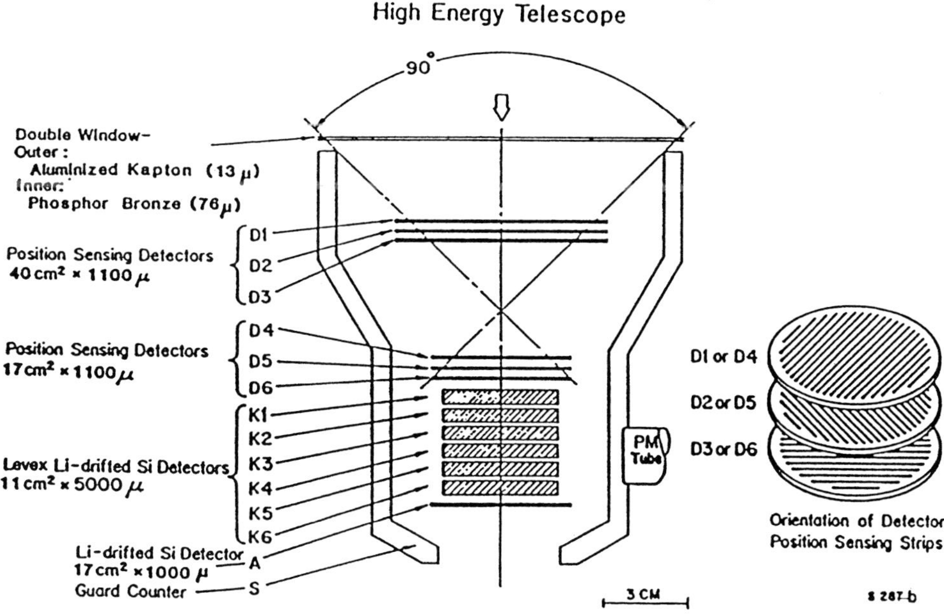

Figure 11. Cross-section schematic of the *Ulysses* High Energy Telescope. With the strips of successive detectors (e.g., D1, D2, D3, or D4, D5, D6) and with successive strips rotated by 60° (for redundancy) the angle of incidence of a cosmic-ray particle is determined with an accuracy better than 1° (Simpson, et al., 1992).

This HET has obtained the highest resolution for isotopes of elements in the element range of hydrogen to nickel that has so far been reported. It will be a challenge for ACE to meet or exceed this isotopic resolution, as it is expected to do.

In addition to instrument resolution, there is a major advance for ACE to study the less abundant, even rare, isotopes due to its huge collection power relative to the high energy telescope on *Ulysses*. Table III illustrates the importance of this difference in isotope collecting power. Table III is a comparison of the less abundant isotope measurements between the *Ulysses* HET and the CRIS instruments estimated for the modulation levels projected in Figure 2 (top panel). Clearly, the CRIS measurements will be a major advance for our understanding of the origin of cosmic-ray matter, especially the less abundant nuclear species.

3.3. The isotopic abundance of Fe, Ni, and Mn derived from the *Ulysses* high energy telescope

To illustrate what we believe to be the present state of isotopic research we have used recent work by Connell and Simpson (1997) for the iron and nickel isotopes, Simpson and Connell (1998) for ^{26}Al, and DuVernois (1996, 1997) for the Mn isotopes.

Fe and Ni.   Figure 12 shows mass histograms for Fe and Ni. The histograms contain a total of 6035 Fe and 284 Ni events. Gaussian peaks were fit to the histograms using the maximum likelihood method. The peak positions for ^{56}Fe and ^{58}Ni were free, but the other peak positions were determined by fitting a single center-to-center spacing parameter for each element. A single fractional σ was fit. A flat background was also fit, giving 1.04 and 0.09 events per bin for Fe and Ni, respectively. Mass resolution varies from 0.277 ± 0.004 amu for ^{56}Fe to 0.29 ± 0.02 amu for ^{58}Ni. The actual distribution of the data is known to have a low mass tail, probably due to neutron changing nuclear interactions within the HET and events stopping in the thin dead layers ($\sim$40 μm) of the K detectors (cf. Figure 11).

For the propagation of the Fe and Ni isotopes we used a weighted slab model of interstellar propagation (e.g., Garcia-Muñoz et al., 1987) using a single exponential pathlength distribution ('leaky-box' model) in which the nuclei are mainly confined to the galaxy. The maximum of the energy dependent average pathlength is $\sim$9.2 g cm^{-2}. This propagation model satisfied the observed B/C and sub Fe/Fe element ratios and employs the most recent nuclear cross sections and estimates of interstellar ionization losses including ionized HII. We note that Fe is a factor of $>10^3$, and Ni > 10 more abundant than the elements beyond Ni. This ensures that, aside from ^{56}Fe spalling into ^{54}Fe and ^{55}Fe, secondary nuclear species produced during interstellar propagation are not major contributors to the observed Fe and Ni isotopes.

TABLE III

Galactic cosmic-ray less abundant isotopes *Ulysses* HET and ACE estimated numbers of events

Symbol	Z	A	U. of C *Ulysses* HET (totals)		ACE/CRIS
			Prime (1991–1995)	Two orbits* (1991–2001)	3 years (1997.5–2000.5)
Be	4	–	950	2695	43 700
Be	4	10	31	88	1 420
Al	13	–	1207	3765	66 800
Al	13	26	82	255	4 530
Cl	17	–	147	451	8 010
Cl	17	36	37	114	2 030
Ar	18	–	345	860	12 100
Ar	18	40	17	42	595
K	19	–	321	733	9 630
K	19	41	69	157	2 060
Ca	20	–	910	2403	36 900
Ca	20	41	70	186	2 850
Ca	20	44	185	488	7 480
Mn	25	–	507	1361	21 100
Mn	25	54	70	189	2 920
Co	27	–	35	99	1 530
Ni	28	–	285	812	13 500
Ni	28	59	8	23	380
Ni	28	60	76	216	3 580
Ni	28	62	11	32	520
Ni	28	64	2	6	95

(1) Rates are based on observed *Ulysses* HET rates, including consistency cuts.

(2) Isotope estimates are based on observed *Ulysses* HET ratios. Statistical uncertainties apply as for the Prime mission.

(3) Projected solar modulation is based upon an average of the last two 22-year solar cycles using Climax neutron monitor data and Φ based on the CHIME Model (Chenette et al., 1994).

(4) ACE CRIS rates based on 80 000 Fe events per year for a normalization factor of 45 times the collecting power of the HET. No spectral correction was made.

*Second orbit has been approved.

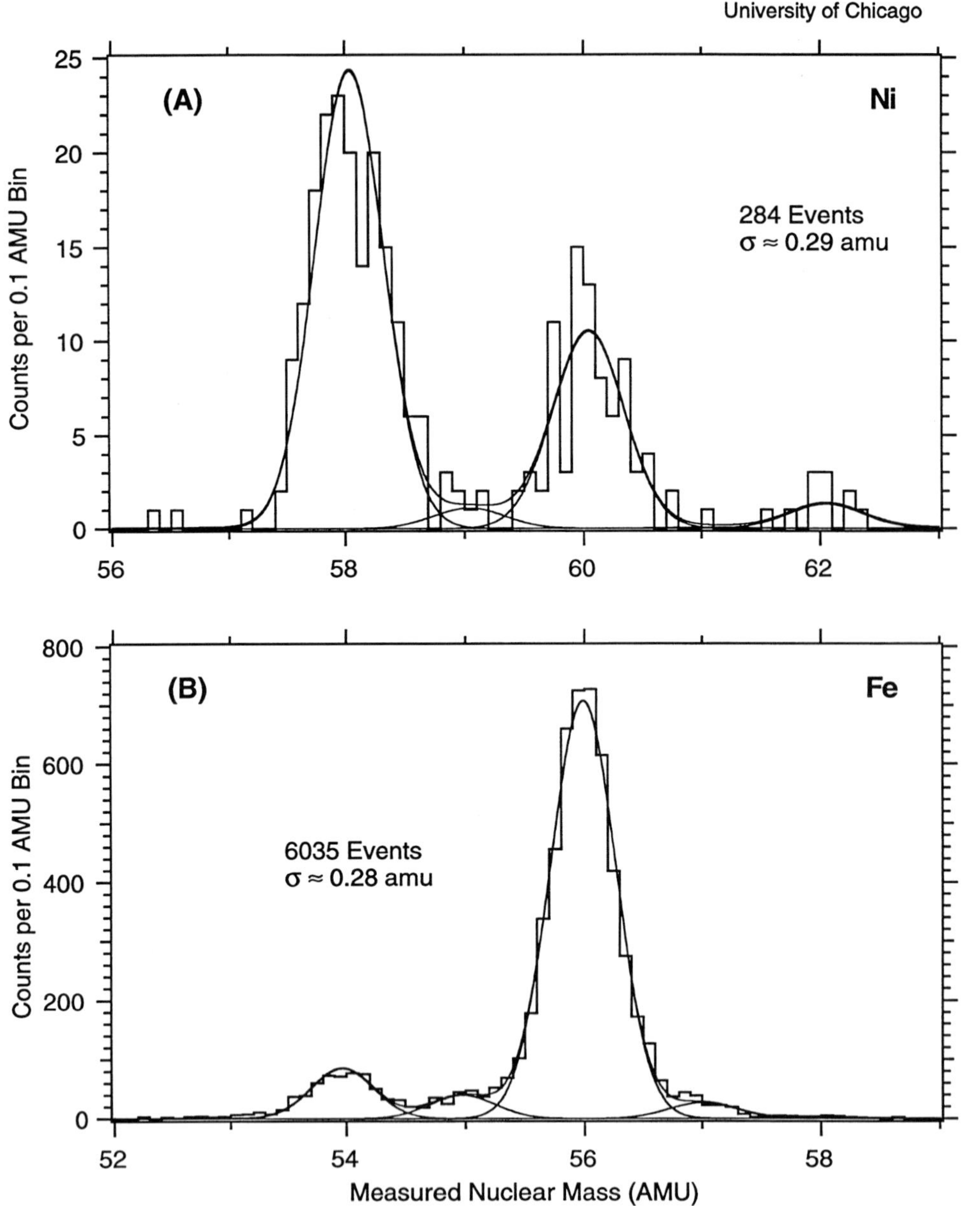

Figure 12. (A) The mass histogram for the nickel isotopes (B) The mass histogram for the iron isotopes. Gaussian peaks are fitted to the histograms (see text, from Connell and Simpson, 1997).

TABLE IV

Ulysses HET Ni and Fe isotopic abundance measurements

	(a) Measured (%)	(b) Source (%)	(c) Solar* (%)	(d) GCRS/SS ratio
^{59}Ni/^{58}Ni	$4.6^{+2.6}_{-2.1}$	$2.60^{+2.1}_{-1.7}$	0	–
^{60}Ni/^{58}Ni	$45.0^{+6.8}_{-6.5}$	$43.2^{+6.7}_{-6.4}$	38.2	1.1 ± 0.2
^{61}Ni/^{58}Ni	<1.6	<1.2	1.7	<0.7
^{62}Ni/^{58}Ni	$5.8^{+2.3}_{-1.9}$	$5.4^{+2.2}_{-1.8}$	5.3	$1.0^{+0.4}_{-0.3}$
^{54}Fe/^{56}Fe	11.4 ± 0.6	9.3 ± 0.6	6.3	1.5 ± 0.1
^{55}Fe/^{56}Fe	5.4 ± 0.4	1.6 ± 0.5	0	–
^{57}Fe/^{56}Fe	$3.9^{+0.35}_{-1.9}$	$3.7^{+0.33}_{-0.36}$	2.3	$1.6^{+0.14}_{-0.16}$
^{58}Fe/^{56}Fe	$0.34^{+0.10}_{-0.14}$	$0.18^{+0.10}_{-0.14}$	0.32	$0.6^{+0.3}_{-0.4}$

* Anders and Grevesse (1989); Cameron (1982).

The cosmic-ray source abundance ratios derived from these calculations are listed in Table IV, column (b). The current accepted values for the corresponding solar system isotopic abundance ratios are shown in column (c). Based on the reviews by Anders and Grevesse (1989) and Cameron (1982) we conclude that, for the Fe and Ni, the solar system ratios, derived mainly from meteorites, probably have fractional errors of the order of $< 2\%$, far smaller than the statistical uncertainties in our galactic source determination.

The principal ratio ^{60}Ni/^{58}Ni clearly shows no significant neutron-rich enhancement which would be predicted by nucleosynthesis models requiring neutron enrichment in Ni (e.g., Arnett, 1996) but is close to its solar system value. The upper limit on the ^{61}Ni/^{58}Ni ratio appears to be inconclusive, or perhaps too low, for source models enriched by Wolf-Rayet nucleosynthesis products (e.g., Prantzos et al., 1987). The ^{59}Ni/^{58}Ni source ratio, where ^{59}Ni is an electron capture nucleus ($\tau = 7.6 \times 10^4$ years), is 1.5σ from 0, which does not rule out a source component of ^{59}Ni. The absence of ^{59}Ni in the cosmic-ray source would indicate a significant time delay between ^{59}Ni synthesis and cosmic-ray acceleration. Since contributions to ^{62}Ni from spallation of the rare, heavier nuclei are negligible, the principal error in the ^{62}Ni/^{58}Ni ratio in Table IV arises from the low statistical weight.

Beyond ^{62}Ni (not shown in Figure 12(A)) are two events near the position of ^{64}Ni. That would be consistent with the solar system ratio of 0.9% of ^{64}Ni/^{58}Ni.

The cosmic-ray source ratio ^{54}Fe/^{56}Fe is approximately 50% higher than the solar system ratio. Part of the excess ^{54}Fe may be the result of our propagation model slightly underestimating the production of secondary nuclides from ^{56}Fe; the 3σ source abundance of ^{55}Fe indicated in Table IV would support this conclusion.

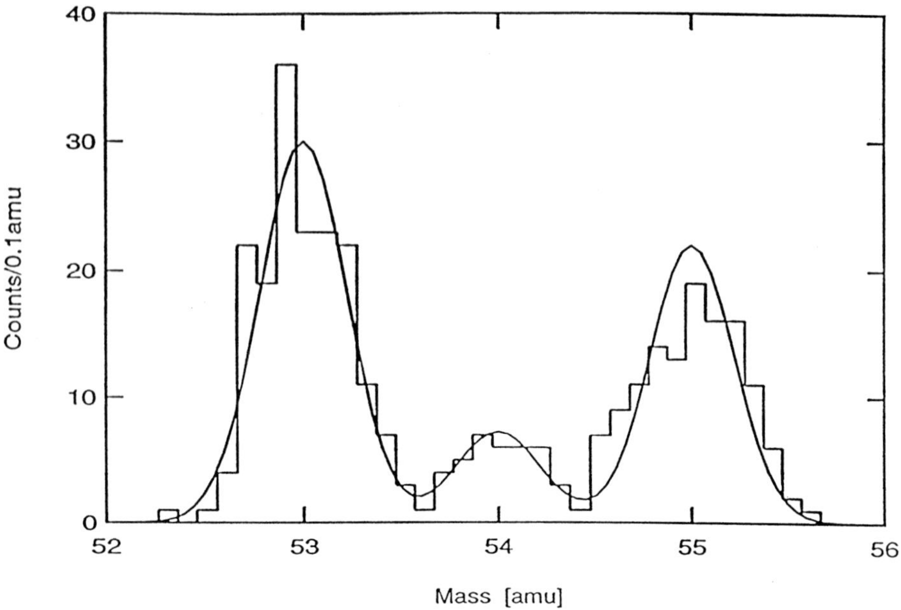

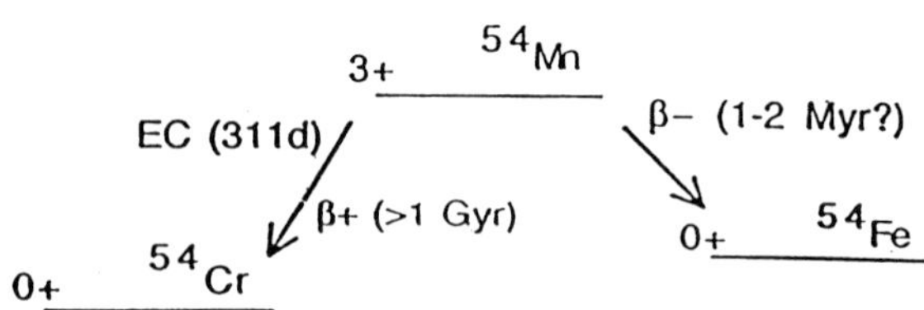

Figure 13. The mass histogram for the Manganese isotopes (from DuVernois, 1997). *Below*: decay modes for ^{54}Mn.

^{55}Fe, which does not exist in the solar system, is a short lived electron capture nuclide ($\tau = 2.68$ years) that would not be present in the cosmic ray source if there were a significant time delay between nucleosynthesis and particle acceleration. If the secondary production in our model were increased to eliminate the source abundance of ^{55}Fe shown in Table IV, the GCRS/SS ratio of ^{54}Fe/^{56}Fe would only be reduced to $\sim 1.3 \pm 0.1$.

^{54}Mn is also a secondary product of ^{56}Fe spallation. As pointed out by Grove et al. (1991), Mn decays via electron capture and β^+ to ^{54}Cr and by β^- to ^{54}Fe (Figure 13). While normally the electron capture channel dominates ($\tau = 312$ days), for cosmic rays (which are fully stripped) the β^- channel to ^{54}Fe is believed to dominate.

We know of no propagation models, including reacceleration models, wherein source abundances could be assumed to be enhanced significantly with heavier isotopes of Fe and Ni, but, after propagation, could be reduced to our observed abundance ratios of ^{58}Fe/^{56}Fe, ^{60}Ni/^{58}Ni or ^{62}Ni/^{58}Ni. From the principal source isotopic ratios of Fe and Ni isotopes in Table IV we conclude that – with the exception of ^{54}Fe/^{56}Fe and ^{57}Fe/^{56}Fe – their source matter is close to solar system values.

Since the interstellar medium isotopic composition appears from present investigations to be more solar system-like than the composition from explosive nucleosynthesis (e.g., Arnett, 1996), present evidence suggests but does not prove (e.g., Leske et al., 1996) that cosmic-ray particle acceleration occurs mainly in the interstellar medium.

The isotopic abundance ratios, especially for Fe and Ni in the solar system, are derived from meteorites and are believed to be measurements of the primordial solar system isotopic abundances some $\sim$4.6 Gyrs ago (Anders and Grevesse, 1989; Cameron, 1982). On the other hand, since the source composition we report here for cosmic-ray matter is modern (i.e., $\sim$10^7 years old), this work suggests that the galactic sources for cosmic rays – at least for Fe and Ni – have not changed dramatically despite over 4.6 Gyrs of galactic chemical evolution. For further details on the Fe and Ni analysis see Connell and Simpson (1997).

Manganese. In the cosmic rays, manganese nuclei are almost entirely secondary, being the spallation products of Fe, Co, and Ni. DuVernois (1996) in his thesis and publication (1997) has obtained the first clear abundance of ^{54}Mn relative to ^{53}Mn and ^{55}Mn from measurements with the *Ulysses* HET. The radioactive isotope ^{54}Mn has a decay lifetime which is in the range of one to two million years (e.g., Wuosmaa et al., 1998; Chan, et al., 1997). Its decay modes are shown in Figure 13.

In cosmic-ray physics ^{54}Mn is a nuclear decay chronometer for determining cosmic-ray propagation in the galactic disk and magnetic halo. The ^{54}Mn decay into ^{54}Fe is also an important correction for ^{54}Fe abundances.

The mass histograms of ^{53}Mn, ^{54}Mn, and ^{55}Mn, are shown in Figure 13 based on the data in Table V. Since ^{55}Mn is a stable isotope it is useful as a reference with respect to ^{54}Mn. For further discussion of the Mn isotopes see DuVernois (1997).

Table VI is a summary of the principal conclusions derived from the Mn, Fe, and Ni cosmic-ray isotopic measurements.

3.4. ^{26}AL AND ITS DECAY IN THE GALAXY

Among the long-lived, radioactive isotopes detectable in the galactic cosmic rays, ^{26}Al is unique, with its decay by electron capture to ^{26}Mg and 1.809 MeV γ-ray emission, making possible the detection of ^{26}Al throughout the galaxy. It is important as an investigative probe for the origin and propagation of cosmic rays, for testing those models of stellar nucleosynthesis that predict injection of ^{26}Al and ^{27}Al into the interstellar medium, and for the cosmochemistry of refractory inclusions in meteorites containing excess ^{26}Mg.

TABLE V

Manganese isotopes observed in the heliosphere by the *Ulysses* HET

Mn isotope	Analyzed events	Spectral correction factor[a]	Isotopic fraction[b]
53...	168	1.019	0.52 ± 0.09
54...	39	0.995	0.12 ± 0.03
55...	124	0.977	0.37 ± 0.07
All...	331	1.00	1.00

[a]Multiplicative factor to raw isotope fraction – corrects for energy interval and spectral shape between isotopes (From DuVernois, 1997).
[b]^{x}Mn/$^{53+54+55}$ Mn.

The half life of ^{26}Al is 7.2×10^5 yr. This becomes 8.7×10^5 yr as a fully-stripped cosmic-ray spallation product in the interstellar medium where electron capture is suppressed. Thus, the radioactive ^{26}Al relative to the stable ^{27}Al provides a measurement of the average density of matter through which the low energy cosmic radiation in galactic magnetic fields propagate that is independent of using radioactive ^{10}Be relative to the stable ^{9}Be analysis to determine this density (e.g., see review by Simpson and Garcia-Muñoz, 1988). From these measurements we obtain a confinement time or 'age' of the cosmic rays.

The 1.809 MeV γ-ray line emission from the galaxy was discovered by Mahoney et al. (1984) and recently observed by the Compton γ-ray Observatory to extend at low intensity throughout the galaxy with intense local regions of emissions (Diehl et al., 1995). These local hotspots probably represent local injections of ^{26}Al from nucleosynthesis in supernovae, red giants and novae (e.g., reviews by Prantzos, 1993; Diehl and Timmes, 1997). Thus, there are two physically different kinds of sources for the 1.809 MeV γ-ray emissions throughout the galaxy: ^{26}Al from cosmic-ray spallation and ^{26}Al from several types of stellar sources.

Simpson and Connell (1998) have reported the results from measurements that clearly resolve ^{26}Al from stable ^{27}Al using the University of Chicago high resolution, high energy particle telescope carried on the *Ulysses* spacecraft.

The mass histograms for ^{26}Al (92-nuclei) and ^{27}Al (1453-nuclei) are shown in Figure 14. The minimum and maximum energies for this analysis were 139 MeV amu^{-1} and 299 MeV amu^{-1}, respectively, with an average energy of $\sim$200 MeV amu^{-1} under the assumption that the spectral shape of these nuclides were identical. The Gaussian peaks were fitted to the histograms using the maximum likelihood method with free peak positions for ^{26}Al and ^{27}Al; a flat background was included in the fit to the data for Figure 14. A single fractional σ was fitted giving a value of 0.17 amu.

TABLE VI

Principal conclusions

The Ni isotopes	The Fe isotopes	The Mn isotopes	Overall
^{60}Ni/^{58}Ni source ratio is close to the solar sytem ratio and shows no neutron-rich enhancement	^{54}Fe/^{56}Fe ratio is ~50% higher than solar system (partly due to spallation of ^{56}Fe, as indicated by the presence of ^{55}Fe (which is not present in the solar system). ^{58}Fe/^{56}Fe ratio is close to the solar system ratio. ^{57}Fe/^{56}Fe is 50% higher than solar system. There are no significant contributions from secondaries to explain this excess.	^{54}Mn life time values now bounded by 1 to 2 Myr. This is a range which does *not* account for the ^{54}Fe/^{56}Fe excess.*	With exception of ^{54}Fe/^{56}Fe and ^{57}Fe/^{56}Fe the source Fe and Ni isotopic abundances are close to their solar system values.

*See also Chan et al. (1997); Wuosmaa et al. (1998).

TABLE VII

Comparison of measurements of ^{26}Al/^{27}Al

'Age' reference	Average ISM energy (MeV amu^{-1})	Observed ^{26}Al/^{27}Al (%)	Average ISM density (atom cm^{-3})	Cosmic ray (MY)*
Wiedenbeck (1983)	~200	$3.6^{+3.7}_{-2.2}$		$9^{+0.26}_{-6.5}$
Lukasiak et al. (1994)	~400	8.3 ± 2.4	$0.52^{+0.26}_{-0.2}$	$13.5^{+8.5}_{-4.5}$
Simpson and Connell (1998)	~600	6.1 ± 0.7	$0.26^{+0.01}_{-0.04}$	19 ± 3

*For homogeneous propagation models.

The measured ratio of ^{26}Al/^{27}Al = 6.1 ± 0.7% for an averaged energy of ~200 MeV amu^{-1} yields an average interstellar density, $\rho = 0.26+0.05/-0.04$ atom cm^{-3}.

The curves in Figure 15 are the predicted ratio of ^{26}Al/^{27}Al as a function of interstellar density and nucleon energy near Earth orbit. With a mean velocity, $v \approx 0.8c$ the corresponding confinement time for the cosmic radiation in the galactic homogeneous model is 19 ± 3 My. This value is consistent with the *Ulysses* ^{10}Be/^{9}Be ratio analysis for cosmic-ray confinement time (e.g., Connell, 1998). Table VII compares these ^{26}Al/^{27}Al results with earlier spacecraft measurements.

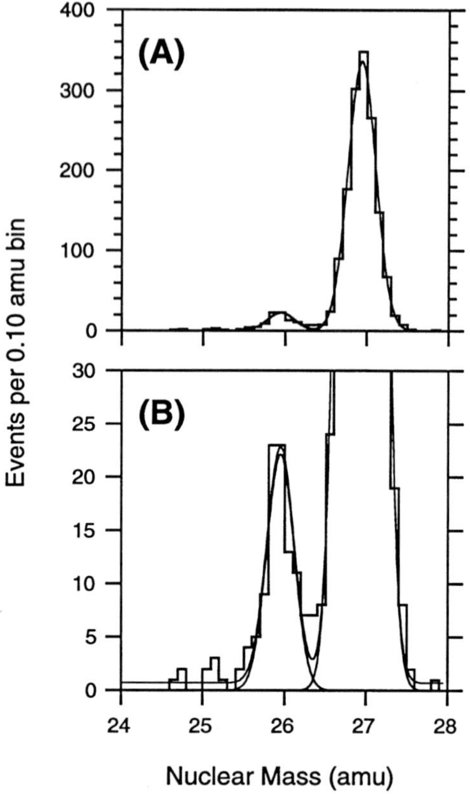

Figure 14. (A) The mass histogram for ^{26}Al and ^{27}Al. (B) Expanded histogram for ^{26}Al. Gaussian peaks are fitted to the histograms (from Simpson and Connell, 1998).

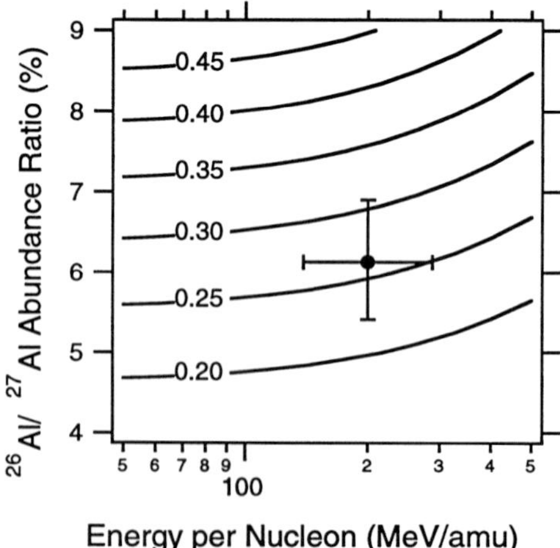

Figure 15. The solid lines are the predicted ratio of ^{26}Al/^{27}Al as a function of interstellar gas density (atom cm^{-3}) and nucleon energy at *Ulysses*. The value derived from the experiment is $0.26 + 0.05/ - 0.04$ atom cm^{-3} under the assumption of 94% H and 6% neutral He, by number, in the interstellar medium.

The average density of matter that cosmic rays have traversed in galactic magnetic fields is significantly less than if they traveled entirely in the galactic disk where the average gas density is 0.7 to 1 atom cm^{-3}. These results support the conclusion of Simpson and Garcia-Muñoz (1988), based on ^{10}Be analysis, that the cosmic-ray nuclei spend part of their confinement time in magnetic fields of a galactic halo.

There is substantial evidence that galaxies containing cosmic-ray electrons with relativistic energies in their galactic planes also have magnetic halos with relativistic electrons, as evidenced by synchrotron emissions tracing out their halo magnetic fields (e.g., Ekers and Sancisi, 1977; Duric et al., 1998, for spiral galaxy NGC 5775). Various mechanisms have been proposed to generate a galactic magnetic halo (e.g., via galactic wind, cosmic ray gas coupling to drag fields from the disk (e.g., Parker, 1991)).

Theoretical cosmic ray diffusion models based on a galactic halo have been proposed by, for example, Preship and Ptuskin (1975), Ginzburg et al. (1980) and most recently in a comprehensive analysis by Ptuskin et al. (1997).

The second class of sources for ^{26}Al in the Galaxy has been predicted by nucleosynthesis theorists in which supernovae, novae or red giants – probably through the mode ^{25}Mg $(p,\gamma)^{26}$Al – resulted in the ejection of ^{26}Al into the interstellar medium. For example, the expelled aluminum from supernovae could have a production mass fraction ratio ^{26}Al/^{27}Al of $\sim 10^{-3}$ (e.g., Truran and Cameron, 1978; Arnett and Wefel, 1978; Woosley and Weaver, 1995; Arnett, 1996). Collectively, the models for supernovae, novae and red giants lead to the prediction that the present day, steady-state mass of ^{26}Al in the interstellar medium is in the range of ~ 0.4 to ~ 2 solar mass units. It is this mass of ^{26}Al that is invoked as the source of the 1.809 MeV γ-ray emission from the galaxy found by Mahoney et al. (1984) and shown by Diehl et al. (1995), to be a diffuse radiation extending throughout the Galaxy with superposed local, intense sources. Since $\lesssim 2$ solar mass units of ^{26}Al would be mixed in an interstellar gas of $\sim 3 \times 10^9$ to $\sim 10^{10}$ solar mass units, there is a huge dilution factor for the ^{26}Al ejecta. However, this suggests that the γ-rays from the decay of cosmic ray ^{26}Al is not observed. For further analysis of these two kinds of sources for ^{26}Al in the Galaxy and cosmic ray tests for them, see Simpson and Connell (1998).

4. Concluding Remarks

Beginning with the first instruments in space in the early 1960's through to the 1990s with the *Ulysses* HET and now the ACE CRIS and SIS instruments, all advances in cosmic-ray isotopic resolution have been instrument-driven. Indeed, the *Ulysses* and ACE isotopic measurement capabilities now exceed in accuracy our knowledge of many of the nuclear and propagation parameters required to determine the isotopic composition of cosmic-ray source abundances in the galaxy. Only a fraction of the required energy-dependent partial and total cross sections have been measured. Silberberg et al. (1997, and references therein) early recognized this problem and determined cross sections based on semi-empirical equations. In the 1980s and early 1990s Webber and collaborators (e.g., Webber et al., 1990) and the TRANSPORT collaboration (e.g., Guzik, 1997; Waddington, 1997) measured selected cross sections mainly using the Bevalac before it was closed. At present there is a critical need for an expanded program of partial and total cross-section measurements that is not being met for lack of access to beam facilities. Waddington (1998) recently has reviewed this problem. It is this writer's estimate that the errors in some of the calculated source abundances may be as high as ± 20–30%. This problem is especially critical for the low abundance isotopes that will be measured by the ACE instruments.

Although this report has focused on the opportunities for heliospheric dynamical phenomena and the composition of the galactic cosmic rays, there are challenges for all the other investigations on ACE – such as for the SIS and SWICS investigations. For example, at the present time the acceleration mechanisms for nucleons in solar energetic particle events are intensely debated (see Ramaty et al., 1996). Currently, solar energetic particle events are divided into two categories: 'impulsive' events occurring near the flare site, and 'gradual' events associated with coronal mass ejections as they propagate outward (e.g., Reames, 1995). A major contribution by ACE would be the search for and measurements of ^{2}H, Li, Be, B in the solar events. These measurements would provide critical information on the coronal depth ('thick' or 'thin' target) and/or time duration of the acceleration process in the corona. The detection of ^{2}H, Li, Be, B also would extend our understanding of the flare-to-flare composition changes (Mogro-Campero and Simpson, 1972; Stone, 1989) occurring among impulsive flares.

Another example is the measurement of pickup ions that become anomalous nuclear components. The SWICS measurements on *Ulysses* have led to a determination of the relative abundance of ^{3}He in the local interstellar cloud (Gloeckler and Geiss, 1996) that is an important constraint on cosmological models (Schramm and Turner, 1996). A further constraint would come from the measurement of the ^{2}H pickup ions by the modified SWICS on ACE when the analysis voltage is raised.

The scientific community is waiting with great expectation for a multitude of discoveries from ACE!

Acknowledgements

The author thanks R. B. McKibben, K. R. Pyle, and J. J. Connell for assistance in the preparation of some of the illustrations and J. Truran for a discussion of nucleosynthesis models. This report was partially supported by NASA-JPL Contract 955432 and the Compton Emeritus Fund.

References

*Review publication useful for graduate student study.

Acuña, M. H. et al.: 1995, in C. T. Russell (ed.), *The Global Geospace Mission*, Kluwer Academic Publishers, Dordrecht, The Netherlands, p. 5.

Althouse, W. E. et al.: 1978, IEEE, *Trans. Geosci. Electrons G.E.* **151**, 1204.

Anders, E. and Grevesse, N.: 1989, *Geochim. Cosmochim. Acta* **53**, 197.

*Arnett, D.: 1996, *Supernovae and Nucleosynthesis*, Princeton University Press, Princeton, N.J.

Arnett, W. D. and Wefel, J. P.: 1978, *Astrophys. J.* **224**, L139.

Baker, D. N. et al.: 1993, *IEEE Trans. on Geo. Sci. & Remote Sensing* **31** (3), 531.

Balogh, A. et al.: 1995, *Science* **268**, 1007.

*Cameron, A. G. W.: 1982, in C. A. Barnes, D. D. Clayton, and D. N. Schram (eds.), *Essays in Nuclear Astrophysics*, Cambridge University Press, Cambridge, p. 23.

Chan, Y. D. et al.: 1997, *Proc. 25th Int. Cosmic Ray Conf.*, *Durban* **4**, 281.

Chenette, D. et al.: 1994, *IEEE Trans Nucl. Sci..* **41**, 2332.

Connell, J. J.: 1998, *Astrophys. J.* **501**, L59.

Connell, J. J. and Simpson, J. A.: 1997, *Astrophys. J.* **475**, L61.

Diehl, R. et al.: 1995, *Astron. Astrophys.* **298**, 445.

Diehl, R. and Timmes, F. X.: 1997, in C. jD. Dermer, M. S. Strickman, and J. D. Kurtess (eds.), *Proc. of the Fourth Compton Symposium*, AIP Conf. Proc. No. 410, AIP, New York, p. 218.

Duric, N., Irwin, J., and Bleomen, H.: 1998, *Astron. Astrophys.* (in press).

DuVernois, M. A.: 1996, Ph.D. Thesis, University of Chicago.

DuVernois, M. A.: 1997, *Astrophys. J.* **481**, 241.

Ekers, R. D. and Sancisi, R.: 1977, *Astron. Astrophys.* **54**, 973.

Fan, C. Y., Gloeckler, G., and Simpson, J. A.: 1966, *Phys. Rev. Letters* **17**, 329.

Fisk, L.: 1996, *J. Geophys. Res.* **101**, 15 547.

Fleck, B., Domingo, V., and Poland, A. (eds.): 1995, *The Soho Mission*, Kluwer Academic Publishers, Dordrecht, The Netherlands.

*Garcia-Muñoz, M. et al.: 1987, *Astrophys. J. Suppl.* **64**, 269.

Ginzburg, V. L., Khazan, Ya. M., and Ptuskin, V. S.: 1980, *Astrophys. Space Sci.* **68**, 295.

Gloeckler, G. and Geiss, J.: 1996, *Nature* **381**, 210.

Gosling, J. J. et al.: 1994, *Geophys. Res. Letters* **21**, 2271.

Greiner, D. E., Bieser, F. S, Heckman, H. N.: 1978, *IEEE Trans. Geosci. Electron.* **GE16**, 163.

Grove, J. E. , Hayes, B. T., Mewaldt, R. A., and Webber, W. R.: 1991, *Astrophys. J.* **377**, 680.

Guzik, T. G. for the TRANSPORT collaboration: 1997, *Proc. 25th Int. Cosmic Ray Conf.*, *Durban* **4**, 337.

*Heber, B., Potgieter, M. S., and Ferrando, P.: 1997, *Adv. Space Res.* **19** (5), 795.

Hunter, S. D. et al.: 1997, *Astrophys. J.* **481**, 205.

King, O. H.: 1982, in C. T. Russell and D. J. Southwood (eds.), *The IMS Sourcebook: Guide to the International Magnetospheric Study Data Analysis*, Am. Geophys. U., p. 10.

Kóta, J. and Jokipii, J. R.: 1995, *Science* **268**, 1024.

Kunow, H. et al.: 1995 in R. G. Marsden (ed.), The High Latitude Heliosphere, Kluwer Academic Publishers, Dordrecht, The Netherlands, p. 397.

Lamport, J. E., Mason, G. M., Perkins, M. A., and Tuzzolino, A. J.: 1976, *Nucl. Instr. Methods* **134**, 71.

Leske, R. A. et al.: 1996, *Space Sci. Rev.* **78**, 149.

Lukasiak, A., McDonald, F. B., and Webber, W. R.: 1994, *Astrophys. J.* **430**, L69.

Lund, N.: 1989, *Cosmic Abundances of Matter*, AIP Conf. No. 183, 111.

Mahoney, W. A., Ling, J. C., Wheaton, W. A., and Jacobson, A. S.: 1984, *Astrophys. J.* **286**, 578.

*McKibben, R. B. et al: 1995, in R. G. Marsden (ed.), *The High Latitude Heliosphere*, Kluwer Academic Publishers, Dordrecht, The Netherlands, p. 403.

McKibben, R. B. et al.: 1998, in R. von Steiger, L. Fisk, and J. Jokipii (eds.), *Cosmic Rays in the Heliosphere*, Kluwer Academic Press, Dordrecht, The Netherlands, pp. 21–32.

*Mewaldt, R. A.: 1989, *AIP Proc.* No. 183, pp. 124–146.

Mewaldt, R. A., Spalding, J. D., Stone, E. C., and Vogt, R. E.: 1981, *Astrophys. J.* **251**, L27.

Meyer, J.-P., Drury, L. O'C., and Ellison, D. C.: 1997, *Astrophys. J.* **487**, 182.

Mogro-Campero, A. and Simpson, J. A.: 1972, *Astrophys. J.* **177**, L37.

Parker, E. N.: 1991, *Proc. 22nd Int. Cosmic Ray Conf., Dublin* **5**, 35.

Perkins, M. A., Kristoff, J. J., Mason, G. M., and Sullivan, J. D.: 1969, *Nucl. Instr. Methods* **68**, 149.

Phillips, J. et al.: 1995, *Science* **268**, 1030.

Pizzo, V.: 1994, *J. Geophys. Res.* **99**, 4185.

Prantzos, N.: 1993, *Astrophys. J.* **405**, L55.

Prantzos, N., Arnould, M., and Arcoragi, J. P.: 1987, *Astrophys. J.* **315**, 209.

Preship, V. L. and Ptuskin, V. S.: 1975, *Astrophys. Space Sci.* **32**, 265.

*Ptuskin, V. S., Völk, H. J., Zirakashvili, V. M., and Brectschwerdt, D.: 1997, *Astron. Astrophys.* **321**, 434.

*Ramaty, R., Mandzhavidge, N., and Hua, X.-M. (eds.): 1996, *AIP Conf. Proc.*, No. 374.

Reames, D. V.: 1995, *Rev. Geophys. Suppl.* **000**, 585.

Schramm, D. and Turner, M.S.: 1996, *Nature* **381**, 193.

Silberberg, R., Tsao, C. H., Barghouty, A. F., and Shapiro, M. M.: 1997, *Proc. 25th Int. Cosmic Ray Conf., Durban* **4**, 321.

Simpson, J. A.: 1971, *Proc. 12th Int. Cosmic Ray Conf., Hobart* **8**, 324.

*Simpson, J. A.: 1983, *Ann. Rev. Nucl. Sci.* **33**, 323.

*Simpson, J. A.: 1998, in R. von Steiger, L. Fisk, and J. Jokipii (eds.), *Cosmic Rays in the Heliosphere*, Kluwer Academic Press, Dordrecht, The Netherlands, pp. 7–22.

Simpson, J. A. and Connell, J. J.: 1998, *Astrophys. J.* **497**, L85.

*Simpson, J. A. and Garcia-Muñoz, M.: 1988, *Space Sci. Rev.* **46**, 205.

Simpson, J. A. et al.: 1992, *Astron. Astrophys. Suppl. Ser.* **92**, 365.

Simpson, J. A. et al.: 1995, *Science* **268**, 1019.

Smith, E. J., Marsden, R. J., and Page, D. E.: 1995, *Science* **268**, 1005.

Stone, E. C.: 1989, *AIP Proc.* No. 183, pp. 72–90.

Stone, E. C. and Wiedenbeck, M. E.: 1979, *Astrophys. J.* **231**, 606.

Truran, J. W.: 1997, private communication.

Truran, J. W. and Cameron, A. G. W.: 1978, *Astrophys. J.* **219**, 226.

Tuzzolino, A. J.: 1988, *Nucl. Instr. Methods Phys. Res.* **A278**, 157.

*Waddington, C. J. (ed.): 1989, 'Cosmic Abundances of Matter', *American Institute of Physics Conf. Proceedings*, No. 183.

Waddington, C. J. for the TRANSPORT Collaboration: 1997, *Proc. 25th Int. Cosmic Ray Conf., Durban* **4**, 337.

*Waddington, C. J.: 1998, *New Astronomy Review*, Elsevier Publishers, Amsterdam (in press).

Webber, W. R., Kish, J. C., and Sherer, D. A.: 1990, *Phys. Rev.* **C41**, 547.

Wiedenbeck, M. E., 1983, *18th Int. Cosmic Ray Conf., Bangalore* **9**, 147.

Woosley, S. E. and Weaver, T. A.: 1995, *Astrophys. J. Suppl.* **101**, 181.

Wuosmaa, A. H. et al.: 1998, *Phys. Rev. Letters* **80**, 2005.

ON THE SLOW SOLAR WIND

L. A. FISK, N. A. SCHWADRON and T. H. ZURBUCHEN
*Department of Atmospheric, Oceanic and Space Sciences, University of Michigan, Ann Arbor,
MI 48109–2143, U.S.A.*

Abstract. A theory for the origin of the slow solar wind is described. Recent papers have demonstrated that magnetic flux moves across coronal holes as a result of the interplay between the differential rotation of the photosphere and the non-radial expansion of the solar wind in more rigidly rotating coronal holes. This flux will be deposited at low latitudes and should reconnect with closed magnetic loops, thereby releasing material from the loops to form the slow solar wind. It is pointed out that this mechanism provides a natural explanation for the charge states of elements observed in the slow solar wind, and for the presence of the First-Ionization Potential, or FIP, effect in the slow wind and its absence in fast wind. Comments are also provided on the role that the ACE mission should have in understanding the slow solar wind.

1. Introduction

The purpose of this paper is to discuss a model for the origin of the slow solar wind – why it exists; how it fits into the larger scheme of coronal structure and evolution; and why it contains compositional differences distinct from fast solar wind. This subject is particularly appropriate for consideration by the ACE mission since ACE will observe primarily the slow wind. ACE will fly near the equatorial plane and at a time of increasing solar activity when high speed flows from the polar regions of the Sun do not readily extend to low latitudes. Moreover, ACE, as we shall discuss, may be uniquely able to make measurements of the slow solar wind which will reveal interesting aspects of the conditions and dynamics in the corona.

We begin by considering the overall picture of fast and slow solar wind in the heliosphere, at least near solar minimum, when the concepts for the overall structure are well developed. At high heliographic latitudes, the polar coronal holes give rise to a fast, $\sim$750 km s^{-1} flow, which is remarkably steady (e.g., Phillips et al., 1995). However, at low latitudes, surrounding the streamer belt, the flow is slower, $\sim$400 km s^{-1}, but also more variable in density and speed, and it exhibits pronounced compositional differences which suggest a different origin from the high speed flow.

The charge composition of the solar wind is frozen-in in the low corona and is thus a measure of coronal electron temperature. Figure 1 is taken from work by von Steiger (1994), who uses *Ulysses* data to show the relative abundance of iron charge states in the fast solar wind from coronal holes and in the slow wind. The solid curve is the equilibrium charge state for a coronal temperature of 1.26

Space Science Reviews **86:** 51–60, 1998.

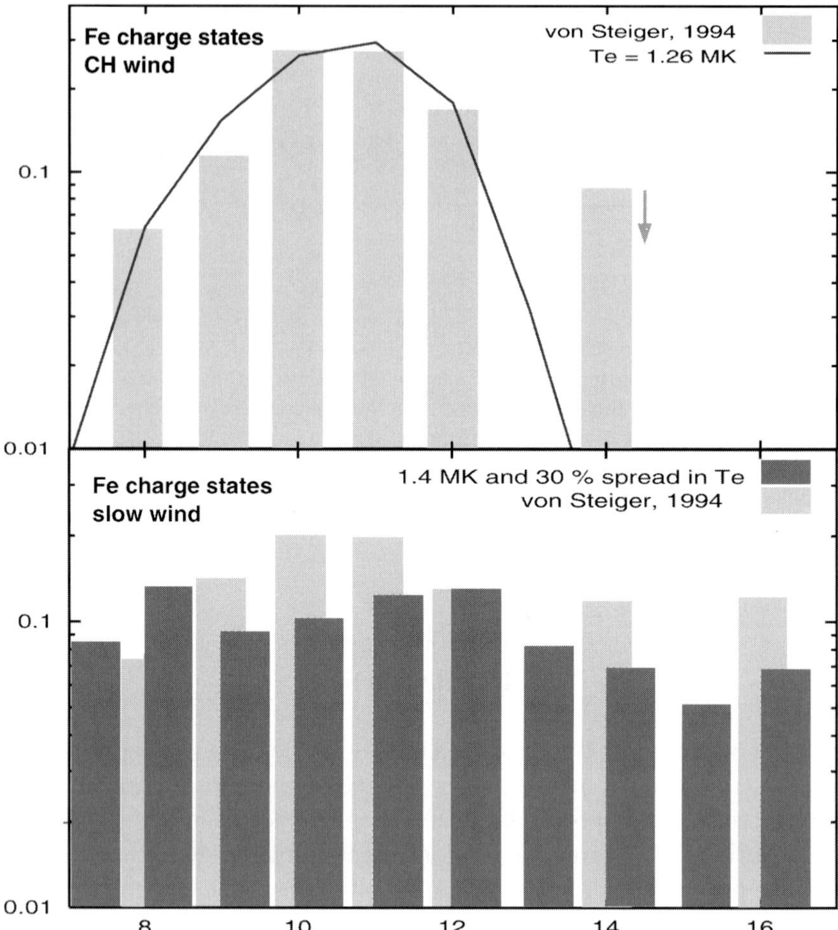

Figure 1. Comparison of charge states of Fe in the slow and coronal hole associated wind. A single uniform freezing-in temperature only applies in case of the fast wind. For slow wind, a temperature distribution of solar wind sources fits the data much better. Figure adapted from von Steiger (1994).

million deg. In the case of the fast wind, a single coronal temperature provides a reasonable fit to the data (note the charge state furthest to the right is an upper limit for the sum of charge states 13 and 14). Of course, looking at the charge states of different elements, which have different freeze-in points, can provide considerable information on the temperature profiles and other coronal parameters (e.g., Ko et al., 1997). However, to lowest order a single coronal temperature will suffice for the fast wind. In contrast, a single temperature will not provide a reasonable fit to the charge states observed in the slow wind. The distribution of charge states is simply too broad.

The second compositional difference between the fast and slow wind, and perhaps the most interesting, is the so-called First Ionization Potential, or FIP effect.

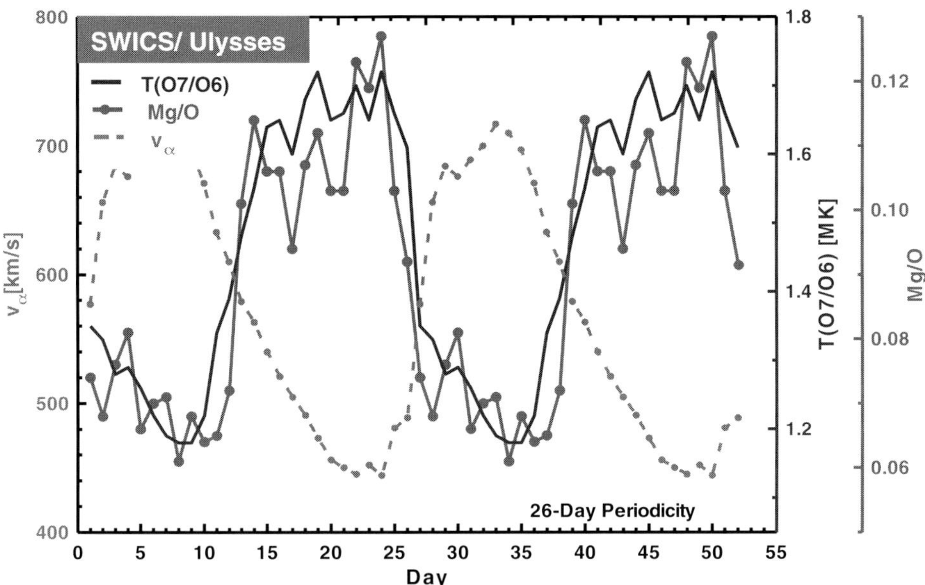

Figure 2. Superposed epoch analysis of *Ulysses* data showing the systematic variation in Mg/O ratio, the solar wind speed, and the coronal temperature inferred from the O^{7+}/O^{6+} ratio during the (effective) solar rotation period. The abrupt transition of freeze-in temperatures and composition indicate a different origin for fast and slow wind. Figure adapted from Geiss et al. (1995).

Figure 2 is from a paper by Geiss et al. (1995), again using *Ulysses* data. Shown here are the results of a superposed epoch analysis showing the solar wind speed, the freeze-in temperature for oxygen, as well as the magnesium to oxygen ratio. Magnesium is a low FIP element; it has a low first ionization potential and is easy to ionize. Clearly, magnesium is enhanced in the slow wind, and the transition is very abrupt, indicating a different origin for the fast and slow wind. When other elements are considered, a very general statement can be made: there is little evidence for a FIP effect in the fast wind, whereas, in the slow wind, there is a FIP effect by a factor of ~4 (Geiss and Bochsler, 1985).

There are several theories which have been developed to explain the FIP effect (see, e.g., summary by Meyer, 1993). However, many have difficulty in explaining the presence of a FIP effect in the slow wind, and its absence in fast wind. Many of the theories assume a quasi-stationary chromospheric layer, with a particular geometry or temperature, which enhances the upward transport of low FIP elements into the corona. The difficulty, however, is that the chromosphere below coronal holes or closed field regions looks remarkably the same, suggesting that mechanisms which are strictly chromospheric should work equally well in fast and slow wind.

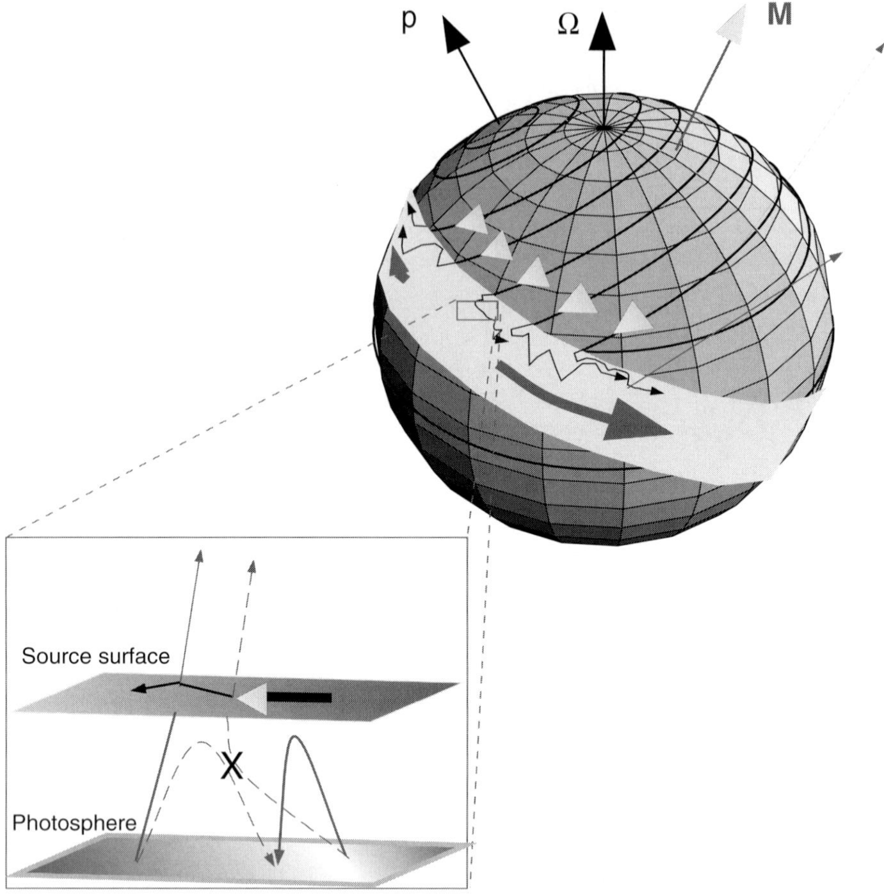

Figure 3. Reconnection scenario as described in text. Shown is the source surface of the solar wind in a frame co-rotating with the equatorial rotation rate. In open magnetic field regions, close to the pole, footpoints move in latitude. They are eventually convected into the band of closed magnetic fields at low latitudes. Due to subsurface reconnection events, a diffusive transport in longitude closes the footpoint curves on the solar wind source surface. For details refer to text.

2. A Theory

Consider a theory for the origin of the slow solar wind. This theory is a natural consequence of the new concept for the heliospheric magnetic field in fast solar wind which was proposed by Fisk (1996), viz., that the footpoints of the heliospheric magnetic field move extensively on the solar wind source surface as a result of the interplay between the differential rotation of the photosphere and the non-radial expansion of the magnetic field through coronal holes, which tend to rotate rigidly at the equatorial rotation rate. We review first this theory and its observational support, and then extend it to describe its consequences for the slow wind. The observational support is described in detail in Zurbuchen et al. (1997).

Consider a frame which co-rotates with the Sun at the equatorial rotation rate. The source surface of the solar wind – the surface beyond which the wind blows radially outward – is illustrated in Figure 3. The polar coronal hole is assumed to be centered on an axis 'M' which is offset from the rotation axis. The polar coronal hole, as does its lower latitude counterparts, is assumed to rotate rigidly at the equatorial rotation rate, i.e., the axis **M** is fixed in this frame. The expansion of the solar wind is taken to be non-radial from a limited region in the polar photosphere to a broader region on the source surface, and is symmetric about **M**. Consider, then, the field line which originates from the heliographic pole. It will penetrate the source surface at the location marked '**p**'. All other field lines, which are anchored in the differentially rotating photosphere, will execute trajectories on the source surface, which are some distorted circular patterns concentric about **p**, and in a direction indicated by the arrows. Clearly, the footpoints of heliospheric magnetic field lines on the source surface execute large excursions in latitude and longitude, which result in, in particular, large excursions in latitude of the heliospheric field. These excursions result in a direct magnetic connection from high to low latitudes which can account for the observation that particles accelerated in Co-Rotating Interaction Regions near the equatorial plane are observed at high latitudes (Fisk, 1996).

In Zurbuchen et al. (1997) observational evidence in support of this theory is presented. First, it is noted that there should be two characteristic frequencies for large-scale fluctuations in the field at high latitudes: 20 days and 34 days. The trajectories in Figure 3 are shown as smooth curves. In practice, the curves should be distorted, with variations in direction and in field magnitude, which move along the trajectories at essentially the differential rotation rate. It can readily be shown that at high latitudes, due to the offset of **p** from the rotation axis, *Ulysses* should observe disturbances on the footpoint trajectories twice, once as the footpoint trajectory crosses the latitude of *Ulysses* on one side of the Sun, and again on the opposite side. The time separation between these observations can be shown to be $\sim$20 days, i.e., a clear periodicity of 20 days in the field observations, which is quite unique to this theory, should be observed. At lower latitudes, where the footpoint trajectories tend to be more closely aligned with a single latitude, the periodicity should be $\sim$34 days, the differential rotation rate. This rate, however, may also be observed at higher latitudes, since the source surface is effectively a surface of constant pressure, and the effects of disturbances at lower latitudes may influence the magnetic field at higher latitudes as the pressure equalizes. Zurbuchen et al. (1997) analyzed magnetic field data from the southern solar pass of *Ulysses* and found that, in fact, two clear periodicities were observed, a strong signal at 20 days, and a somewhat weaker signal at 34 days. Zurbuchen et al. (1997) further analyzed the magnetic field directions seen in the heliosphere at high latitudes and found the clear sinusoidal pattern expected from the motion of the footpoints predicted in Figure 3.

There is clear observational evidence then that the footpoints of heliospheric magnetic field lines are indeed moving across the solar wind source surface. The obvious question to ask, then, is what happens when the field lines encounter low latitude regions, beyond the edge of the coronal holes, or equivalently, when the footpoints of the field lines in the photosphere encounter the coronal hole boundary. Clearly, Maxwell's equations must be satisfied, the divergence of the magnetic field vector must be zero, and the field lines must be continuous. Flux is then being deposited on one side of the Sun at low latitudes, the side where the arrows in Figure 3 are downward in latitude, and depleted on the opposite side. Such a deposition cannot increase indefinitely, and there must be transport of the field through the equatorial region from the side of deposition to the side of depletion, and a steady state achieved. A natural mechanism for such transport is reconnection, as is illustrated conceptually in Figure 3. Underlying the source surface at low latitudes are closed field loops. The field lines can encounter the sides of loops with opposite polarity, reconnect, and by doing so jump from location to location on the Sun, and effectively diffuse around the source surface near the equatorial plane to the opposite side of the Sun. For details concerning this magnetic field transport and reconnection scenario refer to Fisk et al. (1998).

This explanation for the transport of field lines in the solar equatorial region offers an explanation for one of the more puzzling magnetic field observations in the solar wind. There are intervals, sometimes extended, when the magnetic field is observed to be effectively radial near the equatorial plane, despite the strong tendency for the rotation of the Sun to yield a field with a strong azimuthal component. Footpoints diffusing around the Sun near the solar equator, in the direction opposite to that of the solar rotation, will have a reduced azimuthal component and, if the diffusion is sufficient fast, will be effectively radial.

This model for the behavior of the magnetic field at the Sun demands, then, that there are two distinct regions of the solar wind. In coronal holes, the wind is emitted along continuously open magnetic field lines, presumably yielding the fast steady wind. At lower latitudes, however, the field lines, as a result of their reconnection process, are continuously opening the tops of what had been closed loops, and allowing the material to flow outward into the heliosphere. The emission of the slow solar wind is then a sporadic process, dynamically driven when open field lines reconnect with previously closed loops. Such a sporadic origin for the slow wind is not a new concept; however, the driving mechanism, indeed the inevitability of such a process that results from footpoint motions in coronal holes, perhaps is.

Consider then the observations of charge states shown in Figure 1. Coronal loops, of course, have many different sizes, configurations, and particularly temperature. If we assume that the slow solar wind results from the superposition of material from many different loops, the resulting charge states in the observed wind will similarly reflect this spread in temperature. In the darker bars in Figure 1, plotted against the observations of iron charge states in the slow wind, we assume that there is a Gaussian distribution in temperatures, with a 30% spread about a mean temperature of 1.4 MK. Clearly, the fit to the data is much better than that from a single coronal temperature.

Consider also the FIP effect that is observed in the slow solar wind, but which is far less pronounced in the fast wind. Clearly, the obvious mechanism is to take advantage of the loops. Fast wind from coronal holes expands continuously outward; slow wind accumulates in closed loops for hours to days, and then is released into the heliosphere. There are several options here, but one simple mechanism, as is discussed in detail in Schwadron et al. (1998), is to consider the process by which loops are heated. Elements in the corona do not appear to be heated to a uniform temperature, but rather the heating appears to result more in a constant thermal speed, as is frequently observed in the solar wind, at least in regions where the coronal density is not sufficiently high to demand collisional dominance. Such heating will occur, presumably, by the interaction of the ions with MHD turbulence.

Assume, then, that such heating occurs in the larger loops of the corona, which extend to high altitudes and are therefore more likely to open to form the slow solar wind. Assume also that this heating extends downward into the transition layer, where particles are partly ionized. Clearly, it is essential to have the characteristic time for wave heating to be less than the ion collision time, which tends to yield a constant temperature for all ions. As is discussed in Schwadron et al. (1998) such a situation is possible even down into the transition layer. Ions will maintain a high temperature, or equivalently a large scale height throughout the loop. Particles that are neutral will not experience the heating due to MHD turbulence and will have a smaller scale height. Thus, species which are easily ionized – the low FIP elements – will on average have a large scale height throughout the loop. In contrast, the low FIP elements on average will have a smaller scale height at the bottom of the loop. At the top of the loop, then, the density of low-FIP elements is intrinsically larger than that of the high-FIP elements.

This mechanism for producing the FIP effect can be described by a simple set of equations. As is discussed in detail in Schwadron et al. (1998), we assume the loop contains a fixed number of particles with photospheric abundance. The minor ions in the loop, heavier than helium, are heated by waves to a constant thermal speed; hydrogen ions are not heated by waves, and have the usual steep temperature profile between the chromosphere and the corona; and neutrals retain the temperature profile of the hydrogen. The ionization state at the bottom of the loop is determined by both collisional ionization, and by the solar UV and EUV flux. The gas is then assumed to be in hydrostatic equilibrium. The relative composition at the top of the loop, which is placed at 100 000 km, is shown in Figure 4. The predicted ratio of the abundance of a species relative to oxygen, divided by the same ratio in the photosphere, is shown as an open bullet symbol. The error bars indicate qualitatively the natural spread of measurements due to changes in the wave-field (for details refer to Schwadron et al. (1998)). The squared black and white bars are the observations in the fast wind, where there is little or no FIP

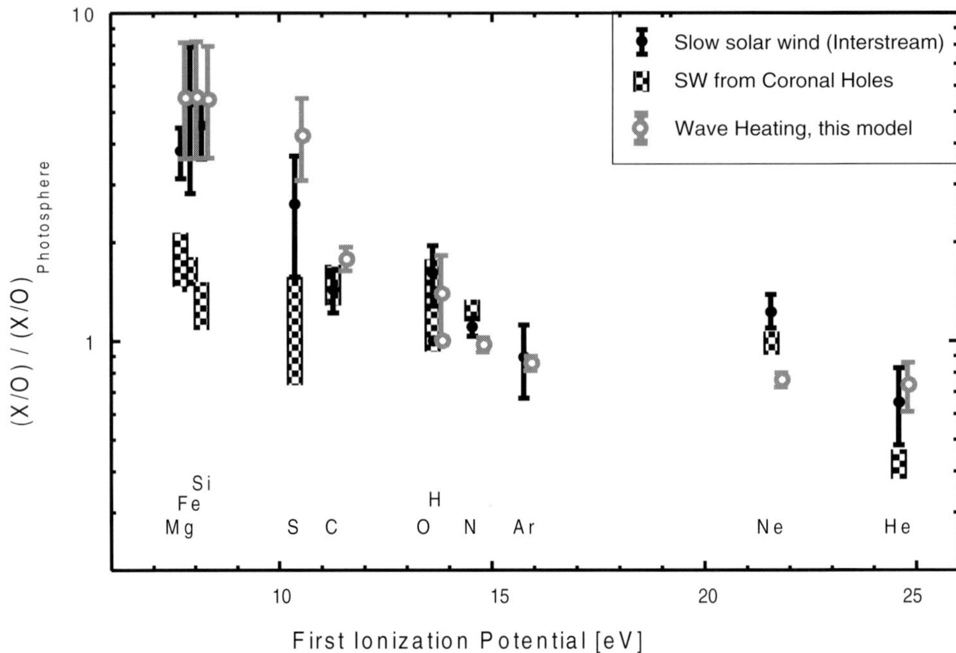

Figure 4. Solar wind abundance ratios, relative to their photospheric values, as a function of the first ionization potential. The measurements are compared with results from a FIP fractionation model described in the text. Figure adapted from von Steiger (1994).

effect. The full symbols with error bars are the observed abundances in the slow wind. Clearly, there is reasonably good agreement for all elements.

Notice, that in order to explain the observed helium abundance, an intermediate heating rate (between heavy ions and protons) has to be assumed. This seems to contradict *in situ* observations of the kinetic properties of solar wind ions, which indicate that helium responds to wave-particle interactions very much in a way similar to heavy ions. However, deep in the corona, wave heating of helium could be less effective than for the minor ions since helium can be a more major constituent of the plasma, with sufficient mass density to effect the properties of the waves.

Finally, we should ask why the wind which is released sporadically from loops is slower. There is no immediate answer here, and further numerical modeling will be required. It may be that the initial density is higher, with a lower final speed resulting. It may be the result of the sudden, almost adiabatic expansion from the previously closed loop, in contrast to the continuous deposition of energy in the steady solar wind.

3. Implications for the ACE Mission

Consider, then, what the solar wind composition instruments SWICS and SWIMS on ACE (Gloeckler et al., 1998) can do to provide information on the origin of the slow solar wind. All of the calculations shown in Figure 1 were averaged over many loops, in part, because the observations with which they were compared were accumulated for hours to days, to acquire sufficient statistics. ACE, however, is in a position to obtain good statistics on shorter time scales. Unlike *Ulysses* which observed the slow solar wind primarily en route to Jupiter, where the density is reduced by the square of the distance, ACE remains at 1 AU and should be able to observe material with sufficient statistics on the scale of $\sim$1 hr. On this time scale the material may originate from a single loop, although we cannot be sure that there is not considerable mixing near the Sun. We may discover that observing on this limited time scale reveals that the observed charge states of the slow solar wind are consistent with a single temperature, characteristic of one loop. Comparing observations from many different time intervals will reveal the variations in the temperatures and conditions in the corona which give rise to the slow wind. Similarly, on a scale of less than 1 hr we should expect considerable variation in the FIP effect in the slow wind, which results from the variations in the wave heating, the altitude where this heating begins, and/or the lifetime of the loops, prior to their being opened to form the slow solar wind.

4. Concluding Remarks

We should remember that the theory described here applies only in the years around solar minimum, when there are well-developed polar coronal holes, across which the field lines move. It is not clear how this mechanism will apply near solar maximum. There are certainly coronal holes on the Sun nearer to solar maximum, but they are short lived, and the concept of field line motion across them, with resulting reconnection in closed loops, may be quite different.

Acknowledgements

The work was supported, in part, by NASA contract NAS5-32626 and NASA/JPL contract 955460. T.H.Z. was supported, in part, by the Swiss National Science Foundation.

References

Fisk, L. A.: 1996, 'Motion of the Footpoints of Heliospheric Magnetic Field Lines at the Sun: Implications for Recurrent Energetic Particle Events at High Heliographic Latitudes', *J. Geophys. Res.* **101**, 15547.

Fisk, L. A., Zurbuchen, T. H., and Schwadron, N. A.: 1998, 'On the Coronal Magnetic Field: Consequences of Large-Scale Motions', *Astrophys. J.*, in press.

Geiss, J. and Bochsler, P.: 1985, 'Ion Composition in the Solar Wind in Relation to Solar Abundances', *Rapports Isotopiques dans le Système Solaire* **213**, Cepadues-Editions, Paris.

Geiss, J., Gloeckler, G., and von Steiger, R.: 1995, 'Origin of the Solar Wind from Composition Data', *Space. Sci. Rev.* **72**, 49.

Gloeckler, G., Bedini, P., Fisk, L. A., Zurbuchen, T. H., Ipavich, F. M., Cain, J., Tums, E. O., Bochsler, P., Fischer, J., Wimmer-Schweingruber, R. F., Geiss, J., and Kallenbach, R.: 1998, 'Investigation of the Composition of Solar and Interstellar Matter Using Solar Wind and Pickup Ion Measurements with SWICS and SWIMS on the ACE Spacecraft', *Space Sci. Rev.* **86**, 497.

Ko, Y.-K., Fisk, L. A., Geiss, J., Gloeckler, G., and Guhathakurta, M.: 1997, 'An Empirical Study of the Electron Temperature and Heavy Ion Velocities in the South Polar Hole', *Solar Phys.* **171**, 245.

Meyer, J.-P.: 1993, 'Element Fractionation at Work in the Solar Atmosphere' in *Origin and Evolution of the Elements*, Cambridge University Press, Cambridge.

Phillips, J. L., Bame, S. J., Feldman, W. C., Goldstein, B. E., Gosling, J. T., Hammond, C. M., McComas, D. J., Neugebauer, M., Scime, E. E., and Suess, S. T. 1995, '*Ulysses* Solar Wind Plasma Observations at High Southerly Latitudes', *Science* **268**, 1030.

Schwadron, N. A., Fisk, L. A., and Zurbuchen, T. H.: 1998, 'Elemental Fractionation in the Slow Solar Wind', *Astrophys. J.* , in press.

von Steiger, R.: 1994, 'Composition of the Solar Wind', Habilitation Thesis, University of Bern.

Zurbuchen, T. H., Fisk, L. A., and Gloeckler, G.: 1996, 'On the Slow Solar Wind (Abstract)', *Eos Trans. AGU* **77**, Fall Meet. Suppl. F563.

Zurbuchen, T. H., Schwadron, N. A., and Fisk, L. A.: 1997, 'Direct Observational Evidence for a Heliospheric Magnetic Field with Large Excursions in Latitude', *J. Geophys. Res.* **102**, 24175.

WIND OBSERVATIONS OF SUPRATHERMAL ELECTRONS IN THE INTERPLANETARY MEDIUM

R. P. LIN

*Space Sciences Laboratory and Physics Department, University of California, Berkeley,
CA 94720–7300, U.S.A.*

Abstract. We review some of the new results for suprathermal electrons obtained with the 3-D Plasma and Energetic Particle Instrument on the WIND spacecraft, which provides high sensitivity electron and ion measurements from solar wind thermal plasma up to $\gtrsim$MeV energies. These results include: (1) the observation of solar impulsive electron events extending down to $\sim$0.5 keV energy; (2) the observation of a turnover at $\sim$12 keV for electrons in a gradual large solar energetic particle (LSEP) event; (3) the detection of a quiet-time population (the 'superhalo') of electrons extending up to $\sim$100 keV energy; and (4) the probing of the magnetic topology and source region for magnetic clouds, using electrons. These unique WIND measurements are highly complementary to the particle composition measurements which will be made by ACE.

1. Introduction

For many years the solar wind thermal plasma has been regarded as a regime essentially independent from the energetic particle populations found in interplanctary space. Because of dynamic range considerations, previous instruments designed to measure the solar wind plasma ions and electrons lack the sensitivity to detect the suprathermal particles from just above solar wind plasma to a few hundred keV, except during highly disturbed times. These suprathermals play a key role in the varied plasma and energetic particle phenomena observed to occur in the interplanetary medium (IPM), and they provide information about their source, whether it is the Sun, the outer heliosphere, or the Earth.

The 3-D Plasma and Energetic Particles Experiment (Lin et al., 1995) on the WIND spacecraft is designed to bridge the gap between solar wind plasma and energetic particle measurements by providing high sensitivity, wide dynamic range, good energy and angular resolution, full 3-D coverage, and high time resolution over the energy range from a few eV to $\gtrsim$300 keV for electrons and $\gtrsim$6 MeV for ions. This is the only instrument operating during the Advanced Composition Explorer (ACE) mission that provides the sensitivity necessary to detect suprathermal electrons from just above the solar wind halo and strahl ($\lesssim$1 keV) to $\sim$30 keV. Observations in this range are essential for studies of particle acceleration at the Sun and in the IPM, for understanding wave-particle interactions in the IPM, and for probing the topology of large structures such as magnetic clouds. The mea-

Space Science Reviews **86:** 61–78, 1998.
© 1998 *Kluwer Academic Publishers. Printed in the Netherlands.*

                    R. P. LIN

surements of this instrument are thus highly complementary to those made by the instruments on the ACE mission.

2. Solar Impulsive Non-relativisitic Electron Events

The steady-state solar wind electron population is dominated by a core with temperature kT $\sim$10 eV, containing $\geq$95% of the plasma density and moving at about the solar wind bulk velocity, plus $\sim$5% in a hot, kT $\sim$80 eV, halo population carrying heat flux outward from the Sun, often in the form of a highly collimated strahl (Feldman et al., 1975). Impulsive non-relativistic electron events were first detected at energies above $\sim$40 keV (Van Allen and Krimigis, 1965; Anderson and Lin, 1966). However, the first high sensitivity measurements of electrons from $\sim$20 keV down to 2 keV, made by the ISEE-3 spacecraft, showed that impulsive acceleration of electrons occurred, on average, several times a day or more during solar maximum (Lin, 1985). Many of these non-relativistic electron events are unaccompanied by reported Hα flares, and many are observed only at energies below $\sim$15 keV.

Accelerated ions are detected above background for some of these electron events but the fluxes are generally very low, indicating the e/p ratio is large. The associated ion emission is primarily at low energies ($\sim$MeV nucl^{-1} and below) and ^{3}He-rich (Reames, von Rosenvinge, and Lin, 1985); that is, the ions have ^{3}He/^{4}He ratios of order unity while the typical ratios for the solar atmosphere or solar wind are a few times 10^{-4}. The soft X-ray (SXR) bursts accompanying these events are typically impulsive, with duration $\lesssim$10 min, so these non-relativistic electron-^{3}He-rich events are sometimes called impulsive solar energetic particle (SEP) events. As these electrons escape, they produce solar and interplanetary type III radio bursts through beam-plasma interactions (see Lin, 1990 for review). Table I, right column, summarizes the properties of these events.

Figure 1 shows a solar impulsive electron event observed by WIND (Lin et al., 1996) over the entire energy range from solar wind to suprathermal particle energies (a few eV to hundreds of keV). The main event beginning at 11:00 UT is easily identified by its velocity dispersion, i.e., the faster electrons arriving earlier, as expected if the electrons of all energies were simultaneously accelerated at the Sun and traveled the same distance along the interplanetary field to reach the spacecraft. The solar event can clearly be identified down to the 0.908 keV and even the 0.624 keV channels. A second, much weaker impulsive electron event is seen beginning about 16:20 UT in the 8.77 keV channel. Another very small event may be starting at $\sim$11:20 UT at $\sim$6 keV energy.

The 3-D angular distributions (Figure 2) show that the electrons in the main event are streaming outward from the Sun with pitch-angle distribution highly peaked, within $\lesssim$30°, along the magnetic field.

TABLE I

Solar energetic particle event characteristics

Characteristic	Large solar energetic particle (LSEP) events	Non-relativistic electron – ^{3}He-rich events
Dominant particle species	$\gtrsim 10$ MeV protons	~ 2–100 keV electrons
Electron to proton ratio	small	large
^{3}He/^{4}He ratio	'normal' solar ($\sim 5 \times 10^{-4}$)	~ 0.1–1 ($\sim 10^2$ to $\gtrsim 10^3$ times normal)
Heavy nuclei	'normal' solar	enhanced abundances of Fe, Mg, Si, S
Ionization states	typical of 1–2 $\times$ 10^6 K normal corona e.g., Fe^{+13})	highly stripped (e.g., Fe^{+20}) typical of $\sim 10^7$ K plasma
Extent in solar longitude	$\gtrsim 100$ deg	tens of degrees
Event rate (at solar maximum)	tens per year	$\gtrsim 10^3$ per year
Flare association	large solar flare (but sometimes missing)	mostly small flares but often no flare
Solar soft X-ray burst	gradual, e-folding decay times > 10 min	impulsive, e-folding decay times < 10 min
Interplanetary association	Coronal Mass Ejection (CME) with fast shock	interplanetary type III radio burst

Figure 3 shows electron differential flux vs energy spectra at different times during, and integrated over, the entire impulsive event, as well as the pre-event solar wind spectrum, with core, halo and 'super-halo' (discussed later). The event spectra have the pre-event spectra subtracted. The primary event spectra progress to lower energies with time; the peak in the spectrum at ~ 5 keV at 12:30–12:40 UT moves to ~ 3 keV at 13:00–13:10 UT and ~ 2 keV at 14:00–14:10 UT. Contributions from the smaller events are also evident, e.g., at 11:45–11:55 UT. The event-integrated spectrum displays a peak at $\lesssim 1$ keV, with significant flux at ~ 0.5 keV. No electrons are detected in the 422 eV channel or below for this impulsive event. Above the peak, the spectrum is similar to those reported previously above ~ 2 keV (Potter et al., 1980), and can be fit to a power-law shape $dJ/dE = AE^{-\delta}$, where E is the electron energy in keV and A and δ are constants. The best fit gives $\delta = 3.0$ from ~ 1 keV to 40 keV, steepening to $\delta = 4.4$ above ~ 40 keV.

Because at coronal temperatures electrons are not gravitationally bound while protons are, an ambipolar electric field (the Pannekoek–Rosseland field (Pannekoek, 1922; Rosseland, 1924)) is set up with a total potential drop of about 1 kV from the base of the corona to 1 AU. This potential varies inversely with distance from Sun center, and it accelerates protons outward and decelerates electrons. Thus, the peak in the spectrum of the electrons just as they escape the corona could be up to

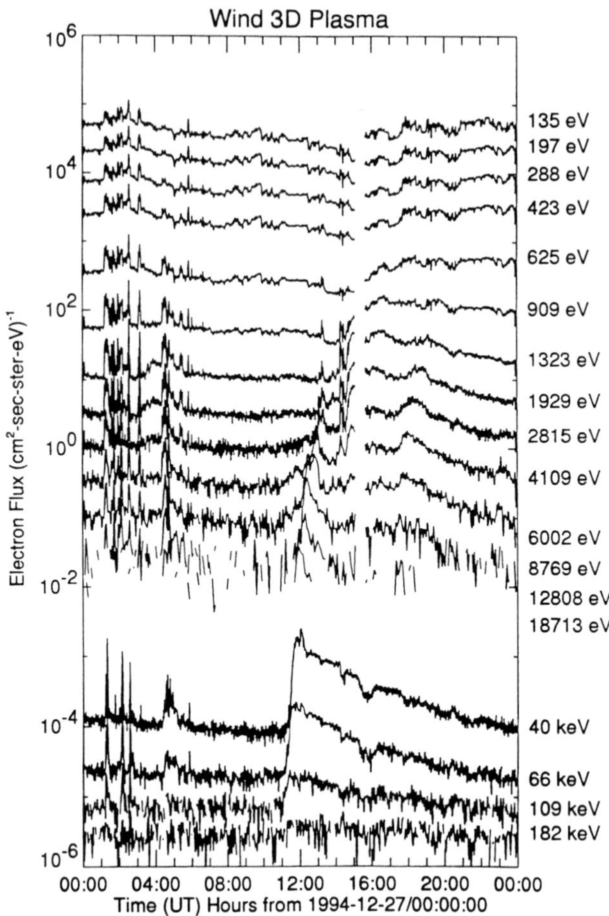

Figure 1. Electron fluxes from $\sim$100 eV to $\gtrsim$100 keV for 27 December 1994. The solar electron event begins at $\sim$11:00 UT at $\sim$100 keV, with velocity dispersion evident down to 624 eV. Two smaller events, at $\sim$11:00 UT and 16:20 UT, are visible below $\sim$6 keV. The dip at $\sim$15:00 UT at low energies is due to the close approach to the Moon, resulting in a plasma shadow (from Lin et al., 1996).

$\sim$1 keV more than measured at 1 AU, e.g., ranging from $\lesssim$1 up to $\sim$2 keV for the event of Figure 1, depending on the height of the acceleration.

The fact that the event spectrum extends down to such low energies indicates that at least some of the electron acceleration must occur high in the corona, since the range of $\sim$keV energy electrons in ionized hydrogen, due to Coulomb collisions, is short compared to the column depth through the corona. Assuming that the initial accelerated electron spectrum is a power law with the same exponent as seen at energies above the peak, the maximum overlying column density can be calculated (Lin, 1974). For a peak at $\sim$1.5 keV, the column density must be less than $\sim$9 $\times$ 10^{17} cm^{-2}. This value implies that the lowest energy electrons must

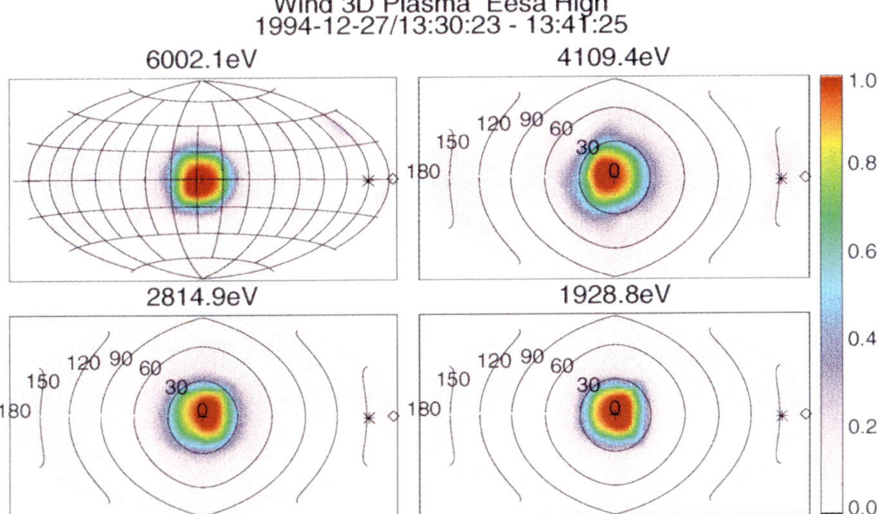

Figure 2. The 3-D angular distribution of the electrons during the impulsive event. the magnetic field direction is centered at the origin and the Sun direction indicated by the asterisk. The angular grid of the measurements (upper left panel) and pitch-angle contours are shown (from Lin et al., 1996).

have been accelerated at altitudes of $\sim 1\ R_\odot$, for the typical active coronal density models (Dulk and McLean, 1978), or $\sim 0.2\ R_\odot$ for the quiet equatorial corona at sunspot minimum (Saito et al., 1977).

A ^{3}He enhancement was detected in solar energetic ions during this electron event (J. Mazur, private communication, 1995). If the ^{3}He originated at the same altitude in the corona as these low energy electrons, the proposed mechanism of acceleration by electromagnetic hydrogen cyclotron waves (Temerin and Roth, 1992) would be ruled out since the magnetic field would be much too weak. However, a previous comparison of the solar flare X-ray spectrum with the escaping electron spectrum over the 2–100 keV energy range (Pan et al., 1983) for an event observed by ISEE-3 suggests that X-ray producing electrons are present over a region which extends from the chromosphere to the high corona, while the lowest energy escaping electrons come only from the high corona. Thus both the ^{3}He and the more energetic electrons might be accelerated in the lower corona where the magnetic field is higher.

Even though the first year of the WIND observation, late 1994 to late 1995, is near solar minimum, tens of impulsive solar electron events have been seen, with many detected down to ~ 0.5 keV. The coronal flare acceleration process thus appears to produce a power-law spectrum extending down to $\lesssim 1.5$ keV or lower, compared to a coronal thermal electron energy of kT ~ 0.1 keV ($\sim 10^6$ K). Integrating over the energy spectrum and over the duration of the event, and assuming that the cone of propagation in the interplanetary medium for the electron event of Figure 4 is $\sim 40°$, we estimate a total energy of $\gtrsim 3 \times 10^{26}$ ergs in escaping

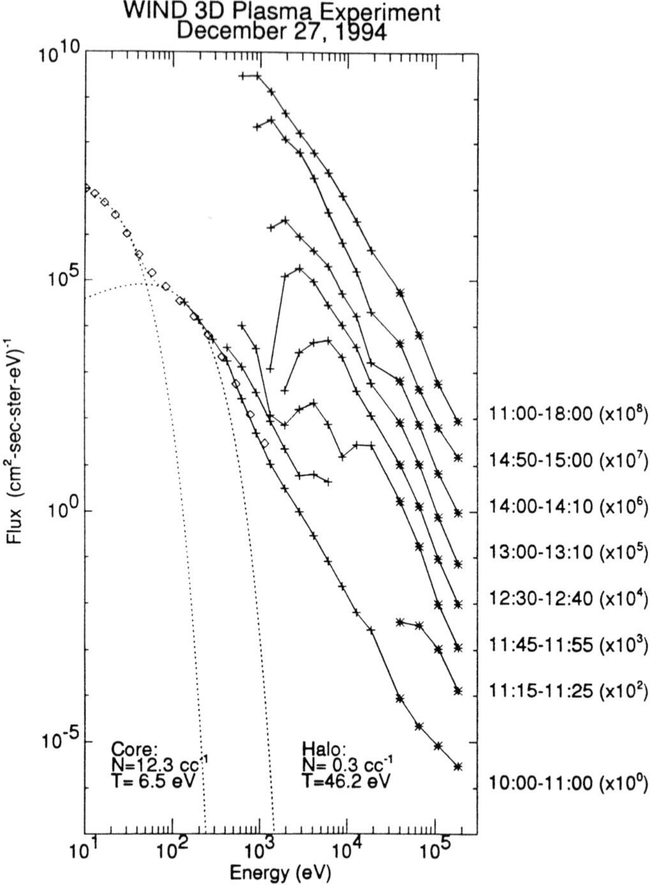

Figure 3. Omnidirectional electron spectra (with pre-event electron fluxes subtracted) are shown for various times during the 27 December 1994 event, and averaged over the entire event. The pre-event electron spectrum shows the solar wind electron core and halo components, as well as a 'super halo' extending to $\gtrsim$100 keV (from Lin et al., 1996).

electrons. Thus, at least that much energy was released in the coronal flare process at the Sun.

3. Large Solar Energetic Particle (LSEP) Events

The other type of solar energetic particle event (actually discovered first) produces significant flux of >10 MeV protons (Table I, left column). These are the largest (hence LSEP) and most energetic events. They usually occur after a large solar flare, and occasionally exhibit acceleration up to relativistic energies. Electrons are also observed, but the fluxes of energetic protons dominate over electrons. Tens of LSEP events are detected per year near solar maximum.

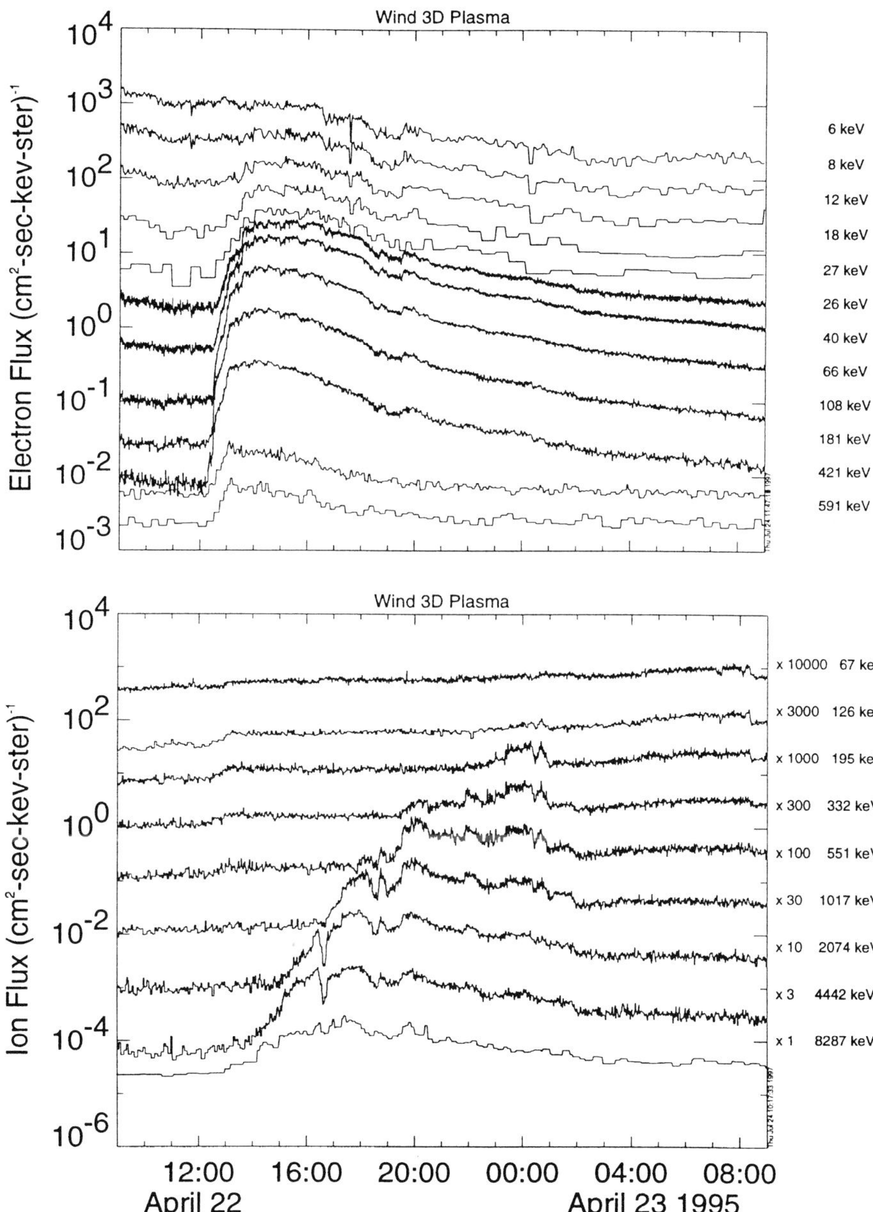

Figure 4. Energetic electron (*upper panel*) and ion (*lower panel*) fluxes for 22–23 April 1995. The energies (and flux factors) are indicated on the right. The gradual (LSEP) event is evident at energies above ∼12 keV in electrons.

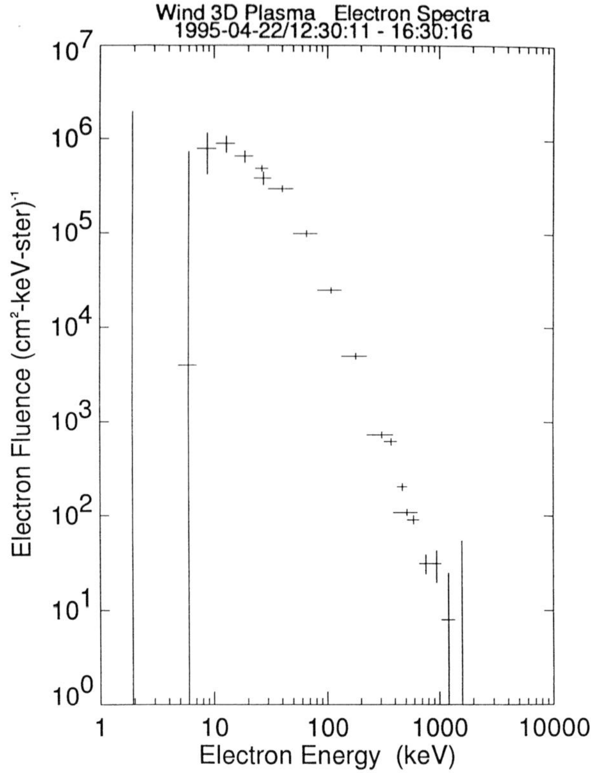

Figure 5. Omnidirectional electron energy spectrum (with pre-event background subtracted) averaged over the first four hours of the event.

The observed ionization states of LSEP particles are typical of a $1-2 \times 10^6$ K plasma, suggesting that these come from the quiescent solar corona and/or solar wind. It is believed that shock waves, from Coronal Mass Ejections (CME) or perhaps flares, propagating over a wide longitude range of the solar corona are the accelerating agent. LSEP events are also called 'gradual' events since they are typically accompanied by flare soft X-ray (SXR) emission of relatively long duration, with e-folding decay times of $\gtrsim$ tens of minutes.

Figure 4 shows the 22 April 1995 SEP event, the first event observed by WIND where energetic solar ions up to $\gtrsim 20$ MeV and electrons up to ~ 1 MeV were detected. A M7.1 SXR burst with two maxima, a sharp impulsive spike at ~ 11:50 UT and a second, much smaller, maximum at ~ 12:30 UT followed by a gradual decay lasting several hours, was detected by GOES. The presence of significant fluxes of > 10 MeV protons and the gradual SXR burst confirm this is an LSEP event, and no enrichment in ^{3}He was detected for this event (J. Mazur, private communication).

Electrons are detected down to ~ 8 keV in this event (Figure 4), but no increase is detected below ~ 6 keV, in contrast to the impulsive event of Figure 1. The electron spectrum for the event (Figure 5) shows a peak at ~ 12 keV, and a rapid falloff below $\sim$8 keV. Above $\sim$12 keV the spectrum fits a double power law with a break around 50 keV, and power-law exponents of $\delta = 1.1$ below and $\delta = 3.1$ above the break. If the peak is due to the electrons traversing overlying material, the source region is at a column depth of $\sim$1 $\times$ 10^{19} cm^{-2}. Thus for quiet coronal density models the electrons would have been accelerated in the chromosphere, while for active corona models the height would be $\sim$0.15 R_o. In contrast, near solar maximum it was found that essentially all LSEP events with gradual SXRs have electron spectra extending down to 2 keV without turnover; the only events with a turnover were those with impulsive SXR bursts (Lin, 1990). The presence of this turnover at relatively high electron energies for this LSEP event suggests that the simple picture of particle acceleration by CME shocks high in the corona, or in the interplanetary medium, may need to be re-examined.

4. Quiet Time Electrons

Unlike ISEE-3, the WIND spacecraft was launched near the minimum in the solar activity cycle. Thus, there are substantial periods free from energetic solar particle events and streams. Figure 6 shows the WIND omnidirectional electron spectrum from $\sim$5 eV to $\sim$100 keV measured during such a quiet period on February 22, 1995. Because of the extremely wide range of electron fluxes, three separate detectors – EESA-L, EESA-H, and SST – are required to measure this spectrum. The Maxwellian core of the solar wind plasma dominates from $\sim$5 to $\sim$50 eV; the solar wind halo takes over from $\sim$100 eV to $\sim$1 keV. The halo is believed to be due to the escape of coronal thermal electrons which have a temperature of $\sim$10^6 K. Note, however, that the halo spectrum departs significantly from isothermal at energies above $\sim$0.7 keV.

A third, much harder component has been discovered in the WIND observations, beginning above $\sim$2 keV and extends to $\gtrsim$100 keV, which we have denoted the 'super-halo'. The spectrum of the 'super-halo' appears to be approximately power-law with exponent δ $\sim$2.5. If this 'super-halo' is solar in origin, it would imply that electrons of such energies must be continuously present at, and escaping from, the Sun.

The angular distribution of these 'super-halo' electrons at very quiet times appears to be nearly isotropic, however, with a very slight anisotropy flowing toward the Sun. Such angular distributions would be expected for sources beyond 1 AU; the near isotropy would be due to mirroring of this incoming population in the stronger magnetic fields of the inner heliosphere. Possibly, acceleration by Co-rotating Interaction Regions (CIRs) beyond 1 AU may be responsible for the superhalo, but a preliminary analysis shows no obvious one-to-one correlation with CIRs or with solar active regions.

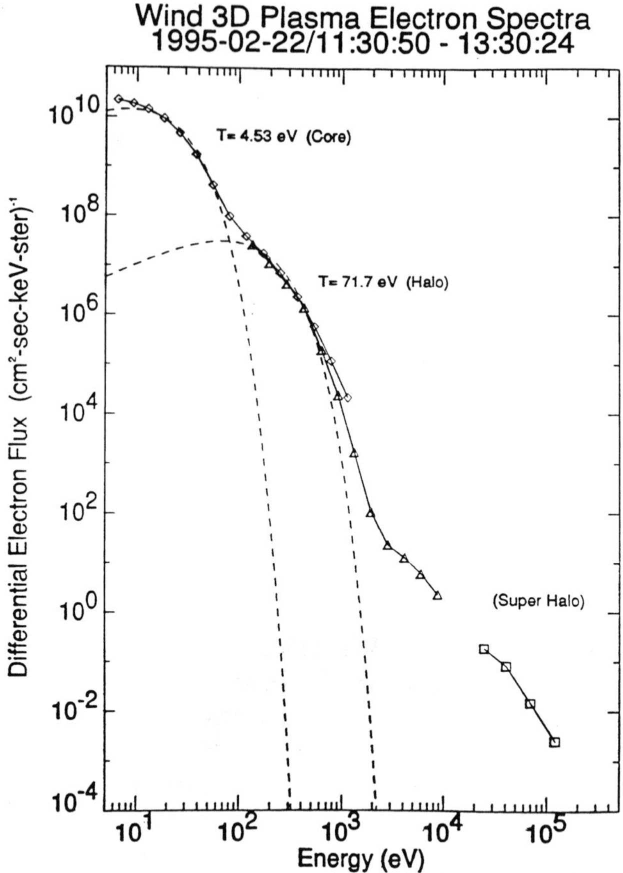

Figure 6. Electron differential flux spectrum from ~5 eV to $\gtrsim$100 keV, measured at a very quiet time, in the absence of any solar particle events, by the WIND 3-D Plasma and Energetic Particle experiment. The diamonds, triangles, and squares indicate the three different detectors used to accommodate the wide range of fluxes over this energy range. The dashed lines give fits to Maxwellians for the solar wind core and halo.

5. Energetic Electrons, Plasma Waves, and Solar Type III Radio Bursts

Observations by the WIND 3DP instrument generally confirm the basic process for generating solar type III radio emission, wherein the velocity dispersion of the escaping electrons forms a beam which generates electron plasma (Langmuir) waves which in turn, produce electromagnetic (radio) emission (see Lin, 1990, for review).

Four solar impulsive electron events were detected by WIND on April 2, 1995 (Ergun et al., 1998). Figure 7(a) (top panel) plots the omnidirectional electron fluxes at 96s resolution as measured by EESAH. The center energies of the channels range from 140 eV to 27.1 keV, with the lowest energy traces on top. Immediately below (Figure 7(b)) are the omnidirectional electron fluxes as measured by the

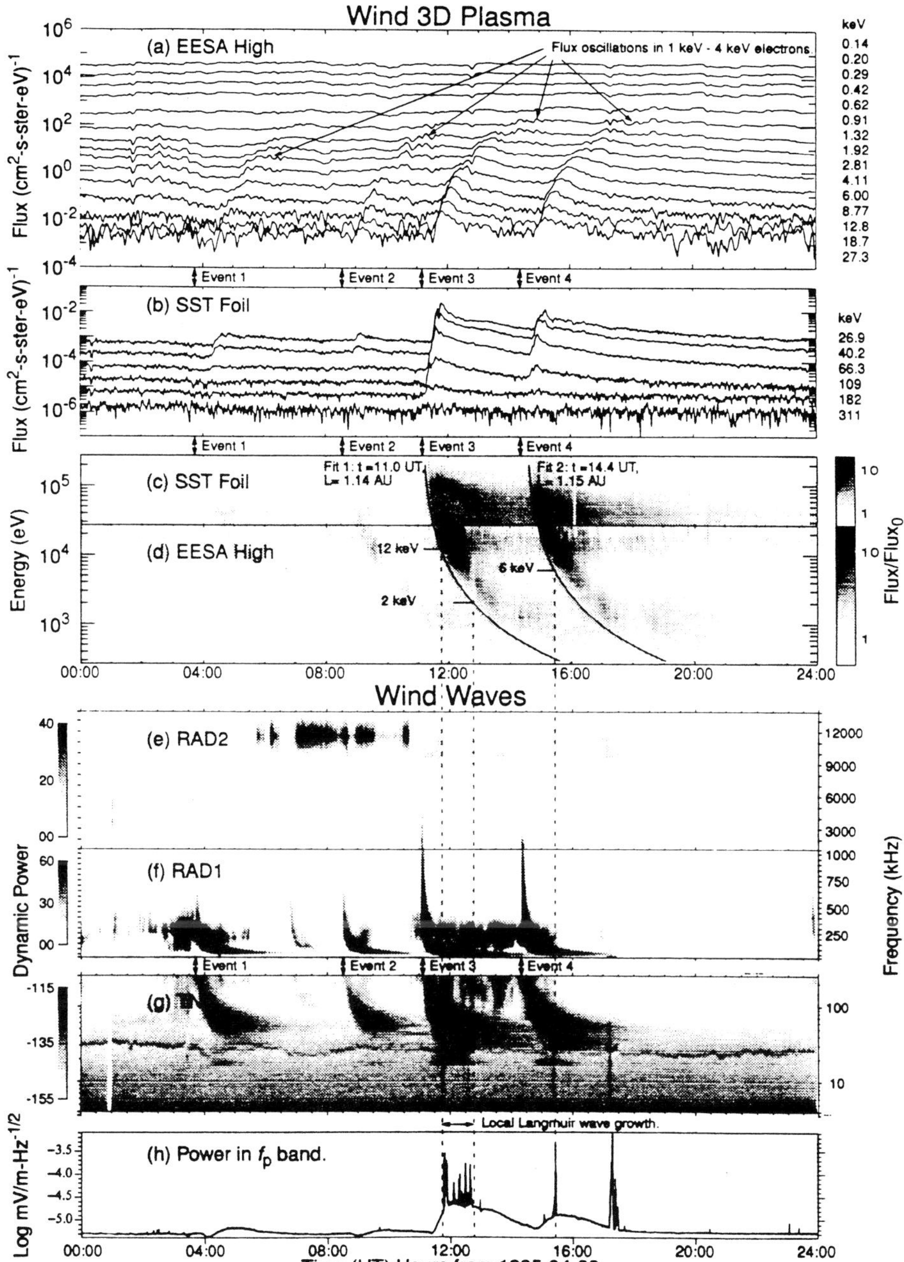

Figure 7. (a) The omnidirectional electron fluxes at 96s resolution as measured by EESA-H on WIND. The center energies are listed on the right. (b) The omnidirectional electron fluxes as measured by the SST- Foil detector. (c) and (d) Spectrograms of the ratio (F/F_0) of electron fluxes (F) to a reference flux (F_0) prior to the first event. Superimposed on the plot are linear fits of the arrival time of the electrons versus their inverse velocity. The event start times are marked on the plot. (e), (f), and (g) Solar type III radio bursts as observed by the WIND WAVES instrument. The RAD2 panel displays the radio emissions from 1 MHz to 14 MHz (linear frequency axis) The TNR (Thermal Noise Receiver) panel displays the frequency range from 4 kHz to 250 kHz. (h) The electric field wave power in the frequency band (19 kHz to 41.5 kHz) encompassing the local Langmuir frequency (from Ergun et al., 1998).

SST-Foil detector. The impulsive electron events can be clearly seen in the electron fluxes, with the strongest beginning at $\sim$11:10 UT with the 182 keV electrons and extending down in energy to $\sim$600 eV.

Persistent feature in all of the events were $\sim$15-min period modulations of the electron fluxes (see Figure 7(a)) that appeared as the low-energy ($\sim$1 keV to $\sim$4 keV) electrons arrived. Similar flux modulations can be seen in several other solar impulsive electron events observed by the WIND satellite and by the ISEE-3 satellite (see Figure 3 of Lin et al., 1981). These modulations showed no measurable velocity dispersion, and the modulations of the second event are accompanied by modulations in the magnitude of the magnetic field with a similar period, suggesting some type of local hydromagnetic instability is occurring.

Figures 7(c) and 7(d) are spectrograms of the ratio (F/F_0) of the event electron fluxes (F) to the pre-event flux (F_0), determined by averaging from $\sim$00:00 UT to $\sim$04:00 UT. Superimposed on the spectrograms are linear fits of the arrival time of the electrons versus their inverse velocity. In all of the events, velocity dispersion of the electrons is consistent with the expected Archimedean spiral length ($\sim$1.15 AU) for the measured solar wind velocity ($\sim$370 km s^{-1}).

All four of the solar impulsive electron events were associated with solar type III radio bursts (Figure 7(e–g)) observed by the WIND WAVES instrument (Bougeret et al., 1995). The radio bursts for events 3 and 4 extended up to 14 MHz, the maximum frequency of the WAVES experiment. The lowest frequency radio electromagnetic emissions appeared to have a cut-off near the local second harmonic. The narrow-banded emissions that persisted throughout the day varying between 20 kHz and 40 kHz (see Figure 7(g)) are locally generated Langmuir waves; the power in Langmuir emissions is shown in Figure 7(h). The series of Langmuir bursts between $\sim$11:45 UT and $\sim$12:45 UT coincide with the arrival of $\sim$12 keV through $\sim$2 keV electron fluxes of event 3. The bursts at $\sim$15:25 UT coincided with the arrival of $\sim$6 keV electron fluxes of event 4.

Although the plasma processes involved in the generation of type III radio bursts are still not understood in detail (see Ergun et al., 1998), these observations confirm that the non-relativistic electrons are the source of the emission, and therefore the radio bursts can be used to track the escaping electrons from the Sun into the interplanetary medium.

6. Magnetic Clouds

Magnetic clouds are a type of CME characterized by relatively strong magnetic fields (low plasma beta) and a smooth rotation of the magnetic field direction over a $\sim$1 day period at 1 AU, consistent with the passage of an approximately force-free helical magnetic flux rope of diameter $\sim$0.25 AU (Burlaga, 1988). They often exhibit CME characteristics such as enriched alpha particle content, low proton temperature and bi-directional electron streaming (Gosling, 1996). The extended periods of southward (and northward) magnetic field provided by the smooth rotation make clouds important for geomagnetic activity. As discussed by Gosling (1975) and MacQueen (1980) CMEs, including magnetic clouds, expel new magnetic flux, still connected to the Sun, into the interplanetary medium, so to maintain the IMF magnitude at a roughly constant level (King, 1979) compensating disconnection of magnetic flux must occur.

Larson et al. (1997) has shown that WIND electron observations of both the continuously outflowing halo at $\sim$0.1–1 keV and impulsive electron events from $\sim$1 to $\sim$$10^2$ keV can be used to trace the magnetic field of a cloud back to the Sun, measure the lengths of the cloud field lines, and determine when magnetic disconnection from the Sun occurs.

The top three panels (a, b, c) of Figure 8 show the magnetic field strength ($|\mathbf{B}|$) and direction (Θ, Φ) for the October 18–20, 1995 magnetic cloud, from the magnetometer experiment on WIND (Lepping et al., 1995). Figures 8(d) and 8(e) show the solar wind speed and density respectively, measured by the 3DP experiment. A shock and magnetosheath-like region precede the cloud. The field rotates smoothly from south to north during the 30 hour passage of the cloud. Lepping et al. (1997) have fit this signature to a force-free flux-rope geometry and estimated the axis of the cloud to be nearly in the ecliptic plane ($\Theta = -12°$), close to the Parker spiral angle ($\Phi = 291°$), and with a right handed helicity. The cloud diameter is estimated to be about 0.27 AU and WIND passed very close to the central axis, ($y_0/R_0 = 0.087$).

Figure 8(f) shows the flux of electrons streaming away from the Sun (135°–180° pitch angles since $\Phi = 291°$) in 17 energy channels logarithmically spaced between $\sim$0.1 and 100 keV, respectively. At energies $\gtrsim$20 keV, several impulsive electron events with rapid rise and slow decay can be seen, for example, at $\sim$21:00 UT October 18, $\sim$06:00, $\sim$11:30 and $\sim$18:00 UT October 19. These events are also evident at lower electron energies, and generally show velocity dispersion.

Figure 8(g) shows a color spectrogram of the ratio F/F_0 of the outward streaming electron flux, F (pitch angles 135°–180°) divided by an average background flux (F_0) in the absence of impulsive events, to illustrate the velocity dispersion in the arrival of energetic electrons impulsively accelerated at the Sun. Figure 8h is a color spectrogram of the radio observations provided by the WAVES instrument on WIND. The numerous fast drift bursts, from high (14 MHz) to low (tens of kHz) frequency, are solar type III radio bursts. produced by fast electrons propagating from near the Sun (a few solar radii) to $\sim$1 AU, as discussed in the preceding section.

Each of the impulsive electron events can be traced back, with the type III radio burst produced by these electrons, to a flare observed by the *Yohkoh* Soft X-ray Telescope within solar active region 7912. Clearly, electrons accelerated in these flares escape outward along open field lines and are detected by the 3-D Plasma experiment if the WIND spacecraft is on a field line connected to the flare site. This tracking by the type III radio bursts of these electron events from $\sim$1 AU back

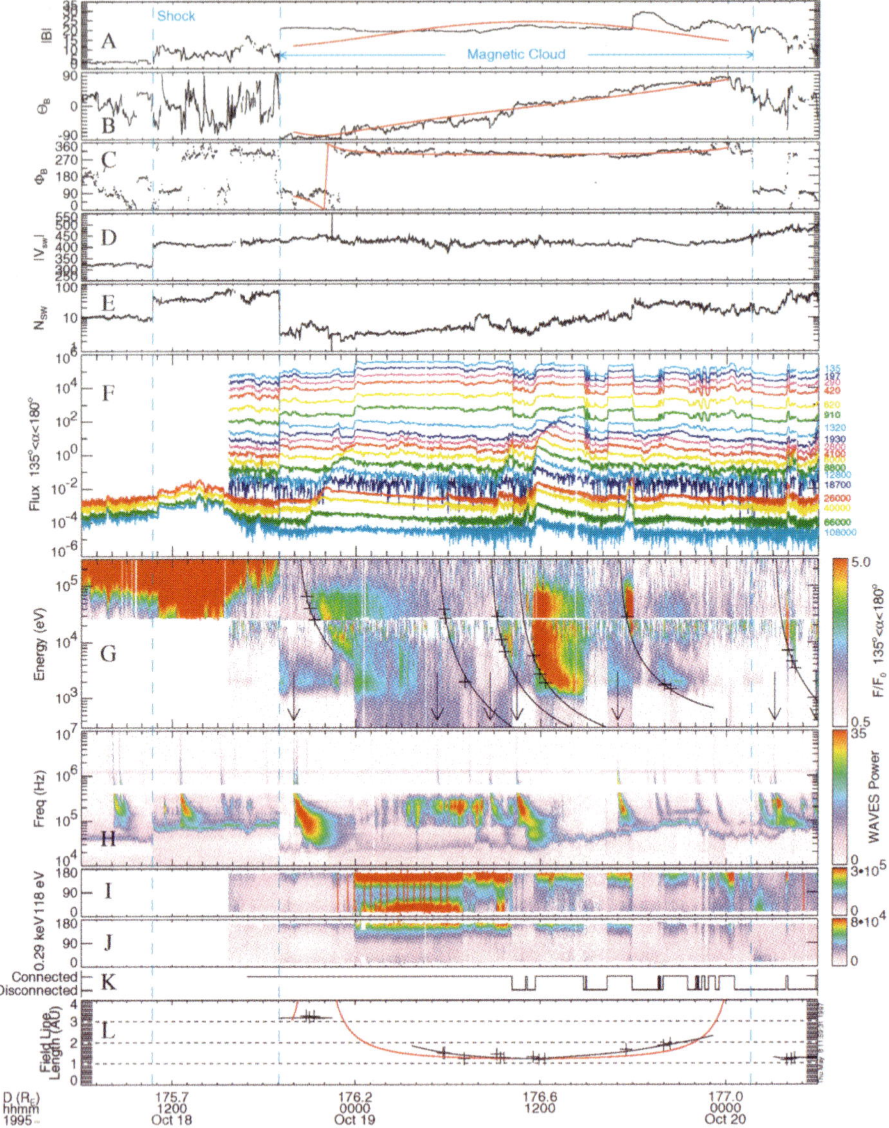

Figure 8. Summary plot of cloud event showing the magnetic field magnitude ($|B|$) and direction (Θ, Φ) in panels (A), (B) and (C). Solar wind speed and density are shown in panels (D) and (E). Panel (F) shows the electron flux for electrons traveling anti-parallel ($135° < a < 180°$) to the magnetic field direction for 18 logarithmically-spaced energy channels ranging in energy from ~ 0.14 to 110 keV. The following spectrogram (G), presents the ratio, F/F_0, of electron flux (F) divided by an average background flux (F_0) measured in the absence of impulsive solar electron events. Panel (H) is a spectrogram of the radio observations from 10 kHz to 10 MHz from the WIND WAVES experiment. Panels (I) and (J) are pitch-angle spectrograms of electrons at 118 eV and 290 eV. Magnetic connection to, or disconnection from, the Sun along the negative leg of the cloud is indicated in (K). From the flare time (arrows, (G)) and the arrival time of electrons as a function of energy, the field line length at each point was determined and is shown in panel (L). The black curve in panel (L) is a smooth curve drawn by hand. The red curves in panels (A), (B), (C) and (I) show the expected values for a model flux rope.

to X-ray flares at the Sun provides the first direct identification of the footprints of the magnetic fields of a cloud, in this case to an active region.

The crosses in Figure 8(g) mark the initial onset (at various energies) of the injected electrons. Using the start of the solar type III burst – identified by the sharp increase in >5 MHz wave power measured by the WAVES experiment – as the electron injection time (arrows) at the Sun, the field line length $L = v_e \Delta_t$ traveled by the electrons was determined and plotted in Figure 8(l). The field line length varies from $\sim$3 AU near the leading edge of the cloud to $\sim$1.2 AU near the center and rises again on the trailing part, qualitatively consistent with a flux rope model (smooth curve drawn in Figure 8(l)) with field lines twisted about a central core. The 3 AU outer field line compared to $\sim$1.2 AU near to the axis implies that it undergoes $\sim$4 turns from the Sun to 1 AU, or $\sim$8 turns for the entire cloud, assuming symmetry. The number of turns should be invariant if the field has not been magnetically reconnected while propagating from the Sun to 1 AU. The implication is that the flux rope was highly coiled at the time of ejection from the Sun.

Figures 8(i) and 8(j) show pitch-angle spectrograms of 118 eV and 289 eV halo electrons. Early in the cloud ($\sim$19:00 UT October 18–07:00 UT October 19), bi-directional streaming of halo electrons is observed, indicating that both ends of the cloud field line are connected back to the solar corona. Later, after $\sim$07:00 UT October 19, the electron pitch-angle distributions are uni-directional, indicating magnetic connection to the Sun on only one leg of the cloud. In addition, the electron fluxes exhibit many abrupt, discontinuous drops in level, occurring simultaneously at all energies (see Figure 8(f)).

The pitch-angle spectrograms of Figures 8(i) and 8(j) identify these as heat flux dropouts (HFDs), which McComas et al. (1989) suggested are due to the disconnection of the IMF from the corona. Almost all the HFDs for this cloud extend up to high energies, and are observed in the impulsive events as well, confirming that these are times when the magnetic field is truly completely disconnected from the Sun, unlike many of McComas et al.'s HFDs which were not dropouts above 2 keV (Lin and Kahler, 1992). Figure 8(k) plots the connection to the Sun. Note that the disconnected regions range from a few minutes to hours long and they are intertwined with connected regions.

As the magnetic cloud passes by, the spacecraft crosses field lines with different connectivity (Figure 8(K)). The disconnections presumably result from magnetic reconnection near the Sun. Since the HFDs are observed simultaneously across all electron energies, the disconnections must have occurred >6 hours earlier (the travel time to 1 AU for the slowest electrons). Furthermore, the flux of energetic (20 keV/nuclon) ions (from the Energetic Particle: Acceleration Composition and Transportation (EPACT) investigation on WIND (von Rosenvinge et al., 1995)) shows coincident dropouts, implying the disconnections occurred as much as $\sim$20 hours before the HFDs are detected at 1 AU.

Burlaga (1991) and Rust (1994) suggest that magnetic clouds are the interplanetary manifestation of solar filaments (prominences on the limb), which are often observed to have helical magnetic topology and to be embedded in CMEs. Gosling et al. (1995) have argued that magnetic clouds are formed by 3-dimensional reconnection of sheared magnetic arcades, and that, in the process, a mix of field lines connected at one end, at both ends and completely disconnected from the Sun can be produced. Alternatively, the cloud could also have emerged from the solar surface and been ejected from the Sun as a fully evolved flux rope with little or no reconnection.

We believe the disconnections probably occurred after the ejection of the cloud from the Sun $\sim$4 days earlier; otherwise the magnetic tension forces might be expected to significantly distort the cloud. As the cloud moves through the interplanetary medium, the legs of the flux rope may become partially disconnected from the solar surface through magnetic reconnection with adjacent field lines. Evidence for this is provided by an interplanetary type III radio storm detected by the WIND WAVES experiment in the 2.0–5.0 MHz range following the ejection of the cloud and prior to its detection at 1 AU. Such storms are believed to be the results of magnetic reconnection of magnetic fields high ($\sim$1 R_s) in the solar corona (Bougeret et al., 1984).

7. Summary

The unique measurements of $\sim$1–30 keV suprathermal electrons provided by the WIND 3-D Plasma and Energetic Particle experiment are highly complementary to the energetic particle composition measurements which will be made from ACE. Together with solar observations from SOHO, Yohkoh, and TRACE, these observations will enable comprehensive studies of the origin of solar energetic particles and the tracing of interplanetary structures back to the Sun.

Acknowledgements

This research was supported in part by NASA grant NAG5–2815.

References

Anderson, K. A. and Lin, R. P.: 1966, 'Observations on the Propagation of Solar Flare Electrons in Interplanetary Space', *Phys. Rev. Letters* **16**, 1121.

Bougeret, J.-L., Kaiser, M. L., Kellogg, P. J., Manning, R., Goetz, K., Monson, S. J., Monge, N., Friel, L., Meetre, C. A., Perche, C., Sitruk, L., and Hoang, S.: 1995, 'WAVES: the Radio and Plasma Wave Investigation on the WIND Spacecraft', *Space Sci. Rev.* **71**, 231.

Bougeret, J.-L., Fainberg, J., and Stone, R. G.: 1984, 'Interplanetary Radio Storms', *Astron. Astrophys.* **141**, 17.

Burlaga, L. F.: 1988, 'Magnetic Clouds and Force-Free Fields with Constant Alpha', *J. Geophys. Res.* **93**, 7217.

Burlaga, L. F.: 1991, 'Magnetic Clouds', in R. Schwenn and E. Marsch (eds), *Physics of the Inner Heliosphere II*, Springer-Verlag, Berlin, p. 1.

Dulk, G. A. and McLean, D. J.: 1978, 'Coronal magnetic fields', *Solar Phys.*, **57**, 279.

Ergun, R. E., Larson, D., Lin, R. P., McFadden, J. P., Carlson, C. W., Anderson, K. A., Muschietti, L., McCarthy, M., Parks, G., Rème, H., Bosqued, J. M., d'Uston, C., Sanderson, T. R., Wenzel, K. P., Kaiser, M., Lepping, R. P., Bale, S. D., Kellogg, P., and Bougeret, J. L.: 1998, 'Wind Spacecraft Observations of Solar Impulsive Electron Events Associated with Solar Type III Radio Bursts', *Astrophys. J.* **503**, 435.

Feldman, W. C., Asbridge, J. R., Bame, S. J., Montgomery, M. D., and Gary, S. P.: 1975, 'Solar Wind Electrons', *J. Geophys. Res.* **80**, 4181.

Gosling, J. T.: 1975, 'Large-Scale Inhomogeneities in the Solar Wind of Solar Origin', *Rev. Geophys.* **13**, 1053.

Gosling, J. T., Birn, J., and Hesse, M.: 1995, 'Three-Dimensional Magnetic Reconnection and the Magnetic Topology of Coronal Mass Ejection Events', *Geophys. Res. Letters* **22** (8), 869.

Gosling, J. T.: 1996, 'Corotating and Transient Solar Wind Flows in Three Dimensions', *Ann. Rev. Astron. Astrophys.* **34**, 35.

King, J. H.: 1979, 'Solar Cycle Variations in IMF Intensity', *J. Geophys. Res.* **84**, 5938.

Larson, D. E., Lin, R. P., McTiernan, J. M., McFadden, J. P., Ergun, R. E., McCarthy, M., Rème, H., Sanderson, T. R., Kaiser, M., Lepping, R. P., and Mazur, J.: 1997, 'Tracing the Topology of the October 18–20, 1995 Magnetic Cloud with ~ 0.1–10^2 keV Electrons', *Geophys. Res. Letters* **24**, 1911.

Lepping, R. P., Acuna, M. H., Burlaga, L. F., Farrell, W. M. et al.: 1995, 'The WIND Magnetic Field Investigation', *Space Sci. Rev.* **71**, 207.

Lepping, R. P., Burlaga, L. F., Szabo, A., Ogilvie, K. W., Mish, W. H., Vassiliadis, D., Lazarus, A. J., Steinberg, J. T., Farrugia, C. J., Janoo, L., and Mariani, F.: 1997, 'The WIND Magnetic Cloud and Events of October 18–20, 1995: Interplanetary Properties and as Triggers for Geomagnetic Activity', *J. Geophys. Res.* **102**, 14049.

Lin, R. P.: 1974, 'Non-Relativstic Solar Electrons', *Space Sci. Rev.* **16**, 189.

Lin, R. P.: 1985, 'Energetic Solar Electrons in the Interplanetary Medium', *Solar Phys.* **100**, 537.

Lin, R. P.: 1990, in E. Priest and V. Krishan (eds), 'Electron Beams and Langmuir Turbulence in Solar Type III Radio Bursts Observed in the Interplanetary Medium', *Basic Plasma Processes on the Sun*, International Astronomical Union, The Netherlands, p. 467.

Lin, R. P. and Kahler, S. W.: 1992, 'Interplanetary Magnetic Field Connection to the Sun During Electron Hear Flux Dropouts in the Solar Wind', *J. Geophys. Res.* **97**, 8203.

Lin, R. P., Potter, D. W., Gurnett, D. A., and Scarf, F. L.: 1981, 'Energetic Electrons and Plasma Waves Associated witha Solar Type III Radio Burst', *Astrophys. J.* **251**, 364.

Lin, R. P., Anderson, K. A., Ashford, S., Carlson, C., Curtis, D., Ergun, R., Larson, D., McFadden, J., McCarthy, M., Parks, G. K., Rème, H., Bosqued, J. M., Coutelier, J., Cotin, F., d'Uston, C., Wenzel, K.-P., Sanderson, T. R., Henrion, J., and Ronnet, J. C.: 1995, 'A Three-Dimensional Plasma and Energetic Particle Investigation for the Wind Spacecraft', *Space Sci. Rev.* **71**, 125.

Lin, R. P., Larson, D., McFadden, J., Carlson, C. W., Ergun, E. R., Anderson, K. A., Ashford, S., McCarthy, M., Parks, G. K., Rème, H., Bosqued, J. M., d'Uston, C., Sanderson, T. R., Wenzel, K.-P.: 1996, 'Observation of an Impulsive Solar Electron Event Extending Down to $\sim$0.5 keV Energy', *Geophys. Res. Letters* **23**, 1211.

MacQueen, R. M.: 1980, 'Coronal Transients: A Summary', *Phil. Trans. R. Soc. London, Ser.* **A297**, 605.

McComas, D. J., Gosling, J. T., Philips, J. L., Bame, S. J., Luhmann, J. G., and Smith, E. J.: 1989, 'Electron Heat Flux Dropouts in the Solar Wind: Evidence for Interplanetary Magnetic Field Reconnection?', *J. Geophys. Res.* **94**, 6907.

Pan, L., Lin, R. P., and Kane, S. R.: 1983, 'Comparisons of Solar Flare X- Ray Producing and Escaping Electrons from $\sim$2 to 100 keV', *Solar Phys.* **91**, 345.

Pannekoek, A.: 1922, 'Ionization in Stellar Atmospheres', *Bull. Astron. Inst. Neth.* **1**, 107.

Potter, D. W., Lin, R. P., and Anderson, K. A.: 1980, 'Impulsive 2–10 keV Solar Electron Events Not Associated with Flares', *Astrophys. J.* **236**, L97.

Reames, D. V., von Rosenvinge, T. T., and Lin, R. P.: 1985, 'Solar ^{3}He-Rich Events and Nonrelativistic Electron Events: A New Association', *Astrophys. J.* **292**, 716.

Rosseland, S.: 1924, 'Electrical State of a Star', *Monthly Notices Roy. Astron. Soc.* **84**, 720.

Rust, D. W.: 1994, 'Spawning and Shedding Helical Magnetic Fields in the Solar Atmosphere', *Geophys. Res. Letters* **21**, 241.

Saito, K., Poland, A. I., and Munro, R. H.: 1977, 'A Study of the Background Corona Near Solar Minimum', *Solar Phys.* **55**, 121.

Temerin, M. and Roth, I: 1992, 'The Production of ^{3}He and Heavy Ion Enrichments in ^{3}He-Rich Flares by Electromagnetic Hydrogen Cyclotron Waves', *Astrophys. J.* **391**, L105.

Van Allen, J. A. and Krimigis, S. M.: 1965, 'Impulsive Emission of $\sim$40 keV Electrons from the Sun', *J. Geophys. Res.* **70**, 5737.

Von Rosenvinge, T. T., Barbier, L. M., Karsch, J., Liberman, R. et al.: 1995, 'The Energetic Particles: Acceleration, Composition, and Transport (EPACT) Investigation on the Wind Spacecraft', *Space Sci. Rev.* **71**, 155.

PARTICLE ACCELERATION IN IMPULSIVE SOLAR FLARES

JAMES A. MILLER

Department of Physics, The University of Alabama in Huntsville, Huntsville, AL 35899, U.S.A.

Abstract. We present the major observationally-derived requirements for a solar flare particle acceleration mechanism, briefly discuss some general electrodynamic constraints that also need to be considered, and suggest a unified electron and ion acceleration theory. This theory consists of two elements: cascading MHD turbulence generated at large scales during the primary flare energy release, which is responsible for the energization of electrons and all ions except ^{3}He, and an electron beam, which excites the waves necessary for ^{3}He acceleration. An issue of special importance for understanding ion acceleration is the convincing measurement of the charge state of Fe, which can be accomplished by the *Advanced Composition Explorer* in the upcoming solar maximum.

1. Introduction

Solar flares are among the most energetic and interesting phenomena in the solar system, releasing up to 10^{32} ergs of energy on time scales of several tens of seconds to several tens of minutes. (For comparison, this amount of energy could be used to push the Moon about 50 km further from the Earth.) Much of this energy is in the form of suprathermal electrons and ions, which remain trapped at the Sun and produce a wide variety of radiations (Ramaty and Murphy, 1987) as well as escape into interplanetary space (Reames, 1990). The radiation from trapped particles consists in general of (1) continuum emission, which ranges from radio and microwave wavelengths to soft ($\sim$1–20 keV) X-rays, hard ($\sim$20–300 keV) X-rays, and finally gamma rays (above $\sim$300 keV), which may have energies in excess of 1 GeV; (2) narrow gamma-ray nuclear de-excitation lines between $\approx$ 4 and 8 MeV; and (3) high-energy neutrons observed in space or by ground-based neutron monitors. The particles that escape into space consist of both electrons and ions, which often have compositions quite different than that of the ambient solar atmosphere. Flares thus present many diagnostics of the particle acceleration mechanism(s), the identification of which is the ultimate goal of flare research. Moreover, flares in fact offer the only opportunity in astrophysics to study the simultaneous energization of both electrons and ions. Hopefully, an understanding of flares with their wealth of diagnostic data will lead to a better understanding of particle acceleration at other sites in the Universe (see, e.g., Li and Miller (1997), where a model developed for flares was applied successfully to galactic black hole accretion disks, or Dermer, et al. (1996), for a discussion of active galactic nuclei).

Primarily as a result of the work of Reames and collaborators, it is now generally accepted that flares are roughly divided into two classes: impulsive and gradual

Space Science Reviews **86:** 79–105, 1998.
© 1998 *Kluwer Academic Publishers. Printed in the Netherlands.*

(see Cliver, (1996) for a refinement of this scheme). Gradual events are large, occur high in the corona, have long-duration soft and hard X-rays and gamma rays, are electron poor, are associated with Type II radio emission and coronal mass ejections (CMEs), and produce energetic ions with coronal abundance ratios. Impulsive events are more compact, occur lower in the corona, produce short-duration radiation, and exhibit dramatic abundance enhancements in the energetic ions. Their ^{3}He/^{4}He ratio is ~ 1, which is a huge increase over the coronal value of about 5×10^{-4}, and they also posses smaller but still significant enhancements of Ne, Mg, Si, and Fe relative to ^{4}He, C, N, and O (e.g., Reames et al., 1994). In addition to these elemental enhancements, Ne and Mg have isotopic enhancements as well (Mason et al., 1994).

The general scenario that initially emerged from these observations is that energetic particles in gradual events are accelerated by a CME-driven shock, while those particles in impulsive events are accelerated by another mechanism(s). However, this picture has been modified by Cliver (1996) and Mandzhavidze and Ramaty (1993), who argue that the particles which remain trapped at the Sun in gradual events have impulsive flare abundances, and that the acceleration mechanism(s) which is responsible for impulsive flares is also responsible for energizing these gradual event trapped particles. In other words, gradual events posses an impulsive flare 'core', which is responsible for the trapped particles (but which does not necessarily imply that impulsive flares cause CMEs, a la the 'solar flare myth' controversy (see Great Debate, 1995)). Some of the particles from the core escape, as in pure impulsive events, and join the particles accelerated by the CME-driven shock. In this paper, we will deal only with impulsive flares (hereafter referred to simply as flares), but point out that the discussion is equally appropriate for those particles in gradual events that were not shock accelerated.

The present canonical solar flare picture is that both electrons and ions are accelerated in and above the ionized coronal region of a magnetic loop, which consists of closed field lines that are anchored at both ends in the denser chromosphere and photosphere. The length of the coronal portion of this loop can vary quite a bit, but typically lies in the 10^8 to 10^9 cm range. The average magnetic field and plasma number density of the loop are not well known, but can be roughly constrained by the modeling of microwave and X-ray emission; they are probably around 100 to 500 G and 10^9 to a few times 10^{10} cm^{-3}, respectively. In addition to the closed magnetic field lines, there are overlying open field lines that are anchored at one end in the photosphere but then extend into interplanetary space. Primarily on the basis of some well-publicized X-ray images from *Yohkoh* (Masuda, 1994), which showed a knot of hard X-ray emission just above the top of the loop seen in soft X-rays, it is now believed that the primary site of particle acceleration lies just above the loop top, in a region permeated by both open and closed field lines. (Hard X-ray rays are an indication of highly suprathermal accelerated electrons, whereas soft X-rays are produced by the thermal plasma; see below.) This acceleration region location and geometry naturally leads to both trapped and escaped particles and further predicts that they belong to the same population, which is in fact confirmed by the observed similarity of the ion compositions in the two types of particles (the composition of the escaped ions can be observed directly, while that of the trapped or interacting ions can be inferred by gamma-ray line spectroscopy (e.g, Murphy et al., 1991)).

After acceleration near the top of or throughout the coronal section of the loop, energetic electrons streaming along the magnetic field lines produce microwaves via gyrosynchrotron emission in the corona, and then hard X-rays and continuum gamma rays via bremsstrahlung emission when they strike the dense chromosphere (or sometimes in the corona, as in the so-called Masuda flares mentioned above). Indeed, it is the observation of two footprints of hard X-ray emission in the chromosphere, separated by some 10^8 to 10^9 cm, that was the initial evidence for a magnetic loop geometry. However, since the Coulomb collision energy loss rate is far larger than that for bremsstrahlung, most (about 99.99%) of the energetic electron energy goes into heating the ambient chromosphere. This heated plasma then radiates at soft X-ray wavelengths and expands into the coronal portion of the loop, in a process often called chromospheric 'evaporation' (Mariska et al., 1993). Soft X-rays will also result from the plasma that is directly heated by the energy release or particle acceleration mechanism. In addition, the coronal electron beam excites electron plasma waves, which then undergo mode conversion into radio waves with a frequency of about the local plasma frequency. If the electrons are on closed field lines, this radio emission takes the form of inverted U or J bursts (e.g., Aschwanden et al., 1992), while electrons on open field lines yield coronal and then interplanetary Type III bursts (e.g., Aschwanden and Benz, 1997). This general sequence of events often has been called the 'electron beam model' in the literature, even though it is essentially an overall semi-quantitative scenario and it lacks any kind of acceleration model for producing the beam; it also implies that ions are not major players in the flare acceleration process or energetics, which is understood not to be the case anymore (see below).

Energetic ions either escape directly along open field lines or, if their energy is above about 1 MeV nucl^{-1}, interact with the ambient nuclei (again in the relatively dense chromosphere) to produce excited nuclei, radioactive nuclei, pions, and neutrons. Accelerated protons and alpha particles striking heavy ambient nuclei produce excited nuclei that subsequently de-excite via gamma-ray line emission. This emission is manifested as the narrow gamma-ray lines around a few MeV. The inverse reactions also yield gamma-ray lines, but these are significantly Doppler broadened as a result of the motion of the heavy incident nucleus. The broad lines are not resolvable but contribute to the nuclear emission in the MeV range. Radioactive nuclei can decay via the emission of positrons. The positrons can also remain trapped in the loop for some time, eventually thermalizing and producing the 511 keV positron annihilation line (e.g., Ramaty, et al., 1987). Lastly, if the accelerated ion energy exceeds ≈ 300 MeV nucl^{-1}, pion production can occur. Charged pions decay ultimately into electrons and positrons, which in turn yield bremsstrahlung and positron annihilation-in-flight radiation (Murphy, et al., 1987). Neutral pions decay into two ≈ 70 MeV gamma-rays, which produce a bump or hardening in the photon spectrum near that energy (Forrest et al., 1986).

From the preceding discussion, it is clear that nowhere else in astrophysics is the wealth and breadth of remote diagnostic data so great. The goal of solar flare research is to deduce the particle acceleration mechanism or mechanisms from these observations. It should be acknowledged at once that, since we will not have *in situ* measurements of solar flares, one cannot prove that a particular model or class of models is or is not the one which is actually operating. The best one can do is evaluate a model based on its ability to account for the greatest amount of solar flare data and its simplicity (Occam's Razor), both in terms of assumptions and implications. However, it should also be acknowledged that since the amount and variety of data is so vast, and flares exhibit so much variation, it is not likely that any theory will account precisely (or nearly so) for every individual observation. How then is one to proceed and make some progress? The approach we use is to distill the observations into a few central requirements (both electron and ion related) that any theory must account for. While glossing over some of the details, we demand that a theory account for all of these main facts simultaneously (i.e., that it unify electron and ion acceleration under the aegis of a single unified model). We specifically avoid as much as possible theories which focus exclusively on one or two observations, since they might not account for any other and eventually lead to a collection of unrelated and possibly contradictory proposed mechanisms and no overall scheme.

In the next section, we present the central requirements for a particle acceleration mechanism. In Section 3, we discuss an attractive quasi-unified model which accounts for these core observations. We summarize our results in Section 4 and discuss how observations with the *Advanced Composition Explorer* can clarify further the issue of solar flare particle acceleration.

2. Central Observations

Recent reviews of solar flare observations can be found in Miller et al. (1997) and Vestrand and Miller (1998), and the reader is referred to these papers and references therein for further information and discussion. We briefly note here that most of the information on the number of interacting electrons and their acceleration time scale is provided by fitting hard X-ray bremsstrahlung emissions, while similar information for the interacting ions is obtained by modeling the nuclear de-excitation line and pion radiation emission. Furthermore, while all of the following features may not be present in a given individual flare, they are routinely observed and are essential aspects that should be accounted for by a successful model. What follows are the most prominent features of particle energization during solar flares.

(1) Electrons are accelerated out of the thermal distribution to ≈ 100 keV on time scales of about 1 s and to ≈ 100 MeV on time scales of a few seconds.

(2) $\approx 10^{36}$ to 10^{37} electrons s^{-1} are accelerated above the hard X-ray producing threshold energy of 20 keV for several tens of seconds, yielding a total electron energy content around 10^{31} ergs.

An energy of 20 keV is typically chosen as the transition from heated electrons to accelerated electrons. Note that for typical flare coronal volumes of $\sim 10^{27}$ cm^3 and densities of $\sim 10^{10}$ cm^{-3}, the electron population of the acceleration region will be entirely depleted in about 1 s. Hence, real-time replenishment must occur, and this in turn is a critical item an acceleration mechanism must be consistent with (but which is usually neglected). While this is the number requirement for the entire flare, there is considerable evidence that electron acceleration is fragmented, occurring in multiple smaller units or volumes called energy release fragments (ERFs) by Machado et al. (1993). Some of this evidence is in the form of spikes of ≈ 400 ms duration in the hard X-ray time profiles of moderate-sized flares. Individual spikes are not present in the time profiles of larger flares since they presumably merge together and form a much more smoothly-varying emission. Hence, point (2) can be amended to include:

(2a) Electron acceleration is fragmented into ERFs, which involve the acceleration of about 5×10^{34} electrons s^{-1} above 20 keV for about 400 ms, for a total energy content of $\approx 5 \times 10^{26}$ ergs.

These are the essential observations as far as electron acceleration is concerned; now consider the ions.

(3) Ions must be accelerated from the thermal distribution to ≈ 100 MeV in about 1 s and to ≈ 1 GeV on time scales of a few seconds.

(4) Protons must be accelerated above approximately 1 MeV at a rate of $\approx 10^{35}$ s^{-1} for several tens of seconds, for a total energy content of $\approx 10^{31}$ ergs.

The number of energetic interacting protons below about an MeV is unknown at present, since these particles do not have a detectable gamma-ray line signature and other diagnostics of their presence (such as the polarization of chromospheric Hα emission) have not been fully investigated yet (e.g., Chambe and Hénoux, 1978; Vogt and Hénoux, 1996). This energization rate is more than an order of magnitude higher than believed prior to 1995, when ^{20}Ne and ^{16}O de-excitation line data became available for 19 flares observed with the *Gamma Ray Spectrometer* on the *Solar Maximum Mission* (Share and Murphy, 1995). The observed fluence ratio of these two lines is a good diagnostic of the proton spectrum between ≈ 1 and 10 MeV, and revealed that the proton spectrum was steeper than previously thought but had the same number of particles above the higher energy of 30 MeV, which in turn drove the total number of accelerated protons to a much higher value (Ramaty et al., 1995). The other ions have an energy content comparable to that in the protons, so that the energy contained in the ions is thus about equal to the energy contained in the electrons. The ion and electron energy contents have been determined for each of 12 flares which have the necessary simultaneous hard X-

TABLE I

Ion abundance ratios

Ratio	Impulsive flares	Gradual flares (Corona)
^{3}He/^{4}He	~1 ($\times 2000$ increase)	~0.0005
^{4}He/O	≈ 46	≈ 55
^{4}He/H	≈ 0.5	≈ 0.1
C/O	≈ 0.436	≈ 0.471
N/O	≈ 0.153	≈ 0.128
Ne/O	≈ 0.416 ($\times 2.8$ increase)	≈ 0.151
Mg/O	≈ 0.413 ($\times 2.0$ increase)	≈ 0.203
Si/O	≈ 0.405 ($\times 2.6$ increase)	≈ 0.155
Fe/O	≈ 1.234 ($\times 8.0$ increase)	≈ 0.155
H/He	~10	~100

ray and nuclear gamma-ray line data, and approximate equipartition is confirmed in each instance (see Figure 3 of Miller et al., 1997). This is also a significant change over the previously-held notion that ions are minor particles or in the noise as far as the acceleration process is concerned, and that the only significant problem was the acceleration of the electrons. Hence,

(5) Electron and ion acceleration are not only simultaneous to within ≈ 1 s, there is also a near equipartition in energy between the two particle species.

Ion abundances yield the last central diagnostic of flare particle acceleration. We show in Table I typical ion abundance ratios for particles $\gtrsim 1$ MeV nucl^{-1} for impulsive flares as well as the gradual events with their associated CMEs (data from Great Debate (1995) and Reames et al. (1994, references therein); the impulsive flare ^{4}He/H ratio is from Share and Murphy (1997)). The abundances for the gradual events are essentially the same as those in the ambient corona, and is evidence for the shock accelerated nature of these particles. The abundances for the impulsive events show enhancements of ^{3}He, Ne, Mg, Si, and Fe relative to C, N, O, and ^{4}He, which in turn are not enhanced relative to one another. The abundance ratios place possibly the most stringent constraints on an acceleration theory. Thus,

(6) The ion abundance ratios need to be similar to those given (in the Impulsive Flares column) in Table I.

3. Theory

Particle acceleration can occur generically through interaction with DC electric fields, shocks, or plasma wave turbulence (by turbulence we mean a superposition of a large number of randomly-phased waves occupying a range of wave numbers, propagation angles, and frequencies). We will briefly discuss the first two and then a particular form of the third in more detail.

Perhaps the simplest way to accelerate particles is by a large-scale DC electric field. Most work in this area focuses on the electrons, which we consider first. In addition to the force due to the electric field, an electron also experiences a Coulomb drag force from the other electrons in the distribution. It is the interplay between these two forces that govern whether or not an electron is accelerated out of the thermal distribution. As the speed of an electron increases from zero, the drag force increases, until reaching a maximum at the electron thermal speed v_{te}. Above the electron thermal speed, this drag force decreases with increasing electron speed. The value of the electric field $\mathcal{E}$ where the drag force at the thermal speed equals the electric field force is called the Dreicer field $\mathcal{E}_D$ (Dreicer, 1960) and is given by $\mathcal{E}_D = (e/4\pi\epsilon_0)(\omega_{pe}/v_{te})^2 \ln \Lambda$ V m^{-1}. Here, e is the magnitude of the electron charge, ω_{pe} is the electron plasma frequency, $\ln \Lambda$ is the Coulomb logarithm, and all quantities are in SI units. The Dreicer electric field for typical flare parameters is $\approx 10^{-4}$ V cm^{-1}.

This simple picture is modified somewhat by Coulomb pitch angle scattering and electron/ion collisions (Fuchs et al., 1986). Neglecting these, for $\mathcal{E} > \mathcal{E}_D$ the electric force exceeds the drag force on all electrons, which will then be freely accelerated to higher energies. Such fields are called super-Dreicer. For $\mathcal{E} < \mathcal{E}_D$, there exists a critical velocity v_c, below which the drag force overcomes the electric force. Above v_c, the situation is reversed. Electrons with speeds $< v_c$ will then be heated, while those with speeds $> v_c$ will be freely accelerated or, as is said, 'run away'. Electric fields greater than $\mathcal{E}_D$ are called super-Dreicer, while those less than $\mathcal{E}_D$ are called sub-Dreicer.

Sub-Dreicer acceleration has been considered in detail for several years, with most of this work being applied to laboratory plasmas (see review by Knoepfel and Spong, 1979). Electron distribution functions have been numerically calculated for a variety of electric field values, taking into account both electron/electron and electron/ion Coulomb energy loss rates and pitch angle scattering (e.g., Wiley et al., 1980; Fuchs et al., 1986). However, a detailed simulation is not necessary in order to derive the basic shape and normalization of the accelerated (or runaway) electron distribution. Imagine an electric field of finite length L permeating a plasma. All along the electric field, electrons with parallel velocity (relative to the electric field) above the critical speed v_c will be accelerated up the field lines. Upon reaching the beginning of the field, the energy of an electron will be proportional to the distance it traveled; since an equal number of electrons were accelerated in each small segment of the loop, the electron energy distribution at the start of the field will simply be a flat distribution extending from about v_c (the electrons here were initially close to the beginning) to the maximum kinematic drop $e\mathcal{E}L$ (the electrons at this energy where initially at the end of the field). Coulomb collisions pitch angle scatter the particles and thus increase the effective perpendicular temperature of the distribution somewhat, but this essentially flat spectrum is confirmed nicely by simulations (see also Moghaddem-Taaheri and Goertz, 1990).

For solar flare applications, the sub-Dreicer field must be very long: taking the field to be of order 10^{-5} V cm^{-1}, the length must be about 3×10^9 cm in order to yield a maximum energy around 10 to 100 keV, which would just be sufficient to account for hard X-ray emission. This field is assumed to exist in the coronal section of the flare loop, extending over its whole length. There are three main problems with this simple picture however:

First, when the electrons with their flat electron energy distribution strike the chromosphere, they will produce hard X-ray bremsstrahlung radiation that has an E_γ^{-s} spectrum (units of photons per unit photon energy E_γ) with $s = 1$; however, the observed spectra have power-law indices s greater than 2 (e.g., Dennis, 1985, 1988). Reconciliation of these results would require a suitable distribution of electric field strengths and/or lengths, such that the superposition of many composite E_γ^{-1} spectra (with different high-energy cutoffs) produce what is observed. There is no proposed physical mechanism for obtaining this critical distribution. An alternative is that given by Benka and Holman (1994), who use a single electric field but employ an *ad hoc* escape term that significantly softens the electron distribution to that needed.

Second, even if the field were taken to be Dreicer, the observed size of flares places a limit of about 1 MeV on the electron energy, which is substantially lower than the maximum observed. This situation can be corrected by invoking anomalous resistivity, which amounts to saying that Coulomb collisions are negligible compared with the scattering rate that results from resonance with waves. If this rate is assumed to be very high, the electric field required to accelerate a thermal electron (the effective Dreicer field) will also become much larger than that for the usual Coulomb collision case. Hence, the electric field could be large ($\approx 10^{-2}$ V cm^{-1}, say) and still be sub-(effective)Dreicer. The problem now is generating suitable (electron plasma? lower hybrid?) waves, in the face of ferocious Landau damping.

Third, the requirement that the magnetic field produced by the streaming accelerated electrons not exceed the coronal magnetic field of 100–500 G yields a severe constraint on the geometry of the current channels carrying the electric fields. To see how this constrains the model, consider that a typical large flare produces a current I due to the accelerated electrons of about 2×10^{18} A. For a flare footpoint area of 10^{18} cm^2 and a coronal magnetic field of 500 G, we find from Ampère's law that the radius l of a cylindrical current channel carrying these electrons cannot be greater than a few meters. This implies that the flaring corona must be filamented in such a way that neighboring current channels have oppositely directed electric fields, so that the self-induced magnetic fields due to the accelerated electrons cancel each other out (or the oppositely-directed currents yield a small net current through an Amperian path). For the above parameters, $\approx 10^{12}$ cylindrical current channels are needed. If the currents are in sheets rather than cylinders, $\approx 10^4$–10^6

sheets are required (Holman, 1995) (but it is not clear how to incorporate a sheet into a cylindrical geometry)

The replenishment problem is similar, in that the coronal portion of the loop needs to be replenished by electrons from the electron-rich (i.e., relatively high density) chromosphere. A single electric field will not admit upflowing co-spatial electrons (electrons cannot be driven both ways by the same field), and so again the current needs to be filamented into oppositely-directed electric fields. (A single electric field could draw thermal electrons from the end opposite to where the accelerated electrons are deposited, but no mechanism of current closure has been suggested, and there is still the magnetic field problem.) An additional argument for filamentation is provided by the fact that current cannot be turned on instantaneously due to the self inductance of the current channels. Taking the current turn-on time to be about an acceleration time of 1 s, Emslie and Hénoux (1995) found that the filamentation requirements are about the same as those given above.

The simultaneous formation of this huge number of large-scale oppositely-directed electric fields, each with an aspect (length to width) ratio of nearly 10^7 and existing for some tens of seconds (comparable to the magnetic diffusion time scale), is a formidable issue. Combined with the facts that electrons with energies ~ 100 MeV cannot be obtained (without anomalous resistivity of unknown origin throughout the loop), the basic electron distribution is wrong (without *ad hoc* escape or a superposition of spectra), and the observed ion abundances do not arise (Holman, 1995), we view sub-Dreicer electric fields as a particularly unattractive candidate for flare particle acceleration, even though they have been extensively employed in the past.

Super-Dreicer electric fields occur within the context of reconnection, and are typically several orders of magnitude larger than the Dreicer field. In a recent incarnation of reconnection, Litvinenko (1996, see also references therein) proposed that the open magnetic field lines above the solar flare loop reconnect and generate an electric field of about 10 V cm^{-1} oriented parallel to the solar surface and normal to the footpoint-to-footpoint line along the loop (actually more like an arcade of loops here). The field is in a current sheet of height $\approx 10^9$ cm, length $\approx 10^9$ cm, and width $\approx 10^4$ cm. Since the field exists over a length of about 10^9 cm, the highest energy electrons could be obtained in principal. However, this kinematic maximum is not realized because of rapid particle escape out of the side of the current sheet resulting from a small transverse magnetic field component. But it is precisely this rapid escape which allows this geometry to bypass the filamentation and replenishment problems encountered by the sub-Dreicer field models. Namely, mass inflow in the sides of the current sheet provides the seed electrons for the acceleration, which gain an energy of about 100 keV before escaping from the sides onto open field lines. Some of these electrons propagate to the chromosphere and generate the hard X-rays, while others escape into space. The crucial point is that there is no electric field along these magnetic field lines, so that a co-spatial return can form from the chromosphere to the current sheet and thus ensure charge neutrality and the constant replenishment of electrons in the sheet. The rapid escape also limits the steady-state current in the sheet to values below those that would generate unrealistically-high magnetic fields.

The geometry of this model is thus quite a bit simpler, and can yield the requisite number of hard X-ray producing electrons. However, it is not likely that energetic protons or the highest energy electrons could be obtained, and no mechanism for the generation of the observed abundance enhancements has been proposed.

The second general mechanism is shock acceleration. As pointed out in Miller et al. (1997), this process has several issues that need to be investigated in the context of solar flares before somewhat firm statements can be made regarding its viability. The only fair thing to say at this time is that it cannot be ruled out. However, the ion abundance enhancements will likely be the biggest problem encountered by this mechanism.

Given the preceding discussion, it is not the case that one of these mechanisms will account for all of the main points outlined in the last section. A combination of them might come close, but this does not seem to be the simplest solution, and so far none of them can deal with the ion abundance enhancements. In the next section we will present a paradigm for a unified acceleration model, that accounts for all the main observations with the exception of the ^{3}He/^{4}He ratio, which is well explained by a second (possibly related) mechanism.

3.1. STOCHASTIC ACCELERATION

The last general mechanism whereby the energy of a particle can be increased is stochastic acceleration. Stochastic acceleration may be broadly defined as any process in which a particle can either gain or lose energy in a short interval of time, but where the particles systematically gain energy over longer times. In its earliest form, stochastic acceleration relied on a collection of magnetic inhomogeneities or 'blobs' (Fermi, 1949; Davis, 1956) to energize the particles. A particle could either make a head-on collision with a blob, be adiabatically reflected, and gain energy, or a trailing collision and lose energy; hence, in any interaction, the energy could either increase or decrease. However, since the probability for a head-on collision is greater (because of the larger relative speed between scatterer and particle), there will be a net increase in energy over long time scales. This type of stochastic acceleration has been applied to solar flares to account for some limited aspects of ion acceleration (e.g., Miller, et al., 1990).

The far richer (and more modern) type of stochastic acceleration relies on plasma waves to do the energization. Here, there is a wide choice not only of plasma wave but specific type of wave-particle interaction. There are many different types of plasma waves (see Swanson, 1989) that could be, and have been, employed, but we will consider in detail just the MHD and EMIC modes, since they are employed in the next sections; other modes, like the magnetic blobs, appear suitable for more limited aspects of the flare acceleration problem.

In a hydrogen plasma, there are two important electromagnetic modes which comprise different branches in the $\omega - k$ plane. They are the Alfvén branch, which exists below the hydrogen gyrofrequency Ω_H, and the fast mode branch, which exists below the electron gyrofrequency Ω_e. For wave frequency $\omega \ll \Omega_H$ (the MHD regime), the Alfvén branch has the dispersion relation $\omega = v_A|k_\parallel|$, while the fast mode branch has the dispersion relation $\omega = v_A k$, where v_A is the Alfvén speed and k and $k_\parallel$ are the magnitude of the wave vector $\mathbf{k}$ and its field-aligned component, respectively. MHD Alfvén waves propagating obliquely with respect to the ambient magnetic field $\mathbf{B}_0$ have a linearly polarized electric field normal to $\mathbf{B}_0$ and a linearly polarized magnetic field normal to both $\mathbf{B}_0$ and $\mathbf{k}$. MHD oblique fast mode waves have a linearly polarized electric field normal to both $\mathbf{B}_0$ and $\mathbf{k}$ and a linearly polarized magnetic field normal to $\mathbf{k}$ and the electric field. The wave magnetic field thus has transverse and compressive components with respect to $\mathbf{B}_0$. In each case the electric field can be decomposed into left- and right-handed components. However, for parallel propagation, all waves on the Alfvén branch are left-handed, while all those on the fast mode branch are right-handed.

When the Alfvén branch approaches Ω_H, the phase speed approaches zero and waves in this regime are called H^+ electromagnetic ion cyclotron (H^+ EMIC) waves. When the fast mode branch passes through Ω_H, the phase speed increases. For $\Omega_H \ll \omega \ll \Omega_e$, the waves are usually called whistlers. As the frequency increases still further, the phase speed approaches zero and whistlers become electromagnetic electron cyclotron waves. In a multi-ion plasma, the dispersion relation below the various ion cyclotron frequencies becomes more complicated, and we refer the reader to Miller and Viñas (1993) and references therein for further details.

The next key issue for understanding stochastic acceleration by waves is the nature of a resonant wave-particle interaction. When the wave amplitude is small (as is usually assumed to be the case when waves are used), stochastic acceleration is a resonant process and occurs when the condition $x \equiv \omega - k_\parallel v_\parallel - \ell\Omega/\gamma = 0$ is satisfied. Here $v_\parallel$ and γ are the parallel particle speed and Lorentz factor, Ω is the cyclotron frequency of the particle, and x is referred to as the frequency mismatch parameter. For harmonic numbers $\ell \neq 0$ (gyroresonance), this equation is a matching condition between the particle's cyclotron frequency and the Doppler-shifted wave frequency in the particle's guiding center frame. It means that the frequency of rotation of the wave electric field is an integer multiple of the frequency of gyration of the particle in that frame and that the sense of rotation of the particle and electric field is the same.

The convention we employ is that Ω is always positive and the sign of ℓ depends upon the sense of rotation of the electric field and the particle in the plasma frame: if both rotate in the same sense (right or left handed) relative to $\mathbf{B}_0$, then $\ell > 0$ (normal Doppler resonance); if the sense of rotation is different, then $\ell < 0$ (anomalous Doppler resonance). Hence, when the resonance condition is satisfied, the particle sees an electric field for a sustained length of time and will either be strongly accelerated or decelerated, depending upon the relative phase of the field and the gyromotion. The most effective gyroresonance is $|\ell| = 1$, and $\ell = +1$ is usually referred to as cyclotron resonance. For $\ell = 0$ the resonance condition specifies matching between the parallel components of the wave phase velocity and particle velocity. This resonance is sometimes referred to as the Landau or Cerenkov resonance.

When a particle is in resonance with a single small-amplitude wave, $v_\parallel$ executes approximate simple harmonic motion about the parallel velocity which exactly satisfies the resonance condition (Karimabadi et al., 1992). There is no energy gain on average. The amplitude of the oscillation is proportional to the square root of the wave amplitude, and the maximum energy gain is small. The frequency ω_b of oscillation, or the bounce frequency, is also proportional to the root of the wave amplitude, and is important for the following reason: If $|x| \leq 2\omega_b$, the particle and wave effectively are in resonance. Hence, the exact resonance condition $x = 0$ does not have to be satisfied in order for a strong wave-particle interaction to occur, which immediately implies that large systematic energy gains in a spectrum of waves are possible.

Consider two neighboring waves, i and $i + 1$, where $i + 1$ will resonate with a particle of higher energy than i will. A particle initially resonant with wave i will periodically gain and lose a small amount of $v_\parallel$. If the gain at some time is large enough to allow it to satisfy $|x| \leq \omega_{b,i+1}$, where $\omega_{b,i+1}$ is the bounce frequency for wave $i + 1$, then the particle will resonate with that wave next. After 'jumping' from one wave to the next in this manner, the particle will have achieved a net gain in energy. If other waves are present that will resonate with even higher energy particles, the particle will continue jumping from resonance to resonance and achieve a maximum energy corresponding to the last resonance present. If the wave spectrum is discrete, then the spacing of waves is critical; if the spectrum is continuous (as is almost certainly the case in flares), however, then resonance overlap will automatically occur. Of course, the particle can also move down the resonance ladder, but over long time scales, there is a net gain in energy and stochastic acceleration is the result. This process can be treated by a momentum diffusion equation, and the diffusion coefficients can be calculated using a convenient Hamiltonian approach found in the work of Karimabadi et al. (1992).

A broadband spectrum of waves is thus typically required in order to stochastically accelerate particles to high energies. The exception is acceleration by resonance overlap in a single large-amplitude wave (Karimabadi et al., 1990). In this interesting process, a particle resonates with the same wave but through many harmonic numbers, and huge energy gains are possible. However, the importance of such acceleration in flare plasmas has not been considered in detail at the present time. We will be considering a spectrum of waves.

While acceleration can result from any harmonic number, there are two specific ones that are most important. The first is $\ell = +1$, or cyclotron resonance. From the resonance condition given above, we readily see that in order for a wave to cyclotron resonate with a low-energy particle (say near the thermal speed), the wave must have a frequency near the cyclotron frequency of the particle. On the other hand, as the particle gains energy, the frequency of the resonant wave can either increase or decrease from the cyclotron frequency. Higher order gyroresonances can sometimes occur as well, but produce an acceleration rate much lower than that for $|\ell| = 1$. The second important resonance is the $\ell = 0$ resonance, which leads to ordinary Landau damping if the resonance is with the wave parallel (with respect to the ambient magnetic field) electric field, or transit-time damping if the resonance is with the wave parallel magnetic field. Transit-time damping can produce huge particle energy gains.

There is one last point that is crucially important for stochastic acceleration in the flare environment: it does not suffer from the replenishment or filamentation issues discussed above for the electric fields, since the acceleration is not directed and a deficit of electrons or protons in the corona will quickly establish an electrostatic field that will draw a co-spatial return current from the electron- and proton-rich chromosphere (e.g., van den Oord, 1990).

The mechanism that we believe is most promising for flare particle acceleration is cascading MHD turbulence. This process is able to account for the salient constraints given in Section 2, is conceptually simple, and unifies electron and ion acceleration within a single overall process. It cannot deal, however, with the ^{3}He/^{4}He ratio. It appears here, and in general, that the ^{3}He/^{4}He enhancement requires a totally separate mechanism which preferentially excites plasma waves in the vicinity of the ^{3}He cyclotron frequency Ω_3, which was the original proposal of Fisk (1978). We consider the ^{3}He issue first.

3.1.1. ^{3}He Acceleration

The best theory from the initial set of theories for this enhancement is that of Fisk (1978), which accounts for the preferential (pre)acceleration of ^{3}He by electrostatic ion cyclotron (EIC) waves in the vicinity of Ω_3. These waves are generated by an electron current, the existence of which is certainly a reasonable assumption for the solar flare environment, and interact with ^{3}He via the cyclotron resonance. However, wave excitation requires a large enhancement of ^{4}He/H in the ambient plasma. Zhang and collaborators (e.g., Zhang, 1995) have proposed variants on this model which bypass the ^{4}He/H issue, but which still lead to small ^{3}He energy gains. A remaining issue for these electron current models (which may or may not be a problem) is one of consistency: An electron current will generate an azimuthal magnetic field, which is not taken into account in the EIC wave dispersion relation. The most promising theory to date was first suggested by Temerin and Roth (1992), who employed an electron beam (as opposed to a current) and argued that it should excite H$^+$ EMIC waves around Ω_3. Furthermore, the maximum energy of the ions is several tens of MeV, so that a secondary acceleration mechanism does not need to be assumed.

Their idea was elaborated upon by Miller and Viñas (1993) and Miller, et al. (1993a, b, 1998). They performed a linear Vlasov stability analysis of a nonrelativistic electron beam, and showed that it will excite H^+ EMIC as well as shear Alfvén waves in a realistic H–He flare plasma. The EMIC waves exist between the H and ^{4}He cyclotron frequencies and typically propagate almost perpendicular to the magnetic field; the shear Alfvén waves exist below the ^{4}He cyclotron frequency and assume the usual $\omega = v_A |k_\parallel|$ dispersion relation at low frequencies (v_A is the Alfvén speed).

An example of this is given in Figure 1, where we show some of the characteristics of the excited EMIC waves. We assumed that the plasma consists of fully ionized H and ^{4}He (^{4}He/H = 0.1) at 3×10^6 K, and contains a 5 keV electron beam (modeled by a drifting Maxwellian distribution with drift speed $0.14c$) whose density is 10% that of the background electrons. The H density was taken to be 10^{10} cm^{-3} and the ambient magnetic field 100 G. At a given propagation angle, waves are typically excited over a range of wave numbers or frequencies. Figure 1(a) shows the maximum frequency ω_m of the excited EMIC waves (in units of the H cyclotron frequency Ω_H) as a function of propagation angle θ with respect to the ambient magnetic field. The dotted line is the ^{3}He cyclotron frequency. For all angles, this frequency lies between the H and ^{4}He cyclotron frequencies. The minimum frequency of the excited waves at all propagation angles is the cutoff frequency for this wave mode, which is about $0.6\Omega_H$. Figure 1(b) shows the maximum growth rate of the waves γ_m as a function of propagation angle; the growth rate is a maximum at $87°$ and hence most of the EMIC waves that are excited will be quasi-perpendicular. Figure 1(c) gives the frequency of the waves with the maximum growth rate. From Figures 1(b) and 1(c) we see that the waves with the highest growth rate, and thus the ones that will be mostly excited, lie almost exactly on the ^{3}He cyclotron frequency. We thus expect ^{3}He to suffer very strong selective cyclotron acceleration from these waves, which is indeed verified using test particle and quasi-linear simulations.

The shear Alfvén waves are excited around the Fe, Ne, Mg, and Si cyclotron frequencies, and will therefore preferentially accelerate these ions by cyclotron resonance. We demonstrate this in Figure 2, where we show the frequency range of the unstable Alfvén waves for a plasma identical to that employed in Figure 1. We give in Figure 2(a) the maximum frequency of the excited Alfvén waves as a function of propagation angle, and show the cyclotron frequencies for the C, N, and O group, the Ne, Mg, and Si group, and Fe by dashed lines. It is clear that no waves are generated over the C, N, and O cyclotron frequencies, so that these waves will suffer no energization from the beam-generated Alfvén waves. However, there are unstable waves around the other cyclotron frequencies for basically all propagation angles, enabling these ions to be accelerated from the background thermal distribution. Figure 2(b) shows the maximum growth rate of the unstable waves, and Figure 2(c) gives the wave frequency at this maximum growth rate. Notice that the waves with the highest growth rate reside almost exactly on the Fe

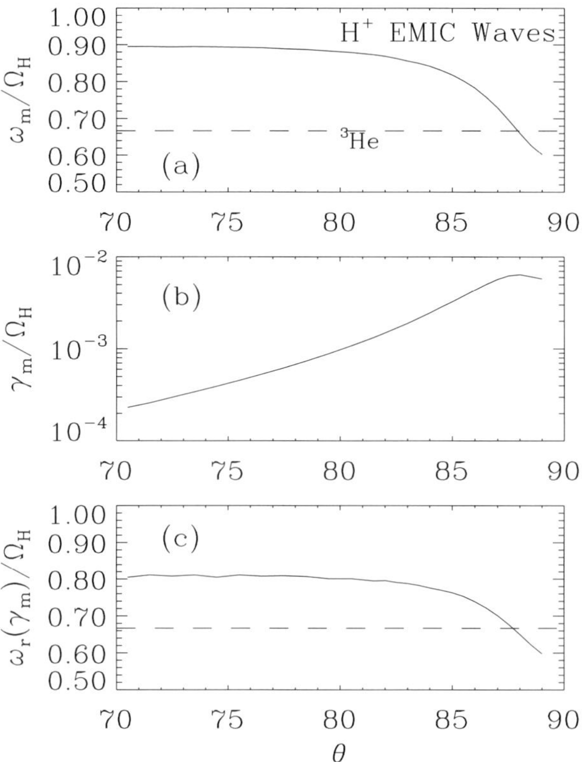

Figure 1. Properties of the H$^+$ EMIC waves excited by a nonrelativistic electron beam in a typical solar flare plasma. (a), (b), and (c) show, respectively, the maximum frequency, the maximum growth rate, and the frequency at the maximum growth rate of the unstable (excited) waves. Strong excitation near the ^{3}He cyclotron frequency (dashed line in (a)) is evident.

cyclotron frequency; we would therefore expect this ion to be enhanced the most in the heavy ion set of Ne, Mg, Si, and Fe (in accordance with observations). While the waves with the maximum growth rate are never on the Ne, Mg, and Si cyclotron frequencies, there are waves excited in this range (as seen in panel (a)); hence, these ions will be accelerated as well, albeit not as strongly as Fe.

Acceleration of both ^{3}He and the heavy ions was examined with a Hamiltonian and quasi-linear analysis, and the maximum ion energies were in the tens of MeV nucleon^{-1} range. A secondary acceleration mechanism is therefore not needed. Moreover, these wave modes are excited even if the electron beam is neutralized by counterstreaming electrons (which may be the case in flares because of the need for replenishment of the coronal acceleration region) and there is no net current.

3.1.2. *Cascading MHD Turbulence*

While the electron-beam model can account nicely for ^{3}He preferential acceleration (as well as the heavy ions), it does not address the acceleration of H, ^{4}He, C, N, and O. It is important to realize that while these ions do not possess any of the

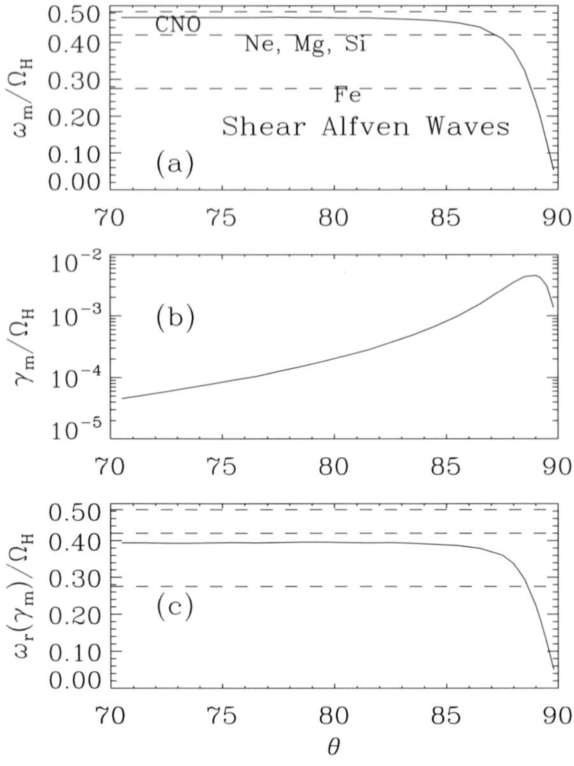

Figure 2. Properties of the shear Alfvén waves excited by a nonrelativistic electron beam in a typical solar flare plasma. (a), (b), and (c) show, respectively, the maximum frequency, the maximum growth rate, and the frequency at the maximum growth rate of the unstable waves. Strong excitation near the Fe cyclotron frequency, and more mild excitation near the Ne, Mg, and Si cyclotron frequencies, is evident. No waves are generated about the C, N, and O cyclotron frequencies. Selective acceleration of the heavy ions (especially Fe) will therefore result.

spectacular enhancements seen in other ions, they are nevertheless accelerated in large numbers. We wish to stay away from either (1) requiring one mechanism for those ions heavier than O, and another mechanism for H, ^{4}He, C, N, and O; or (2) invoking the electron-beam model as a selective pre-acceleration mechanism for the ions heavier than O and then saying that all ions are further accelerated by something else. In either case, the number of parameters is uncomfortably large in our view and the resulting mix of mechanisms will not be compelling. We thus now think that heavy ion acceleration is mostly caused by another process (but the electron-beam generated waves can still contribute).

The heavy ions along with the electrons are accelerated by MHD waves, which can also further accelerate ^{3}He. (We are somewhat loose, for simplicity of nomenclature, in our use of the term MHD here; we are referring to Alfvén and fast mode waves below Ω_H.) Specifically, everything heavier than and including ^{4}He is accelerated by cascading MHD Alfvén waves, electrons are accelerated by cascading

MHD fast mode waves (the 'twin' of the Alfvén waves), and protons are accelerated by mode converted Alfvén waves above the ^{4}He cyclotron frequency. The essential thing to realize is that there are just two main ingredients in this overall scheme: (1) an electron beam at some point in the flare for ^{3}He and (2) MHD waves generated at some large scale for everything else.

The MHD waves accelerate particles in the following scenario:

(1) During the primary flare energy release phase, magnetic free energy is released as the magnetic field undergoes large-scale restructuring, such as relaxation of twisted nonpotential magnetic fields. This is typically thought to be the case, but is a model assumption.

(2) This restructuring leads to the formation of MHD (Alfvén and fast mode) waves of wavelength comparable to the size of the flare. MHD waves should be formed whenever the magnetic field is perturbed on large scales, but this is also an assumption.

At this point, there are long-wavelength, low-frequency waves present. The Alfvén waves are not able to cyclotron resonate with the thermal ions, and so will suffer no dissipation. By virtue of their parallel magnetic field (see above), the fast mode waves can interact via the $\ell = 0$ resonance with super-Alfvénic electrons (of which there are a large number since v_{te} is comparable to v_A) in a process known as transit-time acceleration (Miller, et al., 1996; Miller, 1997; Lenters and Miller, 1998). Transit-time acceleration is essentially resonant Fermi acceleration, and yields an acceleration rate directly proportional to the mean wave number of the fast mode waves. Since this is initially low, the electron acceleration rate is tiny, and so electrons above the threshold of v_A cannot be accelerated against Coulomb collisions. The wave damping rate is very small as well, so that fast mode waves suffer no initial dissipation either.

(3) In the absence of dissipation, the MHD waves cascade to small wavelengths. This is not an assumption since it is well established that cascading will occur via wave steepening (e.g., Zhou and Matthaeus, 1990; Verma et al., 1996). The wave-number region in which they freely cascade with a minimum of dissipation is called the inertial range. The inertial range spans a wide range of wave numbers and the spectral density therein is a power law of form $k^{-5/3}$ for Kolmogorov-like cascading, which is likely the case in flares.

We will now follow the fast mode and Alfvén waves separately.

(4a) As the mean wave-number of the fast mode waves increases as a result of the cascading, the electron acceleration rate increases as well. Eventually, when the waves have cascaded out to sufficiently high wave-numbers, the acceleration rate will exceed the Coulomb energy loss rate above the Alfvén speed v_A. At this point, electrons can be accelerated directly out of the tail and to relativistic energies on subsecond time scales.

(4b) As the shear Alfvén waves cascade from low to high frequency, they are able to cyclotron resonate with ions of progressively lower energy. Eventually, they will be able to resonate with ions in the tail of the thermal distribution, overcome

Coulomb collisions there, and accelerate the ions out of the tail and to relativistic energies (remember that high energies require low frequencies, which are already present in the wave spectrum).

In the flare plasma, the waves will encounter Fe first. Iron will be strongly accelerated but is not abundant enough to damp the waves. Thus, some wave energy will cascade to higher frequencies where it encounters Ne, Mg, and Si (the Ne group; see Figure 2). In the same way, these ions suffer strong acceleration but the wave dissipation is not complete. Some wave energy cascades to reach ^{4}He, C, N, and O (the He group). These ions are sufficiently numerous to totally damp the waves and the cascade ceases. Iron will resonate with the most powerful waves; the Ne group will resonate with waves having less power; and the He group will resonate with the least powerful waves. Hence, Fe should be enhanced more than the Ne group relative to the He group. Since ^{4}He, C, N, and O all have the same cyclotron frequency and behave similarly, they should not be enhanced relative to each other. This necessary layering of the cyclotron frequencies will occur in a plasma of temperature $\approx 3 \times 10^6$ K, which is the initial temperature argued for on this basis by Reames et al. (1994) and Meyer (1996). Hence, cascading Alfvén waves can qualitatively account for heavy ion enhancements.

The last issue is proton acceleration, which should be the easiest (one might think) but which perhaps involves the most complicated process. The problem is that, since the Alfvén wave cascade stops at ^{4}He, none of the initial Alfvén wave energy will directly reach the H cyclotron frequency. Hence, protons will not be cyclotron accelerated out of the tail. There is, however, a natural way around this. Between the ^{4}He and H cyclotron frequencies, there will be only fast mode waves (these are not totally dissipated by the electrons until slightly above $\Omega_{\rm H}$), and these should readily and partially mode convert to Alfvén waves of comparable frequency. It is these mode converted Alfvén waves that can interact with ambient protons and energize some from the tail; the cascading waves at lower frequencies can then accelerate them to relativistic energies. Hence,

(5) Fast mode waves between the ^{4}He and H cyclotron frequencies mode convert to Alfvén waves, which then accelerate the protons.

At present, we are conducting MHD simulations in order to determine this mode conversion efficiency. Without it, we cannot perform a comprehensive simulation of ion and electron acceleration. However, we can illustrate the two main aspects of this scenario: the wave cascading process and heavy ion acceleration.

To illustrate the first, consider MHD wave production and cascading in a H plasma. Here, we are neglecting all ions heavier than H. We assume that the H density and magnetic field of the plasma are 10^{10} cm^{-3} and 500 G, respectively, so that the ambient magnetic field energy density $U_B \approx 10^4$ ergs cm^{-3} and $\Omega_{\rm H} = 4.8 \times 10^6$ s^{-1}. Both the fast mode and Alfvén waves are excited at a wavelength of 10^7 cm and with a volumetric energy production rate of $Q = 8 \times 10^{-9} U_B \Omega_{\rm H} \approx 380$ ergs cm^{-3} s^{-1}. (The results, however, do not depend on the scale at which the waves are generated (see Miller and Roberts, 1995).) The process of particle acceleration is explored with a quasi-linear simulation. Protons and electrons are described by a Fokker–Planck equation that takes into account Coulomb collisions along with wave-particle resonance and acceleration; escape from the acceleration region of scale size L is described by a leaky-box loss term with an escape time scale equal to the transit time across the region L/v. Here, $L = 3 \times 10^8$ cm. The evolution of both wave species is described by a wave diffusion equation that takes into account nonlinear cascading and wave damping by particle acceleration. (For these basic equations, see Miller and Roberts (1995) and Miller et al. (1996); the treatment here is similar to the studies in those papers, except now we treat both electrons and protons at the same time with consistent parameters and include a return current.) In addition, we also model a co-spatial return current by reinjecting the number of particles that escape in a time interval back into the acceleration region in the form of a drifting Maxwellian distribution. In this way, the acceleration region is constantly replenished, as is expected to occur.

In Figure 3 we show the proton and electron energy distributions $N(E)$ (units of particles per unit energy per cm^{-3}) and the Alfvén and fast mode spectral densities W_t (units of wave energy density per unit wave number) for time $t = 0$ to $5 \times 10^5\ T_{\mathrm{H}}$, where $T_{\mathrm{H}} = \Omega_{\mathrm{H}}^{-1}$. Wave injection begins at $t = 0$. During this time interval, the waves cascade through the inertial range and just reach the dissipation range at high k; at this point, they begin accelerating particles and a nonthermal tail develops.

The evolution of the system from $t = 5 \times 10^5$ to $10^6\ T_{\mathrm{H}}$ is given in Figure 4. This time interval is marked by quasi-steady wave cascading and a progressive growth of both the proton and electron nonthermal tails. Note that both types of particle are already accelerated well past an MeV. Equilibrium has not been attained yet, since particle escape from the acceleration region is just beginning to become important.

From $t = 10^6$ to $5 \times 10^6\ T_{\mathrm{H}}$ (about 1 s), the rate of particle escape increases and so does the magnitude of the co-spatial return current. We do not show this interval since the development of the quantities is not significantly different than in Figure 4. In Figure 5, we show the energy distributions and spectral densities for $t = 5 \times 10^6$ to $10^7\ T_{\mathrm{H}}$ (about 2 s). Shortly after $5 \times 10^6\ T_{\mathrm{H}}$, the system reaches equilibrium. At this point, the rate of wave energy density injection equals the rate at which the particles absorb the energy density from resonant wave-particle interactions and the resulting stochastic acceleration, and the rate of particle escape is balanced by the return current that replenishes the region. The acceleration region will remain in this state until the waves are no longer generated at large scales. We have not shown the energy distribution of the escaped particles for the sake of space, but they are sightly harder than the distributions of trapped particles $N(E)$.

We summarize this simulation in Figure 6, where we show the volumetric rate F_e at which particles above a specified energy leave the acceleration region and the time-integrated number density N_e of particles above a specified energy. The escaping flux of both electrons and protons quickly reaches a steady-state value after the beginning of the turbulence injection. Taking the acceleration region vol-

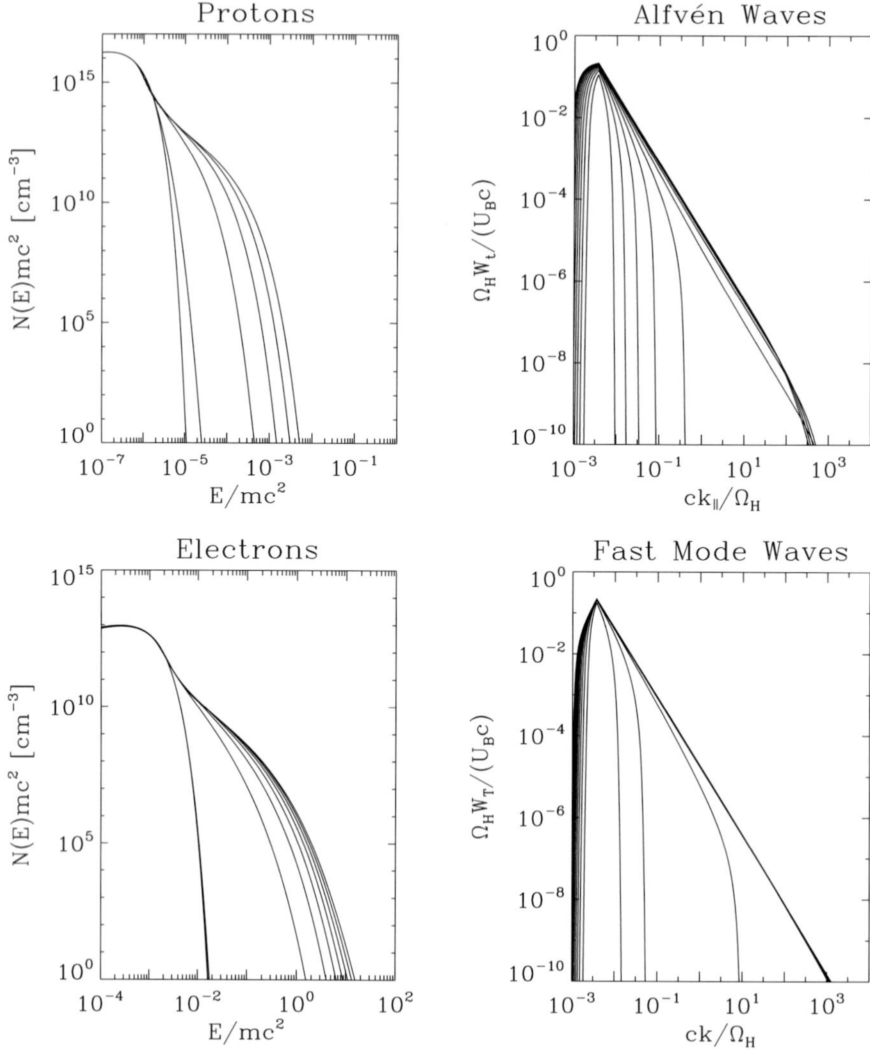

Figure 3. Particle energy distributions $N(E)$ and wave spectral densities W_t for 10 equally spaced times between $t = 0$ to 5×10^5 T_H (about 0.1 s). The leftmost proton and electron distribution are at the initial time (where both are Maxwellians at 3×10^6 K), while those at progressively later times march out to the right. The same for the wave spectral densities. The quantity m refers to either the proton or electron mass. Alfvén waves cascade primarily along the magnetic field, and so $k_\parallel$ is shown, while fast mode waves are assumed to cascade isotropically in wave number.

ume to 10^{27} cm^3, the rate at which protons above 1 MeV escape and thus interact to produce gamma-ray lines is $\approx 7 \times 10^{34}$ s^{-1}, while the rate at which electrons above 20 keV escape and interact to produce hard X-rays is $\approx 2 \times 10^{36}$ s^{-1}. The rates are consistent with those given in Section 2.

While we have shown the results for a single plausible set of parameters, they are very similar for a broad range of acceleration region conditions. Moreover,

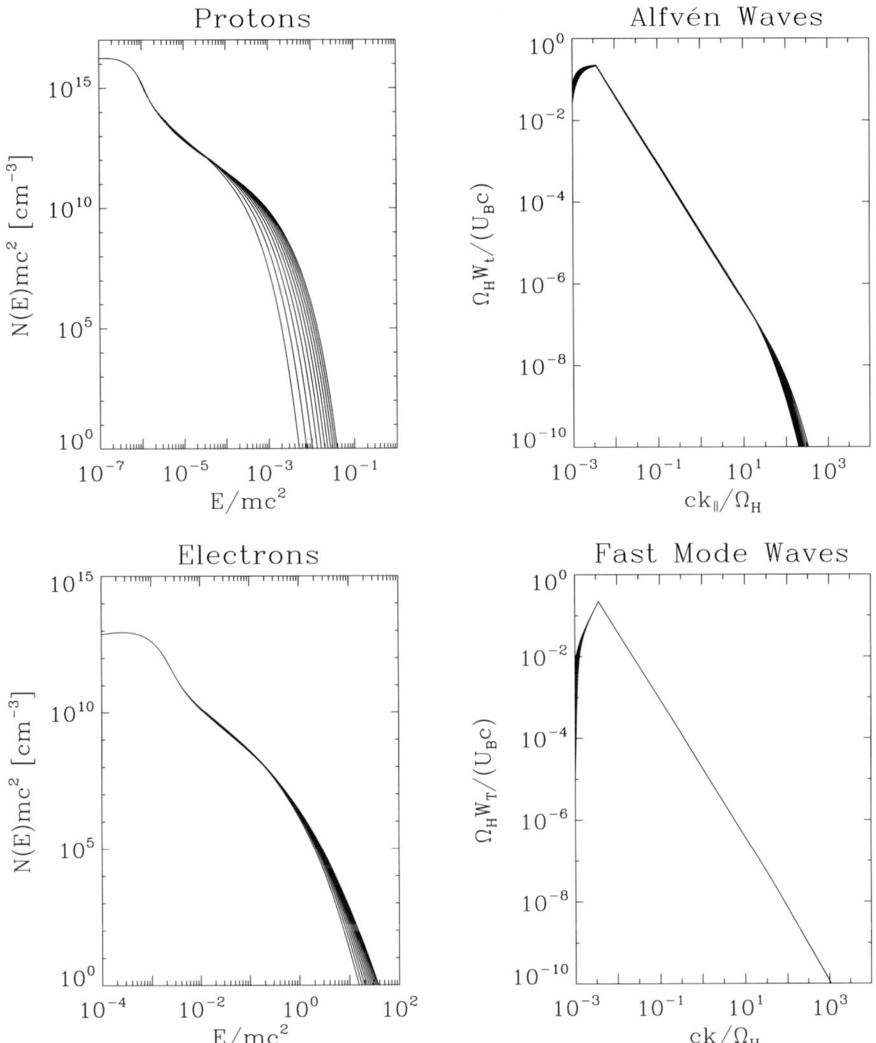

Figure 4. Particle energy distributions $N(E)$ and wave spectral densities W_t for 10 equally spaced times between $t = 5 \times 10^5$ to 10^6 T_{H} (about 0.2 s). The leftmost proton and electron distribution are at the initial time, while those at progressively later times march out to the right. The fast mode spectral density shows no basic change. The Alfvén spectral density at the initial time is the one that extends to highest $k_{\parallel}$; the spectral densities at progressively later times move to slightly lower $k_{\parallel}$ as a result of cyclotron damping by protons.

this model naturally gives equipartition between the energy in accelerated protons and electrons, as required by the observations discussed in Section 2. Hence the cascading MHD turbulence model is easily able to account for the essential features of electron and proton acceleration in flares.

The next general feature of the cascading model that we will illustrate is heavy ion acceleration. Now we will neglect the fast mode waves and the electrons and

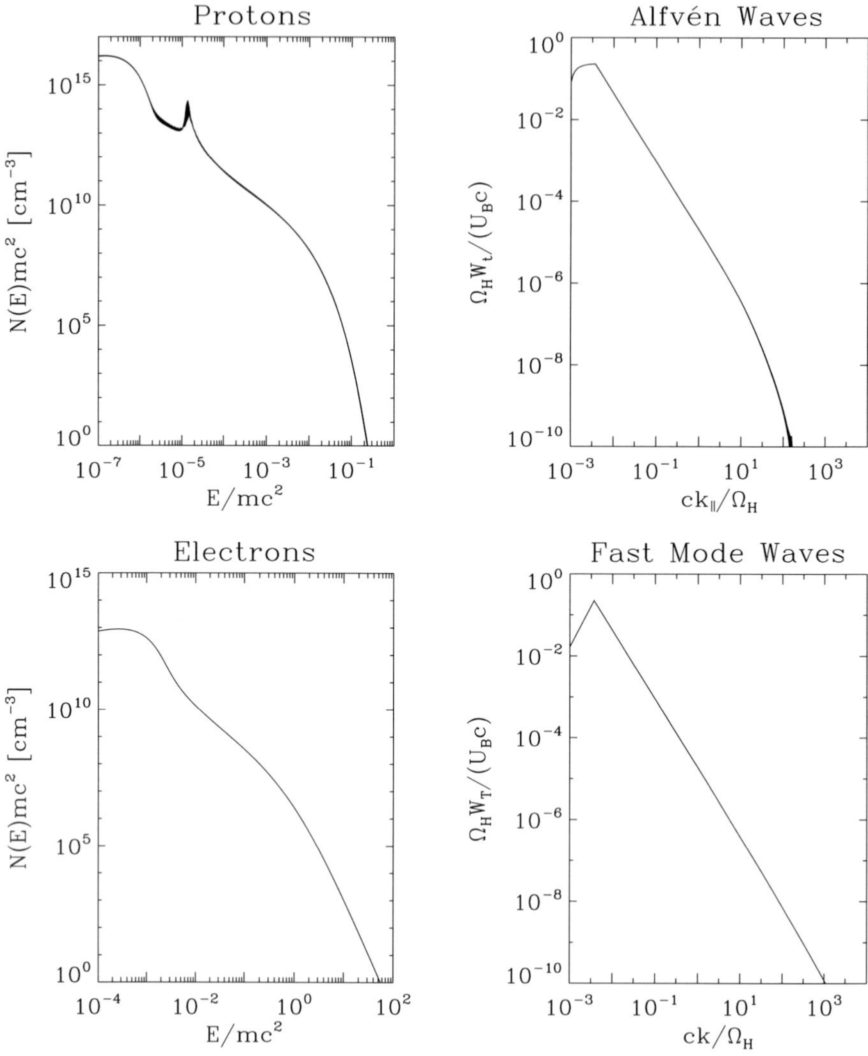

Figure 5. Particle energy distributions $N(E)$ and wave spectral densities W_t for 10 equally spaced times between $t = 5 \times 10^6$ to 10^7 $T_{\rm H}$ (about 2 s). The system is in equilibrium now. The feature in the proton energy distribution at $E \approx 10^{-5}$ mc^2 is due to the return current.

consider just the cascading Alfvén waves. Since the waves will be totally dissipated at ^{4}He, and we have not constrained the mode conversion process discussed above, we also neglect the protons. We quantitatively explore this ion acceleration process with a quasi-linear simulation (described in more detail in Miller and Reames, 1996, 1997) that includes wave cascading, ion escape, cyclotron acceleration, and consequent wave damping. That is, it is almost identical to the treatment above except that we follow more ions species now.

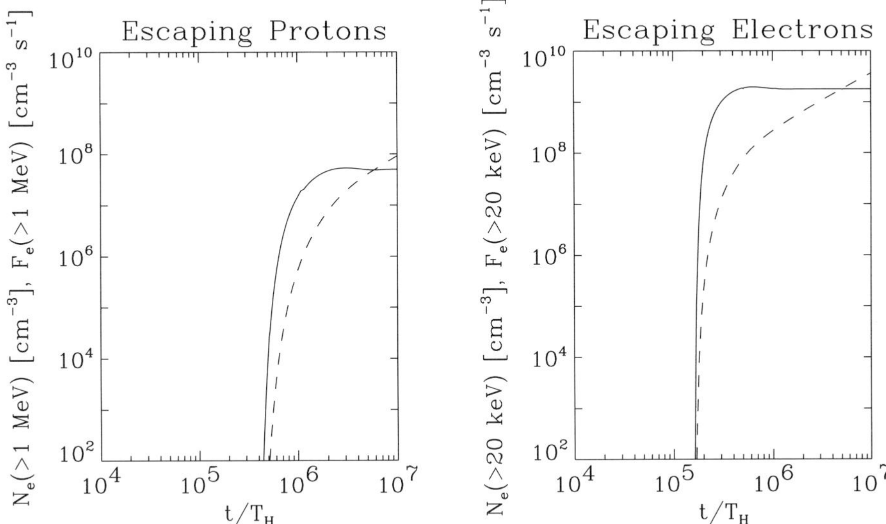

Figure 6. The time-integrated number density N_e of protons above 1 MeV and electrons above 20 keV that have escaped from the acceleration region (*dashed lines*). The volumetric rate at which protons above 1 MeV and electrons above 20 keV escape from the acceleration region (*solid lines*).

Taking parameters (acceleration region length, turbulence energy density injection rate, density, magnetic field) almost identical to those in the electron/proton simulation above, and the ambient relative abundances to be coronal, we find an Fe/O ratio of about 2.5, and Ne/O, Mg/O, and Si/O ratios of about 1.0. The ^{4}He, C, N, and O ions are not enhanced relative to each other. It appears that the length of the acceleration region needs to be near the small end of flare sizes to achieve (approximately) the correct ratios. These ratios are still somewhat too high, but are not extremely different than the values of ≈ 1.2 and ≈ 0.4 given in Table I. Given the relatively small number of model parameters, this appears for now to be an attractive heavy ion acceleration theory.

A few points need to be acknowledged now. While the theory is successful in accounting for the major features of particle acceleration in flares, the resulting particle spectra are not quite immediately consistent with observations. The electron and ion spectra are too hard at low energies and too soft at high energies. However, the size of the acceleration region assumed in the simulations is somewhat small, and could be interpreted as an ERF. Since a flare consists of many ERFs, the distribution that is observed in space or inferred from gamma-ray observations will be a superposition of these elementary spectra. Whether power-law distributions can readily be obtained by reasonable superpositions of these elementary distributions needs to be investigated further.

Second, while the two major components of the overall unified theory work very well separately, they need to be merged next. As discussed above, this requires that we know the rate at which fast mode waves are converted to Alfvén waves between the ^{4}He and H cyclotron frequencies. This is presently being investigated. Third, the formation of a (bump-on-tail) electron beam at some point in the flare needs to be explained. Bump-on-tail distributions readily result from transit effects, but we do not think at present this will be able to account for the stronger beams required for H$^+$ EMIC wave production.

Next, the initial temperature of the plasma was 3×10^6 K, which yields an Fe charge state of about +13. However, it has been observed (Luhn et al., 1987) that the charge state of energetic Fe is about +20. Meyer (1996) has called this result into question, but the convincing measurement of the Fe charge state along with those of the other heavies is crucial to the establishment of this overall acceleration mechanism. This point can be settled with measurements from the *Advanced Composition Explorer.*

Lastly, there is still the underlying ultimate issue of how the MHD waves are generated. A similar formation question confronts every solar acceleration theory, be it stochastic, DC electric field, or shock, and rests upon how energy is released in the first place. Possibly the fairly firm establishment of this or any other model will assist in answering this question.

4. Conclusion

Solar flares place severe constraints on any particle acceleration mechanism as a result of the large number of particles that must be accelerated from thermal to relativistic (even ultrarelativistic) energies on time scales of about a second, as well as the abundance enhancements present in ions heavier than H. We have outlined a unified electron and ion acceleration theory based on cascading MHD waves and electron-beam excited H$^+$ EMIC waves that has been shown to be able to account for the major features of this particle energization.

A couple of theoretical issues still need to be addressed before the model is totally compelling (see above), and the most pressing observational issues are (1) the charge state of Fe and (2) the presence of isotopic abundance enhancements. Observations by the *Advanced Composition Explorer* in the upcoming solar maximum will shed considerable light on both points. Firstly, should charge states consistent with a temperature of about 3×10^6 K be found, this will confirm a model prediction. If the earlier charge states more indicative of a 10 to 20×10^6 K plasma are obtained, this could pose a problem not only for this model but for any ion acceleration mechanism. Either the ambient plasma actually is this hot, in which case a mechanism (much different than those described here) that can strongly differentiate between ions of the same charge-to-mass ratio must be found, or the plasma is cooler and Fe somehow suffers further stripping during or after acceleration (see Meyer, 1996). The first possibility can be settled by examining the charge states of Ne, Mg, and Si, which should be fully stripped at the higher temperature. The second possibility is not really a significant problem, since a mechanism for stripping was proposed by Miller and Viñas (1993); here, the electron beam that excites the H^+ EMIC waves further collisionally ionizes Fe to the observed charge state.

Secondly, the cascading Alfvén wave model predicts isotopic abundance enhancements (of a magnitude as yet undetermined). Specifically, isotopes with a lower charge-to-mass ratio should be enhanced relative to those with a higher ratio, for exactly the same reason that Fe is enhanced relative to C, N, and O: namely, they will encounter waves of higher power and thus be accelerated more strongly. These enhancements need to be calculated and compared with those observed by *ACE*.

A significant amount of progress has been made in understanding particle acceleration in impulsive solar flares, with their wealth and broad array of striking data. Ultimately, one would need *in situ* measurements of the flaring region in order to confirm or rule out an acceleration mechanism. Since this is clearly out of the question, the next best thing is to use a combination of Occam's Razor and remote diagnostic data (such as hard X-rays, gamma-rays, and the energetic particles themselves after they have escaped from the energization region) in order to evaluate proposed models. The model with the minimum of assumptions and with the broadest agreement with the available data wins, at least for a while. But this process is not static and requires constant comparison with the latest data, since some particular data might be the one to invalidate the theory. Only after some time and many passed tests will the theory be considered to be 'the' flare acceleration mechanism. At present, the cascading MHD turbulence model is on the right track, and has gone much further than others. However, the points mentioned above and in the last section need to be addressed next. With energetic particle spectra and charge states from the *Advanced Composition Explorer*, and hard X-ray spectral imaging from the *High Energy Solar Spectroscopic Imager* (HESSI), this and any other model will be able to be tested with unprecedented precision.

References

Aschwanden, M. J., Bastian, T. S., Benz, A. O., and Brosius, J. W.: 1992, *Astrophys. J.* **391**, 380.
Aschwanden, M. J. and Benz, A. O.: 1997, *Astrophys. J.* **480**, 825.
Benka, S. G. and Holman, G. D.: 1994, *Astrophys. J.* **435**, 496.
Chambe, G. and Hénoux, J.-C.: 1978, *Astron. Astrophys.* **80**, 123.
Cliver, E. W.: 1996, in R. Ramaty, N. Mandzhavidze, and X.-M. Hua (eds.), *High Energy Solar Physics*, AIP, New York, p. 45.
Davis, L.: 1956, *Phys. Rev.* **101**, 351.
Dennis, B. R.: 1985, *Solar Phys.* **100**, 465.
Dennis, B. R.: 1988, *Solar Phys.* **118**, 49.
Dermer, C. D., Miller, J. A., and Li, H.: 1996, *Astrophys. J.* **456**, 106.
Dreicer, H.: 1960, *Phys. Rev.* **117**, 329.
Emslie, A. G. and Hénoux, J. C.: 1995, *Astrophys. J.* **446**, 371.
Fermi, E.: 1949, *Phys. Rev.* **75**, 1169.

Fisk, L. A.: 1978, *Astrophys. J.* **224**, 1048.

Forrest, D. J., Vestrand, W. T., Chupp, E. L., Rieger, E., Cooper, J., and Share, G. H.: 1986, *Adv. Space Res.* **6** (6), 115.

Fuchs, V., Cairns, R. A., and Lashmore-Davies, C. N.: 1986, *Phys. Fluids*, **29** 2931.

Great Debates in Space Physics (articles by Miller, J. A., Hudson, H. S., and Reames, D. V.): 1995, *Eos Trans. AGU*, **76** (41), 401.

Holman, G. D.: 1995, *Astrophys. J.* **452**, 451.

Holman, G. D. and Benka, S. G.: 1992, *Astrophys. J.* **400**, L79.

Karimabadi, H., Krauss-Varban, D., and Terasawa, T.: 1992: *J. Geophys. Res.* **97**, 13853.

Karimabadi, H., Akimoto, K., Omidi, N., and Menyuk, C. R.: 1990, *Phys. Fluids* **B2**, 606.

Karimabadi, H., Omidi, N., and Gary, S. P.: 1994: in J. L. Burch and J. H. Waite Jr. (eds.), *Solar System Plasmas in Space and Time (Geophys. Monogr. Series 84)*, AGU, Washington, DC, p. 221.

Knoepfel, H., and Spong, D. A.: 1979, *Nucl. Fusion* **19**, 785.

Lenters, G. and Miller, J. A.: 1998, *Astrophys. J.* **493** 451.

Li, H. and Miller, J. A.: 1997, *Astrophys. J.* **478**, L67.

Litvinenko, Y. E.: 1996, *Astrophys. J.* **462**, 997.

Luhn, A., Klecker, B., Hovestadt, D., and Möbius, E.: 1987, *Astrophys. J.* **317**, 951.

Machado, M. E., Ong, K., Emslie, A. G., Fishman, G. J., Meegan, C., Wilson, R., and Paciesas, W. S.: 1993, *Adv. Space Sci.* **13** (9), 175.

Mandzhavidze, N. and Ramaty, R.: 1993, *Nuc. Phys.* **B33**, 141.

Mason, G. M., Mazur, J. E., and Hamilton, D. C.: 1994, *Astrophys. J.* **425**, 843.

Masuda, S.: 1994, *'Hard X-ray Sources and the Primary Energy Release Site in Solar Flares'*, Ph.D. Thesis, University of Tokyo, Tokyo, Japan.

Mariska, J. T., Doschek, G. A., and Bentley, R. D.: 1993, *Astrophys. J.*, **419** 418.

Meyer, J.-P.: 1996, in R. Ramaty, N. Mandzhavidze, and X.-M. Hua (eds.), *High Energy Solar Physics*, AIP, New York, p. 461.

Miller, J. A.: 1997, *Astrophys. J.*, in press.

Miller, J. A., Cargill, P. J., Emslie, A. G., Holman, G. D., Dennis, B. R., LaRosa, T. N., Winglee, R. M., Benka, S. G., and Tsuneta, S.: 1997, *J. Geophys. Res.* **102**, 14631.

Miller, J. A., Guessoum, N., and Ramaty, R.: 1990: *Astrophys. J.* **361**, 701.

Miller, J. A., LaRosa, T. N., and Moore, R. L.: 1996, *Astrophys. J.* **461**, 445.

Miller, J. A. and Reames, D. V.: 1996, in R. Ramaty, N. Mandzhavidze, and X.-M. Hua (eds.), *High Energy Solar Physics*, AIP, New York, p. 450.

Miller, J. A. and Reames, D. V.: 1997, in *Proc. 25th Int. Cosmic Ray Conf.* **1**, 141.

Miller, J. A. and Roberts, D. A.: 1995, *Astrophys. J.* **452**, 912.

Miller, J. A. and Viñas, A. F.: 1993, *Astrophys. J.* **412**, 386.

Miller, J. A., Viñas, A. F., and Reames, D. V.: 1993a, in *Proc. 23rd Int. Cosmic Ray Conf.* **3**, 13.

Miller, J. A., Viñas, A. F., and Reames, D. V.: 1993b, in *Proc. 23rd Int. Cosmic Ray Conf.* **3**, 17.

Miller, J. A., Viñas, A. F., and Reames, D. V.: 1997, *Astrophys. J.*, in preparation.

Miller, J. A., Viñas, A. F., and Roberts, D. A.: 1998, *Astrophys. J.*, in preparation.

Moghaddam-Taaheri, E. and Goertz, C. K.: 1990: *Astrophys. J.* **352**, 361.

Murphy, R. J., Dermer, C. D., and Ramaty, R.: 1987, *Astrophys. J. Suppl.* **63**, 721.

Murphy, R. J., Ramaty, R., Kozlovsky, B.-Z., and Reames, D. V.: 1991, *Astrophys. J.* **371**, 793.

Ramaty, R., Mandzhavidze, N., Kozlovsky, B.-Z., and Murphy, R. J.: 1995, *Astrophys. J.* **455**, L193.

Ramaty, R. and Murphy, R. J.: 1987, *Space Sci. Rev.* **45**, 213.

Ramaty, R., Murphy, R. J., and Dermer, C. D.: 1987, *Astrophys. J.* **316**, L41.

Reames, D. V.: 1990, *Astrophys. J. Suppl.*, **73** 235.

Reames, D. V., Meyer, J.-P., and von Rosenvinge, T. T.: 1994, *Astrophys. J. Suppl.* **90**, 649.

Temerin, M. and Roth, I.: 1992, *Astrophys. J.* **391**, L105.

Share, G. H. and Murphy, R. J.: 1995, *Astrophys. J.* **452**, 933.

Share, G. H. and Murphy, R. J.: 1997, *Astrophys. J.* **485**, 409.

Swanson, D. G.: 1989, *Plasma Waves*, Academic Press, San Diego.

Van den Oord, G. H. J.: 1990, *Astron. Astrophys.* **234**, 496.

Verma, M. K., et al.: 1996, *J. Geophys. Res.* **101**, 21619.

Vestrand, W. T. and Miller, J. A.: 1998, in B. Haisch, K. Strong, and J. Schmelz (eds.), *The Many Faces of the Sun: The Scientific Results of the Solar Maximum Mission*, Springer-Verlag, New York, in press.

Vogt, E. and Hénoux, J.-C.: 1996, *Solar Phys.* **164**, 345.

Wiley, J. C., Choi, D.-I., and Horton, W.: 1980: *Phys. Fluids* **23**, 2193.

Zhang, T. X.: 1995, *Astrophys. J.* **449**, 916.

Zhou, Y. and Matthaeus, W. H.: 1990, *J. Geophys. Res.* **95**, 14881.

INTERSTELLAR MATTER AND THE BOUNDARY CONDITIONS OF THE HELIOSPHERE

PRISCILLA C. FRISCH

Department of Astronomy and Astrophysics, University of Chicago, Chicago, IL 60637, U.S.A.

Abstract. The interstellar cloud surrounding the solar system regulates the galactic environment of the Sun, and determines the boundary conditions of the heliosphere. Both the Sun and interstellar clouds move through space, so these boundary conditions change with time. Data and theoretical models now support densities in the cloud surrounding the solar system of $n(\mathrm{H}^0) = 0.22 \pm 0.06$ cm^{-3}, and $n(e^-) \sim 0.1$ cm^{-3}, with larger values allowed for $n(\mathrm{H}^0)$ by radiative transfer considerations. *Ulysses* and Extreme Ultraviolet Explorer satellite He^0 data yield a cloud temperature of 6400 K. Nearby interstellar gas appears to be structured and inhomogeneous. The interstellar gas in the Local Fluff cloud complex exhibits elemental abundance patterns in which refractory elements are enhanced over the depleted abundances found in cold disk gas. Within a few parsecs of the Sun, inconclusive evidence for factors of 2–5 variation in Mg$^+$ and Fe$^+$ gas phase abundances is found, providing evidence for variable grain destruction. In principle, photoionization calculations for the surrounding cloud can be compared with elemental abundances found in the pickup ion and anomalous cosmic-ray populations to model cloud properties, including ionization, reference abundances, and radiation field. Observations of the hydrogen pile up at the nose of the heliosphere are consistent with a barely subsonic motion of the heliosphere with respect to the surrounding interstellar cloud. Uncertainties on the velocity vector of the cloud that surrounds the solar system indicate that it is uncertain as to whether the Sun and α Cen are or are not immersed in the same interstellar cloud.

1. Introduction

The physical conditions of the surrounding interstellar cloud establish the boundary conditions of the solar system and heliosphere. The abundances and ionization states of elements in the surrounding interstellar cloud determine the properties of the parent population of the anomalous cosmic-ray and pickup ion components. In addition, the history of the interstellar environment of the heliosphere appears to be partially recorded by radionucleotides such as ^{10}Be and ^{14}C in geologic ice core records (Sonett et al., 1987; Frisch, 1997). Because the solar wind density decreases as R^{-2} ($R =$ distance to Sun), the solar wind and interstellar densities are equal at about 5 AU (the orbit of Jupiter), in the absence of substantial 'filtration'*. Approximately 98% of the diffuse material in the heliosphere is interstellar gas (Gruntman, 1993). Thus, the physical properties of the outer heliosphere are dominated by interstellar matter (ISM). Were the Sun to encounter a high density interstellar cloud, it is anticipated that the physical properties of the inner

---

* 'Filtration' refers to the deflection of interstellar H^0 around the heliopause due to the coupling between interstellar protons and H^0 resulting from charge exchange.

Space Science Reviews **86:** 107–126, 1998.

heliosphere would also be ISM-dominated. Zank and Frisch (1998) have shown that if the space density of the interstellar cloud which surrounds the solar system were increased to ~ 10 cm^{-3}, the properties of the inner heliosphere at the 1 AU position of the Earth would be dramatically altered.

The accuracy with which the physical properties of the surrounding cloud can be derived from observations of stars within a few parsecs of the Sun (1 pc $\sim$ 200 000 AU) depends on the homogeneity and physical parameters of nearby ISM. Observations of nearby stars gives sightlines which probe the ensemble of nearby clouds constituting the 'Local Fluff' cloud complex. Conclusions based on observations of nearby stars, however, must be qualified by the absence of detailed data pertaining to the small-scale structure of the local ISM (LISM). More distant cold diffuse interstellar gas is highly structured, replete with dense ($\sim 10^4$–10^5 cm^{-3}), small (20–200 AU) inclusions occupying in some cases less than 1% of the cloud volume (Frail et al., 1994; Falgarone et al., 1997; Falgarone and Puget, 1995; Heiles, 1997). Small-scale structures are ubiquitous in interstellar gas, and individual velocity components exhibiting column densities as low as $N(\mathrm{H}^0) \sim 3 \times 10^{18}$ cm^{-3} are found in cold clouds (Frail et al., 1994; Heiles, 1997). The presence of dense low column density wisps near the Sun is allowed by currently available data.

The Sun has a peculiar motion with respect to the 'Local Standard of Rest' (LSR*); the Sun moves through the LSR with a velocity $V \sim 16.5$ km s^{-1} towards the apex direction $l = 53°$, $b = +25°$ (Mihalas and Binney, 1981). Uncertainties on the relative solar-LSR motion appear to be less than 3 km s^{-1} and $\pm 5°$. This motion corresponds to ~ 17 pc per million years. Note that the solar path is tilted by $\sim 25°$ with respect to the galactic plane. The Sun oscillates about the galactic plane, crossing the plane every 33 Myr, reaching a maximum distance from the plane of ~ 77 pc. The last galactic plane 'crossing' was about 21 Myr ago (Bash, 1986). This amplitude of oscillation can be compared to scale heights on the order of ~ 50–80 pc for cold H$_2$ and CO, ~ 100 pc for cold H^0 and infrared cirrus, ~ 250 pc for warm H^0, and ~ 1 kpc for warm H$^+$ (the 'Reynolds Layer').

There are three time scales of interest in understanding the environmental history of the solar galactic milieu – $\sim 10^6$ years, $\sim 10^5$ years, and $\sim 10^4$ years. Prior to entering into the Local Fluff complex of interstellar clouds, the Sun traveled through a region of the galaxy between the Orion spiral arm and the spiral arm spur known as the Local Arm. On the order of a million years ago, the Sun was displaced ~ 17 pc in the anti-apex direction, which is towards the present day location of the junction of the borders of the constellations of Columba, Lepus and Canis Major. The motions of the Sun and surrounding interstellar cloud with respect to interstellar matter within 500 pc, projected onto the plane, are illustrated in Figure 1. Note that the velocity vectors of the Sun and interstellar cloud sur-

---

* The LSR is the velocity frame of reference in which the vector motions of a group of nearby comparison stars are minimized. Stars in the LSR corotate around the galactic center with a velocity of ~ 250 km s^{-1}.

rounding the solar system are nearly perpendicular in the LSR, implying that the surrounding cloud complex is sweeping past the Sun (see Section 2). When the morphology of the Local Fluff complex is considered, it is apparent that sometime during the past $\sim$200 000 years the Sun appears to have emerged from a region of space with virtually no interstellar matter (densities $n(H^0) < 0.0005$ cm^{-3}, $n(e^-) < 0.02$ cm^{-3}) and entered the Local Fluff complex of clouds (average densities $n(H^0) \sim 0.1$ cm^{-3}) outflowing from the Scorpius-Centaurus Association of star-forming regions. One model for the morphology of the cloud surrounding the solar system predicts that sometime within the past 10 000 years, and possibly within the past 2000 years, the Sun appears to have entered the interstellar cloud in which it is currently situated (Frisch, 1994, 1997). The cloud surrounding the solar system will be called here the 'Local Interstellar Cloud' (LIC*).

2. Velocity, Magnetic Field

The velocity of the cloud surrounding the solar system provides the sole criteria for selecting LIC absorption features in nearby stars, and therefore for deducing the structure of this cloud.

The velocity of the cloud feeding neutral gas into the solar system is found from backscattered solar He^0 584 A radiation (e.g., Flynn et al., 1998), direct measurements of inflowing He^0 atoms by *Ulysses* (Witte et al., 1993, 1996; Witte, 1998, private communication), and measurements of pickup ions (Moebius, 1996). In the rest frame of the Sun, the downwind direction of the velocity vector found from He^0 backscattered radiation is $V = 26.4 \pm 1.5$ km s^{-1}, $\lambda = 76.0° \pm 0.4°$, $\beta = -5.4° \pm 0.6°$ (ecliptic coordinates), with $T = 6900 \pm 600$ K (Flynn et al., 1998). This corresponds to a heliocentric upwind vector in galactic coordinates of $l = 4.1° \pm 0.8°$, $b = +14.8° \pm 0.7°$, $V = -26.4 \pm 1.5$ km s^{-1}. The *Ulysses* neutral gas detector yields a downwind vector of $V = 25.3 \pm 0.5$ km s^{-1}, $\lambda = 74.8° \pm 0.5°$, $\beta = -5.7° \pm 0.2°$, $T = 6100 \pm 300$ K. The *Ulysses* vector corresponds to a heliocentric upwind direction in galactic coordinates of $l = 3.8°$, $b = +16.0°$.

These determinations of the cloud velocity from measurements of interstellar gas within the solar system can be compared with the values derived from observations of interstellar absorption lines towards the nearest stars. The velocity vector of the surrounding cloud has also been determined by fitting the velocities of interstellar absorption line components in a limited number of nearby stars. The fits by Bertin et al. (1993a) give an upwind direction $V = -25.7$ km s^{-1}, $l = 5.9°$, $b = 16.7°$, which is consistent with the *Ulysses* He^0 observations. Uncertainties on this velocity are hard to ascertain due to the small number of stars on which the

<hr>

*This cloud surrounding the solar system is also referred to as the 'surrounding interstellar cloud', or SIC, which unambiguously defines the cloud feeding interstellar matter into the solar system. For the sake of uniformity of notation, however, the term LIC is used here.

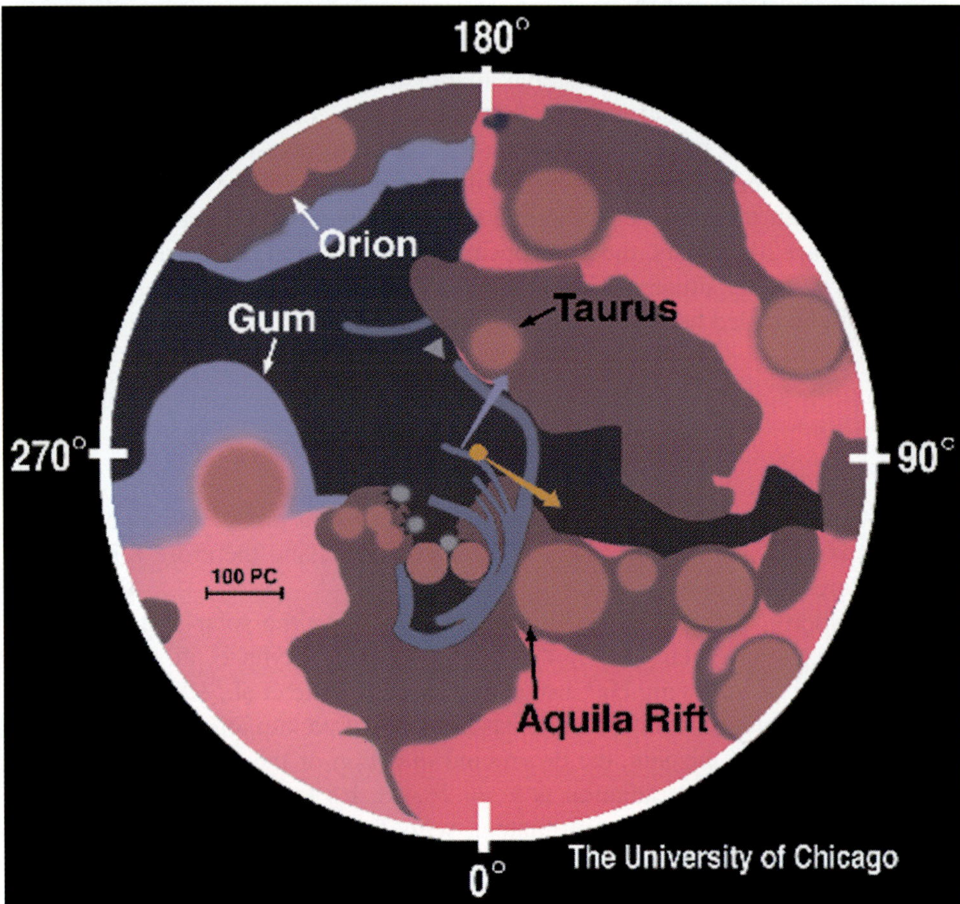

Figure 1. The distribution of interstellar molecular clouds (traced by the CO 1->0 115 GHz rotational transition) and diffuse gas (traced by $E(B - V)$ color excess due to the reddening of starlight by interstellar dust) within 500 pc of the Sun are shown. The round circles are molecular clouds, and the shaded material is diffuse gas. The horizontal bar (lower left) illustrates a distance of 100 pc. Interstellar matter is shown projected onto the galactic plane, and the plot is labeled with galactic longitudes. The distribution of nearby interstellar matter is associated with the local galactic feature known as 'Gould's Belt', which is tilted by about 15–20° with respect to the galactic plane. ISM towards Orion is over 15° below the plane, while the Scorpius–Centaurus material (longitudes 300°–0°) is about 15–20° above the plane. Also illustrated are the space motions of the Sun and local interstellar gas, which are nearly perpendicular in the LSR velocity frame. The three asterisks are three subgroups of the Scorpius–Centaurus Association. The three-sided star is the Geminga Pulsar. The arc towards Orion represents the Orion's Cloak supernova remnant shell. The other arcs are illustrative of superbubble shells from star formation in the Scorpius–Centaurus Association subgroups. The smallest (i.e., greatest curvature) shell feature represents the Loop I supernova remnant.

vector derivation is based, but will be assumed to be ± 1.5 km s^{-1} or $\sim \frac{1}{2}$ of the typical spectral resolution of the data.

Removing solar motion from these three velocity vectors gives an upwind direction for the LIC cloud in the LSR of $V = -18.7 \pm 0.6$ km s^{-1} from the direction $l = 327.3° \pm 1.4°$, $b = 0.3° \pm 1.0°$ (Frisch, 1995). The *Ulysses* and EUVE data He^0 data indicate a cloud temperature of $T = 6400$ K. As a curiosity, this LSR direction is $\sim 10°$ from the 1.3 pc distant star α Cen. In essence, the Local Fluff gas is sweeping down on the solar system from the vicinity of α Cen.

The small difference in the magnitudes and directions of the *Ulysses* versus backscatter vectors introduces a significant uncertainty concerning the structure of the Local Fluff and the location of the Sun with respect to the boundary of the surrounding interstellar cloud. The uncertainties on the velocity vectors as defined by the Flynn et al. versus *Ulysses* versus Bertin et al. ('AG') vectors do not allow us to make a definitive statement as to whether or not the Sun and the nearby star α Cen are immersed in the same interstellar cloud. The Flynn et al. vector projected in the direction of the star α Cen ($d = 1.3$ pc, $l = 315.8°$, $b = -0.7°$) predicts a heliocentric velocity of -16.9 ± 1.0 km s^{-1} for the LIC cloud. The *Ulysses* flow vector predicts a velocity of -16.2 ± 1.0 km s^{-1}. The Bertin et al. AG flow vector predicts a velocity of -15.7 ± 1.0 km s^{-1}. The small differences between these three predictions (~ 1.2 km s^{-1}) may seem academic; but, when compared to observational data, these differences have consequences regarding the location of the Sun with respect to the edge of the interstellar cloud in which it is situated. The velocities of the Mg$^+$ and Fe$^+$ absorption lines observed in this direction are -17.5 to -18.2 km s^{-1} with assumed uncertainties on the individual velocities of ± 1.0 km s^{-1} (corresponding to $\frac{1}{3}$ of the nominal spectral resolution with which the observations were acquired, Lallement et al., 1995; Linsky and Wood, 1997)*. As noted by Lallement et al. (1995), if the AG velocity vector is correct, less than 2.5×10^{16} cm^{-2} column density of H is found between the Sun and the location of the cloud edge in this direction. This corresponds to a distance to the cloud edge of less than 0.03 pc (6000 AU) for n(H)=0.3 cm^{-2} (see Section 3). In contrast, if the Flynn et al. vector is correct, the uncertainties allow the Fe$^+$ and Mg$^+$ absorption features to be formed in the surrounding cloud, so that $\log(N(\mathrm{Fe}^+))$=12.36 cm^{-2} towards α Cen (Lallement et al., 1995). Using $\log(N(\mathrm{Fe}^+)/N(\mathrm{H}^0)) \sim -5.71$ in the surrounding cloud (Table I) then indicates that the cloud edge is at or beyond α Cen. Since the star α Cen is located in the true upwind direction (i.e., in the LSR), if the AG vector is correct the Sun would thus exit the cloud now surrounding the solar system within the next 2000 years, while if the Flynn et al. vector is correct the Sun will not exit the current cloud for over 70 000 years. These data allow, but do not require, the Local Fluff gas to be a structured or turbulent medium on the ≤ 5000 AU scale. Evidently reducing these uncertainties in the velocity of the cloud

---

* These data were acquired with the Hubble Space Telescope GHRS instrument at a nominal spectral resolution of 3.3 km s^{-1}.

around the solar system will enable us to understand interactions between the Sun and interstellar gas over the next 100 000 years.

The strength of the interstellar magnetic field in the cloud surrounding the solar system has not been directly measured. A strength of $\sim$1.6 μG has been found to be consistent with observations of polarization and pulsar dispersion measures (Frisch, 1991), while estimates of the heliosphere confinement pressure give an upper limit $B < 3$–4 μG (Gloeckler et al., 1997). Ultimately, three-dimensional MHD heliosphere models, in comparison with *Voyager* data, may provide the best estimate for the magnetic field strength in the cloud surrounding the solar system. The presence of a local magnetic field is confirmed by observations of the polarization of light from nearby stars ($d < 35$ pc) in the galactic center hemisphere (Tinbergen, 1982; Frisch, 1991). *If* the cloud surface is perpendicular to the direction of gas flow, then the direction of the magnetic field is found to be approximately parallel to the LIC cloud surface (Frisch, 1996). The pile up of H^0 at the heliosphere nose is consistent with an interstellar magnetic field strength of 3 μG (see Section 7).

3. Neutral Density

The quantities of interest in constraining the heliosphere are the absolute neutral and ion space densities (cm^{-3}), which can be derived from column densities (cm^{-2}) only if cloud length is known. The *average* density of nearby gas can be inferred from observations of H^0 and other elements seen in absorption. For example, combining H^0 column densities for the six clouds detected towards the four nearby stars α Cen (1.3 pc), α CMa (2.7 pc), α CMi (3.5 pc) and α Aql (5 pc), gives an average density $\langle n(H^0) \rangle = 0.08$–0.11 cm^{-3} for interstellar gas within 3–5 pc of the Sun. This value can be compared with $\langle n(H^0) \rangle \sim 0.2$ cm^{-3} for the surrounding cloud, suggesting a filling factor of $\sim$50%[*]. Data for α Cen and α CMa are given in Table I. The hydrogen column density for α Aql has been inferred from Ca^+ absorption lines using the conversion factor $N(Ca^+)/N(H) \sim 10^{-8}$, based on $N(Ca^+) \sim 10^{10}$ cm^{-2} and $N(H^0 + H^+) = 10^{18}$ cm^{-2} towards η UMa (which samples a low column density sightline through the Local Fluff complex, Frisch and Welty, 1998). With this conversion factor, $N(Ca^+) \sim 1.7 \times 10^{10}$ observed towards α Aql, gives $N(H) = 1.7 \times 10^{18}$ cm^{-2}. (The depletion variations of Ca^+, discussed in Section 5, will affect this conclusion.) The average density $\langle n(H^0) \rangle \sim 0.10\,cm^{-3}$ can be used as a guide when inferring the overall morphology of interstellar clouds in the Local Fluff complex.

The absolute space density of the LIC can be inferred from observations of pickup ions combined with the H^0/He^0 ratio in the surrounding cloud (e.g., Gloeckler et al., 1997; Vallerga, 1996; Frisch and Slavin, 1996). An average value

---

[*] The 'filling factor' of a cloud complex is the fraction of space occupied by the gas in the complex.

$N(H^0)/N(He^0) \sim 14$ is found from observations of white dwarf stars (Dupuis et al., 1995; Frisch, 1995; Vallerga, 1996, also see Section 6). EUVE observations of the two lowest column density white-dwarf stars GD 71 ($l = 192°$, $b = -5°$, $N(H^0) = 8.7 \pm 0.7 \times 10^{17}$ cm^{-2}) and HZ 43 ($l = 54°$, $b = 84°$, $N(H^0) = 6.3 \pm 1.6 \times 10^{17}$ cm^{-2}) give $N(H^0)/N(He^0)$ ratios of 12.1 and 15.8, respectively, over the sightlines to these stars (Vallerga, 1996), a difference which may reflect local variations in ionization levels. In principle hydrogen ionization rises towards cloud exterior more rapidly than helium ionization because of the low column densities (Slavin, 1989; Cheng and Bruhweiler, 1990; Vallerga, 1996; Slavin and Frisch, 1998a, b) so the actual ratio $N(H^0)/N(He^0)$ at the solar location will be larger than the sightline averaged values measured for white-dwarf stars. Observations by *Ulysses* of interstellar helium within the heliosphere yield $n(He^0) \sim 0.016\pm0.002$ cm^{-3} (Witte et al., 1993, 1996; Witte, 1998, private communication). Combining the *Ulysses* and EUVE data gives a lower limit on the neutral density in the interstellar cloud outside of the solar system of $n(H^0) = 0.22\pm0.06$ cm^{-3}, with larger values allowed by the above radiative transfer considerations. Both hydrogen and helium are ionized in the surrounding cloud, however, so the total space density (neutral plus ions) will be larger (see Section 5).

Observations of Lα backscattered radiation from interstellar hydrogen in the inner heliosphere yield a value $n(H^0) = 0.15 \pm 0.05$ cm^{-3} for the interstellar H^0 density inside of the heliopause (Quemerais et al., 1995). Combined with the LIC value of $n(H^0)$ above, this gives $\sim$30% filtration of H^0 at the heliopause.

4. Structure

The largest data set on the Local Fluff complex is derived from observations of the optical Ca$^+$ H and K lines. Although globally calcium has highly variable depletions onto interstellar dust grains in the ISM, Ca$^+$ observations constitute the most complete set of available data and therefore provide a first look at the structure and morphology of nearby gas. The rest of this section rests on the assumption that calcium depletions and ionization are relatively constant across the clouds in the Local Fluff cloud complex. This premise may not be true, but it offers an insight into the distribution of ISM within 30 pc of the Sun. Because of the variation of both calcium depletion and ionization in interstellar gas in general, this distance estimate is useful mainly to provide a qualitative picture of the distribution of Local Fluff gas.

About 80 individual Ca$^+$ velocity components have been observed that can be attributed to the Local Fluff complex (Frisch and Welty, 1998). The overall morphology of the Local Fluff complex is shown in Figure 2, where Ca$^+$ column densities ($N(Ca^+)$) are plotted versus galactic longitude and latitude for stars sampling the Local Fluff complex. (Data are from Bertin et al., 1993b; Crawford, 1994; Crawford and Dunkin, 1995; Crawford et al., 1997; Frisch, 1997; Lallement et al.,

                    PRISCILLA C. FRISCH

1986, 1994, 1995; Lallement and Bertin, 1992; Vallerga et al., 1993; Welty et al., 1996; Frisch and Welty, 1998.) The well-known asymmetry in the distribution of Local Fluff complex between the galactic center and anti-center hemispheres is apparent. In the absence of data on non-refractory elements in a given sightline, the distance to the 'edge' of the Local Fluff cloud complex can be guesstimated using the somewhat uncertain relation for the Local Fluff d (pc) $\sim 3 \times 10^{-10} N(\mathrm{Ca}^+)$, for $\langle n \rangle \sim 0.1$ cm^{-3}. The sightline to the star τ^3 Eri ($l = 214°$, $b = -60°$, $N(\mathrm{Ca}^+) = 5 \times 10^{10}$ cm^{-2}) shows that the Local Fluff complex extends into the southern galactic hemisphere. The low column densities ($N(\mathrm{Ca}^+) < 10^{10}$ cm^{-2}, Vallerga et al., 1993; Frisch and Welty, 1998) towards the three high-latitude stars σ Leo ($l = 253°$, $b = 60°$), η UMa ($l = 101°$, $b = 65°$), and α^2 CVn ($l = 117°$, $b = 79°$) show the Sun is 'above' most of the mass of the Local Fluff complex (in the sense of 'above' the galactic plane). The cloud in front of τ^3 Eri may be an extension of the interstellar cloud complex seen towards α Gru, δ Vel, ϵ Gru, α Hyi, and ι Cen and other nearby stars (with data from Crawford and Dunkin, 1995; Lallement et al., 1986; Crawford et al., 1997; Frisch and Welty, 1998). The location of the upwind cloud in the LSR (which is the galactic center hemisphere of the sky) coincides with the interstellar dust patch which polarizes the light of nearby stars (Tinbergen, 1982; Frisch, 1991). This cloud has been identified as part of the leading edge of the 4 Myr old superbubble shell expanding from the last epoch of star formation in the Scorpius-Centaurus association (Frisch, 1995).

The Local Fluff complex is structured and inhomogeneous. At 3 km s^{-1} resolution, about one interstellar absorption line velocity component is seen per 1–2 parsecs of sightline in the direction of the stars α CMa at 2.7 pc and α Aql at 5 pc (Lallement et al., 1994, 1995), but the actual filling factor of the gas is not known since the densities of LISM clouds other than the LIC are unknown. Globally, cold H$°$ clouds and cold molecular (H$_2$, CO) clouds fill $\sim 10^{-3}$–10^{-4} of cloud volume, evidently revealing structure consistent with a Kolmogorov spectrum, including features on ≤ 100 AU scale sizes. Ionization also appears to be inhomogeneous, based on both the variations in the white dwarf $N(\mathrm{H}^0)/N(\mathrm{He}^0)$ ratios and factors of 2–4 variation in LIC electron densities (Section 5 and Table I). Other possible evidence for local structure on subparsec scales are the absence of gas at the LIC velocity in the Fe$^+$ and Mg$^+$ lines towards α Cen (1.3 pc) (Section 2) and the possible small-scale structure (≤ 0.1 pc) revealed by the finite time width of the ^{10}Be spikes in the antarctic ice core record (Frisch, 1997). The upper limits on Ca$^+$ column density towards λ Aql ($l = 30°$, $b = -6°$, Vallerga et al., 1993) indicate either the LIC cloud surface is closer than 1.2 pc or that the calcium is doubly ionized in this direction.

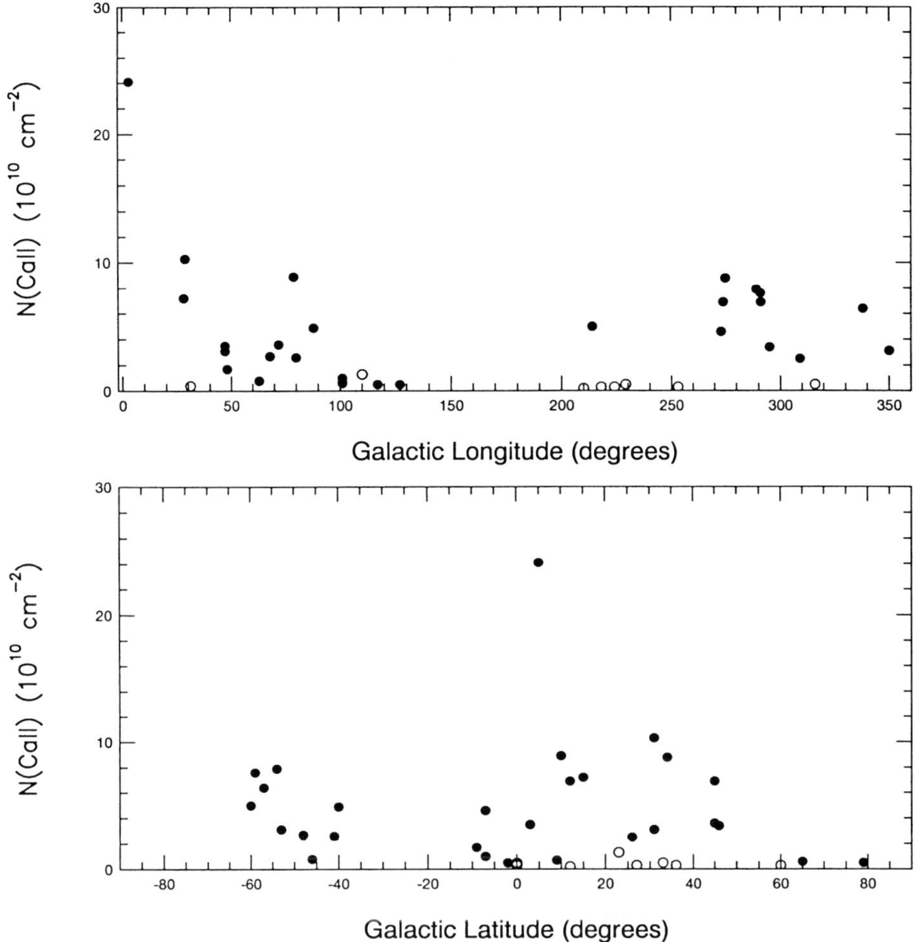

Figure 2. Overall morphology of Local Fluff complex of interstellar clouds based on Ca^+ observations of nearby stars. Total Ca^+ column densities for stars sampling the Local Fluff complex are plotted versus galactic longitude and latitude. The open circles are upper limits. With the exception of τ^3 Eri, at $l = 214°$, $b = -60°$, $d = 17$ pc, the section of the galaxy between $l = 130°$ and $270°$ is relatively devoid of nearby interstellar gas. (Data are from Bertin et al., 1993b; Crawford, 1994; Crawford and Dunkin, 1995; Crawford et al., 1997; Frisch, 1997: Lallement et al., 1986, 1994, 1995; Lallement and Bertin, 1992; Vallerga et al., 1993; Welty et al., 1996.)

5. Ionization

The relative ionizations of elements in interstellar gas at the solar location are a sensitive function of the relative number of EUV (200–900 Å) versus FUV (900–1500 Å) photons, which in turn depends on the mix of different sources of ionizing radiation and radiative transfer in the Local Fluff (e.g., Slavin, 1989; Reynolds, 1991; Cheng and Bruhweiler, 1990; Vallerga and Welsh, 1995). Locally, hydrogen ionization is dominated by the ultraviolet flux from the star ϵ CMa, a luminous

B2 II star located in a direction with very low column densities. The radiation field which ionizes He, Ne, and other species with ionization potentials greater than 13.6 eV is dominated by white dwarf stars, the diffuse soft X-ray flux, and radiation from a predicted conductive interface between the Local Fluff complex and the nominal hot ($\sim 10^6$) plasma filling the adjacent Local Bubble interior (Vallerga, 1998; Slavin and Frisch, 1998a; Slavin, 1989).

The question of which methods to use in calculating interstellar electron densities needs to be put in the context of global ISM studies. Different estimates of interstellar ionization equilibrium towards denser clouds do not produce consistent results. Welty et al. (1998) have observed a $n \sim 10$ cm^{-3} cloud towards the star 23 Ori and found that thirteen different ratios of neutral-to-first ionization states yield electron density values that span a range of a factor of 25. Welty attributes this range to a combination of unidentified physical processes (such as charge exchange) and uncertain rate constants. In light of these generic ISM studies, for the LIC a variety of approaches are needed to estimate the electron density.

The electron density in the LIC cloud has been found with several methods, some of which provide cloud-averaged values versus values at the solar location. Based on the results outlined below, it appears that $n(e^-) \sim 0.1$ cm^{-3} in the LIC, but more work is needed. These methods include:

– Observations of absorption lines of trace ions in the LIC cloud as seen towards nearby stars. These data give electron density values that are averaged over the cloud segment being sampled.

– Theoretical predictions of the LIC ionization at the solar location based on a radiation field which is the sum of measurements of ultraviolet and extreme ultraviolet fluxes of point sources such as white dwarf and other hot stars near the Sun. These calculations include radiative transfer effects.

– Theoretical predictions of the LIC ionization, including radiative transfer effects, at the solar location based on a radiation field which is the sum of both a diffuse radiation field (from scattered starlight, plasma in the local hot bubble, and emission from the predicted conductive interface on the LIC cloud) and measurements of ultraviolet and extreme ultraviolet fluxes from point sources such as white dwarf and hot stars near the Sun.

– Measurements of H°, He°, and He$^+$ in nearby white dwarf stars, which when compared to the nominal reference abundance ratio $N(\text{H}°)/N(\text{He}°) = 10/1$ give an estimate of the LISM ionization. This approach gives electron densities averaged over the cloud segment being sampled.

– Theoretical models of the hydrogen wall of the heliosphere, which has properties dominated by charge exchange between interstellar protons and hydrogen atoms piled up at the heliopause. Proton and electron densities are assumed to be equal.

Absorption line data.   The LIC electron density has been inferred from the ratio Mg$^+$/Mg^0 and C$^+$ fine-structure lines (Frisch et al., 1990; Lallement et al.,

1994; Lallement and Ferlet, 1996; Gry et al., 1995; Frisch, 1994; Lallement, 1996; Wood and Linsky, 1997). Electron densities found towards nearby stars are 0.09 (+0.23, −0.07) cm^{-3} towards ϵ CMa (LIC cloud only), 0.11 (+0.12, −0.06) towards α Aur, $\sim$0.24 towards η UMa, 0.31±0.20 towards δ Cas, and 0.2–0.4 cm^{-3} towards α CMa (see Table I). The large value seen towards α CMa may indicate a cloud interface between the LIC cloud (T$\sim$10^4 K) and adjacent plasma ($T \sim 10^6$ K, Frisch, 1994; Bertin et al., 1995). The convergence between the theoretical models (Table II) and the $n(e^-)$ values from ϵ CMa Mg$^+$/Mg^0 and α Aur C$^+$ fine-structure data suggest that the LIC value is $n(e^-) \approx 0.1$ cm^{-3}. In principle, electron densities can also be inferred from optical observations of Na^0 and Ca$^+$, but as these species are subordinate ionization states of elements with variable depletions in the ISM, and trace ISM at different temperatures, the resulting electron densities are highly uncertain. The range $n(e^-) = 0.15 \pm 0.11$ cm^{-3} is found from Na^0 data (Lallement and Ferlet, 1996; Lallement, 1996).

Stellar radiation field.   Predicted ionization levels of the LISM depend on whether the radiation field estimates are based on point sources, such as white dwarf and hot stars, or also emission from diffuse hot gas. Including stellar fluxes alone, e.g., ϵ CMa and white dwarf stars, hydrogen is predicted to be $\sim$13% ionized and He $\sim$4.3% ionized at the solar location (Vallerga, 1996, 1998), yielding $n(e^-) \sim 0.03$ cm^{-3}.

Diffuse and stellar radiation field.   Diffuse EUV emission is expected from the Local Bubble 10^6 K plasma and the conductive interface between the LIC and plasma (Slavin, 1989). As part of an ongoing project, we (Slavin and Frisch, 1998a, b; Frisch and Slavin, 1996) are modeling the ionization balance in the LIC using a radiative transfer ionization which includes both diffuse and point source EUV and FUV emission, yielding a self-consistent solution for cloud heating and cooling. The predictions of this model are sensitive to assumptions about the input radiation field and cloud abundances. These results have been calculated using parts per million (PPM) values for He, C, N, O and Ne of 1×10^5, 191, 85, 646, and 123, respectively, based on LIC abundances towards ϵ CMa (Gry and Dupin, 1995). In this particular model hydrogen and helium ionizations are predicted to be $\sim$23% and $\sim$36%, respectively, giving $n(e^-) \sim 0.08$ cm^{-3} if $n(\mathrm{H}) = 0.3$ cm^{-2}. Table II shows that these results for He, O, N are consistent with PUI data providing O and N are filtered at the heliopause by $\sim$30%. (Alternatively, they could have lower intrinsic abundances.) However, Ne predictions agree with the PUI data only if Ne^0 increases by a factor of $\sim$2.5. Since Ne is only $\sim$20% neutral at the solar location in this particular model, small changes in the extreme ultraviolet radiation field yield larger changes in the fraction of neutral Ne. Since He has about the same ionization potential, however, this change would create problems in understanding the He data. The model used for the Table II predictions yields predictions of S^0/O$^0 = 2.8 \times 10^{-6}$ and Mg^0/O$^0 = 2.2 \times 10^{-5}$. These predictions are highly sensitive to the diffuse radiation field in the 200–1200 Å region which has not been directly measured. These results represent one possible model, but better understandings of the radiation field, ISM reference abundances, and cloud geometry are needed before the problem can be considered solved.

White dwarf stars.   The two white dwarf stars with the lowest column densities are GD 71 and HZ 43. EUVE observations of He^0 and He^+ toward the white dwarf star HZ 43 (a high latitude white dwarf at 60 pc which samples the Local Fluff complex) yield an upper limit on the electron density of $n(e^-) < 0.14$ cm^{-3} for $n(H^0 + H^+) = 0.15$–0.34 cm^{-3}, after comparing $N(H^0)$, $N(He^0)$ and $N(He^+)$ with the expected cosmic ratio $N(H)/N(He) = 10/1$ and heliospheric He data (Vallerga, 1996).

Hydrogen wall.   The observed velocity distribution of interstellar H^0 decelerated by charge exchange with interstellar protons at the heliopause can be reproduced theoretically in the direction of α Cen for the case where the interstellar gas confining the heliosphere has plasma density $n(e^-) = 0.1$ cm^{-3} (see Section 7).

6. Abundances

In principle *in situ* measurements of pickup ions and anomalous cosmic rays can help resolve an active debate as to whether solar or B-star abundances (which are $\sim$ 70% solar) apply as the correct reference abundance for the interstellar medium*. Since He is primordial, the correct reference ratio H/He ratio should $\sim$10/1, regardless of whether solar or B-star abundances apply in the ISM. Elements that will be sensitive as to whether the correct ISM reference abundances are solar versus B-star are Ne, Ar, N, and S, none of which are thought to be depleted onto interstellar dust grains. Therefore, when LIC ionization and the filtration of elements at the heliopause are properly understood, measurements of N/He, O/He, Ne/He, and Ar/He (for example) will constrain the reference abundances of the LIC cloud.

The gas phase abundances of the neutral atomic constituents in the LIC are a function of both the ionization and depletion patterns of the gas. Abundant elements in the ISM such as C, S, Mg, Fe, and Si, have first ionization potentials (FIP) less than the 13.6 eV ionization potential of H^0, so they will be predominantly ionized in the LIC. Thus, low-FIP elements will be underrepresented in the pickup ion population when compared to cosmic abundances. The relative abundances of common low-FIP elements, based on observations typical of warm and cold gas in interstellar clouds, are shown in Table III. Absorption lines from C^0, Si^0, Fe^0, S^0 are weak in warm low column density ISM.

* The abundances of C, N, O, and other elements in the Sun are $\sim$50% higher than abundances in B-stars, and it is an open question as to whether B-star or solar abundances apply to the ISM (Snow and Witt, 1996; Savage and Sembach, 1996).

TABLE I

Abundances in directions of nearby stars[1]

Star	l, b, r	V_{obs}/V_{LIC} (km s^{-1})	Log $N(H^0)$ (cm^{-2})	$n(H^0)$ (cm^{-3})	$n(e^-)$ (10^4 K)	Log T (cm^{-2})	Log $N(Fe^+)$ (cm^{-2})	Log $N(Mg^+)$ (cm^{-2})	$N(O^0)$	$\delta(Fe^+)/\delta(Mg^+)/\delta(O^0)$
α Cen[2]	316°, −1°, 1.3 pc	−18/−15.7	17.6–18.0	0.1–0.2		5.4 ± 0.5	12.36 ± 0.07			∼−0.95/∼−0.68/
α CMa[3]	227°, −9°, 2.7 pc	19/19.5	17.23 ± 0.17	(0.04)[12]	0.2–0.4	7.6 ± 3.0	11.93 ± 0.07	12.20 ± 0.08		−0.98/−0.78/
ϵ CMa[4]	240°, −11°, 187 pc	17/15.6	17.30[8]		0.09 (+0.23, −0.07)	7.2 ± 2.0	12.13	12.48	14.15	−0.68/−0.40/−0.02
α CMi[5]	214°, +13°, 3.5 pc	21/19.5	17.88 ± 0.05	0.07		6.9 ± 0.38	12.05 ± 0.02			−1.34/−1.10/
α Aur[6]	163°, +5°, 12 pc	22/22.0	18.4 ± 0.05	0.05	0.11 (+0.12, −0.06)	7.0 ± 0.9	12.49 ± 0.03	12.83 ± 0.02		−1.26/−0.99/
δ Cas[7]	127°, −2°, 27 pc	12.3/12.9			0.30 ± 0.20		12.81 ± 0.09			
G191-B2B[9]	156°, +8°, 44 pc	20/20.1	18.27	0.01		∼7	12.48 ± 0.17	12.72 ± 0.14	14.63	−1.30/−1.13/−0.51
η UMa[10]	101°, +65°, 42 pc	−3.2/ − 4.8	17.85	0.005	(0.24)[11]	(2.7)[11]	12.39	12.70	14.37	−0.97/−0.73/−0.45

[1]Depletions calculated with respect to H^0 only. The depletion of element X is Log[X/H] − Log[X/H]$_{solar}$, where solar abundances with respect to hydrogen are: Log[O] = −3.13, Log[Fe] = −4.49, Log[Mg] = −4.42 (Savage and Sembach, 1996). For α CMa and ϵ CMa, values are given only for the LIC component.

[2]Piskunov et al., 1997; Lallement et al., 1995; Crawford, 1994; Linsky and Wood, 1997

[3]Lallement et al., 1994; Lallement et al., 1995; Frisch, 1994. LIC component only.

[4]Gry et al., 1995; Gry and Dupin, 1995. LIC component only.

[5]Piskunov et al., 1997.

[6]Piskunov et al., 1997; Lallement et al., 1995.

[7]Lallement et al., 1995.

[8]LIC component only; Vallerga (1996) gives log $N(H^0)$ = 17.90 cm^{-2} for the total neutral column density towards this star, based on EUVE data.

[9]Lemoine et al., 1996; Lallement et al., 1995.

[10]From Frisch, 1998; Frisch and Welty, 1998; Frisch and York, 1991, including preliminary results based on HST GHRS data.

[11]The electron density is based on the T = 2700 K temperature at which $n(e^-)$ derived from Mg^0/Mg$^+$ and C^{+*}/C$^+$ ratios are equal.

[12]Including the H column density of the second, blue-shifted cloud gives a total column density of $N(H^0)$ = 17.53 cm^{-3} (Bertin et al., 1994).

TABLE II

Photoionization predictions vs pickup ion data*

Ratio	Cloudy[1] predictions	Cloudy with 30% $O°$, $N°$ filtration at heliopause	Pickup ions[2]	Anomalous cosmic rays[3]	Assumed LIC ratio (no ionization)
He^0/O^0	125	178	196 (+70, −50)	13.3	155
N^0/O^0	0.12	0.12	0.118 (+0.070, −0.048)	0.15	0.13
Ne^0/O^0	0.05	0.08	0.147 (+0.056, −0.049)	0.067	0.19
He^0/Ne^0	2330	2330	1330 (+570, −366)	200	813
C^0/He^0	1.2×10^{-6}		<0.08	1.8×10^{-5}	0.0026

*The photoionization code predictions were made with the photoionization code CLOUDY (Ferland, 1996). The photoionization code predictions are sensitive to the uncertain diffuse radiation field in the 200 Å–1200 Å region. The assumed LIC abundances are based on Gry and Dupin (1995) observations towards ε CMa.

[1] These predictions are from Slavin and Frisch (1998a, b) and Frisch and Slavin (1996), using LIC reference abundances (see text).

[2] Geiss et al. (1994), Gloeckler (1996, private communication).

[3] Cummings and Stone (1996).

TABLE III

Gas-phase neutral elements in sample of interstellar clouds

Ratio	Log(ratio)/H^0)	Source	Reference
O^0/H^0	−3.19	ε CMa LIC	Gry and Dupin (1995)
Mg^0/H^0	−7.45	ε CMa LIC	Gry and Dupin (1995)
D^0/H^0	−5.12	λ Sco warm cloud	York (1983)
N^0/H^0	−4.14	λ Sco warm cloud	York (1983)
O^0/H^0	−3.37	λ Sco warm cloud	York (1983)
Ar^0/H^0	−5.81	λ Sco warm cloud	York (1983)
Na^0/H^0	−8.06	ζ Oph warm cloud	Morton (1975)
Mg^0/H^0	−8.01	ζ Oph warm cloud	Savage et al. (1992)
$C^0/(H^0+H_2)$	−5.77	ζ Oph cold cloud	Morton (1975)
$Si^0/(H^0+H_2)$	≤−8.54	ζ Oph cold cloud	Morton (1975)
$Fe^0/(H^0+H_2)$	−9.6	ζ Oph cold cloud	Morton (1975)
$S^0/(H^0+H_2)$	−7.19	ζ Oph cold cloud	Federman and Cardelli (1995)
C^+/H^0	−3.85	global ISM	Cardelli et al. (1996)
N^0/H^0	−4.12	global ISM	Meyer et al. (1998a)
O^0/H^0	−3.50	global ISM	Meyer et al. (1998b)

Most elements are not present in solar abundances in interstellar gas, a phenomena explained by invoking the depletion of the missing elements onto dust grains which are mixed into the interstellar gas. Recently arguments have been made that the correct reference abundance for the interstellar medium should be B-star abundances, at about 70% solar (e.g., Snow and Witt, 1996), reducing the amount of ISM tied up in grain mass.

Typical gas-to-dust mass ratios of 100:1 are found in the global ISM. Depletion patterns depend on cloud type, and nearby interstellar gas shows enhanced abundances with respect to the abundance patterns seen in distant cold clouds (Frisch, 1981). Generally elements with high condensation temperatures (e.g., Si, Mg, Mn, Fe, Cr, Ni, Ca, Co, Ti) are the most depleted in cold clouds. In warm interstellar clouds these refractory elements show less depletion, i.e., have higher gas phase abundances (e.g., see Savage and Sembach, 1996). Warm clouds in the galactic halo have similar properties as warm disk clouds (e.g., Fitzpatrick, 1996). The abundances of Fe and Mg in Table I indicate LISM gas-phase abundance variations of factors of 2–5, suggesting variable grain destruction. In the cold cloud towards the star ζ Oph, 99.4% of Fe and 97% of Mg are depleted onto dust grains. In contrast, $\sim$90% of Fe and $\sim$85% of Mg are depleted onto dust grains in the LIC (Table I). These enhanced abundances seen in local gas were the basis for my original conclusion that the interstellar gas surrounding the solar system had been processed through a supernova shock front which had partially destroyed embedded dust grains (Frisch, 1981).

Figure 3 shows gas phase abundances of the most common elements, where in this case the solar abundances have been used as the reference abundance for interstellar gas. The gas phase abundances in both warm and cloud clouds are shown. Elements with FIPs less than 13.6 eV will be mainly ionized, whereas O, N, Ar, Ne, He will be primarily neutral when hydrogen is neutral. Also shown are uncertainties in Mg, Fe, Si, and O abundances in the LIC. Predicted ratios $Mg^0/O^0 = 2.2 \times 10^{-5}$ and $S^0/O^0 = 2.8 \times 10^{-6}$ are deduced from the radiative transfer model with LIC abundances discussed in Section 5 and Table II. Thus, Mg^0 and S^0 are candidates for positive detections in the pickup ion population.

Table I also summarizes Fe and Mg abundances towards several nearby stars. A puzzling pattern emerges from the data in Table I. The stars α CMa, ϵ CMa, and α Cen, appear to have enhanced Mg^+/H^0 and Fe^+/H^0 abundances in the foreground interstellar gas when compared to the other stars in the table, a property which may be partly attributed to ionization effects which are not included in the depletion estimates. Of the stars listed in Table I, these three stars sample clouds in the most compact region of space around the solar system. For stars sampling more distant regions, uncertainties could be introduced by unresolved velocity structure in saturated lines, but the good correlation between Fe^+ and Mg^+ suggests this is not the case (Frisch et al., 1998). Alternatively, this variation could be evidence for abundance variations in the gas-phase ISM over the longer sightlines within 20 pc of the Sun (e.g., Lallement et al., 1995; Piskunov et al., 1997; Vallerga and

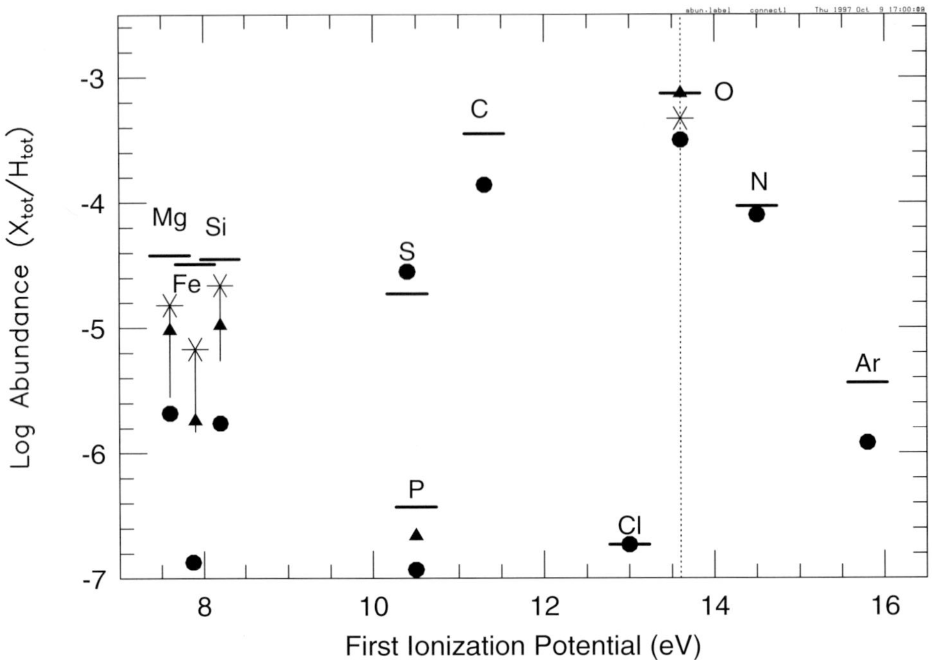

Figure 3. The abundances of the most common elements present in interstellar gas are shown plotted against FIP. X$_{tot}$ is the total abundance of element X, while H$_{tot}$ is the total abundance of H. The solar abundances (short line), and typical abundances in warm clouds (triangles) and cold clouds (circle) are shown, based on observations of ζ Oph. (The apparent overabundance of S in cold gas is probably the result of measurement uncertainties.) The vertical bars represent the range of abundances found for the LIC (Table I); the asterisks are the ϵ CMa LIC values. Since over twelve interstellar absorption components are seen towards ζ Oph, these abundances are guides only to the differences found in abundance patterns. Not shown is the ratio Na/H = -6.64 in the cold cloud towards ζ Oph, where the ionization potential of Na is 5.1 eV. Elements with FIPs less than 13.6 eV (the hydrogen ionization potential) will be ionized in neutral interstellar clouds. Both abundance patterns and ionization must be considered when estimating the abundances of neutrals in interstellar gas. These abundances, and the reference abundances, are taken from Savage and Sembach (1996), York (1983), and Table I.

Welsh, 1995; Frisch, 1995). This variation corresponds to a factor of $\sim$4 variation in gas phase abundances, but much smaller variations in dust mass due to the highly depleted nature of refractories.

7. Hydrogen 'Wall' at Heliosphere Nose

The relative motions of the Sun and surrounding interstellar cloud result in a pile up of interstellar H^0 at the nose of the heliosphere caused by the charge exchange coupling of interstellar H^0 and H$^+$ (e.g., Baranov et al., 1971; Baranov and Malama, 1993; Gayley et al., 1997, and references therein). This pile up leaves a signature in the Lyα absorption line which can, in principle, be distinguished and modeled in the spectrum of nearby stars (Linsky and Wood, 1997; Frisch et al., 1996). Gayley et al. (1997) have modeled the Lyα due to the pile up in the spectrum of α Cen, and found that a barely subsonic heliosphere model with a Mach number of 0.9 provided the best fit. The model parameters used in this fit were $n(\mathrm{H}^0) = 0.14$ cm^{-3}, $n(e^-) = 0.1$ cm^{-3}, $T = 7600$ K, and $V = -26$ km s^{-1}. For this set of parameters, if the excess velocity (over the sound speed) is the Alfvén speed resulting from an interstellar magnetic field, then a magnetic field strength of $B = 3.1$ μG is required to achieve a Mach number of 0.9. This gives sound and Alfvén speeds of 10.5 and 20.9 km s^{-1}, respectively. Although the full range of parameter space has not yet been explored for the modeling process, supersonic and subsonic models corresponding to Mach numbers 1.5 and 0.7, or magnetic field strengths of 2 and 5 μG, were found to provide unacceptable fits to the observed Lyα profile. Based on this exercise, evidently the properties of astrospheres may be used to determine the pressure of the interstellar medium at the location of a star, as predicted earlier by Frisch (1993).

8. Concluding Remarks

One new conclusion presented here is that in principle the *in situ* pickup ion data can help resolve the outstanding question of whether the correct reference abundances for the LIC are given by solar versus B-star abundances.

A second new result is that the uncertainties on the LIC velocity vector indicate that it is not yet clear whether the Sun and α Cen are immersed in the same interstellar cloud.

Based on the discussions in this paper, the best values for LIC properties are given by $n(\mathrm{H}^0) = 0.22 \pm 0.06$ cm^{-3}, $n(e^-) = n(\mathrm{H}^+) = 0.1$ cm^{-3}, $T = 6900$ K and a relative Sun-cloud velocity of 25.8 ± 0.8 km s^{-1}. However, radiative transfer considerations in the LIC suggest that the quoted neutral density is a lower limit. Ulysses and EUVE observations of He^0 indicate a cloud temperature of $T = 6400$ K. The magnetic field strength is weakly constrained to be in the range of 2–3 μG. Models of the Lyα absorption line towards α Cen are consistent with an Alfvén velocity of 20.9 km s^{-1}, which in turn is consistent with an interstellar magnetic field of 3 μG in the absence of additional unknown contributions to the interstellar pressure. *Ulysses* and EUVE observations of interstellar He^0 within the solar system give an upwind direction for the 'wind' of interstellar gas through the solar system, in the rest frame of the Sun, of $V = -25.9 \pm 0.6$ km s^{-1} arriving from the galactic direction $l = 4.0° \pm 0.2°$, $b = 15.4° \pm 0.6°$. Removing solar motion from this vector gives an upwind direction for the LIC cloud in the LSR of $V = -18.7 \pm 0.6$ km s^{-1} arriving from the direction $l = 327.3° \pm 1.4°$, $b = 0.3° \pm 1.0°$.

Through a combination of observations and theory, uncertainties in the LIC electron density are narrowing. Radiative transfer in the sightlines towards nearby stars require that cloud models must be combined with data in order to deduce properties at the cloud location. Radiative transfer models of ionization in the LISM show interesting results, but additional understanding of the input radiation fields is needed. The Local Fluff complex is structured and inhomogeneous. Striking progress would be made in understanding this structure if interstellar absorption lines could be observed at resolutions of ~ 1 km s^{-1} in the ultraviolet. The most glaring uncertainty is the absence of detailed knowledge about the interstellar magnetic field. Many of the most abundant elements in the LIC are ionized and densities of neutral atoms with FIPs less than 13.6 eV are typically down by 1–3 orders of magnitude from the dominant ions. The current approach of trying to understand the interaction of the ISM with the heliopause, from both the outside in and the inside out, is finally bearing fruit.

Acknowledgements

I would like to thank John Vallerga and a second 'anonymous' referee for thoughtful comments which have improved the quality of this paper. I would also like to thank Dan Welty and Adolf Witt for helpful discussions. I gratefully acknowledge the support of NASA grants NAG 5–6188 and NAG 5–6405.

References

Baranov, V. B., Krasnobaev, K. I., and Kulikovsky, A.: 1971, *Soviet Phys. Dokl.* **15**, 791.

Baranov, V. B. and Malama, Y. G.: 1993, *J. Geophys. Rev.* **98**, 15 157.

Bash, F.: 1986, in R. Smoluchowski, J. N. Bahcall, and M. Matthews, (eds.), *The Galaxy and the Solar System*, University of Arizona Press, Tucson, AZ, p. 83.

Bertin, P., Lallement, R., Ferlet, R., and Vidal-Madjar, A.: 1993a, *J. Geophys. Rev.* **98**, 15 193.

Bertin, P., Lallement, R., Ferlet, R., and Vidal-Madjar, A.: 1993b, *Astron. Astrophys.* **278**, 549.

Bertin, P., Vidal-Madjar, A., Lallement, R., Ferlet, R., Lemoine, M.: 1995, *Astron. Astrophys.* **302**, 889.

Cardelli, J. A., Meyer, D. M., Jura, M., and Savage, B. D.: 1996, *Astrophys. J.* **467**, 344.

Cheng, K.-P. and Bruhweiler, F. C.: 1990, *Astrophys. J.* **364**, 573.

Cummings, A. C. and Stone, E. C.: 1996, *Space Sci. Rev.* **78**, 117.

Crawford, I. A.: 1994, *The Observatory* **114**, 288.

Crawford, I. A. and Dunkin, S. K.: 1995, *Monthly Notices Roy. Astron. Soc.* **273**, 219.

Crawford, I. A., Craig, N., and Welsh, B. Y.: 1997, *Astron. Astrophys.* **317**, 889.

Dupuis, J., Vennes, S., Pradhan, A. K., and Thejll, P.: 1995, *Astrophys. J.* **455**, 574.

Falgarone, E.: 1997, *IAU Colloq.* **166**.

Falgarone, E. and Puget, J.-L.: 1995, *Astron. Astrophys.* **293**, 840.

Federman, S. R. and Cardelli, J. A.: 1995, *Astrophys. J.* **452**, 269.

Ferland, G. J.: 1996, *Hazy, a Brief Introduction to Cloudy*, University of Kentucky, Department of Physics and Astronomy Internal Report.

Fitzpatrick, E. L.: 1996, *Astrophys. J.* **473**, L55.

Flynn, B., Vallerga, J., Dalaudier, F., and Gladstone, G. R.: 1998, *J. Geophys. Rev.*, in press.

Frail, D. A., Weisberg, J. M., Cordes, J. M., and Mathers, C.: 1994, *Astrophys. J.* **436**, 144.

Frisch, P. C.: 1981, *Nature* **293**, 377.

Frisch, P. C.: 1991, in S. Grzedzielski and D. E. Page (eds.), 'Physics of the Outer Heliosphere', *COSPAR Colloquia Series*, Vol. 1, p. 10.

Frisch, P. C.: 1993, *Astrophys. J.* **407**, 198.

Frisch, P. C.: 1994, *Science* **265**, 1423.

Frisch, P. C.: 1996, *Space Sci. Rev.* **72**, 499.

Frisch, P. C.: 1997, *Science*, submitted.

Frisch, P. C.: 1998, *Astrophys. J.*, in preparation.

Frisch, P. C. and Slavin, J. D.: 1996, *Space Sci. Rev.* **78**, 223.

Frisch, P. C. and Welty, D. G., :1998, in preparation.

Frisch, P. C. and York, D. G.: 1991, in *Extreme Ultraviolet Astronomy*, Pergamon Press, London, pp. 322–332.

Frisch, P. C., Welty, D. E., York, D. G., and Fowler, J.: 1990, *Astrophys. J.* **357**, 514.

Frisch, P. C., Welty, D. E., Pauls, H. L., Williams, L. L., and Zank, G. P.: 1996, *Bull. Am. Astron. Soc.* **28**, 760.

Frisch, P., Dorschner, J., Geiss, J., Greenberg, M., Gruen, E., Landgraf, M., Hoppe, P., Jones, A., Kraetschmer, W., Linde, T., Morfill, G., Reach, W., Slavin, J., Svestka, J., Witt, A., and Zank, G.: 1998, *Astrophys. J.*, in preparation.

Gayley, K. G., Zank, G. P., Pauls, H. L., Frisch, P. C., and Welty, D. E.: 1997, *Astrophys. J.* **487**, 259.

Geiss, J., Gloeckler, G., Mall, U., von Steiger, R., Galvin, A. B., and Ogilvie, K. W.: 1994, *Astron. Astrophys.* **282**, 924.

Gloeckler, G., Fisk, L. A., and Geiss, J.: 1997, *Nature* **386**, 374.

Gruntman, M. A.: 1993, *Planetary Space Sci.* **41**, 307.

Gry, C., Lemonon, L., Vidal-Madjar, A., Lemoine, M., and Ferlet, R.: 1995, *Astron. Astrophys.* **302**, 497.

Gry, C. and Dupin, O.: 1995, in P. Benvenuti, F. D. Macchetto, and E. J. Schreier (eds.), Paris, France. Science with the Hubble Space Telescope – II.

Heiles, C.: 1997, *Astrophys. J.* **481**, 193.

Lallement, R.: 1996, *Space Sci. Rev.* **78**, 361.

Lallement, R. and Ferlet, R.: 1996, *Astron. Astrophys.*, in press.

Lallement, R. and Bertin, P.: 1992, *Astron. Astrophys.* **266**, 479.

Lallement, R., Vidal-Madjar, A., and Ferlet, R.: 1986, *Astron. Astrophys.* **168**, 225.

Lallement, R., Bertin, P., Ferlet, R., Vidal-Madjar, A., and Bertaux, J. L.: 1994, *Astron. Astrophys.* **286**, 898.

Lallement, R., Bertin, P., Ferlet, R., Vidal-Madjar, A., and Bertaux, J. L.: 1995, *Astron. Astrophys.* **304**, 461.

Lemoine, M., Vidal-Madjar, A., Ferlet, R., Bertin, P., Gry, C., and Lallement, R.: 1996, *Astron. Astrophys.* **308**, 601.

Linsky, J. L., and Wood, B. E.: 1997, *Astrophys. J.* **463**, 254.

Meyer, D. M., Cardelli, J. A., and Sofia, U. J.: 1998a, *Astrophys. J.*, in press.

Meyer, D. M., Jura, M., and Cardelli, J. A.: 1998b, *Astrophys. J.*, in press.

Mihalas, D. and Binney, J.: 1981, *Galactic Astronomy Structure and Kinematics*, W. H. Freeman & Co., San Francisco.

Moebius, E.: 1996, *Space Sci. Rev.* **78**, 375.

Morton, D. C.: 1975, *Astrophys. J.* **197**, 85.

Piskunov, N., Wood, B. E., Linsky, J. L., Dempsey, R. C., and Ayres, T. R.: 1997, *Astrophys. J.* **474**, 315.

Quemerais, E., Bertaux, J. L., Sandel, B. R., and Lallement, R.: 1996, *Astron. Astrophys.* **308**, 279.

Reynolds, R. J.: 1991, in S. Grzedzielski and D. E. Page (eds.),'Physics of the Outer Heliosphere', *COSPAR Colloquia Series*, Vol. 1., p. 101.

Savage, B. D. and Sembach, K. R.: 1996, *Ann. Rev. Astron. Astrophys.* **34**, 279.

Savage, B. D., Cardelli, J. A., and Sofia, U. J.: 1992, *Astrophys. J.* **401**, 706.

Slavin, J. D.: 1989, *Astrophys. J.* **346**, 718.

Slavin, J. D. and Frisch, P. C.: 1998a, in Breitschwerdt and Freyberg (eds.), 'Local Bubble and Beyond', *IAU Colloq.* **166**.

Slavin, J. D. and Frisch, P. C.: 1998b, in preparation.

Snow, T. P. and Witt, A. N.: 1996, *Astrophys. J.* **468**, L65.

Sonett, C. P., Morfill, G. E., and Jokipii, J. R.: 1987, *Nature* **330**, 458.

Tinbergen, J.: 1982, *Astron. Astrophys.* **105**, 53.

Vallerga, J.: 1996, *Space Sci. Rev.* **78**, 277.

Vallerga, J.: 1998, *Astrophys. J.* **497**, 921.

Vallerga, J. V. and Welsh, B. Y.: 1995, *Astrophys. J.* **444**, 702.

Vallerga, J. V., Vedder, P. W., Craig, N., and Welsh, B. Y.: 1993, *Astrophys. J.* **411**, 729.

Welty, D. E., Hobbs, L. M., Lauroesch, J., Morton, D. C., Spitzer, L., York, D. G.: 1998, in preparation.

Welty, D. E., Morton, D. C., and Hobbs, L. M.: 1996, *Astrophys. J.* **106**, 533.

Witte, M., Rosenbauer, H., Banaskiewicz, M., and Fahr, H.: 1993, *Adv. Space Res.* **13**, 121.

Witte, M., Banaskiewicz, M., Rosenbauer, H.: 1996, *Adv. Space Res.* **78**, 289.

Wood, B. E., Linsky, J. L.: 1997, *Astrophys. J.* **474**, 39.

York, D. G.: 1983, *Astrophys. J.* **264**, 172.

Zank, G. and Frisch, P. C.: 1998, *Astrophys. J.*, in preparation.

INTERSTELLAR AND INNER SOURCE PICKUP IONS OBSERVED WITH SWICS ON *ULYSSES*

GEORGE GLOECKLER

Department of Physics and IPST, University of Maryland, College Park, MD 20742, U.S.A. and Department of Atmospheric, Oceanic and Space Sciences, University of Michigan, Ann Arbor, MI 48109, U.S.A.

JOHANNES GEISS

International Space Science Institute, Hallerstrasse 6, CH-3012 Bern, Switzerland

Abstract. Many species of pickup ions, both of interstellar origin and from an inner, distributed source have been discovered using data from the Solar Wind Ion Composition Spectrometer (SWICS) on *Ulysses*. Velocity distribution functions of these ions were measured for the first time over heliocentric distances between 1.35 and 5.4 AU, both at high and low latitudes, and in the disturbed slow solar wind as well as the steady fast wind of the polar coronal holes. This has given us the first glance at plasma properties of suprathermal ions in various solar wind flows, and is enabling us to study the chemical and, in the case of He, the isotopic composition of the local interstellar cloud. Among the new findings are (a) the surprisingly weak pitch-angle scattering of low rigidity, suprathermal ions leading to strongly anisotropic velocity distributions in radial magnetic fields, (b) the efficient injection and consequent acceleration of pickup ions, especially He^+ and H^+, in the turbulent solar wind, and (c) the discovery of a new extended source releasing carbon, oxygen, nitrogen and possibly other atoms and molecules in the inner solar system. Pickup ion measurements are now used to study the characteristics of the local interstellar cloud (LIC) and, in particular, to determine accurately the abundance of atomic H, He, N, O, and Ne, the isotopes of He and Ne, as well as the ionization fractions of H and He in the LIC. Pickup ion observations allow us to infer the location of the termination shock and, in combination with measurements of anomalous cosmic rays, to investigate termination shock acceleration mechanisms.

1. Introduction

Ions in the expanding corona are virtually all of solar origin. With increasing distance from the Sun, however, non-solar ions are gradually mixed in, so that when the solar wind approaches the termination shock (estimated to be between $\sim$85 AU and 110 AU in the direction of motion of the Sun through the interstellar cloud, the solar apex), their relative abundance is estimated to have reached the order of 10%. These non-solar ions originate from extended sources such as the interstellar gas, interstellar grains and interplanetary grains, as well as from local sources such as comets or planetary atmospheres as is illustrated in Figure 1.

At relatively unperturbed solar wind conditions, two distinct types of ion velocity distributions are observed in interplanetary space. (1) The dominant solar wind ions of solar origin are found in a characteristically narrow Mach angle that cor-

Space Science Reviews **86**: 127–159, 1998.

Heliospheric Pickup Ions

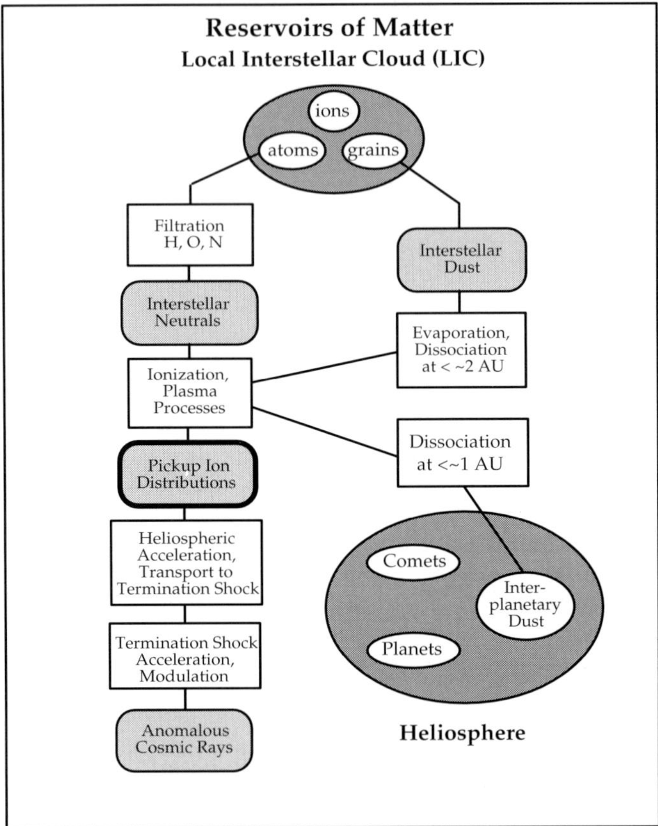

Figure 1. Schematic representation of sources and processes leading to the creation of pickup ions in the heliosphere and their subsequent transport and acceleration to form Anomalous Cosmic Rays. In addition to the interstellar source, a new extended inner source (interstellar and interplanetary dust) inside the heliosphere has been identified using SWICS/*Ulysses* data. The large shaded ovals represent source regions of matter shown in small ovals. Physical processes are shown in clear boxes. These processes produce the particle populations shown in the lightly shaded boxes.

responds to Mach numbers of typically 10 to 20. (2) Less abundant are the pickup ions which have broad suprathermal distribution functions that in the undisturbed solar wind show a sharp decrease at the upper velocity limit of twice the solar wind speed (Möbius et al., 1985; Gloeckler et al., 1993). At heliocentric distances of several AU and sufficiently far from local sources, their number densities are of the order of 10^{-3} relative to the number densities of the solar wind ions. All pickup ions that have been identified up to now are of non-solar origin.

For presenting and comparing different pickup ion populations, it is advantageous to relate V_{ion}, the speed of an incoming ion to V_{sw}, the solar wind bulk speed at that time. We will use

$$W = V_{\text{ion}}/V_{\text{sw}}\,, \tag{1}$$

for the relative speed of an ion in the reference frame of the Sun (and the spacecraft). Then

$$\mathbf{w} = (\mathbf{V}_{\text{ion}} - \mathbf{V}_{\text{sw}})/V_{\text{sw}}, \tag{2}$$

relates the relative velocity, $\mathbf{w}$, in the reference frame of the solar wind to the ion velocity, $\mathbf{V}_{\text{ion}}$, in the spacecraft frame.

Three criteria are available for identifying pickup ion populations and determining their sources of origin:

(1) *Charge states.*   Solar wind ions of heavy elements have high charges reflecting the typical coronal temperature of $\sim 10^6$ K. Pickup ions, on the other hand, are predominately singly charged. The reason for this is simple. Pickup ions, created as singly charged, are rapidly convected by the solar wind from regions of high EUV and solar wind flux that caused their ionization to distances where further ionization becomes negligible. Another way of stating this is that the time it takes to ionize an atom (by combination of photoionization by solar EUV, charge exchange with solar wind ions, and ionization by suprathermal electrons) is long compared to the time it takes to convect the pickup ion from regions of high to low ionization, a distance of several AU. The only important exception is He which also has a relatively large cross section for double ionization. Therefore, the vast majority of pickup ions produced from atoms or molecules outside magnetospheres remains singly charged. Inside magnetospheres, where ion residence lifetimes are long and electron energies, and hence ionization rates, are high, ions produced from neutral atmospheres may carry two or more charges (Young et al., 1977; Gloeckler et al., 1985; McNutt et al., 1981; Geiss et al., 1992), and thus, multiply charged ions of planetary origin may be found in the wake of planets. Measurements of ionization states of heavy ions is one of the easiest ways to separate and identify low-charge state, heavy pickup ions from the far more abundant, highly charged solar wind ions.

(2) *Spatial distributions in the heliosphere.*   Ions produced from the interstellar gas have a very characteristic spatial distribution in the heliosphere. Their number density depends on both the angle of the observer with respect to the direction of the solar apex, as well as on the ionization rate which determines the distance from the Sun of their maximum abundance (e.g., Thomas, 1978; Holzer, 1989). The density of ions of local origin (e.g., comets, Jupiter, etc.), on the other hand, decrease with distance from the source (cf., Neugebauer et al., 1987; Geiss et al., 1994b; Ogilvie et al., 1995). However, local sources of neutrals, located directly between the Sun and the point of observation, can be detected even at large distances from the source as was recently demonstrated by Grünwaldt et al. (1997) who, with an

instrument near Earth, recorded pickup ions produced from neutrals originating from Venus.

(3) *Velocity distributions.* In the solar wind flowing through the heliosphere, pickup ions are produced with $W \ll 1$ from slowly moving atoms or molecules, such as the interstellar gas atoms, cometary constituents or molecules, and atoms and ions liberated from grains by evaporation or sputtering. These ions are immediately picked up by the electromagnetic field of the solar wind, and rapidly convected outward. Pitchangle scattering and or variations in the magnetic field ($\mathbf{B}$) directions are then required to spread the ions into shell-like distributions in $\mathbf{w}$ and, in particular, change W of some of the ions from less than to greater than 1. When the average direction of $\mathbf{B}$ is nearly perpendicular to V_{sw}, i.e., for distances greater than about a few AU at low heliocentric latitude, distributions showing little anisotropy between $W < 1$ and $W > 1$ are expected and are indeed observed (Gloeckler et al., 1993, 1997). On the other hand, at high latitudes or very close to the Sun where the B field is oriented more radially, the velocity distribution of pickup ions is found to be highly anisotropic with little scattering into the $W > 1$ phase-space hemisphere (Gloeckler et al., 1995). In either case, pickup ions are adiabatically cooled in the expanding solar wind. Thus, when moving away from the region where they are created, the shell-like distribution of these pickup ions shrinks to lower w values. All these processes tend to produce a pickup ion velocity distribution in the solar wind frame that is relatively flat to $w_c = V_{sw}$ with a sharp cutoff beyond w_c. These highly suprathermal pickup ion distributions are distinctly different from the thermal spectra of solar wind ions. For protons and $^4\mathrm{He}^{++}$ this difference in the velocity distributions provides the only means of separating pickup hydrogen and $^4\mathrm{He}^{++}$ from solar wind protons and alpha particles respectively.

Using one or more of the three criteria given above, the following pickup ions have been discovered: $^1\mathrm{H}^+$ (Gloeckler et al., 1993), $^4\mathrm{He}^+$ (Möbius et al., 1985; Gloeckler et al., 1993), $^3\mathrm{He}^+$ (Gloeckler and Geiss, 1996), $^4\mathrm{He}^{++}$ (Gloeckler et al., 1997), N^+, O^+, and Ne^+ (Geiss et al., 1994a). The dependence of their abundances on solar distance and latitude, as well as the shape of their velocity distributions show that the majority of the pickup ions of these species in the heliosphere as a whole is produced from the neutral interstellar gas as it flows through the solar system.

C^+ ions are also found at all solar latitudes and distances visited by *Ulysses* (cf., Figures 13 and 14). However, their distribution in space and their phase-space density spectra differ radically from pickup ions produced from the interstellar gas. Geiss et al. (1995, 1996) have shown that the majority of the C^+ ions as well as a fraction of O^+ and N^+ are produced by an 'inner source' that is located at solar distances below a few AU. These authors presented evidence showing that at medium and high solar latitudes, molecules sublimating from interstellar grains that have penetrated into the inner heliosphere constitute a significant component of this inner source. In Sections 3 and 4 we present new data and relate these new results to our earlier work, in order to further characterize the inner source.

2. Hydrogen and Helium Pickup Ions

The orbit of *Ulysses* and the low background capabilities of the SWICS instrument (Gloeckler et al., 1992) made it possible to discover pickup hydrogen (Gloeckler et al., 1993) whose existence was postulated in the early 1970s (e.g., Blum and Fahr, 1970; Axford, 1972, and references therein; Vasyliunas and Siscoe, 1976), and whose detection was anticipated with the launches of the deep space missions Pioneers 10 and 11 and Voyagers 1 and 2. The exceptionally low background of SWICS and the long time periods that *Ulysses* spent in the fast, steady and quiet solar wind from the polar coronal holes enabled us to identify the uncommon $^4\mathrm{He}^{++}$ pickup ions and from these measurements deduce the abundance of He in the local interstellar cloud (Gloeckler et al., 1997). With SWICS we also discovered the extremely rare pickup $^3\mathrm{He}^+$ and used its measured abundance to place a new lower limit on the amount of missing matter in the universe (Gloeckler and Geiss, 1996).

2.1. Pickup Ions Observed in the Quiet, High-Speed Solar Wind of the Polar Coronal Hole

2.1.1. *Protons*
Pickup hydrogen ions are distinguishable from the far more abundant solar wind protons by their distinctly different velocity distributions as displayed in Figure 2. The proton phase-space density shown was accumulated over a 100-day period (10 April to 19 July 1994) when *Ulysses* was at an average heliocentric distance of 3.0 AU and at a high latitude ($-66°$) in the undisturbed and fast (785 km s^{-1}) solar wind from the south polar coronal hole. The solar wind proton distribution around $W = 1$ is narrow and well represented by a kappa function in the solar wind frame:

$$f_{\mathrm{sw}}(w) = f_0[1 + (w/\theta)^2/\kappa]^{-(\kappa+1)} , \tag{3}$$

$$\theta = V_{\mathrm{th}}[1 - (1.5/\kappa)]^{1/2} , \tag{4}$$

The fit (curve labeled Solar Wind) to the solar wind peak around $W = 1$ is the kappa distribution (Equation (3)) transformed to the spacecraft frame using Equation (2), and integrated over the instrument view angles (see Gloeckler, 1996, for details). Best values for the density, thermal speed, V_{th}, and κ were 0.30 cm^{-3}, 32.3 km s^{-1} and 6.3, respectively. We note in passing that a maxwellian distribution fails to fit the solar wind proton distribution we observe.

The speed distribution of pickup hydrogen, on the other hand, is broad and extends to $W = 2$ at which point the phase-space density drops by about three orders of magnitude. The fact that the two distributions are quite distinct implies

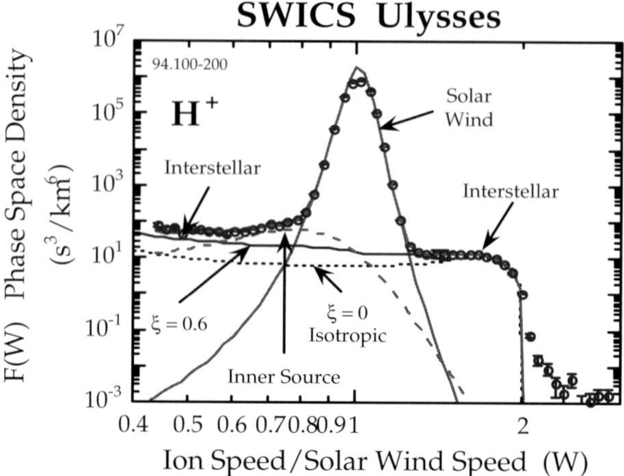

Figure 2. Phase-space density of H$^+$ versus W, the ion speed in the spacecraft frame of reference divided by the solar wind speed. This time averaged spectrum was observed with SWICS during a 100-day period in 1994 when *Ulysses* was in the steady high speed ($\sim$785 km s^{-1}) stream of the southern polar coronal hole. The average heliocentric distance and latitude were 3.0 AU and $-66°$ respectively, and the average direction of the magnetic field almost radial (165°). The two highest density points closest to $W = 1$ fall below the kappa curve fit (labeled Solar Wind) because instrument deadtime saturated the proton count rate used here. We note that the solar wind proton number density is computed from another rate (start detector rate) which does not suffer from these deadtime effects. In these and subsequent plots the phase-space density, $F_{m,m/q}(W)$, is computed from the differential energy per charge flux of the ion with mass (m) and mass/charge (m/q) in the standard fashion using a calculated value for the isotropic geometrical factor. Model distribution functions are then computed, first in the solar-wind frame of reference and then transformed to the spacecraft frame, integrating over the SWICS view angles using the measured directional geometrical factor to obtain the predicted count rate for each of the 64 W intervals. These count rates are then converted to a predicted phase-space density for each W step using the calculated isotropic geometrical factor of the instrument. (See Gloeckler, 1996 for further details.)

that the interaction between the solar wind and pickup ions is sufficiently weak so that no indication of thermalization of the pickup ion populations is apparent. The very sharp drop in phase-space density at $W = 2$ indicates little energy diffusion, at least in the quiet solar wind.

The dotted curve labeled $\xi = 0$ represents the predicted phase-space density of pickup protons using the 'hot model' of Thomas (1978) for the spatial distribution of hydrogen atoms in the heliosphere. We use the Vasyliunas and Siscoe (1976) equation, $f_{iso}(R, \Theta, w)$, derived under the assumption of rapid pitchangle scattering and hence isotropy (see also Gloeckler, 1996) for computing the phase-space density in the solar wind frame of reference of the resulting pickup protons at the location (R, Θ) of *Ulysses*, with R the heliocentric radial distance in AU and Θ the angle between the direction of motion of the Sun relative to the interstellar cloud and the Sun–*Ulysses* line. Finally, we make the coordinate transformation to the spacecraft frame and integrate over the view directions of SWICS. The parameter values used for the density, temperature and flow speed of atomic hydrogen at the termination shock were 0.115 cm^{-3} (Gloeckler et al., 1997), 7000 K, and 20 km s^{-1} (Lallement, 1996; Geiss and Witte, 1996) respectively. The loss rate ($\beta_{\text{Hloss}} = 5.5 \times 10^{-7}$ s^{-1}) and the ratio of radiation pressure to the gravitational force (μ) were taken from the best fit to the spatial distribution of the pickup H$^+$ measured between 1.5 and 3.3 AU (see Figure 6). The total production rate of pickup protons from atomic hydrogen required to match the observations for W above $\sim$1.7 was $\beta_{\text{Hprod}} = 2.25 \times 10^{-7}$ s^{-1}. This value is considerably below the measured value of the production rate (3.5×10^{-7} s^{-1}) due to charge exchange with solar wind protons alone. Furthermore, it is clear that the model distribution, based on the assumption of rapid pitchangle scattering, fails to account for the observed spectral shape. In fact, no isotropic (in the solar wind frame) velocity distribution of any kind would fit the observed spectrum both below and above $W = 1$. Following Gloeckler et al. (1995) and Fisk et al. (1997) we therefore conclude that fast pitchangle scattering leading to isotropic phase-space density distributions in the solar wind frame does not occur during the time period and solar wind conditions of these observations.

The solid curve labeled $\xi = 0.6$ uses a modified form of the Vasyliunas and Siscoe (1976) equation. Instead of fast scattering at all pitchangles we assume a strong suppression of scattering at 90° pitchangle, but allow rapid scattering for all pitchangles less than as well as greater than 90° (Fisk et al., 1997). The modified form of the Vasyliunas and Siscoe (1976) equation we use is $f(R, \Theta, w, \theta) = f_{\text{iso}}(R, \Theta, w)\,(1+\xi(w))$ for $\cos(\theta) < 1$ and $f(R, \Theta, w, \theta) = f_{\text{iso}}(R, \Theta, w) \times (1 - \xi(w))$ for $\cos(\theta) > 1$, where θ is the angle between the velocity, $\mathbf{w}$, of the pickup ion and the direction of the average magnetic field $\mathbf{B}$ taken to point inward. In the present case of nearly radial (173°) average magnetic field, pickup ions created at $W \ll 1$ stay at speeds below $W = 1$ with only a small fraction scattering past 90° into the speed range above $W = 1$. To approximate the effects of spatial transport of the pickup ion population, we assume that $\xi(w) = \xi \exp(-(1 - w^{3/2})R\phi/V_{\text{sw}})$ and treat ξ and ϕ as free parameters. The degree of anisotropy, ξ, related to the probability of scattering through 90°, is defined in terms of the ratio, ρ, of the density in the <90° ($W < 1$) to that in the >90° pitchangle ($W > 1$) hemisphere of phase-space, $\rho = (1 + \xi)/(1 - \xi)$.

The $\xi = 0.6$ model distribution fits the observed pickup proton speed spectrum above $W = 1.25$ quite well. The parameters used to compute this model distribution are the same as those for the $\xi = 0$ model curve, except for β_{Hprod}, the total production rate and μ. To obtain the best match to the observed distribution above $W = 1.25$ required $\mu = 1.17$ and $\beta_{\text{Hprod}} = 4.95 \times 10^{-7}$ s^{-1}. The contributions to the total production rate are primarily from charge exchange with solar wind protons (measured to be 3.5×10^{-7} s^{-1}) with the rest coming from ionization by solar UV and electron impact (Rucinski et al., 1996). For the anisotropy parameters we chose $\xi = 0.6$ and $\phi = 800$ (km s^{-1}) AU^{-1}. Using Equation (2) of Fisk et al. (1997) an anisotropy $\xi = 0.6$ indicates a large ($\sim$1.2 AU) mean free path.

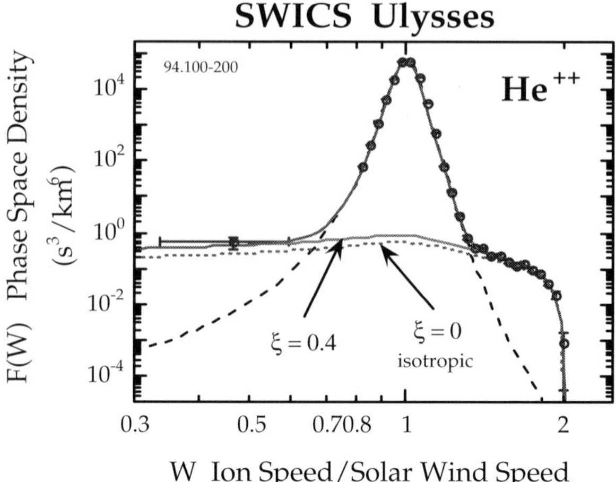

Figure 3. Same as Figure 2 for He^{++}. This time averaged spectrum was taken during the same time period as the proton distribution of Figure 2. Pickup He^{++} phase-space densities were used to derive the absolute interstellar helium number density of 0.0153 cm^{-3} because its production rate (almost entirely by double charge exchange with solar wind alpha particles) was measured by SWICS. This method eliminates instrumental systematic error because the interstellar He density is then essentially proportional to the ratio of the pickup He^{++} flux (at W close to 2) and the solar wind He^{++} flux (at $W = 1$). Because of the lower count rate no saturation of the rate near $W = 1$ occurs.

While the $\xi = 0.6$ fit matches the observed spectrum above $W = 1.25$ extremely well it falls below the measured distribution for $W < 0.85$. When data from the 45° Sun sector is excluded (as was done in Gloeckler et al., 1995) the observed phase-space density below $W \sim 0.62$ falls right on the $\xi = 0.6$ curve. To account for the relatively large contributions to the total density below $W = 0.8$, especially in the Sun sector, we postulate the existence of an inner source of pickup protons as indicated by the dashed curve. The shape of this 'inner source' proton spectrum is similar to that of the inner source carbon and oxygen distribution discussed in Section 3 and shown in Figures 11 and 13 below.

2.1.2. *Doubly Charged Helium-4*

In Figure 3 we show the ^{4}He^{++} speed distribution averaged over the same 100-day period in 1994 used for protons of Figure 2 above. The distribution of alpha particles is very similar to that of H$^+$, showing the clear separation of the solar wind alphas from interstellar pickup ^{4}He^{++}, which is produced almost entirely by double charge exchange of atomic helium with solar wind alpha particles (Gloeckler et al., 1997; Gloeckler, 1996; Rucinski et al., 1998). The distribution function of pickup ^{4}He^{++} (solid curve) is again anisotropic with $\phi = 800$ (km s^{-1}) AU^{-1} and $\xi \approx 0.4$, which corresponds to a mean free path of $\sim$1 AU using Equation (2) of Fisk et al. (1997). To obtain the model pickup ^{4}He^{++} distribution function we used the hot model of Thomas (1978) with a neutral interstellar helium density of 0.0153 cm^{-3}

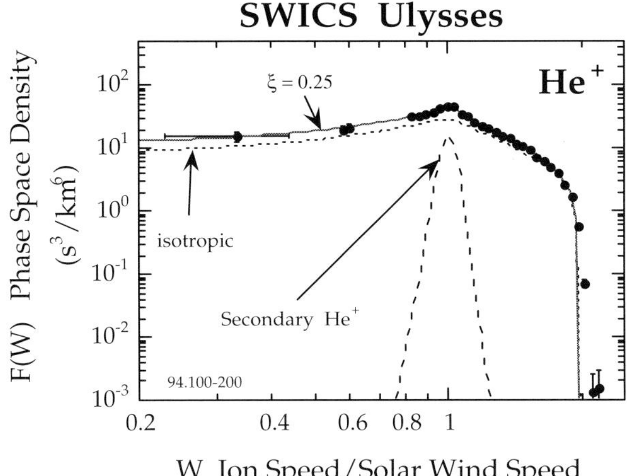

Figure 4. Same as Figure 3 for He$^+$. Although pickup He$^+$ is about 30 times more abundant than He^{++}, it could not be used to compute the interstellar helium density because its rate of production, almost entirely by solar UV, was not measured during the time of these observations. See text for further details and explanation of the various model curves.

(Gloeckler et al., 1997), a temperature of 7000 K (Witte et al., 1993, 1996) and a loss rate of 0.55×10^{-7} s^{-1} (see Figure 4; Rucinski et al., 1996). Ninety percent of total production rate comes from double charge exchange of solar wind alphas with interstellar atomic He and the rest from charge exchange with pickup ^{4}He$^+$ and photoionization (Gloeckler et al., 1997). Thus, the dominant portion of the total production rate of 0.0215×10^{-7} s^{-1} used here is known reasonably well from the product of the measured solar wind alpha particle flux (adjusted to 1 AU) and the double charge exchange cross section of 1.9×10^{-16} cm^2 (Rucinski et al., 1998) at an energy corresponding to the $\sim$800 km s^{-1} wind. Best values for the solar wind alpha particle density, thermal speed, V_{th} and κ were 0.0145 cm^{-3}, 36 km s^{-1}, and 5.2, respectively.

2.1.3. *Singly Charged Helium-4*

Interstellar ^{4}He$^+$ pickup ions are the most abundant suprathermal ions at 1 AU and comparable in density to pickup protons at $\sim$5 AU. They were the first of the interstellar pickup ions to be discovered (Möbius et al., 1985), although the detailed spectrum of these ions over a wide velocity range was obtained only recently with the SWICS/*Ulysses* instrument (Gloeckler et al., 1993). Figure 4 shows the ^{4}He$^+$ phase-space density as a function of W, the speed of ^{4}He$^+$ divided by the solar wind ^{4}He^{++} speed. The time period for this average spectrum was the same as used for the two previous figures. SWICS samples almost all of the relevant phase-space in the case of pickup ^{4}He$^+$. We show for the first time the ^{4}He$^+$ phase-space density over the entire interval of speeds between $\sim 0.2 < W < \sim 2$. The small gaps around $W = 0.5$ and $W = 0.7$ result from spillover of solar wind protons and alpha particles for which no corrections have as yet been made. To obtain the ^{4}He$^+$ densities in the important speed interval around $W = 1$ we carefully subtracted known contributions from the measured fluxes of solar wind Si^{+7}, Mg^{+6}, Fe^{+14}, as well as O^{+6} and Fe^{+12}. The solid and dotted curves are the predicted spectra using the same parameters as for the computation of the ^{4}He^{++} pickup ion distribution, except for the production rate. The dotted curve, which assumes an isotropic distribution, does not fit the measured distribution. The solid curve, which fits the data over the entire speed range, uses an anisotropy factor $\xi = 0.25$, corresponding to a mean free path of $\sim$0.8 AU, and $\phi = 800$ (km s^{-1}) AU^{-1}. The total production rate required to match the data was 0.8×10^{-7} s^{-1}. We note the major part of the production rate of ^{4}He$^+$ is due to photoionization. The EUV photon flux was not measured during the present observation period.

In the region around $W \approx 1$ significant contributions to the production rate of pickup ^{4}He$^+$ can also come from electron impact ionization close ($<$1 AU) to the Sun (e.g., Rucinski et al., 1996, 1998). In this same W interval there are also contributions from 'secondary' ^{4}He$^+$, produced primarily in the charge exchange reaction ($\sigma = 8.5 \times 10^{-16}$ cm^2, Rucinski et al., 1998) of solar wind alpha particles with atomic hydrogen. In this reaction ^{4}He$^+$ is produced with the same velocity distribution as solar wind ^{4}He^{++}, as well as a small fraction of pickup hydrogen. For the time period used in this analysis we calculate the production rate and flux of secondary ^{4}He$^+$ to be 0.078×10^{-7} s^{-1} and 300 cm^{-2} s^{-1}, respectively using the measured solar wind alpha parameters and the atomic hydrogen spatial distribution deduced from our pickup proton measurements. The dashed curve in Figure 4 represents the calculated secondary ^{4}He$^+$ velocity distribution whose shape is taken to be the same as that of the measured solar wind alpha particles of Figure 3. Adding the secondary ^{4}He$^+$ to our model fit did not entirely account for the small peak in the ^{4}He$^+$ phase-space density around $W = 1$. To obtain the excellent fit (solid curve) to the observed spectrum over the entire range of speeds required additional production of ^{4}He$^+$ by electron impact ionization. The production rate we used was $\beta_{\mathrm{el}} = (0.16 \times 10^{-7}) R^{-2.3}$ s^{-1}, with R the heliocentric distance in AU.

2.1.4. *Singly Charged Helium-3*

The rarest of the light-mass pickup ions observed is ^{3}He$^+$. Long accumulation times were required to positively identify this uncommon pickup ion and measure its abundance relative to ^{4}He$^+$. The mass/charge distribution of ions with masses between 1.75 and 4.6 amu and $1.6 < W < 2.0$ measured by SWICS during the 40-month period (July 1992 to November 1995) of high-speed solar wind is given in the top panel of Figure 5. The triple coincidence requirement reduced residual background to about 10% of the ^{3}He$^+$ counts (less than one count in three years). While limited by counting statistics, the peak due to ^{3}He$^+$ is clearly visible at mass per charge around 3, between the peaks at mass/charge 2 and 4 from the far more abundant pickup ^{4}He^{++} and ^{4}He$^+$, respectively.

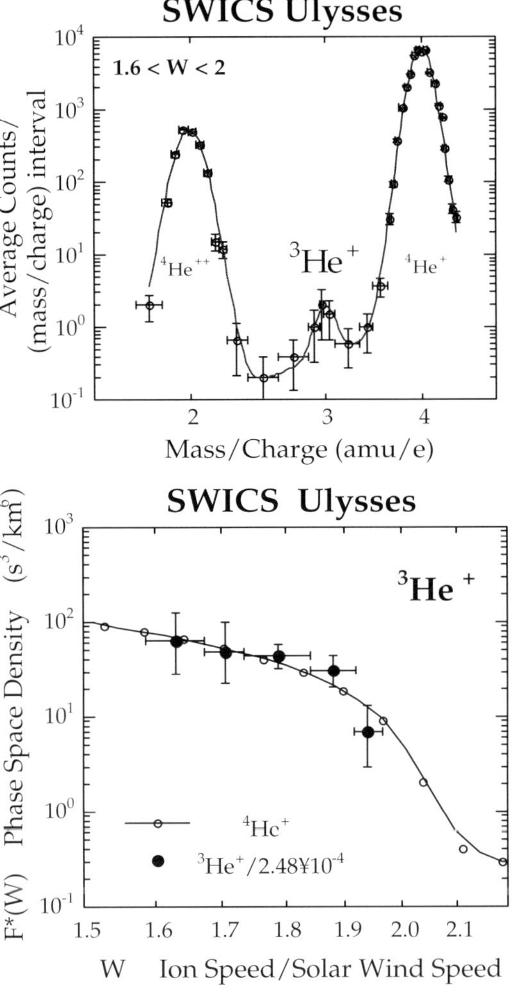

Figure 5. Top panel: plot of the average triple-coincidence counts of ions with masses between 1.75 and 4.6 amu and with normalized speed W between 1.6 and 2 per mass/charge (m/q) interval versus the mean m/q of each interval. Triple coincidence is required to determine both the mass and the mass/charge of the ions (see Gloeckler et al., 1992) and to reduce the background sufficiently to detect these extremely rare ions. In regions of low count rate the m/q intervals were increased to contain at least one count. The peak at m/q of $\sim$3 is due to ^{3}He$^+$. Note that ^{3}He$^+$ is well resolved from the far more abundant pickup ^{4}He$^+$ to the right, and pickup ^{4}He^{++} to the left. The time interval between the detection of adjacent ^{3}He$^+$ followed a Poisson distribution with a mean time interval of 121 days. Because of the extremely low fluxes of ^{3}He$^+$, data were taken during a 40-month period when the solar wind speed exceeded 750 km s^{-1} for maximum detection efficiency. *Bottom panel*: phase-space density of ^{4}He$^+$ and ^{3}He$^+$ (divided by 2.48×10^{-4}) versus W (ion speed/solar wind speed). Despite the large statistical error for ^{3}He$^+$, the similarity of the two spectral shapes is obvious. No ^{3}He$^+$ counts were detected above $W = 2$ consistent with the expected cutoff in density. Below $W \sim 1.5$ the probability of ^{3}He$^+$ producing triple coincidence becomes extremely small since such ions have insufficient energy to trigger the 40 keV threshold solid-state detector.

The phase-space density of $^3\mathrm{He}^+$ (divided by 2.48×10^{-4}) is compared to that of $^4\mathrm{He}^+$ in the bottom panel of Figure 5. Within the statistical errors the two spectra are comparable as expected, further strengthening the case for the correct identification of interstellar $^3\mathrm{He}$. We also note that $^3\mathrm{He}^+$ was not detected above $W = 2$ indicating the characteristic cutoff of a pickup ion spectrum. Based on the results shown in Figure 5, Gloeckler and Geiss (1996, 1998) were able to place a new lower limit on the amount of missing matter in the early Universe, and concluded that the amount of $^3\mathrm{He}$ production by stars was less than predicted by some models of stellar chemical evolution.

2.1.5. *Spatial Distribution of Atomic Hydrogen*

The spatial distribution of an atomic species from which the corresponding pickup ions are created depends on the loss rate, β_{loss}, for removing neutrals through ionization, and on μ, the ratio of the force on the atom due to radiation pressure to the gravitational force. The loss rate and μ can be obtained from the shape of the distribution function of the pickup ion species between $W \sim 1.3$ and 2, as was discussed above, or from measurements of the gradual change (gradient) of the pickup ion fluxes with heliocentric distance R at a given angle Θ with respect to the direction of the solar apex. The gradient of the ratio of pickup hydrogen to pickup $^4\mathrm{He}^+$ is shown in Figure 6 which includes data for the time periods of Figures 2–4. Assuming $\mu = 1$ for protons and $\mu = 0$ for He, the best fit to the observed gradient is with β_{loss} (hydrogen) $= (5.5 \pm 1.0) \times 10^{-7} \text{ s}^{-1}$. This same value for the loss rate was used to derive the model pickup proton distribution shown in Figure 2.

2.2. ACCELERATED PICKUP IONS IN THE DISTURBED SOLAR WIND INSIDE A CO-ROTATING INTERACTION REGION

Ion velocity distributions observed in the turbulent solar wind, such as the wind inside co-rotating interaction regions (CIRs) have a distinctly different character, showing unmistakable evidence for heating and acceleration. This is best illustrated for $^4\mathrm{He}^+$, a pure pickup ion species. In Figure 7 we compare the phase-space density of $^4\mathrm{He}^+$ measured by SWICS during a $\sim$2-day period immediately behind the reverse shock of a CIR (28 and 29 December 1992), to the distribution functions observed during a $\sim$3-day period immediately upstream of the reverse shock of that CIR, and in the quiet polar coronal hole wind (from Figure 4), respectively. We have removed the $(V_{\mathrm{sw}}/V_{\mathrm{swCIR}})^4$ dependence of the phase-space density on the solar wind speed (e.g., Vasyliunas and Siscoe, 1976) by multiplying the preshock spectrum by 2.97 and the quiet distributions by 6.24. The difference between the CIR and quiet time spectra is most striking. While the quiet time distribution has a sharp cutoff at $W = 2$ indicating negligible energy diffusion and acceleration, the CIR spectrum has a pronounced high-velocity tail showing unmistakable acceleration above $W = \sim 1.5$. We conclude, in agreement with Gloeckler et al. (1994) and Schwadron et al. (1996), that in the turbulent solar wind behind the

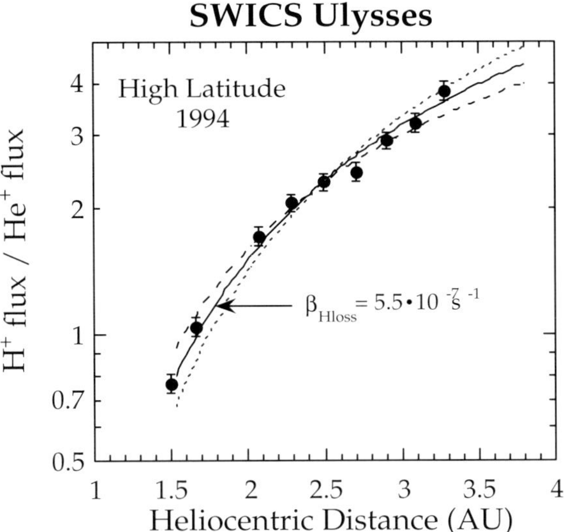

Figure 6. Ratio of H^+ to $^4He^+$ pickup ion fluxes versus heliocentric range. The flux ratio is obtained from the measured phase-space densities of H^+ and He^+ for W between 1.75 and 2. Each data point is a one-month average. During the entire time period of these observations *Ulysses* was in the high speed stream of the southern polar coronal hole. The best fit to the data is the solid curve which was computed assuming μ, the ratio of radiation to gravitational forces on hydrogen to be 1, using a hydrogen loss rate of 5.5×10^{-7} s^{-1}, a He loss rate (Rucinski et al., 1996) of 0.55×10^{-7} s^{-1}, and a H/He density ratio at the termination shock of 7.5. The dashed and dotted curves were calculated using H loss rates of 4.5×10^{-7} s^{-1} and 6.5×10^{-7} s^{-1}, respectively, but keeping the other parameters the same as before.

shock, pickup $^4He^+$ ions are easily accelerated. Thus, the once puzzling ubiquitous presence of energetic (0.4–0.6 MeV amu^{-1}) $^4He^+$ measured during a $\sim$1.5-year period in 1978–1979 at 1 AU by Hovestadt et al. (1984), can now be quite naturally explained to be interstellar pickup $^4He^+$ that were accelerated in CIRs at $\sim$3 to 6 AU and then transported back to 1 AU. The distribution function immediately upstream of the reverse shock shows a modest density enhancement (tail) above the cutoff at $W = 2$. This tail probably represents ions accelerated by the turbulence within the CIR that have gained enough energy to make their way back into the upstream region. It is this tail population upstream of the shock that may be further accelerated to higher energy by the shock mechanism.

In the top panel of Figure 8 we compare the spectral shapes of H^+, He^+, and He^{++} measured downstream of the forward shock (FS) with those observed in the downstream region of the reverse shock (RS) in the late December 1992 CIR. The normalization factors applied to the reverse shock data are shown in the figure in parentheses. There are several remarkable features apparent from these data. First, we note that more He^+ than He^{++} is accelerated even though solar wind alpha particles are at least a factor of 10^3 more abundant than pickup He^+. Second, the spectral shapes in the high speed ($W > 2.4$) tail region behind the reverse shock

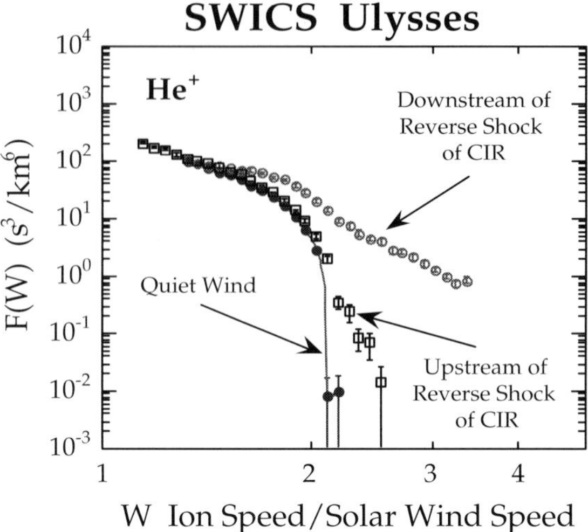

Figure 7. Phase-space density versus W of ^{4}He$^+$ (a) during a 40-hour period at the end of 1992 (December 28, 07:47 to December 29, 23:47) behind (downstream) the reverse shock (RS) of a co-rotating interaction region (CIR), (b) during a $\sim$3-day period immediately upstream of the same reverse shock (December 29, 23:47 to January 1, 1993, 00:59), and (c) during the extended quiet time period in the high speed solar wind (from Figure 4). *Ulysses* was at $-22.6°$ and 5.06 AU from the Sun in late December, 1992. In the region downstream of the RS the average solar wind He bulk speed was 498 km s^{-1}, while upstream of the RS the average speed was 654 km s^{-1}. The spectrum downstream of the reverse shock has a pronounced high-energy tail above $W \sim 1.5$ showing clear evidence of acceleration of pickup ^{4}He$^+$.

are identical (within experimental uncertainties) to those behind the forward shock. This is even more remarkable because the two shocks are so different. The FS is weaker ($M_s = 1.55$) than the RS ($M_s = 2.38$) and one is quasi-perpendicular while the other quasi-parallel (Balogh et al., 1995). Our observation that the acceleration process produces spectra of identical shapes despite the large difference in the shock parameters suggests that the acceleration is not due to the shocks themselves but rather to the turbulence in the downstream regions of the shocks (Gloeckler et al., 1994, see also Schwadron et al., 1996). It is evident from the solar wind phase-space densities that the solar wind behind the reverse shock is heated more than the wind downstream of the weaker forward shock. This heating supplies more suprathermal solar wind ions for injection into the acceleration process downstream of the RS than in the region behind the forward shock. The fact that four times more He^{++} and only 1.7 times more H$^+$ and He$^+$ is accelerated behind the RS compared to the FS implies that a large percentage of pickup protons is accelerated in the FS. The heating of solar wind protons behind the forward shock does not provide enough suprathermal solar wind protons to completely overwhelm the pickup protons that appear as the hump in the FS spectrum (open circles).

In the bottom panel of Figure 8 we compare the phase-space density of protons (open squares) in the downstream region of the CIR reverse shock (same time

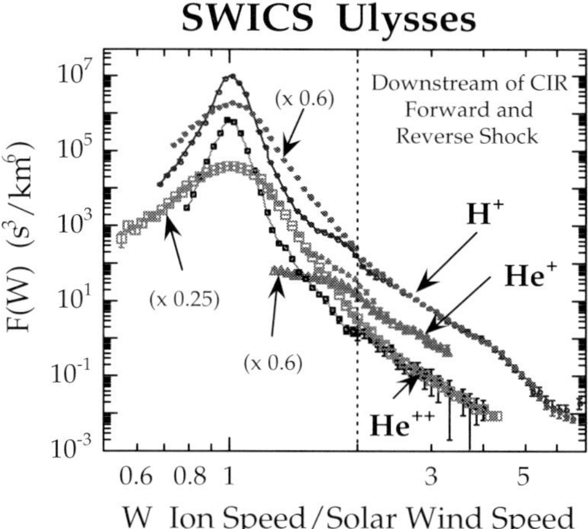

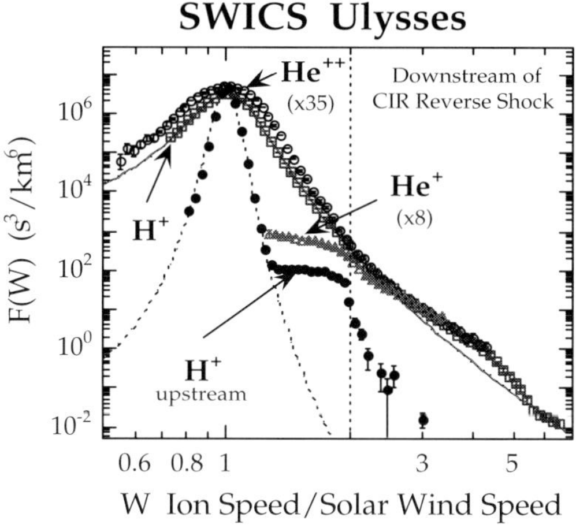

Figure 8. Top panel: distribution functions of H^+, He^+ and He^{++} downstream of the forward shock (open circles, small triangles and small squares respectively) and downstream of the reverse shock (filled circles, large triangles and large squares respectively) of the late December 1992 CIR. The time periods used for averaging the downstream FS spectra was from December 26, 19:41 to December 28, 07:47. For the downstream reverse shock (RS) distributions the time interval is the same as was used in Figure 7. The downstream RS distributions have been multiplied by the factors shown in parenthesis in order to match the respective spectra in the high velocity region above $W = 2.4$. Notice the remarkable similarity in the spectral shapes of all three ion species in the high velocity tail region. *Bottom panel*: velocity distributions of H^+ (squares), He^+ (triangles), and He^{++} (open circles) in the region downstream of the RS. The accumulation times were the same as those used for the distributions in Figure 7. The He^+ and He^{++} phase-space densities were multiplied by 8 and 35, respectively, in order to match the proton spectrum in the high velocity tail above $W \approx 2.5$. For comparison we also show the proton distribution (solid circles) in the upstream RS region (same time period as used for the He^+ upstream RS spectrum of Figure 7). We note that (a) the tails (above $W \approx 2.5$) of all three ion species have the same shape in the downstream region of the shock, and that the shape is not a simple power law as predicted by standard shock acceleration models, (b) the He^+/He^{++} ratio in the tail is $\sim$4.4, and (c) the $H^+/(He^+ + He^{++})$ ratio in the tail is about 6.5. The upstream RS distribution shows little evidence of acceleration.

period as in Figure 7) with that of He$^+$ (triangles) of Figure 7 (multiplied by a factor of 8) and He^{++} (multiplied by 35). These normalization factors were chosen so that all three velocity distributions overlapped in the tail region above $W \sim$ 2.4. For reference we also show the distribution function of protons (filled circles) measured in the upstream region during the same time period as in Figure 7. In comparing the upstream to the downstream H$^+$ distribution we note that the solar wind proton distribution in the downstream region behind the reverse shock is very broad and non-maxwellian. The solid curve is a kappa function fit (Equation (3)) to protons with $V_{\text{th}} = 67.2$ km s^{-1} and an extremely low value of $\kappa = 3.1$. This is to be contrasted with the upstream solar wind proton distribution (dotted curve) which is much colder ($V_{\text{th}} = 29$ km s^{-1}) and has a weaker tail ($\kappa = 4.9$). The pickup hydrogen distribution, quite obvious in the upstream spectrum is completely obscured by the over-ten-times-more-abundant heated solar wind protons above $W \sim 1.5$ in the downstream region. We therefore conclude that, in the case of strong heating illustrated here, mostly solar wind protons are injected for further acceleration to higher energies. On the other hand, in instances where less heating occurs (as shown to be the case downstream of the FS in this CIR and as is often the case behind forward shocks of CIRs in general) the protons in the high-velocity tail will be a mixture of solar wind and pickup protons (Gloeckler et al., 1994).

The speed distributions of all three species in the tail region have identical shapes above $W \sim 2.4$. This, combined with the fact that the FS and RS spectra also have the same shapes (see left-hand panel of Figure 8), implies that the acceleration mechanism depends primarily on the ion speed. In addition, the spectral shapes are not simple power laws, as would be predicted by standard shock acceleration models, providing further evidence that the high velocity tail distributions we observe are not produced by a simple shock acceleration mechanism.

3. Pickup Ions with Masses Heavier than Helium

Heavy pickup ions (O$^+$, N$^+$, and Ne$^+$) of interstellar origin were discovered (Geiss et al., 1994a) using SWICS/*Ulysses* data at distances beyond several AU. The presence of singly charged heavy ions of non-solar origin is clearly evident from Figure 9 which shows a mass/charge (m/q) histogram of ions with speeds above 1.4 times the solar wind speed ($W > 1.4$) in the mass/charge range above 10 amu e^{-1}. The prominent peak at $m/q = 16$ is due to singly charged interstellar pickup oxygen, with smaller peaks at m/q of 14 and 20 corresponding respectively to pickup N$^+$ and Ne$^+$. In particular, C$^+$ ($m/q = 12$) is not observed in this velocity range, placing a rough upper limit on C$^+$/O$^+$ of 0.05. These observations are consistent with an interstellar origin for these three heavy pickup ion species. Atomic O, N, and Ne penetrate deep into the solar system before they become pickup ions. Carbon, on the other hand, is mostly ionized in the local interstellar cloud (e.g.,

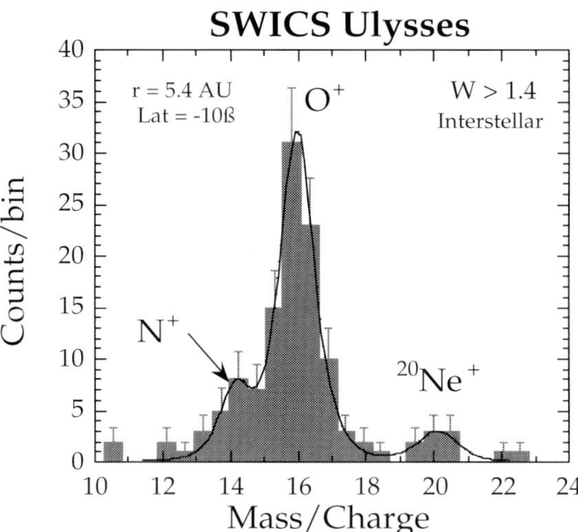

Figure 9. Plot of the double coincidence counts of ions with normalized speed W above 1.4 per mass/charge (m/q) intervals versus the mean m/q of each interval. These SWICS data were accumulated during a 135-day period in 1992 at an average distance of 5.4 AU and average helio-latitude of $-10°$ (also see Geiss et al., 1994a). The average solar wind speed was 442 km s^{-1}. In addition to the prominent peak at mass/charge 16 (^{16}O$^+$), ^{20}Ne$^+$ and ^{14}N$^+$ are also clearly observed. The speed (W) selection criteria exclude solar wind heavy ions whose distribution peaks around $W = 1$.

Frisch, this issue). Thus, a large fraction of interstellar C is excluded from the heliosphere.

Conclusive evidence for interstellar pickup oxygen comes from the velocity distribution function of O$^+$ shown in Figure 10. These SWICS data (and those shown in Figure 9) were accumulated during a 135-day period in 1992 at a distance of 5.4 AU and an average heliolatitude of $-10°$. During this time period the solar wind speed remained relatively steady at 442 ± 36 km s^{-1}. The distribution has the characteristic cutoff at $W = 2$, and the flat shape of the spectrum is similar to that of pickup hydrogen. Both the sharp cutoff and the flatness of the distribution function indicate a source location beyond 5.4 AU, the heliocentric distance of *Ulysses* when these data were obtained. A Jovian origin was ruled out by Geiss et al. (1994b). The smooth dotted curve is a model fit for the interstellar oxygen using 12×10^{-7} s^{-1} for the loss and production rates of O (see Rucinski et al., 1996) and the same anisotropy factor as for hydrogen ($\xi = 0.6$). The agreement with the observed spectrum above $W = 1.3$ is excellent. We find the atomic oxygen number density at the termination shock to be 7.8×10^{-5} cm^{-3}.

For W between 0.8 and ~ 1.2 additional O$^+$, beyond that expected from the interstellar gas source is clearly evident. This extra component of pickup O$^+$ has most likely the same origin as pickup C$^+$ discovered by Geiss et al. (1995). This hypothesis is confirmed by comparing, in Figure 11, the pickup carbon spectrum observed during this period with the pickup oxygen spectrum from which the in-

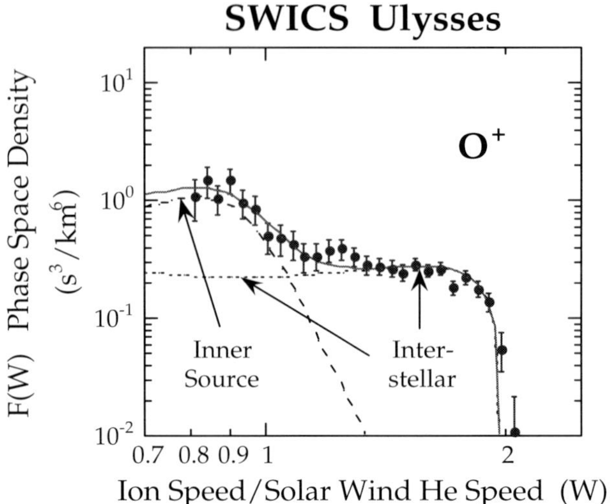

Figure 10. Phase-space density of O$^+$ versus W averaged over a 135 days from 17 February to 1 July 1992 when *Ulysses* was still near the ecliptic plane. The average solar wind speed was relatively steady at 442 ± 36 km s^{-1} and the distance from the Sun remained close to 5.4 AU during the entire time period of these observations. The dotted curve is the model calculation of interstellar pickup O$^+$ assuming an anisotropy factor $\xi = 0.6$ (see text). However, because the average magnetic field was nearly perpendicular (112°) to the solar wind flow direction the distribution function would appear to be isotropic in the spacecraft frame with nearly equal densities above and below $W = 1$. While the agreement between the model and data is excellent above W of about 1.3 a significant amount of O$^+$ from another source is evident below $W \approx 1.2$. A hypothetical inner source O$^+$ spectrum, represented by the dashed curve, combined with the interstellar O$^+$ curve would match the observed distribution. Below W of ~0.8 significant corrections for background from low-charge-state heavy solar wind ions such as Fe^{+6} are required. These corrections have not yet been applied to the distribution function of heavy pickup ions shown here. Note the similarity in the shapes of the inner source spectrum of O$^+$ shown here and that of H$^+$ given by the dashed curve of Figure 2.

terstellar pickup O$^+$ (the dotted model curve) has been subtracted. The two spectra (O$^+$ and C$^+$) are nearly identical within experimental uncertainties, and both reach maximum phase-space density of a few s^3 km^{-6} at W of about 0.85. This suggests a common source for these two heavy pickup ions, a source that supplies almost as much oxygen as carbon. Because the phase-space densities increase with decreasing W to peak just below $W = 1$, the source for these pickup C$^+$ and O$^+$ must be located much closer to the Sun than the 5 AU distance of *Ulysses* during these observations. Peaked distributions would be expected to be the result of adiabatic cooling of ions picked up far away from where they are detected (Vasyliunas and Siscoe, 1976; Gloeckler et al., 1993; Gloeckler, 1996; Fisk et al., 1997).

We have begun to investigate the characteristics of this inner source (or inner sources), located between *Ulysses* and the Sun, by measuring the velocity distributions of heavy pickup ions at different positions of the spacecraft. The mass/charge histogram of heavy, singly charged ions with W between 0.8 and 1.0 observed

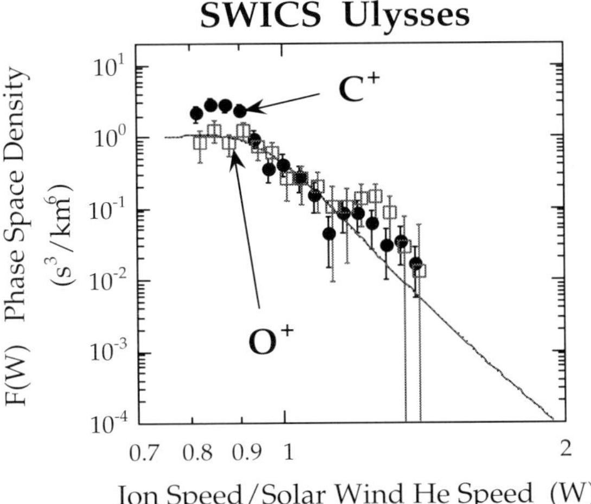

Figure 11. Same as Figure 10 except for O$^+$, from which the interstellar contribution represented by the dotted curve of Figure 10 has been subtracted, and C$^+$. Within the statistical errors the two spectra are about the same. The somewhat higher density of C$^+$ at $W \sim 0.85$ may indicate a higher production of pickup carbon compared to oxygen (Rucinski et al., 1996) from a source that produces comparable densities of atomic C and O. The smooth curve is the inner source spectrum represented by the dashed curve of Figure 10. Integrating the inner source spectrum of C$^+$ (assumed to be the same as for O$^+$) and combining this with the O$^+$ spectrum of Figure 10 gives a total C$^+$/O$^+$ ratio of 0.003 above $W \approx 1.4$.

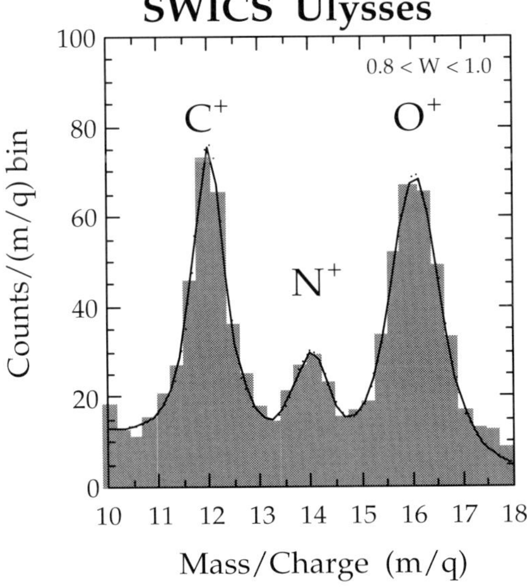

Figure 12. Plot of the double coincidence counts per mass/charge (m/q) interval of ions with W between 0.8 and 1.0 versus the mean m/q of each interval. These SWICS data were accumulated during all of 1994 at an average distance of 2.8 AU from the Sun and average helio-latitude of $-65°$. The average solar wind helium speed was 784 km s^{-1}. In addition to the ^{16}O$^+$ and ^{14}N$^+$, ^{12}C$^+$ is now observed at a level comparable to ^{16}O$^+$.

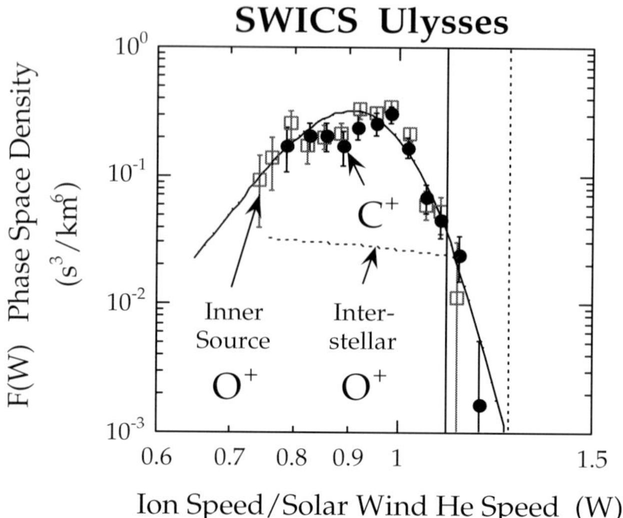

Figure 13. Phase-space density of C^+ and O^+ versus W during all of 1994 when *Ulysses* was at high latitudes at an average distance of 2.8 AU. The vertical solid and dashed lines indicate the average upper W cutoff for O^+ and C^+, respectively, above which the SWICS instrument upper energy per charge limit of 60 keV e^{-1} prevents detection of these singly charged heavy pickup ions in the high speed solar wind. The dashed curve is the expected contribution of interstellar O^+. It is evident that in the high speed stream interstellar O^+ is entirely obscured by the far more abundant O^+ from the inner source. The solid curve represents a hypothetical inner source spectrum. The shape of the inner source distribution viewed here from high latitudes is somewhat different from that observed in the ecliptic plane (Figure 11). This difference is most likely due to different spatial profiles of this distributed inner source in and out of the ecliptic. See text for further discussion.

during 1994, when *Ulysses* was in the high-speed solar wind and at high latitudes, is shown in Figure 12. In contrast to what was seen at low latitude and high W (cf., Figure 9), C^+ is now present at a level comparable to O^+. A peak at $m/q = 14$ is also clearly visibly. It is unlikely that this peak is due predominantly to solar wind Fe^{+4} because no significant amount of the more abundant Fe^{+5} is observed at $m/q = 11.2$.

The distribution functions of C^+ and O^+ for this one-year period, when *Ulysses* was at high latitudes, are shown in Figure 13. Because of the high solar wind speed ($\langle V_{sw} \rangle = 780$ km s^{-1}) the SWICS energy per charge upper limit of 60 keV e^{-1} prevented detection of C^+ and O^+ beyond W of 1.26 and 1.1, respectively (dotted and solid vertical lines). This instrumental limit excludes interstellar O^+ which otherwise would be visible beyond W $\sim$ 1.1 as is indicated by the model curve (labeled Interstellar O^+). The velocity distributions of C^+ and O^+ in the observable W range are identical within experimental errors but have rather complicated shapes. The observed phase-space densities decrease modestly with decreasing W below $\sim$0.95 but fall rapidly above W $\sim$ 0.95. Concentration of density around W close to 1 again suggests that the source of these heavy pickup ions again lies well

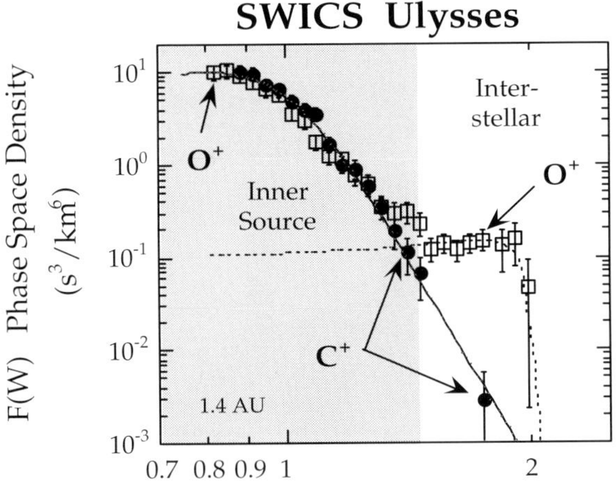

Figure 14. Average phase-space density of C⁺ and O⁺ versus W during the fast pole-to-pole pass of *Ulysses* through the ecliptic plane (5 February to 29 March 1995). Only data for slow solar wind speed (<550 km s⁻¹) were included. The average speed for the selected data was 444 km s⁻¹ and the average distance of *Ulysses* was 1.4 AU from the Sun. The dashed curve represents the expected contribution of interstellar O⁺. Despite the dominance of the inner source at W less than about 1.5 interstellar pickup oxygen is still seen above $W = 1.5$. The solid curve represents a hypothetical inner source phase-space density distribution. Above $W \approx 1.5$ little if any C⁺ appears to be of interstellar origin. Notice the similarity in the shape of the inner source spectrum shown here to that given in Figure 11.

within the average radial distance of *Ulysses* of 2.8 AU during 1994 when these data were taken.

The rapid pole-to-pole passage of *Ulysses* in 1995 provided the opportunity to observe pickup ion distributions when *Ulysses* was again in the ecliptic plane but now only 1.4 AU away from the Sun. Because much of this time the solar wind speed was well below ~600 km s⁻¹, the distribution functions of C⁺ and O⁺ over the entire speed range up to $W = 2$ could be measured. The C⁺ and O⁺ spectra we observed are shown in Figure 14. The mass/charge histogram for heavy ions with W between 0.9 and 1.5 for this same time period is given in Figure 15. The distribution functions of C⁺ and O⁺ are again identical (within errors) and the phase-space density is decreasing with increasing W from a maximum of ~10 s³ km⁻⁶ at $W \sim 0.85$. The decrease in the density is not as rapid as for the spectra of Figure 13, when *Ulysses* was at high latitude and about twice as far away from the Sun, indicating an inner source close to the Sun extending not much beyond 1 AU. For O⁺ the spectrum becomes flat for $W >\sim 1.5$ with a cutoff at $W = 2$. The phase-space density in this high W interval is entirely consistent with oxygen of interstellar origin and an atomic oxygen density of 7.8×10^{-5} cm⁻³ at the termination shock. Thus, as close as 1.5 AU and perhaps even at 1 AU where the ACE spacecraft is positioned, observations of interstellar pickup oxygen should

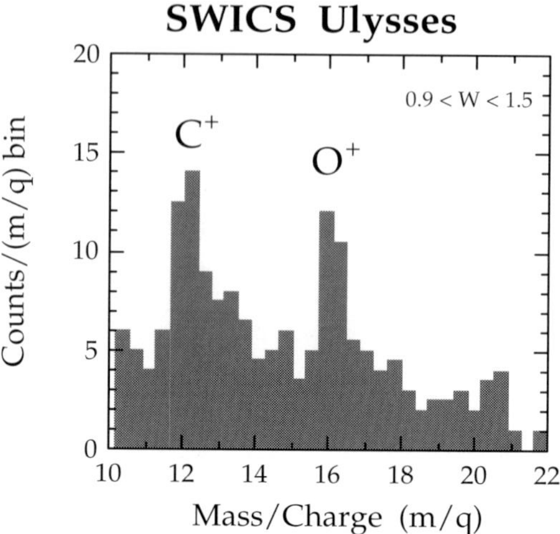

Figure 15. Plot of the double coincidence counts per mass/charge (m/q) interval of ions with W between 0.9 and 1.5 versus the mean m/q of each interval. These SWICS data were accumulated for the same period and selection criteria as given in the caption to Figure 14. $^{16}O^+$ and $^{12}C^+$ are observed at comparable levels. Other elements or molecules may be present but cannot be resolved in this data set.

be possible since it appears to be not entirely obscured by the O^+ from the inner source.

The distribution functions of the inner source pickup oxygen and carbon are complicated and not well understood. If one accepts adiabatic energy losses then the peaked distributions indicate a source located predominately upstream of *Ulysses*, very likely within a few AU from the Sun. The phase-space densities of both O^+ and C^+ peak consistently somewhat below $W = 1$. We cannot find any instrumental or analysis effects that would cause the peaks of these distributions to shift to slightly below $W = 1$. One way to interpret the inner source spectra is to postulate weak scattering at all pitchangles in regions close to the Sun (i.e., <1 AU) where, in the primarily radial magnetic field, most of these pickup ions are created with average speeds of W less than 1. Adiabatic cooling would still take place due to conservation of the first adiabatic invariant in the diverging magnetic field. The cooling, however, would be primarily perpendicular to the magnetic field. While this could account in a qualitative manner for the peaked spectra of the inner source ions, much work still remains to be done to fully understand these observations and to make firm inferences concerning the spatial distribution of the inner source material.

4. Discussion and Conclusions

The capability to measure mass/charge, mass, and speed of an ion, and the low background of the SWICS instrument combined with the trajectory of *Ulysses* make it possible to distinguish ions originating from the Sun, the interstellar gas, and the newly discovered inner source (cf., Geiss et al., 1995, 1996). The singly charged suprathermal interstellar pickup ions can be easily identified and separated from the far more abundant solar wind ions which are multiply charged and have relatively narrow velocity distributions peaking at the solar wind bulk speed. This has enabled us to study pickup ions to an extent never possible before, and to draw numerous conclusions regarding their origin and behavior, and the characteristics of the regions of their origin.

4.1. INTERACTION OF PICKUP IONS WITH THE SOLAR WIND

The velocity distributions of interstellar pickup hydrogen and helium observed in the steady and relatively unperturbed solar wind of the polar coronal hole provide clear evidence for lack of significant thermalization and energy diffusion. However, while adiabatic cooling is unmistakable, the strongly anisotropic distributions in the predominantly radial field at high latitudes indicate weak pitch-angle scattering. To reconcile these two apparently conflicting results one can imagine two extreme cases: (a) pitch-angle scattering is suppressed only at 90° pitch-angles (Fisk et al., 1997) or (b) pitch-angle scattering is weak at all pitch-angles and cooling results only perpendicular to the magnetic field (Möbius, 1996). In either case modeling of the expected evolution of the pickup ion distribution function is complicated and has as yet not been accomplished. In a magnetic field predominately perpendicular to the solar wind flow direction, nearly isotropic distributions (i.e., equal density below and above $W = 1$) are observed (Gloeckler et al., 1997) and are expected in either of the two cases. Fisk et al. (1997) find that weak pitch-angle scattering at 90° is most consistent with SWICS observations of interstellar pickup hydrogen.

A consequence of weak pitch-angle scattering is the observed correlation between fluctuations of the phase-space densities of pickup hydrogen and helium, both integrated between $W = 1.6$ and 2.0, shown in Figure 16 (Fisk et al., 1997). This initially surprising correlation is quite likely due to the variation of the mean free path for scattering through 90° which controls and modulates the densities of both pickup H^+ and He^+ in the $W > \sim 1.6$ range. Calculations of the dependence of the fractional density fluctuations, $\Delta n/n$, on scattering mean free path (λ) variations, $\Delta\lambda/\lambda$, show that, in order to be consistent with observations, a large mean free path ($\lambda \sim 1$ AU) is required (Fisk et al., 1997) for reasonable $\Delta\lambda/\lambda$ ($<\sim 1$).

We have interpreted interstellar pickup ion spectra assuming weak pitch-angle scattering only at 90° pitch-angle, and strong scattering on either side of 90°. It is then possible to model H^+ and He^+ and He^{++} distribution functions and obtain in a self-consistent fashion the respective ionization loss rates for atomic H and He in

     GEORGE GLOECKLER AND JOHANNES GEISS

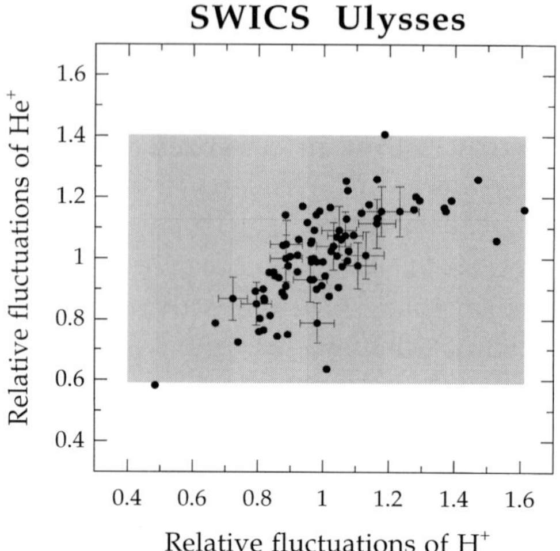

Figure 16. The fractional density fluctuations of interstellar pickup H$^+$ versus those of interstellar pickup He$^+$. Each data point represents a 2-day average of the respective integrated phase-space density from $W = 1.6$ to 2.0 divided by the corresponding 30-day smoothed density. These observations were made between January 1 and June 30, 1994 when *Ulysses* was in the high speed stream (770 to 790 km s^{-1}) of the southern polar coronal hole. The helio-centric distance of *Ulysses* ranged from 2.88 to 3.8 AU during this period.

the heliosphere as well as atomic abundances at the heliospheric termination shock. The ionization rates we determined from the shapes of the pickup ion spectra were found to be consistent with independently derived rates (Rucinski et al., 1996), and our values for the atomic H and He densities agreed well with those derived from UV backscatter observations (Quemerais et al., 1996) for H and from direct detection of neutral He on *Ulysses* (Witte et al., 1993, 1996).

In the disturbed solar wind, as is the case in co-rotating interaction regions between the forward and reverse shocks, the distribution functions of pickup as well as solar wind ions become broader with pronounced tails extending to high energies (cf., Figures 7 and 8). Clearly, pickup H$^+$ and He$^+$ are easily accelerated in these regions. Less certain is the proportion of pickup as opposed to solar wind ions that is accelerated. Just how much of each of these two populations is accelerated in CIRs will very likely depend on the relative fraction of these populations in the suprathermal range, perhaps around $W \approx 2$. Since the solar wind generally is heated more behind the reverse shock than it is behind the forward shock (Wimmer-Schweingruber et al., 1997) one would expect that ions accelerated at the reverse shock are predominantly solar wind ions (see Figure 8). On the other hand, when the solar wind speed is high, pickup ion distributions extend to higher speeds, and at twice the solar wind speed they may again have densities comparable to or higher than those of solar wind ions heated to these higher speeds. It seems to be the case that in CIRs at $\sim$3–5 AU, the heated solar wind distributions for H, He, C, and O and the corresponding pickup ion distributions have roughly comparable densities near $W \sim 2$ and therefore relatively small changes in either the solar wind thermal speed or its bulk speed will affect the proportion of solar wind versus pickup ions accelerated in CIRs. Observations of the larger relative abundance of H and He of energetic particles in CIRs compared to that in the solar wind (Gloeckler et al., 1979; Reames et al., 1991) support our conclusion that pickup ions form a significant part of the seed population of energetic particles accelerated in these turbulent regions.

The inner source could also contribute substantially to the energetic particles accelerated in co-rotating interaction regions (cf., Geiss et al., 1995, 1996). The average C/O ratio of CIR accelerated particles is higher than the solar wind ratio (Gloeckler et al., 1979; Reames et al., 1991; Fränz et al., 1995) but close to the C/O ratio of about 1 observed in the inner source. However, we would again expect that the proportion of inner source pickup ions versus solar wind C and O that is accelerated will be very sensitive to the thermal and bulk speed of the ambient solar wind.

To what extent heliospheric acceleration of pickup ions, as shown here, contributes to the production of anomalous cosmic rays (ACR) remains an open question. At the very least some pre-acceleration of pickup ions inside the heliosphere is required in order to compensate for adiabatic cooling. In the absence of this acceleration, pickup ions would be adiabatically cooled, as is observed, and at the termination shock their phase-space density would be more concentrated around $W = 1$, with smaller density near the cutoff at $W = 2$ and very few pickup ions beyond the $W = 2$ cutoff. At the other extreme, heliospheric acceleration may occur not only in the inner (~ 3 to ~ 10 AU) heliosphere as is observed, but also at large distances from the Sun at latitudes not yet explored. One possibility, suggested by Fisk (1996) is acceleration even at distances far beyond ~ 10 AU at the interface between the low speed flow in the ecliptic and the high speed flow from the polar coronal holes at higher latitudes. Presumably, pickup ions in the turbulence associated with this interface would be accelerated as efficiently as in CIRs. Finally, we note that because little atomic C exists in the local interstellar cloud (Frisch, this issue) only a small amount of interstellar carbon is available for acceleration inside the termination shock. Thus, much of the ACR carbon may come from the inner source (Geiss et al., 1996), requiring then substantial heliospheric acceleration to overcome the drastic adiabatic cooling of inner source pickup carbon.

4.2. COMPOSITION OF THE LOCAL INTERSTELLAR CLOUD (LIC)

The absolute number densities of atomic H, He, N, O, and Ne in the interstellar gas flowing through the heliosphere can be obtained without much ambiguity from measurements of the phase-space densities of the corresponding pickup ions. For H and He there are no significant contributions from possible inner sources of these atoms, and even for N and O the interstellar gas ions can be well separated from inner source ions (cf., Geiss et al., 1995, 1996) as is best demonstrated by the phase-space density spectra (cf., Figures 10 and 14). Thus, at distances greater than 5 AU, less than a few percent of the O^+ ions in the velocity range $W > 1.4$ come from the inner source (see Figures 9 and 11). Since the interstellar ion abundances of N^+, O^+ and Ne^+ given by Geiss et al. (1994a) were essentially based on ions with $W > 1.3$ near *Ulysses'* aphelion, no significant corrections for an inner source contribution are needed for the interstellar gas abundances reported there, even though these abundances were obtained before the inner source was discovered.

We have used our measurements of the velocity distributions of pickup ions (such as shown in Figures 2, 3, 5, and 10) to derive the atomic number densities of ^{1}H, ^{4}He, ^{3}He, ^{14}N, ^{16}O, and ^{20}Ne at the termination shock ($\sim$100 AU). The ratios (relative to ^{4}He) and the number densities are given in columns 2 and 3, respectively, of Table I. The ^{1}H and ^{4}He values are from Gloeckler et al. (1997), the ^{3}He abundance is taken from Gloeckler and Geiss (1998). The interstellar heavy element abundances are derived from SWICS measurements of O^+ reported here (cf., Figures 9 and 10) and from similar distribution functions of N^+ and Ne^+ (not shown). The errors given next to the values include estimates of systematic errors in the ionization rates of the respective species. These systematic errors are in general the largest source of uncertainty.

In the local interstellar cloud (LIC) matter exists in the form of atoms, ions and grains (or dust). Only atoms and very low-charged grains can enter the heliosphere as indicated in Figure 1. For H, O, and N there is a reduction in the density of atoms reaching the termination shock compared to their abundance in the LIC due to charge-exchange ionization with the slow moving (compared to neutrals) interstellar protons in the filtration region (roughly 100–200 AU wide) just beyond the heliopause (Holzer, 1989; Baranov and Malama, 1995; Gloeckler et al., 1997). Because of the small charge-exchange cross section for noble gases, He and Ne are not affected by this process. Using SWICS measurements of pickup ions (H^+ and He^{++}) Gloeckler et al., (1997) have found the amount of reduction (filtration) for hydrogen and from this computed the fractional ionization for H and He in the LIC. The derived densities in the LIC of atoms and ions are given in columns 4 and 5 of Table I.

The O/He ratio at the termination shock derived from the phase-space density data shown in Figures 10 and 2 falls well within the error limits of our earlier value. The uncertainties of the present ratio are smaller, because of the availability of a larger data base of slow solar wind measurements at large solar distance, and a better knowledge of the ionization rates. Geiss et al. (1994a) included in their error limits large uncertainties in the production and loss rates for oxygen.

The depletion of atomic oxygen in the heliosphere caused by filtration in the interaction region beyond the heliopause can be estimated by comparing our oxygen density at the termination shock (7.8×10^{-5} cm^{-3}) with the atomic oxygen density in the LIC. The LIC density of O is derived by combining the LIC HI/OI ratio, determined to be 2090 by Linsky et al. (1995), with our estimate of the atomic hy-

TABLE I

Densities of atoms at the heliospheric termination shock ($\sim$85–110 AU) and densities of atoms and ions in the local interstellar cloud (LIC) of elements and isotopes measured as pickup ions with SWICS on *Ulysses*

Isotope	Termination shock		Local Interstellar Cloud (LIC)				Solar system abundance ratio
	Ratio	Density (cm^{-3})	Density*			Ratio	
			Atoms	Ions	Total		
^{1}H	7.5	0.115 ± 0.025	0.20	0.043	0.243	10	10
^{4}He	1.000	0.0153 ± 0.0018	0.0153	0.0090	0.0243	1	1
^{3}He	2.48×10^{-4}	$(3.8 \pm 1.0) \times 10^{-6}$	3.8×10^{-6}	2.2×10^{-6}	6.0×10^{-6}	2.5×10^{-4}	–
^{14}N	0.6×10^{-3}	$(0.92 \pm 0.28) \times 10^{-5}$	1.0×10^{-5}	2.2×10^{-6}	1.2×10^{-5}	5.1×10^{-4}	1.12×10^{-3}
^{16}O	5.1×10^{-3}	$(7.8 \pm 1.4) \times 10^{-5}$	9.5×10^{-5}	2.1×10^{-5}	1.2×10^{-4}	4.8×10^{-3}	8.51×10^{-3}
^{20}Ne	0.75×10^{-3}	$(1.15 \pm 0.25) \times 10^{-5}$	1.15×10^{-5}	6.7×10^{-6}	1.8×10^{-5}	7.5×10^{-4}	1.23×10^{-3}

*For nitrogen and oxygen we asumed the same ionization fraction was measured for hydrogen (0.18). For Ne a the same ionization fraction as determined for He (0.37) was assumed.

drogen density of 0.20 cm^{-3} in the local cloud (column 4 of Table I). We arrive at an $\sim$18% loss of atomic oxygen by filtration using the estimated LIC atomic oxygen density of 9.5×10^{-5} cm^{-3} and our measured oxygen density at the termination shock. The loss of H by filtration is about 42% (Gloeckler et al., 1997). Thus, the observed filtration losses of H and O are consistent with the theoretical predictions of Geiss et al. (1994b) who showed that under all conditions the filtration loss caused by charge exchange is smaller for O than for H. Since the charge-exchange cross section between nitrogen and hydrogen is smaller than between oxygen and hydrogen, and since the ionization energy of N is larger than that of O, the loss by filtration of N should be even smaller than that of O. For the LIC abundance given in Table I we therefore assume a small N loss (10%) due to filtration.

The ionization fractions of the high FIP (first ionization potential) N, O, and Ne are not well known, but believed to be far less than for low FIP elements (e.g., C, Mg, Fe) which are highly ionized (e.g., Frisch, this issue). We have assumed the same ionization fraction for N and O as for hydrogen (0.18) and the same ion fraction of Ne as for He (0.37) to compute the ion densities and total densities of N, O, and Ne in the Local Interstellar Cloud. In comparing the abundance ratios (relative to ^{4}He) of elements measured in the LIC (column 7 of Table I) to those in the protosolar nebula given in column 8 (Anders and Grevesse, 1988), we find that nitrogen and oxygen are underabundant in the LIC by about a factor of 2 and Ne by about 1.6. The abundance of N, O and other elements in B-stars is about 50% lower than in the Sun (e.g., Frisch, this issue). Our results suggest that the composition in the LIC may be better represented by B-star rather than solar abundances. Using B-star rather than solar abundances our results further suggest that about 15% of the total oxygen abundance in the LIC may be in the form of grains.

4.3. LOCATION OF THE HELIOSPHERIC TERMINATION SHOCK

Observations of pickup ions were used by Gloeckler et al. (1997) to estimate the minimum distance to the termination shock. Assuming pressure balance at the heliopause, measurements of atomic densities and ion fractions in the LIC were used to calculate the position of the termination shock as a function of the magnetic field strength in the LIC. A value of 1.6 μG for the magnetic field (Frisch, 1996) places the shock between 75 and 95 AU. Lack of detection of the shock by Voyager 1 at its present location close to 70 AU restricts the field to an upper limit of $\sim$4 μG, a value insufficient to pressure balance the hot large cloud in which the LIC is embedded. To provide the required balance Gloeckler et al.,(1997) suggested that either the magnetic field in the LIC is inhomogeneous or that there is a significant and, as yet undetected, non-thermal component to the pressure in the local cloud.

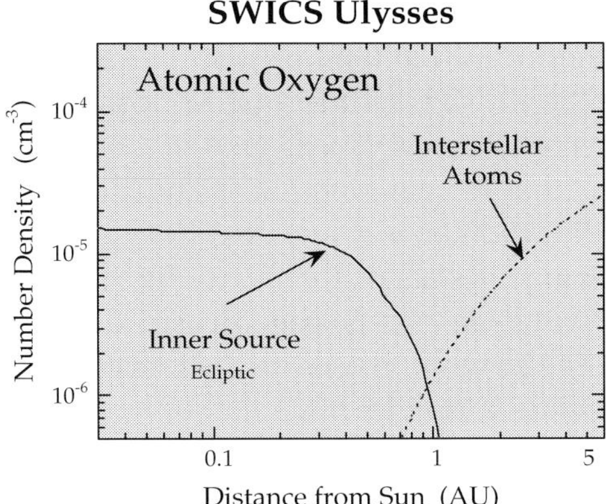

Figure 17. Dependence of the density of atomic oxygen from the inner and interstellar sources on radial distance. The dotted curve, representing the interstellar atomic oxygen, has been computed with the same ionization rates used for the model fit of O^+ shown in Figure 10. The solid curve represents a sketch of the radial density profile of the inner source in the ecliptic plane. This spatial distribution is roughly consistent with the spectral shape of the inner source oxygen shown in Figure 14. Statistical uncertainties of the data used and poor knowledge of the evolution of pickup ion distributions do not allow us to exclude more complex radial profiles for the inner source.

4.4. THE EXTENDED INNER SOURCE OF PICKUP IONS

Densities of pickup ions produced by the inner source are expected to decrease approximately with r^2 at heliocentric distances sufficiently beyond the source region. The distance of maximum density of interstellar gas pickup ions, on the other hand, scales roughly with their ionization rate. For H^+, N^+, and O^+ this maximum is at or beyond the aphelion distance of *Ulysses* (see Figure 6). There is also a striking difference in the velocity distributions of the two pickup ion populations, as is demonstrated in Figures 10, 11, and 14. Once the solar wind has left the region of the inner source, no new pickup ions from this source are created and the velocity distributions of these ions become progressively narrower with increasing distance due to adiabatic cooling, and therefore, the cutoff speed moves away from $W = 2$ toward $W = 1$. Closer to the Sun and for $W < \sim 1.5$, the contribution by the inner source to the total pickup ion population is substantial, as the relatively high abundance of C^+ in Figure 14 indicates. Nevertheless, the total production in the heliosphere of inner source ions is much smaller than the production of ions from the interstellar gas (cf., Table 1 in Geiss et al., 1996).

The nature of the inner source of pickup ions is at present not so readily specified. This source seems to produce ions over a range of solar distances and over virtually all solar latitudes. In Figure 17 we show a sketch of the approximate density profile for atomic oxygen from the inner source viewed from the ecliptic at 1.4 AU (solid curve) as well as the spatial density profile of the interstellar source (dotted curve). Our determination of the inner source radial profile (from the distribution function shown in Figure 14) is subject to uncertainties due to our poor knowledge of evolution of the pickup ion distributions. Nevertheless, this density profile can be used to estimate the total production rate of oxygen from the inner source to be $\sim 7 \times 10^6$ g s^{-1}. The production rate for carbon is comparable since roughly equal amounts of C$^+$ and O$^+$ are found in the inner source (cf., Figures 13 and 14). This suggests that the contribution of the inner source to the production of anomalous cosmic-ray carbon (at a rate of $\sim 10^7$ g s^{-1} estimated by Geiss et al., 1996) may be important.

Present observations of the phase-space densities of carbon and oxygen (cf., Figures 11, 13, and 14) cannot be used to rule out the existence of more than one inner source of pickup ions. Our observations suggest that the extended inner source is likely to be complex, with different source materials and different ion production mechanisms contributing. In our judgment, the most important of these are:

(1) *Thermal release of molecules from interstellar grains passing through the heliosphere.* The source strength is not expected to depend strongly on latitude. The radial dependence of densities shows that for C$^+$ and O$^+$, the mechanism of release cannot be temperature-independent sputtering.

(2) *Thermal release of molecules from interplanetary grains.* This source is expected to be concentrated near the ecliptic plane.

(3) *Small comets.* Whereas medium and large comets would appear to SWICS/ *Ulysses* as point sources (cf., Geiss et al., 1996), a large number of small comets ('mini-comets') could not be distinguished from a truly distributed source.

(4) *Collisions of grains.* Grün et al. (1985) have shown that, inside the orbit of earth, collisions become the main loss mechanism for grains on Poynting–Robertson trajectories. Due to solar radiation pressure, small fragments created by these collisions will assume hyperbolic orbits (beta meteoroids) and escape rapidly, but some of the material will evaporate locally and subsequently ionize, forming pickup ions.

(5) *Solar wind ions absorbed by grains and desorbed as atoms and molecules.* Pickup ions are then created from this solar material through re-ionization by solar EUV and solar wind charge exchange.

All these sources depend on the existence of small solid objects in the solar wind, upstream of *Ulysses*. The initial velocity distributions of pickup ions created from these sources are virtually the same, but they evolve with distance from the point of origin by pitch-angle scattering, adiabatic cooling and perhaps other processes. The composition of sources 1 to 4 reflects the composition of components in the solids. However, interstellar and interplanetary grains are likely to have very inhomogenous compositions, so that even a single grain population could produce a complex source pattern.

Source 5 creates pickup ions of solar origin. The underlying mechanism has been discussed by various authors (e.g., Banks, 1971; Fahr et al., 1981; Gruntman, 1996) in the context of producing atomic and molecular hydrogen, as well as H$^+$ and H$_2$$^+$ pickup ions. This mechanism will of course also produce heavier pickup ions, such as C^+ and O^+ and Ne^+. The source strength will depend on the grain density in the upstream region. Grains on Poynting-Robertson trajectories will be more effective than grains on transient orbits (interstellar grains, beta-meteoroids), because the latter spend only a relatively short time near perihelion where solar wind fluxes and temperatures are high.

In order to quantify the relative contributions of these five sources, a comprehensive investigation of the phase-space densities of H^+, He^+, C^+, N^+, O^+, Ne^+, and other species as a function of heliospheric latitude, longitude and distance from the Sun is needed. This can be achieved by using the data of SWICS/*Ulysses*, supplemented by data from STICS/WIND, CELIAS/SOHO and especially SWICS/ACE. It is also essential to understand far better than we do now the evolution of the pickup ion distribution functions and the scattering properties of these ions. Such work is now in progress. Comparing the results with measured and inferred densities of interstellar and interplanetary grains, and with observations relevant to the possible existence of mini-comets (cf., Brandt et al., 1996) should enable us to better specify the relative contributions to the inner source.

Acknowledgments

We are very grateful to the many individuals (see Gloeckler et al., 1992) at the University of Maryland, the University of Bern, the Max-Planck-Institute für Aeronomie and the Technical University of Braunschweig for their many outstanding contributions to the success of the SWICS experiment on *Ulysses* during the many years of its development phase. We wish to thank Fred Ipavich, Fritz Gliem, and Wolfgang Rieck for developing and implementing the on-board data reduction algorithms which enabled high data recovery despite the limited telemetry. Of particular benefit have been the many illuminating discussions with Len Fisk, Nathan Schwadron and Thomas Zurbuchen concerning evolution of the distribution functions of pickup ions, with Daniel Rucinski regarding ionization rates of interstellar atoms in the heliosphere and with Pricilla Frisch and Vlad Izmodenov on abundances in the LIC. We thank Christine Gloeckler for her help with data reduction. This work was supported in part by NASA/JPL contract 955460 (GG) and by the International Space Science Institute and the Swiss National Science Foundation (JG).

References

Anders, E. and Grevesse, N.: 1988, *Geochim. Cosmochim. Acta* **53**, 197.
Axford, W. I.: 1972, in C. P. Sonett, P. J. Coleman, Jr., and J. M. Wilcox (eds.), *Proc. Solar Wind Conf.*, Asilomar, California, 1971, NASA SP-308, p. 609.
Banks, P. M.: 1971, *J. Geophys. Res.* **76**, 4341.

Balogh, A., Gonzalez-Esparza, J. A., Forsyth, R. J., Burton, M. E., Goldstein, B. E., Smith, E. J., and Bame, S. J.: 1995, *Space Sci. Rev.* **72**, 171.

Baranov, V. B. and Malama, Y. G.: 1995, *J. Geophys. Res.* **100**, 755.

Blum, P. W. and Fahr, H. J.: 1970, *Astron. Astrophys.* **4**, 280.

Brandt, J. C., A'Hearn, M. F., Randall, C. E., Schleicher, D. G., Shoemaker, E. M., and Stewart, A. I. F.: 1996, *Earth, Moon and Planets* **72**, 243.

Fahr, H. J., Ripken, H. W., and Lay, G.: 1981, *Astron. Astrophys.* **102**, 359.

Fisk, L. A.: 1996, *Space Sci. Rev.* **78**, 129.

Fisk, L. A., Schwadron, N. A., and Gloeckler, G.: 1997, *Geophys. Res. Letters* **24**, 93.

Fränz, M., Keppler, E., Krupp, N., Rouss, M. K., and Blake, J. B.: 1995, *Space Sci. Rev.* **72**, 339.

Frisch, P. C.: 1996 *ISSI Workshop, Space Sci. Revs.* **78**, 213.

Frisch, P. C.: 1998, *Space Sci. Rev.* **86**, 107.

Geiss, J. and Witte, M.: 1996, *Space Sci. Rev.* **78**, 229.

Geiss, J., Gloeckler, G., Balsiger H., Fisk, L. A., Galvin, A. B., Gliem, F., Hamilton, D. C., Ipavich, F. M., Livi, S., Mall, U., Ogilvie, K. W., von Steiger, R., and Wilken, B.: 1992, *Science* **257**, 1535.

Geiss, J., Gloeckler, G., Mall, U., von Steiger, R., Galvin, A. B., and Ogilvie, K. W.: 1994a, *Astron. Astrophys.* **282**, 924.

Geiss, J., Gloeckler, G., and Mall, U.: 1994b, *Astron. Astrophys.* **282**, 933.

Geiss, J., Gloeckler, G., Fisk, L. A., and von Steiger, R.: 1995, *J. Geophys. Res.* **100**, 23 373.

Geiss, J., Gloeckler, G., and von Steiger, R.: 1996, *Space Sci. Rev.* **78**, 43.

Gloeckler, G.: 1996, *Space Sci. Rev.* **78**, 335.

Gloeckler, G. and Geiss, J.: 1996, *Nature* **381**, 210.

Gloeckler, G. and Geiss, J.: 1998, *Space Sci. Rev.* **84**, 275.

Gloeckler, G., Hovestadt, D., and Fisk, L. A.: 1979, *Astrophys. J.* **230**, L191.

Gloeckler, G., Wilken, B., Stüdemann, W., Ipavich, F. M., Hovestadt, D., Hamilton, D. C., and Kremser, G.: 1985, *Geophys. Res. Letters* **12**, 325.

Gloeckler, G., Geiss, J., Balsiger, H., Bedini, P., Cain, J. C., Fischer, J., Fisk, L. A., Galvin, A. B., Gliem, F., Hamilton, D. C., Hollweg, J. V., Ipavich, F. M., Joss, R., Livi, S., Lundgren, R., Mall, U., McKenzie, J. F., Ogilvie, K. W., Ottens, F., Rieck, W., Tums, E. O., von Steiger, R., Weiss, W., and Wilken, B.: 1992, *Astron. Astrophys. Suppl. Ser.* **92**, 267.

Gloeckler, G., Geiss, J., Balsiger, H., Fisk, L. A., Galvin, A. B., Ipavich, F. M., Ogilvie, K. W., von Steiger, R., and Wilken, B.: 1993, *Science* **261**, 70.

Gloeckler, G., Geiss, J., Roelof, E. C., Fisk, L. A., Ipavich, F. M., Ogilvie, K. W., Lanzerotti, L. J., von Steiger, R., and Wilken, B.: 1994, *J. Geophys. Res.* **99**, 17 637.

Gloeckler, G., Schwadron, N. A., Fisk, L. A., and Geiss, J.: 1995, *Geophys. Res. Letters* **22**, 2665.

Gloeckler, G., Fisk, L. A., and Geiss, J: 1997, *Nature* **386**, 374.

Grün, E., Zook, H. A., Fechtig, H., and Giese, R. H.: 1985, in R. H. Giese and P. Lamy (eds.), *Mass Input Into and Output from the Meteoritic Complex in Properties and Interactions of Interplanetary Dust*, D. Reidel Publ. Co., Dordrecht, Holland, pp. 411–430.

Gruntman, M.: 1996, *J. Geophys. Res.* **101**, 15 555.

Grünwaldt, H., Neugebauer, M., Hilchenbach, M., Bochsler, P., Hovestadt, D., Bürgi, A., Ipavich, F. M., Reiche, K.-U., Axford, W. I., Balsiger, H., Galvin, A. B., Geiss, J., Gliem, F., Gloeckler, G., Hsieh, K. C., Kallenbach, R., Klecker, B., Livi, S., Lee, M. A., Managadze, G. G., Marsch, E., Möbius, E., Scholer, M., Verigin, M. I., Wilken, B., and Wurz, P.: 1997, *Geophys. Res. Letters* **24**, 1163.

Holzer, T. E.: 1989, *Ann. Rev. Astron. Astrophys.* **27**, 199.

Hovestadt, D., Klecker, B., Gloeckler, G., Ipavich, F. M., and Scholer, M: 1984, *Astrophys. J.* **282**, L39.

Lallement, R.: 1996, *Space Sci. Rev.* **78**, 361.

Linsky, J. L., Diplas, A., Wood, B. E., Brown, A., Ayres, T. R., and Savage, B. D.: 1995, *Astrophys. J.* **451**, 335.

McNutt, R. L., Belcher, J. W., and Bridge, H. S.: 1981, *J. Geophys. Res.* **86**, 8319.

Möbius, E.: 1996, *Space Sci. Rev.* **78**, 375.

Möbius, E., Hovestadt, D., Klecker, B., Scholer, M., Gloeckler, G., and Ipavich, F. M.: 1985, *Nature* **318**, 426.

Neugebauer, M., Lazarus, A. J., Altwegg, K., Balsiger, H., Schwenn, R., Shelley, E. G., and Ungstrup, E.: 1987, *Astron. Astrophys.* **187**, 21.

Ogilvie, K. W., Gloeckler, G., and Geiss, J.: 1995, *Astron. Astrophys.* **299**, 925.

Quemerais, E., Bertaux, J.-L., Sandel, B. R., and Lallement, R.: 1996, *Astron. Astrophys.* **290**, 961.

Reames, D. V., Richardson, I. G., and Barbier, L. M.: 1991, *Astrophys. J.* **382**, L43.

Rucinski, D., Cummings, A. C., Gloeckler, G., Lazarus, A. J., Möbius, E. and Witte, M.: 1996, *Space Sci. Rev.* **78**, 361.

Rucinski, D., Bzowski, M., and Fahr, H.-J.: 1998, *Astron. Astrophys.*, in press.

Schwadron, N. A., Fisk, L. A., and Gloeckler, G.: 1996, *Geophys. Res. Letters* **23**, 2871.

Thomas, G. E.: 1978, *Ann. Rev. Earth Planetary Sci.* **6**, 173.

Vasyliunas, V. M. and Siscoe, G. L.: 1976, *J. Geophys. Res.* **81**, 1247.

Wimmer-Schweingruber, R. F., von Steiger, R., and Paerli, R.: 1997, *J. Geophys. Res.* **102**, 17 407.

Witte, M., Banaszkiewicz, M., and Rosenbauer, H.: 1996, *Space Sci. Rev.* **78**, 289.

Witte, M., Rosenbauer, H., Banaszkiewicz, M., and Fahr, H. J.: 1993, *Adv. Space Res.* **13** (6), 121.

Young, D. T., Geiss, J., Balsiger, H., Eberhardt, P., Ghielmetti, A., and Rosenbauer, H.: 1977, *Geophys. Res. Letters* **4**, 561.

INSIGHTS INTO COSMIC-RAY ACCELERATION FROM THE STUDY OF ANOMALOUS COSMIC RAYS

J.R. JOKIPII

The University of Arizona, Tucson, AZ 85721, U.S.A.

Abstract. Cosmic rays from many sources and in many locations exhibit similar, inverse-power-law energy spectra, which suggests a common origin for most cosmic rays. Diffusive shock acceleration appears at present to be this common accelerator. Hence, anomalous cosmic rays, thought to be accelerated at the solar-wind termination shock, provide a relatively accessible laboratory for the study of the mechanism of cosmic-ray acceleration. Observations showing a transition from singly-charged anomalous cosmic-ray oxygen to multiply-charged at an energy of some 250 MeV support the picture of acceleration at the quasi-perpendicular termination shock. Such acceleration may be important in other sources, as well. The basic physics of this acceleration process is discussed in some detail.

1. Introduction

Cosmic rays, in a variety of sources and settings, exhibit energy spectra which are power laws to a good approximation, and whose spectral indices are very similar, being generally in the range -2 to -3 (see Figure 1 for the spectrum observed at Earth). Diffusive shock acceleration is generally thought to be the mechanism for the acceleration of the bulk of these cosmic rays. This is in no small part because the mechanism has the virtue of naturally producing an energy spectrum which is quite close to the spectrum observed, and in many cases the small discrepancy is plausibly explained in terms of transport effects.

However, there are still outstanding questions regarding the scenario for shock acceleration. For example, the highest energy attainable by the particles in a given shock depends on a number of assumptions. The 'strict upper energy limit' in supernova shocks has been set by some at about 10^{14}–10^{15} eV (e.g., Lagage and Cesarsky, 1983) for quasi-parallel shocks, which is somewhat lower than the observed knee in the spectrum at about 2×10^{15} eV. Utilizing the fact that the supernova shock is quasiperpendicular over much of its area can, in principle, bridge the gap. But quasiperpendicular shocks remain not well understood, and injection of low-energy or thermal particles remains a difficulty.

Space Science Reviews **86:** 161–178, 1998.

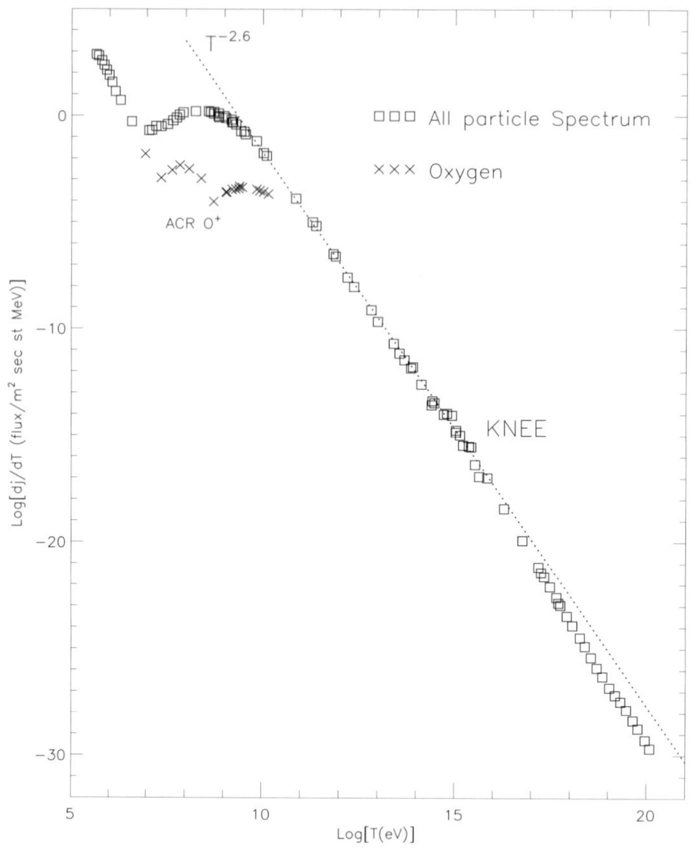

Figure 1. The cosmic-ray energy spectrum observed at Earth, ranging from the lowest energies, dominated by heliospheric particles, to the very highest energies, which may be extragalactic. Anomalous oxygen produces the peak at some 10^8 eV.

2. The Transport Equation

Astrophysical plasmas are turbulent and this results in scattering of fast charged particles, which drives the pitch-angle distribution toward isotropy. If this scattering occurs faster than other processes which produce anisotropy, the anisotropy will be small, and we may describe the particle motion as diffusive.

The general diffusive theory of the transport of energetic particles was developed about thirty years ago (see, e.g., Parker, 1965; Axford, 1965; Gleeson and Axford, 1967; Jokipii and Parker, 1970; general reviews were written by Jokipii, 1971; Völk, 1975). The equation for the pitch-angle-averaged distribution function $f(\mathbf{r}, p, t)$ as a function of position $\mathbf{r}$, particle momentum p, and time t, in a fluid moving with velocity $\mathbf{U}(\mathbf{r}, t)$ may be written as

$$\frac{\partial f}{\partial t} = \frac{\partial}{\partial x_i}\left[\kappa_{ij}^{(S)}\frac{\partial f}{\partial x_j}\right] - \mathbf{U}\cdot\nabla f - \mathbf{V_d}\cdot\nabla f + \frac{1}{3}\nabla\cdot\mathbf{U}\left[\frac{\partial f}{\partial\ln n(p)}\right] + Q, \quad (1)$$

where the successive terms on the right-hand side correspond to diffusion, convection, particle drift, adiabatic cooling or heating and any local source Q. Here, for particles of speed w, momentum p, and charge q, and scattering time somewhat greater than the gyroperiod, the drift velocity is

$$\mathbf{V_d} = \frac{pcw}{3q} \; \nabla \times \left[\frac{\mathbf{B}}{B^2} \right], \tag{2}$$

where c is the speed of light and $\kappa_{ij}^{(S)}$ is the symmetric part of the diffusion tensor. In general, we may write $\kappa_{ij}^{(S)}$ in terms of the magnetic field components B_i and $\kappa_{\parallel}$ and $\kappa_{\perp}$, the diffusion coefficients parallel and perpendicular to it, as

$$\kappa_{ij}^{(S)} = \kappa_{\perp}\delta_{ij} - \frac{\left(\kappa_{\perp} - \kappa_{\parallel}\right) B_i B_j}{B^2}. \tag{3}$$

Associated with the solution to the pitch-angle-averaged distribution function f is a streaming flux $\mathbf{S}$, which may be written in terms of f as

$$S_i = -\kappa_{ij}\frac{\partial f}{\partial x_j} - \frac{1}{3}U_i\frac{\partial f}{\partial \ln(p)}. \tag{4}$$

with an associated anisotropy magnitude

$$\delta = 3 \mid \mathbf{S} \mid /(wf). \tag{5}$$

Equations (1)–(5) are very general. They appear to be a reasonable approximation subject to only two physical constraints. First, there must be a sufficient level of magnetic irregularities or turbulence that the distribution function is kept nearly isotropic. Second, the speed w of the individual particles must be significantly greater than the convection speed U. Hence, the transport equation can be readily applied to models in which complicated transport effects such as drifts occur at the same time as shock acceleration. The divergence of the plasma flow velocity is singular at a shock, as is the drift velocity in general, but these effects are readily included in Equation (1) in the form of jump conditions connecting upstream and downstream solutions, as discussed below. In this context, therefore, acceleration is but one important aspect of general diffusive transport theory.

In general, the energy change of cosmic rays in a collisionless space plasma must be caused by the electric field associated with the fluid motions, so $\dot{T} = q\mathbf{w} \cdot \mathbf{E} = -q\mathbf{w} \cdot [\mathbf{U} \times \mathbf{B}/c]$. In the transport equation (1), the energy change term is the term proportional to $\nabla \cdot \mathbf{U}$, so the effects of the electric field in changing the particle's energy must all be contained in this term, even though the electric field does not appear. That this is in fact true may be illustrated by considering the expression for energy change of a particle moving without scattering in a large-scale electromagnetic field in a hydromagnetic fluid with velocity $\mathbf{U}$ (Kota, 1979; Jokipii, 1979; Webb et al., 1981), for an isotropic distribution of particles,

$$\frac{\partial T}{\partial t} = q\langle\mathbf{V}_d\rangle \cdot \mathbf{E} + \frac{pw}{2B}\langle\sin^2\theta\rangle\frac{\partial B}{\partial t} = \frac{1}{3}T\nabla \cdot \mathbf{U}_{\perp}\frac{d \ln T}{d \ln p}, \tag{6}$$

where the brackets denote an average over the distribution, and $\mathbf{U}_\perp$ is the component of $\mathbf{U}$ normal to $\mathbf{B}$. In addition, $\nabla \cdot \mathbf{U}_{\mathrm{par}}$ will be added to $\nabla \cdot \mathbf{U}_\perp$ from a consideration of the scattering needed to maintain near isotropy. This simple, intuitive, result states that the energy gain caused by drifting in the electric field, or the change of $\mathbf{B}$, with time, is simply part of the net energy change given by the more-general expression $\nabla \cdot \mathbf{U}$ term in Equation (1).

Now, we must also remember that particles may also move in the direction of $\mathbf{E}$ as a consequence of diffusion, down a density gradient having a component in the direction of $\mathbf{E}$, for example. That this is quite different from the drift motion may be seen by a simple thought experiment. Consider the system shown in Figure 2, top, in which $\mathbf{B}$ is uniform and normal to a uniform $\mathbf{U}$, and the density gradient of the particle distribution, ∇f, is normal to both. Clearly, the transport equation gives no energy change in this case, since $\nabla \cdot \mathbf{U} = 0$. But the particles will move down the density gradient, along (or opposite to) $\mathbf{E}$. What happens to offset the energy change resulting from the finite $\mathbf{w} \cdot \mathbf{E}$? The answer becomes apparent from a careful consideration of the particle trajectories in the magnetic field. In order to move across the magnetic field, the particles must be scattered in predominantly overtaking (or head-on) collisions with the scattering centers, and hence will suffer a net loss (or gain) of energy, which may be shown to precisely offset the energy change due to the large scale $-\mathbf{U} \times \mathbf{B}/\mathbf{c}$ electric field. Hence diffusion in the electric field results in zero net energy change, which is consistent with the transport equation. The bottom part of Figure 2 shows the case where a compression of the flow results in a gradient B drift in the direction of the magnetic field.

These considerations show that the energy changes due to the electric fields carried with the flow are indeed completely contained in the $\nabla \cdot \mathbf{U}$ term in the transport equation.

3. Diffusive Shock Acceleration

We next consider the solution of the transport equation in a standard, planar shock configuration. We work in the shock frame, with the shock at $x = 0$, and take the magnetic field to be weak enough that the flow is unaffected by it. The $x - z$ plane is chosen to contain $\mathbf{B}$. The upstream and downstream quantities are given the subscripts 1 and 2, respectively. We denote the strength of the shock by the ratio of upstream to downstream velocity $r_{sh} = V_1/V_2 = \rho_2/\rho_1$, where ρ is the density. The normal magnetic field is unchanged at the shock and the transverse field increases by r_{sh}, $B_{z2}/B_{z1} = r_{sh}$.

The solution to the transport equation may be obtained by solving it in the upstream regions and relating the solutions across the shock by the jump condition (obtained by integrating across the shock)

$$\left[\kappa_{xx} \frac{\partial f}{\partial x} + \frac{V_x}{3} \frac{\partial f}{\partial \ln p} - \frac{pcw}{3q} \frac{B_z}{B^2} \frac{\partial f}{\partial y} \right]_1^2 = Q_*. \tag{7}$$

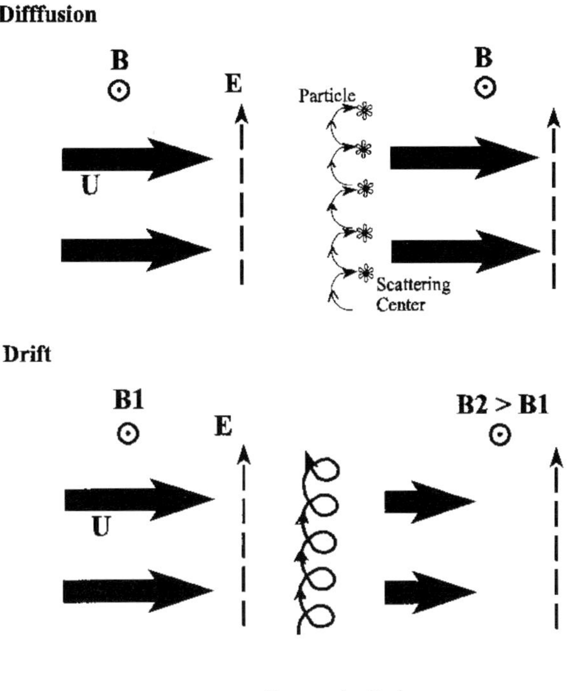

Figure 2. Schematic illustration of particle motion across a magnetic field convected at velocity **U**, indicated by the arrows. The top portion illustrates the case for uniform flow, where the cross-field motion is the result of scattering. The energy loss due to overtaking collisions compensates for the effect of the electric field, resulting in zero net energy change. In the lower figure, the slowing of the flow results in a gradient in |**B**|. The resulting drift motions are in the direction to result in energy gain.

where Q_* is that part of the source which is concentrated at the shock, and where κ_{xx}, the coefficient of diffusion normal to the shock face, may be written, in terms of the angle θ between the magnetic field and the x direction, as $\kappa_{xx} = \kappa_\parallel \cos^2 (\theta) + \kappa_\perp \sin^2 (\theta)$.

The first two terms on the left yield the standard jump condition, and the third term gives the effects of the gradient and curvature drifts at the shock front. The drift term vanishes in planar, one-dimensional systems, since $\partial f / \partial y$ is zero. However, for the reasons discussed above, even though the drifts do not appear explicitly for purely one-dimensional shock acceleration, they play an extremely important role in understanding the physics of particle acceleration in all but purely parallel shocks.

Consider now the interpretation of energy change in the presence of drift. Assume a perpendicular shock, in which the average field on both sides of the shock are parallel to the shock front. Then, remembering that the electric field in this frame is $\mathbf{E} = -\mathbf{V} \times \mathbf{B}/c$, one may show that the mean energy gain of a particle is related to the displacement along the shock front by (Jokipii, 1982)

$$dT = q E_y \, dy. \tag{8}$$

where T is kinetic energy. This is clearly a special case of Equation (6). As shown in the discussion of Equation (6), diffusion in the y-direction will spread the particles in the y direction, without causing further energy change in the process.

In the limit that f is independent of y, z and time, the solution to Equation (1) is the usual time-independent solution (obtained by many authors). In this case, the energy spectrum is found to be $f \propto p^q$, where $q = -3r_{sh}/(r_{sh} - 1)$, independent of the diffusion coefficient. Hence, any one-dimensional shock produces the same spectrum in a steady-state, independent of the angle of the magnetic field or the drifts.

In order to study the acceleration *rate*, we must go to time-dependent solutions. The time for acceleration of particles from an initial momentum p_0 to a momentum p may be written as (see, e.g., Forman and Morfill, 1979; Drury, 1983)

$$\tau_a = \frac{3}{(V_1 - V_2)} \int_{p_0}^{p} \left\{ \frac{\kappa_1}{V_1} + \frac{\kappa_2}{V_2} \right\} \frac{dp}{p} \frac{1}{p} \frac{dp}{d\tau_a} = \frac{3r_{sh}[\kappa_1 + r_{sh}\kappa_2]}{V_{sh}^2(r_{sh} - 1)}. \tag{9}$$

where κ_1 (κ_2) $= \kappa_{xx}$ as a function of momentum p in the upstream (downstream) region. The solution at any time is the power law given above, up to a gradual cutoff at a momentum near that defined in Equation (9). We see that the rate of acceleration will depend on the direction of the magnetic field relative to the shock normal, and on the functional form of the dependence of $\kappa_\parallel$ and $\kappa_\perp$ on particle energy. There is no general agreement on the value of $\kappa_\perp$ for particle transport in a turbulent magnetic field, so it is often treated as a free parameter. Alternatively one may take the result from standard kinetic theory, which corresponds to ordinary collisions (see, e.g., Axford, 1965), and this will also be done here. Let the mean free path parallel to the magnetic field, $\lambda_\parallel$, be a constant factor $\eta(> 1)$ times the gyroradius r_g, so that $\kappa_\parallel = \eta r_g w/3$. Then

$$\kappa_\perp = \frac{\kappa_\parallel}{\{1 + \eta^2\}}. \tag{10}$$

which is generally *less* than the Bohm limit $r_g w/3$. This corresponds, approximately, to a particle being shifted one gyroradius normal to the magnetic field in one scattering mean free path. It is clear that, for a quasiperpendicular shock, the acceleration rate can be enhanced considerably over the standard values. In the limit that the shock has an angle of 90° the acceleration rate is increased by a factor of $1 + \eta^2$, which can be large. Physically, the reason for the increased acceleration rate is that the particle can drift along the shock face, effectively colliding with it many times in one scattering mean free path.

4. Maximum Energy

A number of authors have estimated the maximum energy attainable in diffusive shock acceleration. In particular, the case of a supernova shock has received a great deal of attention. Lagage and Cesarsky (1983) presented a careful analysis of this case and concluded that the *maximum* energy was of the order of 10^{14}–10^{15} eV, and that in reality the value was probably lower. Their approach was based on arguing that the parallel diffusive mean free path must be greater than the gyroradius r_g . Hence $\kappa_\parallel > r_g c/3$. The absolute maximum energy would then be obtained by substituting this 'Bohm' value in Equation (6). However, as we see from Equation (7), for quasi-perpendicular shocks, κ_{xx} is always *less* than the Bohm value, and this result is inapplicable.

The analysis in Section 4 placed no limit on $\lambda_\parallel/r_g$ and thus placed no limit on the energy gain, which increases as the square of the ratio $\lambda_\parallel/r_g$. Of course, other factors must act to limit the attainable acceleration rate.

First (Jokipii, 1982), the finite size L_{sh} of the shock (transverse to the shock propagation direction) results in a limit to the distance a particle can drift along the shock face. For a quasi-perpendicular shock, if $\kappa_\perp << \kappa_\parallel$, the energy gain is of the order of the potential energy gained in the $\mathbf{V} \times \mathbf{B}$ electric field. The finite size of the shock then limits the energy gain to a value of the order of the potential energy gain available in drifting along the shock face $\Delta T \leq q V_{sh} B L_{sh}/c$.

Second (Jokipii, 1987), if $\lambda_\parallel$ were to be too large, the distribution function would become highly anisotropic, and the diffusion approximation would no longer be valid. Consider, for simplicity, a perpendicular shock. As a particle interacts with the shock and drifts along its face, it gains energy (mainly in the perpendicular velocity). At the same time it is being engulfed by the shock at the shock velocity V_d. One may estimate that the perpendicular energy of the particle will increase by a factor of the order of the shock ratio r in this process. This will produce a substantial anisotropy, and hence the particle must be scattered in the time required to drift through the shock to maintain near isotropy. This leads to the criterion for this special case (Jokipii, 1987, 1991).

$$\eta \ll \frac{w}{V_{sh}} \qquad \text{or, more generally,} \qquad \kappa_\perp >> r_g U_1. \tag{11}$$

It is also instructive to evaluate the streaming anisotropies in terms of the upstream gradients. They come out to be of order $r_g U_1/\kappa_\perp$. Again, the requirement that the anisotropy be substantially less than unity leads to the result in (11), the extra factor of 3 being of no consequence. In addition, we should guarantee that the particles can diffuse upstream ahead of the shock, which requires that the 'diffusion velocity' be equal to or of order V_{sh}. But the gradient length scale cannot be less than the particle gyroradius, which then leads to the same upper limit on η.

Note that Equation (11) becomes a quite stringent constraint for low-energy particles, where w may not be much larger than V_{sh}. Hence, these considerations lead to the expectation that low-energy injection occurs more readily when the shock is quasi-parallel, as suggested previously by other authors (e.g., Pesses et al., 1981).

The maximum possible energy will be attained, for any given situation, if $\lambda_{\parallel}/r_g$ is equal to a fraction $\epsilon\,(\approx 0.1)$ times V_{sh}/w. If the time available for acceleration is denoted by τ_a, then integrating Equation (6) yields the new estimate for the maximum attainable momentum, for the optimal diffusion coefficient,

$$p_{\max} = \frac{\epsilon}{c} \int_0^{\tau_a} \frac{(r-1)}{2r} V_{sh} q B \, d\tau. \tag{12}$$

Note the interesting fact that if the shock ratio r does not change, the maximum energy is proportional to the magnetic flux swept up by the shock, $\Phi = \int V_{sh} B \, d\tau$. In any actual shock, the maximum energy could be considerably smaller.

Equation (12) can be applied to a variety of situations to give an estimate of the maximum energies attainable, although they may be unattainable for other reasons. In actual fact, the maximum energy will be decided in any given case by the competition between the various factors in any given shock. In any given situation we must solve the full transport equation, including all loss processes to adequately estimate the spectrum. Nonetheless, it appears that perpendicular shocks should be capable of accelerating particles to higher energies than parallel shocks.

5. Applications

5.1. ANOMALOUS COSMIC RAYS

The anomalous cosmic rays (ACR) are composed of fluxes of helium, nitrogen, oxygen, neon, protons and low levels of carbon (Klecker, 1995), which are observed in the inner heliosphere to be enhanced at energies ranging from 20 MeV to perhaps 300 MeV. The observed spectra of anomalous cosmic-ray oxygen, helium and protons are shown in Figure 3 (left panel).

Fisk et al. (1974) suggested that the anomalous component was the result of heliospheric acceleration (by some unspecified mechanism) of freshly-ionized interstellar particles. Interstellar neutral atoms which stream into the solar system have a probability of being ionized, either by solar ultraviolet radiation, or by charge exchange with the solar wind and are subsequently accelerated. This explains very nicely the composition of the anomalous component, since only initially-neutral particles can participate in the process.

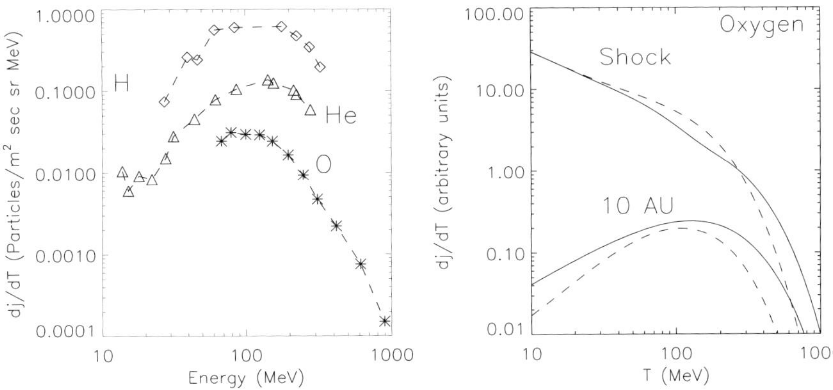

Figure 3. Left: Energy spectra of anomalous oxygen, hydrogen and helium observed near 21 AU in 1985 (Cummings et al., 1987). *Right*: plot of the computed energy spectra of anomalous oxygen near the heliospheric equator, for two radii, and for the two signs of the heliospheric magnetic field. The solid lines correspond to the case where the northern heliospheric magnetic field is directed outward $A > 0$), corresponding to the present sunspot minimum. The dashed lines are the same for $A < 0$, corresponding to the 1986 sunspot minimum.

Pesses et al. (1981) pointed out that many features of the anomalous component could be explained if the acceleration of the ACR occurs at the termination shock of the solar wind, by the mechanism of diffusive shock acceleration. Jokipii (1986) presented results from a quantitative two-dimensional numerical simulation of this model, in which the full transport equation was solved. This model showed that (a) energetic, singly-charged particles could be accelerated very efficiently at the termination shock, and (b) the essential observed features of the spectrum and spatial gradients could be explained very naturally in terms of this model. Because of the small charge and consequent high magnetic rigidity of the anomalous cosmic rays, gradient and curvature drifts of the particles, both along the face of the shock and in the solar wind play a major role in producing the observed spectrum and intensity gradients. Other aspects of this problem have been discussed by le Roux et al. (1996), Chalov et al. (1997), and Savolopulos and Quenby (1997).

It has been established observationally that the acceleration of anomalous oxygen to ≈ 200 MeV must occur in less than a few years because otherwise further electron loss would occur (see, e.g., Adams and Leising, 1991; Jokipii, 1992). Jokipii (1992) showed, further, that diffusive shock acceleration can accelerate particles to these energies in the required time, whereas other mechanisms in the weak fields of the outer heliosphere take much longer. Mewaldt et al. (1996) reported observations showing a transition from singly- to multiply-charged anomalous cosmic-ray oxygen at energies above some 250 MeV, and Jokipii (1996) presented model results showing that this was consistent with the picture outlined above.

Therefore, diffusive shock acceleration at the solar wind termination shock seems to be the acceleration mechanism of choice for ACR. Recently, Giacalone and Jokipii (1996) suggested that the initial acceleration of pickup ions is difficult at the termination shock, and have suggested that the initial acceleration occurs in the inner heliosphere (Giacalone et al., 1997).

At Arizona, we have developed a numerical code which solves the Parker equation in a 2-dimensional model heliosphere. We first specify the heliospheric configuration, which is reasonably well understood, and extrapolation to the termination shock probably does not introduce major uncertainties. The flow and magnetic field beyond the shock is uncertain, and we adopt a simple configuration which contains the basic physics.

The solar wind velocity is taken to be radial out to a spherical shock at a heliocentric radius $r = R_{sh}$ at which it drops by a factor of 4 (for a strong shock) and then decreases as $1/r^2$ (small Mach number flow) out to an outer boundary R_b, where the energetic particles are presumed to escape. Typically, R_{sh} is taken to be some 90–120 AU and R_b somewhat larger.

During the years around each sunspot minimum, the interplanetary magnetic field is organized into two hemispheres separated by a thin, nearly equatorial current sheet, across which the field reverses direction. In each hemisphere the field is approximately a classical Parker Archimedean spiral. The field direction in each hemisphere alternates with the 11-year sunspot cycle, so that during the 1975 minimum, the northern field was directed outward from the sun (conventionally denoted as $A > 0$), but in 1965 the northern field pointed inward ($A < 0$). This field is assumed to continue beyond the termination shock, with the spiral angle reflecting the local V_w.

There is now evidence that the *polar* magnetic field differs considerably from the Parker spiral (Jokipii and Kóta, 1989) and Jokipii et al. (1995), so the polar field is modified in our simulations. The structure for the years near sunspot maximum is not simple, so the following discussion is most relevant during the period around sunspot minimum.

The numerical model follows the acceleration and spatial motions of low-energy, singly-charged particles injected into the solar wind and then accelerated at the termination shock. The characteristic time for these processes to approach a steady state is found to be 2–3 years except for the highest-energy ACR.

From Equation (8) we note that if the scattering frequency is significantly less than the gyrofrequency, the energy ΔT gained by a particle having electric charge Ze, at the quasi-perpendicular termination shock, is approximately Ze times the electrostatic potential energy $\Delta\phi$ gained in drifting along the shock face from the pole to the equator:

$$(\Delta\phi)_{\max} = B_r r^2 \Omega_\odot / c$$

$$\approx 240 \text{ MeV/unit charge,} \tag{13}$$

where c is the speed of light and the numerical value results from using a radial magnetic field of 3.5 nT at a radius of 1 AU and a solar rotational angular velocity

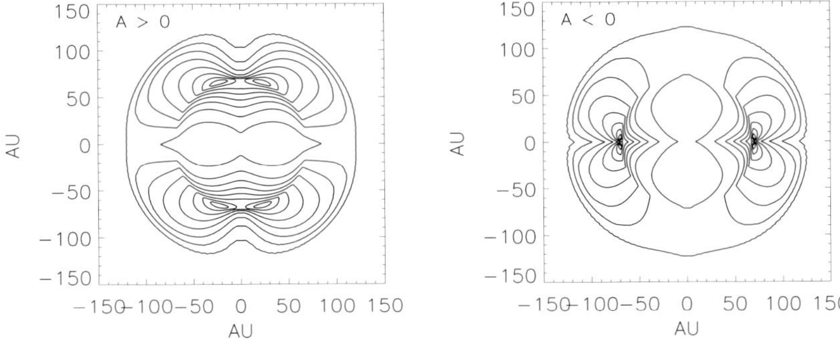

Figure 4. Equal intensity contours computed for 8 MeV nucl^{-1} anomalous oxygen, as a function of position in a meridional plane, as computed for the model parameters discussed in the text. The left panel is for $A > 0$ and the right panel is for $A < 0$.

$\Omega_{\odot} = 2.9 \times 10^{-6}$ s^{-1}. The cutoff need not be steep, since some particles will be scattered back and forth, receiving more or less energy than T_c. Because of this, singly-charged particles should be accelerated to lower energies than multiply-charged ones. Singly-charged particles accelerated at the termination shock would then have a spectrum which exhibits a decrease above an energy between 200 and 300 MeV. Multiply-charged particles could gain much more. Following the discussion of the transport equation at the end of section 2, we find that some particles can gain more than this amount because they *diffuse* back along the direction of drift, and can pick up part of the potential energy again. This diffusion occurs without the particle losing the potential energy it has gained in drifting.

5.2. ACR SIMULATION RESULTS.

Model results are presented for the case where the low-energy particles are injected uniformly at all latitudes at the shock. The computed spectra are shown in Figure 3 (right). Note that the relative normalization of the two spectra may be changed by variation of the parameters. The computed energy spectra of anomalous protons, helium, neon, etc are in excellent agreement with the observed spectra (see Figure 3 (left)).

Figure 4 illustrates contour plots of the intensity of the modeled anomalous oxygen as a function of heliocentric radius and polar angle, in a solar meridional plane. We see that the intensity increases with radius, much as does that of the galactic cosmic rays, out to the termination shock (at a radius of 60 AU). Beyond, the intensity decreases out to the outer boundary of the heliosphere. Along the shock, the *maximum* intensity occurs at a latitude which shifts as the sign of the magnetic field changes. If $A > 0$, the particles drift toward the pole along the shock face and then inward and down from the poles to the current sheet, the intensity maximum is near the poles. For $A < 0$, the drifts are in the opposite direction, and the maximum shifts to the equator.

There have been numerous measurements of the radial and latitudinal gradients of the anomalous component (McKibben et al., 1979; McDonald and Lal, 1986; Cummings et al., 1987). In particular it is found that in the inner solar system the radial gradient of the anomalous oxygen is as high as 16% per AU, whereas in the outer solar system it is 3% per AU. The magnitude of the radial gradient computed in the model is quite consistent with that observed, both in absolute magnitude and in radial variation. It is a general feature of the simulations of ACR and galactic cosmic rays that, for identical parameters, the latitudinal gradients of the ACR are several times larger than those for galactic particles. This is in large part due to the large latitudinal gradient induced by the acceleration at the shock.

A robust prediction of the models is that the latitudinal gradient near the current sheet, near sunspot minimum, should change sign in alternate sunspot minima. In particular, during the sunspot minimum near 1975, the intensity of both galactic cosmic rays and the anomalous component should *increase* away from the current sheet, whereas during the 1986 sunspot minimum the cosmic rays should *decrease* away from the current sheet. Near sunspot minimum, when the current sheet is nearly flat, the effects of these drifts are expected to be the most important. In 1977, Pioneer observed a positive gradient of anomalous helium away from the current sheet (McKibben et al., 1979), as predicted. Observations carried out during sunspot minimum in 1984 and 1985, when the current sheet tilt went below the latitude of the Voyager 1 spacecraft, found that the sign of the latitudinal gradient changed from being positive to negative (Cummings and Stone, 1987), just as predicted by the theory.

6. The Charge State of Anomalous Cosmic Rays

Adams and Leising (1991) showed that the charge state of the anomalous oxygen at ≈ 10 MeV nucl^{-1} implied that the source was less than 0.2 pc away, if they came on a straight line. Jokipii (1992) pointed that this result implied an upper limit, of the order of 4 years, on the time for acceleration of these particles. Mewaldt et al. (1996) reported that at energies above some 16 MeV nucl^{-1}, multiply-charged ACR oxygen atoms are observed and they become dominant at higher energies. As they pointed out, this has important implications for our picture of ACR. Among other things it gives support to the discussion leading up to Equation (2). Singly-charged oxygen can readily reach energies of about $\Delta\phi_{max}$, or some 15–20 MeV nucl^{-1}. In the process, some of these are stripped of one or more further electrons, which can then reach energies of several times $\Delta\phi_{max}$. These higher-energy particles are the multiply-charged particles reported by Mewaldt et al. (1996).

These conclusions are consistent with the simulation models. Jokipii (1996) discussed the acceleration, ionization and transport of anomalous cosmic-ray oxygen, extending previous simulation codes to include all eight non-zero charge states.

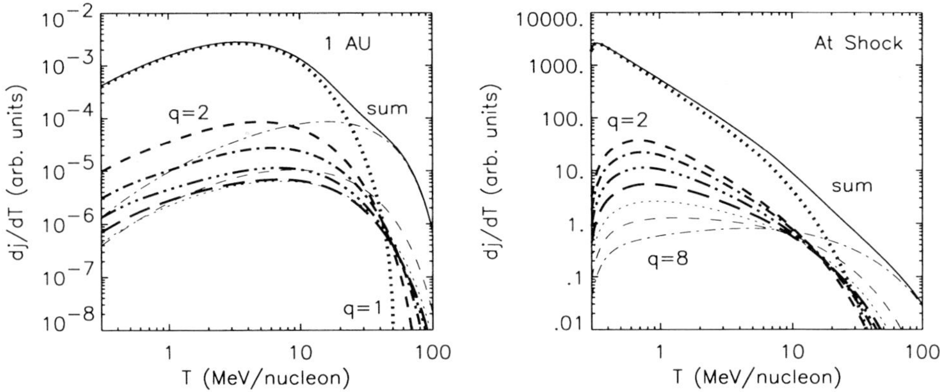

Figure 5. The computed spectra of the 8 charge states of anomalous oxygen, and their sum, as a function of energy, at two different radii, near the heliospheric equator, for nominal parameters.

The ratio of multiply-charged to singly-charged anomalous cosmic rays in the inner heliosphere is found to increase rapidly above an energy T_c of about 20 MeV nucl^{-1} consistent with expectation based on Equation (13). These simulation results, illustrated in Figure 5 are found to agree well with the SAMPEX observations reported by Mewaldt et al. (1996), as put forth in their Figure 5. The cross sections used in the simulations are not well known, so detailed agreement is not expected. The increase in multiply-charged ACR oxygen above some 250 MeV energy is, however, robust and insensitive to the details of the cross sections.

6.1. THE ACCELERATION OF PICKUP IONS

The above discussion has summarized the present status of the acceleration of ACRs to the highest energies observed, in excess of 1 GeV. Diffusive acceleration at the termination shock of the solar wind seems to be the most-likely mechanism for this acceleration.

Not yet fully resolved is the question of how previously unaccelerated pickup ions are accelerated to energies large enough to be further accelerated at the termination shock. This 'injection problem' stems from the fact that the non-accelerated pickup ions have speeds that are less than or equal to the convection speed of the solar wind and hence are not mobile enough to encounter the nearly-perpendicular shock the many times necessary to gain significant energy. It has been shown (Jokipii and Giacalone, 1996; Giacalone et al., 1997) that propagating shocks more readily accelerate low-energy pickup ions than shocks that stand in the solar wind. Consequently, they are more natural injectors of particles and may represent the first stage in a two-step process in accelerating pickup ions to ACR energies.

To examine the consequences of pre-acceleration at propagating shocks, Jokipii and Giacalone (1996) considered a modification of the model discussed previously. Rather than inject the particles at the shock, they considered a source in

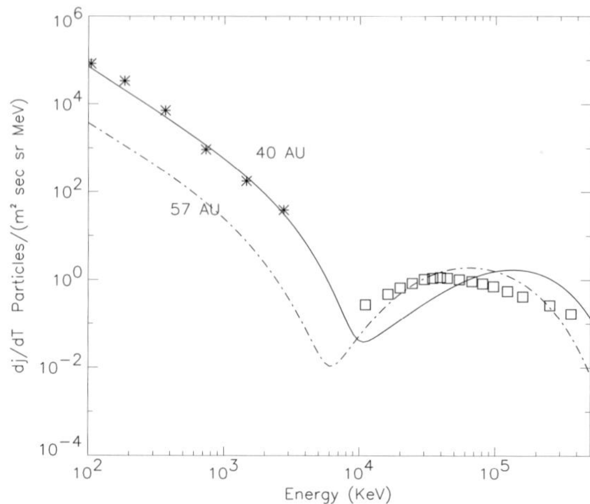

Figure 6. Comparison of the computed energy spectra of anomalous hydrogen, near the equator and at 20 and 40 AU, for the parameters discussed in the text. The overall normalization is chosen to fit the low-energy points. The higher energy points are in good agreement with the data.

the inner heliosphere distributed in radius and latitude to represent acceleration by Co-rotating Interaction Regions (CIRs).

They assumed an interplanetary-shock-accelerated source of energetic particles of the form

$$Q(r, \theta, p, t) = p^{-\gamma} \exp\left[-\frac{p}{p_{\text{cut}}} \right] \exp\left[-\frac{(r - r_0)^2}{L_c^2} \right] \times$$

$$\times \exp\left[-\frac{(\theta - \pi/2)^2}{\theta_0^2} \right], \tag{14}$$

where r, θ are spherical coordinates.

Equation (1) was solved numerically with (14) as a source until a steady-state was reached. The resulting spectra were normalized using Voyager 2 observations at 57 AU.

The parameters of the source are $\gamma = 5$, p_{cut} corresponds to an energy of 6 MeV, $r_0 = 10$ AU, $L_c = 6$ AU, and $\theta_0 = 0.4$ rad. The remaining simulation parameters are: $r_{sh} = 100$ AU, $r_b = 140$ AU $\kappa_\parallel = 1.5 \times 10^{22} P^{1/2}\beta$ cm^2 s^{-1} (where P is the particle rigidity in GV, and β is the ratio of the particle speed to the speed of light) and $\kappa_\perp = 0.015\kappa_\parallel$. The particle energy range considered in the simulation was from 50 keV to 1.7 GeV. The results are shown in Figure 6. The stars are the Voyager data at low energies, used to normalize the simulation. The dot-dashed line shows the simulated spectrum at 57 AU. The square data points are Voyager observations reported by Christian et al. (1995) which were taken when Voyager was at 57 AU.

We see that the model, normalized to the 40 AU observations at $\sim$ 5 MeV, are of the same order of magnitude as the higher-energy observations of anomalous cosmic rays obtained near 57 AU. In fact, the simulated fluxes are somewhat higher than the observations. This leads us to conclude that the model scenario is plausible. Note that relatively minor adjustments to the input parameters may lead to an improved comparison between the simulation results and the data. However, this is not our intention. The main point that is illustrated in Figure 6 is that low-energy proton fluxes that are clearly associated with recurrent structures found in the inner heliosphere, are high enough to account for the observed fluxes of anomalous cosmic-ray protons. Therefore, it is possible, and perhaps likely, that pre-acceleration of anomalous cosmic rays in the inner heliosphere plays a major role in producing the observed flux.

7. Blast wave

We may apply the above conclusions reached from studying ACR to elucidate the problem of acceleration in the velocity and magnetic field generated by a supernova blast wave. Approximate the supernova by a point explosion occurring in a uniform plasma having a number density N_o particles per cm^3, in which is imbedded a frozen-in, initially-uniform magnetic field $B_0 \hat{e}_z$,which fixes the polar axis of a spherical coordinate system. B_0 is taken to be small enough that it has a negligible effect on the dynamics of the gas. Clearly, around the 'equator' the shock is perpendicular, and, based on the above analysis, we expect more-rapid acceleration in the equatorial regions. This expectation has been born out by detailed numerical calculations described below.

Initially, however, let us estimate the attainable energies by applying the equivalent of Equation (13) to the perpendicular shock around the equator. Thus, if the shock remains strong out to a radius $R_{10} \times 10$ pc, and the ambient magnetic field is $B_6 \times 10^{-6}$ G, the maximum energy for particles of charge q would be

$$(pc)_{\max} \approx \epsilon R_{10} B_6 Z \times 3.4 \times 10^{15} \text{ eV}. \tag{15}$$

Thus, taking $\epsilon = 0.1$, and considering only the quasi-perpendicular shock at the equator, a supernova expanding to a radius of 30 pc in a 5×10^{-6} G magnetic field would be able to accelerate protons to nearly 10^{16} eV, substantially higher than the maximum energies obtained in previous calculations. ϵ could probably be taken to be somewhat larger.

However, the actual situation is more complicated, since the shock is perpendicular only in a small area near the equator and the more-rapid parallel transport along the magnetic field in the non-equatorial regions will certainly act to make the maximum energy less than that given in Equation (15). A full calculation is required to determine the actual spectrum.

8. Discussion and Summary

Anomalous cosmic rays can be understood as the consequence of acceleration of interstellar pickup ions, from their original energy of a few keV to more than 1 GeV, at the termination shock of the solar wind. Models which incorporate such acceleration in a global transport model can account naturally for the observed energy spectrum, charge state, composition and spatial distribution. This acceleration can proceed rapidly enough that the particles can remain singly charged from their initial ionization up to energies of the order of 200 MeV, including the bulk of the ACR. This model also accounts naturally for the rapid change to predominantly multiply-charged ACR oxygen at energies above 300 MeV, as observed by SAMPEX. However, the direct acceleration of pickup ions at the termination shock seems to be more difficult. Acceleration of these low-energy particles at propagating shocks in the inner heliosphere is found to be much easier. We have shown that a recently-proposed scenario of initial acceleration (or 'pre-acceleration') at propagating or co-rotating shocks in the inner solar system is consistent with observations of both low-energy particles and $\approx$ 100 MeV anomalous cosmic rays in the inner heliosphere.

Acknowledgements

This work was supported by NASA/JPL as part of an interdisciplinary investigation on the *Ulysses* mission, JPL contract number 960843. It was also supported, in part by the National Science Foundation under Grant ATM 9616547 and by the National Aeronautics and Space Administration under grants NAG 2251 and NAGW 1931.

References

Adams, J. H. and Leising, M. D.: 1991, 'Maximum Distance to the Acceleration Site of the Anomalous Component of Cosmic Rays', *Proc. 22nd Int. Cosmic Ray Conf.*, Paper SH4.2.8.

Axford, W. I.: 1965, 'The Modulation of Galactic Cosmic Rays in the Interplanetary Medium', *Planetary Space. Sci.* **13**, 115.

Chalov, S. V., Fahr, H. J., and Ismodenov, V. T.: 1997, 'Spectra of Energized Pick-Up Ions Upstream of the Two-Dimensional Heliospheric Termination Shock. II. Acceleration by Alfvénic Fluctuations', *Astron. Astrophys.* **320**, 659.

Christian, E. R., Cummings, A. C., and Stone, E. C.: 1995, 'Observations of Anomalous Cosmic-Ray Hydrogen from the Voyager Spacecraft', *Astrophys. J.* **446**, L105.

Cummings, A. C. and Stone, E. C.: 1987, 'Energy Spectra of Anomalous Cosmic-Ray Oxygen During 1977–1987', *Proc. 20th Int. Cosmic Ray Conf., Moscow* **3**, 421.

Cummings, A. C., Stone, E. C., and Webber, W. R.: 1987, 'Latitudinal and Radial Gradients of Anomalous and Galactic Cosmic Rays in the Outer Heliosphere', *Geophys. Res. Letters* **14**, 174.

Decker, R. B.: 1988, 'Computer Modeling of Test Particle Acceleration at Oblique Shocks', *Space Sci. Rev.* **48**, 195.

Drury, L.: 1983, 'An Introduction to the Theory of Diffusive Shock Acceleration of Energetic Particles in Tenuous Plasmas', *Rept. Prog. Phys.* **46**, 973.

Fisk, L. A., Kozlovsky, B., and Ramaty, R.: 1974, 'An Interpretation of the Observed Oxygen and Nitrogen Enhancements in Low-Energy Cosmic Rays', *Astrophys. J.* **190**, L35.

Forman, M. A. and Morfill, G.: 1979, 'Time-Dependent Acceleration of Solar Wind Plasma to MeV Energies at Co-rotating Interplanetary Shocks', *Proc. 16th Int. Cosmic Ray Conf., Kyoto* **5**, 328.

Giacalone, J. and Jokipii, J. R.: 1996, 'Perpendicular Transport in Shock Acceleration', *J. Geophys. Res.* **101**, 11 095.

Giacalone, J., Jokipii, J. R., Decker, R. B., Krimigis, S. M., Scholer, M., and Kucharek, H.: 1997, 'The Preacceleration of Anomalous Cosmic Rays in the Inner Heliosphere', *Astrophys. J.*, in press.

Gleeson, L. and Axford, W. I.: 1967, 'Cosmic Rays in the Interplanetary Medium', *Astrophys. J.* **149**, L115.

Gloeckler, G., Geiss, J., Balsiger, H., Fisk, L. A., Galvin, A. B., Ipavich, F. M., Ogilvie, K. W., von Steiger, R., and Wilken, B.: 1993, 'Detection of Interstellar Pick-Up Hydrogen in the Solar System', *Science* **261**, 70.

Gloeckler, G., Roelof, E. C., Ogilvie, K. W., and Berdichevsky, D. B.: 1995, 'Proton Phase Space Densities (0.5 eV $< E_p < 5$ MeV) at Mid-Latitudes from Ulysses SWICS/HI-SCALE Measurements', *Space Sci. Rev.* **72**, 321.

Jokipii, J. R.: 1971, 'Propagation of Cosmic Rays in the Solar Wind', *Rev. Geophys. Space Phys.* **9**, 27.

Jokipii, J. R.: 1979, 'Rapporteur Paper for Session mg-1 Modulation Theory', *Proc. 16th Int. Cosmic Ray Conf., Kyoto* **14**, 175.

Jokipii, J. R.: 1982, 'Particle Drift, Diffusion, and Acceleration at Shocks', *Astrophys. J.* **255**, 716.

Jokipii, J. R.: 1986, 'Particle Acceleration at a Termination Shock 1. Application to the Solar Wind and Anomalous Component', *J. Geophys. Res.* **91**, 2929.

Jokipii, J. R.: 1987, 'Rate of Energy Gain and Maximum Energy in Diffusive Shock Acceleration', *Astrophys. J.* **313**, 842.

Jokipii, J. R.: 1991, in G. P. Zank and T. K. Gaisser (eds), 'Diffusive Shock Acceleration: Acceleration Rate, Magnetic-Field Direction and the Diffusive Limit', *Particle Acceleration in Cosmic Plasmas*, Am. Inst. of Phys., College Park, Md., pp. 137–147.

Jokipii, J. R.: 1992, 'Constraints on the Acceleration of Anomalous Cosmic Rays', *Astrophys. J.* **393**, L41.

Jokipii, J. R.: 1993, 'Particle Drifts for a Finite Scattering Rate', *Proc. Int. Cosmic Ray Conf., Calgary* **3**, 497.

Jokipii, J. R.: 1996, 'Theory of Multiply Charged Anomalous Cosmic Rays', *Astrophys. J.* **466**, L47.

Jokipii, J. R. and Giacalone, J.: 1996, in R. von Steiger, R. Lallemont, and M. A. Lee (eds), 'The Acceleration of Pickup Ions', *The Heliosphere in the Local Interstellar Medium*, Kluwer Academic Publishers, Dordrecht, Holland, pp. 137-148.

Jokipii, J. R. and Kóta, J.: 1989, 'The Polar Heliospheric Magnetic Field', *Geophys. Res. Letters* **16**, 1.

Jokipii, J. R. and Parker, E. N.: 1970, 'On the Convection, Diffusion and Adiabatic Deceleration of Cosmic Rays in the Solar Wind', *Astrophys. J.* **160**, 735.

Jokipii, J. R., Kóta, J., Giacalone, J., Horbury, T. S., and Smith, E. J.: 1995, 'Interpretation and Consequences of Large-Scale Magnetic Variances Observed at High Heliographic Latitude', *Geophys. Res. Letters* **22**, 3385.

Klecker, B.: 1995, 'The Anomalous Component of Cosmic Rays in the 3-D Heliosphere', *Space. Sci. Rev.* **72**, 419.

Kóta, J.: 1979, 'Drift, the Essential Process in Losing Energy', *Proc. 16th Int. Cosmic Ray Conf., Kyoto* **3**, 13.

Lagage, P. O. and Cesarsky, C. J.: 1983, 'Acceleration of Cosmic Rays', *Astron. Astrphys.* **118**, 223.

le Roux, J. A., Potgieter, M. S., and Ptuskin, V. S.: 1996, 'A Transport Model for the Diffusive Shock Acceleration and Modulation of Anomalous Cosmic Rays in the Heliosphere', *J. Geophys. Res.* **101**, 4791.

McDonald, F. B. and Lal, N.: 1986, 'Variation of Cosmic Rays with Heliolatitude in the Outer Heliosphere', *Geophys. Res. Letters* **13**, 781.

McKibben, R. B, Pyle, K. R., and Simpson, J. A.: 1979, 'The Solar Latitude and Radial Dependence of the Anomalous Cosmic-Ray Helium', *Astrophys. J.* **227**, L147.

Mewaldt, R. A., Selesnick, R. S., Cummings, J. R., Stone, E. C., and von Rosenvinge, T. T.: 1996, 'Evidence for Multiply Charged Anomalous Cosmic Rays', *Astrophys. J.* **466**, L43.

Parker, E. N.: 1965, 'The Passage of Energetic Charged Particles through Interplanetary Space', *Planetary Space Sci.* **13**, 9.

Pesses, M. E., Jokipii, J. R., and Eichler, D.: 1981, 'Cosmic Ray Drift, Shock-Wave Acceleration, and the Anomalous Component of Cosmic Rays', *Astrophys. J.* **246**, L85.

Savopolus, M. and Quenby, J. J.: 1997, 'The Source of Anomalous He Acceleration and the Determination of the Interplanetary Transport Coefficient', Proc. 25th Int. Cosmic Ray Conf., Durban **2**, 325.

Vasyliunas, V. M. and Siscoe, G. L.: 1976, 'On the Flux and the Energy Spectrum of Interstellar Ions in the Solar System', *J. Geophys. Res.* **81**, 1247.

Volk, Heinrich J.: 1975, 'Cosmic Ray Propagation in Interplanetary Space', *Rev. Geophys. Space Phys.* **13**, 547.

Webb, G. M., Martinic, N. J., and Moraal, H.: 1981, 'Scatter Free Propagation and Drifts of Cosmic Rays in the Heliosphere', *Proc. 17th Int. Cosmic Ray Conf., Kyoto*, Paper SH5.3-9.

A COSMIC-RAY COMPOSITION CONTROLLED BY VOLATILITY AND *A/Q* RATIO. SNR SHOCK ACCELERATION OF GAS AND DUST

JEAN-PAUL MEYER

Service d'Astrophysique, DSM/DAPNIA, CEA/Saclay, 91191 Gif-sur-Yvette, France

LUKE O'C. DRURY

Dublin Institute for Advanced Studies, School of Cosmic Physics, 5 Merrion Square, Dublin 2,
Ireland

DONALD C. ELLISON

Department of Physics, North Carolina State University, Box 8202, Raleigh NC 27695, U.S.A.

Abstract. The composition of Galactic Cosmic Ray Sources (GCRS) shows the following features: (i) an enhancement of the refractory elements relative to the volatile ones, and (ii) an enhancement of the heavier *volatile* elements relative to the lighter ones; this mass dependence should reflect a mass-to-charge (A/Q) dependence of the acceleration efficiency; among the refractory elements, there is only a very weak enhancement of heavier species, or *none at all*. We consider it fortuitous that the GCRS composition resembles that of the solar corona, which is biased according to first ionization potential. In a companion paper by Ellison et al. (1998, this issue), this GCRS composition is interpreted in terms of a supernova shock wave acceleration of interstellar and/or circumstellar (e.g., ^{22}Ne-rich Wolf–Rayet wind) gas phase and especially dust grain material. These two papers summarize and complement the content of two papers that recently appeared in *Astrophys. J.* (Meyer et al., 1997; Ellison et al., 1997).

1. Introduction

The elemental and isotopic composition of Galactic Cosmic-Ray (GCR) constitute major clue to their origin. It was long believed that the GCR nuclei originate in newly processed supernova (SN) ejecta, the SN explosion being held responsible for both their general heavy element enhancement relative to H and He, and their acceleration. In the 1970s, however, it appeared that the *detailed* GCR source (GCRS) composition anomalies (Section 2) did not seem to be controlled by nucleosynthetic processes (Section 3). By contrast, they seemed controlled by *atomic*, rather than nuclear, parameters, such as the first ionization potential (FIP), the lower-FIP elements being systematically in excess relative to the higher-FIP ones (Section 3). This remark, associated with the discovery of an *extremely similar* FIP-bias in the composition of the solar corona, solar wind, and solar energetic particles (SEPs), and with the *lack* of a depletion in GCRs of the refractory elements locked in grains in most of the interstellar medium (ISM), suggested a source medium for the GCR nuclei: the (grain free!) coronal material of later-type stars possessing a cool, neutral H chromosphere, in which an ion-neutral separation can possibly

Space Science Reviews **86:** 179–201, 1998.

take place. It has been conjectured that GCRs consisted of stellar energetic particles with frozen-in coronal composition (similar to SEPs), first injected at MeV energies by stellar activity, and later reaccelerated to GeV and TeV energies by passing SNR shock waves (Section 3). This view required a two-stage acceleration process, in two separate sites. In addition, the presence of a ^{22}Ne excess in GCRs further required the presence of a totally unrelated second GCR component, presumably originating in Wolf–Rayet (WR) wind material (Sections 2 and 3). This scenario also had difficulty accounting for the fairly large spread of the enhancements among the high-FIP elements, and in particular for the low abundances of H, He, and N.

On the other hand, it has been noted long ago that the FIP of the various elements and the volatility of the chemical compounds they form are correlated: Typically, low-FIP elements (metals) form refractory compounds, while high-FIP elements (hydrogen, non-metals, noble gases) form volatile compounds or do not condense at all. Therefore, the apparent ordering of the GCRS composition in terms of FIP could as well reflect an actual ordering in terms of volatility (Section 4)! Such an ordering would imply an enhanced acceleration of those elements locked in dust grains in the ISM, as compared to those in the gas phase. Models for a preferential acceleration of ISM grain destruction products by SNR shock waves were actually explored in the early 1980s (Section 4).

To remove the ambiguity and choose between FIP and volatility as the parameter governing the GCRS composition, the behavior of those elements which are exceptions to the above general correlation must be considered: low-FIP volatile elements, and high-FIP refractories. In Section 5, we find nine such appropriate 'clue elements', whose GCRS abundance is reasonably well determined. Out of these, five are found equally consistent with both FIP and volatility; the four other ones are clearly discrepant with FIP and suggest volatility as the relevant parameter.

When analyzed in terms of volatility (Section 6), the GCRS abundances of *all* elements seem remarkably well organized in terms of the combined effects of: (i) volatility, the more refractory elements being in excess relative to the more volatile ones, and (ii) mass, most probably reflecting an effect of the mass-to-charge ratio A/Q, the more massive volatiles being in excess relative to the less massive ones; this description, in particular, accounts for the low GCRS abundances of H, He, and N, although the H/He ratio does not follow the above rule. This mass effect is much weaker or absent among the refractory elements. A similar combination of an ordering in terms of FIP and of mass would not account for the data as satisfactorily. Our conclusions are summarized in Section 7.

In a companion paper by Ellison et al. (1998, this issue), this behavior is interpreted in terms of an A/Q dependent acceleration of interstellar and/or circumstellar volatile gas-phase elements by smoothed SNR shock waves, and of a preferential acceleration of entire dust grains followed by their sputtering, accounting for the roughly mass-independent excess of refractory elements. In this scenario, the acceleration takes place in a single step and at a single site; the ^{22}Ne excess is also naturally accounted for, since higher mass stars produce SNR shocks which accelerate their own ^{22}Ne-rich pre-SN WR wind material.

This paper and Ellison et al. (1998, this issue) summarize and complement the content of two papers that recently appeared in *Astrophys. J.* (Meyer et al., 1997; Ellison et al., 1997; hereafter Papers I and II, respectively). The common figures have been slightly revised (cf., Table I).

2. The Empirical GCR Source Composition

The GCRS elemental abundances, the solar ones, and the GCRS/solar abundance ratios are given in Table I, normalized to H, and the GCRS/solar ratios have been plotted versus various parameters in Figures 1, 3, 4 and 5. The GCRS abundances for elements up to Ni are mainly based on the HEAO-C2 abundances of Engelmann et al. (1990), the review by Ferrando (1993), and the recent *Ulysses* data of Du Vernois and Thayer (1996); those for the 'Ultra-Heavy' (UH) elements with $Z > 30$ by Binns et al. (1989) and Binns (1995). The reference solar elemental abundances are essentially taken from Grevesse et al. (1996) (meteoritic determination adopted preferentially; for noble gases, see Table I) and the isotopic ratios from Anders and Grevesse (1989).

Two remarks apply for UH elements. First, the observations for $Z > 60$ have a limited charge resolution, forcing us to deal only with *groups* of elements: especially the 'Pt-group' elements with $Z = 74$–80 (hereafter 'Pt') and the 'Pb-group' elements with $Z = 81$–83 (hereafter 'Pb'). Second, the current data suggest GCRS excesses by factors of ~ 2 for many elements with $Z \gtrsim 40$ relative to Fe. This applies, in particular, for the comparatively abundant 'Pt' elements, as well as for the secondaries produced by their spallation in the $Z \sim 61$–73 range. By contrast, the rarer 'Pb' elements do *not* seem enhanced relative to Fe. Actually, in view of the increase of the total nuclear *destruction* cross-sections σ_{destr} with mass, the derived source abundances of UH elements *relative to Fe* are very sensitive to the propagation conditions (see Paper I); so, the source 'Pt'/Fe and 'Pb'/Fe ratios cannot be precisely determined. By contrast, the low source 'Pb'/'Pt' ratio is rather firmly established. In Figures 1, 3, 4 and 5, the error bars for those UH elements whose source abundance *relative to Fe* might be affected by such poorly known systematic errors are shown dashed, and with a '?' sign.

The abundances of dominant H and He pose specific problems. The source He/O has been assessed to lie in the range 19 ± 4, i.e., 0.145 ± 0.030 times solar, based on Webber and McDonald's (1994) determination of this ratio, and on a comparison of the He fluxes obtained by Webber et al. (1987) and Seo et al. (1991) with Engelmann et al.'s (1990) observed O fluxes. As for the H/He ratio, it is mainly based on the data by Webber et al. (1987) and Seo et al. (1991). Altogether, these data suggest a local interstellar H/He ratio around 23 ± 5 at a given energy/nucleon in the ~ 5 to 30 GeV nucl^{-1} range which, with a rigidity-dependent escape length

JEAN-PAUL MEYER ET AL.

TABLE I

Element properties and abundances

Z	Elem.	Notes	FIP[a]	T_c[b]	GCRS[c]		Solar[c]		GCRS/solar[c]	
1	H	d	13.6	<400	$\equiv 1.00 \times 10^6$	41	$\equiv 1.00 \times 10^6$	3	$\equiv 1.00$	41
2	He	d	24.6	<400	0.069×10^6	24	0.098×10^6	8	0.70	26
6	C	b, e	11.3	<400	3000.	3	355.	12	8.45	12
7	N		14.5	<400	137.	31	93.	17	1.47	37
8	O	b	13.6	<400	3720.	1	741.	17	5.02	17
10	^{20}Ne	f, g	21.6	<400	282.	6	112.	30	2.52	31
11	Na		5.1	970	21.0	51	2.09	5	10.0	51
12	Mg		7.6	1340	734.	3	38.0	2	19.3	4
13	Al		6.0	1650	54.0	21	3.09	2	17.5	21
14	Si		8.2	1310	707.	1	36.3	2	19.5	2
15	P		10.5	1150	4.92	58	0.339	10	14.5	60
16	S		10.4	650	92.4	7	15.8	10	5.85	12
18	Ar	f	15.8	<400	15.2	34	3.31	30	4.59	48
20	Ca		6.1	1520	42.0	17	2.24	2	18.8	17
26	Fe		7.9	1340	713.	2	31.6	2	22.6	3
27	Co	h	7.9	1350	1.28	37	0.0813	2	15.7	37
28	Ni		7.6	1350	40.2	4	1.78	2	22.6	5
29	Cu		7.7	1040	0.489	25	0.0195	10	25.1	27
30	Zn		9.4	660	0.445	18	0.0468	10	9.50	21
31	Ga		6.0	920	424×10^{-4}	51	13.5×10^{-4}	5	31.4	51
32	Ge		7.9	830	542×10^{-4}	19	42.7×10^{-4}	10	12.7	22
34	Se		9.8	680	205×10^{-4}	25	24.0×10^{-4}	5	8.53	26
36	Kr	i, f	14.0	<400	83.0×10^{-4}	56 ?	17.0×10^{-4}	40	4.88	75 ?
38	Sr		5.7	>1400	240×10^{-4}	29	8.32×10^{-4}	5	28.8	30
40	Zr		6.8	1780	80.1×10^{-4}	38	4.07×10^{-4}	5	19.7	39
42	Mo	i	7.1	1610	49.1×10^{-4}	76 ?	0.933×10^{-4}	5	52.6	76 ?
54	Xe	i, f	12.1	<400	18.3×10^{-4}	82 ?	1.70×10^{-4}	40	10.8	99 ?
56	Ba	i	5.2	>1400	50.0×10^{-4}	36 ?	1.66×10^{-4}	5	30.1	37 ?
58	Ce	i	5.5	1500	22.7×10^{-4}	59 ?	0.427×10^{-4}	5	53.2	59 ?
74–80	'Pt'	i, j	~9.0	~1560	50.8×10^{-4}	117 ?	1.27×10^{-4}	10	40.0	118 ?
81–83	'Pb'	i, j	~7.4	~500	25.6×10^{-4}	121 ?	1.28×10^{-4}	10	20.0	122 ?

[a]Element First Ionization Potential, in eV.

[b]Element condensation temperature T_c, in Kelvin. It is the calculated 50% condensation temperature of the dominant solid compound formed by the element in a 10^{-4} atm gas with a solar composition (mainly from Wasson, 1985; Section 4). It describes beautifully the condensation of the elements in the various types of carbonaceous chondrite meteorites (Paper I). Special care must be taken for O and C. Oxygen can be significantly condensed at higher temperatures, though always much less than 50 %. As for C, it can be largely condensed at higher temperatures for non-solar, C-rich gas compositions (see Section 4).

TABLE I

Continued

---

[c]For each entry, we give the central value V, on a geometric scale, and the (upward) error e, in percent. The adopted error bar extends from $V \cdot (1 + e/100)$ down to $V / (1 + e/100)$, so that e is equivalent to a 'dex' error on a log scale. For instance, the GCRS/Solar value for 'Pb' lies between 9.0 and 44.4. The error on each GCRS/Solar ratio is the quadratic sum of the GCRS and the solar logarithmic errors. The main sources for the GCRS and Solar values are given in Section 2, and more details can be found in Paper I (Sections 2.1 and 4).

[d]The 41% and 24% errors on the GCRS H and He abundances refer to their abundances relative to the heavies (Oxygen). The error on the GCRS He/H ratio is only $\sim$31% (Section 2; Paper I, Section 2.1). That on the GCRS/Solar He/H ratio is therefore $\sim$ 33%.

[e]Based on Wolf–Rayet nucleosynthesis considerations only, we estimate a GCRS non-Wolf–Rayet C abundance of 818, with a $\sim$53% error, i.e., a non-Wolf–Rayet C GCRS/solar ratio of 2.30, with a $\sim$ 55% error. Since we have not considered the likely preferential acceleration of a fraction of the C locked in grains in the C-rich WC wind material, this value might well be an overestimate (Section 6; Paper I, Appendix).

[f]For noble gases, for which we have no conventional, meteoritic or quiet photospheric data, the error bars given by Grevesse et al. (1996) might be too small. For Ne and Ar, based on the available local galactic (H II regions, hot stars, H I gas) and solar 'near-photospheric' data (gas emerging from the photosphere, compact flare gas, chromospheric gas; also possibly, with some caution, coronal gas, solar wind and energetic particles), we estimate a 30% error bar (e.g., Meyer, 1989, 1993; Feldman, 1992; Ramaty et al., 1995; Widing 1997; and references therein). For Kr and Xe, we conservatively adopt an error bar of 40% (see Anders and Grevesse, 1989).

[g]For Ne, both the GCRS and the solar values refer to the sole isotope ^{20}Ne, in view of the large Wolf–Rayet contribution to ^{22}Ne.

[h]For clarity, Co is not plotted in the figures. While its behavior is consistent with those of Fe and Ni, plotting its large error bar would just blur the picture without bringing any new information.

[i]The '?' sign indicates a possible additional, systematic error on the GCRS composition, due to uncertainties on the effect of GCR propagation, which are difficult to assess (Section 2; Paper I, Sections 2.1 and 4) . In the figures, the error bars for these elements are shown dashed and with a '?' sign.

[j]The 117% and 121% errors on the GCRS 'Pt' and 'Pb' abundances refer to their abundances relative to the much lighter nuclei (Fe). The error on the GCRS 'Pb'/'Pt' ratio is only $\sim$83% (Sections 2 and 5; Paper I, Sections 2.1 and 4). That on the GCRS/Solar 'Pb'/'Pt' ratio is therefore $\sim$86%.

from the Galaxy $\propto R^{-0.6}$ applying above $\sim$ 4 GeV nucl^{-1}, corresponds to a $2^{0.6}$ times smaller source H/He ratio of 15 ± 4. This correction is not sensitive to a possible reacceleration of GCRs while propagating in the Galaxy (see Paper I). All in all, with $(H/He)_\odot = 10$, H seems slightly enhanced relative to He at a fixed energy/nucleon in the range we consider, by a factor of 1.5 ± 0.4.

Key determinations of GCRS *isotopic* ratios will be found, e.g., in Leske (1993), Leske and Wiedenbeck (1993), Ferrando (1993, 1994), Lukasiak et al. (1994, 1997), Shibata (1996), Du Vernois et al. (1996a), Connell and Simpson (1997), Webber et al. (1997).

3. Significance of the GCRS Composition: the Current Wisdom

For a long time, it was generally accepted that GCRs originate in newly processed SN ejecta. This view was very tantalizing since GCRs are *globally* enriched in heavy elements, while supernovae synthesizes heavy elements, disperse them, have ample energy available for acceleration and are actually observed to accelerate electrons. It has, however, become clear that the *detailed* GCRS composition is inconsistent with a predominant selection of the elements according to specific nucleosynthesis processes and, more particularly, with what could be expected from SN nucleosynthesis (Arnould, 1984; Meyer, 1985b, 1988). Large anomalies are, indeed, found in the GCRS relative abundances of ^{20}Ne, Mg, and Al, all made by C-burning, as well as in those of Si, S, Ar, and Ca, all made by O- and Si-burning (e.g., Meyer, 1988, Figure 2). No such anomalies are found in existing SN nucleosynthesis calculations (Woosley and Weaver, 1995; Timmes et al., 1995; Arnett, 1995). By contrast, Mg, Al (C-burning), Si and Ca (O- and Si-burning), and Fe and Ni (e-process) are found in solar proportions in GCRSs (to within $\sim 20\%$), while these nucleosynthesis calculations commonly yield deviations of these ratios by factors of ~ 2 for specific types of SNae. Further, the elements through the first and the second r- and s-process peaks show no specific abundance anomaly of either r- or of s-nuclei in the GCRSs. This observed *lack* of a s-element deficiency definitely implies that GCRs do *not* predominantly originate in SN processed material, since no type of SN synthesize s-nuclei (Prantzos et al. 1993). In this context, an interpretation of the low GCRS 'Pb'/'Pt' ratio in terms of an excess of the *third* r-process peak elements seems unlikely. Finally, most GCRS isotope ratios are found consistent with solar; and the isotopic ^{59}Co/^{57}Co ratio indicates the absence of freshly synthesized Fe peak nuclei. So, most of the GCRS material seems to be of the 'solar mix' type as regards nucleosynthesis (with one exception discussed below).

Early in the 1970s, it was noted that the *detailed* GCRS heavy element composition anomalies could be rather well organized in terms of an atomic physics parameter, the FIP, which controls the tendency of an element to be neutral or ionized in a gas at $\sim 10^4$ K (Cassé and Goret, 1978, and references therein; Meyer, 1985b; Silberberg and Tsao, 1990). As shown in Figure 1, heavy elements with FIP $\lesssim 8.5$ eV ('low-FIP') are typically enhanced by a factor of ~ 5 relative to those with FIP $\gtrsim 11$ eV ('high-FIP').

The low temperature required for the parent gas ($\sim 10^4$ K), together with the lack of a depletion in GCRs of the refractory elements locked in grains in virtually all but the hottest interstellar medium (Cassé et al., 1975; Dwek, 1979), first suggested that the GCR nuclei did not originate in the ISM, but in stellar surfaces (Meyer et al., 1979; Meyer, 1985b). This conclusion, of course, assumed implicitly that only gas-phase atoms of the ISM could be accelerated.

In addition, it was found that the entire solar coronal gas (outside coronal holes), the slow solar wind, and the solar energetic particles (SEP) have a FIP-biased com-

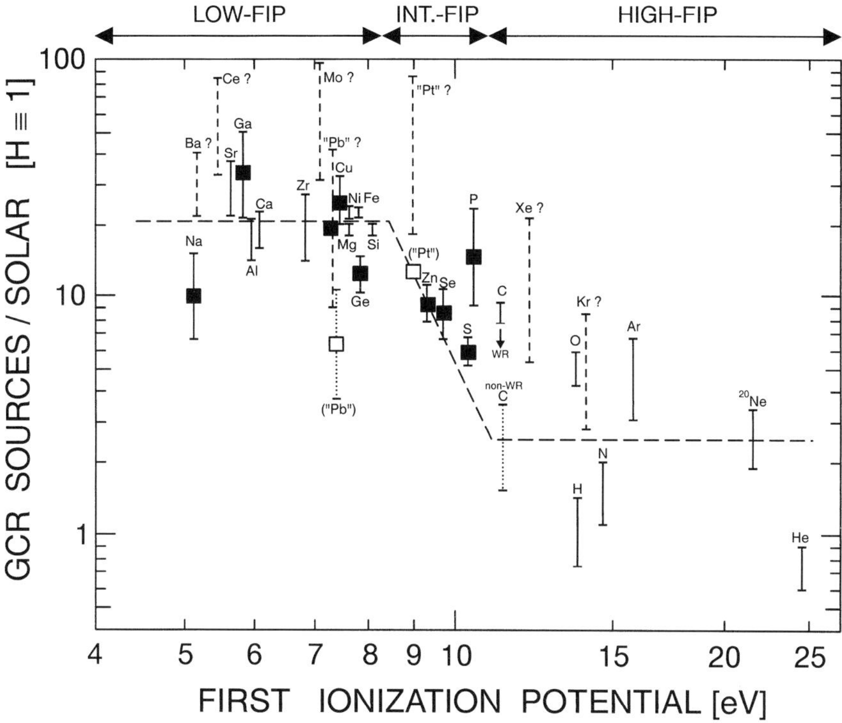

Figure 1. The GCRS to solar abundance ratios in the $\sim$ GeV range, plotted versus FIP (Section 2), normalized to H at a given energy/nucleon. Those elements which can serve as clues to distinguish between FIP and volatility as the parameter governing the GCRS composition are emphasized by a solid square (abundances discussed in Section 5). The point for C is plotted as an upper limit since its total source abundance includes a specific ^{12}C contribute associated with the ^{22}Ne-rich component from WR stars. We propose a tentative estimate of the C *non*-WR source abundance (*not* considering any preferential acceleration of C relative to ^{22}Ne in the WR component); WR stars may also contribute to O, but this contribution is probably insignificant (Section 6). Also, we have plotted the isotopic ^{20}Ne abundance for Ne. We have marked by a dashed bar and a '?' sign those ultra-heavy elements whose source abundance *relative to Fe* is quite uncertain (Sections 2 and 5). This is, in particular, the case for the crucial 'Pt'- and 'Pb'-group elements. But the source 'Pb'/'Pt' ratio is much better determined. To visualize it, we have also plotted the 'Pb' point relative to the 'Pt' point *arbitrarily* placed where it would fit assuming the standard FIP pattern (open squares). The figure defines three classes of elements, according to their FIP (Section 3).

position, much like that of GCRSs (e.g., Meyer, 1985a, b, 1993, 1996; Breneman and Stone, 1985; Feldman, 1992; Geiss et al., 1995; Garrard and Stone, 1993; Reames, 1995). This similarity between the GCRS and the solar coronal composition has further suggested that the GCR nuclei had been first extracted from stellar atmospheres. More precisely, the GCR particles probably originated in the coronae of ordinary F to M stars possessing, like the Sun a predominantly neutral-H chromosphere at $\sim$ 7000 K, in which, in some yet debated way, ionized heavies may be separated from neutral ones, and rise preferentially into the corona (Meyer,

1985b; review by Hénoux, 1995). FIP effects have been recently observed in the coronae of some later type stars (e.g., Laming et al., 1996; Drake et al., 1997; Singh et al., 1996).

Stellar flare activity is, however, energetically unable to accelerate the bulk of the (GeV and TeV) GCRs in the Galaxy. It may, however, accelerate some of the 'FIP-biased' coronal material to low (MeV) energies, just as the Sun accelerates SEPs, thus providing a suprathermal component with 'frozen' coronal composition. These 'injected' MeV particles must then be later *preferentially* accelerated by more powerful SNR shock waves, which boost them to the GeV and TeV energies of GCRs (Meyer, 1985b); so, this scenario requires two separate acceleration stages in two separate sites.

Further, there is one exception to the lack of signature of specific nucleosynthetic processes in the GCRS composition: the GCRS ^{22}Ne/^{20}Ne ratio is definitely enhanced and there are hints for depleted ^{13}C/^{12}C and ^{18}O/^{16}O ratios. This ^{22}Ne excess, together with the high GCRS elemental C/O ratio, suggests the presence of a pure He-burning material component in GCRs, whose most likely origin is WR wind material (Meyer, 1981a, 1985b; Cassé and Paul, 1982; Maeder, 1983; Prantzos et al., 1987; Maeder and Meynet, 1993; Paper I). Associated with the WR ^{22}Ne excess, we indeed expect a ^{12}C excess, whose contribution can be reliably assessed, and is certainly large. A WR contribution to ^{16}O is, in principle, also possible; but it is probably insignificant, as indicated in particular by the *lack* of a ^{26,26}Mg excess in the GCRS (Paper I, Note Added in Proof). In any case, the FIP/later-type-star scenario for the bulk of the GCRs requires an additional, entirely unrelated source for the ^{22}Ne-C-rich component.

Figure 1 shows the correlation of the GCRS abundance enhancements with FIP. Obviously, FIP does roughly order the data; but there is a lot of scatter around the correlation. In particular, H, He and N are deficient relative to the other high-FIP elements. We have plotted the point for C as an upper limit, in view of the expected, specific ^{12}C contribution associated with the WR ^{22}Ne-rich component (Section 6), and we propose an estimate of the *non*-WR C-source abundance (see Paper I; the WR contribution to ^{16}O is probably negligible). Other elements deviate from the correlation. The source abundance of Na seems low and that of P, high, at least based on the currently best available spallation cross sections for these largely secondary species (Section 5). The UH element data show a larger scatter, and a general trend towards larger enhancements for $Z \gtrsim 40$. This trend may be real, or due to an improper account of the interstellar propagation (Section 2). An important exception is Ge, which is reliably determined to be low as compared to Fe, with exactly the same FIP value (Section 5). Finally, while the 'Pt' and 'Pb' abundances relative to Fe are poorly determined (Section 2), the more reliable low 'Pb'/'Pt' ratio also conflicts with a FIP ordering.

4. Element Volatility versus FIP: Acceleration of Grain Material?

For most elements, the volatility of the chemical compounds the element can form and its FIP are correlated. Figure 2 shows a cross plot of the element condensation temperature T_c versus its FIP, for all elements, updated from Meyer (1981b). This temperature T_c, given in Table I, is the calculated 50% condensation temperature of the dominant solid compound formed by the element in a 10^{-4} atm gas with solar composition, taken mainly from Wasson (1985). The lower the T_c, the higher the volatility of the element. Figure 2 shows that FIP and T_c are, indeed, anti-correlated for the majority of the elements.

Therefore, the apparent correlation of the GCRS abundances with FIP could just *mimic* an actual correlation with the element volatility. With this viewpoint, the refractory (low-FIP!) elements, those generally locked in grains in the ISM, would be overabundant in GCRs. This would imply *a preferential acceleration of grain destruction products*, presumably in SNR shocks. One nice point with such a scenario is that the same SNR shocks could destroy the grains and fully accelerate the particles to their final GeV and TeV energies. We are back to a one-step acceleration process in a single site.

Now, it is believed that SNR shocks accelerate mainly external interstellar or circumstellar material, not the SN ejecta themselves (see Paper II). It is therefore no surprise that we find little trace of SN nucleosynthesis in the GCRS composition (Section 3), in spite of the key role played by SNR shocks in accelerating the particles! This applies to both Type I and Type II SNae.

What does this external material consist of? Around Type I and lower mass Type II SNae, this material ought to be ordinary interstellar material (ISM), with roughly solar composition (gas+grain). Its grains consist of old ISM grains. But for massive SN progenitors, earlier wind ejections may be important, so that the shocks may accelerate the progenitor's own wind material. Its grains will then consist of newly formed grains, presumably with different properties. The highest mass, WR star progenitors, have been stripped off by huge winds, to the point where their He-burning layers have been tapped and their winds are highly enriched in He-burning products. So, another extremely nice feature of this type of scenario is that it may *account in a natural, continuous way* for the observed excess of He-burning material (^{22}Ne, ^{12}C, possibly ^{16}O excesses) in GCR sources.

Note that, in this context, the logic of the 'grain constraint' earlier put forward to *exclude* the ordinary ISM as a possible source of the GCR ions is completely reversed: the ISM was excluded as a possible source of the GCRs, based on the implicit assumption that ions had to be accelerated out of the ISM *gas-phase*, which is depleted of its low-FIP, refractory elements locked in grains. Here we turn the argument around, considering the opposite possibility of a preferential acceleration of this very material locked in grains!

Such a preferential acceleration of grain material in SNR shocks has been considered in the early 1980s. Epstein (1980) first introduced the concept considered in

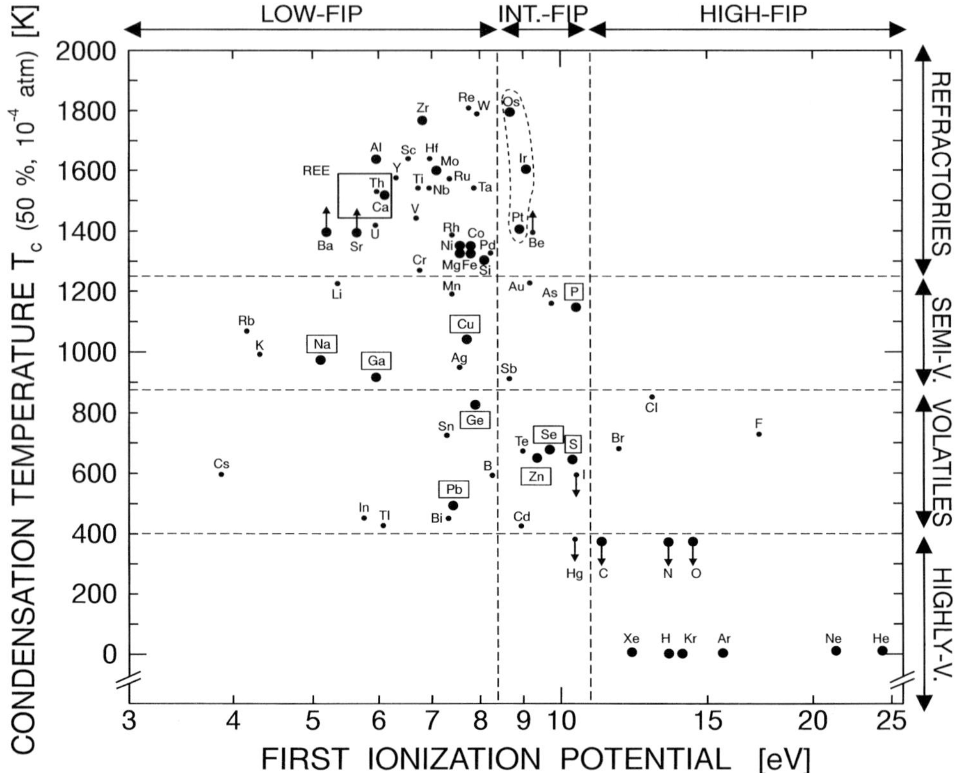

CONDENSATION TEMPERATURE T_c (50 %, 10^{-4} atm) [K]

FIRST IONIZATION POTENTIAL [eV]

Figure 2. Cross plot of the condensation temperature T_c of each chemical element, versus the element's FIP (Section 4). 'REE' stands for 'rare earth elements'. We define four classes of volatility (Section 4; Paper I). The general anti-correlation between FIP and T_c is conspicuous, most lower-FIP elements being refractory and higher-FIP ones, volatiles. The elements whose GCRS abundance is reasonably well determined are denoted by big, solid dots. Among them, the elements lying *outside* the main FIP – T_c correlation, which can currently serve as clues to distinguish between FIP and T_c as the parameter governing the GCRS composition, are framed (marked by solid squares in Figures 1 and 3).

the present work: an acceleration of the entire grain, followed by grain sputtering and by a first-order Fermi re-acceleration of the suprathermal grain destruction products to GeV and TeV energies. Cesarsky and Bibring (1981) and Bibring and Cesarsky (1981), on the other hand, suggested that grains freely cross the shock, so that they acquire a bulk speed relative to the ambient shocked gas equal to the shock speed. When these grains undergo strong sputtering in the downstream region, the sputtered ions are thus injected with the shock speed, relative to the ambient gas, and preferentially *stochastically* accelerated to GeV and TeV energies. This, we will not pursue, because of the well-known problems with stochastic acceleration of cosmic rays (see Paper I). Further, Meyer (1981b), Tarafdar and Apparao (1981), and Sakurai (1995, and references therein) have provided analyses of the GCRS composition data in the light of the condensation of elements into grains.

In Paper I we have discussed in detail the significance of the condensation temperature T_c and shown, in particular, that T_c describes very nicely the condensation of the elements in the various types of carbonaceous chondrite (CC) meteorites. On this basis, we have defined four groups of elements: (i) 'refractory' elements with $T_c > 1250$ K, consisting of all refractory metals, including Mg, Si, and Fe, which are predominantly condensed in all types of CCs; (ii) 'semi-volatile' elements with 1250 K $> T_c > 875$ K, still entirely condensed in type-I CCs, but with quite T_c-dependent condensation in other types; (iii) 'volatile' elements with 875 K $> T_c > 400$ K, also entirely condensed in type-I CCs, but only weakly condensed in other types; (iv) 'highly-volatile' elements with 400 K $> T_c$, which are not predominantly (<50 %) condensed even in type-I CCs, and include H, C, N, O, and noble gases.

Among the 'highly volatile' elements, H and noble gases can never be significantly condensed in any astrophysical context. The situation is more subtle for C, N, and O. In particular, a very significant fraction of C may be condensed in solid carbon compounds, if the condensation takes place in a C-rich atmosphere ($C/O > 1$), such as Red Giants and WC-type WR wind material. Carbon, indeed, behaves as a highly volatile element when entirely trapped in gaseous CO molecules at high temperature; this is the case whenever $C/O < 1$, and in particular for a solar composition ($C/O = 0.48$), but it behaves as a highly refractory element when CO formation is hindered for lack of a sufficient amount of O in the medium ($C/O > 1$). As for O, the trapping of a major fraction of it is possible only in environments in which H_2O ice mantles survive, which is not the case in the situations we will consider (SN shocks). The amount of O trapped in sturdy organic refractory mantles seems negligible (Schutte, 1988; Schutte and Greenberg, 1988). However, some 15 to 20% of the O is certainly trapped in silicates, the principal constituents of the grain cores, throughout the ISM (Meyer, 1989, scaled to $O/Si = 20.4$ from Grevesse et al., 1996). By contrast, N is probably never significantly trapped in grains (e.g., Mason, 1979; Wasson and Kallemeyn, 1988; Sharp and Wasserburg, 1995; Mathis, 1996). So, while formally 'highly volatile elements' as defined above, never get >50% condensed out of a solar-type gas, C and O (but not N!) do get partly condensed in warm astrophysical environments. These remarks will be important when interpreting the C and O excesses relative to the GCRS/solar vs A correlation for other 'highly volatile' elements (Figure 4). These excesses will have to be interpreted in terms of, not only WR star nucleosynthesis, but also of preferential acceleration of some of the C and O trapped in grain material (Section 6).

The relevance of T_c to the fraction of condensed material can also be tested by the observed depletion of the various elements in the ISM gas phase (e.g., Savage and Sembach, 1996; see Paper I). There is a general trend for an increasing condensation in grains for elements with increasing T_c, but with a very large spread of the gas phase depletions for any given T_c, and with possible problems with the solar abundance normalization of the observations (e.g., Mathis, 1996; Dwek,

1997). The spread might result from a slow chemical reprocessing governing grain destruction and growth in the ISM, largely independent of T_c, subsequent to the primary grain condensation phase in cooling stellar ejecta and winds, which could be more closely controlled by T_c. So, newly formed grains in stellar wind envelopes might have a composition different from old ISM grains, and more closely controlled by T_c (regarding P in this context, see Section 5).

5. Observational Evidence for the Relevance of Volatility

To distinguish between FIP and volatility as the key factor governing the GCRS composition, we have to consider the GCRS abundances of those few elements which do *not* fit into the general correlation between FIP and volatility shown in Figure 2 (Meyer, 1981b): (i) low-FIP volatile elements, which should have solar abundances relative to the standard, refractory low-FIP elements if FIP is relevant, and should be comparatively depleted if T_c is relevant; (ii) high-FIP refractory elements, which should be depleted relative to these same low-FIP elements if FIP is relevant, and should have solar abundances if T_c is relevant.

In Figure 2 we have singled out those elements for which we currently have reasonably accurate GCRS abundances (large, solid dots). Among them, $_{11}$Na, $_{15}$P, $_{16}$S, $_{29}$Cu, $_{30}$Zn, $_{31}$Ga, $_{32}$Ge, $_{34}$Se, $_{82}$Pb lie outside or only marginally within the FIP $-$ T_c correlation, and are therefore the 'clue elements' of interest in this context (framed in Figure 2). We now examine the relevance of each of these elements, mainly based on the studies by Engelmann et al. (1990), Ferrando (1993), Du Vernois and Thayer (1996), Du Vernois et al. (1996b) for $Z \leq 30$, and by Lund (1984), Fowler et al. (1987), Binns et al. (1981, 1989), Binns (1995), Clinton and Waddington (1993), and Waddington (1996) for UH elements (for more specifics, see Paper I).

5.1. SODIUM

In spite of its very low FIP of 5.1 eV, Na is a rather volatile element ($T_c = 970$ K; Figure 2). Na seems deficient by a factor of 2.0 ± 0.8 relative to Si. This points towards a GCRS composition controlled by volatility. The measured Na abundance in arriving GCRs is absolutely foolproof. However, Na is a predominantly secondary element in these observed GCRs, so that the determination of its source abundance is very sensitive to spallation cross-section errors; while the large error bar on its adopted source abundance is based on a conservative estimate of these errors, it might still be not entirely definitive.

5.2. SULPHUR, ZINC, AND SELENIUM

These three elements have neighboring, 'intermediate' values of FIP $\sim$ 10 eV, and very similar values of $T_c \sim$ 660 K (Figure 2), which make them full-fledged volatile elements. Their source abundances are reliably determined. That of S is very accurately known. We have good data on Zn from both the C2 and the C3 instruments on board the HEAO-3 spacecraft, which converge on a source Zn/Fe ratio of 0.43 ± 0.08 times solar. Finally, Se is well measured by both the Ariel and the HEAO-C3 instruments. The secondary fraction of all three elements is small, and the source Zn/Fe and Se/Fe ratios are not much affected by interstellar propagation (similar σ_{destr}). The GCRS abundances of these elements is well interpreted in terms of FIP (Figure 1). In terms of volatility, the lower GCRS abundance of S, as compared to Zn and Se, may, at first, seem disturbing; it will later be interpreted as a mass effect.

5.3. PHOSPHORUS

The FIP of P, another 'intermediate-FIP' element, is 10.5 eV, i.e., virtually the same as that of S. But, while S is a full-fledged volatile ($T_c = 650$ K), P is a rather refractory semi-volatile element ($T_c = 1150$ K; Figure 2). While S is depleted by a factor of 3.4 ± 0.5 relative to Si, P seems depleted by a factor of 1.5 ± 0.7 only (consistent with being undepleted), so that the P/S ratio is enhanced relative to Solar by a factor of 3.0 ± 1.6. Thus, the high P/S ratio represents another hint in favor of volatility controlling the GCRS composition. There are, however, two caveats. First, even more so than Na, P is a predominantly secondary element in the observed GCRs, so that the determination of its source abundance is very sensitive to spallation cross-section errors. Second, while P, a siderophile element, behaves as a rather refractory element in CCs (where the fractionation seems well controlled by T_c), in the ISM gasphase it seems much less depleted than other elements with comparable values of T_c, actually hardly more than S (see Paper I). This difference might be due to a slow chemical reprocessing of the grains in the ISM (Section 4); if this were the case, P could be much more condensed in the grains recently formed in pre-SN stellar winds than in the general ISM.

5.4. COPPER AND GALLIUM

Cu and Ga are low-FIP (7.7 and 6.0 eV), semi-volatile elements ($T_c = 1040$ and 920 K; Figure 2). We have, unfortunately, only one observation of these odd-Z elements in GCRs, by the HEAO-C2 experiment, in which these elements seem well resolved. It yields Cu/Fe $= 1.14 \pm 0.25$ and Ga/Fe $= 1.51 \pm 0.59$ times solar. These values are consistent with FIP as the relevant parameter.

5.5. GERMANIUM

Ge has virtually the same FIP as Fe (7.9 eV), but Ge is a volatile element ($T_c =$ 830 K), while Fe is refractory ($T_c = $ 1340 K; Figure 2). The C2 and the C3 instruments on board the HEAO-3 spacecraft have yielded independent, consistent GCRS Ge/Fe ratios, both significantly lower than solar. All in all, they lead to a GCRS Ge/Fe $= 0.57 \pm 0.10$ times solar. In both instruments, the charge resolution is appropriate to safely observe Ge, and possible systematic errors are limited. The errors are predominantly statistical. The *source* Ge/Fe ratio is close to the measured one (within 13%), and the error due to interstellar propagation is insignificant. Therefore, the low abundance of Ge strongly argues in favor of volatility controlling the GCRS source composition.

5.6. LEAD

The 'Pb'-group elements essentially consist of low-FIP, volatile, *s*-elements, and the 'Pt'-group elements of intermediate-FIP, refractory, *r*-elements (Section 2; Figure 2). Since source abundances *relative to Fe* are very sensitive to propagation (Section 2; Figure 1), we will deal only with the 'Pb'/'Pt' ratio. According to standard calculations, the source 'Pb'/'Pt' ratio is estimated to be $\sim$ 1.65 times higher than the observed one; considering various sources of error, this factor could actually lie anywhere between $\sim$ 1.3 and $\sim$ 2.6. Our current knowledge of the 'Pb'/'Pt' ratio comes essentially from the Ariel-6 and the HEAO-C3 spacecraft experiments, which seem confirmed by the preliminary LDEF data. Both sets of data yield low 'Pb'/'Pt' ratios, in the range $\sim$ 3.9 $\pm$ 1.1 times lower than solar, resulting in a source 'Pb'/'Pt' ratio $\sim$ 2.4 $\pm$ 1.3 times lower than solar. Now, 'Pb' elements have low-FIPs, and 'Pt' elements mostly intermediate-FIPs (Figure 2); based on a plain FIP-biased solar source composition, one would therefore expect a source 'Pb'/'Pt' ratio slightly *higher* than solar, by a factor of $\sim$ 1.6. So, the actual source 'Pb'/'Pt' ratio is $\sim$ 3.9 $\pm$ 2.0 times lower than would be expected, based on a FIP-biased solar source composition (see Figure 1). An interpretation of this low ratio in terms of an excess of the *third r*-peak Pt-group elements seems unlikely, in view of the lack of a similar excess of the lighter *r*-nuclei (Section 3). The other possible interpretation is that 'Pb' is depleted relative to 'Pt' because 'Pb' elements are very volatile ($T_c \sim$ 500 K) while 'Pt' elements are refractory ($T_c \sim$ 1400 to 1800 K; Figure 2).

5.7. TO SUMMARIZE

One very dependable indicator, Ge, a rather solid one, Pb, and two less foolproof but still significant indicators, Na and P, point to volatility, not FIP, as the relevant parameter governing the GCRS composition. The other five indicators, S, Zn, Se, Cu, and Ga are consistent with either the FIP picture or volatility. In terms of volatility, the low S/Zn,Se ratio may seem problematic, and semi-volatile Cu and

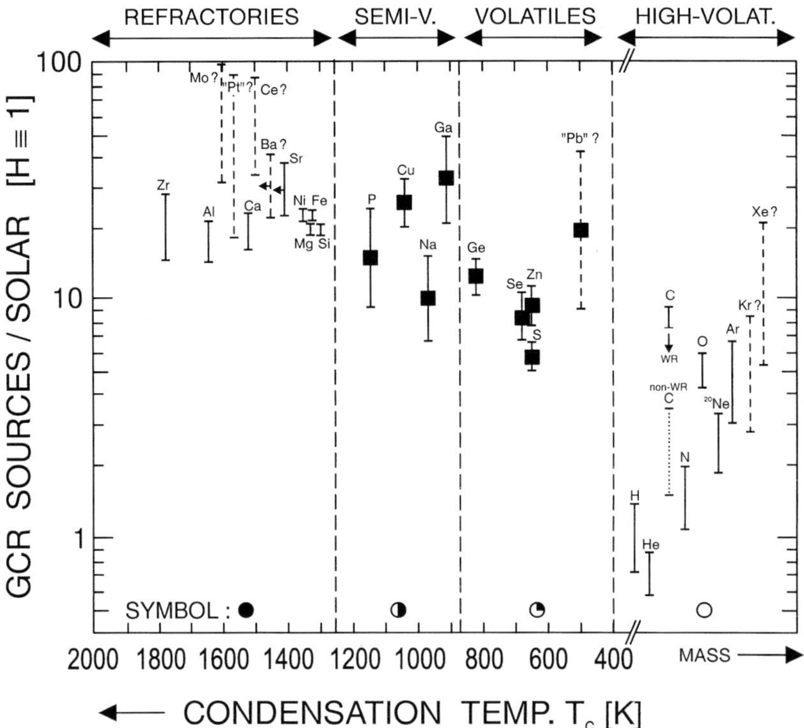

Figure 3. The same GCRS to Solar abundance ratios as in Figure 1, this time plotted versus condensation temperature T_c (Section 6). See Figure 1 caption. Note that our 'clue elements' (solid squares, as in Figure 1) are all those in the two intermediate classes of volatility (cf., Figure 2). Clearly, the enhancements progressively decrease with decreasing T_c, throughout the four classes of volatilities. For the 'highly-volatile' elements, T_c does not make sense, and we have plotted these elements simply in order of increasing mass; their enhancements seem a monotonic function of the mass, except for H, C, and O (Section 6; see Figure 4 caption). The large spread in the enhancements of the elements in the two intermediate classes of volatility will also be interpreted as a mass effect (Figure 4).

Ga seem rather high. In Section 6, these apparent difficulties will be interpreted in terms of a mass effect.

6. Interpretation: a CGRS Compsition Controlled by Volatility and Mass-to-Charge Ratio

In Figure 3, we show the same GCRS overabundances relative to solar as in Figure 1, but this time versus condensation temperature T_c. For the 'highly volatile' elements with $T_c < 400$ K, T_c has no physical relevance, and we have just ordered these elements by mass. Two conclusions can be drawn from Figure 3: (i) The enhancement of the 'refractory' elements relative to the 'highly volatile" ones is

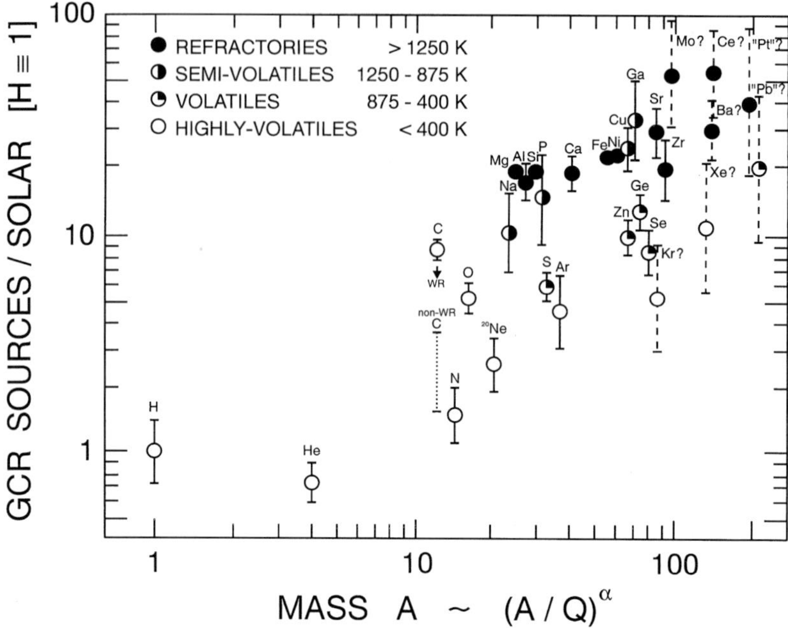

Figure 4. The same GCRS to solar abundance ratios as in Figures 1 and 3, this time plotted versus element atomic mass number A, for the elements in the four classes of volatility (Section 6). See Figure 1 caption. Clearly, the more refractory elements are generally more enhanced than the more volatile ones. For the 'highly-volatile' elements, the GCRS/Solar enhancements seem roughly $\propto A^{0.8\pm0.2}$, except for H (high thermal speed), C (excess from WR stars; recall that our non-WR C estimate may still be an overestimate, if a significant fraction of the C is locked in grains in the C-rich WC wind material, and hence preferentially accelerated relative to ^{22}Ne), and O (partly locked in grains), which are discussed in Sections 4 and 6. This apparent correlation of the enhancements with A most likely reflects a correlation with A/Q in the source gas, which is a roughly monotonic function of A in all actual ionization situations. For the 'refractory' elements, by contrast, there is only a very weak increase of the enhancements with mass A, if any. The elements in the two intermediate classes of volatility show intermediate behaviors. These contrasting behaviors will be interpreted in terms of the volatile elements being accelerated as individual ions directly out of the gas-phase, while the refractory elements are first accelerated as constituents of entire grains.

obvious. The two intermediate classes, 'semi-volatiles' and 'volatiles', tend to show intermediate overabundances, but with quite a large scatter. (ii) The overabundances of the 'highly volatile' elements seem to be an increasing function of their mass. H, C, and O form exceptions to this rule, to be discussed below.

This behavior of the 'highly volatile' elements leads us to suspect that the overabundances of the elements in the other classes of volatility might also be correlated with mass. We therefore plot in Figure 4 the same overabundance versus element mass A, distinguishing the elements in the four classes of volatility. Figure 4 contains the essential conclusions of this paper:

(i) The 'refractory' elements are globally enhanced relative to the 'highly volatile' ones.

(ii) There is only a very weak mass dependence of the 'refractory' element overabundances, or none at all. For instance, the GCRS Fe/Mg ratio is close to solar, enhanced by 20% at the most. The current ultra-heavy abundance estimates relative to Fe suggest modest excesses of most elements with $Z > 40$, but these estimates need confirmation (Sections 2 and 5).

(iii) By contrast, the overabundances of most 'highly volatile' elements are a strongly increasing function of their atomic mass, roughly $\propto A^{0.8\pm0.2}$.

(iv) C and O, are totally out of the correlation for other 'highly volatile' elements. But these are precisely the two 'highly volatile' elements which may be *partly* trapped in grains (see discussion in Section 4); in addition, we expect a major specific contribution to C from WR stars (Section 3). In Paper I, this WR contribution has been roughly estimated for C, based on the WR nucleosynthetic yields only, i.e., assuming that all the C lies in the gas-phase in the WR wind. We have not considered a possible preferential acceleration of a significant fraction of the C, which is partly locked in the grains formed in WC wind material, in which C/O > 1. So, our assessment of the WR contribution to C may be an underestimate, and therefore that of the non-WR abundance of C, plotted in Figures 1, 3, 4 and 5, an overestimate. As such, it fits quite well with the trend given by the neighboring elements in Figure 4. As for O, its WR contribution is probably insignificant. But some 15 to 20% of the O is trapped in grain silicates in the general ISM. Comparing the GCRS abundances of the neighboring mass 'highly volatile' N and ^{20}Ne with those of the refractories in Figure 4, we estimate that the grain material gets accelerated ~ 10 times as efficiently as the gas-phase O (recall that N is not significantly locked in grains, Section 4). With 15 to 20% of the O in grains getting 10-fold preferentially accelerated, the O excess by a factor of ~ 3 is readily accounted for within the framework of our model.

(v) Another exception is H, which is slightly enhanced relative to He (at a given energy/nucleon, the relevant parameter for acceleration), hence much less depleted than expected based on the pattern for He and heavier elements. It is shown in Paper II that low Mach number shocks, with shock speed comparable to the H thermal speed, can accelerate both H and heavy ions more efficiently than He.

(vi) Regarding the two intermediate classes of volatility, they show intermediate overabundances and fit beautifully into the picture. In particular, the low Na, P/Cu, Ga ratios in the 'semi-volatile' group, the low S/Zn, Se ratio and the high 'Pb' abundance (if confirmed) among the 'volatile' one (Figure 3), are now readily interpreted in terms of a mass effect.

(vii) With the current errors, it is, of course, not possible to know whether the 'volatile' elements behave significantly differently from the 'highly volatile' ones, or not. In the future, it may be interesting to investigate more precisely the compared behaviours of S and O, both of which get slightly trapped in grains.

Of course, the mass A is not a physical parameter capable of governing by itself the acceleration efficiency for the various elements. The observed rough mass dependence of GCRS overabundances of the more volatile elements most likely just reflects an actual dependence on A/Q, i.e., a rigidity dependence of the acceleration efficiency (Paper II). In any ionization situation, indeed, A/Q is, by and large, a monotonically increasing function of A. Clearly, the appropriate abscissa scale in Figure 4 would have been A/Q, rather than A. This plot, however, would have required a precise knowledge of the ionization states for all elements in the source gas.

Qualitatively, we can say that the accelerated gas cannot be a purely collisionally ionized gas around $\sim 10^4$ K, since in such a gas Ne and He, for example, would be entirely neutral, hence not accelerated. It could be hot $\sim 10^6$ K gas, in which grains have been somehow preserved, in which case the charge states Q are a rather smooth function of A, very roughly described by $A/Q \approx A^{0.4}$ (Arnaud and Rothenflug, 1985). It might also be $\sim 10^4$ K gas photoionized by stellar UV radiation, in which case most elements will have charges of $Q = +1$ or $+2$, so that $A/Q = (0.5 \text{ to } 1) \times A$.

Note that A/Q-dependent abundance enhancements more or less similar to those observed among the GCR volatiles are found in several heliospheric accelerated particle populations: (i) There is clear evidence for A/Q-enhancements among the particles accelerated at the Earth's bow shock, which are fully consistent with the smoothed shock acceleration theory developed in Paper II for GCRs (e.g., Ellison et al., 1990). (ii) The same is true for the 'anomalous cosmic rays', accelerated by the solar wind termination shock (Cummings and Stone, 1996). (iii) *Gradual* SEP events also show smooth A/Q enhancements. Here, however, heavier, higher-A/Q elements, while most frequently enhanced, are also sometimes depleted relative to lighter ones. So, shock smoothing, which produces essentially heavier element enhancements (Paper II), is certainly not always the dominant factor shaping the SEP composition (Meyer, 1985a; Breneman and Stone, 1985; Cane et al., 1991; Garrard and Stone, 1993; Reames, 1995). (iv) In *impulsive*, ^{3}He-rich SEP events, the heavy element enhancements have quite different, specific characteristics, which are to be explained in terms of resonant wave acceleration (e.g., Reames et al., 1994; Steinacker et al., 1997; Roth and Temerin, 1997, and references therein).

7. Summary

The GCRS composition is best described (Figure 4) in terms of (i) an enhancement of the refractory elements relative to the volatile ones, and (ii) among the volatile elements, an enhancement of the heavier elements relative to the lighter ones; this mass dependence most likely reflects a mass-to-charge (A/Q) dependence of the acceleration efficiency; H, C, and O, however, do not fit well into the pattern for the other 'highly-volatile' elements (noble gases and N); these deviations are discussed below. Among the refractory elements, by contrast, there is only a very weak enhancement of heavier species, or none at all.

This conclusion is based on an analysis of the detailed GCRS composition, in terms of both FIP and volatility. One may, however, object that, when discussing the spread of the GCRS overabundances versus FIP (Figure 1), we did not consider a possible superimposed mass (or A/Q) dependent fractionation (as we did in the context of volatility!). After all, the mass dependence of the volatile element overabundances could as well be interpreted in terms of a mass dependence of the high-FIP element overabundances! So, one may wonder whether a combination of a FIP and of a mass fractionation could not account for the data, just as well the volatility-mass combination! To clarify this point, we have replotted, in Figure 5, the GCRS overabundances versus mass, but with the elements sorted, not according to their volatility, but according to their FIP. So, Figure 5 is identical to Figure 4, except for the symbols describing the sorting of the elements. As could be expected, the general pattern in Figure 5 still seems by and large ordered: FIP and volatility are, indeed, largely correlated, and the mass effect among the 'highly-volatiles' cannot be distinguished from a mass effect among the high-FIP elements! However, when looking specifically at the 'clue elements' (Section 5), it can be seen that definitely Ge, most likely 'Pb' (relative to 'Pt'), and probably Na and P do not fit into the general pattern set by the other elements. More precisely, the GCRS Na/Mg, P/S, Ge/Fe, and 'Pb'/'Pt' ratios between elements of comparable FIP and mass, but widely different volatilities, cannot be interpreted in terms of a combined FIP and mass fractionation. In addition, both the marked difference between the mass dependences of the high- and of the low-FIP element overabundances, and the singular behavior of O, may be understood in terms of the trapping of elements in grains (see below and Section 6); in a FIP context, they would remain to be interpreted without resorting to this trapping in grains.

So, we consider the similarity between the volatility-biased GCRS composition and the FIP-biased composition of the solar corona, wind and energetic particles as fortuitous! Note that this similarity is not complete: crucial elements such as Na and P do seem to behave differently in GCR sources and SEPs, where they clearly follow the FIP pattern (Garrard and Stone, 1993; Reames, 1995).

To confirm or disprove these views, new determinations of the GCRS abundance of all low-FIP volatile and high-FIP *moderately* volatile elements (in the lower left and upper right parts of Figure 2) would be essential. In addition to the key elements already studied: $_{11}$Na, $_{15}$P, $_{16}$S, $_{29}$Cu, $_{30}$Zn, $_{31}$Ga, $_{32}$Ge, $_{34}$Se, $_{82}$Pb, the following elements, whose GCRS abundances may become accessible in the future, can serve as clues: $_{17}$Cl, $_{19}$K, $_{25}$Mn, $_{33}$As, $_{35}$Br, $_{37}$Rb, $_{47}$Ag, $_{48}$Cd, $_{49}$In, $_{50}$Sn, $_{51}$Sb, $_{52}$Te, $_{55}$Cs, $_{79}$Au, $_{81}$Tl, and $_{83}$Bi.

This GCRS composition can be understood in terms of an acceleration of interstellar and/or circumstellar gas and dust material by SNR shocks (Ellison et al., 1997, 1998). SNR shocks, smoothed by the feedback pressure of the very accelerated particles, preferentially inject and accelerate the higher rigidity ions. Among the ISM gas-phase volatile elements, they therefore enhance the higher A/Q, i.e., the more massive elements. The same shocks treat the dust grains as extremely

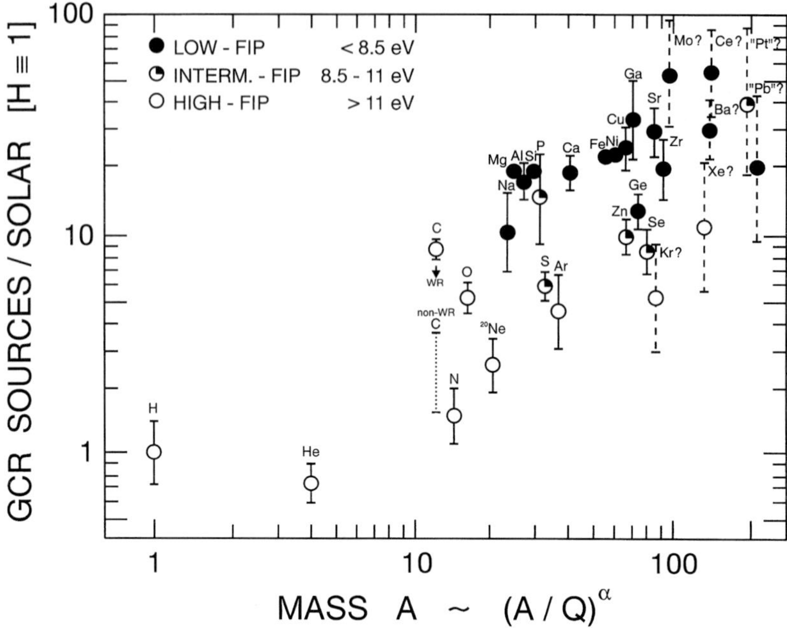

Figure 5. Same as Figure 4, but with the elements sorted, not according to their volatility, but according to their FIP. It shows that a combination of a FIP and of a mass fractionation would *not* order the data as well as the volatility-mass combination (Section 7). As could be expected, the general picture is not too different from that of Figure 4, with the high-FIP elements taking the role of the 'highly-volatiles' and the low-FIP elements that of the 'refractories'. FIP and volatility are, indeed, largely correlated (Section 4; Figure 2), and a mass effect among the 'highly-volatiles' cannot be distinguished from a mass effect among the high-FIP elements! However, contrary to Figure 4, four 'clue elements' (Section 5) do not fit well into the general pattern set by the other elements: definitely Ge, most likely 'Pb' (relative to 'Pt'), and probably Na and P. More precisely, the GCRS Na/Mg, P/S, Ge/Fe, and 'Pb'/'Pt' ratios between elements of comparable FIP and mass, but widely different volatilities, cannot be interpreted in terms of a combined FIP and mass fractionation (Section 5). In a FIP context, in addition, both the marked difference between the mass dependences of the high- and the low-FIP element overabundances, and the singular behavior of O, would remain to be interpreted without resorting to a trapping of elements in grains.

high A/Q 'ions' and accelerate them very efficiently to $\sim$ 0.1 MeV nucl^{-1} energies, where friction and sputtering become important. The sputtered ions form a population of $\sim$ 0.1 MeV nucl^{-1} refractory element ions, which can be further accelerated by the shock, and for which the *crucial* early acceleration phases have taken place while the ion was a constituent of an entire grain, hence independent of its own individual mass. Therefore, both the strong mass dependence of the abundance enhancements among the volatile elements, and the *lack* of such a mass dependence among the refractories may be understood consistently.

In such a picture, the GCR ions are accelerated in a single step in a single site, contrary to the earlier models in terms of FIP. The low GCRS abundances of the 'highly-volatile' H, He, and N is readily interpreted in terms of their low mass.

Contrary to FIP models, this picture also accounts naturally for the presence of ^{22}Ne and ^{12}C excesses in GCRs, since the shocks associated with the most massive SNae accelerate their own pre-SN ^{22}Ne-^{12}C-rich WR wind material (in which C is, in addition, partly locked in grains). The O excess is well understood in terms of its partial ($\sim 20\,\%$) locking in grain silicates in the ISM. As for H, its excess relative to heavier 'highly-volatile' species might be understood in terms of its higher thermal speed ($v_{\mathrm{th}} \propto A^{-1/2}$), which may become comparable to the shock speed in weak SNR shocks.

Acknowledgements

D. Ellison and L. Drury wish to acknowledge the hospitality of the Service d'Astrophysique, Centre d'Etudes de Saclay where much of this work was carried out. L. Drury's visit was supported by the Commission of the European Communities under contract ERBCHRXCT940604, and D. Ellison was supported, in part, by the NASA Space Physics Theory Program. The authors also wish to thank Monique Arnaud, Jean Ballet, Bob Binns, Michel Cassé, Anne Decourchelle, Mike Du Vernois, Jean-Jacques Engelmann, Philippe Ferrando, Martin Israel, Frank McDonald, Georges Meynet, Renaud Papoular, Etienne Parizot, Vladimir Ptuskin, Don Reames, Charles Ryter, Blair Savage, Eun-Suk Seo, Christopher Sharp, Aimé Soutoul, and Bill Webber for valuable discussions regarding various aspects of this paper. We would also like to thank the referee, whose appropriate remarks have helped to significantly improve this paper.

References

Anders, E. and Grevesse, N.: 1989, *Geochim. Cosmochim. Acta* **53**, 197.
Arnaud, M. and Rothenflug, R.: 1985, *Astron. Astrophys. Suppl.* **60**, 425.
Arnett, D.: 1995, *Ann. Rev. Astron. Astrophys.* **33**, 115.
Arnould, M.: 1984, *Adv. Space Res.* **4** (2–3), 45.
Bibring, J. P. and Cesarsky, C. J.: 1981, *Proc. 17th Int. Cosmic-Ray Conf., Paris* **2**, 289.
Binns, W. R.: 1995, *Adv. Space Res.* **15** (6), 29.
Binns, W. R., Fickle, R. K., Garrard, T. L., Israel, M. H., Klarmann, J., Stone, E. C., and Waddington, C. J.: 1981, *Astrophys. J.* **247**, L118.
Binns, W. R., Garrard, T. L., Gibner, P. S., Israel, M. H., Kertzman, M. P., Klarmann, J., Newport, B. J., Stone, E. C., and Waddington, C. J.: 1989, *Astrophys. J.* **346**, 997.
Breneman, H. H. and Stone, E. C.: 1985, *Astrophys. J.* **299**, L57.
Cane, H. V., Reames, D. V., and von Rosenvinge, T. T.: 1991, *Astrophys. J.* **373**, 675.
Cassé, M., and Goret, P.: 1978, *Astrophys. J.* **221**, 703.
Cassé, M. and Paul, J. A.: 1982, *Astrophys. J.* **258**, 860.
Cassé, M., Goret, P., and Cesarsky, C. J.: 1975, *Proc. 14th Int. Cosmic-Ray Conf., Munich* **2**, 646.
Cesarsky, C. J. and Bibring, J. P.: 1981, in G. Setti, G. Spada, and A. W. Wolfendale (eds.), 'Origin of Cosmic Rays', *IAU Symp.* **94**, 361.
Clinton, R. R. and Waddington, C. J.: 1993, *Astrophys. J.* **403**, 644.

Connell, J. J. and Simpson, J. A.: 1997, *Astrophys. J.* **475**, L61.

Cummings, A. C. and Stone, E. C.: 1996, *Space Sci. Rev.* **78**, 117.

Drake, J. J., Laming, J. M., and Widing, K. G.: 1997, *Astrophys. J.* **478**, 403.

Du Vernois, M. A. and Thayer, M. R.: 1996, *Astrophys. J.* **465**, 982.

Du Vernois, M. A., Garcia-Munoz, M., Pyle, K. R., Simpson, J. A., and Thayer, M. R.: 1996a, *Astrophys. J.* **466**, 457.

Du Vernois, M. A., Simpson, J. A., and Thayer, M. R.: 1996b, *Astron. Astrophys.* **316**, 555.

Dwek, E.: 1979, Orange Aid Preprint OAP-570, Caltech.

Dwek, E.: 1997, *Astrophys. J.* **484**, 779.

Ellison, D. C., Drury, L. O'C., and Meyer, J. P.: 1997, *Astrophys. J.* **487**, 197 (Paper II).

Ellison, D. C., Drury, L. O'C., and Meyer, J. P.: 1998, *Space Sci. Rev.* **86**, 203.

Ellison, D. C., Möbius, E., and Paschmann, G.: 1990, *Astrophys. J.* **352**, 376.

Engelmann, J. J., Ferrando, P., Soutoul, A., Goret, P., Juliusson, E., Koch-Miramond, L., Lund, N., Masse, P., Peters, B., Petrou, N., and Rasmussen, I. L.: 1990, *Astron. Astrophys.* **233**, 96.

Epstein, R. I.: 1980, *Monthly Notices Royal Astron. Soc.* **193**, 723.

Feldman, U.: 1992, *Physica Scripta* **46**, 202.

Ferrando, P.: 1993, *J. Phys. G: Nucl. Part. Phys.* **19**, S53.

Ferrando, P.: 1994, in D. A. Leahy, R. B. Hicks, and D. Venkatesan (eds.), *Proc. 23rd Int. Cosmic-Ray Conf., Calgary, Invited, Rapporteur and Highlight Papers*, World Scientific, Singapore, p. 279.

Fowler, P. H., Walker, R. N. F., Masheder, M. R. W., Moses, R. T., Worley, A., and Gay, A. M.: 1987, *Astrophys. J.* **314**, 739.

Garrard, T. L. and Stone, E. C.: 1993, *Proc. 23rd Int. Cosmic-Ray Conf., Calgary* **3**, 384.

Geiss, J., Gloeckler, G., and von Steiger, R.: 1995, *Space Sci. Rev.* **72**, 49.

Grevesse, N., Noels, A., and Sauval, A. J.: 1996, in S. S. Holt and G. Sonneborn (eds.), *Cosmic Abundances*, ASP Conf. Series, Vol. 99 (*Astron. Soc. Pacific*), p. 117.

Hénoux, J. C.: 1995, *Adv. Space Res.* **15** (7), 23.

Laming, J. M., Drake, J. J., and Widing, K. G.: 1996, *Astrophys. J.* **462**, 948.

Leske, R. A.: 1993, *Astrophys. J.* **405**, 567.

Leske, R. A. and Wiedenbeck, M. E.: 1993, *Proc. 23rd Int. Cosmic-Ray Conf., Calgary* **1**, 571.

Lukasiak, A., Ferrando, P., McDonald, F. B., and Webber, W. R.: 1994, *Astrophys. J.* **426**, 366.

Lukasiak, A., McDonald, F. B., Webber, W. R., and Ferrando, P.: 1997, *Adv. Space Res.* **19**, 747.

Lund, N.: 1984, *Adv. Space Res.* **4** (2–3), 5.

Maeder, A.: 1983, *Astron. Astrophys.* **120**, 130.

Maeder, A. and Meynet, G.: 1993, *Astron. Astrophys.* **278**, 406.

Mason, B.: 1979, Data of Geochemistry/Cosmochemistry/Meteorites, Geological Survey Professional Paper No. 440-B-1, U.S. Govt. Printing Office.

Mathis, J. S. 1996, *Astrophys. J.* **472**, 643.

Meyer, J. P.: 1981a, *Proc. 17th Int. Cosmic-Ray Conf., Paris*, **2**, 265.

Meyer, J. P.: 1981b, *Proc. 17th Int. Cosmic-Ray Conf., Paris*, **2**, 281.

Meyer, J. P.: 1985a, *Astrophys. J. Suppl.* **57**, 151.

Meyer, J. P.: 1985b, *Astrophys. J. Suppl.* **57**, 173.

Meyer, J. P.: 1988, in G. J. Mathews (ed.), *Origin and Distribution of the Elements*, World Scientific, Singapore, p. 310.

Meyer, J. P.: 1989, in C. J. Waddington (ed.), *Cosmic Abundances of Matter*, AIP Conf. Proc., Vol. 183, The Amer. Inst. of Physics, p. 245.

Meyer, J. P.: 1993, in N. Prantzos, E. Vangioni-Flam, and M. Cassé (eds.), *Origin and Evolution of the Elements*, Cambridge University Press, Cambridge, p. 26; or *Adv. Space Res.* **13** (9), 377.

Meyer, J. P.: 1996, in J. Trân Thanh Vân, L. Celnikier, H. C. Trung and S. Vauclair (eds.), *The Sun and Beyond*, Editions Frontières, Gif-sur-Yvette, p. 27; or in S. S. Holt and G. Sonneborn (eds.), *Cosmic Abundances*, ASP Conf. Series Vol. 99 (*Astron. Soc. Pacific*), p. 127.

Meyer, J. P., Cassé, M., and Reeves, H.: 1979, *Proc. 16th Int. Cosmic Ray Conf., Kyoto* **12**, 108.

Meyer, J. P., Drury, L. O'C., and Ellison, D. C.: 1997, *Astrophys. J.* **487**, 182 (Paper I).

Prantzos, N., Arnould, M., and Arcoragi, J. P.: 1987, *Astrophys. J.* **315**, 209.

Prantzos, N., Cassé, M., and Vangioni-Flam, E.: 1993, in N. Prantzos, E. Vangioni-Flam, and M. Cassé (eds.), *Origin and Evolution of the Elements*, Cambridge University Press, Cambridge, p. 156.

Ramaty, R., Mandzhavidze, N., Kozlovsky, B., and Murphy, R. J.: 1995, *Astrophys. J.* **455**, L193.

Reames, D. V.: 1995, *Adv. Space Res.* **15** (7), 41.

Reames, D. V., Meyer, J. P., and von Rosenvinge, T. T.: 1994, *Astrophys. J. Suppl.* **90**, 649.

Roth, I. and Temerin, M.: 1997, *Astrophys. J.* **477**, 940.

Sakurai, K.: 1995, *Adv. Space Res.* **15** (6), 59.

Savage, B. D. and Sembach, K. R.: 1996, *Ann. Rev Astron. Astrophys.* **34**, 279.

Schutte, W. A.: 1988, 'The Evolution of Interstellar Organic Grain Mantles', Ph.D. Thesis, University of Leiden.

Schutte, W. A. and Greenberg, J. M.: 1988, in M. E. Bailey and D. A. Williams (eds.), *Dust in the Universe*, Cambridge University Press, Cambridge, p. 403.

Seo, E. S., Ormes, J. F., Streitmatter, R. E., Stochaj, S. J., Jones, W. V., Stephens, S. A., and Bowen, T.: 1991, *Astrophys. J.* **378**, 763.

Sharp, C. M. and Wasserburg, G. J.: 1995, *Geochim. Cosmochim. Acta* **59**, 1633.

Shibata, T.: 1996, in N. Iucci and E. Lamanna (eds.), 'Proc. 24th Int. Cosmic-Ray Conf., Rome, Invited, Rapporteur and Highlight Papers', Società Italiana di Fisica, *Nuovo Cimento* **19C**, 713.

Silberberg, R. and Tsao, C. H.: 1990, *Astrophys. J.* **352**, L49.

Singh, K. P., White, N. E., and Drake, S. A.: 1996, *Astrophys. J.* **456**, 766.

Steinacker, J., Meyer, J. P., Steinacker, A., and Reames, D. V.: 1997, *Astrophys. J.* **476**, 403.

Tarafdar, S. P. and Apparao, K. M. V.: 1981, *Astrophys. Space Sci.* **77**, 521.

Timmes, F. X., Woosley, S. E., and Weaver, T. A.: 1995, *Astrophys. J. Suppl.* **98**, 617.

Waddington, C. J.: 1996, *Astrophys. J.* **470**, 1218.

Wasson, J. T.: 1985, *Meteorites: Their Record of Early Solar System History*, Freeman, New York.

Wasson, J. T. and Kallemeyn, G. W.: 1988, *Phil. Trans. R. Soc. London* **A325**, 535.

Webber, W. R. and McDonald, F. B.: 1994, *Astrophys. J.* **435**, 464.

Webber, W. R., Golden, R. L., and Stephens, S. A.: 1987, *Proc. 20th Int. Cosmic Ray Conf., Moscow* **1**, 325.

Webber, W. R., Lukasiak, A., and McDonald, F. B.: 1997, *Astrophys. J.* **476**, 766.

Widing, K. G.: 1997, *Astrophys. J.* **480**, 400.

Woosley, S. E. and Weaver, T. A.: 1995, *Astrophys. J. Suppl.* **101**, 181.

COSMIC RAYS FROM SUPERNOVA REMNANTS: A BRIEF DESCRIPTION OF THE SHOCK ACCELERATION OF GAS AND DUST

DONALD C. ELLISON

Department of Physics, North Carolina University, Box 8202, Raleigh NC 27695, U.S.A.

LUKE O'C. DRURY

Dublin Institute for Advanced Studies, School of Cosmic Physics, 5 Merrion Square, Dublin 2, Ireland

JEAN-PAUL MEYER

Service d'Astrophysique, CEA/DSM/DAPNIA, Centre d'Etudes de Saclay, 91191 Gif-sur-Yvette, France

Abstract. We summarize our model of galactic cosmic-ray (GCR) origin and acceleration, wherein a mixture of interstellar and/or circumstellar gas and dust is accelerated by a supernova remnant (SNR) blast wave. A detailed analysis of observed GCR abundances (Meyer et al., 1997), combined with the knowledge that many refractory elements known to be locked in grains in the interstellar medium (ISM) are abundant in cosmic rays, has lead us to revive an old suggestion (Epstein, 1980) that charged dust grains can be shock accelerated. Here, we outline results (presented more completely in Ellison et al., 1997) from a nonlinear shock model which includes (i) the direct acceleration of interstellar gas-phase ions, (ii) a simplified model for the direct acceleration of weakly charged grains to ~ 100 keV amu^{-1} energies, simultaneously with the acceleration of the gas ions, (iii) the energy losses of grains colliding with the ambient gas, (iv) the sputtering of grains, and (v) the simultaneous acceleration of the sputtered ions to TeV energies. We show that the model produces GCR source abundance enhancements of the volatile, gas-phase elements, which are an increasing function of mass, as well as a net, mass independent, enhancement of the refractory, grain elements over protons, consistent with cosmic-ray observations. The GCR ^{22}Ne and C excesses may also be accounted for in terms of the acceleration of ^{22}Ne-C-enriched pre-SN Wolf–Rayet star wind material surrounding the most massive supernovae. The O excess seen in cosmic rays probably cannot be interpreted in terms of W–R star nucleosynthesis, but is easily accounted for in our model since 15 to 20% of O is trapped in grain cores and this O will be preferentially accelerated. We have expanded the parameter range explored in Ellison et al. (1997) to lower shock speeds and higher maximum cosmic-ray energies and find similar fits to the H/He ratio and the cosmic-ray source spectra.

1. Introduction

The galactic cosmic-ray source (GCRS) composition is characterized by a general overabundance of heavier elements relative to H and He, and by a fine structure among the heavy elements. This fine structure is primarily governed by atomic, not nuclear, physics; in particular, it does not at all correspond to fresh supernova nucleosynthesis products; (there exists, however, a ^{22}Ne, ^{12}C, ^{16}O-rich component in GCRs, suggesting the acceleration of some Wolf–Rayet wind material). The

Space Science Reviews **86**: 203–224, 1998.
© 1998 *Kluwer Academic Publishers. Printed in the Netherlands.*

data indicate that the relevant atomic physics parameter could be either the first ionization potential (FIP), which controls the neutral or ionized state of each element in a $\sim 10^4$ K gas, or the element volatility (i.e., its condensation temperature, T_c), which controls the element's ability to condense into solid compounds. For most elements, values of FIP and T_c are anti-correlated, so it is not easy to tell which of these two parameters shapes the GCRS composition. In any case, *either* the easily ionized low-FIP elements, *or* the easily condensed high-T_c elements, are found enhanced by a factor of order 5 relative to the other heavy elements (and $\sim$30 relative to H).

Based on a few elements which do *not* fit in the general anti-correlation between FIP and T_c, it has now become apparent that all four key ratios Na/Mg, P/S, Ge/Fe, Pb/Pt point towards volatility, not FIP, as the controlling parameter for GCR composition (see Meyer et al., 1997, for complete details). In fact, all of the GCRS composition data are remarkably well ordered in terms of two specific behaviors for the volatile and the refractory elements: (i) among the volatile elements, the abundance enhancements strongly increase with element mass A (A is the number of nucleons); only hydrogen does not entirely fit into the pattern. We believe this reflects an increase of the acceleration efficiency with the element mass-to-charge ratio A/Q, i.e., with its rigidity at a given velocity; in any ionization model, indeed, A/Q is a roughly monotonically increasing function of the mass A (Q is the charge number). The low GCRS abundances of H, He, and N can be interpreted in this framework. The somewhat odd behavior of H can also be modeled if sufficiently low Mach shocks produce the bulk of the GCRs (see Ellison et al., 1997 for details). (ii) The refractory elements are all enhanced relative to volatiles; but this enhancement is approximately the same for all refractories, i.e., it has *little or no* mass-dependence (see Figure 1).

It is quite clear from UV, IR, and visible observations that the refractory elements are largely locked into solid dust grains in most of the interstellar medium (ISM) (e.g., Cardelli, 1994; Sembach and Savage, 1996; Savage and Sembach, 1996), as well as in supernova ejecta (e.g., Lucy et al., 1989, 1991; Dwek et al., 1992), and in stellar, and especially Wolf–Rayet star, wind envelopes (e.g., Bode, 1988; Gehrz, 1991; van der Hucht and Hidayat, 1991; van der Hucht and Williams, 1995). Clearly, the GCRS composition features suggest a preferential acceleration of those elements locked in grains (refractories), relative to the gas-phase elements (volatiles). This leads to the idea of a preferential acceleration of grain erosion products in supernova shocks, as first suggested by Epstein (1980). In this paper, we present a summary of the work contained in Ellison et al. (1997) where we re-examine grain acceleration in the light of the above, specific composition features, and of modern nonlinear (i.e., smoothed) shock acceleration theories. We have extended that work slightly by adding an example with a slower shock speed and a higher maximum cosmic-ray energy.

First, if particles with the same energy per nucleon are considered, smoothed shock models predict an increased particle acceleration efficiency for increased

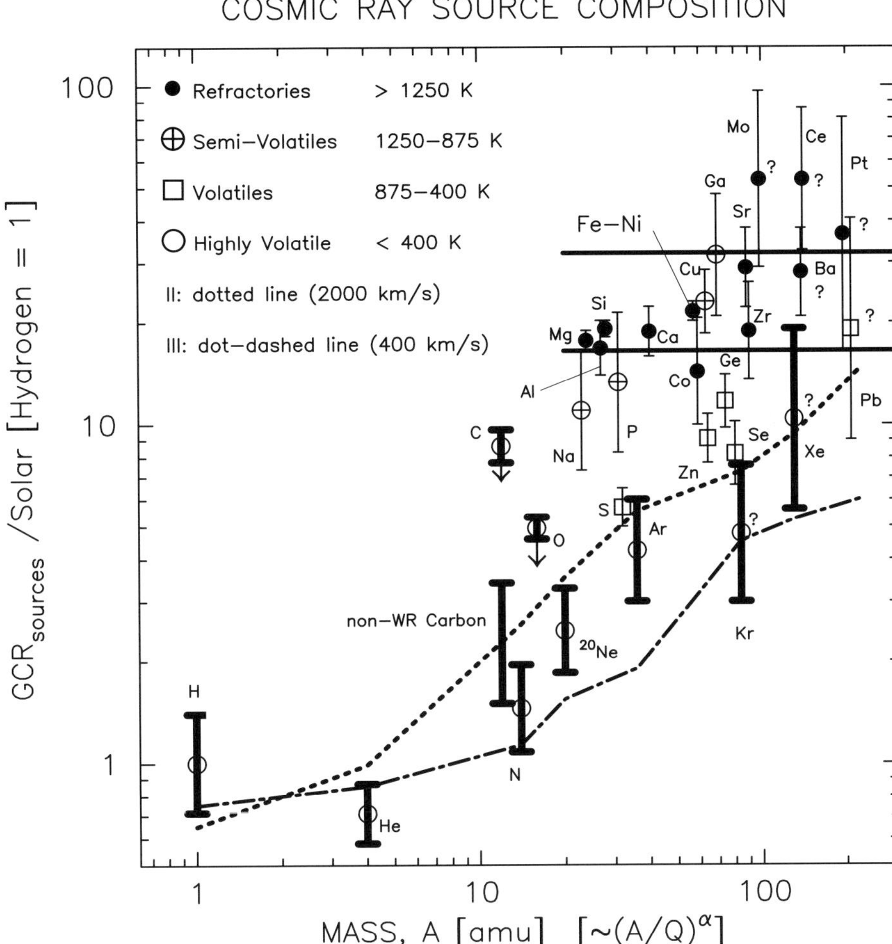

Figure 1. Galactic cosmic-ray source abundance relative to solar abundance versus atomic mass number. All values are measured relative to cosmic-ray hydrogen at a given energy per nucleon. The elements are divided, on the basis of condensation temperature, into refractories, semi-volatile, volatile, and highly volatile groups. The refractories are essentially completely locked in grains in the ISM, while the highly volatile elements are gaseous. The arrows on carbon and oxygen indicate that these elements have an additional source; ^{22}Ne and C from Wolf–Rayet wind material and O from the 15% to 20% that is locked in grains. Our estimate for the non-W-R contribution of carbon is labeled. Our predictions for the abundances of volatile elements from a high Mach number shock model are shown as a dotted line, and for a lower Mach number model with a dot-dashed line. The horizontal solid lines on the right side of the plot are limits on our predicted abundance of iron and other refractory elements. The label on the abscissa [$\sim (A/Q)^\alpha$, where α is some unspecified constant] is a reminder that, for most ionization models, A/Q is a roughly monotonically increasing function of the mass. We note that the abundances of Kr, Xe, Mo, Ba, Ce, Pt, and Pb relative to Fe may contain systematic errors which are difficult to evaluate (we indicate this with a '?' to the right of the point). For a complete discussion of the observations, see Meyer, Drury, and Ellison (1997).

particle rigidity, i.e., A/Q ratio, because higher rigidity ions can diffuse further back upstream of the shock than low A/Q particles at the same energy per nucleon. This assumes, of course, that the diffusion coefficient is an increasing function of rigidity. Hence, high A/Q particles 'see' a larger velocity difference and are more easily injected to suprathermal energies (e.g., Eichler, 1979, 1984; Ellison and Eichler, 1984; Jones and Ellison, 1991). This effect fits qualitatively with the mass dependence of the volatile, gas-phase element enhancements. Second, weakly charged grains can behave as extremely high A/Q ions, and thus get very efficiently injected and accelerated provided they obtain similar energies per nucleon to protons when first shock heated. Third, if the refractory elements are, at the early crucial stage, accelerated as part of grains, their own A/Q plays no role in this acceleration stage, so that the approximate mass *in*dependence of their GCRS enhancements is not surprising.

In this paper, we sketch a calculation of the expected source composition and spectral shapes of the GCRs, using a Monte Carlo model of cosmic-ray acceleration at SNR shocks (e.g., Ellison et al., 1997) including both interstellar gas and 'grains'. The shock model includes nonlinear effects from shock smoothing and a parameterized description of injection from the thermal background for any ion species or grain. It yields both the spectral shapes and absolute abundances of various species of cosmic rays. To this we have added a simple model of grain deceleration, sputtering of individual atoms off grains, and acceleration of sputtered ions to cosmic-ray energies, yielding a first principles estimate of the refractory element/hydrogen ratio in cosmic rays. Despite the approximations that must be made for such a complex calculation, we find *excellent agreement with observations* for both the spectral shape and the relative abundances of the various nuclear components, at least above a few GeV, where solar modulation is no longer important, and below the observed 'knee' in the GCR spectrum and the Lagage and Cesarsky (1983) limit at $\sim 10^{14-15}$ eV (also Prishchep and Ptuskin, 1981). We believe this is the first attempt to simultaneously and self-consistently address the intensity and shape of the major cosmic-ray components using a mixture of interstellar gas and dust.

2. Model Assumptions

For a complete description of the model assumptions, see Ellison et al. (1997). Briefly, we assume the forward shocks in the Sedov phase of supernova remnants are responsible for producing the bulk of the cosmic rays. These shocks accelerate mainly ambient interstellar material via diffusive shock acceleration (see Drury, 1983; Völk, 1984; Blandford and Eichler, 1987; Berezhko and Krymsky, 1988; Jones and Ellison, 1991, for reviews). While test-particle shock acceleration produces power-law spectral shapes in momentum, nonlinear acceleration, where the reaction of the accelerated particles causes the shock to be smooth, produces spectra which are no longer exact power laws, and different ion species are treated differently (e.g., Eichler, 1984; Ellison and Eichler, 1984). These nonlinear effects produce different spectral shapes (in the range where all particles are not fully relativistic) and different injection and acceleration efficiencies for different ion species, with the ion rigidity and thermal speed becoming the important distinguishing parameters.

We define the rigidity of species α as

$$R = \frac{cp}{Q_\alpha e} = \frac{m_p c^2}{e} \left(\frac{A}{Q}\right)_\alpha \beta\gamma \tag{1}$$

(with units of Volts in the SI system), where c is the velocity of light *in vacuo*, e is the electronic charge, m_p is the proton mass, $m_\alpha = A_\alpha m_p$ is the rest mass of species α with A_α nucleons, Q_α is its charge number, β is the particle's v/c and γ is its Lorentz factor. In a magnetic field, B, the gyroradius of the particle in SI units is $r_g = p/(Q_\alpha eB) = R/(cB)$, and we assume in all that follows that the mean free path parallel to the mean magnetic field, λ, is given by $\lambda = \eta r_g$, where η is a constant independent of ion species, particle energy, or position relative to the shock. The so-called strong scattering Bohm limit obtains when $\eta \sim 1$.

Thus the corresponding spatial diffusion coefficient parallel to the mean magnetic field will be

$$\kappa_\alpha = \frac{\lambda(R)v}{3}\,, \tag{2}$$

and if the shock speed is V_{sk}, the diffusion length in the precursor is $L_{D,\alpha} \sim \kappa_\alpha/V_{sk} \propto \lambda_\alpha v_\alpha$, and the ratio of diffusion lengths of species α to protons, at fixed energy per nucleon, is simply

$$\frac{L_{D,\alpha}}{L_{D,p}} \propto \left(\frac{A}{Q}\right)_\alpha . \tag{3}$$

If the shock is smooth rather than discontinuous (cf., Figure 2), as will always be the case in nonlinear shock acceleration if κ is an increasing function of energy, ions with large A/Q will diffuse further upstream than protons of the same energy per nucleon (provided both are nonrelativistic). These ions will be turned around against a more rapid upstream flow, will receive a larger energy boost for each shock crossing, and will be accelerated more efficiently and obtain a flatter spectrum than protons. Recent specific observational support for this effect in anomalous cosmic rays has been reported by Cummings and Stone (1996). While the differences in A/Q may be small for atoms, grains can have huge A/Q's ($\sim 10^{4-8}$). Any differences in intensities of the spectra of various species, acquired when they were nonrelativistic, persist to the highest energies obtained.

It is important to note that this conclusion is based on our assumption of a steady state. The longer diffusion lengths of heavy ions also result in longer acceleration

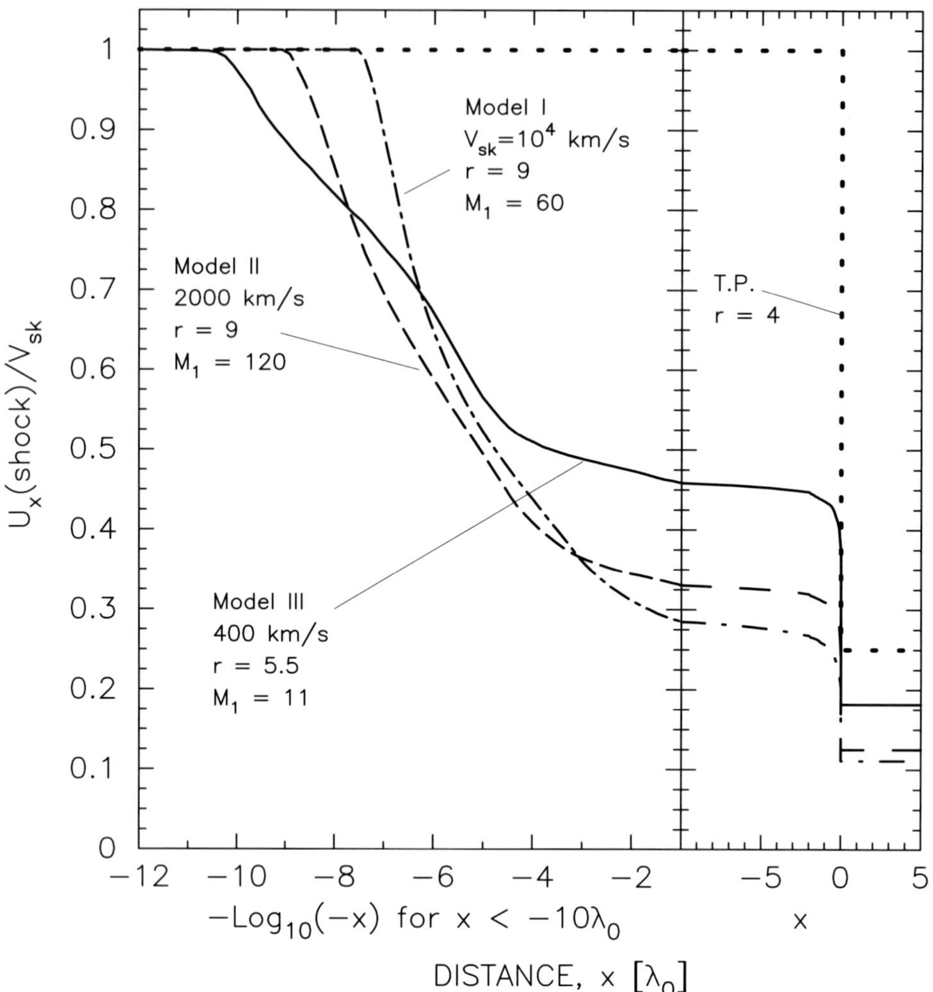

Figure 2. Average bulk flow speed in the shock frame, $U_x(x)$, versus distance (i.e., the shock structure) obtained from the Monte Carlo model. The distance scale is logarithmic for $x < -10\lambda_0$ and linear for $x > -10\lambda_0$. The vertical scale is in units of the far upstream speed, V_{sk}, and $\lambda_0 = \eta r_{g1}$, where r_{g1} is the gyroradius of a far upstream proton with a speed equal to the shock speed, which varies for each model. The compression ratios, r, in these nonlinear models depend on the fraction of pressure carried by relativistic particles and on the amount of energy escaping at the upstream free escape boundary (FEB) and are always greater than the standard Rankine–Hugoniot value. The dotted line shows a test-particle profile of a shock with $r = 4$.

times to a given energy per nucleon. If the shock has a finite age, the acceleration of heavy ions, and particularly grains, may cutoff before protons.

2.1. ACCELERATION OF DUST GRAINS

The basic idea of Epstein (1980) was that dust grains could behave like ions of very large mass to charge ratio, thus large rigidity, and should therefore be relatively efficiently accelerated to velocities where they are eroded by sputtering. The sputtered grain material will have the velocity of the parent grain which can be well above thermal. If the sputtering occurs in the upstream region the sputtered products (i.e., refractory elements) will be carried back into the shock and further accelerated to cosmic-ray energies with higher net efficiency than the gas-phase thermal protons and volatile element ions. Of course, material which is sputtered from the grains downstream of the shock is mainly carried away and lost from the system because the sputtered ions are, on average, many more mean free paths downstream from the shock then the parent grain.

Small grains in a plasma will be charged by a number of processes all of which are uncertain. However, what one can say with certainty is that a grain will only be uncharged very briefly, if at all, and that in general the potential of the grain will be of order 10 to 100 V. It follows that if the grain potential is ϕ, the charge on a spherical grain is of order $q \sim 4\pi\epsilon_0 a\phi$, where a is a characteristic size of the grain. In terms of electronic charges this gives numerically

$$Q_{\mathrm{G}} = \frac{q}{e} \simeq 700 \left(\frac{a}{10^{-7}\ \mathrm{m}}\right)\left(\frac{\phi}{10\ \mathrm{V}}\right). \tag{4}$$

The number of atoms in the grain will be of order $(a/10^{-10}\ \mathrm{m})^3$, or 10^9 for a 10^{-7} m size grain. If μ is the mean atomic weight of the grain atoms, the entire grain 'atomic weight', A_{G}, is $\mu(a/10^{-10}\ \mathrm{m})^3$. Thus the effective A/Q value for a grain is very large, of order

$$\left(\frac{A}{Q}\right)_{\mathrm{G}} \simeq 1.4 \times 10^6 \mu \left(\frac{a}{10^{-7}\ \mathrm{m}}\right)^2 \left(\frac{\phi}{10\ \mathrm{V}}\right)^{-1}. \tag{5}$$

If the dust grain has velocity $\beta_{\mathrm{G}}c$ with $\beta_{\mathrm{G}} \ll 1$, then the magnetic rigidity of the grain is

$$R = \frac{cp}{q} = \frac{A_{\mathrm{G}} m_p \beta_{\mathrm{G}} c^2}{Q_{\mathrm{G}} e} \simeq 10^9 \beta_{\mathrm{G}} \left(\frac{A}{Q}\right)_{\mathrm{G}},$$

$$V \simeq 1.3 \times 10^{15} \beta_{\mathrm{G}} \mu \left(\frac{a}{10^{-7}\ \mathrm{m}}\right)^2 \left(\frac{\phi}{10\ \mathrm{V}}\right)^{-1}\ \mathrm{V}. \tag{6}$$

Ultraviolet and optical extinction measurements indicate that the grain size distribution is quite broad extending from very small grains to an upper cutoff at $\sim 0.25\ \mu$m (e.g., Mathis e al., 1977). The amount of total grain mass in particles with radii a or less, $M_G(< a)$, goes roughly as $a^{1/2}$, so that nearly half of the total grain mass is in a relatively small range of sizes around 0.1 μm.

In general, supernova remnant shocks have velocities in the range 30 to 3000 km s^{-1} (e.g., Reynolds, 1988). Let us consider a high Mach number shock of velocity 400 km s^{-1} which overtakes a dust grain in a typical interstellar hydrogen density, $n_H \sim 1$ cm^{-3}. Relative to the post-shock gas, the grain will have a velocity of ~ 300 km s^{-1} or $\beta_G \sim 10^{-3}$, and thus a rigidity of about 10^{14} V if it is 0.1 μm in size, is charged to a surface potential of 10 V, and is made of material with $\mu \sim 56.$*

We assume that SNR shocks are capable of accelerating *protons* to energies of order 10^{14-15} eV. If this is the case, three conditions must be met; (i) the magnetic field near the shock has to contain structures capable of scattering protons of rigidities up to 10^{14-15} V, (ii) the shock radius (i.e., the size of the acceleration region) must be considerably larger than a $\sim 10^{14-15}$ eV proton gyroradius, and (iii) the age of the remnant must be greater than the acceleration time to $\sim 10^{14-15}$ eV. Our fundamental assumption is: if relativistic protons of energy $> 10^{14}$ eV are being efficiently scattered and accelerated (i.e., are being scattered nearly elastically and isotropically in the local plasma frame), *then so should dust grains with the same rigidity*. There is however one vital difference. The dust grains, far from being relativistic, only have a velocity of order the shock velocity, at least initially. Since we also assume that the mean free path depends only on rigidity, not velocity, the diffusion coefficient of the grains is smaller than that of the relativistic protons with the same rigidity by a factor of the grain β, typically 10^{-3}. While we believe this assumption is reasonable, particularly since the Alfvén speed should be considerably less than the grain speed in all cases we consider, it has not been proven and is, we believe, the cause for greatest concern in the model we present.

Using standard estimates for the shock acceleration time scale and the frictional loss time scale (see Ellison et al., 1997), we estimate the maximum energy per nucleon grains can acquire in the shock:

$$\left(\frac{E}{A}\right)_{G,\max} \simeq 100\eta^{-2/3} \left(\frac{V_{sk}}{400\ \text{km s}^{-1}}\right)^{4/3} \left(\frac{a}{10^{-7}\ \text{m}}\right)^{-2/3} \times$$

$$\times \left(\frac{n_H}{1\ \text{cm}^{-3}}\right)^{-2/3} \left(\frac{\phi}{10\ \text{V}}\right)^{2/3} \left(\frac{B}{3\ \mu\text{G}}\right)^{2/3}\ \text{keV} . \tag{7}$$

This yields $(E/A)_{G,\max} \simeq 100$ keV, well above thermal energies. Of course this does not mean that all grains are accelerated by this amount, in fact, a distribution extending upwards from thermal energies will result with only a small fraction of

---

* While the value $\mu = 56$ only applies to pure iron grains, we have chosen it for simplicity. For silicate grains containing Mg, Si, Fe, and O, $\mu \sim 20$ to 30, but this factor of about two difference does not seriously influence the results that follow.

grains obtaining the cutoff energy $(E/A)_{G,\max}$. In addition, of course, the shock must be large enough and old enough for acceleration to these velocities to occur.

2.2. GRAIN SPUTTERING AND INJECTION OF SPUTTERED MATERIAL

The acceleration of grains has significant implications for grain erosion by sputtering. At the energies indicated in Equation (7), the sputtering process is quite uncertain and grains may even become transparent (e.g., Dwek, 1987), but we assume that roughly 0.5 to 1% of collisions with ambient gas atoms result in the sputtering of an atom from the grain surface. We compare the grain-destruction time scale for collisional sputtering with the momentum-loss time scale for direct collisions with the same gas atoms and conclude that the acceleration time scales are always shorter than the destruction time scales. This means that the grain has time to diffuse back and forth between both sides of the shock before being destroyed. Thus, some of the sputtering must occur while the grain is in the upstream region ahead of the shock. These sputtered particles can then be advected into the shock as a seed population of energetic ions which can then be accelerated with high efficiency.

When an ion is sputtered off a grain, it is unlikely to be fully stripped, especially if of high nuclear charge. On ejection from the grain it will carry some electrons with it, and there will also be electron exchange with the atoms of the background plasma. We expect the ions to have an effective charge Q^* of at most $+3$ and those elements with high first ionization potentials actually have a significant chance of becoming neutral atoms. Such neutrals, if formed, are no longer trapped by the magnetic field and move in a straight line until they again become ionized on a time scale short compared to other relevant times. During acceleration, partially charged ions will lose electrons and eventually become fully stripped. The rate of stripping will influence the acceleration efficiency, but we have not yet included this in our model. In Figure 4 (discussed below) we quantify this effect somewhat by showing two extreme charge states for sputtered iron.

2.3. REFRACTORY ELEMENT INJECTION RATE

We can now estimate the suprathermal refractory element injection rate resulting from the sputtering of grains in the upstream region, followed by the advection of the resulting suprathermal ions into the shock. The key ideas are (i) that the sputtered ions have the same velocity as the parent grain, and (ii) that the maximum probability for sputtering occurs when the acceleration time scale and the direct collisional momentum loss time scales coincide, i.e., near $(E/A)_{G,\max}$. The sputtered ions are injected at superthermal energies.

After some calculation, we find that the total injection rate of sputtered ions per unit surface area, q_{sput}, is

$$q_{\mathrm{sput}} \sim O(10^{-4}) n_G A_G V_{sk} . \tag{8}$$

The quantity $n_G A_G V_{sk}$ is the flux of nucleons contained in grain material coming from far upstream. This result can be interpreted as saying that, with a probability of order 10^{-4}, an atom in a dust grain will be sputtered as a suprathermal ion while in the upstream region and be convected back to the shock without major energy loss (at least in a 10^4 K gas). The resulting suprathermal ions typically have velocities of order 10 times the downstream thermal proton velocity and, in a 10^4 K gas without ongoing photoionization, have a charge of $Q^* \leq +3$. In a 10^6 K gas, the mean charge can be higher (i.e., $\sim +9$ for Fe) and energy losses for the sputtered ions may be significant.

In view of the rather crude nature of this estimate it is remarkable that the answer appears close to what is required by the GCR composition observations. As is well known, if one simply assumes that the accelerated protons have a p^{-2} power-law spectrum from a few times thermal energy to an upper cut-off at around 10^{14} eV, the condition that the total energy flux in accelerated protons out of the shock cannot exceed the mechanical power in, implies that only about 1 in 10^4 of the incident thermal protons can become part of the cosmic-ray proton power law. The coincidence of this figure with the estimate for the ion sputtering probability suggests that the resultant accelerated cosmic-ray composition will be fairly close to the average chemical composition of the interstellar medium. But, since the sputtered ions are injected at a velocity about a decade higher than the protons, the refractory elements should show an enhancement which is also of order 10 (see Figure 3.4 in Jones and Ellison, 1991).

2.4. DETAILED ASSUMPTIONS OF MONTE CARLO SHOCK MODEL

In the Monte Carlo technique used here, we model a plane, parallel, steady-state shock, and mimic the curvature of a real SNR shock by placing a free escape boundary (FEB) at some distance, d_{FEB}, upstream from the shock. Shocked particles reaching the FEB are lost from the system, thus truncating the acceleration process. Our steady-state assumption precludes a description of the overall dynamics of the SNR explosion; instead, we use standard Sedov estimates for SNR shock radii, speeds, and ages where these are required.

The Monte Carlo model makes the same assumption for the scattering mean free path as made above, i.e.,

$$\lambda = \eta r_g , \tag{9}$$

where η is a constant independent of particle species, energy, or position. All lengths are measured in units of $\lambda_0 = \eta r_{g1}$, where $r_{g1} = m_p V_{sk}/(eB_1)$ is the gyroradius of a far upstream proton with a speed equal to the shock speed. We further assume that all particles scatter elastically and isotropically in the local plasma frame. By assuming that the scattering is elastic against a massive background, we model a situation where particles scatter against waves which are frozen in the plasma.

As mentioned above, we further assume the SNR shocks in question are capable of accelerating *protons* to energies on the order of $E_{p,\max} \sim 10^{14-15}$ eV. This limit is imposed by the observed constancy of the energetic proton spectral shape up to those energies (Shibata, 1995); cutoffs above this energy can be interpreted in terms of either the finite size of the shock acceleration region (e.g., Berezhko et al., 1996), or the finite age of the remnant (e.g., Prishchep and Ptuskin, 1981; Lagage and Cesarsky, 1983), depending on the parameters. For our models here, we assume a finite shock size limits proton acceleration. If the waves responsible for scattering high energy particles are self-generated, the upstream diffusion length of the highest energy particles currently in the system will define the turbulent foreshock region. Energetic particles backstreaming to the limits of the foreshock region will leave the system truncating the acceleration.

In terms of the actual SNR environment, the maximum energy depends on three parameters, d_{FEB} which is some measure of the shock radius or age, η, and the magnitude of the upstream magnetic field, B_1. As long as we confine ourselves to parallel shocks, these three parameters can be combined into one. if, for example, we take d_{FEB} to be some fraction f of the shock radius R_{sk}, and set this distance equal to the upstream diffusion length, i.e., $f R_{sk} = \kappa_1 / V_{sk}$, we get for highly relativistic particles:

$$\frac{\eta r_g c}{3 V_{sk}} = f R_{sk} , \tag{10}$$

and replacing R_{sk} and V_{sk} with their Sedov values at a remnant age, t_{SNR}, we have

$$\left(\frac{E}{A}\right)_{\max} \simeq 2 \times 10^{14} \left(\frac{Q}{A}\right) \left(\frac{f B_1}{\eta\, 3\, \mu G}\right) \left(\frac{n_{\mathrm{H}}}{1\ \mathrm{cm}^{-3}}\right)^{-1/3} \times$$

$$\times \left(\frac{E_{SN}}{10^{51}\ \mathrm{erg}}\right)^{1/3} \left(\frac{V_{sk}}{10^3\ \mathrm{km\ s}^{-1}}\right)^{1/3}\ \mathrm{eV} . \tag{11}$$

Values of $(E/A)_{\max} \sim 10^{14}(Q/A)$ eV can be obtained for $f \sim 0.3$, $\eta \sim 1$ (i.e., the Bohm limit), and $B_1 \sim 3 \times 10^{-6}$ G over a fairly wide range of t_{SNR}. In the examples presented here, we arbitrary set the parameter $f B_1 / \eta$ so that a maximum proton energy of $\sim 10^{14}$ eV is obtained in all cases.

3. Numerical Results

We first produce nonlinear solutions for the shock structure including the acceleration of protons and He^{+2} to energies $\sim 10^{14}$ eV. Both species are included self-consistently and contribute to the smoothing of the shock. Helium is injected far upstream from the shock at 'cosmic' abundance, i.e., $n_{\mathrm{He}}/n_{\mathrm{H}} = 0.1$. Once the shock structure has been determined, we accelerate other gases and grains as test particles in the smooth shock, including the slowing and sputtering of these grains from direct collisions with the ambient gas. Once the grains have been accelerated, we determine the rate at which sputtered ions are injected into the shock and reaccelerated (as test particles) as described above.

3.1. NONLINEAR SHOCK MODELS

We have tested several SNR models, including one with a high shock speed, $V_{sk} = 10^4$ km s^{-1}, typical of a young SNR at the end of the free expansion (or ballistic) stage (e.g., Drury et al., 1994) (Model I), one with an intermediate speed (i.e., $V_{sk} = 2000$ km s^{-1}, Model II), and one with a slow speed typical of older, slower remnants in the Sedov phase ($V_{sk} = 400$ km s^{-1}, Model III). These three models span a wide parameter range and show the essential effects for all but the lowest Mach number SNR shocks in the ISM.

Our solutions are obtained by iteration and the technique is described in detail in Ellison et al. (1990). In Figure 2 we show the gas flow speed in the shock frame, versus distance from the shock (i.e., the shock structure or profile, $U_x(x)$), determined by our Monte Carlo technique for models I, II, and III. Notice that the distance is plotted with a logarithmic scale for $x < -10\lambda_0$ and a linear scale for $x > -10\lambda_0$. The shock is smoothed on the diffusion length scale $\sim \kappa / V_{sk}$ of the highest energy particles in the system. Despite this extreme smoothing, a distinct subshock persists with an abrupt transition to the downstream state occurring in about one thermal ion gyroradius. While the three cases shown differ in details, they are qualitatively the same and result in similar particle spectra as discussed below.

In analyzing Figure 2, it is essential to realize that the distance unit λ_0 is proportional to V_{sk}. Since approximately the same maximum energy is obtained in each case, the precursor length in real units scales essentially as $1/V_{sk}$. Another important point to notice in comparing the nonlinear solutions to the test particle one (shown as a dotted line in Figure 2), is that the overall compression ratio is well above four in the nonlinear cases (e.g., Ellison and Eichler, 1984; Jones and Ellison, 1991).

In Figure 3 we show differential flux spectra for models I, II, and III. The spectra are calculated in the shock frame, at a position downstream from the shock, and measured in energy per nucleon. The light solid lines are the proton spectra and the light dashed lines are the He^{+2} spectra. The proton spectra are normalized to one thermal proton injected far upstream per cm^2 per sec and thermal helium is injected far upstream with $n_{He}/n_H = 0.1$. We will discuss the grain spectra (heavy solid and dotted lines) below.

The result of shock smoothing is seen in the H$^+$ and He^{+2} spectra; below $\sim Am_p c^2$ ($\sim 10^6$ keV A^{-1}) the spectra curve slightly upward as the particles get more energetic. This comes about because as the particles increase in energy, they develop a longer diffusion length and 'feel' a stronger compression ratio. Around

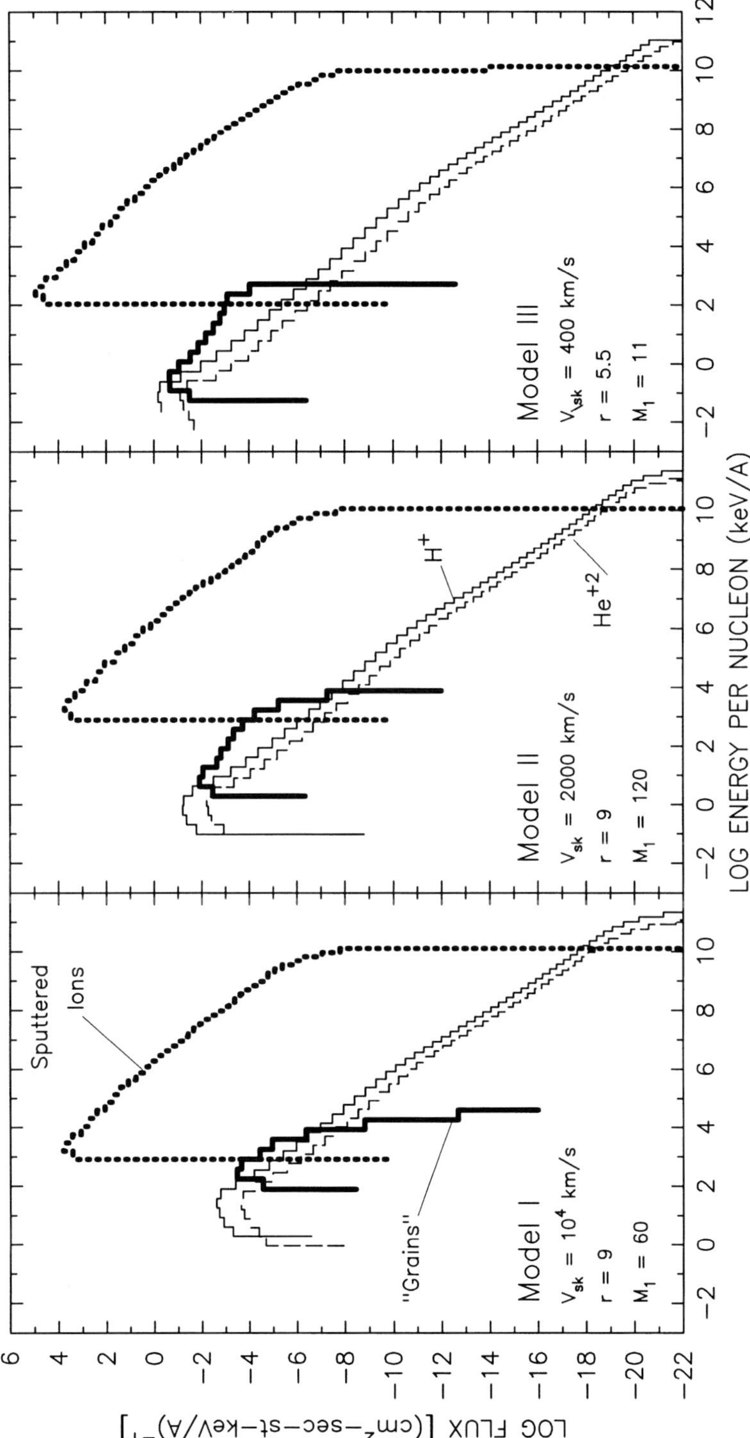

Figure 3. Differential flux spectra in energy per nucleon obtained from shocks with $V_{sk} = 10^4$ km s^{-1} (Model I), $V_{sk} = 2000$ km s^{-1} (Model II), and $V_{sk} = 400$ km s^{-1} (Model III). The light solid curves are the proton spectra, the light dashed curves are He^{+2} spectra, the heavy solid curves are 'Fe grain' spectra, and the heavy dotted lines are Fe ions sputtered off the grains. In all models, the far upstream proton flux is normalized to one particle per cm^2 per sec, $n_{He}/n_H = 0.1$ far upstream from the shock, and the grains are test particles and are injected far upstream with the same number density as protons. At energies below the falloff produced by frictional losses, grains experience a large enhancement over protons at least in models II and III. All spectra here and elsewhere are calculated in the shock frame at a position downstream from the shock.

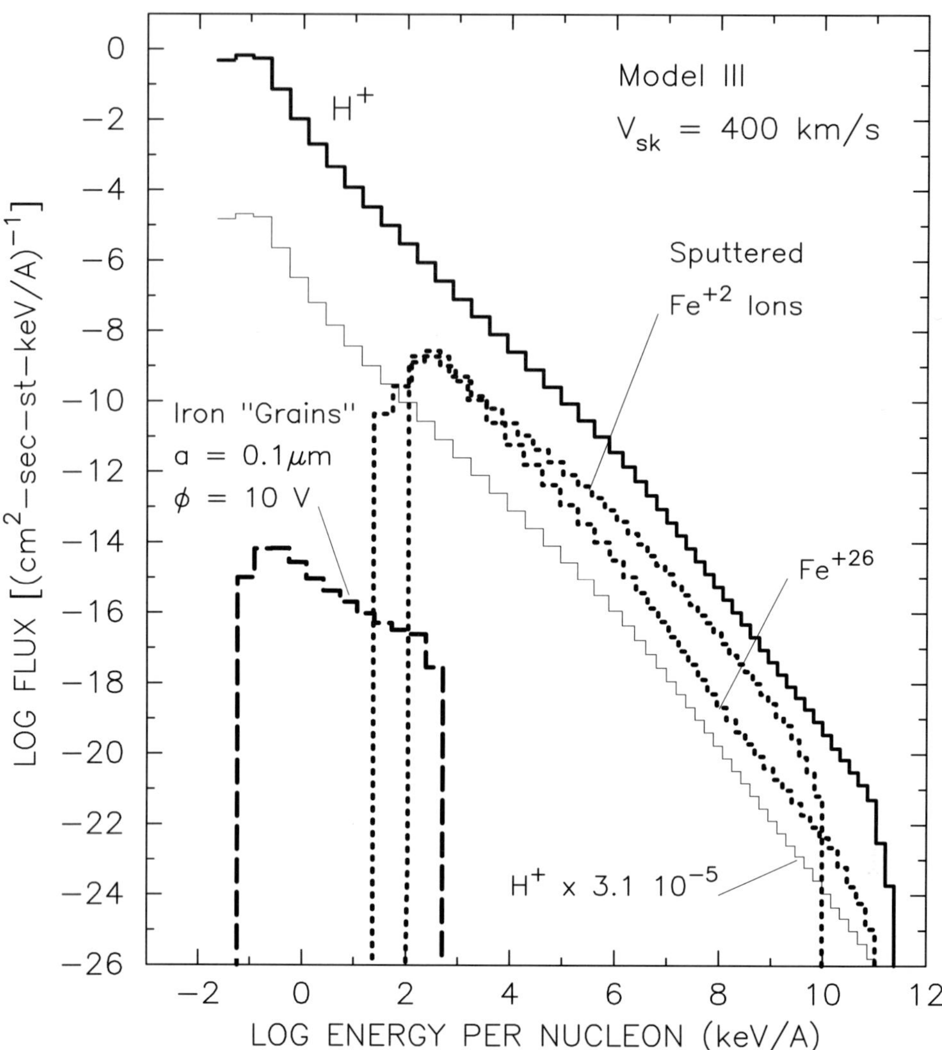

Figure 4. The heavy solid line shows the same proton spectrum as shown in the right-hand panel of Figure 3. The light solid line is this proton spectrum multiplied by 3.1×10^{-5}, i.e., scaled to the solar Fe/H ratio. The dashed line shows the grain spectrum normalized to the cosmic abundance of iron, i.e., the far upstream grain number density, $n_G = 3.1 \times 10^{-5} n_H/10^9$ cm^{-3}, where there are 10^9 iron atoms per grain. The two dotted lines show the spectra of the accelerated sputtered iron ions assuming constant charge states of $+2$ and $+26$.

$\sim Am_p c^2$, kinematic effects cause a steepening as the particles become relativistic, but above $\sim Am_p c^2$, the upward curvature begins again. Since the highest energy particles diffuse across the full density jump, r, just below the turnover caused by the FEB the spectra develop slopes not too different from that expected from test-particle Fermi acceleration, i.e., $dJ/dE \propto E^{-\sigma}$, where $\sigma = (r + 2)/(r - 1)$ for relativistic energies.

3.2. ACCELERATION OF INTERSTELLAR GASES

Once we assume that all ion species obey Equation (9), smooth shocks produce injection and acceleration efficiencies which are increasing functions of A/Q. If, in addition, all elements start with similar charge states in the ISM, the abundance of heavy ions relative to protons will be an increasing function of mass. Because of this, we believe the increase of abundance with mass seen for gas phase elements in cosmic rays (Figure 1) is a fairly clear signature of particle acceleration in smooth shocks (e.g., Eichler, 1979; Ellison, 1981).

To obtain specific estimates of the abundance enhancements, we have calculated the spectra for a number of gases, all injected with the same number density far upstream from the shock and all (except H^{+1}) with a charge of $+2$, which remains unchanged during acceleration. A charge state of $+2$ might occur in the cool ISM ($\sim 10^4$ K) subjected to UV or X-ray photoionization. The abundance ratios determined from these runs are compared to the gas-phase element cosmic-ray observations in Figure 1. The dotted line is from Model II ($V_{sk} = 2000$ km s^{-1}) and the dot-dashed line is from Model III ($V_{sk} = 400$ km s^{-1}). Despite the uncertainties involved for the charge state and other approximations, it is clear that the shock model does an excellent job of reproducing the abundances of gas-phase elements in cosmic rays. The general increase with mass is reproduced as is the magnitude of the abundance enhancement relative to hydrogen (we note that a similar relationship was obtained by Ellison, 1981). Even the fact that the observed H/He ratio is actually more in cosmic rays than in the Sun can be naturally accounted for if the shocks producing the bulk of the cosmic rays are of sufficiently low Mach numbers (see Ellison et al., 1997).

We also note that exceptions to the general increase of abundance with mass may occur, as with carbon, oxygen, and ^{22}Ne, if an additional source of material is present (i.e., Wolf–Rayet stars for C and ^{22}Ne), or some fraction of the element is locked in grains (15% to 20% of O is in grains; see Meyer et al., 1997). In such a case, the abundance will lie *above* our predictions. Our estimate for the non-Wolf–Rayet abundance of cosmic-ray carbon is indicated in Figure 1.

3.3. GRAIN SPUTTERING AND ABUNDANCES OF REFRACTORY COSMIC RAYS

Having determined abundances for cosmic-ray gas-phase elements, we now determine the abundance of cosmic-ray refractory elements from ions sputtered off grains. Clearly this is a very complicated problem since grains come in many sizes, with varied compositions, and largely unknown structures. Our aim here is to obtain quantitative results by making simple, straightforward assumptions and approximations for grain and shock properties.

In our model, the same shock which accelerates interstellar gases will simultaneously accelerate dust grains. In Figure 3, the heavy solid lines are test-particle 'grain' spectra for grains with our standard parameters, i.e., $a = 10^{-7}$ m, $\phi = 10$ V, and $\mu = 56$ (yielding $A/Q \simeq 8 \times 10^7$). The grains have been accelerated in the smooth shocks shown in Figure 2. They have undergone losses from collisions with the gas and the high energy turnover reflects the situation where the loss time approximately equals the acceleration time. Note that we take $B_1 = 3 \times 10^{-6}$ G and $n_H = 1$ cm^{-3} here and in all of the examples below. The test-particle grains were injected with $n_G = n_H$ and must be scaled by the actual ambient thermal grain density to obtain the absolute normalization (i.e., $n_G/n_H \sim 3 \times 10^{-14}$).

The enhancement effect for large A/Q particles is clearly seen in the grain spectra. While the grains were injected with the same number density as protons, they obtained a much flatter spectrum at low energies resulting in a substantial enhancement before their spectra cut off. As they accelerate, the grains sputter and ions sputtered off in the upstream region will be further accelerated upon convecting into the shock. We include, in the Monte Carlo simulation, a direct determination of this injection process.

The acceleration of grains and sputtered products is an extremely complicated process with a number of factors influencing the final cosmic-ray abundance. These factors include: the size, charge, and mean molecular weight of the grain, the background gas density, the collision and sputtering rate, any losses sputtered ions experience before being further accelerated, and the charge state of the sputtered ion including charge stripping before and during acceleration. In addition, shock properties, such as the Mach number, shock age, geometry, ambient magnetic field strength, and maximum proton energy obtained, will modify the resultant cosmic-ray abundance. Since many of these factors are poorly known, it is impossible to precisely predict the refractory cosmic-ray abundance. However, we have investigated many of these factors and believe the results presented here are quite robust.

We now make specific assumptions for the shock and grain properties in order to obtain a direct estimate of the cosmic-ray abundance of iron. Our predictions, of course, will depend on the particular assumptions we make. We assume that *all* iron which ends up as cosmic rays originates in grains. Using our 400 km s^{-1} shock Model III, we inject and accelerate grains with $a = 0.1$ μm, $\phi = 10$ V, $\mu = 56$, assuming $B_1 = 3 \times 10^{-6}$ G, and $n_H = 1$ cm^{-3}. We next assume that all upstream sputtered iron ions are swept back into the shock without significant energy losses. Iron sputtered downstream is lost. Furthermore, we assume that in the 100–1000 years or so they spend convecting back to the shock they become fully stripped. The Fe^{+26} ions are then re-accelerated in the same shock which accelerated the grains.

Our results are shown in Figure 4 and compared to the proton spectrum (light and heavy solid lines). The accelerated grains are shown with a dashed line, and sputtered Fe ions are shown with dotted lines. The test-particle grains are injected far upstream with solar (i.e., 'cosmic') abundance, that is, the total number of Fe *atoms* in the grains is $\sim 3.1 \times 10^{-5}$ times the number of H atoms. This means that the ratio of far upstream number densities is $n_G Fe/n_H \sim 3 \times 10^{-14}$ since, for $a = 0.1\mu$m, there are $\sim 10^9$ Fe atoms per grain. In order to indicate the difference the sputtered ion charge state makes, we show Fe spectra for Fe^{+2} (i.e., without stripping), as well as Fe^{+26}. In the actual case, there will be a mixture of charge states since sputtered ions formed close to the shock will be convected back before much charge stripping can occur. These ions will be accelerated more efficiently than the Fe^{+26} ions, but fully stripped ions obtain a higher maximum energy per nucleon in the finite size shock. On the other hand, sputtered ions will experience some ionization energy losses before encountering the shock which will cause them to be accelerated less efficiently. In want of a more complete model, we compare the Fe^{+26} spectrum with no upstream losses to the observations in Figure 5.

In any event, it is clear from this figure that a large enhancement of iron over protons occurs. Between ~ 10 and 100 GeV A^{-1}, the sputtered Fe^{+2} ions stand about a factor of 10^3 above their solar abundance relative to hydrogen and the Fe^{+26} stands about a factor of 20 above (compare the light solid proton line, which has been divided by the cosmic abundance of Fe, to the heavy dotted lines). In the next section we compare these predictions, both normalization and spectral shape, to observed cosmic-ray spectra.

3.4. SPECTRAL COMPARISONS

In Figure 5 we compare our proton, helium, and Fe^{+26} spectra to cosmic-ray observations. The data in Figure 5 is adapted from the compilation of cosmic-ray observations presented by Shibata (1995). Both data and model spectra have been multiplied by $(E/A)^{2.5}$ to flatten the steep spectra. This comparison uses Model III ($V_{sk} = 400$ km s^{-1}) and our results for protons are shown as a solid line and those for He^{+2} with a dashed line. All model source spectra have been multiplied by $R^{-0.65}$ to correct for rigidity dependent escape from the galaxy and helium and iron have been corrected for nuclear destruction. The rigidity dependence we have chosen, $\delta = 0.65$, is close to the best value, $\delta = 0.6$, adopted by Engelmann et al. (1990) and Shibata (1995).

Several features are evident in Figure 5. First of all, solar modulation is not included in our model and its effect shows in the data which fall off somewhat below the kinematic turnover near 1 GeV nucl^{-1}. Second, our model matches the gross features of the spectra extremely well above the energy where modulation is important and below where our model spectra fall off from the effects of our adopted finite size SNR shock. We must caution that the fit to the slope depends strongly on the value for δ we choose and should not be over interpreted. Cosmic rays undoubtedly come from a number of supernovae of varying sizes, Mach numbers, etc., and a model such as ours of a single shock is not intended to model the observations in complete detail. Having said this, we wish to emphasize that while we have adjusted the overall normalization of the *proton* spectrum to match the observations, there is *no adjustment of the relative normalization* between protons and helium. Our 400 km s^{-1} shock model, using the cosmic abundance of helium, reproduces the observed relative fluxes fairly accurately.

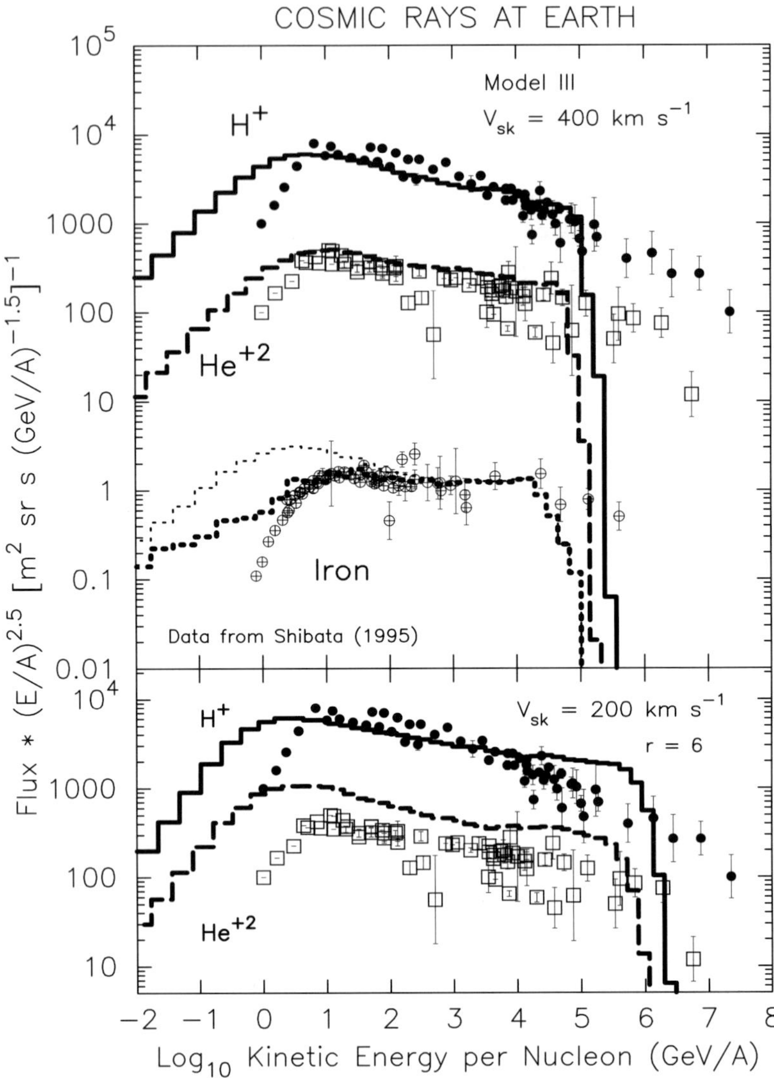

Figure 5. Cosmic-ray spectra in energy per nucleon. The cosmic-ray data are from the compilation of Shibata (1995) and have been multiplied by $(E/A)^{2.5}$ to produce nearly flat spectra above ~ 1 GeV A^{-1}. Note that in addition to being multiplied by $(E/A)^{2.5}$, the source model curves (protons, solid line; He^{+2}, dashed line; Fe^{+26}, dotted lines) have been multiplied by $R^{-0.65}$ to mimic rigidity dependent escape from the galaxy and corrected for nuclear destruction during propagation. In the model, both helium and iron are injected at solar abundance, i.e., $n_{He}/n_H = 0.1$ and $n_{Fe}/n_H = 3.1 \times 10^{-5}$. The normalization of the model proton spectrum is varied to match the observations but the relative normalization of helium to hydrogen and iron to hydrogen is fixed by the model. The light dotted line shows the iron spectrum without correction for nuclear destruction. The bottom panel shows model H and He spectra from our 200 km s^{-1} shock example ($r \simeq 6$, $M_s \simeq 7$) with a maximum proton energy extending to $\sim 10^{15}$ eV. As in the top panel, the overall normalization is varied to match the observations but the relative normalization of helium to hydrogen is fixed by the model. The difference in the He/H ratio between the two models is mainly the result of the poor statistics for the 200 km s^{-1} example.

The heavy dotted line in Figure 5 shows the Fe^{+26} curve from Figure 4 multiplied by $(E/A)^{2.5} \times R^{-0.65}$ and corrected for nuclear destruction. Once these corrections are made, the normalization relative to protons is exactly that shown in Figure 4, i.e., the iron is injected out of a medium with solar abundance $n_{Fe} = 3.1 \times 10^{-5} n_H$ and the model sets the abundance relative to hydrogen. Clearly, the excellent match to the observed normalization is somewhat fortuitous considering the many uncertainties in the model. However, we have chosen parameters we feel to be realistic and believe our model is quite robust; we do not expect that the fundamental result that cosmic rays that originate from grain material are enhanced will change as refinements are made. The light dotted line shows the Fe^{+26} spectrum without the nuclear destruction correction which tends to remove the otherwise distinctive curvature expected from nonlinear shock acceleration. We predict that this curvature is real and sufficiently accurate cosmic-ray observations may reveal it.

Our prediction for the source cosmic-ray iron abundance is included in Figure 1. To give some indication of the errors intrinsic to our calculation, we show two horizontal lines on the right side of the plot. The lower line is the ratio, Fe^{+26}/H^+, taken at 100 GeV A^{-1} in Figure 4, and the upper line is taken at 10 GeV A^{-1}. The match is excellent, but we again emphasize that many uncertainties remain in the model that are not taken into account by the spread in the horizontal lines. The same enhancements should apply roughly for the other refractory elements in Figure 1 since, in the crucial early acceleration phase, they are all accelerated, not as individual ions, but as constituents of the same grains.

3.4.1. *Example with Slower Shock Speed and Higher E_{max}*

To investigate further how the shock speed and the maximum particle energy influence our results, we have calculated the shock structure and spectra for a model with $V_{sk} = 200$ km s^{-1} and a maximum energy of $\sim 10^{15}$ eV (about an order of magnitude higher than the previous examples). These parameters require long computer runs and our statistics are not as good as in our $V_{sk} = 400$ km s^{-1} model. We compare the H^+ and He^{+2} spectra to observations in the lower panel of Figure 5. All corrections are identical to those described for Model III.

Except for the higher maximum energies, the results for the 400 and 200 km s^{-1} models are essentially the same with the differences seen in Figure 5 being largely statistical. Most important, the overall slopes of the H^+ and He^{+2} spectra are still slightly flatter than the observations (even with our use of $R^{-0.65}$ to correct for rigidity dependent escape from the Galaxy). While we have reduced the shock speed from 400 to 200 km s^{-1}, the Mach number for our 200 km s^{-1} example is still ~ 7 and the overall compression ratio from our self-consistent solution is $r \simeq 6$, approximately the same as in the 400 km s^{-1} example. Depending on the galactic propagation models, shocks with lower Mach numbers may be needed to explain the overall spectral shape. The difference in the H/He ratio between the two models is, we believe, largely the result of the poor statistics for the 200 km s^{-1}

model. We have calculated the abundance of Fe from grains in an identical fashion to that described above and find a strong enhancement similar to that shown in Figure 4 and the upper panel of Figure 5.

While considerably longer computer runs are needed to investigate lower Mach number shocks and improve statistics, we believe the most important conclusion from this additional model is that factor of two changes in shock speed and factor of 10 changes in the maximum energy produce only small changes in our basic results.

4. Conclusions

The analysis of Meyer et al. (1997) strongly suggests that cosmic-ray source material consists mainly of two ISM components: volatile elements from the gas-phase, and refractory elements from dust grains. Relative to solar abundances, the abundances of the volatile elements are found to be a strongly increasing function of mass with the possible exception of hydrogen. In contrast, the abundances of the refractory elements are systematically higher, but with little or no mass dependence, allowing a clear separation of these components in the data. The elements which are likely to be *partly* locked in grains in the ISM show up midway between these two groups. Here, we briefly outlined our theory for how this comes about (for a more complete discussion, see Ellison et al., 1997). The essential point is that, if one assumes that all charged particles of a given magnetic rigidity (ion and charged dust grain) act identically as they interact with the turbulent magnetic field, a single modified shock will accelerate gas and dust simultaneously with the observed abundance variations.

Standard nonlinear shock acceleration theory predicts that ions are accelerated directly to cosmic-ray energies in the smoothed shock, which produces an enhancement of high mass/charge (i.e., A/Q) elements. Since heavier elements always tend to have higher A/Q ratios, the mass dependence seen in the volatile cosmic-ray abundances is a clear signature of acceleration by smoothed shocks. If the weakly charged, massive grains (i.e., large A/Q particles) act as protons of the same rigidity, these grains will be very efficiently accelerated by the same shocks, although up to far lower energy per nucleon (~ 100 keV nucl^{-1}) than the gas ions, due to friction and to the limited age and size of the SNR. A simple model of the sputtering of these grains upstream from the shock, and the acceleration of the sputtered ions to TeV cosmic-ray energies, then yields the relative abundances of refractory elements to volatile ones. This is found to agree with observations, including the lack of a significant mass dependence of the refractory element abundances, as observed.

We believe that this is the first attempt to give a detailed description of grain and gas acceleration in nonlinear shocks. Specifically, our most important predictions are that: (i) All gas-phase elements will have abundances which lie in the range given by the dotted and dot-dashed lines shown in Figure 1. Based on the observed ^{22}Ne excess, we expect an additional source of carbon from the acceleration of ^{12}C-enriched Wolf–Rayet wind material affected by He-burning nucleosynthesis, causing C to lie *above* the line by the amount contributed by this additional source. Our estimate for the non-Wolf–Rayet carbon contribution is labeled in Figure 1. (ii) All refractory elements should end up with abundances independent of condensation temperature and mass, and not too far from our prediction for iron. (iii) The observed O excess in cosmic rays is the result of the fact that a significant fraction of oxygen (15 to 20%) is locked in grains and this fraction will be accelerated more efficiently than the gaseous O. (iv) No significant amount of SN ejecta material is being accelerated. (v) The shock acceleration of interstellar grains will produce grain speeds relative to the background plasma considerably greater ($\beta_G \sim 0.01$ for the fastest ones) than is generally assumed.

It is important to note that, if our model is correct, the long-standing belief that cosmic rays accelerated to TeV energies by SNR shock waves were first injected to MeV energies with appropriate coronal gas composition by later-type stars (so called FIP effect models), is rejected. Not only is our one-step, one-site model much simpler than any scenario based on stellar injection processes, which involve at least two acceleration stages in two unrelated sites, but we can naturally account for the relative deficiency of the two most abundant species, hydrogen and helium, and can account, without recourse to a different type of source, for the ^{22}Ne–C–O excess observed in galactic cosmic rays.

We have extended the work of Ellison et al. (1997) slightly here by including a model with a slower shock speed (200 km s^{-1}) and a higher maximum cosmic-ray energy ($\sim 10^{15}$ eV). If the standard corrections for galactic propagation are correct, quite slow shocks are required to produce the implied steep source spectra. We find that our 200 km s^{-1} model gives essentially the same H$^+$ and He^{+2} spectra (within our statistical errors) as our previous 400 km s^{-1} model (except, of course, they now extend to higher energies). Shocks with even lower Mach numbers may be required to match the steep source spectra implied by standard propagation models.

Acknowledgements

D. Ellison and L. Drury wish to acknowledge the hospitality of the Service d'Astrophysique, Centre d'Etudes de Saclay where much of this work was carried out. L. Drury's visit was supported by the Commission of the European Communities under contract ERBCHRXCT940604, and D. Ellison was supported, in part, by the NASA Space Physics Theory Program. The authors also thank T. Shibata for kindly furnishing recent cosmic-ray data, and K. Borkowski, S. Reynolds, and D. Reames for helpful discussions.

References

Berezhko, E. G. and Krymsky, G. F.: 1988, *Usp. Fiz. Nauk* **154**, 49 (English translation, *Soviet Phys. Usp.* **31**, 27).

Berezhko, E. G., Elshin, V. K., and Ksenofontov, L. T.: 1996, *JETP* **82**, 1.

Blandford, R. D. and Eichler, D.: 1987, *Physics Reports* **154**, 1.

Bode, M. F.: 1988, in M. E. Bailey and D. A. Williams (eds.), *Dust in the Universe*, Cambridge University Press, Cambridge, p. 73.

Cardelli, J. A.: 1994, *Science* **265**, 209.

Cummings, A. C. and Stone, E. C.: 1996, *Space Sci. Rev.* **78**, 117.

Drury, L. O'C.: 1983, *Rep. Prog. Phys.* **46**, 973.

Drury, L. O'C., Aharonian, F. A., and Völk, H. J.: 1994, *Astron. Astrophys.* **287**, 959.

Dwek, E.: 1987, *Astrophys. J.* **322**, 812.

Dwek, E., Moseley, S. H., Glaccum, W., Graham, J. R., Loewenstein, R. F., Silverberg, R. F., and Smith, R. K.: 1992, *Astrophys. J.* **389**, L21.

Eichler, D.: 1979, *Astrophys. J.* **232**, 106.

Eichler, D.: 1984, *Astrophys. J.* **277**, 429.

Ellison, D. C.: 1981, Ph.D. Thesis, The Catholic University of America.

Ellison, D. C. and Eichler, D.: 1984, *Astrophys. J.* **286**, 691.

Ellison, D. C., Drury, L. O'C., and Meyer, J. P.: 1997, *Astrophys. J.*, in press.

Ellison, D. C., Jones, F. C., and Reynolds, S. P.: 1990, *Astrophys. J.* **360**, 702.

Engelmann, J. J., Ferrando, P., Soutoul, A., Goret, P., Juliusson, E., Koch-Miramond, L., Lund, N., Masse, P., Peters, B., Petrou, N., and Rasmussen, I. L.: 1990, *Astron. Astrophys.* **233**, 96.

Epstein, R. I.: 1980, *Monthly Notices Royal Astron. Soc.* **193**, 723.

Gehrz, R. D.: 1991, in L. J. Allamandola and A. G. G. M. Tielens (eds.), 'Interstellar Dust', *IAU Symp.* **135**, 445.

Jones, F. C. and Ellison, D. C.: 1991, *Space Sci. Rev.* **58**, 259.

Lagage, P. O. and Cesarsky, C. J.: 1983, *Astron. Astrophys. J.* **125**, 249.

Lucy, L. B., Danziger, I. J., Gouiffes, C., and Bouchet, P.: 1989, in S. E. Woosley (ed.), *Supernovae*, Springer-Verlag, New York, p. 82.

Lucy, L. B., Danziger, I. J., Gouiffes, C., and Bouchet, P.: 1991, in G. Tenorio-Tagle, M. Moles, and J. Melnick (eds.), 'Structure and Dynamics of the Interstellar Medium', *IAU Colloq.* **120**, 164.

Mathis, J. S., Rumpl, W., and Nordsieck, K. H.: 1977, *Astrophys. J.* **217**, 425.

Meyer, J. P., Drury, L. O'C., and Ellison, D. C.: 1997, *Astrophys. J.*, in press.

Prishchep, V. L. and Ptuskin, V. S.: 1981, *Astron. Zh.* **58**, 779 (English translation *Soviet Astron.* **25**, 446).

Reynolds, S. P. 1988, in G. L. Verschuur and K. I. Kellermann (eds.), *Galactic and Extragalactic Radio Astronomy*, Springer-Verlag, Berlin, p. 439.

Savage, B. D. and Sembach, K. R.: 1996, *Ann. Rev. Astron. Astrophys.*, in press.

Sembach, K. R. and Savage, B. D.: 1996, *Astrophys. J.* **457**, 211.

Shibata, T.: 1995, *24th Int. Cosmic Ray Conf., Rome*, Invited, Rapporteurs and Highlight Papers, p. 713.

Van der Hucht, K. A. and Hidayat, B. (eds.): 1991, 'Wolf–Rayet Stars and Interrelations with Other Massive Stars in the Galaxy', *IAU Symp.* **143**.

Van der Hucht, K. A. and Williams, P. M. (eds.): 1995, 'Wolf–Rayet Stars: Binaries, Colliding Winds, Evolution', *IAU Symp.* **163**.

Völk, H. J.: 1984, in Tran Than Van (ed.), *High Energy Astrophysics, Proc. 19th Rencontre de Moriond*, Editions Frontières, Gif-sur-Yvette, p. 281.

COSMIC-RAY CLOCKS

VLADIMIR S. PTUSKIN

*Institute for Terrestrial Magnetism, Ionosphere and Radio Wave Propagation (IZMIRAN) of the
Russian Academy of Sciences, Troitsk, Moscow region 142092, Russia*

AIMÉ SOUTOUL

Service d'Astrophysique, DAPNIA/SAp, CE Saclay, 91191 Gif-sur-Yvette, Cedex, France

Abstract. Secondary radioactive isotopes that are used for the determination of cosmic-ray age
have relatively short decay lifetimes. The measured abundance of these isotopes at low energies
is representative of the cosmic-ray diffusion and the gas distribution in a region of a few hundred
parsecs around the Sun. We show how to determine the local cosmic-ray diffusion coefficient in the
Galaxy using the data on decaying cosmic-ray nuclei. Calculated surviving fractions of decaying
secondary isotopes in diffusion and leaky box models are presented.

1. Introduction

Leaving their cosmic-ray sources and wandering in the interstellar medium, primary relativistic nuclei experience fragmentation in interstellar gas and give rise to secondary nuclei in cosmic rays. The content of different stable and radioactive secondary isotopes reflects the conditions of cosmic-ray propagation in the Galaxy.

Bradt and Peters (1948) put forward nuclear fragmentation as the explanation for the presence of a considerable number of nuclei of rare elements among the cosmic rays. The observed abundance of secondary cosmic-ray nuclei gives $X \sim 10\ \mathrm{g\ cm^{-2}}$ for the value of the mean thickness of matter traversed by particles with energy about $1\ \mathrm{GeV\ nucl^{-1}}$ in the interstellar gas. This value of X is of the order of magnitude of the nuclear destruction pathlength for the relativistic medium-mass nucleus. This coincidence and the large dispersion of the matter thickness allow one to obtain simultaneously information on cosmic-ray source composition and information on the conditions of cosmic-ray propagation in the interstellar medium.

Another significant circumstance is the presence of a number of radioactive isotopes in cosmic rays with a decay lifetime that is comparable to the age of cosmic rays in the Galaxy. Hayakawa et al. (1958) realized that the abundance of radioactive secondary isotopes could be used as a 'clock' to determine cosmic-ray age. The isotope ^{10}Be with a decay lifetime at rest 2.3×10^6 yr is most frequently used for this procedure. The list of other isotopes includes ^{26}Al (1.3×10^6 yr), ^{36}Cl (4.3×10^5 yr), ^{54}Mn ($\sim 10^6$ yr), and ^{14}C (8.2×10^3 yr). The decay lifetime of completely ionized ^{54}Mn is not yet well determined.

Space Science Reviews **86**: 225–238, 1998.
© 1998 *Kluwer Academic Publishers. Printed in the Netherlands.*

Below we discuss the transport of energetic radioactive nuclei in the Galaxy and present an interpretation of available data. A strong motivation for this work comes from the good agreement between satellite measurements of the ^{10}Be abundance in low energy cosmic rays (Garcia-Muñoz et al., 1977; Wiedenbeck and Greiner, 1980; Simpson and Garcia-Muñoz, 1988; Lukasiak et al., 1994a), the availability of satellite data on ^{26}Al (Wiedenbeck, 1983; Lukasiak et al., 1994b), ^{36}Cl (Leske and Wiedenbeck, 1993), and ^{54}Mn (Leske, 1993; Lukasiak et al., 1995), soon to be released results on radioactive isotopes from the *Ulysses* experiment, and, primarily, those from the Cosmic-Ray Isotope Spectrometer (CRIS) experiment on board the Advanced Composition Explorer (ACE) spacecraft. CRIS has a collection power about 100 times greater than existing instruments. All the foregoing stimulate the development of improved models of cosmic-ray propagation in the Galaxy.

2. Models of Cosmic-Ray Propagation in the Galaxy

The semi-empirical models of cosmic-ray propagation serve a purpose of ordering numerous data on cosmic rays. The models provide a basis for the interpretation of radio-astronomical, X-ray, and gamma-ray measurements.

The homogeneous leaky-box model is a well accepted approximation for the description of the propagation of stable nuclei. Hundreds of isotopes are included in the calculations of nuclear fragmentation and transformation of energetic nuclei in the course of their interaction with interstellar gas. Leakage from the Galaxy, described by the escape length X, and ionization losses accompany these processes. The value $X = 14.4\beta$ g cm^{-2}, for particles with magnetic rigidity $R \leq R_c$, where $R_c = 4.4$ GV, and $X = 35.7\beta R^{-0.62}$ g cm^{-2}, for $R > R_c$ (Engelmann et al., 1990) found from the HEAO-3 data on B/C ratio up to 20 GeV nucl^{-1} is still in use. Here $\beta = v/c$ is the ratio of the particle velocity v to the velocity of light c.

Success in the interpretation of observations in the frameworks of the leaky box model may appear surprising. Radio and gamma-ray observations definitely show very nonhomogeneous distributions of cosmic-ray protons and electrons in our and other galaxies (e.g., Berezinskii et al. 1990), whereas homogeneity is the most essential feature of the leaky box model. These observations, confirmed by the theory of motion of charged relativistic particles in interstellar magnetic fields, show that diffusion is a good approximation for the description of cosmic-ray propagation in the Galaxy. Diffusion in random and regular galactic magnetic fields may be accompanied by advection of cosmic rays under the action of convective motions of interstellar gas and, possibly, by galactic wind flow.

We know however that solutions of diffusion transport equations may be in some cases approximated by the leaky box formulas. Diffusion and nuclear fragmentation of stable nuclei in a flat-halo galaxy with free escape of cosmic rays at the boundaries and negligibly small cosmic-ray density in intergalactic space (Ginzburg and Ptuskin, 1976) is an important example. It is assumed that cosmic-ray sources and interstellar gas are concentrated in a region which is thin compared with the size of the galactic cosmic-ray halo. The relation between parameters of the diffusion and equivalent leaky box model in this case is given for an observer in the galactic disk by the following equation

$$X_m = \frac{v\mu H}{2D} ,$$ (1)

where μ is the total surface gas density of the galactic disk, H is the scale height of the cosmic-ray halo, and D is the cosmic-ray diffusion coefficient. The equivalence holds for not very heavy nuclei with total cross sections $\sigma \ll mH/(hX)$, where m is the mean mass of atoms in the interstellar gas, and $h \ll H$ is the characteristic height of the gas disk. The above limitation on the value of σ means that nuclear fragmentation of an energetic nucleus is weak during one diffusive crossing of the gas disk.

Convection of cosmic rays may work in the Galaxy in addition to the diffusion (Jokipii, 1976). Both the real motion of the interstellar medium (the galactic wind) and the flux of hydromagnetic waves in a medium at rest can result in convective transport of energetic particles. Assuming that the convection velocity u is constant and directed away from the galactic disk, one can find the following asymptotic expressions for the escape length X (Jones, 1979; Prishchep and Ptuskin, 1979; Freedman et al., 1980):

$$X \approx \frac{3}{2(\gamma + 2)} \frac{v\mu}{u} \quad \text{at} \quad uh \ll D \ll uH ,$$ (2)

and

$$X \approx \frac{v\mu H}{2D} \quad \text{at} \quad D \gg uH .$$ (3)

Here

$$\gamma = -\frac{E + 2m_N c^2}{E + m_N c^2} \frac{d \ln J(E)}{d \ln E} ,$$

$J(E)$ is the differential cosmic-ray intensity, E is the kinetic energy per nucleon, m_N is the nucleon mass. Equations (2) and (3) reproduce the observations if $D \propto v R^{0.62}$ and $D(R_c) \sim uH$. Equations (2) and (3) describe the low-energy and high-energy regimes of cosmic-ray transport, respectively. In the last case the convection is not important and Equation (3) coincides with Equation (1). At low energies ($R < R_c$) diffusion transport dominates over convection inside the boundary layer with the thickness D/u adjacent to the galactic disk. Convection dominates over diffusion above this layer. An energetic particle which gets to the convection zone has almost no chance to return back to the galactic disk. Thus, in this context the situation for an observer in the disk is approximately the same as for the pure diffusion model with the effective size of the halo $H_{\mathrm{eff}} \sim D/u$ at $R < R_c$ and $H_{\mathrm{eff}} \sim H$ at $R > R_c$. Equation (2) together with the observed value of $X \approx 15\,\mathrm{g\,cm^{-2}}$ at $R < 4.4\,\mathrm{GV}$ determines the convection velocity $u \approx 2.7 \times 10^6\,\mathrm{cm\,s^{-1}}$ (at $\mu = 2.3 \times 10^{-3}\,\mathrm{g\,cm^{-2}}$, $\gamma = 0.5$).

The actual realization of the leaky box model is possible if cosmic-ray streaming from the Galaxy is accompanied by the excitation of hydromagnetic waves. This streaming instability may work efficiently only above some critical height $h_w \sim 0.5\,\mathrm{kpc}$ above the galactic midplane since the considerable density of neutrals in the interstellar gas below this height suppresses the development of instability. A high level of turbulence outside this region leads then to strong scattering of cosmic-ray particles and makes them return back to the internal zone. The probability of not being scattered by the wave barrier is small as v_A/v, where v_A is the Alfvén velocity. Thus it might be that cosmic rays are well confined in a real flat leaky box with a thickness of about 1 kpc and with an effective escape length $X \propto v\mu/v_A$. Such is the case of the cosmic-ray propagation at low energies ($E < 2\,\mathrm{GeV\,nucl^{-1}}$) in the galactic wind model suggested by Ptuskin et al. (1997). In this self-consistent model, the pressure of cosmic rays generated in the galactic disk determines the structure of the galactic wind flow and the cosmic-ray streaming instability balanced by the nonlinear Landau damping on thermal ions determines the level of turbulence. The entire wind flow extends over a few hundred kpc from the galactic disk. The transport coefficients for cosmic rays, the diffusion coefficient and the convective velocity, are not prescribed but are calculated under this approach. Cosmic-ray particles move freely (formally, with very large diffusion coefficient $D_t \gg h_w v_A \approx 5 \times 10^{28}\,\mathrm{cm^2\,s^{-1}}$) in the internal zone $|z| \leq h_w$. The wave barrier bounding this zone is not effective for particles with energies higher than $2\,\mathrm{GeV\,nucl^{-1}}$ and their propagation may be described as some combination of diffusion and convection. At high energies and for an observer at the disk, this model is almost equivalent to a pure diffusion model with absorbing boundaries moving apart and with scaling $X \propto v R^{-0.62}$ for the escape length.

Figure 1 schematically depicts three types of galactic models of cosmic-ray propagation discussed above.

In the following, we will consider the transport of low-energy decaying nuclei in two distinct simple models: the pure diffusion model without reflecting boundaries and the standard leaky box model (which actually may present a limiting case of a diffusion model with reflecting boundaries and fast diffusion between them). The results for the diffusion-convection model would be close to the pure diffusion model with the effective size of the halo of the order of D/u; see Berezinskii et al. (1990), Webber et al. (1992), Bloemen et al. (1993) for discussion about different aspects of cosmic-ray propagation in galactic wind models where convective transport is essential.

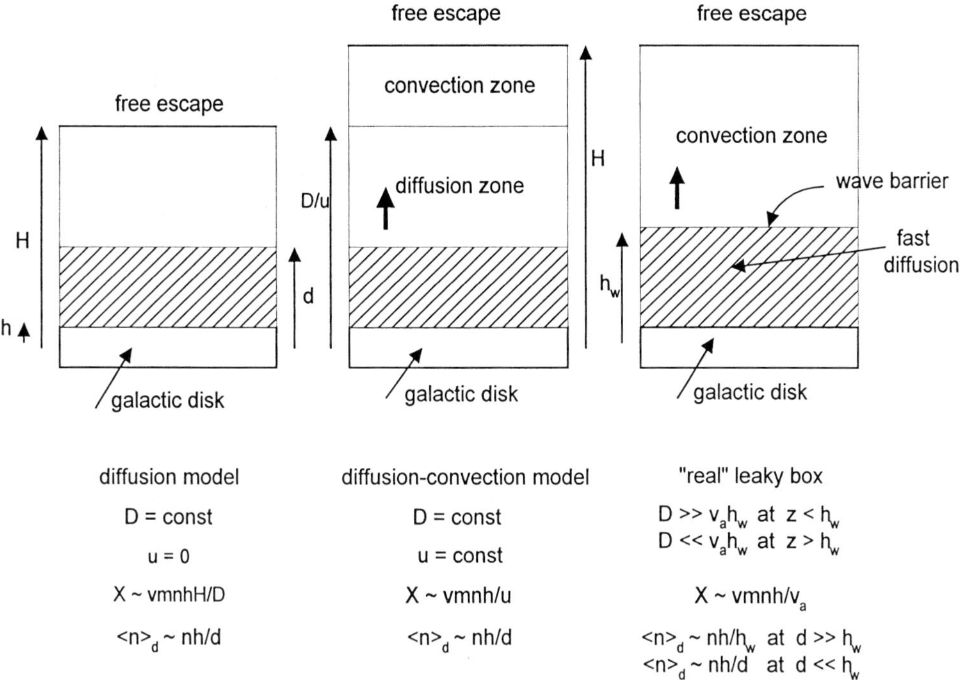

Figure 1. An illustration of cosmic-ray propagation in the Galaxy in different models. Cosmic-ray sources and interstellar gas (with gas density n) are concentrated in the galactic disk $|z| \leq h$. Distribution of radioactive nuclei in the galactic halo is shown by the hatched area. Approximate expressions for the escape length X and the gas density $\langle n \rangle_d$ averaged over the volume occupied by the decaying nuclei are indicated. (The case $d = (D\tau)^{1/2} \gg h_w$ is shown in the picture for the 'real' leaky box model. The hatched area would be limited by $|z| \leq d$ if $d \ll h_w$.)

3. Transport of Radioactive Nuclei

The interpretation of observations of radioactive nuclei is rather sensitive to the model of cosmic-ray propagation in the Galaxy. In particular, the diffusion model and the leaky box model turn out to not be equivalent (Prishchep and Ptuskin, 1975; Ginzburg and Ptuskin, 1976). The difference is mainly due to the inhomogeneous interstellar gas distribution where the secondary nuclei are produced. Slow diffusion and relatively fast decay of radioactive secondaries may make their spatial distribution strongly inhomogeneous whereas the leaky box model assumes a uniform distribution for all cosmic-ray species. The well-studied effect of this kind is caused by the most prominent feature in the large-scale gas distribution in the Galaxy: the concentration of a large fraction of mass of interstellar gas in a relatively thin gas disk (Prishchep and Ptuskin, 1975; Ginzburg and Ptuskin, 1976; Prishchep and Ptuskin, 1979; Freedman et al., 1980; Berezinskii et al., 1990; Ptuskin and Soutoul, 1991; Webber et al., 1992; Bloemen et al., 1993), see Figure 1. The spatial scale which is important here is of order $d = (D\tau)^{1/2}$, where

D is the cosmic-ray diffusion coefficient and τ is the lifetime for decay (including relativistic factor). The typical value of D at energies less than 2 GeV nucl^{-1} is about 3×10^{28} cm^2 s^{-1} and the value of d may range from 50 to 500 pc for the isotopes listed above. The variations of the gas density are essentially on these scales and must be incorporated in the calculations of the surviving fraction in the diffusion model. The appropriate procedure is described below following the approach suggested by Ptuskin and Soutoul (1998).

The steady-state transport equation for secondary radioactive nuclei that describes their diffusion, nuclear fragmentation and decay has the form:

$$-\nabla D \nabla N_2 + n v \sigma_2 N_2 + \frac{N_2}{\tau} + \frac{\partial}{\partial E}\left(\left(\frac{dE}{dt}\right)_{\mathrm{ion}} N_2\right) = n v \sigma_{12} N_1 \, . \tag{4}$$

Here N_2 and N_1 are the number densities of primary and secondary nuclei, n is the interstellar gas density, σ_2 and σ_{12} are the total and production cross sections for secondaries, and the term with $(dE/dt)_{\mathrm{ion}} < 0$ describes the ionization energy losses. The term for decay in the left-hand side of Equation (4) is larger than the term for fragmentation and the last is larger than the term for ionization losses for the radioactive isotopes listed above and at the energies under consideration, so that the interstellar medium is only weakly nontransparent against fragmentation during the typical time interval spent by these isotopes in the Galaxy. We shall consider the solution of Equation (4) in an unbounded medium assuming that the size of the galactic cosmic-ray halo is relatively large compared to d. In these circumstances the following approximate solution of Equation (4) may be used:

$$N_2(\mathbf{r}) = v \sigma_{12} \iiint d^3 r_0 n(\mathbf{r}_0) N_1(\mathbf{r}_0) G(\mathbf{r}, \mathbf{r}_0, \tau_m) \, , \tag{5}$$

where the Green function $G(\mathbf{r}, \mathbf{r}_0, \tau_m)$ obeys the equation

$$-\nabla D \nabla G + \frac{G}{\tau_m} = \delta^3(\mathbf{r} - \mathbf{r}_0) \tag{6}$$

and

$$\tau_m = \frac{\tau}{1 + v(\sigma_2 + \sigma_{2,\mathrm{ion}}) \iiint d^3 r_0 n(\mathbf{r}_0) G(\mathbf{r}, \mathbf{r}_0, \tau)} \, . \tag{7}$$

The effect of ionization energy losses is taken into account by applying the correction to the cross section of nuclear fragmentation (the values of correction $\sigma_{2,\mathrm{ion}}$ for different isotopes are given by Ptuskin and Soutoul (1998); this correction depends on the slope of the cosmic-ray energy spectrum). The quantity τ_m approximately describes the combined effect of radioactive decay with lifetime τ and nuclear fragmentation in the interstellar gas with density averaged with the corresponding weight over the volume where radioactive nuclei can diffuse to reach the observer at $\mathbf{r}$.

The presence of 10% of He in interstellar gas was taken into account in the actual calculations of nuclear fragmentation and ionization energy losses. The density of primary nuclei N_1 was assumed to be constant on the scale d, i.e., $N_1(\mathbf{r}) \approx N_1(\mathbf{r}_0)$ in the integral in Equation (5). This assumption is based on the radio and gamma-ray observations, see, e.g., Berezinskii et al. (1990). It was anticipated also that cosmic-ray diffusion is isotropic and does not depend on position, i.e., the scalar $D = $ const. This assumption may be justified only by the lack of detailed knowledge about the structure of the galactic magnetic field since the magnetic field makes cosmic-ray diffusion anisotropic.

The assumption $D = $ const. allows simple analytic solutions of Equation (6) which transforms now into the Helmholtz equation. For example one has

$$G(\mathbf{r}, \mathbf{r}_0, \tau_m) = \frac{1}{4\pi D |\mathbf{r} - \mathbf{r}_0|} \exp\left(-\frac{|\mathbf{r} - \mathbf{r}_0|}{\sqrt{D\tau_m}} \right) \tag{8}$$

in three dimensions;

$$G(z, z_0, \tau_m) = \frac{1}{2} \sqrt{\frac{\tau_m}{D}} \exp\left(-\frac{|z - z_0|}{\sqrt{D\tau_m}} \right) \tag{9}$$

in one dimension.

The choice of the spatial gas distribution for the study of cosmic-ray propagation within a few hundred parsecs from the Sun is not an easy one. Observations show complicated gas structures with possible traces of supernova explosions during the last 10^6 years (Paresce, 1984; Bochkharev, 1987; Cox and Reynolds, 1987; Dickey and Lockman, 1990; Frisch, 1995). In particular, the Sun is located in a cloud (the Local Fluff) with diameter a few parsecs. The local fluff is inside a low-density cavity (the Local Bubble) which is separated from another hot bubble circumscribed by Loop I by a wall of hydrogen. The size of this system is about 300 pc.

A simplified model of gas distribution may be chosen as follows. There are three gas layers with exponential profiles: $n(z) = \Sigma n_a \exp(-|z|/h_a)$, $a = 1, 2, 3$. The component with parameters $n_1 = 0.45$ cm^{-3}, $h_1 = 130$ pc represents the smeared out contribution of small numerous neutral clouds, the component $n_2 = 0.21$ cm^{-3}, $h_2 = 200$ pc represents the more extended warm medium, and the component $n_3 = 0.025$, $h_3 = 1$ kpc represents the ionized hot gas. In addition we take into account the individual molecular clouds inside a circle of 1 kpc around the Sun as presented by Dame et al. (1987). There is a uniform molecular gas distribution with a surface density 1.3 $M_\odot$ pc^{-2} ($M_\odot$ is the solar mass) beyond 1 kpc.

The outlined gas distribution must be corrected for the above mentioned HI Hole in the vicinity of the Sun. It can be done by withdrawing all atomic hydrogen described as the components 1 and 2 from a cylindrical cavity with diameter 400 pc and full height 260 pc and replacing it with a radial dependent gas distribution built from the absorption map presented by Paresce (1984).

The description of cosmic-ray transport in the framework of the leaky box approximation is much simpler than in the diffusion model. One has the following equation for cosmic-ray number density of a decaying secondary isotope in the leaky box model (compare this equation with Equation (4)):

$$\frac{N_2}{T} + n_{lb} v \sigma_2 N_2 + \frac{N_2}{\tau} + \frac{\partial}{\partial E}\left(\left(\frac{dE}{dt}\right)_{ion} N_n\right) = n_{lb} v \sigma_{12} N_1 \,. \tag{10}$$

Here T is the escape time of cosmic rays, n_{lb} is the mean gas density. The escape length X introduced above is defined as $X = m n_{lb} v T$.

The measured abundance of radioactive isotopes in cosmic rays may be conveniently expressed through the surviving fraction s defined as follows:

$$s = N_2(\tau)/N_2(\tau = \infty) \,. \tag{11}$$

Here $N_2(\tau = \infty)$ is the density for the isotope considered to be stable. It is clear that the value of the surviving fraction is limited by the inequality $s < 1$.

As discussed earlier, the empirical leaky box model successfully reproduces the abundance of stable secondary isotopes. Thus the value of $N_2(\tau = \infty)$ may be calculated from Equation (10) at $\tau = \infty$. The flat halo diffusion model provides the same results with the relation between parameters of these models given by Equation (1).

4. Results of Calculations and Discussion

The calculations of the surviving fractions for ^{10}Be, ^{26}Al, ^{36}Cl, and ^{14}C as functions of the diffusion coefficient D are presented in the four panels of Figure 2 (Ptuskin and Soutoul, 1998). Together with the basic model for the gas distribution with the HI Hole described above (labeled 'WH' in Figure 2) we show the results of the calculation (labeled 'NH' in Figure 2) with uniform gas surface density without the HI Hole but with the three HI layers and the molecular clouds as described above. Also, the results of the calculations for two simplified models, one with a single exponential gas layer (labeled 'SL' in Figure 2) with the height $h = 100$ pc, and another with an infinitely thin gas layer (labeled 'δ' in Figure 2) are presented. In the last two cases the total gas surface density is the same as in the model 'NH'. The calculations are performed at the particle energy 0.4 GeV nucl^{-1} in the interstellar medium and assuming that $X = 15\beta$ g cm^{-2}. Shaded areas in Figure 2 indicate the present range of cosmic-ray observations when expressed in terms of surviving fractions. Corresponding references to experimental work were given in the Introduction.

The combined measurements from IMP 7–8 (Garcia-Muñoz et al., 1977; Simpson and Garcia-Muñoz, 1988), ISEE-3 (Wiedenbeck and Greiner, 1980), and Voyager 1–2 (Lukasiak et al., 1994a) give the value of $s(^{10}$Be$) = 0.21 \pm 0.04$ for the

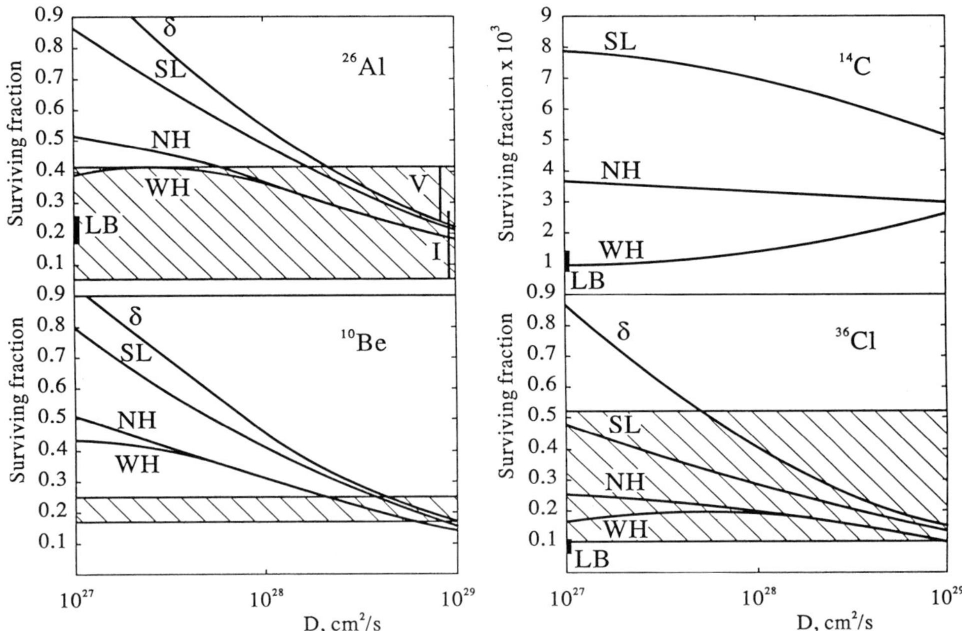

Figure 2. The surviving fractions of ^{10}Be, ^{26}Al, ^{36}Cl, ^{14}C in the bottom left panel, the top left panel, the bottom right panel and the top right panel respectively as functions of the diffusion coefficient for cosmic rays. Symbols of the curve: δ delta disk, **SL** single layer, **WH** with HI hole, **NH** without HI hole (see text for details). Hatched areas: one sigma limits of the surviving fractions from combined satellite observations, see references in the text. Thick vertical bars **LB** on the ordinate axis for ^{26}Al, ^{36}Cl and ^{14}C show the calculated values of surviving fractions in the leaky box model based on fitting the ^{10}Be observations. (These values do not depend on the diffusion coefficient.)

surviving fraction of ^{10}Be isotope. Figure 2 shows that the corresponding diffusion coefficient derived from the WH and NH curves is equal to $D = (3.4(+2.8, -1.4)) \times 10^{28}$ cm^2 s^{-1}. The use of models with very simple interstellar gas distributions such as the one-layer and δ-layer approximation may be too crude. Ignoring the local HI hole does not affect this relatively large value of D. Taking the HI hole into account is important for rapidly decaying ^{14}C as is seen from Figure 2. The ^{10}Be data imply the value $n_{lb} = (0.29(+0.07, -0.07))$ nuclei cm^{-3} when interpreted in the framework of the leaky box model.

The above value of the diffusion coefficient is in agreement with that accepted in the flat halo diffusion model with relatively large halo (Ginzburg et al., 1980; Berezinskii et al., 1990; Ptuskin and Soutoul, 1990; Ptuskin, 1996). It is also in conformity with the results of investigation on the diffusion of electrons with very high energies about 10^3 GeV when some increase of the diffusion coefficient with energy is taken into account (see Dorman et al. (1984) and Nishimura et al. (1995)). The very high energy electrons, observed at the Earth, may come only from local galactic sources (presumably, from the local supernovae remnants) since their lifetime relative to the synchrotron and inverse Compton energy losses is of the order 3×10^5 yr.

Using the preceding value of the diffusion coefficient and Equation (1), one can determine the value of $H = (4.9(+4, -2))$ kpc. This value is not in disagreement with the radioastronomical observations (Beuermann et al., 1985) which indicate the presence of a thick nonthermal galactic radio disk with the full equivalent width $(3.6(+0.4, -0.4))$ kpc. Our simple model implies that the density of relativistic stable nuclei is halved at somewhat larger height about 2.5 kpc above the galactic midplane. This difference seems natural since the nonthermal radio emissivity is determined by the density of relativistic electrons that decreases with height more rapidly than stable nuclei because of energy losses. Also, the magnetic field that appears in the expression for the emissivity is decreasing with height.

The characteristic time of cosmic-ray diffusion from the Galaxy is estimated in the one-dimensional approximation as $H^2/2D = (1.1(+1.2, -0.5)) \times 10^8$ yr while it is only $T = X/(vmn_{lb}) = (3.3(+1.1, -0.6)) \times 10^7$ yr in the leaky box model. We emphasize that the time T may be related to the particle escape from the internal zone $|z| \leq h_w$ which represents only a small part of the region of cosmic-ray propagation in the Galaxy if the internal region is bounded by the wave barrier as was discussed above in Section 2.

The available data on ^{26}Al and ^{36}Cl shown in Figure 2 are still not sufficiently accurate, and they are not in disagreement with the ^{10}Be data. The Voyager data on ^{26}Al, which has the highest statistics, agrees better with the diffusion model than with the leaky box model, although with still marginal statistical significance (Lukasiak et al., 1994b). A similar tendency is present in the ISSE-3 data on ^{36}Cl (Leske and Wiedenbeck, 1993; Ferrando, 1994).

There is as yet no data on the ^{14}C isotope in cosmic rays. The short decay time of this isotope and correspondingly very small expected surviving fraction demand a new generation of instruments with large collecting power such as on ACE (Stone et al., 1998). Figure 2 shows that because of this short lifetime ^{14}C is rather sensitive to the local interstellar gas distribution. The splitting between the 'NH' (no halo) and 'WH' (with halo) models is quite marked especially at small diffusion coefficients. So measurements of ^{14}C abundance are needed to clearly demonstrate the effect of the HI hole. Such a demonstration would not be just one more sophisticated confirmation of this well-established feature of the local gas distribution around the Sun. The point is that the radioactive isotopes carry information about galactic gas distribution i.e. on a time scale of a few thousand years for ^{14}C and of 5×10^5–5×10^6 years for the other ones. One can expect considerable changes of the structure of the local gas over these periods mainly because of star evolution and supernovae activity (see, e.g., Gehrels and Chen, 1993). Note that the value of density of the very local interstellar medium is determined with some uncertainty.

The anisotropic diffusion of strongly magnetized cosmic-ray particles probably makes the interpretation of cosmic-ray data more complicated than discussed so far. Particles diffuse mainly along magnetic field lines and, depending on unknown field geometry, may spend more time in dense or, on the contrary, in rarefied regions than calculated with the isotropic diffusion model.

Another critical point is our assumption of a constant distribution of primaries. Local supernovae remnants like Geminga and Loop I could have produced variations of cosmic rays in the past correlated with considerable distortions of the galactic gas distribution and the magnetic field structure. The supernovae outburst which 3×10^4–3×10^5 years ago gave rise to the gamma and X-ray pulsar Geminga was probably responsible for the variation of cosmic ray intensity by a factor of $\sim$2 (Raisbeck et al., 1987; Konstantinov et al., 1991; Dorman et al., 1983; Sonnet et al., 1987; Ramadurai, 1995), for the formation of a local gas bubble with extremely low density (Gerhels and Chen, 1993), with the orientation of the local magnetic field being along the bubble shell (Frisch, 1995). This local source could also explain the observed cosmic-ray anisotropy at 10^{12}–10^{14} eV and the observed intensity of very high-energy electrons at about 10^{12} eV (Dorman et al., 1984; Nishimura et al., 1995). The elaboration of appropriate nonlinear models may become desirable as new data on decaying isotopes are available.

At present we cannot unambiguously distinguish between the diffusion model and the leaky box model. These models when normalized to the same abundance of stable secondaries and to the same surviving fraction of ^{10}Be give different surviving fractions for other isotopes. In Figure 2 the predicted values of the surviving fractions for ^{26}Al, ^{36}Cl, ^{14}C in the leaky box model are indicated under the assumption that the diffusion and the leaky box models are fitted to the observed surviving fraction of ^{10}Be and the B/C ratio. The differences between models are rather small since the ratios of respective decay times to destruction times are not very different. The case of ^{36}Cl looks promising in connection with the *Ulysses* and ACE missions. There is a one-to-one correspondence between the values of the surviving fractions of the isotopes at each value of the diffusion coefficient. In Figure 3 the surviving fraction of ^{36}Cl is shown as a function of that of ^{10}Be. The full curves are for values of the diffusion coefficient between 10^{27} and 10^{29} cm^2 s^{-1}. The dashed part of the curve labeled WH is for values of the diffusion coefficient between 10^{26} and 10^{27} cm^2 s^{-1}.

The measurements of even a single radioactive isotope, for example ^{10}Be, in an extended energy range is very useful because the Lorenz factor changes the decay time of a relativistically moving isotope. For the proposed ISOMAX experiment at about 4 GeV nucl^{-1} (Streitmatter et al., 1993) the diffusion model (the leaky box model) predicts $s(^{10}$Be$) = 0.45$ (0.58) assuming that $s(^{10}$Be$) = 0.21$ at 0.4 GeV nucl^{-1}.

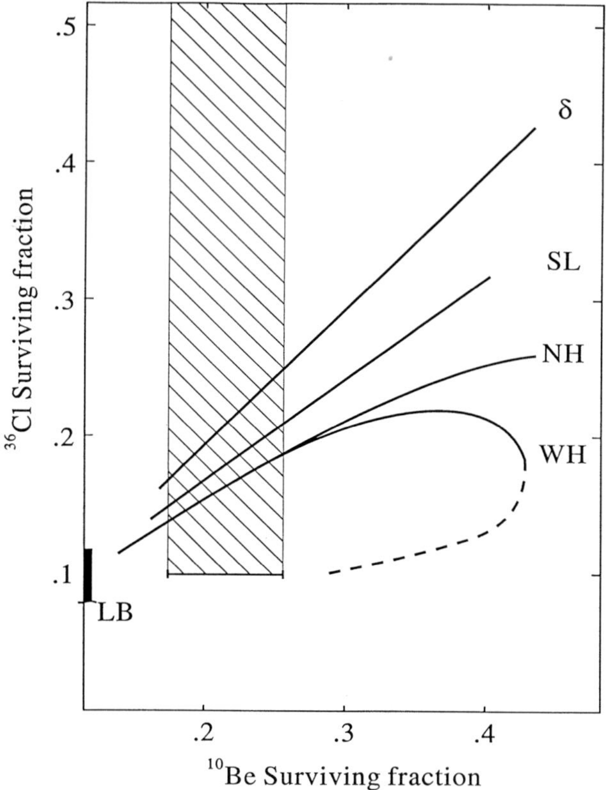

Figure 3. The surviving fraction of ^{36}Cl as a function of the surviving fraction of ^{10}Be. Symbols as in Figure 2. Full curves: values of the diffusion coefficient between 10^{27} and 10^{28} cm^2 s^{-1}. Dashed curves: values of the diffusion coefficient between 10^{26} and 10^{27} cm^2 s^{-1}.

5. Conclusion

The surviving fraction of rapidly decaying radioactive secondary isotopes in cosmic rays is determined by the value of the density of the interstellar gas in a sphere with a radius of the order of $(D\tau)^{1/2}$ from where the isotopes can reach us without significant decay. This result generalizes the conclusion made in the framework of the homogeneous leaky box model that the measurement of the content of radioactive secondary nuclei is actually an experiment on the determination of the gas density in the system once the abundance of stable secondaries is known. This is also in agreement with previous studies of diffusion models with simplified gas distribution when different constant gas densities in the disk and in the halo were assumed. Our present investigation takes into account complicated distribution of the local interstellar gas.

At a given gas distribution and under the assumption that diffusion determines cosmic-ray transport, the abundance of radioactive secondaries ^{10}Be, ^{26}Al, ^{36}Cl, and ^{14}C at low energies offers information on the cosmic-ray diffusion coefficient in the vicinity of a few hundred parsecs around the solar system. The surviving fraction $s = 0.21 \pm 0.04$ of ^{10}Be allows us to find the value of the cosmic-ray diffusion coefficient $D = (3.4(+2.8, -1.4)) \times 10^{28}$ cm^2 s^{-1} at energy 0.4 GeV nucl^{-1} in the interstellar medium. Under the additional assumptions that the diffusion coefficient is constant over the entire Galaxy and that cosmic rays escape freely at some distance H above the galactic plane, one can determine the value of $H = (4.9(+4, -2))$ kpc, which is the size of the cosmic-ray halo in this model (or the size of the 'boundary layer' in the diffusion-convection model). The corresponding diffusion exit time is $H^2/2D = (1.1(+1.2, -0.5)) \times 10^8$ yr. The leaky-box gas density which corresponds to the ^{10}Be abundance indicated above is equal to $n_{lb} = (0.29(+0.07, -0.07))=$ nucleon cm^{-3} and the leaky-box leakage time is equal to $X/(vmn_{lb}) = (3.3(+1.1, -0.6)) \times 10^7$ yr which is small compared with the prediction of the diffusion model but may refer not to the entire Galaxy but to some thin region $|z| \leq h_w$, $h_w \sim 0.5$ kpc adjacent to the galactic midplane and bounded by the wave barrier.

Some essential simplifying assumptions were made in the consideration of cosmic-ray diffusion: isotropic diffusion tensor and a constant primary cosmic-ray density within a few hundred parsecs around the Sun, together with steady-state conditions in the local interstellar medium on a time scale of 10^6 years.

The uncertainty about the choice between the models of cosmic-ray propagation in the Galaxy and the validity of some of the simplifying assumptions should be significantly clarified with the help of the CRIS experiment on cosmic-ray isotopes aboard the ACE spacecraft.

Acknowledgements

V.S.P. is grateful to the Organizers of the ACE Science Workshop at the California Institute of Technology for kind hospitality and sponsorship.

References

Berezinskii, V. S., Bulanov, S. V., Dogiel, V. A., Ginzburg, V. L., and Ptuskin V.S.: 1990, *Astrophysics of Cosmic Rays*, North Holland, Amsterdam.

Beuermann, K., Kanbach, G., and Berkhuisjen, E. M.: 1985, *Astron. Astrophys.* **153**, 17.

Bloemen, J. B. G. M., Dogiel, V. A., Dorman, V. L., and Ptuskin, V. S.: 1993, *Astron. Astrophys.* **267**, 372.

Bochkarev, N. G.: 1987, *Astropys. Space Sci.* **138**, 229.

Bradt, H. L. and Peters, B.: 1948, *Phys. Rev.* **74**, 1828.

Cox, D. P. and Reynolds, R. J.: 1987, *Ann. Rev. Astron. Astrophys.* **25**, 303.

Dame, T. M., Ungerechts, H., Cohen, R. S. et al.: 1987, *Astrophys. J.* **322**, 706.

Dickey, J. M. and Lockman, F. J.: 1990, *Ann. Rev. Astron. Astrophys.* **28**, 215.0

Dorman, L. I., Ghosh, A., and Ptuskin, V. S.: 1984, *Soviet Astron. Letters* **10**, 345.

Engelmann, J. J. et al.: 1990, *Astron. Astrophys.* **233**, 96.

Ferrando, P.: 1994, in D. A. Leahy et al. (eds), '*Proc. of the XXIII Int. Cosmic Ray Conf., Invited, Rapporteur and Highlight Papers*, World Scientific, Singapore, p. 279.

Freedman, I., Giler, M., Kearsey, S., and Osborne, J. L.: 1980, *Astron. Astrophys.* **82**, 110.

Frisch, P. C.: 1995, *Space. Sci. Rev.* **72**, 499.

Garcia-Muñoz, M., Mason, G. M., and Simpson, J. A.: 1977, *Astrophys. J.* **217**, 859.

Gehrels, N. and Chen, W.: 1993, *Nature* **361**, 706.

Ginzburg, V. L. and Ptuskin, V. S.: 1976, *Rev. Mod. Phys.* **48**, 161.

Ginzburg, V. L., Khazan, Ya. M., and Ptuskin, V. S.: 1980, *Astrophys. Space Sci.* **68**, 295.

Hayakawa, S., Ito, K., and Terashima, Y.: 1958, *Progress Theor. Phys. Suppl.* **6**, 1.

Jokipii, J. R.: 1976, *Astrophys. J.* **208**, 900.

Jones, F. C.: 1979, *Astrophys. J.* **229**, 747.

Leske, R. A.: 1993, *Astrophys. J.* **405**, 583.

Leske, R. A. and Weidenbeck, M. E.: 1993, *23rd Int. Cosmic Ray Conf., Calgary* **1**, 571.

Lukasiak, A., Ferrando, P., McDonald, F. B., and Webber, W. R.:1994a, *Astrophys. J.* **423**, 426.

Lukasiak, A., McDonald, F. B., and Webber, W. R.: 1994b, *Astrophys. J.* **430**, L69.

Lukasiak, A., McDonald, F. B., Webber, W. R., and Ferrando, P.: 1995, *24th Int. Cosmic Ray Conf., Rome* **2**, 576.

Konstantinov, G. N., Kocharov, G. E., and Levchenko, E. V.: 1991, *Soviet. Astron. Letters* **16**, 343.

Nishimura, J., Kobayashi, T., Komori, Y., and Tateyama, N.: 1995, *24th Int. Cosmic Ray Conf., Rome* **3**, 29.

Paresce, F.: 1984, *Astron. J.* **89**, 1022.

Prishchep, V. L. and Ptuskin, V. S.: 1975, *Astrophys. Space Sci.* **32**, 265.

Prishchep, V. L. and Ptuskin, V. S.: 1979, *16th Int. Cosmic Ray Conf., Kyoto* **2**, 137.

Ptuskin, V. S.: 1996, *Nuovo Cimento* **19C**, 755.

Ptuskin, V. S. and Soutoul, A.: 1990, *Astron. Astrophys* **237**, 445.

Ptuskin, V. S. and Soutoul, A.: 1991, *22nd Int. Cosmic Ray Conf., Adelaide* **2**, 197.

Ptuskin, V. S. and Soutoul, A.: 1998, *Astron. Astrophys.* **337**, 859.

Ptuskin, V. S., Volk, H. J., Zirakashvili, V. N., and Breitschwerdt, D.: 1997, *Astron. Astrophys.* **321**, 434.

Raisbeck, G. M. et al.: 1987, *Nature* **326**, 273.

Ramadurai, S.: 1995, *Adv. Space Res.* **15**, 41.

Simpson, J. A. and Garcia-Muñoz, M.: 1988, *Space Sci. Rev.* **46**, 205.

Soutoul, A., Ferrando, P., and Webber, W. R.: 1990, *22nd Int. Cosmic Ray Conf., Adelaide* **3** , 337.

Sonnet, C. P., Morfill, G. E., and Jokipii, J. R.: 1987, *Nature* **330**, 458.

Stone, E. C. et al.: 1989, *Particles Astrophysics*, AIP Conference Proceedings **203**, 48.0

Streitmatter, R. E. et al.: 1993, *23rd Int. Cosmic Ray Conf., Calgary* **2**, 623.

Webber, W. R., Lee, M. A., and Gupta, M.: 1992, *Astrophys. J.* **390**, 96.

Wiedenbeck, M. E.: 1983, *18th Int. Cosmic Ray Conf., Bangalore* **9**, 147.

Wiedenbeck, M. E. and Greiner, D. E.: 1980, *Astrophys. J.* **239**, L139.

WHAT ARE THE LIMITS FOR ACE GALACTIC COSMIC-RAY ISOTOPE STUDIES?

W. R. WEBBER

Astronomy Department 4500, NMSU, Las Cruces, NM 88003–8001, U.S.A.

Abstract. The CRIS experiment on ACE, with its excellent charge and mass resolution and a geometrical factor $\sim 10\times$ that of any previous experiment, holds the promise of rewriting the book on galactic cosmic-ray abundance studies. Translating these measurements into precise cosmic-ray source abundances and using these measurements to determine more accurately the propagation history of cosmic rays is a different matter, however. In many important cases these studies will be limited by the accuracy of the nuclear cross-sections that determine how the cosmic-ray composition is modified as it traverses the interstellar matter. In this paper we will discuss these cross-sections and how well they are known as a function of the energy and the charge and mass of the cosmic-ray nuclei. This will then be used to discuss what new limits can be expected on several contemporary problems of interest in cosmic rays from the CRIS measurements.

1. Introduction

The CRIS experiment on ACE has the potential to make giant strides in the study of galactic cosmic rays. The combination of high statistical accuracy and excellent charge and mass resolution will provide a new and precise insight into the source composition of cosmic rays and the related nucleosynthesis as well as specific details related to the propagation of cosmic rays through interstellar (IS) matter. This study, the detailed composition of cosmic rays at their source, has been a foremost part of cosmic-ray research since their discovery. But the CRIS experiment still has significant limitations in what it can hope to achieve in this area. These limitations, which mainly have to do with cross-section uncertainties, are discussed in this article. The modern era of cosmic-ray charge and isotope studies really began in the 1970s when both balloon borne instruments and those on the IMP, HEAO, and ISEE-3 spacecraft provided measurements of sufficient charge resolution, mass resolution and statistics to identify individual charges and separate isotopes of the same charge and to begin the study of cosmic-ray composition at their sources. These studies seemed to reveal a pattern of compositional differences in the cosmic-ray sources (CRS) that was significantly different than that found in matter in the solar system. This data also stimulated a renewed effort to measure cross-sections for it was soon realized that many of these compositional differences could simply be due to uncertainties in the cross-sections used in the interstellar propagation calculations. This cross-section effort provided an important step forward in our understanding of the effects of interstellar propagation, which

Space Science Reviews **86**: 239–256, 1998.

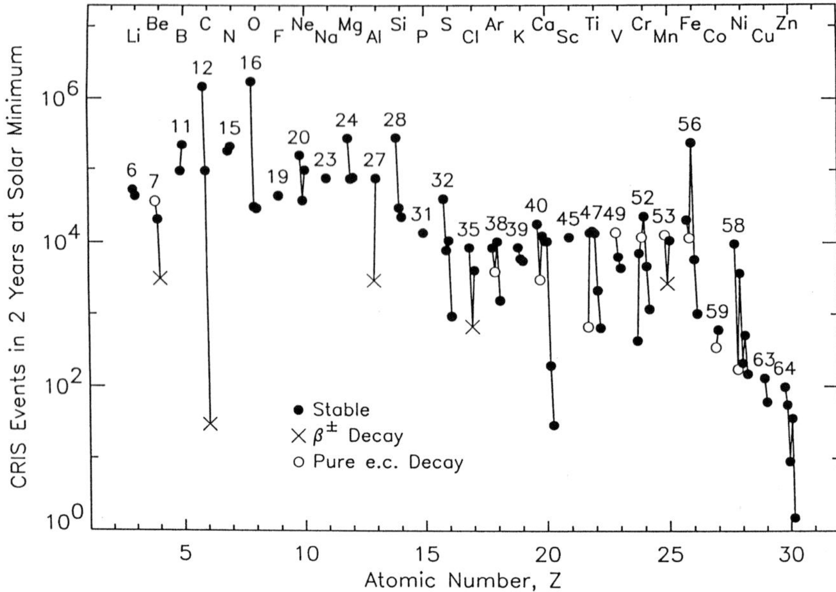

Figure 1. Number of events of each isotope expected for 2 years of operation for the ACE isotope telescope under sunspot minimum modulation conditions (Stone et al., 1998).

along with greatly improved charge and isotope data from the Voyager and *Ulysses* spacecraft in recent years have led to a new picture of the differences between the solar system and cosmic-ray source composition thus setting the stage for the CRIS experiment.

Basically these compositional differences can be considered in two classes. The first is related to charge compositional differences that seem to be a function of the first ionization potential of the elements involved. In this case elements with a first ionization potential $\geq 10\,\mathrm{eV}$ seem to be underabundant in the CRS by a factor ~ 4 in much the same manner that they are underabundant in the hot solar corona relative to the photosphere. Volatility effects related to matter in interstellar dust grains could also play a role in these compositional differences (see discussion by Meyer et al., 1997). The second class of differences, superimposed on the first, seems to be related to possible nucleosynthesis effects in the cosmic-ray sources themselves. Specifically the isotopes ^{4}He, ^{13}C, ^{14}N and possibly ^{32}S appear to be underabundant in the CRS relative to the solar system whereas the isotopes ^{12}C and ^{22}Ne are overabundant. These abundance differences suggest a Wolf–Rayet contribution to the CRS through the helium burning reaction ^{14}N$(\alpha, \gamma)^{18}$F$(\beta^{+}, \nu)^{18}$O$(\alpha, \gamma)^{22}$Ne which effectively depletes ^{4}He and ^{14}N and enhances ^{22}Ne. This is known to be an important phase of nucleosynthesis in these giant pre-supernova stars. Just how much of this material finds its way into the nearby IS medium and how it is mixed with other possible source components is a key question regarding the origin of cosmic rays and one in which the ACE data will play an important role.

2. The ACE Data and What Is Involved In Its Interpretation

Figure 1, taken from the CRIS instrument paper (Stone et al., 1998), shows the number of events of each isotope to be expected after 2 years of operation under sunspot minimum modulation conditions. These statistics are very conservatively an order of magnitude greater than the sum of all previous results including those from the Voyager and *Ulysses* spacecraft which provide the backbone of our current understanding of cosmic-ray charge and isotopic composition. Is it really possible to fully utilize this giant advance in statistics? The initial reaction might well be – how is this possible given the well-known uncertainties in galactic propagation models and the effects of solar modulation – not to mention the uncertainties in the interstellar fragmentation cross-sections themselves? Fortunately there is an approach that circumvents some of the uncertainties noted above. This is the secondary tracer approach to the determination of the galactic CRS isotopic abundances and is due to Stone and Wiedenbeck (1979). This formalism derives the CRS isotopic abundances from the observed abundances near the Earth using a purely secondary nuclide as a tracer of the amount of secondary production during propagation and the effects of solar modulation. In essence the propagation and modulation effects can be normalized out, so that source abundances can be derived that are reasonably independent of detailed propagation models or solar modulation theories. The resultant errors in these source abundances are then due mainly to the statistical errors in the observed abundances and the uncertainties in the secondary production cross-sections themselves. The CRIS measurements will reduce these statistical uncertainties to a level where the cross-section uncertainties will dominate for most isotopes. So the determination of a particular source isotopic abundance essentially depends on the fraction of that particular isotope which results from secondary production in the Galaxy. The larger this fraction is, the more difficult this calculation is and the more dependent it is on the cross-section uncertainties. For example, if only 3% of a particular isotope observed at Earth is due to the source (the other 97% being produced during propagation) and the cross-section accuracy itself is $\pm 3\%$, which is the best that can be hoped for from today's measurements, then the source abundance of this isotope can only be determined to within $\sim \pm 100\%$.

2.1. STABLE (OR UNSTABLE) SOURCE ISOTOPES

In Table I we present a list of cosmic-ray isotopes classified into categories that depend on the fraction of the isotope observed at the Earth that is from the CRS. This calculation is based on a Leaky-Box propagation model using the latest available cross-sections as of February 1997. The largest group of isotopes – classified as EASIEST – have at least 40% of their composition at the Earth from the CRS. Most of these isotopes now have well-determined source abundances (within $\sim \pm 10\%$) thanks to the present level of accuracy of the cross-section measurements and the

TABLE I

Source fraction of various isotopes at the Earth ($E = 260$ MeV nucl^{-1}, $\phi = 700$ MV)

Easiest			Harder			Hardest			Impossible		
	% source	CTS		% source	CTS		% source	CTS		% source	CTS
^{14}N	37	4500	^{18}O	12	600	^{13}C	~3 (12)	1800	^{11}B	~0	6000
^{22}Ne	51	1800	^{23}Na	18	1800	^{15}N	~2	4500	^{17}O	~1.5	600
^{25}Mg	52	1600	^{31}P	20	300	^{33}S	~3	250	^{19}F	<1	600
^{26}Mg	60	1800	^{34}S	13	300	^{35}Cl	~3	220	^{21}Ne	~1	600
^{27}Al	39	1800	^{36}Ar	30	180	^{44}Ca	~3	360	^{41}K	~1	200
^{29}Si	48	600	^{38}Ar	12	250	^{48}Ti	~3	400			
^{30}Si	41	500	^{39}K	10	250						
^{32}S	76	1000	^{52}Cr	18	550						
^{40}Ca	>90	600	^{55}Mn	29	400						
^{54}Fe	69	900	^{57}Fe	>90	450						
^{58}Ni	>90	360	^{58}Fe	75	75						
^{60}Ni	>90	150	^{62}Ni	>90	18						
			^{64}Ni	>90	8						

CTS = total events estimated to have been observed by both Voyager and *Ulysses* through the end of 1996.

statistics of the Voyager and *Ulysses* measurements (e.g., Webber et al., 1997; Connell and Simpson, 1997). The CRIS data will make some improvements in the source abundances of these isotopes but the major impact of the CRIS measurements will come in the determination of the source abundances of the isotopes labelled HARDER and HARDEST in columns 2 and 3 of Table I. Those isotopes labelled HARDER include isotopes whose source fraction of arriving nuclei lies in the range 10–30% and isotopes such as ^{57}Fe that have a large source fraction but are difficult to resolve including isotopes whose statistics at the present time are limited. The CRIS data will make an immediate impact on our understanding of the source abundances of these latter isotopes. For the other isotopes in this category the CRIS statistics will be so good that the source abundance accuracies will be almost completely determined by the cross-section uncertainties themselves. For these isotopes it is important to know what fraction of a particular isotope is produced by what heavier primary and how accurately this cross-section is known. In Table II we show examples of this production for the isotopes ^{18}O, ^{23}Na, ^{31}P, ^{39}K, and ^{52}Cr in hydrogen targets at $\sim$600 MeV nucl^{-1} from available measurements (e.g., Webber et al., 1990). In each case one element is usually a dominant producer of the secondary isotope in question and the combined uncertainty can be estimated by assuming that the errors are random. If the cross-section from this dominant element to the secondary isotope can be measured more accurately then the source abundance determination can be correspondingly improved. Where one dominant isotope is not present, e.g., ^{18}O, the problem is more difficult.

For those isotopes in the HARDEST category the source fraction is the order of the cross-section uncertainties themselves, that is 2–5%, so the limits on the source abundance are very uncertain. We show examples in Table III of the production of two isotopes in this category, ^{13}C and ^{15}N, from heavier elements. For these isotopes it seems that the best that can be hoped for, given the present accuracy of the cross-sections, is that the source abundance can be determined to a factor of $\sim$2. ^{13}C is an interesting example. If this cosmic-ray isotope had the normal ^{13}C/^{12}C solar system ratio of 0.011 then the arriving source fraction of ^{13}C would be $\sim$12% of the total ^{13}C. Already the most recent Voyager measurements (Webber et al., 1996) have shown that the ^{13}C/^{12}C ratio in the CRS appears to be underabundant by a factor $\sim$4. The improved CRIS statistics on this isotope and on ^{15}N may not lead to a better source abundance estimate, however, because of the limitations of the cross-sections.

The isotopes/charges listed in the IMPOSSIBLE category are really the tracer isotopes discussed by Stone and Wiedenbeck (1979). The statistical accuracy of all of these isotopes from the ACE experiment will be such that their abundance can be compared with the propagation predictions to $\sim$3% or less, the estimated best possible uncertainty in the cross-sections themselves. It may be that, if the predictions and measurements of these isotopes/charges consistently agree to less than 3%, one may gain further confidence in the cross-sections and their limitations.

TABLE II

Decayed cross-sections at 600 MeV nucl^{-1} (hydrogen target)

Transition	σ (mb)	$\Delta\sigma$ (%)	Fractional contribution
^{18}O			
^{19}F $\rightarrow$	50.5	8	6.8
^{20}Ne $\rightarrow$	25.4	3	20.8
^{21}Ne $\rightarrow$	20.5	8	2.8
^{22}Ne $\rightarrow$	19.6	5	8.8
^{23}Ne $\rightarrow$	16.0	6	3.8
^{24}Mg $\rightarrow$	16.3	4	22.0
^{25}Mg $\rightarrow$	20.2	5	5.4
^{26}Mg $\rightarrow$	19.0	5	5.3
^{27}Al $\rightarrow$	15.5	5	3.8
^{28}Si $\rightarrow$	12.0	4	15.6
^{23}Na			
^{24}Mg $\rightarrow$	72.0	4	57.0
^{25}Mg $\rightarrow$	51.5	5	8.4
^{26}Mg $\rightarrow$	36.0	6	5.9
^{27}Al $\rightarrow$	32.0	5	5.3
^{28}Si $\rightarrow$	24.1	5	16.9
^{32}S $\rightarrow$	22.6	5	2.5
^{31}P			
^{32}S $\rightarrow$	70.2	3	42.8
^{33}S $\rightarrow$	59.7	6	6.4
^{34}S $\rightarrow$	39.3	10	6.0
^{35}Cl $\rightarrow$	39.5	6	3.6
^{38}Ar $\rightarrow$	36.8	5	4.5
^{39}K $\rightarrow$	26.2	10	3.1
^{40}Ca $\rightarrow$	27.0	3	6.7
^{56}Fe $\rightarrow$	4.9	7	14.6
^{39}K			
^{40}Ca $\rightarrow$	46.6	3	12.1
$^{42-44}$Ca $\rightarrow$ (Avg)	40.0	10	14.9
^{45}Sc $\rightarrow$	32.5	10	4.3
$^{46-48}$Ti $\rightarrow$ (Avg)	28.0	10	12.9
^{49}V $\rightarrow$	25.0	10	3.8
$^{51-52}$Cr $\rightarrow$	19.5	10	6.4
$^{53-55}$Mn $\rightarrow$ (Avg)	16.5	10	4.5
^{54}Fe $\rightarrow$	12.0	10	3.6
^{56}Fe $\rightarrow$	10.5	4	33.6
^{52}Cr			
^{53}Mn $\rightarrow$	90	10	3.8
^{55}Mn $\rightarrow$	49	10	2.1
^{54}Fe $\rightarrow$	54	10	9.2
^{56}Fe $\rightarrow$	57	3	80.3

TABLE III

Transition	σ (mb)	$\Delta\sigma$ (%)	Fractional contribution
^{13}C			
^{14}N $\rightarrow$	16.0	3	4.3
^{15}N $\rightarrow$	41.0	3	8.5
^{16}O $\rightarrow$	22.8	3	54.3
^{20}Ne $\rightarrow$	19.9	4	4.2
^{22}Ne $\rightarrow$	22.0	4	2.2
^{24}Mg $\rightarrow$	12.2	4	3.8
^{26}Mg $\rightarrow$	25.0	5	1.9
^{28}Si $\rightarrow$	9.0	4	2.6
^{15}N			
^{16}O $\rightarrow$	66.5	3	73.7
^{20}Ne $\rightarrow$	40.4	4	4.0
^{22}Ne $\rightarrow$	36.5	3	2.0
^{24}Mg $\rightarrow$	22.4	4	3.3
^{26}Mg $\rightarrow$	30.5	5	1.0
^{28}Si $\rightarrow$	16.2	4	2.2

Along with the 36 or so isotopes in Table I there is also the case of ^{59}Ni. This isotope will generally decay by K-capture with a half-life $\sim 7 \times 10^4$ yr. However, after acceleration with no K electrons present it is essentially stable. This fact has been suggested by Casse and Soutoul (1978), to provide a chronometer between the time of production of ^{59}Ni in Fe-peak nucleosynthesis and its acceleration later after the SN explosion, perhaps in the material of the SNR. If there is no decay of ^{59}Ni (implying rapid acceleration) then $\sim 75\%$ of all ^{59}Ni observed in cosmic rays should come from the source (the remaining 25% being secondary production from ^{60}Ni during propagation). If, however ^{59}Ni has completely decayed to ^{59}Co (implying a long time between production and acceleration) then $\sim 80\%$ of the ^{59}Co observed in cosmic rays should be from the source (the remaining 20% being secondary production from ^{60}Ni during propagation). Already the Voyager data shows too much ^{59}Co to be accounted for by secondary production alone indicating that most of the ^{59}Ni has indeed decayed (Lukasiak et al., 1997). This is confirmed by *Ulysses* data (Connell and Simpson, 1997) which show only a small ^{59}Ni presence, just about that to be expected from ^{60}Ni production only. These conclusions are based on ~ 50 total ^{59}Co and ^{59}Ni counts. With ~ 1000 counts expected from CRIS,

TABLE IV

Radioactive isotopes (100–250 MeV nucl^{-1})

	τ (10^6 yr)	Measured f	Implied n	Calc f for $n = 0.3$	ℓ (pc)	CTS		CTS
^{10}Be	1.6	0.21 ± 0.04	0.29 ± 0.08	0.225	300	$\sim$50	$\rightarrow ^{10}$B	1000
^{14}C	6×10^{-3}	–	–	0.006	20	$\sim$2	$\rightarrow ^{14}$N	4500
^{26}Al	0.9	0.23 ± 0.07	0.37 ± 0.12	0.170	230	150	$\rightarrow ^{26}$Mg	1800
^{36}Cl	0.3	0.13 ± 0.13	0.48 ± 0.48	0.070	130	$\sim$15	$\rightarrow ^{36}$Ar	200
^{54}Mn	1.2	0.32 ± 0.06	–	0.420	260	$\sim$100	$\rightarrow ^{54}$Fe	900

$\tau = \beta$ decay half life.

f = surviving fraction as determined in a Leaky Box propagation model.

n = density in H atoms cm^{-3} as determined in a Leaky Box propagation model.

R = characteristic distance travelled in β decay half life.

CTS = see Table I.

this decay can be examined to a much higher level of precision, perhaps leading to a finite (small) number for the fraction of ^{59}Ni surviving.

2.2. SECONDARY RADIOACTIVE DECAY ISOTOPES

There are 5 radioactive secondary isotopes with half lives between 6×10^3 and 2×10^6 yr that provide unique information on the propagation of cosmic rays. These isotopes and the current status, as of May 1997, of the measurements are listed in Table IV. The present data, which effectively measure the fraction of the radioactive isotopes that survive, f, is very clearly statistics limited, therefore the CRIS measurements will provide a major improvement. We comment briefly on the current situation and potential CRIS improvements below.

^{10}*Be.* Approximately 20% of this isotope survives – based on a sum of $\sim$50 ^{10}Be counts recorded by the IMP-7 and -8, ISEE-3 and Voyager experiments (Lukasiak et al., 1994a). Given the total grammage traversed of $\sim$6.5 g cm^{-2} as determined by the measured B/C ratio at this energy, this surviving fraction translates into an average matter density of 0.29 ± 0.08 H atoms cm^{-3} for a Leaky-Box propagation model as indicated in Table IV. During the ^{10}Be half life of 1.6×10^6 yr cosmic rays will thus diffuse a distance $\ell = (D\tau)^{1/2}$ which for a diffusion coefficient $= 2 \times 10^{28}$ cm^2 s^{-1}, gives $\ell = 10^{21}$ cm or $\sim$300 pc. So in a sense, the matter density derived from the ^{10}Be surviving fraction is an average over a region of this radius in the galaxy surrounding the Sun.

^{14}*C.* This is a particularly interesting isotope which to our knowledge has not yet been observed in earlier experiments including Voyager and *Ulysses*. Using relative cross-sections, the ^{14}C production (most of which comes from ^{15}N and ^{16}O) should be $\sim$0.15 of that of ^{13}C, which when taken with the estimated surviving fraction of 0.6% (see Table IV) suggests that for every 1500 ^{13}C nuclei one ^{14}C should be observed. So for the same IS density as implied by the ^{10}Be measurements $\sim$1–C^{14} count should now have been seen by Voyager and $\sim$2 counts by *Ulysses*. Of course ^{14}C is sampling over a smaller distance ℓ than ^{10}Be and, if the density is less in this more local region, the surviving fraction of ^{14}C will be correspondingly less. With perhaps 100 ^{14}C recorded by CRIS, this density should be quite well determined.

^{26}Al.   This isotope provides very similar information to ^{10}Be since its surviving fraction for a given IS density is similar and the characteristic radius ℓ, is also similar although smaller. However the current measurements (e.g., Lukasiak et al., 1994b) give a slightly higher value for the surviving fraction of this isotope as compared with ^{10}Be (at less than the 1σ level of significance, however) implying a higher density. If this higher density is real it could be evidence for propagation from the matter disk of thickness 100 pc into the halo as revealed by the different scales of ℓ for the two isotopes.

^{36}Cl.   This isotope is very interesting because its surviving fraction is much less than either ^{10}Be or ^{26}Al because of its short half life, a fact which leads to characteristic radius, ℓ, closer to the thickness of the matter disk itself. As a result one might expect to find a higher matter density from the ^{36}Cl measurements. Indeed the present data from Voyager hints at this but with very limited statistics (4 ^{36}Cl events, giving a surviving fraction of $\sim$0.12). Here also CRIS data should greatly improve the present situation.

^{54}Mn.   Data from both Voyager (Lukasiak et al., 1997) and *Ulysses* (DuVernois, 1997) suggest a surviving fraction $\sim$0.3 for this isotope. The half life of this isotope for β decay has not yet been measured in the laboratory but it can be estimated to be $\sim$1.2 $\times$ 10^6 yr from the measured surviving fraction above – assuming the same IS density of $\sim$0.3 cm^{-3} as obtained from the average of the ^{10}Be and ^{26}Al measurements. The CRIS data will greatly improve our knowledge of the surviving fraction of ^{54}Mn; but since this isotope appears to have a lifetime bracketed by ^{26}Al and ^{10}Be, it will probably provide no propagation information not already obtained by these two other β-decay isotopes.

2.3. SECONDARY K-CAPTURE DECAY ISOTOPES

There are several isotopes in the charge range from Be to Ni that decay only by capturing a K shell electron. Under normal conditions (K electrons present) these lifetimes range from $\sim$0.1 yr to $\sim$10^6 yr. The cosmic rays, once accelerated, are stripped of their K-electrons and as a result cannot decay. However, there is a probability that these nuclei can pick up a K-electron during their traversal of interstellar matter. This attachment probability is strongly energy dependent ($\sim E^{-2}$) and can become significant at energies less than a few hundred MeV nucl^{-1}. Once this attachment occurs, several of the K-capture isotopes with relatively short half-lives (less than a few years) can now decay. The effective lifetime for decay depends on the attachment probability and can be calculated to an accuracy $\pm20\%$. This lifetime is essentially independent of the density if it is >0.1 cm^{-3}. A list of these isotopes is given in Table V. The probability that one of these isotopes decays essentially depends on the probability of attachment. Calculations of this attachment have been made by several authors (e.g., Letaw et al., 1984). These results are shown in Table V where the attachment probability has been converted into an effective lifetime in column 1. This effective lifetime is shown at 3 interstellar energies, 350, 500, and 650 MeV nucl^{-1} for $n = 0.3$. The energy dependence as well as the charge dependence of the lifetime can be seen – e.g., at 500 MeV nucl^{-1} these lifetimes range from $\sim$7.5 to $\sim$75 $\times 10^6$ yr as the charge decreases from 27 to 18. This is longer than the β-decay lifetimes discussed earlier – but still significant in terms of the amount of decay that might occur as shown by the surviving fraction given in column 2 of Table V for an IS energy of 500 MeV nucl^{-1}. It is seen that this decay probability is strongly energy dependent and can result in surviving fractions that change from $\sim$0.75 at 350 MeV nucl^{-1} to 0.93 at 650 MeV nucl^{-1} for ^{51}Cr, for example. If we believe that the production cross-sections for these isotopes during IS propagation are known accurately enough, this energy dependence may be used to determine if the energy at the time of attachment is the same as that measured now; the enhanced K-capture decay at the possibly lower attachment energies being manifested in a reduced intensity of that isotope as compared to predictions made for the presently measured IS energy. This effect has been suggested as a possible measure of interstellar energy gain, a so-called reacceleration after an initial pulse of acceleration.

Previous attempts to measure possible K-capture decay have largely been inconclusive due to both uncertainties in the data and the cross-sections. But recently it has been claimed that this decay has been observed by studying the mass distribution of V isotopes using results from the ISEE-3 and Voyager spacecraft (Soutoul et al., 1998). In effect the isotope ^{49}V is found to be underabundant by $\sim$25% corresponding to the K-capture decay of ^{49}V and the isotope ^{51}V is found to be overabundant corresponding to the decay of $\sim$25% of the K-capture isotope ^{51}Cr. The combined mass histogram from ISEE-3 and Voyager for V is shown in Figure 2. This illustrates the current state of isotopic resolution of the heavier elements where the mass fractions must be deconvolved from the observed mass distribution (both *Ulysses* and CRIS should provide resolved mass histograms). The Voyager result, which is based on the weighted average mass fractions of ^{49}V and ^{51}V obtained from the individual mass histograms, suggests that, because the percent decay is more than that expected at the current interstellar energy, some reacceleration has indeed occurred, but because there is expected to be interplanetary energy loss of the same magnitude due to solar modulation, the possible energy gain is partially masked by the assumed interplanetary energy loss and its uncertainty.

TABLE V

K-capture isotopes

	τ (10^6 yr) 350–500–650 MeV nucl^{-1}	f 500 MeV nucl^{-1}	CTS		CTS
^{37}Ar	$40 - 75 - 90$	0.96	200	$\rightarrow ^{37}$Cl	35
^{44}Ti	$18 - 30 - 48$	0.93	20	$\rightarrow ^{44}$Ca	330
^{49}V	$12 - 22 - 36$	0.90	250	$\rightarrow ^{49}$Ti	200
^{51}Cr	$9 - 18 - 30$	0.88	400	$\rightarrow ^{51}$V	180
^{54}Mn	$7.5 - 13 - 24$	0.85	100	$\rightarrow ^{54}$Cr	80
^{55}Fe	$6 - 10 - 20$	0.82	300	$\rightarrow ^{55}$Mn	400
^{57}Co	$4.5 - 7.5 - 16$	0.78	30	$\rightarrow ^{57}$Fe	450

f = surviving fraction calculated in a Leaky Box model for the effective lifetime listed.
CTS = see Table I.

Making a very simple estimate for this interplanetary energy loss, the above authors conclude that the energy gain between attachment and observation implied by this data could be as much as $\sim$130 MeV nucl^{-1} or about 25% of the current interstellar energy of 500 MeV nucl^{-1}.

The CRIS experiment with its excellent mass resolution and greatly increased statistics is made to order for the study of this problem. The surviving fraction of all of the K-capture isotopes listed in Table V should be measured to an accuracy of a few percent thus enabling the pattern of decay to be determined for the various K-capture isotopes. These observations may allow the solar modulation energy loss to be separated from the IS energy gain. One limitation of the CRIS data may be the level of solar modulation at the start of observations. Under sunspot minimum conditions the minimum modulation corresponds to an average interplanetary energy loss $\sim$200–250 MeV nucl^{-1} so the CRIS observations, which will be mainly in the range 200–300 MeV nucl^{-1} at the Earth, will correspond to IS energies -400-550 MeV nucl^{-1}. At these energies the surviving fraction should be larger than that observed by Voyager where the effective IS energy was $\sim$400 MeV nucl^{-1} (because of the lower solar modulation). As a result the measurement of the decay might be more difficult. But this observation by CRIS has the potential for answering two of the most important questions in cosmic rays today: (1) How much IS reacceleration is occurring? and (2) Is the interplanetary energy loss predicted by theories of solar modulation actually occurring?

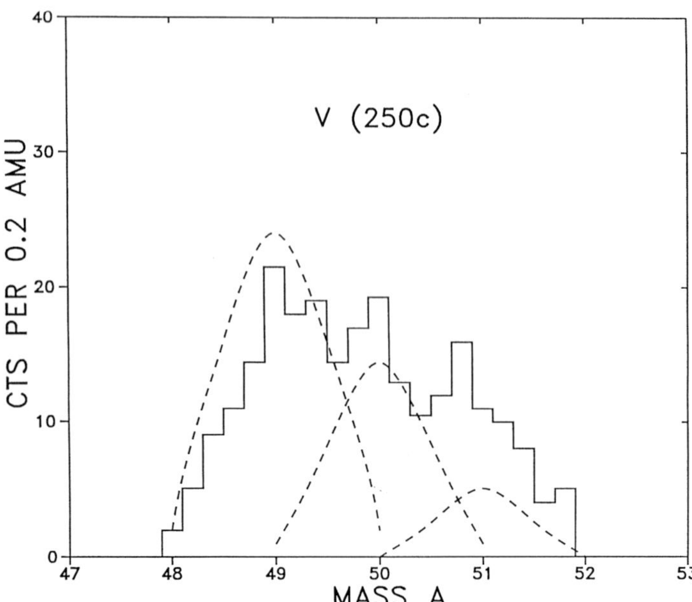

Figure 2. Combined data from ISEE-3 (Leske: 1993) and Voyager (Lukasiak et al., 1997) on the isotopic composition of V nuclei. (Both experiments have essentially the same mass resolution therefore the data can be simply added.) Predictions of the number of events from a Leaky Box propagation model assuming the same mass resolution shown as a dashed line. The excess of ^{51}V from ^{51}Cr decay and the deficiency of ^{49}V, $\sim$25% of which has decayed to ^{49}Ti, are clearly seen.

3. The Fragmentation Cross-Sections – Their Accuracy and Range of Coverage in Charge and Mass

It is clear from the previous discussion that, without fragmentation cross-sections of comparable accuracy, the interpretation of the high quality cosmic-ray data from spacecraft like Voyager or *Ulysses*, to say nothing of the CRIS experiment, would be very difficult at best. The measurements of the cross-sections themselves, carried out at various accelerators around the world, have generally been done in relative obscurity and with minimal financial support. From the point of view of the nuclear physicists, who are the main users of the accelerators, the measurement of cross-sections to an accuracy better than a factor $\sim$2 is unnecessary. It is easy to see why they have this point of view. Nuclear physics programs used to calculate cross-sections give predictions which themselves are apparently only accurate to within a factor of $\sim$2 (see, for example, Tull et al., 1998) rather than the few percent necessary for cosmic-ray propagation calculations. And cosmic-ray physicists (and funding agencies) who need and use the cross-section data seem only to realize its importance when they attempt to interpret their data.

In spite of these problems we now have a large data base of fragmentation cross-sections of quite high accuracy, limited more in energy range rather than charge and mass coverage. Measurements of cross-sections specifically for cosmic-ray propagation problems had their origin in France where several groups began measuring cross-sections (e.g., Raisbeck and Yiou, 1971; Fontes et al., 1971; Perron, 1976) following the pioneering work on cosmic-ray propagation by Meneguzzi et al. (1971), which demonstrated the need for nuclear cross-sections in order to extrapolate the charge and isotopic information obtained at the Earth back to the cosmic ray sources. Most of these cross-section measurements were made using radio-chemical techniques and were therefore restricted to radioactive secondaries, generally not the most abundant and useful products to describe the global propagation effects (not so of course for ^{10}Be and ^{26}Al). The measurements were made using proton beams up to $\sim$25 GeV. In total $\sim$100 individual isotopic cross-sections or more were measured during this period and those at energies above a few GeV still represent the best data base of high energy nuclear cross-sections available.

With the advent of more sophisticated balloon and spacecraft measurements the limitations of this cross-section data base for detailed propagation studies soon became evident. The author and his colleagues began a program in the early 1980s to measure cross-sections using beams of heavy nuclei available at the BEVALAC. This program culminated with the publication of 4 papers summarizing these results and presenting a new parametric formula for calculating cross-sections in 1990 (Webber et al., 1990a–d). Overall during a 7-yr period, over 40 individual beams ranging in charge from He to Ni and in energy from $\sim$300 to 1700 MeV nucl^{-1} were used in this study. The cross-sections for hydrogen targets were obtained using a CH$_2$–C target subtraction and, late in the program, results were obtained using He targets as well. Overall several hundred charge changing cross-sections were measured at different energies. For the most important nuclei such as ^{12}C, ^{16}O, ^{28}Si, and ^{56}Fe up to 6 energies from $\sim$300 to 1700 MeV nucl^{-1} were used. Over 500 isotopic cross-sections were measured at energies from 400 to 1100 MeV nucl^{-1} – these were presented as a series of $\sim$300 cross-sections from 13 beam charges at a single energy of 600 MeV nucl^{-1}. This truncation in energy could be done because of the weak energy dependence of the mass fractions of the individual isotopes that was observed. This (restricted energy) coverage is shown in Figure 3. This overall effort concentrated on beams of the most abundant cosmic-ray nuclei at the source and covered more than $\sim$90% of the arriving cosmic-ray $Z \geq 3$ nuclei by abundance. The experimental accuracy was estimated to be as good as $\pm$3% for the most abundant isotopes near the center of the individual mass distributions – this accuracy being dominated by systematic uncertainties.

Still more work needed to be done, however, and to extend this effort the Transport collaboration was formed. This international collaboration made measurements again at the BEVALAC in 1990 and 1991. The emphasis here was on neutron-rich isotope beams such as ^{22}Ne, ^{26}Mg, and ^{40}Ar to help in understanding the systematics of secondary fragmentation as well as to make measurements at several energies to study the energy dependence. The coverage of this program is shown

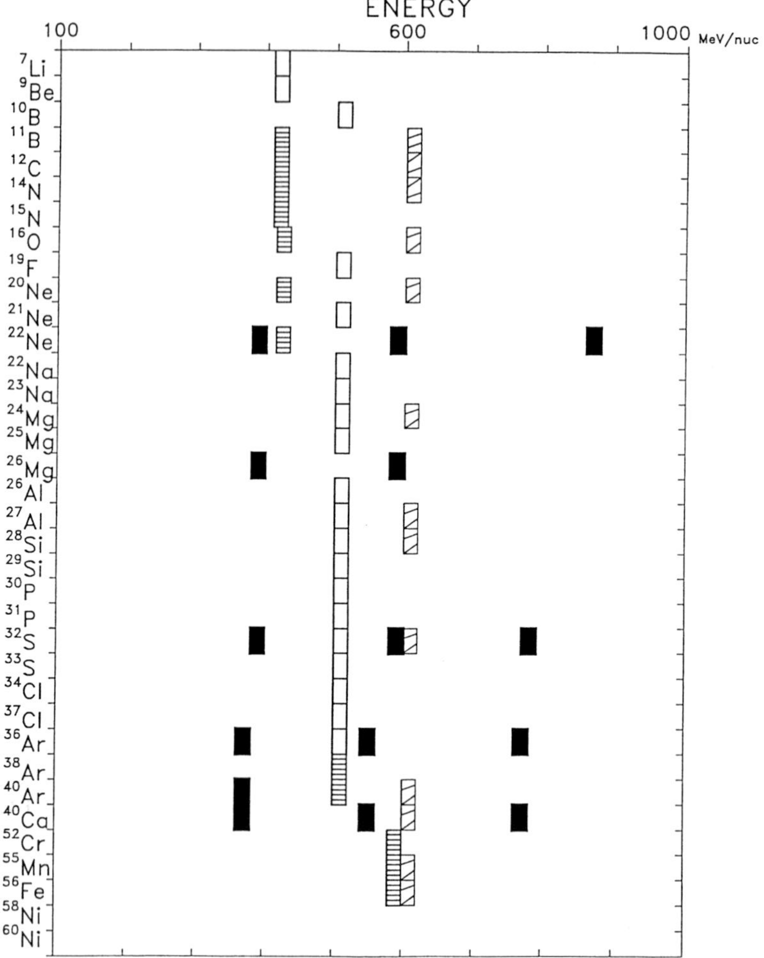

Figure 3. Beams and energies for which the isotopic cross-sections of secondary products have been measured in (1) The Webber, Kish and Schrier (WKS) measurements (diagonal); (2) the Transport measurements (solid); and (3) the SATURNE measurements (open and horizontal).

by the solid bars in Figure 3. These data are just now appearing in the published literature (Chen et al., 1994, 1997a, b; Knott et al., 1996, 1997). The isotopic cross-sections now available cover 5 charges at 3 separate energies. (The charge-changing cross-sections include 7 charges). Over $\sim$500 isotopic cross-sections were measured of which $\sim$150 are from 5 new beams at an energy $\sim$600 MeV nucl^{-1}. The experimental accuracy for the most abundant isotopes near the center of the individual mass distributions was estimated to be as good as $\pm$5%. Unfortunately this program was cut short with the closure of the BELAVAC in 1993.

Anticipating this closure the author and his French colleagues reassembled the telescope used for the WKS cross-section studies in the 1980s at the BELAVAC and made new runs at the SATURNE accelerator in France in 1993 and 1994 (and also in 1996 using ^{3}He and ^{4}He beams) before this accelerator also shut down in 1996. These runs were designed to fill in the gaps still remaining from previous runs – recognizing again the limited energy range available at SATURNE. A series of runs using ^{11}B, ^{12}C, ^{14}N, ^{15}N, ^{16}O, ^{20}Ne, and ^{22}Ne beams were made (at one energy - 450 MeV nucl^{-1}) to complete our understanding of low-Z fragmentation of all the important nuclei in the cosmic-ray beam – this time using a liquid hydrogen target to compare with earlier measurements using a CH_2–C subtraction. These measurements for isotopes again had a best accuracy estimated to be $\pm 3\%$. Similar runs were also made later with a ^{56}Fe beam at ~ 600 MeV nucl^{-1}. These results are being submitted to *Astrophys. J.* These runs are shown as the horizontal bars in Figure 3.

This program also included two long runs in which ^{40}Ar and ^{56}Fe beams were fragmented within the beam line using a 'thin' CH_2 'target' with the individual fragments from these interactions then being focused on the primary hydrogen target and the cosmic-ray telescope according to their A/Z ratio. For the ^{40}Ar runs two separate A/Z tunes were used – one with an A/Z ratio = 2.00 giving effectively simultaneous beams of ^{36}Ar, ^{32}S, ^{30}P, ^{28}Si, ^{26}Al, ^{24}Mg ... (all the way to ^{10}B) – 14 beams in all. The second A/Z tune was =2.06 and was designed for ^{27}Al but included ^{33}S, ^{31}P, ^{29}Si, ^{25}Mg ... down to ^{19}F - 8 beams in all. For the ^{56}Fe runs two separate A/Z tunes were also used – one with an A/Z ratio = 2.20 giving beams of ^{55}Mn, ^{52}Cr, ^{50}V, ^{48}Ti, ^{46}Sc and ^{40}Ca ... and an A/Z ratio = 2.078 giving beams of ^{54}Fe, ^{52}Mn, ^{50}Cr, ^{48}V, ^{46}Ti, ^{44}Sc, and ^{42}Ca .... This coverage at an energy ~ 500–600 MeV nucl^{-1} is shown by the open bars in Figure 3. The results from these runs have also been analyzed and are being submitted to *Phys. Rev. C.* The isotopic cross-sections from these runs are slightly less accurate than other runs – the accuracy of the isotopic cross-sections from the best of these multiple beam runs being ± 6–8%. This is due mainly to some contamination of the beams with adjacent isotopes and to limited statistics for some beams.

In total then the cross-section measurements made to date include almost all significant isotopes from A = 7 to 58 – comprising $\sim 99\%$ of the incoming cosmic-ray beam. The coverage in energy is less good – concentrated around ~ 600 MeV nucl^{-1} but including significant measurements in the range 300–1000 MeV nucl^{-1} and up to ~ 1700 MeV nucl^{-1}. The need for an accurate formula to describe these cross-sections for propagation studies is now greatly lessened – except for a description of the energy dependence which becomes more uncertain above ~ 15 GeV nucl^{-1}. This is perhaps not as important for the CRIS measurements at relatively low energies – but for higher-energy measurements including the escape length dependence of cosmic rays from the Galaxy, the energy dependence of the cross-sections becomes critical.

So how accurate are these cross-section measurements anyhow? Note that, in all of the measurements using heavy nuclei beams, the limitation appears to be related to systematic uncertainties rather than simple statistics. As a result it is difficult

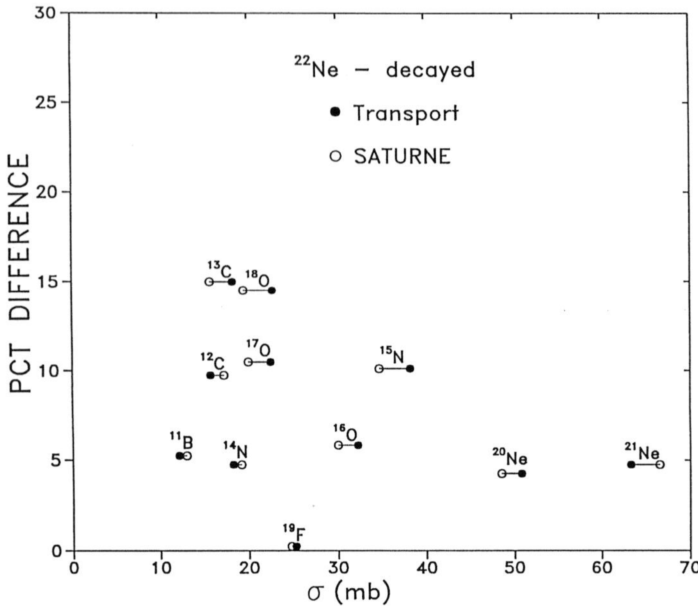

Figure 4. Percent difference in secondary isotopic cross-sections from ^{22}Ne beams interacting in H targets as measured by Transport and at SATURNE.

to evaluate the absolute errors. Perhaps the best way to evaluate the errors is to compare different sets of measurements. It is possible to compare the original WKS measurements with the more recent SATURNE measurements using a H target for the beams ^{12}C, ^{14}N, ^{16}O, ^{20}Ne, ^{24}Mg, ^{27}Al, ^{28}Si, ^{32}S, and ^{56}Fe, 9 beams providing $\sim$200 isotopic cross-sections. It is also possible to compare either the WKS or SATURNE measurements with the Transport measurements for the following beams, ^{22}Ne, ^{32}S, ^{36}Ar, and ^{40}Ca. These comparisons may be made in several ways. Figure 4 shows a comparison of the cross-sections measured at SATURNE and by Transport for ^{22}Ne fragmentation. The larger cross-sections show consistencies $\sim$5% or less, increasing to a maximum $\sim$15% for the smaller cross-sections. These differences are consistent with the quoted errors in the Transport measurements.

Figure 5 shows a comparison of the cross-sections into ^{13}C, ^{15}N, and ^{18}O isotopes from ^{14}N, ^{16}O, ^{20}Ne, and ^{24}Mg beams (WKS – SATURNE comparison) and ^{22}Ne beams (SATURNE – Transport comparison). These show a similar pattern of differences roughly consistent with the quoted errors on the individual measurements added in quadrature.

We have also compared all of the cross-sections measured at SATURNE and the WKS measurements for a total of $\sim$200 isotopic cross-sections from the 9 beams from ^{12}C to ^{56}Fe listed above. We simply take the sum of the ratio of the percent difference of the individual measurements over the percent error in the poorest measurement (which is the WKS measurements for the ^{12}C, ^{14}N, ^{16}O, ^{22}Ne, and ^{56}Fe beams and the SATURNE multiple beams for the ^{24}Mg, ^{27}Al, ^{28}Si, and

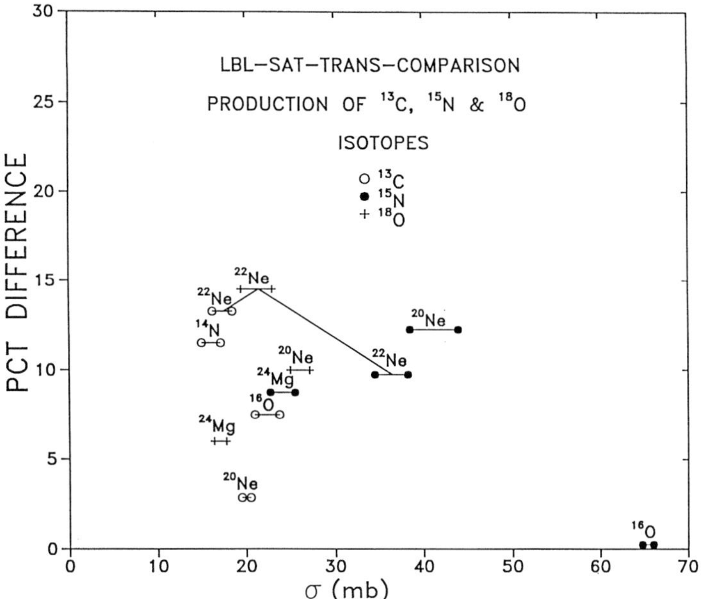

Figure 5. Percent difference in isotopic cross-sections into ^{13}C, ^{15}N and ^{18}O from ^{14}N, ^{16}O, ^{20}Ne and ^{24}Mg beams (WKS – SATURNE comparison) and ^{22}Ne beams (SATURNE – Transport comparison).

^{32}S beam comparisons). For the first 5 beams noted above the average differences of this ratio for 102 individual isotopic cross-sections is 1.49, for the second 4 beams this average is 1.48 for 85 individual isotopic cross-sections. If one set of measurements was precisely accurate, this ratio should be 1.0, representing the typical error of the poorest measurements. If both sets of measurements have the same error, this ratio should be 1.41. We also made this same comparison between the Transport measurements using ^{22}Ne, ^{32}S, ^{36}Ar, and ^{40}Ca beams and either the WKS or SATURNE cross-sections measured using the same beams. This ratio was 1.52 for 113 individual isotopic cross-sections.

So overall this comparison shows that the extensive set of recent cross-section measurements are consistent with one another within the quoted experimental errors, which are $\pm 3\%$ for isotopes with $\sigma > 10$ mb which are near the center of the mass distribution for the WKS and SATURNE measurements and $\pm 5\%$ for the most accurate Transport and SATURNE multiple beam measurements.

Acknowledgements

The author appreciates being invited to participate in the workshop on the capabilities of the ACE spacecraft. It has been a rewarding experience to examine the current situation regarding cross-sections and how they might affect the interpretation of the excellent data to be expected. It is a pleasure to see that the effort that has gone into cross-section measurements is about to be challenged by the interpretation of data of even greater precision.

References

Casse, M. and Soutoul, A.: 1978, *Astrophys. J.* **200**, L75.

Chen, C. X. et al.: 1994, *Phys. Rev.* **C49**, 3200.

Chen, C. X. et al.: 1997a, *Astrophys. J.* **479**, 504.

Chen, C. X. et al.: 1997, *Phys. Rev.* **C56**, 1536.

Connell, J. J. and Simpson, J. A.: 1997, *Astrophys. J.* **475**, 261.

DuVernois, M. A.: 1997, *Astrophys. J.* **481**, 241.

Fontes, P., Perron, C., Lestrinquez, J., Yiou, F., and Bernas, R.: 1971, *Nucl. Phys.* **165**, 405.

Knott, C. N. et al.: 1996, *Phys. Rev.* **C53**, 347.

Knott, C. N. et al.: 1997, *Phys. Rev.* **C56**, 398.

Leske, R. A.: 1993, *Astrophys. J.* **405**, 567.

Letaw, J. R., Silberberg, R., and Tsao, C. H.: 1984, *Astrophys. J. Suppl.* **56**, 369.

Lukasiak, A., Ferrando, P., McDonald, F. B., and Webber, W. R.: 1994a, *Astrophys. J.* **423**, 426.

Lukasiak, A., McDonald, F. B., and Webber, W. R.: 1994b, *Astrophys. J.* **430**, L69.

Lukasiak, A., McDonald, F. B., and Webber, W. R.: 1997, *Astrophys. J.* **488**, 454.

Lukasiak, A., McDonald, F. B., Webber, W. R., and Ferrando, P.: 1997, *Adv. Space Res.* **19,** 747.

Meneguzzi, M., Auduoze, J., and Reeves, H.: 1971, *Astron. Astrophys.* **15**, 337.

Meyer, J. P., Drury, L., and Ellison, D. C.: 1997, *Astrophys. J.* **487**, 182.

Perron, C.: 1976, *Phys. Rev.* **C14**, 1108.

Raisbeck, G. M. and Yiou, F.: 1971, *Phys. Rev.* **A4**, 1858.

Soutoul, A. et al.: 1998, *Astron. Astrophys.*, in press.

Stone, E. C. and Wiedenbeck, M. E.: 1979, *Astrophys. J.* **231**, 606.

Stone, E. C. et al.: 1998, *Space Sci. Rev.* **86**, 285.

Tull, C. E. et al.: 1998, *Phys. Rev. C*, submitted.

Webber, W. R., Kish, J. C., and Schrier, D. A.: 1990a–d, *Phys. Rev.* **C41**, 520, 533, 547, 566.

Webber, W. R., Lukasiak, A., McDonald, F. B., and Ferrando, P.: 1996, *Astrophys. J.* **457**, 435.

Webber, W. R., Lukasiak, A., and McDonald, F. B.: 1997, *Astrophys. J.* **476,** 766.

ACE SPACECRAFT

M. C. CHIU, U. I. VON-MEHLEM, C. E. WILLEY, T. M. BETENBAUGH,
J. J. MAYNARD, J. A. KREIN, R. F. CONDE, W. T. GRAY, J. W. HUNT, JR.,
L. E. MOSHER, M. G. McCULLOUGH, P. E. PANNETON, J. P. STAIGER and
E. H. RODBERG
The Johns Hopkins University Applied Physics Laboratory, Laurel, Maryland, U.S.A.

Abstract. The Johns Hopkins University Applied Physics Laboratory (JHU/APL) was responsible for the design and fabrication of the ACE spacecraft to accommodate the ACE Mission requirements and for the integration, test, and launch support for the entire ACE Observatory. The primary ACE Mission includes a significant number of science instruments – nine – whose diverse requirements had to be factored into the overall spacecraft bus design. Secondary missions for monitoring space weather and measuring launch vibration environments were also accommodated within the spacecraft design. Substantial coordination and cooperation were required between the spacecraft and instrument engineers, and all requirements were met. Overall, the spacecraft was kept as simple as possible in meeting requirements to achieve a highly reliable and low-cost design.

1. Introduction

The ACE spacecraft accommodates a total of ten instruments; nine scientific instruments for the primary mission and one engineering instrument for a secondary mission.

– Solar Energetic Particle Ionic Charge Analyzer (SEPICA) – University of New Hampshire, Max-Planck Institute for Extraterrestrial Physics.

– Ultra Low Energy Isotope Spectrometer (ULEIS) – University of Maryland, The Johns Hopkins University Applied Physics Laboratory (JHU/APL).

– Solar Wind Ion Mass Spectrometer (SWIMS) – University of Maryland, University of Bern (Switzerland).

– Solar Wind Ion Composition Spectrometer (SWICS) – University of Maryland, University of Bern (Switzerland).

– Solar Wind Electron, Proton and Alpha Monitor (SWEPAM)–Los Alamos National Laboratory (includes SWEPAME, the electron component, and SWEPAMI, the ion component).

– Electron, Proton and Alpha Monitor (EPAM) – JHU/APL.

– Magnetic Field Experiment (MAG) – University of Delaware Bartol Research Institute, National Aeronautics and Space Administration (NASA) Goddard Space Flight Center.

– Solar Isotope Spectrometer (SIS) – California Institute of Technology, Goddard Space Flight Center, Jet Propulsion Laboratory.

 *Space Science Reviews* **86**: 257–284, 1998.
© 1998 *Kluwer Academic Publishers. Printed in the Netherlands.*

– Cosmic Ray Isotope Spectrometer (CRIS) – California Institute of Technology, Goddard Space Flight Center, Jet Propulsion Laboratory, Washington University (St. Louis).

– Spacecraft Loads and Acoustics Measurements (SLAM) – Goddard Space Flight Center.

Three of these science instruments, SEPICA, SWICS, and SWIMS, share a data processing unit, designated S3DPU. Nine of the instruments (all but SLAM) support the primary ACE science mission. Four of them – MAG, EPAM, SWEPAM, and SIS – also provide data for the Real Time Solar Wind (RTSW) experiment to monitor space weather (e.g., solar flares and geomagnetic storms). The RTSW experiment is a secondary mission on ACE funded by the National Oceanographic and Atmospheric Administration (NOAA). The ACE spacecraft collects and packages the RTSW data from the four instruments for low-rate transmission during the times when the primary mission data are not being transmitted to the NASA Deep Space Network (DSN). It is anticipated that the RTSW data will be transmitted in this manner 21 hours per day. NOAA is responsible for operating the ground system for the collection and distribution of RTSW data for solar weather monitoring. The RTSW experiment on ACE was implemented at very low cost by using only the spacecraft resources needed for the primary mission. The SLAM instrument is another secondary mission incorporated during the ACE spacecraft development. SLAM is an experiment conducted by Goddard Space Flight Center Code 740 to measure the vibration environment within the spacecraft structure during the first 5 min of launch (before fairing jettison). The flight hardware is self-contained relative to other spacecraft resources; it has its own power supply, transmitter, processor, and sensors. Data from this experiment will go into a database to be used in predicting vibration loads for future spacecraft structural designs.

2. Spacecraft System Description

Figure 1 shows a photograph of the completed ACE Observatory during ground testing. The major technical parameters of the ACE spacecraft are provided in Table I.

2.1. MECHANICAL DESCRIPTION

The ACE structure design was driven primarily by the field-of-view requirements for the instruments, antennas, and attitude components and by the need for a spinning spacecraft to be mass balanced. Further complications were presented by the need for individual thermal radiators on the instruments and structure, which needed views to space, and by handling, access, clearance, and harness routing issues.

An orbital view of the spacecraft is shown in Figure 2, and an exploded view is shown in Figure 3. The primary structural components of ACE consist of ten

TABLE I

Technical parameters of the ACE spacecraft

Attitude	*Star tracker*
	– 20° × 20° field of view
	– +0.1 to +4.5 sensitivity range
	– 30 arc sec (1σ) total random error
	– 1 to 5 simultaneous stars
	Sun sensor
	– ±64° field of view
	– Sun angle is gray-coded in 0.5° increments
	– ±0.02° short-term repeatability of most significant bit
	Observatory orientation
	– Known (after the fact) to ±0.7°; stable to ±0.5°
	Pointing at Earth
	– Angle between the observatory Z axis and the Earth – observatory line is within ±3°, to stay within the required beamwidth of the high-gain antenna
	– Angle between the Sun–Earth line and the observatory – Earth line is ≥ 5°, to limit the solar noise contribution to the receiving system noise temperature
Maneuvering capability	*Tanks*
	– Four tanks with total of 195 kg of hydrazine fuel expelled in blow-down (97% efficiency), providing mission average specific impulse of 216–221 s at 10–21 °C
	Thrusters
	– Four axial and six radial thrusters
	Spin rate
	– Maintained at 5 ± 0.1 rpm to meet science requirements
	Maneuvers
	– All maneuvers except onboard autonomy performed under ground control
Communications	*Downlink data rate*
	– 434 bps (low rate and NOAA)
	– 6944 kbps (real-time transmission)
	– 76 384 kbps (recorder playback interleaved with real-time data)
	– *Uplink data rate*
	– 1000 bps
	RF frequencies
	– 2097.9806 MHz uplink
	– 2278.35 MHz downlink
Data storage	1.073 Gbit solid-state memory per recorder capacity beginning of life

TABLE I

Continued

Ground contact	DSN network 26 m (primary), 34 m (backup)
	Telemetry designed to be compatible with CCSDS
	Contact nominally once/d (requirement is for recorder capacity to support one missed contact)
Spacecraft safing (see safing section)	*Autonomy*
	C&DH autonomy:
	– during launch to switch to redundant shunt regulator in case of an analog shunt short
	– post launch vehicle separation to turn transmitter off in case of a problem
	– spacecraft and instrument health monitoring
	– abort thruster firing in case of maneuver or attitude problems
	– protection against commands which result in an improper spacecraft configuration
	– support for other autonomous spacecraft actions such as switching recorders when one is full
	Power Subsystem autonomy:
	– shunt regulator switched from primary to redundant for bus voltage over or under conditions
	Spacecraft power bus health monitor
	– load shed and switching to redundant shunt regulator, if required
	Watchdog timers
	– RF antenna switching
	– redundant thruster firing timers
	Reset to restore critical parameters
	– C&DH hardware reset and software boot and intitialization C&DH last-resort timer
	– power subsystem processor reset
Observatory power	280285 W nominal; 385 W peak (136 W payload, 249 W spacecraft) Solar Array
	support >425 W (observatory budget) at 28 V for 5 years
	Launch power 59 W (12-Ah 18-cell NiCd battery supplies 200 W-h to loads)
Observatory wet mass	771 756.54 kg (launch vehicle can support 785 kg)
Structure	Decks and Panels: Honeycomb with aluminum alloy facesheet
	Support structure: Longerons and rails aluminum alloy
	Corner Brackets: Machined titanium

TABLE I

Continued

Mechanisms	Four deployable solar panels, each 86.4×149.9 cm
	Two deployable magnetometer booms (magnetometer sensors on ends of boom)
Thermal control	Thermostatically controlled heaters, instrument-specific radiators, and
	observatory radiators are used
Orbit	Halo orbit about L_1
	$A_Y = 264\,071$ km
	$A_Z = 157\,406$ km

Note: C&DH = command and data handling, CCSDS = Consultative Committee on Space Data Systems, DSN = Deep Space Network

Figure 1. Photograph of the ACE spacecraft during ground testing (solar arrays manually deployed).

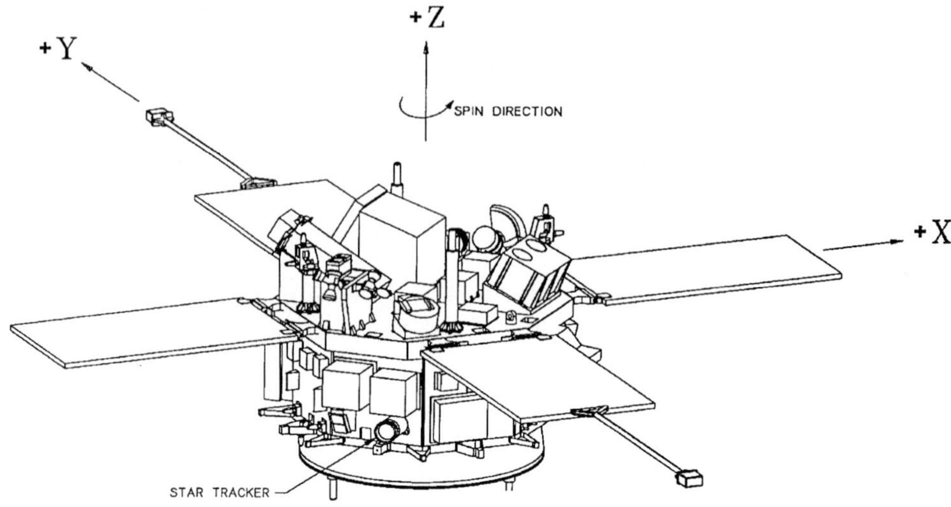

Figure 2. Line drawing of the ACE Observatory in on-orbit configuration.

aluminum honeycomb panels that form a 142.2×76.2 cm closed octagon supported by an internal aluminum and titanium frame. The Observatory Attach Fitting is a 22.9-cm-high aluminum cylinder that attaches the ACE primary structure to a 5624 Delta Payload Fitting with a clampband. The observatory mass, with fuel, is 756.5 kg. Most of the payload instruments (SEPICA, SIS, SWICS, SWEPAME, SWEPAMI, ULEIS, EPAM, and S3DPU) are mounted to the top ($+Z$) deck facing the sun. The $+Z$ deck is isolated from the rest of the structure via buttons (ultem material) and titanium brackets in order to meet instrument temperature requirements. The CRIS and SWIMS instruments and most spacecraft subsystems are mounted to the side decks. The lower ($-Z$) deck houses most of the RF subsystem and the SLAM instrument.

The propulsion subsystem consists primarily of four conospheric titanium tanks and ten thrusters. Thrusters were mounted and located to minimize plume heating effects on the solar arrays and nearby instruments. All propulsion components were integrated and welded together on the $-Z$ deck. This eliminated the need for field joints on the fuel lines and facilitated the final assembly of the system with the rest of the spacecraft primary structure. The tanks were filled with 195 kg of hydrazine fuel and mounted so the primary structural load path was almost directly into the Orbital Attach Fitting. Once the structure assembly was complete, access to the propulsion subsystem was limited. Fill and drain valves and electrical interconnects were all accessible from the outside of the $-X$ side panel.

Four 86.4×149.9 cm deployable aluminum honeycomb solar panels are hinged from the $\pm X$ and $\pm Y$ sides of the $+Z$ deck and are restrained to the $-Z$ deck with pin puller mechanisms during launch. A 152.4-cm titanium boom attaches a magnetometer to the $+Z$ end of each of the $\pm Y$ solar panels. For launch, the $+Z$ end of each boom is restrained to the $\pm Y$ solar panels with a pin puller mechanism.

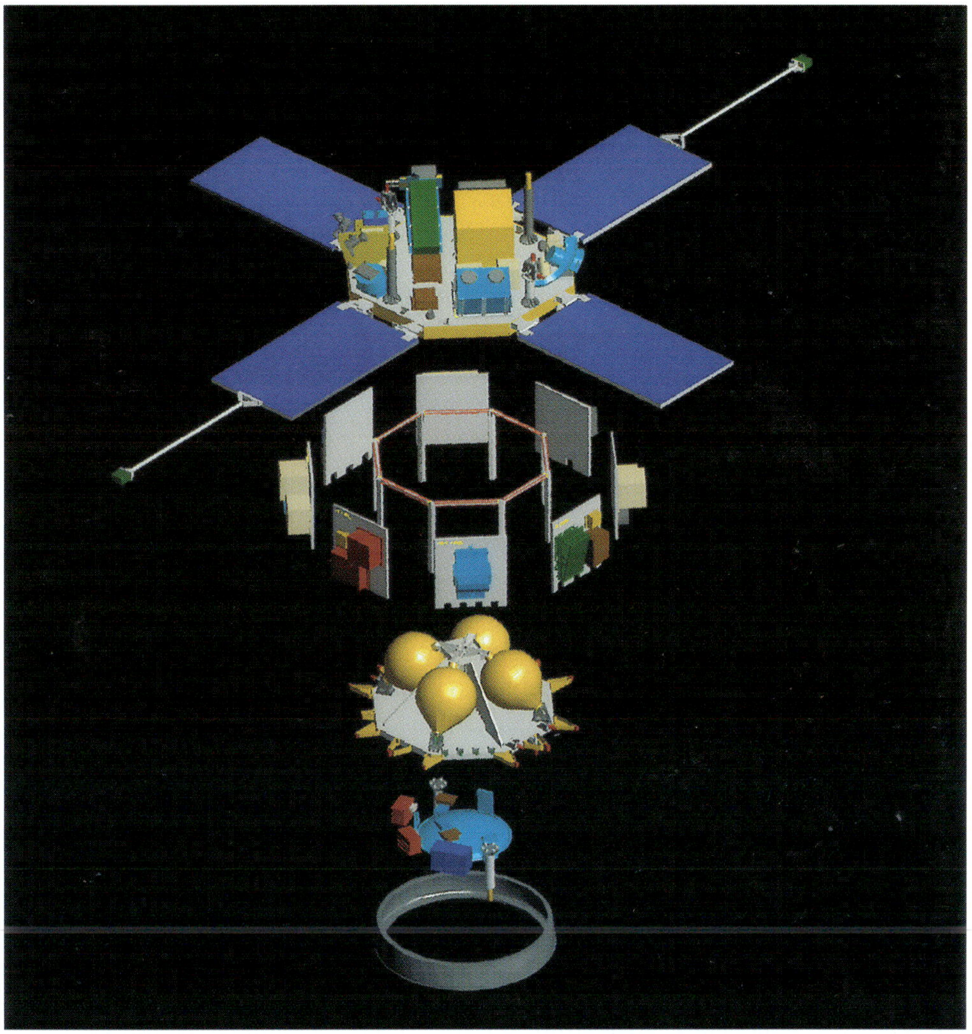

Figure 3. Exploded view of the ACE spacecraft.

Pyrotechnic devices actuate the pin pullers, which release the solar panels and booms. Preloaded torsion springs deploy and center the solar panels and booms in their appropriate positions.

To avoid dynamic coupling of the spacecraft with low-frequency Delta II launch vehicle modes, the primary structure had to be designed with fundamental frequencies above 35 Hz in the thrust (Z) direction and 12 Hz in the lateral (X, Y) directions when mounted to a Delta 5624 Payload Attach Fitting. The spacecraft primary structural modes were measured as about 100 Hz in the thrust direction and 40 Hz in the lateral directions.

The ACE Observatory was designed for launch on a Delta II 7920. The Delta II 7920 flight events produce loads from steady-state and dynamic environments. The steady-state environment produces a maximum thrust acceleration at the end of the first-stage burn or main engine cut-off. The dynamic environments produced by the Delta II are sinusoidal vibration, acoustics, and shock. The sinusoidal environment is a result of liftoff transients, oscillations before main engine cut-off, and engine ignition and shutdown. A spacecraft random vibration environment is generated by launch vehicle acoustics. The ACE Observatory was designed to withstand the loads from these environments. A protoflight test program (flight qualification levels at acceptance-level duration) was used to verify the observatory strength and stiffness. Vibration (sine and random), acoustic, and shock (clampband separation) testing was performed on the fully flight configured observatory.

2.2. THERMAL CONTROL DESCRIPTION

The thermal design of the ACE Observatory was created through the joint efforts of the spacecraft and instrument thermal engineers. Early in the program, it was decided that most of the instruments should be thermally isolated from the spacecraft. Uncertainty in the instrument schedules, together with stringent and differing instrument temperature requirements, made thermal isolation the best path to success. Instruments and instrument components whose temperature requirements could be managed by the spacecraft (SWEPAME, SWEPAMI, S3DPU, ULEIS data processing unit, ULEIS analog electronics) were allowed to thermally conduct their heat to the spacecraft.

The observatory uses a combination of multilayer insulation (MLI), thermal radiators, and thermostatically controlled heater circuits to meet its thermal design requirements, without the need for more active (and expensive) methods of moving heat such as louvers and heat pipes. Aluminum doubler plates (additional aluminum material) from 0.15 to 0.32 cm thick were added where necessary to enhance heat conduction away from an area. Where required, instruments were thermally isolated at the mounting interface using a combination of bushings made of insulating material (ultem), titanium mounting hardware, and MLI blankets.

Most of the MLI blankets on the ACE Observatory were fabricated using an outer layer of white Beta Cloth (a type of thermal insulating material) with an embedded 10.2×15.2 cm graphite weave. The Beta Cloth provides durability and the desired optical properties while minimizing specular reflections. Although a conductive spacecraft was not a requirement for the ACE program, the graphite weave was incorporated into the Beta Cloth to provide some protection from electrostatic discharge.

In some areas, special requirements dictated the use of alternative materials for the MLI outer layer. Thruster plume impingement is possible on the $+X$ side panel in the area of the battery and on the Observatory Attach Fitting below the $+X$ and $-X$ aft axial thrusters. The MLI blankets were hardened to increased temperature by replacing the Beta Cloth with five layers of embossed 0.25-mil aluminized Kapton and a 3-mil aluminized Kapton outer layer. Specular reflections are not a problem in these areas. In addition to the special requirements imposed by the spacecraft, the SWIMS and SWEPAME instruments had special requirements that dictated the use of more electrically conductive MLI outer layer materials. ITO-coated aluminized Kapton was used for SWIMS and for the $+X + Y$ side panel below SWEPAME. The alternative material was allowed only on the aperture face of SWEPAME to avoid problems with specular reflections.

The forward $(+Z)$ deck of the observatory is covered by thermal insulation to shield it from the sun as much as possible. Heat is rejected from the forward deck via radiators attached to each of the eight deck edges. The silvered Teflon radiators face radially away from the axis of symmetry of the observatory to maximize heat rejection.

The aft $(-Z)$ deck of the observatory serves as the mounting platform for the internal components of the propulsion system. In addition, components of the RF subsystem and the SLAM instrument are mounted to the space side of the aft deck. The Observatory Attach Fitting is bolted to the aft deck and is covered with silvered Teflon to serve as a thermal radiator for the side panels and the aft deck. Except for the RF-radiating surfaces of the two low-gain antenna towers and the high-gain antenna dish, the aft deck is protected from a direct view toward space by an MLI blanket.

The spacecraft side panels are used primarily for mounting spacecraft components. However, the $+X - Y$ and $-X + Y$ panels are used for the isolated mounting of the SWIMS and CRIS instruments, respectively. The spacecraft thermal design minimizes heat exchange between the side panels and space by covering the side panels with MLI blankets that extend from the bottom of the forward deck radiators to the Observatory Attach Fitting mounted to the aft deck. For side panels with instruments, the spacecraft blankets provide radiative isolation between the instrument and side panel. Sun sensor and star scanner apertures, thruster nozzles, and umbilical connectors remain free from MLI blockage.

The spacecraft radiators were oversized during design and fabrication. The effective radiator sizes were then tailored using MLI after correlation between the results of the observatory-level thermal vacuum test and the observatory thermal model.

On orbit, the ACE spacecraft thermal environment is expected to show little or no short-term variation. The environment will vary as the sun angle changes from a nominal minimum (4°) to a nominal maximum (20°). However, sun angle variation will occur on a scale of months and will be corrected periodically via the propulsion system. The actual maximum sun angle will be controlled to less than 20° to keep the high-gain antenna pointed at the DSN. The additional control provides greater margin in the spacecraft hot-case thermal analysis, where 20° was used as a worst-case assumption. The spin of the spacecraft (5 rpm) will dampen any short-term azimuthal variations.

The observatory thermal analysis includes worst-case variations in all thermal design parameters, including incident solar flux, sun angle, MLI effectiveness, optical properties of thermal radiators and other external surfaces, and minimum voltage on heaters. Therefore, the analysis results bracket all possible temperature variations over the entire mission life.

The spacecraft thermal control system is robust and versatile. A primary requirement of the thermal control system is to minimize peak heater power while providing adequate support of observatory temperatures. Survival, operational, and interface heaters manage the resources to meet the observatory's thermal requirements.

Survival heater circuits provide control when instruments are off and when most spacecraft components are either off or in low-power states. Operational heater circuits provide control during normal observatory power dissipation states. To minimize peak heater power, interface heaters are used to replace part of the dissipation of a component placed in a nonoperational state while the operational heaters are active. Interface heaters are used for all isolated instruments and for some higher-power spacecraft components. Without interface heaters, larger operational heater circuits would have been needed to accommodate the occasional lower-power modes. The result would have been higher peak operational heater power.

The extensive use of thermal components to distribute heater power about the observatory provides added reliability to the design. In general, heater circuits consist of multiple thermostats, each with multiple small heaters. The failure of an individual heater element is not likely to cause significant problems. Many of the survival thermostats have the same temperature set points as the operational thermostats due to the overriding requirement that hydrazine in the fuel lines not be allowed to freeze. Therefore, in many cases, the survival heaters can provide additional backup capabilities.

2.3. SPACECRAFT SUBSYSTEM DESCRIPTIONS

The spacecraft subsystems are described in the following paragraphs. A simplified block diagram is provided in Figure 4.

2.3.1. *Command and Data Handling Subsystem*

The ACE command and data handling (C&DH) subsystem provides capabilities for ground and stored commanding; onboard autonomy and safing; collection, formatting, and storage of science and engineering data; switching of spacecraft power and energizing of pyrotechnic devices; thruster firing control; and generation and distribution of sun pulses and spin clocks. The C&DH subsystem consists of redundant C&DH components, two 1-Gbit solid-state recorders (SSRs), a power switching component, and an ordnance fire component. Special design features are included to avoid any single-point mission failures and to allow continuous

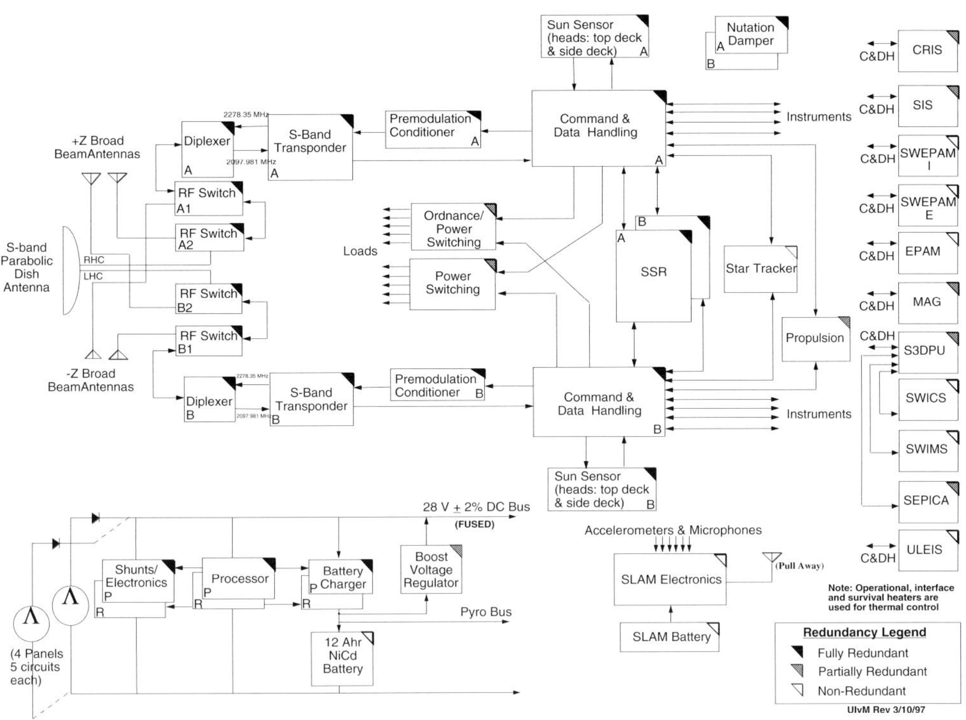

Figure 4. System block diagram of the ACE Observatory.

operation through solar flares. Major requirements of the C&DH subsystem are shown in Table II.

Each C&DH component processes and executes commands, formats downlink and stored telemetry frames, controls thrusters, and generates sun pulses and spin clocks. Uplinked commands are received in Consultative Committee Space Data Systems (CCSDS) – compatible telecommand transfer frames at a rate of 1000 bits per second (bps). Transfer frames are accepted and ground notification is performed using a modified CCSDS Command Operation Procedure-1 command protocol. Commands can be executed immediately, stored and executed at a specific spacecraft time, executed in response to an out-of-limit spacecraft health parameter, or stored as a command macro. Blocks of stored commands are executed in response to discrete fault indications from the power subsystem and also when the spacecraft has separated from the launch vehicle. Antennas are automatically reconfigured if no valid commands are received in a period that can be set from 0.5 to 96 hours by ground command. All command outputs include design features to prevent single failures from causing false commands to be sent. Data commands transfer 1 to 512 bytes of user data to other onboard subsystems at a rate of 1200 bps. Logic pulse commands provide mode switching pulses to other subsystems. Both data and logic pulse command signal outputs are edge-

TABLE II

Major requirements of the C&DH subsystem

Major requirement	Impact
No single point mission failures	Redundant C&DH components; include numerous watchdogs and safety circuits to prevent anomalous operation; redundant SSRs; redundant control of relays
Continuous operation through solar flares	Hardened logic; triple voting FPGAs; EDAC on memory
Interface with heritage instruments and fixed data rate to each instrument	Point-to-point discrete interfaces for science data collection
CCSDS-compatible downlink	Package minor frames in CCSDS Packet-Virtual Channel Transfer Frame; Reed Solomon and convolutional encoding
Time-tag data to allow correlation to 0.1 s with Universal Time	Use of Temperature-Compensated Crystal Oscillator
Perform thruster firing, meet safety requirements	Design tolerates two or more failures before improper thruster activation
Store 50 hours of science data (missed ground contact)	Redundant 1-Gbit SSRs
Downlink real-time science with SSR playback	Interleave Virtual Channel 1 real-time Science frames with Virtual Channel 2 SSR playback frames containing Virtual Channel 4 recorded frames
Respond to uplink and stored commands	Execute real-time commands and store commands in time-tagged and block bins
Reconfigure antennas if no ground contact C&DH component includes RF watchdog timer to periodically reconfigure antennas in absence of valid real-time commands	
Respond to out-of-limit telemetry and fault conditions	Evaluate autonomy rule bins once per second; respond to discrete fault indication signals from power subsystem
Generate and distribute sun pulse and spin clock signals to instruments	Use sun angle information to generate signals

Note: C&DH = command and data handling, CCSDS = Consultative Committee on Space Data Systems, EDAC = error detection and correction, FPGA = field-programmable gate array; SSR = solid-state recorder.

rate controlled to prevent electromagnetic interference problems. Table III shows the number and type of each command interface.

Each C&DH component can operate in Full Collection mode (collect all data and generate sun pulses and spin clocks), Reduced Collection mode (collect all data except science), or Disabled Collection mode (collect no data). One C&DH

TABLE III

C&DH subsystem user commands

Command type	Number provided	Comment
Data command	16	1 to 512 bytes of user data
Logic pulse	16	40-ms pulse
Relay	101	2-A and 10-A latching relays; 2-A and 5-A nonlatching relays

Note: C&DH = command and data handling.

TABLE IV

C&DH subsystem telemetry channels

Telemetry channel type	Number provided	Comment
Serial digital	16	Includes major and minor frame pulses, clock, read-out gate
0–5 V single-ended analog	62	
0–50 mV differential analog	62	Used primarily for measuring currents
AD590 temperature transducer	62	Measures temperatures in range of $-60\,^\circ$ to $+100\,^\circ$C
PT103 temperature transducer	62	Measures temperatures in range of $-100\,^\circ$ to $+150\,^\circ$C
Digital telltale-logic	32	Used primarily for RF transponder telemetry
Digital telltale-switch	16	Used primarily to detect the state of switches

Note: C&DH = command and data handling.

component is designated the active C&DH and operates in Full Collection mode. The other C&DH component is designated the inactive C&DH and operates in Reduced Collection mode. Digital and analog telemetry data are collected with six types of telemetry channels (Table IV) from onboard subsystems at a single rate and format. Data are collected in a major/minor frame structure at a rate of 1 minor frame per second, with 16 minor frames per major frame. The collected data are stored in the Data Collection Buffer resident in C&DH component memory. The Data Collection Buffer is the source of data in all telemetry formats and for autonomy (i.e., checking out-of-limit values of spacecraft telemetry). Both C&DH components can simultaneously limit-check spacecraft telemetry data.

Three requirements – the need to be compatible with several existing instruments, the need to simplify the C&DH – instrument interface by allocating each instrument a continuous fixed science data rate, and the need to have a CCSDS-compatible downlink – led to the development of a hybrid telemetry format scheme that combines a traditional major/minor frame structure with a modern CCSDS packet/transfer frame structure. Each telemetry frame consists of a Virtual Channel Transfer Frame containing one telemetry packet. Each packet contains a single telemetry minor frame in its data field. Real-time data can be formatted into one of ten minor frame formats for downlinking or recording (Table V). The Downlink and Record formats are independently selected. Data can be downlinked at 434 bps (low-rate engineering data or real-time solar wind data), 6944 bps (science and engineering data), or 76 384 bps (real-time and SSR playback data) while another format is being recorded at 6944 bps. During normal operations, data are recorded in the science and engineering format at 6944 bps while real-time solar wind data are downlinked to NOAA ground stations at 434 bps for 21 hours per day. For 3 hours per day, real-time data are interleaved with SSR playback data at a composite rate of 76 384 bps and transmitted to the NASA DSN. Eleven frames are downlinked per second: one real-time frame and ten SSR playback frames. Downlink frames are encoded with the CCSDS standard Reed Solomon (Interleave = 4) and convolutional coding.

Each C&DH component contains a temperature-compensated crystal oscillator that is the source of all C&DH telemetry timing. A 32-bit time count is provided in each minor frame. The precision of the oscillator permits correlation of onboard data with Universal Time (UT) to at least 100 ms accuracy. Each C&DH component utilizes sun angle data from the attitude determination and control (ADandC) subsystem to generate sun pulse and spin clock signals that are distributed to instruments. For each revolution of the spacecraft, 16 384 spin clock pulses are generated. All science data can be correlated to telemetry synchronizing signals (major and minor frame pulses) that are generated at known times or to the sun pulse/spin clock signals. Each sun pulse is tagged with the current spacecraft time.

The two SSRs each provide 1 Gbit of storage. They are designed to operate with a less than 10^{-7} bit error rate after 26 hours of continuous solar flare. (This radiation level significantly exceeds the worst-case scenarios ever anticipated during the ACE mission.) In operation, data are recorded at an average rate of 6944 bps and reproduced at an average rate of 68 640 bps. Data are reproduced in forward order. Under normal operation, the observatory will be under ground control once every 24 hours. However, the observatory is designed for one missed contact, with autonomous collection and storage of data for up to 48 hours. If an SSR fills up with data, the active C&DH component will use the autonomy to start recording data on the second SSR.

Each C&DH component has interfaces to the Power Switching and ordnance fire components to allow the selection and activation of relays. Each includes latching relays for providing switched power to spacecraft subsystems and nonlatching relays for providing pulses. The latching relays include dual coils for redundant control. Redundant nonlatching relays are used. The ordnance fire component also includes interfaces to each C&DH component for firing thrusters.

TABLE V

C&DH telemetry formats

Downlink and record telemetry formats	Rate (bps)	Description
Attitude determination and control	6944	Contains all housekeeping and ADandC data; limited science data; repeats every second
Science	6944	Contains all science data; housekeeping data repeats every 16 s
C&DH memory dump	6944	Replaces housekeeping data in science format with C&DH memory dump data
C&DH bin dump	6944	Replaces housekeeping data in science format with C&DH bin dump data
Recorder test pattern	6944	Data field contains pseudorandom pattern to load SSR with known data
Low-rate housekeeping	434	Contains all housekeeping data, repeats every 16 s
Low-rate C&DH memory dump	434	Identical to low-rate housekeeping except includes C&DH memory dump data
Low-rate C&DH bin dump	434	Identical to low-rate housekeeping except includes C&DH bin dump data
Low-rate attitude determination and control	434	Contains all ADandC data and most housekeeping data
Real-time solar wind	434	Includes science data for real-time evaluation of solar wind

Note: ADandC = attitude determination and control, C&DH = command and data handling, SSR = solid-state recorder.

Each C&DH component controls the firing of thrusters in the propulsion subsystem. Thrusters can be fired for a single specific time duration (axial firing) or they can be pulsed during multiple spacecraft spins (sector-based firing). Thrusters are organized into Top Deck Thruster Fire and Bottom Deck Thruster Fire groups. Each C&DH component controls Thruster Select relays in the ordnance fire component. The Thruster Select relays allow thrusters in each group to be selected for firing; Thruster Arm relays enable firing to take place. When thruster firing is to begin, a C&DH component activates a Top Deck or a Bottom Deck Thruster Fire signal to the ordnance fire component for the duration of the firing. The thruster control circuitry is designed so that at least two failures would have to occur before any thruster would inadvertently fire. Independent watchdog timers are used to detect and halt thruster firing that is not halted by the primary control mechanism. Three commands must be successfully executed in order to begin thruster firing.

The C&DH component was designed and fabricated at JHU/APL. It utilizes a Harris RTX2010 processor executing the FORTH language. The RTX2010 is fabricated in a CMOS/SOS process that is exceptionally hard to single-event upsets (SEUs), making it suitable for operation through solar flares. Code is stored in electrically erasable/programmable read-only memory (EEPROM) and downloaded into random-access memory (RAM) for execution. EEPROM can be reloaded on the ground, and the RAM can be patched in flight. Both RAM and EEPROM utilize error detection and correction (EDAC) circuitry to correct single errors and detect double errors. Digital logic is implemented with radiation-hardened Harris HCS series logic, radiation-tolerant National FACT series logic, and Actel field-programmable gate arrays (FPGAs). The FPGAs are designed with triple voting cells to minimize the probability of SEUs. The RTX2010 is programmed in FORTH using a FORTH kernel and cross-compiler developed at JHU/APL and previously used on the Freja and Near Earth Asteroid Rendezvous (NEAR) spacecraft.

The SSRs were designed and fabricated at SEAKR Corporation. They utilize IBM 16-Mbit dynamic random access memories (DRAMs) for storage and a Harris 80C85 microprocessor for control. Error detection and correction is done on 16-bit words and allows for correction of single errors and detection of double errors. Failed memory segments are automatically mapped out.

The power switching and ordnance fire components were designed and fabricated at JHU/APL. They are implemented with a modular design having redundant relay coil driver cards and application-specific relay cards. They operate directly off the spacecraft power bus. The design was previously used on the Ballistic Missile Defense Organization's Midcourse Space Experiment spacecraft and on the Near Earth Asteroid Rendezvous spacecraft.

2.3.2. *RF Communications Subsystem*

The primary function of the RF communications subsystem is to serve as the observatory terminus for radio communications between the observatory and the NASA Deep Space Network of Earth stations. A secondary function is to transmit downlink data in real time, at 434 bps, to Earth stations supporting the NOAA Real Time Solar Wind project. The system is designed to receive uplink commands and transmit downlink telemetry data concurrently with coherent ranging. The system operates at 2097.9806 MHz for the uplink and 2278.35 MHz for the downlink.

The system consists of two identical (redundant) and independent communications subsystems and a single high-gain, dual-polarized parabolic reflector antenna. Each communications subsystem consists of a transponder (transmitter and receiver), diplexer, coaxial switching network, and two broad-beam antennas. There is no cross strapping between RF subsystems. The coaxial switching network is used to connect a given transponder to an aft $(-Z)$ or forward $(+Z)$ broad-beam antenna or to the aft high-gain parabolic antenna. Watchdog timers, implemented in software within the C&DH subsystem, are designed to switch the broad-beam antennas if no uplink spacecraft commands are received within a preset time. The timers provide a means to recover spacecraft communications in the event of a communications system or attitude anomaly. Both receivers are operated continuously, but only one transmitter is to be powered and one antenna energized at any given time.

Antennas

The purpose of the broad-beam antennas is to provide hemispherical uplink coverage forward and aft of the spacecraft. The broad-beam antennas also provide sufficient gain in the region $\pm 34°$ off boresight for the transmission of 434 bps data to a DSN 26-m ground station. The broad-beam antennas provide right circularly polarized coverage (gain) over a hemisphere. The gain of the broad-beam antennas over the hemisphere is greater than -7 dBic at uplink frequencies, -8 dBic at downlink frequencies. By configuring the RF system so that one receiver is connected to a forward broad-beam antenna and the other to an aft broad-beam antenna, omnidirectional coverage of the spacecraft can be achieved. At angles less than $34°$ off boresight, the broad-beam antenna gain is -1 dBic at uplink frequencies, 0 dBic at downlink frequencies.

The purpose of the high-gain antenna is to support DSN uplink commands and the transmission of downlink data at 76 384 or 6944 kilobits per second (kbps). The high-gain antenna is also used to transmit 434-bps data to NOAA Real Time Solar Wind sites. The specified gain of the high-gain antenna at $4.25°$ off boresight is greater than 18 dBic at the uplink frequency, 19 dBic at the downlink frequency. The high-gain antenna is pointed by ground commands to the spacecraft attitude subsystem. The antenna must be pointed within $\pm 3°$ of the spacecraft–Earth line in order to have sufficient gain to achieve the signal-to-noise ratios required for the data links. On orbit, the attitude of the observatory must be adjusted periodically to keep the Earth within the $\pm 3°$ beam width of the antenna.

Transponder

The transponder consists of a receiver and transmitter that are diplexed onto a single line feeding the antenna. The receiver is a phase-locked, dual conversion, superheterodyne type with a detector and command demodulator. The transmitter is a phase-shift-keyed type with an RF output power of 5 W minimum. The output spectrum consists of a residual carrier with data sidebands.

Communications modes

The transponder receiver and transmitter may be operated independently (noncoherently) or coherently, where the downlink frequency is derived from the uplink signal frequency. The coherent mode permits the measurement of two-way Doppler. The functional modes of operation are command reception, telemetry transmission, and ranging.

Commands reception

The uplink command data rate is 1000 bps. The pulse code modulation is NRZ-L. The command data are phase-shift-keyed onto a 16-kHz subcarrier and then phase modulated onto the 2097.9806-MHz RF carrier.

Telemetry transmission

There are three selectable downlink data rates: 434 bps, 6944 bps, and 76 384 bps. The pulse code modulation is bi-phase-L. The downlink data streams are encoded by the C&DH subsystem before they are routed to the transmitter. The coding consists of a convolutional code $(7, \frac{1}{2})$ concatenated with a (248 217) Reed-Solomon code. The data stream is directly phase modulated onto the 2278.35-MHz downlink carrier, producing a residual carrier and data sidebands. The modulation index is selectable (high/low) by ground command. The low modulation index is required for transmitting the 434-bps data, and the high modulation index is used for the 6944-bps and 76 384-bps data rates.

Ranging

The system is capable of simultaneous command reception, telemetry transmission, and ranging. However, a noninterfering signal structure, such as that produced by the DSN Sequential Ranging Assembly (SRA), is required. The ranging clock frequency is approximately 512 kHz; the lower frequency codes are expected to be component numbers 6 through 17.

2.3.3. *Attitude Determination and Control Subsystem*

The ADandC subsystem was designed to minimize spacecraft cost and complexity while maximizing reliability and mission success. It utilizes the inherent gyroscopic stability of a spinning spacecraft for attitude control coupled with telemetered sun sensor and star scanner data for determining attitude on the ground. The ADandC subsystem consists of a solid-state star scanner, a redundant sun sensor system (which acts as an on-orbit backup to the star scanner), two fluid-filled ring nutation dampers, the ten thrusters of the propulsion system, and the command capability of the C&DH subsystem. It is an extremely simple system that has proven itself on a variety of other missions.

Each redundant sun sensor system from the Adcole Corporation consists of two sun angle sensors and an associated electronics box. Each sensor digitally encodes to an 8-bit value, the sun angle in nominal 0.5° increments over a field of view of ±64° in each of two orthogonal axes. One sun angle sensor of a pair is located on the +Z deck of the spacecraft with its normal parallel to the +Z axis; the other is mounted on the side deck with its normal canted 125° away from the spacecraft +Z axis. Unless there is an attitude anomaly, the sun will always shine on the top-deck sun angle sensor. The sun sensor electronics forwards to the C&DH subsystem the two encoded 8-bit sun angles from the illuminated sun angle sensor and two identification bits indicating which of the two sensors is providing

TABLE VI

Observatory instantaneous attitude error budget

Parameter	Angular error (deg)
Star scanner accuracy	0.025 (3σ)
Star scanner mechanical mounting (total)	0.023
Side panel thermal distortion	negligible
Principal axis misalignment (max)	0.20
Residual nutation (max)	0.25
Total (RSS)	0.498

the data. The C&DH subsystem records the sun angle data for inclusion in the downlinked telemetry and both generates a sun-crossing pulse and initializes a sector clock based on the transition of the most significant bit of one of the two sun angle values from the illuminated sensor.

The observatory attitude is determined on the ground by combining the telemetered sun angle data with high-accuracy data provided by a Ball Aerospace Systems Division CT-632 solid-state star scanner. The CT-632 star scanner is a star tracker modified to operate at the ACE spin rate of $30° \ s^{-1}$, which is two orders of magnitude greater than nominal star tracker angular rates, with no significant impact on the error budget (1σ error of 30 arc sec) (Radovich 1995). Time delay integration is used to accumulate the star image signal on the charge-coupled device (CCD) so that standard star tracker image processing algorithms can be used to determine star centroids and magnitudes. The data from the star scanner are collected by the C&DH subsystem for telemetering to the ground, where they are combined with the sun angle data to determine the attitude of the observatory.

The requirement for attitude knowledge is $\pm 0.7°$ after the fact, with a goal of $\pm 0.5°$ for the magnetometer. The spacecraft components were assigned budgets for attitude errors, which are given in Table VI.

Attitude control is achieved by the inherent passive, gyroscopic stability of a major-axis spinning spacecraft. Two 0.46-m-diameter hoops filled with an ethylene-glycol solution provide purposeful energy dissipation to damp nutational motion. Open-loop, ground commanded firings of the hydrazine thrusters are used to precess the observatory spin axis to follow the nominal $1° \ day^{-1}$ apparent motion of the Earth and Sun and to adjust the spin rate as needed. Operationally the spin axis is precessed once every 5 or 6 days, whereas spin-rate adjustments will be rare.

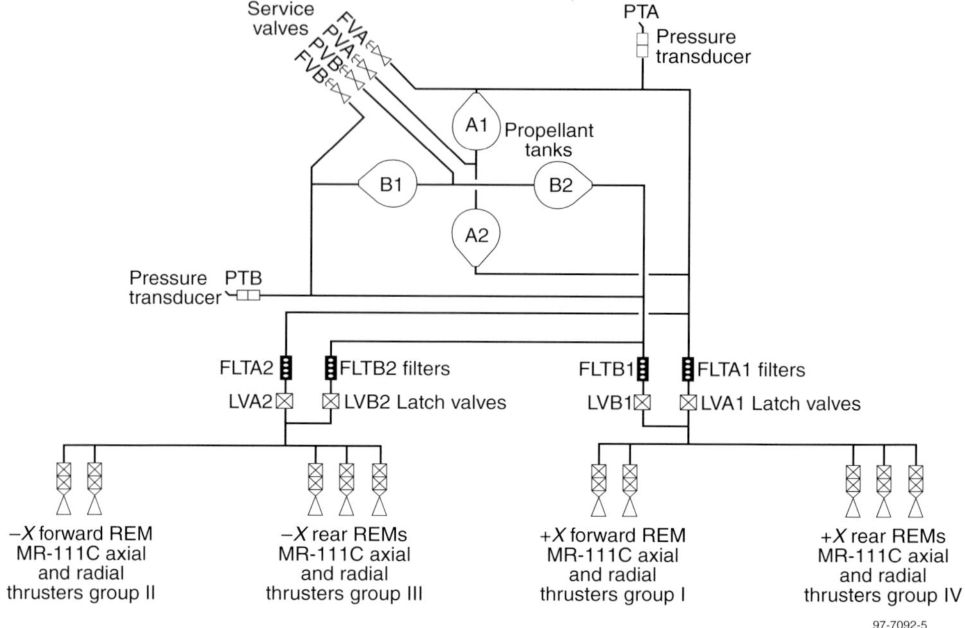

Figure 5. Schematic of the propulsion subsystem, illustrating the cross strapping between the A and B side tanks, service valves, and thrusters (PV = pressurant valve; FV = fill valve; REM = reaction engine module).

2.3.4. *Propulsion Subsystem*

The ACE propulsion subsystem will correct launch vehicle dispersion errors, inject the spacecraft into the L_1 halo orbit, adjust the orbit, adjust spin axis pointing, and maintain a 5 rpm spin rate. The subsystem is a hydrazine blowdown unit that uses nitrogen gas as the pressurant and is made up of four fuel tanks, four axial thrusters for velocity control along the spin axis, and six radial thrusters for spin plane velocity control and spin rate control, as well as filters, pressure transducers, latch valves, service valves, heaters, plumbing, and structure. Figure 5 presents a schematic of the subsystem and illustrates the cross strapping between the A and B side tanks, service valves, and thrusters. Figure 6 shows the location and function of the ten thrusters.

The four propellant tanks are 65.1-m^3 titanium cone spheres with a total capacity of 195 kg, which provides for a mission lifetime of 5 years with considerable margin. The tanks are initially pressurized to 305 pounds per square inch absolute (psia) at 21°C, and they blow down to 91 psia when the propellant is fully expended. The ten thrusters each provide 4.4 N nominal thrust and, combined with the mission duty cycle, provide a mission average specific impulse of 216 s minimum.

The pressure transducers have a 0–500 psia range and are individually powered through relays. The latch valves are flown in the open position and are normally used only for post-loading shipment at Kennedy Space Center. They could be

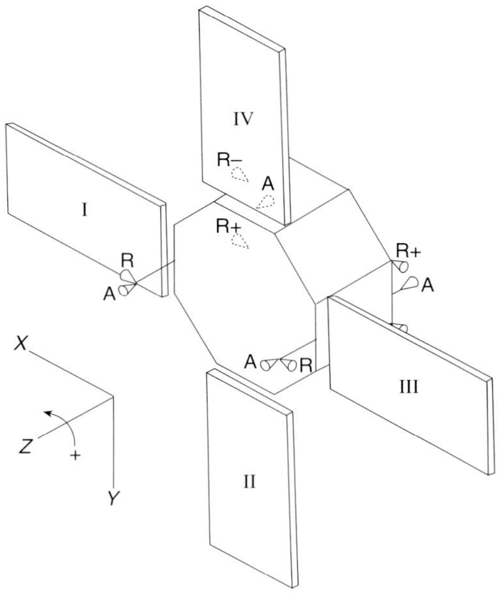

<table>
<tr>
  <th rowspan="2">Thrusters</th>
  <th colspan="4">Velocity Control</th>
  <th rowspan="2" colspan="4">Spin Axis Pointing</th>
  <th colspan="2">Spin Rate</th>
</tr>
<tr>
  <th>+Z</th><th>−Z</th><th colspan="2">X−Y plane</th>
  <th>Increase</th><th>Decrease</th>
</tr>
<tr><td>IA</td><td></td><td>X</td><td></td><td></td><td>X</td><td></td><td></td><td></td><td></td><td></td></tr>
<tr><td>IR</td><td></td><td></td><td>X</td><td></td><td></td><td></td><td>X</td><td></td><td></td><td></td></tr>
<tr><td>IVA</td><td>X</td><td></td><td></td><td></td><td></td><td>X</td><td></td><td></td><td></td><td></td></tr>
<tr><td>IVR+</td><td></td><td></td><td>X</td><td></td><td></td><td></td><td></td><td>X</td><td>X</td><td></td></tr>
<tr><td>IVR−</td><td></td><td></td><td>X</td><td></td><td></td><td></td><td></td><td>X</td><td></td><td>X</td></tr>
<tr><td>IIA</td><td></td><td>X</td><td></td><td></td><td></td><td>X</td><td></td><td></td><td></td><td></td></tr>
<tr><td>IIR</td><td></td><td></td><td></td><td>X</td><td></td><td></td><td></td><td>X</td><td></td><td></td></tr>
<tr><td>IIIA</td><td>X</td><td></td><td></td><td></td><td>X</td><td></td><td></td><td></td><td></td><td></td></tr>
<tr><td>IIIR+</td><td></td><td></td><td></td><td>X</td><td></td><td></td><td>X</td><td></td><td>X</td><td></td></tr>
<tr><td>IIIR−</td><td></td><td></td><td></td><td>X</td><td></td><td></td><td>X</td><td></td><td></td><td>X</td></tr>
</table>

97-7092-6

Figure 6. Location and function of the ten thrusters.

used to isolate a leaking thruster or propellant tank if such a failure were detected during the mission. Tank, line, and thruster valve heaters are thermostatically controlled and maintain the propellant components comfortably above the 0°C hydrazine freezing point. Thruster catalyst bed heaters are relay controlled and preheat the bed prior to thruster firings to prevent cold-start degradation. The subsystem plumbing is all welded and sized to minimize any differential flow pressure drop that might cause a spin imbalance. Orifices are used in the latch valves and at the entrance to the pressure transducers to minimize pressure surge and water hammer effects. Filters upstream of each latch valve and at the inlet of each thruster remove any potential contaminating particulate before it reaches the component.

The propulsion subsystem was designed and fabricated by Primex Corporation (formerly Olin Aeorospace) and shipped to JHU/APL for integration with the spacecraft after thermal blanket installation. It was environmentally tested at protoflight levels using water to simulate propellant and using mass simulators for the rest of the spacecraft subsystems.

2.3.5. *Power Subsystem*

The ACE electrical power subsystem provides 444 W at spacecraft end of life (EOL), with EOL defined as 5 years. The electrical power load budget is 425 W peak. The low-magnetic-emission, fixed planar silicon solar array is connected directly to the 28 V $\pm 2\%$ shunt-regulated bus. The electrical power subsystem also contains an 18-cell 12 A-h NiCd battery booster capable of supporting a 165-W launch load. This subsystem is one-fault tolerant with its autonomously controlled redundant shunt regulator and cross-strapped redundant battery chargers.

Solar array

TECSTAR was contracted to fabricate and lay down five strings, each 89 solar cells long, on each of four aluminum honeycomb substrates. The $86.4 \times 149.9 \times 3.2$ cm substrates, insulated with 3-mil Kapton, were provided by Applied Aerospace Structures Corporation. The 3.9×6.3 cm, 15.1% efficient silicon cells have a diffused boron back surface field/reflector and are covered with 12-mil, ceria-doped microsheet. The cells and covers have anti-reflective coatings, and the covers also have ultraviolet reflective coatings. Strings are back-wired for magnetic emissions below 0.1 nT at the magnetometer, which is housed on a boom extending 150 cm from the panel edge. Silver interconnects, which are nonstandard, were used to achieve low magnetic emissions. TECSTAR provided a coupon (i.e., a sample) and a qualification panel with accelerated thermal cycle testing to validate the silver interconnects.

The solar arrays were designed for several worst-case operating conditions. Maximum temperature expected is 58°C while under thruster plume impingement of the ultrapure hydrazine propellant. Conservative degradation estimates project enough EOL power to tolerate one string failure with no effect on spacecraft performance. A solar cell patch located on each panel supports an experiment to measure the degradation over mission life. Strings also have bypass diodes to protect against shadows cast by spacecraft instruments, antennas, and thrusters. Pyrotechnic-actuated, spring-loaded hinges deploy the panels after spacecraft separation from the Delta II.

Battery

JHU/APL designed and built the 18-cell stack with redundant thermostatic controlled heaters and current, temperature, and full- and half-stack differential voltage monitors. The thermal design will keep the battery between 0° and 25°C. The cells for the flight battery are a SAFT Gates Aerospace Batteries (GAB, Gainsville,

Florida) 12AB28 standard profile design (4.6 × 3.0 × 0.9 cm, 536 g). Although aged for 8 years in dry storage at Gainsville, cells successfully activated in France perform well (typically 15 A-h).

Electronics

The electronics consists of six boxes and a shunt resistor dissipater assembly. Each solar panel has a digital shunt box nearby containing five metal oxide semiconductor field-effect transistors (MOSFETs) and blocking diodes for low-dissipation shorting of each string. The panel current and the voltage of each string are monitored. Fuses, which help filter the regulated bus, protect the bus from a shorting failure of either blocking diodes or capacitors.

The power subsystem dissipater electronics contains redundant linear transistors, the booster, and redundant battery chargers. The 90% efficient booster can be configured in flight to provide partial battery reconditioning. The battery chargers provide closed-loop current and voltage limiting. All boxes are constructed with a combination of semi-rigid flex printed circuit boards and heat sink subassemblies. The traces are sized for 70-A fuse blowing capability and 15-A continuous power to the bus with less than a 2-V drop from the solar panel connector to the subsystem dissipater electronics output.

The power subsystem control electronics contain redundant hybrid power converters from Interpoint, shunt regulation electronics, and a circuit that will switch from the primary to the redundant side in response to bus under- or overvoltage. It also contains redundant processors and low battery voltage and low bus voltage sensors. The 80C85RH-based processors digitize all electrical power subsystem telemetry for transmission via cross-strapped serial digital links to the redundant spacecraft telemetry systems. The processors also decode commands cross-strapped from the redundant spacecraft command systems. Digital-to-analog converters allow adjustment of the bus voltage set point (±2%), battery charge limits (0 to 1.5 A), and temperature-compensated battery voltage limits.

3. Environmental Design Drivers

As a result of specific ACE science and mission concerns, three environmental factors were given significant attention during the design and fabrication of the ACE Observatory. By addressing these factors from the beginning of the design process, goals and requirements were achieved with relatively little additional cost.

3.1. MAGNETIC CLEANLINESS

The observatory has two magnetometer sensors, one at the end of each magnetometer boom. Although ACE had no magnetic requirement, there were magnetic goals. Magnetic cleanliness precautions were undertaken to the extent practical (i.e., without driving the cost), to limit the magnetic fields from the spacecraft components that could radiate to the sensors. The goal was to achieve a spacecraft static magnetic field at the magnetometer sensors of less than 0.1 nT. The goal for AC interference at the sensor locations was less than 0.001 nT over a frequency range of 0 to 10 Hz and the specific frequencies of 15 kHz ($\pm$200 Hz), 30 kHz ($\pm$200 Hz), and 60 kHz ($\pm$200 Hz). These goals were achieved as follows:

– The spacecraft battery was wired to reduce the magnetic field and to minimize magnetic loops, and the battery was demagnetized and located as far as practical from the magnetometer sensors on the $+X$ side panel (sensors are at $\pm Y$ axis).

– Solar panels were back-wired for compensation. Magnetic loops at the end of the panels were minimized.

– Twisted pair wire was used for power lines; a single-point ground system was used; and grounding was designed to minimize ground loops, all to the extent possible.

– Magnetic materials used in spacecraft components were reviewed on a global basis, and a random selection of typical parts was tested.

– Boom material is titanium; hinges and materials close to sensors are nonmagnetic; and de-permed tools were used.

– The magnetometer team used 'sniffing' devices to map the magnetic emissions of selected suspect spacecraft components and parts (e.g., analog shunt resistive strips and latch valves). As a result, propulsion subsystem latch valves and latch valves for the SEPICA instrument were shielded with Met-glass.

– The ordnance fire component was compensated with external magnets after pre-integration tests.

– Nonmagnetic connectors were used for spacecraft components as much as possible.

– The magnetic field generated by the shaker tables used for spacecraft components and the observatory was measured (both with and without the degauss coils), and, where required, degauss coils were used to reduce the large fields generated by the exciter.

– Instruments and spacecraft subsystems were screened prior to integration to reduce the possibility that compensation of the entire spacecraft would be required.

– No observatory-level compensation was performed after preflight static field tests.

Two preliminary results of the static test prior to launch indicated that if the source locations are neglected, and if the conservative assumption is made that the source of the field is at the center of the spacecraft, a worst-case field of 2.3 nT at each flight sensor is predicted. If the sources have been correctly identified, based in part on the static test and on pre-integration screening results, then the in-flight field components are only $\frac{1}{10}$ this value. The preliminary results of the AC test revealed no AC signals at a level that could significantly affect the magnetic field measurements.

The magnetometer sensors are flown in 'twin design,' with one sensor on each boom, as opposed to 'dual design,' with two sensors on a long boom on the same side of the spacecraft, so that the actual field on orbit is folded into the magnetometer sensor and electronics calibration. The actual field will never be known but, from data for the early phase of the mission, is estimated to be <1 nT at the sensors.

3.2. RADIATION ENVIRONMENT

Spacecraft components had to be designed to meet the radiation requirements as follows. Parts were required to survive a total ionizing dose of 15 krad (Si) without part failure (with spot shielding as required). The 15 krad (Si) requirement was based on a 5-year mission goal using the projected radiation exposure for the mission with a 95% level of confidence and assuming a nominal shielding of 75-mil aluminum. Since ACE could operate for 4 years at solar maximum, the components had to withstand a total proton fluence of approximately 2^{11} particles cm^{-2} for protons with energies above 4 MeV. Components had to be immune to latch-up; parts susceptible to single-event latch-up with linear energy transfer threshold less than 120 MeV cm^2 mg^{-1} could not be used in spacecraft components without latch-up mitigation techniques. Parts susceptible to SEU could not be used in critical components if these could cause mission-critical failures. In parts of noncritical components, SEU could not compromise spacecraft health or mission performance. System-level SEU effects were considered, so that upsets did not cause uncorrectable errors that would affect system performance. SEU-susceptible parts used for data storage memory had to use appropriate error detection and correction techniques so that during exposure to maximum particle flux (as experienced during a worst-case solar particle event), the amount of data lost would not cause a violation of the component specification requirements.

3.3. CONTAMINATION CONTROL

ACE instruments are susceptible to hydrocarbons, fluoro-hydrocarbons, particles, and humidity. To satisfy instrument requirements and permit a comfortable working environment, the instruments were purged and the environment kept at class 100 000 clean room level (this level requires only the use of white coats). Chemicals and compounds used were carefully screened. Nitrogen purge with 99.995% or better purity at a temperature of $15°$ to $25°C$ was supplied to the instruments through a manifold at a pressure of 3.0 pounds per square inch (psi) nominal (5.0 psi maximum) with a flow rate of 5.9 cm^3 min^{-1} total. Each instrument used a restrictor to obtain its required flow rate. The purge remained connected (with short interruptions of minutes' duration) until launch. The sun sensor and star scanner optical portions were covered unless ground support equipment (stimulators or sun simulator) was attached.

3.4. SPACECRAFT SAFING

Spacecraft features are used to 'safe' the spacecraft when problems are recognized onboard or when spacecraft actions, independent of ground intervention, are required. These features are implemented with C&DH 'autonomy rules,' built-in power subsystem autonomy, dedicated lines between the power subsystem and the C&DH for power bus fault indications, C&DH RF watchdog timers, and C&DH and power subsystem resets.

The C&DH autonomy rules (called 'autonomy' here) are widely used on the spacecraft to put an instrument, a spacecraft component, or the whole observatory (spacecraft and instruments) in a 'safe' configuration when certain criteria are met. The criteria used to determine the need for action, and the actions are predefined but can be changed. A component parameter such as a current, which requires monitoring, is collected by the C&DH. The value of the parameter is compared (=, >, <, etc.) to a preselected threshold, for example, the maximum expected current for that component. Based on the result, the C&DH autonomy may issue commands such as unpowering the component if the current is too high. Below are examples where autonomy is used on the spacecraft for all phases of the mission: launch, postlaunch vehicle separation, and on-station operation.

- Launch
  - Autonomy switches from the prime to the redundant power subsystem shunt regulator in the unlikely case of an analog shunt short.
- Postlaunch vehicle separation activities
  - After the solar panels are automatically deployed, autonomy controls the safe configuration of the spacecraft. For example, if the solar panels are supplying power (indicating successful panel deployment), the transmitter is commanded to radiate, thereby facilitating first ground contact for the ground operators. Otherwise, the transmitter is commanded to standby, thereby conserving power until ground intervention can determine the cause of the problem.
- Health monitoring
  - When instrument and spacecraft health indicators, such as currents, exceed predefined limits, autonomy takes action, such as unpowering a component, until the cause of the problem can be determined by mission operations.
  - For redundant thermostats / heaters, autonomy removes a failed thermostat / heater of a redundant set but keeps the other powered.
  - During battery reconditioning, autonomy stops battery reconditioning if battery health indicators exceeds limits.
- Maneuver or attitude problems
  - Autonomy aborts thruster firing and closes the ULEIS instrument shutter in case of a spacecraft attitude problem. This problem is indicated by the Sun angle exceeding a predefined limit or by the side panel Sun sensors sensing the Sun.

- Thruster firing abort commands are issued by autonomy after a maneuver if, because of a problem, the maneuver is not completed with the normal process.

- Protection against commands resulting in improper spacecraft configurations

  - The autonomy feature can be used to protect against possible command errors. Two examples follow. Should both transmit/receive chains be erroneously connected to the top-deck broad-beam antennas, autonomy is used to configured the active side to the bottom-deck antenna (which is oriented toward Earth). If both transmitters are configured to radiate, autonomy is used to configure the inactive side transmitter to standby.

- Support for other autonomous spacecraft actions

  - After an RF watchdog timer time-out, which indicates abnormal operation, the C&DH commands the active transmitter to stand by (prior to reconfiguring the broad-beam antennas). Autonomy is used to command the transmitter to radiate and to select the 'attitude determination and control' format for the downlink, thereby minimizing the time required for the ground station to acquire the spacecraft and to reconfigure the downlink.

  - Under certain conditions, e.g., power shedding, both transmitters may be turned off by the C&DH. Autonomy is used to power the interface heater thermostat to keep the bottom deck at an acceptable temperature for the other components on the deck.

  - When one recorder is filled (for example, when a ground contact where the recorder would be dumped is missed), autonomy is used to start recording on the other recorder so data loss is minimized.

The power subsystem autonomy feature is used to detect bus regulation failures indicated by overvoltage (30 V) or undervoltage (26 V) conditions. For either condition, the power subsystem autonomously switches control from the primary to the redundant shunt regulator in <10 ms. Note that in the load shed sequence described in the next paragraph, the C&DH autonomy takes 90.0 to 181.8 ms to switch to the redundant shunt regulator.

The power subsystem monitors the health of the spacecraft power bus and detects instrument or spacecraft hardware fault conditions that cause the bus to fall below specified limits. Dedicated lines between the power subsystem and the C&DH are used to indicate faults to the C&DH. The latter autonomously executes a predetermined sequence of commands to shed loads. The default sequence, loaded at launch, sheds all noncritical loads and switches control from the primary to the redundant power subsystem shunt regulator. Graduated sequences, which shed loads depending on the severity of the failure, will be used later in the mission. The graduated sequences attempt to preserve telemetry data history and minimize changes to the spacecraft configuration as long as possible to help the ground operators determine the cause of the fault. Note that the C&DH load shed and the power subsystem autonomy can be individually disabled, if necessary

The C&DH RF watchdog timer is used to place the spacecraft in a communications safe mode when a valid command has not been received for a specified time. When the specified time has expired and the RF watchdog timer in the active C&DH has timed out, the active C&DH sets the RF antenna switches in its chain to connect the transmit/receive chain on its side to the bottom-deck broad-beam antenna. When the RF watchdog timer in the inactive C&DH times out, it sets the RF antenna switches in its chain to connect the transmit/receive chain on its side to the top-deck broad-beam antenna, thereby providing hemispheric uplink coverage to facilitate reception of a command. At the next time-out, this configuration is reversed, with the active side connected to the top-deck antenna and the inactive side connected to the bottom-deck antenna. The C&DH toggles between these two configurations for each time-out until a valid command is received, which resets the timer.

Each C&DH has a two independent timers used to halt thruster firing. If one timer fails during a maneuver, the other will still time out and stop the firing.

The C&DH has four sources that can generate a reset: last resort timer, software watchdog timer, error detection and correction double bit error, and C&DH circuit bus time-out error. These result in C&DH hardware reset and software boot and initialization. Critical parameters are restored either to their pre-reset value (if two of the three backup copies of the parameter match) or their default value (if these do not match). The power subsystem processor resets whenever the power processor watchdog timer is not reset within 2 s. Again, majority voting is used to restore critical parameters to their pre-reset value or default value.

Acknowledgements

The authors would like to acknowledge the contributions of the many individuals involved with the ACE spacecraft development at JHU/APL and its subcontractors. We would also like to acknowledge and express our appreciation of the many individuals within the ACE Team organizations that also contributed so much to the success of the ACE spacecraft. These organizations include GSFC/NASA and their subcontractors, Caltech, and all of the science and instrument team members in the various universities and government agencies involved with ACE.

Reference

Radovich, Mark A., Jr.: 1995, 'Modifications of a Mosaic Focal Plane Star Tracker for Scanning Application,' *SPIE Int. Symp. Aerospace/Defense Sensing and Control and Dual Use Photonics,* Orlando, Florida.

THE COSMIC-RAY ISOTOPE SPECTROMETER FOR THE ADVANCED COMPOSITION EXPLORER

E. C. STONE, C. M. S. COHEN, W. R. COOK, A. C. CUMMINGS, B. GAULD,
B. KECMAN, R. A. LESKE, R. A. MEWALDT and M. R. THAYER
California Institute of Technology, Pasadena, CA 91125, U.S.A.

B. L. DOUGHERTY*, R. L. GRUMM, B. D. MILLIKEN[†], R. G. RADOCINSKI and
M. E. WIEDENBECK
Jet Propulsion Laboratory, Pasadena, CA 91109, U.S.A.

E. R. CHRISTIAN, S. SHUMAN, H. TREXEL and T. T. VON ROSENVINGE
NASA/Goddard Space Flight Center, Greenbelt, MD 20771, U.S.A.

W. R. BINNS, D. J. CRARY[‡], P. DOWKONTT, J. EPSTEIN, P. L. HINK,
J. KLARMANN, M. LIJOWSKI and M. A. OLEVITCH
Washington University, St. Louis, MO 63130, U.S.A.

Abstract. The Cosmic-Ray Isotope Spectrometer is designed to cover the highest decade of the Advanced Composition Explorer's energy interval, from ~50 to ~500 MeV nucl^{-1}, with isotopic resolution for elements from $Z \simeq 2$ to $Z \simeq 30$. The nuclei detected in this energy interval are predominantly cosmic rays originating in our Galaxy. This sample of galactic matter can be used to investigate the nucleosynthesis of the parent material, as well as fractionation, acceleration, and transport processes that these particles undergo in the Galaxy and in the interplanetary medium.

Charge and mass identification with CRIS is based on multiple measurements of dE/dx and total energy in stacks of silicon detectors, and trajectory measurements in a scintillating optical fiber trajectory (SOFT) hodoscope. The instrument has a geometrical factor of ~r250 cm^2 sr for isotope measurements, and should accumulate $\sim 5 \times 10^6$ stopping heavy nuclei ($Z > 2$) in two years of data collection under solar minimum conditions.

1. Introduction

The Advanced Composition Explorer (ACE) will perform comprehensive studies of the elemental, isotopic, and ionic charge state composition of energetic nuclei from H to Zn ($1 \leq Z \leq 30$) from a vantage point located $\sim 1.5 \times 10^6$ km sunward of Earth, near the L1 libration point. These observations will address a wide range of questions concerning the origin, acceleration and transport of energetic nuclei from solar, interplanetary, and galactic sources. ACE includes six high-resolution

---

* Present address: Mail Stop 264–33, California Institute of Technology, Pasadena, CA 91125, U.S.A.

[†] Present address: Lucent Technologies, Naperville, IL 60566, U.S.A.

[‡] Present address: Pacific-Sierra Research Corp., Warminster, PA 18947, U.S.A.

 *Space Science Reviews* **96**: 285–356, 1998.
© 1998 *Kluwer Academic Publishers. Printed in the Netherlands.*

spectrometers and three interplanetary monitoring instruments that together will measure composition at energies ranging from <1 keV nucl^{-1} (characteristic of the solar wind) to ~500 MeV nucl^{-1} (characteristic of galactic cosmic radiation). The Cosmic-Ray Isotope Spectrometer (CRIS), one of these six spectrometers, is designed to measure the elemental and isotopic composition over the energy range from ~50 to ~500 MeV nucl^{-1} with excellent mass resolution and collecting power.

The primary objective of the ACE mission is to determine and compare the elemental and isotopic composition of several distinct samples of matter, including the solar corona, the interplanetary medium, the local interstellar medium, and galactic matter. ACE will study matter from the Sun by measuring solar wind (SW) and solar energetic particles (SEPs). Matter from the local interstellar medium (LISM) will be observed at energies of ~1 keV nucl^{-1} as interstellar neutral particles enter the heliosphere and are ionized to become solar wind 'pickup ions'. Some of these ions are also observable as 'anomalous cosmic rays' (ACRs) once they have been convected out to the solar wind termination shock and accelerated to energies of ~1 to 100 MeV nucl^{-1}. Galactic cosmic rays (GCRs) provide a sample of matter that enters the heliosphere from other regions of the Galaxy.

CRIS is the primary instrument responsible for measuring galactic cosmic-ray composition on ACE. Table I summarizes the essential features of this instrument. The remainder of the paper presents a more detailed discussion of the objectives, the design, and the measured performance of CRIS.

2. Scientific Objectives

Cosmic rays consist of energetic electrons and nuclei which are a direct sample of material from beyond the solar system. The energy density of cosmic rays is comparable to the energy densities of interstellar gas, interstellar magnetic fields, and starlight. Thus the cosmic rays are an important part of our astrophysical surroundings. It is generally believed that supernovae provide the energy carried by cosmic rays, but many details of where and how the cosmic rays are accelerated to high energies remain unknown. Unlike photons, which propagate on straight lines from their sources, cosmic rays are charged and hence spiral around the galactic magnetic field lines. As a consequence, their arrival directions tell us nothing about their sources. Instead, essential clues to the origin and history of cosmic rays may be found in the relative abundances of elements and of isotopes. Cosmic rays consist primarily of hydrogen nuclei (~90% by number), with helium making up the bulk of the remaining particles. Heavier nuclei only account for ~1%. Nonetheless, most of the available information about cosmic-ray origin comes from the elemental and isotopic composition of the heavier cosmic rays.

CRIS will investigate many of the same processes as the other isotope spectrometers on ACE. The galactic cosmic-ray source (GCRS) composition for the

TABLE I

Summary of CRIS characteristics

Characteristic	Value	Details[a]
Measurement objective	Elemental and isotopic composition of galactic cosmic rays	§2
Measurement technique	Multiple-ΔE vs residual energy plus trajectory	§3.1
Sensor system		
Energy loss measurements	Si(Li) detectors (3 mm thick $\times$ 10 cm diameter, some with active guard rings) in 4 stacks of 15 detectors each	§4.2, T5
Trajectory measurements	Scintillating Optical Fiber Trajectory (SOFT) hodoscope, position resolution $<$ 100 μm r.m.s., 7.2 cm lever arm	§4.1
Charge interval		§8.1, F18
Primary interval	$4 \leq Z \leq 28$	
Extended interval	$1 \leq Z \leq 30$	
Energy interval for mass analysis		§8.1, F18
O	60–280 MeV nucl^{-1}	
Si	85–400 MeV nucl^{-1}	
Fe	115–570 MeV nucl^{-1}	
Field of view		F6
Primary FOV [b]	45° half angle	
Full FOV	70° half angle	
Geometrical factor (for $\theta < 45°$)	$\sim$250 cm^2 sr	§8.2, F19
Event yields (solar minimum; $\theta < 45°$)		§8.2, F20
Be	3.0×10^4 yr^{-1}	
O	8.9×10^5 yr^{-1}	
Si	1.7×10^5 yr^{-1}	
Fe	1.4×10^5 yr^{-1}	
Mass resolution (r.m.s.)		§8.3, §9.1, F21, F23, §A
O	$\lesssim$ 0.15 amu	
Fe	$\lesssim$ 0.25 amu	
Resource utilization		§7
Dimensions ($l \times w \times h$)	53.3 cm $\times$ 43.8 cm $\times$ 23.5 cm	
Mass	29.2 kg	
Instrument power	12 W	
Bit rate	464 bits s^{-1}	

[a]Sections (§), Figures (F), and Tables (T) containing additional details.
[b]The definition of a 'primary' field of view is somewhat arbitrary. In optimizing the instrument design, 45° was treated as a practical maximum angle.

dominant heavy elements bears a remarkable resemblance to the pattern of abundances found in solar system material, although more subtle differences between these populations of matter are of great interest (see Section 2.1). Both cosmic rays and the solar corona/solar wind show evidence of having undergone chemical alteration, and the resulting fractionation patterns are strikingly similar. Furthermore, it is widely thought that diffusive shock acceleration is responsible for energizing galactic cosmic rays, just as it is for accelerating anomalous cosmic rays and for producing energetic particle events in interplanetary space. Thus the physical phenomena addressed by CRIS are in many respects the same as those studied with other ACE spectrometers, but on larger spatial, temporal, and energy scales. The remainder of this section briefly outlines some of the specific measurements CRIS is designed to make.

It has long been recognized that in cosmic rays the observed abundance ratio (Li+Be+B)/(C+N+O) exceeds the value found in solar system material by a factor of about 10^5, and that this must be a measure of how much material cosmic rays have traversed since they were accelerated. Specifically, cosmic rays consist of primary nuclei, that is, nuclei such as C and O which reach Earth without undergoing fragmentation, and secondary nuclei, such as Li, Be, and B, which are the products of fragmentation reactions in the interstellar medium.

Primary cosmic rays must be products of stellar nucleosynthesis, but it is not known what types of stars are primarily responsible for producing this material, or whether the acceleration of the cosmic rays has any direct association with the stellar sources responsible for the nucleosynthesis of the accelerated particles. For example, supernovae explosions may accelerate material from the parent stars, or perhaps supernova shocks accelerate material which was previously ejected into interstellar space by other supernovae explosions and stellar winds. In any case, successful models of how cosmic rays propagate will both account for secondary production and provide information on the spatial distribution of sources. Using such models, it is possible to determine the GCRS abundances and gain information about the nature of the sources themselves (e.g., see Stone and Wiedenbeck, 1979). Hence high-resolution measurements by CRIS of cosmic-ray isotopes will provide vital clues to the astrophysics of our Galaxy.

In addition to stable isotopes, the cosmic rays contain long-lived radioactive nuclides of either primary or secondary origin. Table II lists the decay modes and half lives of the radioactive species with atomic number $Z \leq 30$ that should be observable. The observed abundances of these 'clock' isotopes can be used for establishing various time scales related to the origin of cosmic rays.

There are still important gaps in our understanding of the origin of galactic cosmic rays, both in terms of the source of the parent material from which they are derived and in terms of the details of how and where they are accelerated to relativistic energies. Narrowing the range of possible explanations for the origin and propagation history of galactic cosmic rays is the observational goal of CRIS.

TABLE II

Radioactive nuclides

Nuclide	Decay Mode	Partial Half life[a]
^{7}Be	ec	53 days
^{10}Be	β^{-}	1.5 Myr
^{14}C	β^{-}	5730 yr
^{26}Al	β^{+}	0.87 Myr
	ec	4.0 Myr
^{36}Cl	β^{-}	0.30 Myr
^{37}Ar	ec	35 days
^{41}Ca	ec	0.10 Myr
^{44}Ti	ec	49 yr
^{49}V	ec	330 days
^{51}Cr	ec	28 days
^{53}Mn	ec	3.7 Myr
^{54}Mn	ec	312 days
	β^{+}	400 Myr
	β^{-}	0.8 Myr[b]
^{55}Fe	ec	2.7 yr
^{56}Ni	ec	6.1 days
^{56}Ni	β^{+}	$\gtrsim 1$ yr
^{57}Co	ec	270 days
^{59}Ni	ec	76000 yr

[a]Laboratory half lives.
[b]Estimated value (Zaerpoor et al., 1997).

There are several distinct groups of isotopes that one can take advantage of to address various aspects of this problem:

− Primary isotopes: these isotopes reveal the conditions of nucleosynthesis in the cosmic-ray sources. Models of cosmic-ray propagation, based in large part on the other isotope groups, are essential for determining the abundances of the primary isotopes. As will be discussed below, the primary isotopes have undergone some type of chemical fractionation. It is of major interest to understand how and where this has occurred.

− Acceleration delay clocks: primary isotopes which decay by electron capture can provide information about the elapsed time between nucleosynthesis and particle acceleration.

− Propagation clocks: secondary isotopes which decay by $\beta^{\pm}$ emission can measure the average time between the production of these particles and their escape from the Galaxy (Ptuskin and Soutoul, 1998).

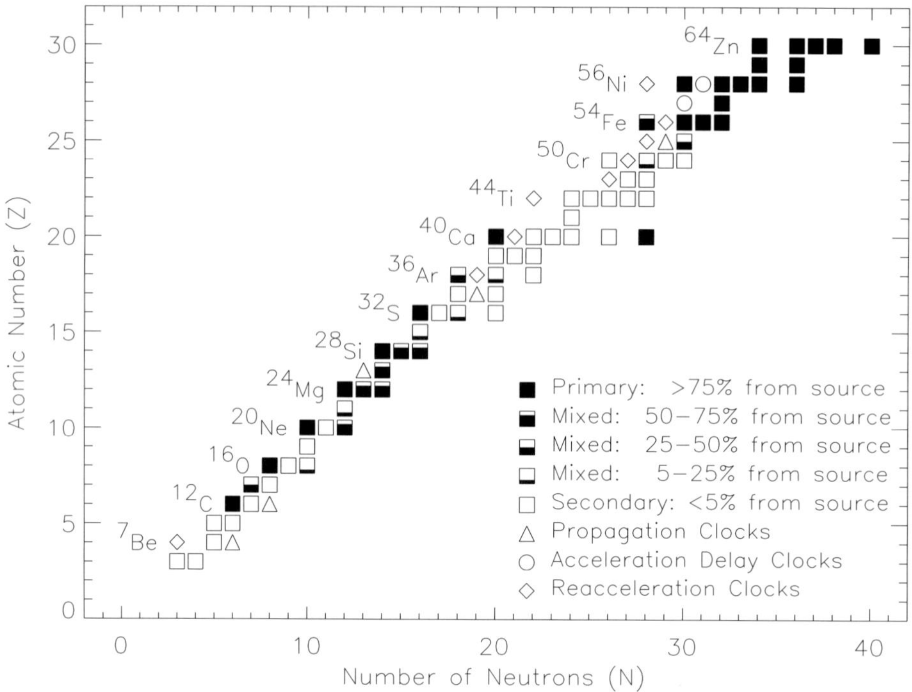

Figure 1. Chart showing the atomic number *Z* and neutron number *N* for nuclides present in galactic cosmic rays. The symbols used for the various nuclides indicate the types of studies to which they are expected to contribute, as discussed in the text. In some cases these assignments are uncertain, particularly for the rarer nuclides.

– Reacceleration clocks: secondary isotopes which decay solely by electron capture can do so only if their velocities have, at some time, been much lower than the velocities at which they are observed. Hence they are potentially probes of energy changing processes that occur during propagation in the Galaxy.

Figure 1 shows which isotopes are in these different groups. Examples of each are discussed in the following paragraphs.

2.1. PRIMARY ISOTOPES

Ratios of individual GCRS element abundances to the corresponding solar system elemental abundances are ordered by first ionization potential (FIP): elements with FIP below ~10 eV are about 5 times more abundant in the galactic cosmic-ray sources than are the higher FIP elements (see Figure 2). A similar FIP effect is well known for solar energetic particles (e.g., Meyer, 1985; Stone, 1989) and is presumably related to the fact that ions, being charged, are transported from the photosphere into the corona more efficiently than neutrals. A model which could perhaps account for the FIP effect in galactic cosmic rays would have the source ejecting its outer envelope for a long period with a FIP selection effect (as the

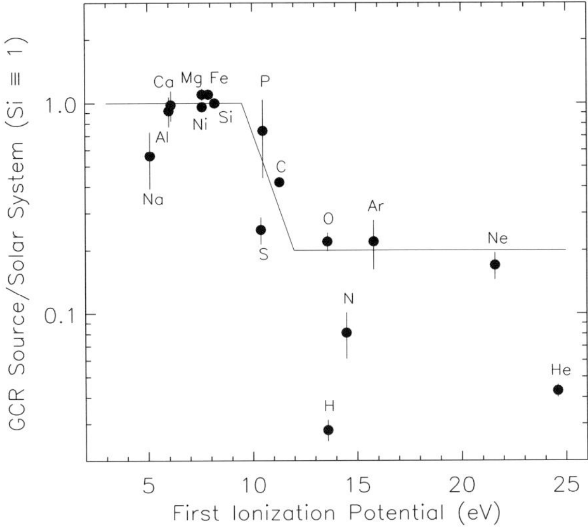

Figure 2. Ratio of galactic cosmic-ray source abundance to solar system abundance versus the element first ionization potential (FIP), normalized to 1 for Si. Cosmic-ray values are from Lund (1989) and are based primarily on data from the HEAO-3-C2 experiment (see Engelmann et al., 1990). More recent data at lower energies from *Ulysses* (DuVernois and Thayer, 1996) are in generally good agreement with these values. The solid line indicates the general trend of the heavy element data. Cosmic-ray hydrogen and helium are distinctly underabundant, even compared to the other high-FIP elements.

sun does) followed by a supernova shock which sweeps up and accelerates this material. Recently it has been claimed (Meyer et al., 1998; Ellison et al., 1998) that the FIP effect is due to non-volatiles being accelerated as grains, while the volatiles are accelerated as individual nuclei (i.e., they propose that the FIP effect is misnamed and really is a volatility effect). In any case, it appears that the cosmic rays have undergone a chemical fractionation process, the origins of which are uncertain.

The GCRS abundances should reveal conditions in the sources themselves. Elemental abundances can probe the chemical processes which fractionate the material prior to acceleration. Such processes should have little, if any, influence on the relative abundances of the isotopes of an element. Particle acceleration, however, is likely to depend on the particle's rigidity, and therefore on its mass-to-charge ratio. Nevertheless, such effects may average out when one observes the composition resulting from a large number of acceleration events, as has been seen in solar energetic particle data (Breneman and Stone, 1985; Stone, 1989). Thus it is thought that the isotopic composition of the GCRS should serve primarily as an indicator of the nuclear processing that the material has undergone.

The abscissa in Figure 3 shows, for different elements, the ratios of low abundance isotopes of each element to the corresponding dominant isotope. The ordinate of Figure 3 shows for each such ratio the GCRS deduced value divided by

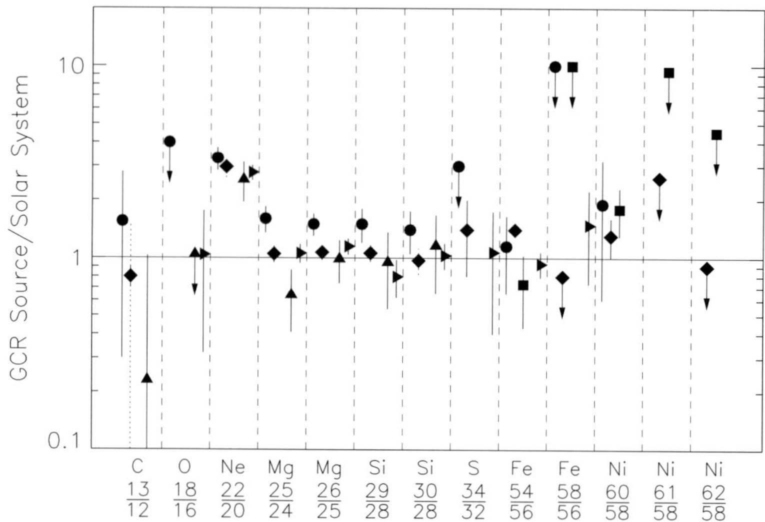

Figure 3. Observed values of various galactic cosmic-ray source isotopic abundance ratios divided by the corresponding solar system ratios. The labels along the abscissa indicate the element and the mass numbers of the isotopes involved in the plotted ratios. Note that in some cases only upper limits are presently available. Symbols: circle – data from prior to 1989, summarized by Mewaldt (1989); diamond – *Ulysses* data from Connell and Simpson (1995, 1997a), Thayer (1997); square – ISEE-3 data from Leske and Wiedenbeck (1993), Leske (1993); triangle pointing up – CRRES data from DuVernois et al. (1996); triangle pointing right – Voyager data from Lukasiak et al. (1997), Webber et al. (1997). The *Ulysses* points are shown with two error bars – the short, solid bars indicate the contribution from the statistical errors in the measurements while the longer, dashed bars give an estimate of the overall uncertainty in the source abundances when propagation errors are also taken into account. For isotopes with large secondary components such as ^{13}C the propagation errors dominate.

the solar system value. With the exception of Ne, most of the ordinate values are very near to unity, i.e., the GCRS isotopic compositions are very similar to those of the solar system. On the other hand, the existing measurements can be greatly improved upon by CRIS and real differences may emerge which are currently obscured by large uncertainties. The Ne anomaly has been known for some time (Fisher et al., 1976). It has been proposed by Cassé and Paul (1982) that Wolf–Rayet stars contribute approximately 2% of all cosmic-ray source material (and $\sim$25% of heavy nuclei), but that the high ^{22}Ne/^{20}Ne ratio ($\sim$120) in the material ejected from Wolf–Rayet stars could account for the ^{22}Ne excess in the GCRS.

2.2. ACCELERATION DELAY CLOCKS

The observed correlation of the fractionation pattern observed in cosmic-ray elemental abundances with atomic properties (Figure 2) suggests that the source material experiences some time at temperature $\lesssim$ 10 eV after nucleosynthesis and before acceleration. If cosmic rays were produced from grains formed in the ejecta of a supernova explosion this time could be relatively short. On the other hand, if the accelerated material were injected from the coronas of flare stars the time delay could be a significant fraction of the age of the Galaxy. Thus a determination of the time delay between synthesis and acceleration could prove decisive in distinguishing among models for the origin of cosmic rays.

It has been proposed (Soutoul et al., 1978) that this time delay can be established using observations of the abundances of radioactive nuclides which are produced in supernova explosions and can decay only by electron capture. In a low-temperature environment these species will decay, but once they are accelerated they are rapidly stripped of their electrons and become effectively stable. The most promising candidates are ^{59}Ni and ^{57}Co, which have half lives $\sim 10^5$ years and ~ 1 year, respectively, if they are not stripped of their electrons (see Table II).

There are already indications that there may have been a significant delay ($>10^5$ years) between cosmic-ray nucleosynthesis and acceleration. For example, Connell and Simpson (1995) have reported a ^{59}Ni/^{58}Ni ratio which suggests the absence of ^{59}Ni in the cosmic-ray source. The isotope ^{59}Co, which is produced by the decay of ^{59}Ni, can also serve as an indicator of this decay. Leske (1993) has also reported a significant delay, based on observations of ^{59}Co and an upper limit on ^{59}Ni.

Because the isotopes of mass 57 and 59 have low abundances and, in the cases of ^{59}Ni and ^{57}Fe, are adjacent to nuclides that are much more abundant, measurements with excellent mass resolution and large collecting power are needed to definitively determine whether the cosmic-ray source material has experienced a long time delay prior to acceleration. CRIS is expected to provide measurements of the necessary quality.

2.3. PROPAGATION CLOCKS

Propagation clocks are secondary isotopes which decay by $\beta^{\pm}$ emission. These provide a measure of the cosmic-ray 'confinement time' (or 'escape time'), the average time which passes between the production of cosmic rays and their escape from the Galaxy. A convenient parameter to use in discussing the decay of propagation clocks is the 'surviving fraction', the ratio of the observed abundance to the abundance expected if the nuclei did not have time to decay. This quantity is plotted in Figure 4 as a function of the half life of the nuclide, based on a standard, homogeneous 'leaky-box' model of cosmic-ray propagation.

The confinement time for cosmic rays in the Galaxy has been estimated from the surviving fraction of ^{10}Be, which decays by β^- emission with a half life of 1.5×10^6 years. The result is only about 1–2×10^7 years (Wiedenbeck and Greiner, 1980; Garcia-Muñoz et al., 1977). The corresponding mean gas density in the confinement volume, ~ 0.3 atoms cm^{-3}, is less than commonly accepted values of the mean interstellar gas density in the galactic disk and it has therefore been argued that cosmic rays must spend a significant amount of time outside of the disk. The short confinement time places heavy requirements on the energy output of the

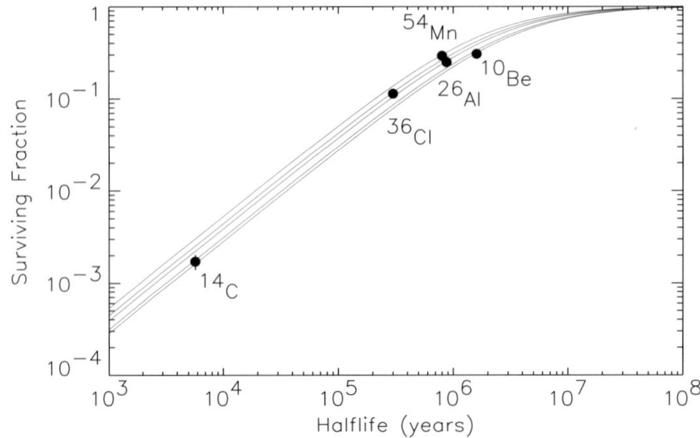

Figure 4. Expected surviving fraction of various radioactive secondary nuclides as a function of their half lives. The calculated surviving fractions are based on a leaky box model of cosmic-ray propagation with the confinement time deduced from the measured ^{10}Be/^{9}Be ratio. The different curves, corresponding to different nuclear masses (plotted for $A = 10$, 14, 26, 36, and 54 from bottom to top), are due to differences in the mean path length for nuclear fragmentation loss for different species. In these calculations time dilation, ionization energy loss, and solar modulation effects were neglected.

sources of cosmic rays ($\sim 10^{41}$ ergs s^{-1} in our Galaxy (Berezinski et al., 1990)) and also means that the GCR source material could be significantly younger than solar system matter.

The isotope ^{10}Be predominantly measures the confinement time for C, N, and O. It is important to study other such clocks to determine whether or not other elements (e.g., those near the iron peak), might have different histories. Possibilities include ^{26}Al, ^{36}Cl, and ^{54}Mn, although the case of ^{54}Mn is difficult because its half life for β^- decay has not been directly measured (Zaerpoor et al., 1997).

Recently, *Ulysses* observations of 92 ^{26}Al nuclei have yielded a confinement time of 16 ± 3 million years (Connell and Simpson, 1997b). CRIS is expected to observe about 3000 ^{10}Be, 3000 ^{26}Al, and 1000 ^{36}Cl in two years (see Section 8.2). With measurements having good statistical accuracy for several clock isotopes over a range of half lives it may also be possible to test more realistic models of cosmic-ray propagation (Ptuskin and Soutoul, 1998), since such data will provide some sensitivity to the distribution of confinement times in the Galaxy, in addition to providing the mean of this distribution.

The isotope ^{14}C is another potentially interesting secondary which decays by β^- emission. Its half life is very short in comparison with the other clock isotopes discussed above, only 5730 years. Secondary ^{14}C produced by fragmentation of heavier nuclei in the interstellar medium will be observable only if it is produced in the immediate neighborhood of the solar system. Based on a homogeneous leaky-box model of cosmic-ray propagation with the confinement time deduced from the ^{10}Be abundance, one expects a ^{14}C surviving fraction of ~ 1–2×10^{-3}

(see Figure 4). With the CRIS instrument's large geometrical factor it should be possible to observe $\sim$30 ^{14}C nuclei in 2 years under these conditions. However, if the interstellar density in the region sampled during the 5730 yr half life of ^{14}C were significantly lower or higher than the average density in the confinement volume, the number of ^{14}C nuclei produced in this local volume would be correspondingly modified. By comparing the density deduced from the ^{14}C measurement with astronomical measurements of the density distribution in the local interstellar medium, it may be possible to derive a limit on the cosmic-ray diffusion coefficient in the local interstellar medium, since this parameter affects the size of the region sampled by the ^{14}C clock.

2.4. REACCELERATION CLOCKS

The abundances of nuclides that can decay only by electron capture are sensitive to the density of the medium in which the particles propagate and to the particle energy at which the propagation occurs. At high energies, cosmic rays spend essentially all their time fully stripped of their orbital electrons, so the electron capture mode of decay is no longer possible and these particles are effectively stable. At somewhat lower energies, $\sim$100 MeV nucl^{-1} and below, the particles will occasionally attach an electron from the surrounding medium and retain it until the electron is either stripped back off in a collision with an atom in the interstellar gas or until the nucleus captures it and decays. In a high-density medium, the time between stripping collisions is short and the nuclei attain an equilibrium charge-state distribution with decay losses depending on the fraction of this distribution which is not fully stripped and on the electron capture half life. At low densities, nuclei which attach an electron are likely to decay before it can be stripped off, so the decay rate for these particles depends primarily on their electron attachment cross section. Thus abundances of electron capture nuclei can, in principle, provide information about the interstellar density in the propagation medium (see, for example, Letaw et al., 1993, and references therein) which can complement that obtained from the propagation clocks discussed above.

The cross sections for electron stripping and, particularly, for electron attachment are very strong functions of the velocity of the nucleus. If, for example, the electron capture nuclide ^{55}Fe ($T_{1/2} = 2.6$ yr in the laboratory) is produced by fragmentation of ^{56}Fe at 500 MeV nucl^{-1} and propagates at essentially this same energy for $\sim$10^7 years, there is negligible chance that it will decay. However, if the ^{55}Fe is produced at 50 MeV nucl^{-1} and propagates at this low energy until reaccelerated by an interstellar shock, there is a significant chance that it will decay before the reacceleration. It has been proposed (Letaw et al., 1985) that electron capture nuclides can be used to probe changes in cosmic-ray energies during propagation.

A variety of secondary isotopes that can decay only by electron capture are available for such investigations. These are listed in Table II, along with their laboratory half lives. (With only a single electron attached half lives are a factor

~2 longer.) The interpretation of the abundances of electron capture secondaries will be complicated by the joint dependence on density and particle energy, and also by the energy changes that occur due to solar modulation as the particles enter the heliosphere. With precise measurements of the abundances of a range of electron capture nuclides it may be possible, for the first time, to extract some of the information carried by this class of cosmic rays.

3. Design Requirements and Instrumental Approach

If CRIS is to meet the objectives discussed in Section 2 and make a significant contribution beyond previous experiments, it must be capable of precisely measuring even the rarest cosmic-ray nuclei between Be ($Z = 4$) and Ni ($Z = 28$). The separation of rare nuclides from adjacent abundant species (for example, ^{57}Fe from ^{56}Fe) demands a mass resolution $\lesssim 0.25$ amu. In addition, the measurement of rare isotopes demands sufficient collecting power to accumulate a statistically significant sample of these particles over the duration of the ACE mission. The necessary geometrical factor, $\gtrsim 100$ cm^2 sr (see Section 8.2), is considerably larger than that of previous cosmic-ray isotope spectrometers flown in space.

The sensor approach that was adopted for CRIS, using multiple measurements of dE/dx vs total energy (see Section 3.1), has been developed and proven over the course of a large number of space missions. During the ACE Definition Study a critical evaluation was made of instrumental requirements for achieving the needed 0.25 amu mass resolution. This process was greatly aided by the heritage of analyses and accelerator data accumulated from earlier isotope spectrometers, such as those flown on ISEE-3, CRRES, *Ulysses*, SAMPEX, and WIND. The joint requirements for precise mass determination and large collecting power influenced essentially all aspects of the CRIS instrument design, placing requirements on the sensors, the electronics, and the mechanical construction. They also necessitated an extensive program of accelerator tests and calibrations, both of individual sensor elements and of the completed instrument.

3.1. MASS AND CHARGE ANALYSIS TECHNIQUE

In CRIS, the charge and mass of detected particles are identified based on the energies they deposit in coming to rest in a stack of silicon solid-state detectors. When a particle of charge Z, mass M, and kinetic energy E penetrates an amount of matter L and emerges with residual kinetic energy E', the change in its range must equal the thickness penetrated:

$$\mathcal{R}_{Z,M}\left(E/M\right) - \mathcal{R}_{Z,M}\left(E'/M\right) = L\,, \tag{1}$$

where the function $\mathcal{R}_{Z,M}$ is available in tabulated form (see, for example, Hubert et al., 1990). Thus from measurements of E', $\Delta E = E - E'$, and L this fundamental

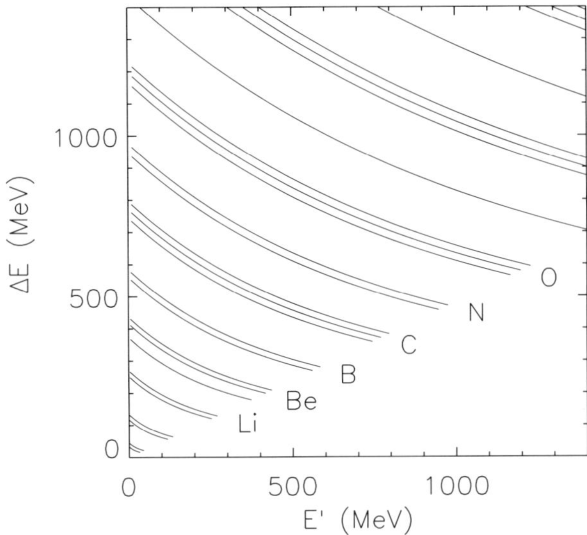

Figure 5. Calculated relationship between energy loss (ΔE) and residual energy (E') for nuclei penetrating a 6 mm silicon detector and stopping in the following 6 mm detector. Each curve corresponds to a different stable or long-lived nuclide. Selected elements are labeled. The locations of the response tracks also depend on the angle of incidence. An approximate correction for this dependence can be made by dividing both ΔE and E' by $(\sec\theta)^{1/a}$ where $a \simeq 1.7$.

equation of the 'ΔE–E' technique' provides information about a particle's charge and mass. An iterative technique for solving this equation is discussed by Stone et al. (1998). Figure 5 illustrates the ΔE vs E' relationship expected for isotopes of several elements when the energy loss is measured in a 6 mm thick layer of silicon.

Further insight into this method can be obtained by introducing an approximate power-law form for the range–energy function, $\mathcal{R}_{Z,M}(E/M) \simeq k(M/Z^2)(E/M)^a$. This form, with $a \simeq 1.7$, is a reasonable approximation to the range–energy relation over much of the energy interval of interest for CRIS. Substituting the power-law form of the range–energy relation into Equation (1) and solving for mass one obtains:

$$M \simeq \left(k/Z^2 L\right)^{1/(a-1)} \left(E^a - E'^a\right)^{1/(a-1)} .$$ (2)

Similarly, by assuming a nominal mass-to-charge ratio typical of the elements of interest, $M/Z \simeq 2 + \epsilon$, one can also solve for the charge:

$$Z \simeq \left(\frac{k}{L\,(2+\epsilon)^{(a-1)}}\right)^{1/(a+1)} \left(E^a - E'^a\right)^{1/(a+1)} .$$ (3)

From these equations one can readily show that the spacing between 'tracks' for adjacent elements on the ΔE–E' plot is larger than the spacing between adjacent isotopes of an element by a factor $(2+\epsilon)(a+1)/(a-1) \simeq 8$. For elements with

$Z \leq 30$ the number of stable or long-lived isotopes is small enough that one never encounters an ambiguity in the identification of charge and mass. In practice, one first calculates the particle's charge. A histogram of the resulting values for a population of particles yields well-separated element peaks. Identification of these peaks gives the integer charge values needed for the mass calculation.

The approximate expressions for charge and mass (Equations (2) and (3)) are also helpful for assessing the importance of various sources of uncertainty in ΔE, E', and L. For example, since $M \propto L^{1/(a-1)}$, one sees that a fractional error σ_L/L in the penetrated thickness contributes a fractional uncertainty in the mass $(\sigma_M/M)_L = (\sigma_L/L)/(a-1)$. For iron, this implies that to keep the contribution to the mass uncertainty from thickness errors to less than 0.1 amu, one needs to determine the thickness to $\sim$0.12% or better. Since CRIS is designed to measure particles incident over a wide range of angles, the value of L depends not only on the detector thickness (denoted by L_0), but also the angle of incidence, θ: $L = L_0 \sec \theta$. To provide the necessary determination of θ, CRIS includes a sensor system capable of precisely measuring particle trajectories. Furthermore, the detector thicknesses are not perfectly uniform and it is necessary to apply maps of the measured thickness as a function of location on the detector. The trajectory system also provides the event-by-event location information needed to make these mapping corrections.

In CRIS there is the additional complication that the detectors have significant surface 'dead layers' in which energy lost by a particle will not contribute to the measured pulse height (see Section 4.2). Equation (1) can be generalized to take these dead layers into account (see Appendix A). Maps of measured dead layer thickness are used to correct for the spatial dependence of this quantity.

Even when the quantities ΔE, E', L_0, and θ are well determined there are important contributions to the mass resolution caused by several physical sources of fluctuations occurring as particles slow in matter. Most important among these are Bohr/Landau fluctuations of the energy loss in the ΔE detector (Rossi, 1952) and multiple Coulomb scattering (Barnett, 1996), which causes the particle's path to deviate from the straight line derived from the trajectory measurements. One strives for an instrument design in which the mass resolution is dominated by these fundamental contributions.

Appendix A discusses the various contributions to the mass resolution, and Section 8.3 presents the overall mass resolution calculated for selected elements using the parameters of the CRIS instrument.

3.2. DETAILED REQUIREMENTS

The development of a CRIS 'error budget' that would make it possible to achieve the required overall mass resolution $\lesssim 0.25$ amu led to specific design requirements on a number of the subsystems of CRIS. These included requirements on (1) the gain, linearity, noise, stability, and dynamic range of the pulse-height analysis electronics; (2) the leakage current, thickness uniformity, and dead layers of the solid-state detectors; (3) the position and angular resolution of the trajectory system; and (4) the alignment and stability of the mechanical design. In addition to these design requirements, it became clear that to achieve optimal mass resolution required mapping the thickness and dead layers of each of the individual solid-state detectors to an accuracy of a few μm, and it required accelerator calibrations of the completed instrument.

Table III summarizes the requirements that were adopted and that guided the design and development of the CRIS instrument. The following sections describe the instrument that was designed to meet these requirements, and present the results of calibrations and calculations used to demonstrate that the intended performance has been achieved.

4. Sensor System

The CRIS sensor system consists of a scintillating optical fiber trajectory (SOFT) hodoscope for measuring the location and direction of incidence of detected nuclei, and four silicon solid-state detector telescopes for measuring the energy loss signatures of these particles. Figure 6 schematically shows the configuration of the instrument. Figure 7 contains a photograph of the fully assembled CRIS instrument. The SOFT hodoscope is described in Section 4.1 and the silicon detector telescopes are discussed in Section 4.2.

4.1. THE SCINTILLATING OPTICAL FIBER TRAJECTORY (SOFT) SYSTEM

The SOFT system (Figure 6) consists of a hodoscope composed of three xy scintillating fiber planes (six fiber layers) and a trigger detector composed of a single fiber plane (two fiber layers). The hodoscope and trigger fibers are coupled to an image intensifier which is then coupled to a charge-coupled device (CCD) for hodoscope readout, and to photodiodes to obtain trigger pulses. The SOFT detector system has two fully redundant image-intensified CCD camera systems viewing opposite ends of the fibers. Only one of these camera systems is operated at any given time because of power and bit rate limitations. Figure 8 is a photograph showing the internal layout of the SOFT system.

4.1.1. *Scintillating Fiber Hodoscope*
The scintillating fibers consist of a polystyrene core doped with scintillation dyes (BPBD and DPOPOP, emission peak 430 nm (Davis et al., 1989)) surrounded by an acrylic cladding. These fibers, fabricated by Washington University, have a 200 μm square cross section which includes a 10 μm cladding thickness on each side. The cladding of the hodoscope fibers is coated with a black ink (called 'extramural absorber' or EMA) to prevent optical coupling between fibers. The fibers are

TABLE III

CRIS design requirements

Parameter	Requirement
Trajectory system	
Zenith angle resolution	$<0.1°$
Relative position resolution	<0.13 mm r.m.s.
Absolute position resolution	<1 mm r.m.s.
Detection efficiency	
$Z = 4$ (Be)	$>50\%$
$Z \geq 8$	$>90\%$
Silicon detectors	
Active thickness	3.0 ± 0.1 mm
Thickness variation	<60 μm
Dead layer thickness	$\lesssim 50$ μm
Depletion voltage	<250 V
Leakage current (20 °C)	<15 μA
Pulse-height analyzer electronics	
Dynamic range	$>700{:}1$
Quantization resolution	12 bits or better
Nonlinearity	$<0.05\%$
Gain temperature coefficient	<200 ppm/°C
Offset temperature coefficient	<200 ppm/°C
Ballistic deficit	$\lesssim 0.2\%$
Noise (r.m.s., for guards)	<50 keV
Power (per pha)	<100 mW
Mechanical construction	
Lateral position tolerance	
Trajectory system	0.25 mm plane-to-plane
	1 mm with respect to Si detectors
Silicon detectors	1 mm detector-to-detector
Orientation co-planarity	
Trajectory system	$0.1°$
Silicon detectors	$0.1°$
Detector azimuthal orientation	
Trajectory system	$0.1°$ plane-to-plane
	$1°$ with respect to Si detectors
Silicon detectors	$1°$ detector-to-detector

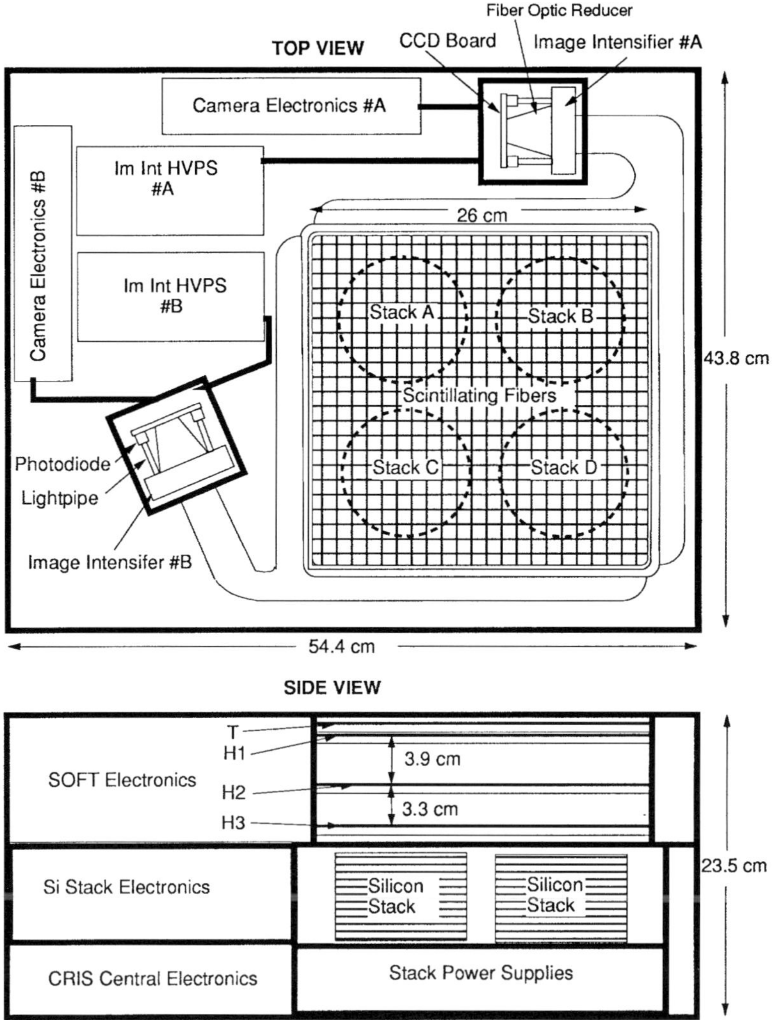

Figure 6. CRIS instrument cross section. The side view shows the fiber hodoscope which consists of three hodoscope planes (H1, H2, H3) and one trigger plane (T) which are located above the four stacks of silicon detectors. The top view shows the fiber readout which consists of two image intensified CCDs at either end of the fibers and the four stacks of silicon detectors.

bonded together with an elastomeric adhesive (Uralane 5753). Each of the three hodoscope planes is composed of two layers of orthogonally crossed fibers. The fiber planes are constructed by mounting one layer (or ribbon) of fibers, which is 26 cm wide and about 125 cm long, onto a 25 μm thick Kapton substrate which has been stretched on a square cross-section frame with outside dimensions of 28 cm$\times$28 cm. A second fiber layer is then mounted orthogonal to the first layer to form an xy plane. Uralane adhesive is used to bond the first layer to the Kapton substrate and the second layer to the first layer. The second layer also has a thin

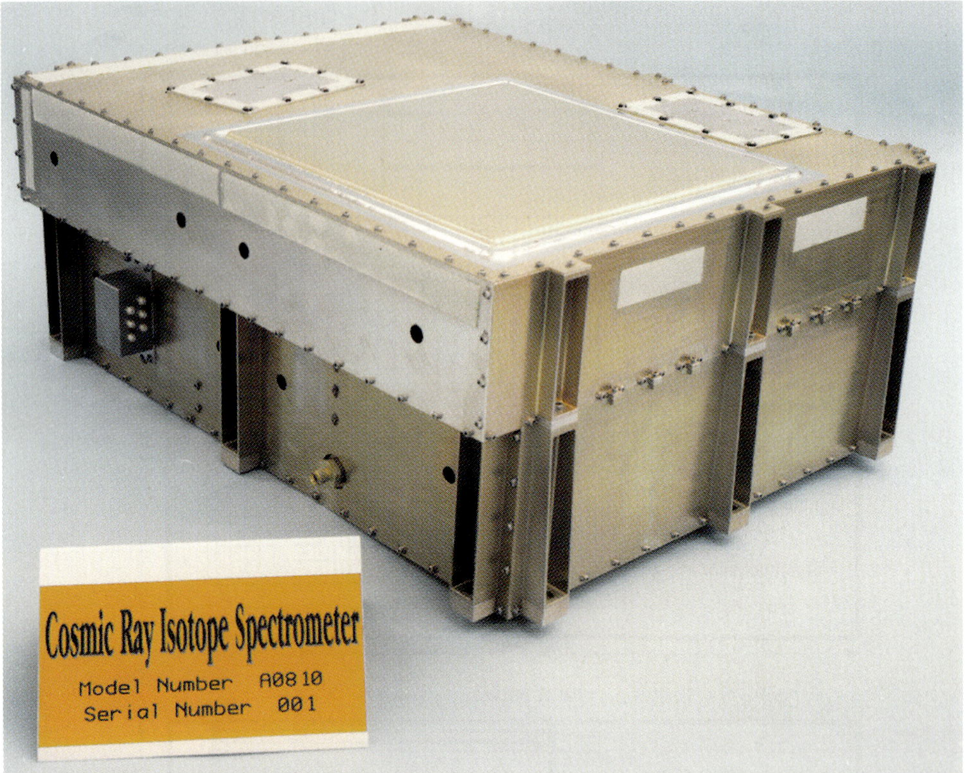

Figure 7. Fully assembled CRIS instrument consists of two boxes bolted together. The upper box contains the SOFT system, while the lower box contains the Si(Li) detector stacks with their pulse-height analysis electronics, as well as the main CRIS control electronics. The large window on the top of the SOFT box is the CRIS entrance aperture. The two smaller patches are thermal radiators for cooling the two CCD cameras.

coating of adhesive on top which completes the encapsulation of the fibers. The crossed-fiber planes form an active region 26 cm×26 cm centered on the frame and Kapton substrate, with approximately a 50 cm length of fibers on either end of each layer to light pipe the photons to the two image intensifier assemblies (Figures 6 and 8).

The spacings between the hodoscope planes are 3.9 cm and 3.3 cm for H1–H2 and H2–H3, respectively. (This unequal plane spacing is used to resolve trajectory ambiguities which can occur for low-Z nuclei due to electron 'hopping' in the intensifier, as discussed in Section 9.3.) Each of the four 26 cm wide hodoscope fiber outputs for a plane is split into 11 'tabs' with width $\sim$2.4 cm of contiguous fibers (about 120 fibers). The tabs are then stacked together with a spacer of thickness 50 μm placed between each tab at the end of the ribbon. The tabs with spacers are bonded together forming four rectangular outputs for each plane with dimensions $\sim$0.3 cm× $\sim$2.4 cm.

Figure 8. Photograph showing a top view of the SOFT detector before integration with the rest of the CRIS instrument. The fiber plane which is visible is the trigger plane. The hodoscope planes are below that and are not visible. The fiber outputs are routed to two image intensifier/CCD assemblies, one at the upper right and one at the lower left of the photo. Thermal radiators are attached to the tops of the assemblies in flight (not shown in this photo, see Figure 7) to cool the image intensifiers and CCD arrays. The high-voltage power supplies for the image intensifiers are located to the upper left of the fiber plane. The readout electronics for each camera are in the black boxes in the upper left of the photograph.

Each of the two sets of six fiber layer outputs (denoted H1x, H1y, H2x, H2y, H3x, and H3y) is routed to one of the image intensified CCD cameras. The six outputs are stacked as illustrated in Figure 9, and clamped together forming a block which is then cut and polished and coupled to the image intensifier using Dow-Corning 93-500 adhesive with thickness $\sim$50 μm. This thickness was necessary to eliminate flashes in the image intensifier caused by electrostatic buildup between the plastic fibers and the image intensifier faceplate. The bundle of fibers running from the active area to the image intensifier was then encapsulated with Uralane.

The trigger fiber plane is fabricated in the same way as the hodoscope planes with the exception that the fibers are not coated with EMA so that maximum sensitivity can be obtained. The trigger fiber outputs are formatted similarly to the hodoscope fibers with the x-layer outputs being placed above and the y-layer outputs placed below the hodoscope outputs on the face of the image intensifier.

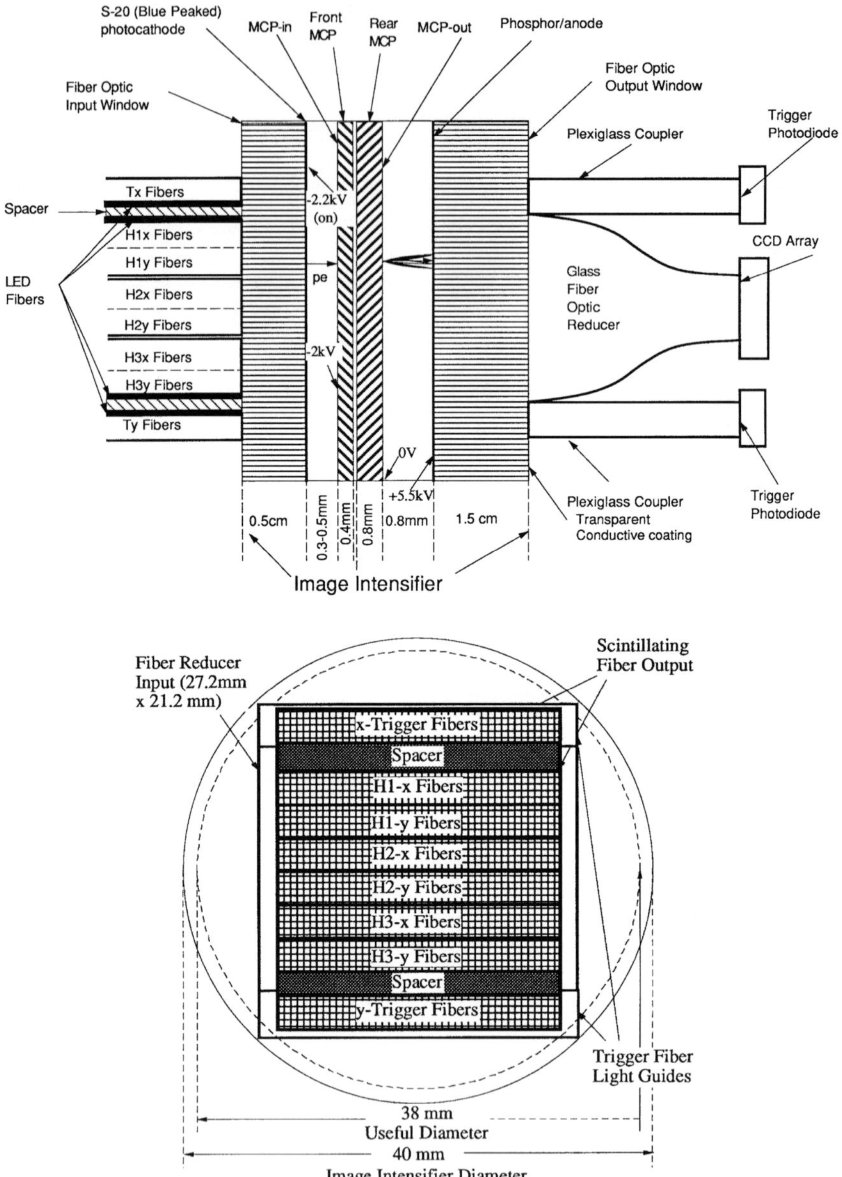

Figure 9. Schematic view of SOFT image intensified CCD camera. The top section of the figure shows a side view of the fiber readout. The scintillating fibers are shown on the left of the image intensifier and are proximity focused onto the fiber-optic face plate of the image intensifier. At the output of the image intensifier the hodoscope fibers (H1–H3) are fiber-optic reduced onto the CCD for readout. The trigger fibers (Tx and Ty) are interfaced with a Plexiglass coupler to photodiodes which produce signals used in the CRIS coincidence determination. The bottom section of the figure shows a face view of the fibers as they are formatted onto the image intensifier. The hodoscope fibers are formatted into the center of the image intensifier and the trigger fibers are spaced off at the top and bottom from the hodoscope fibers so that their output is clearly separated. The fiber reducer input (output of the image intensifier), which proximity focuses the image onto the CCD, and the trigger fiber light guides are also shown.

TABLE IV

Materials in cosmic-ray beam path[a]

Item	Material	Composition	Thickness[b] (mm)	Areal density[b] (g cm^{-2})
MLI thermal blanket	Multiple material types			0.083
Entry window	Aluminum	Al	0.254	0.069
Trigger fiber plane	Polystyrene	C_8H_8	0.324	0.033
	Acrylic	$C_5H_8O_2$	0.076	0.009
	Uralane 5753		0.175	0.018
	Kapton	$C_{22}H_{10}O_5N_2$	0.025	0.004
H1 fiber plane	— same as trigger fiber plane —			0.064
H2 fiber plane	— same as trigger fiber plane —			0.064
H3 fiber plane	— same as trigger fiber plane—			0.064
Exit window	Aluminum	Al	0.127	0.034
			Total:	0.442

[a]Above the silicon detector stacks.

[b]Average over active area.

Spacers are placed between the hodoscope and trigger fiber outputs as shown in Figure 9 to clearly separate the readout of hodoscope and trigger fibers.

Four extra tabs of fibers which are not part of the hodoscope or trigger planes are placed above and below the trigger fiber spacers (for each camera) and are attached to LEDs. These are used for functional testing and alignment checks of the system. Five groups of several fibers in each of the four tabs are attached to the LEDs.

Table IV lists the areal density of materials in SOFT including the thermal insulation and entry window dome as well as the thin aluminum sheet separating the fiber detector from the silicon stack detectors. The SOFT hodoscope is a self-contained assembly which can be tested independent of the rest of the CRIS instrument.

4.1.2. *Image Intensifier Assembly*

A cross-sectional view of the image intensifier assembly is shown in Figure 9. The image intensifiers are 40 mm diameter, dual microchannel plate (MCP), gateable devices (Photek Model MCP-340S). The rear MCP is double thickness (0.8 mm) so that the gain provided by the tube is equivalent to that for a three-MCP intensifier. The tubes are each powered by a high-voltage power supply (Model 2404) made by K&M Electronics. The intensifiers have fiber optic faceplates on the input and output. The blue peaked, S-20 photocathodes have a sensitivity of $\sim$50 mA W^{-1} at 450 nm wavelength (roughly 15% quantum efficiency).

The intensifiers exhibit about 20 to 30 dark counts cm^{-2} s^{-1} at room temperature and a much smaller rate at temperatures of $\lesssim$0 °C. The vacuum side of the fiber optic output window is coated with P-20 phosphor and a thin film aluminum anode is deposited over the phosphor. The phosphor has an e^{-1} decay time of about 50 μs, with a low level tail extending for about 1 ms. The external face of the output window is coated with a thin, optically transparent metallic layer which is grounded to prevent corona from the anode. The internal structure of the image intensifier was ruggedized to ensure that it would survive the vibration and shock levels expected in the ACE launch environment.

The voltages used to power the intensifier are indicated in Figure 9. The MCP-out is held at signal ground, the anode is at +5.5 kV, and the MCP-in is variable from zero to -2 kV. The cathode is held at a voltage with respect to the MCP-in of -200 V when it is gated on and a voltage of +40 V when it is gated off. The gating, which can be accomplished in ~ 1 μs, is used to turn off the image intensifier during event readout. The variable MCP-in voltage allows the overall image intensifier gain to be controlled during flight up to a maximum photon gain of $\sim 2 \times 10^6$. The region of the image intensifier output corresponding to the hodoscope fiber input is coupled onto the large end of a 34.5 mm to 11 mm (diagonal dimension, 3:4 aspect ratio) fiber-optic reducer* with Dow-Corning 93-500 adhesive. The reducer output is coupled with Uralane adhesive to a fiber optic window installed on the CCD. The Thompson TH-7866 CCD is a 244×550 array with individual pixel size 16 μm (width) $\times$ 27 μm (height). A single fiber projects onto an area 4 pixels wide by 2.4 pixels high, after image reduction by the fiber optic reducer. The regions of the image intensifier output corresponding to the two trigger fiber inputs are coupled to Hamamatsu S-3590-01 photodiodes using acrylic light guides. The CCD and photodiodes are mounted on a circular printed circuit board located at the back of the image intensifier assembly.

The image intensifier assembly is thermally isolated from the rest of the SOFT detector by multi-layer insulation (MLI) blankets and thermal standoffs at all mounting points. The top of each assembly is attached to a radiator (Figure 7; two small rectangular plates on top of CRIS) which views space and is designed to provide cooling to ~ -20 °C in flight. From accelerator tests it has been shown that by operating at ~ 0 °C and shuttering the camera as described in Section 5.3 (~ 100 ms shutter speed) there is negligible background from thermal photoelectron emission ($\ll 1$ photoelectron frame^{-1}) and CCD dark current build-up. Heaters are provided to maintain the temperature above -25 °C, thus minimizing thermally induced stresses in the assembly. The outer housing of the image intensifier is 1 cm thick aluminum to reduce radiation damage to the CCD.

4.2. Silicon detector telescopes

Energy losses of particles detected in CRIS are measured in detector 'telescopes' consisting of stacks of 15 silicon detectors. As indicated in Figure 6, CRIS contains four detector telescopes to achieve a large collecting area and to provide redun-

---

* The SOFT fiber-optic reducer was produced by Schott Fiber Optics.

Figure 10. Photographs of CRIS stack detectors. A single groove detector is on the left, and a double groove is on the right. The silicon wafers are 10 cm in diameter and 3 mm thick. The annular region between grooves of the double-groove detector serves as an active guard region for identifying particles which exit or enter through the sides of the stack.

dancy. The individual detectors are made from silicon wafers 10 cm in diameter and 3 mm thick. These thick wafers are processed using lithium compensation technology (Allbritton et al., 1996) to produce a large volume of material with effective impurity concentrations approaching those of intrinsic silicon. The lithium-drifted silicon detectors, denoted Si(Li), that are produced from this compensated material can be thick (several millimeters) yet be fully depleted using relatively modest bias voltages. All CRIS detectors use -400 V bias, applied to the ungrooved surface.

The Si(Li) detectors in CRIS are made in two different designs, as shown in Figure 10. In one design a single groove, concentric with the wafer, has been cut into one of the surfaces. This groove separates the central active region of 68 cm^2 area from an inactive outer ring which is used for physically mounting the detector. The other design has two concentric grooves. The outer groove is identical to those in the single-groove devices, while the inner groove separates the active region of the detector into a central region of 57 cm^2 area surrounded by an annular 'guard ring'.

Each telescope contains four single-groove detectors and 11 double-groove devices arranged as shown in Figure 11. The designations used for each of the detectors are indicated in this figure. Events best suited for mass analysis involve nuclei which enter the telescope through the top of E1, penetrate the central regions of a number of subsequent detectors, and come to rest somewhere between E2 and E8, inclusive. For nuclei which exit from the telescope before stopping, the total energy

 E. C. STONE ET AL.

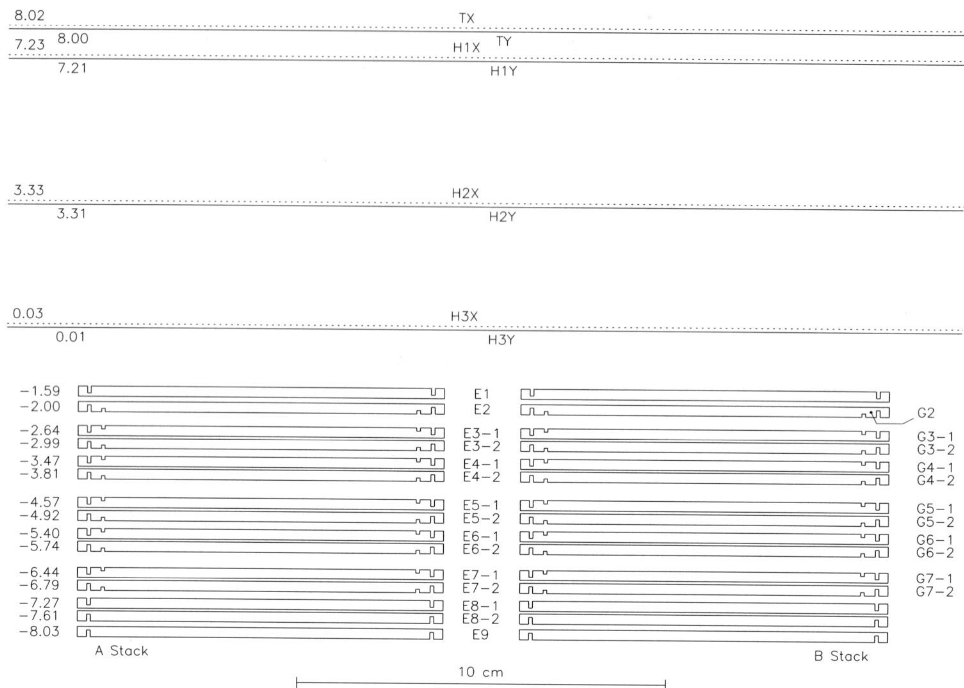

Figure 11. Cross section of the CRIS sensor system in a plane passing through the centers of the A and B detector stacks. The designations used for each sensor element are indicated: TX and TY are the SOFT trigger planes, H1X through H3Y are the SOFT hodoscope planes, E1 through E9 are the central regions of the Si(Li) detectors, and G2 through G7-2 are the guard rings of the double-groove detectors. The dimensions indicated on the left side of the figure show the locations of the top surface of each sensor, measured in cm from an origin chosen near the bottom of the SOFT system.

is not directly measured and the particle identification is less accurate. Particles which exit through the back of the telescope are indicated by an energy loss in E9, while those which exit through the sides are flagged by a signal in one or more of the guard regions. The guards G2 through G7 together with the E9 detector serve as an active anticoincidence which can flag a large fraction of exiting particles, allowing them to be discarded and thus preserving the limited telemetry for higher priority events. Two different guard threshold settings are implemented. For nuclei heavier than helium a high-level guard signal is used to veto an event on board, but a low-level guard signal is simply included in the event telemetry for processing on the ground. This approach assures that a small component of cross-talk between center and guard regions of a detector will not cause the loss of good events.

A fraction of the nuclei which exit through the sides of a stack pass through the gaps between the individual guards and are not identified on board. These events, which mainly occur at wide angles, are processed as though they were due to stopping particles, but are subsequently identified on the ground and discarded based on their trajectory and their pattern of energy losses in the penetrated detectors.

4.2.1. *Si(Li) Detector Fabrication*

Fabrication of Si(Li) detectors began with p-type silicon crystals having resistivity in the range $\sim$1–3 kΩ cm. Topsil Semiconductor Materials produced silicon ingots having the necessary characteristics by controlled gas-phase doping of the material with boron during the float-zone refining process used to purify the silicon and grow the crystal ($\langle 111 \rangle$ orientation). The resulting silicon rod was subsequently machined down to 10 cm diameter and cut into slices. After lapping to the specified thickness, the silicon wafers were delivered to the Lawrence Berkeley National Laboratory (LBNL) for detector fabrication.

At LBNL the silicon wafers were processed into large area diodes. Lithium was vacuum deposited on one surface and diffused into the top several hundred microns of the silicon, forming a thick n^+ contact on the p-type wafer. Grooves of width $\sim$1 mm and depth $\sim$1.5 mm were then ultrasonically cut into the lithiated surface to define the boundaries of the central active region and the guard ring. (A passivating layer of photoresist was later applied in these grooves.)

Acceptor impurities in the crystal were compensated by drifting lithium through the wafer, a process which was performed under carefully controlled temperature, voltage, and current conditions. The lithiated surface was then lapped down to reduce the thickness of the dead layer. The sheet resistance of this surface was monitored and the thinning was stopped when this value had increased to $\sim$20 Ω per square. Metallic contacts were vacuum deposited on the detector faces: aluminum on the junction (grooved) surface and gold on the ohmic (ungrooved) surface. Subsequently, additional metallization (chromium plus gold) was added in selected areas over the aluminum to reduce the resistance and improve the reliability of the electrical contacts to the detector (see Section 6).

A small notch was cut into the edge of each detector in a fixed orientation relative to the crystal axes in order to facilitate aligning and locking the detectors in their modules, as discussed in Section 6. This careful alignment was motivated by observations made during accelerator calibrations of degraded resolution for particles with certain angles of incidence relative to the crystal axes.

Further details of the Si(Li) detector design and fabrication can be found in Allbritton et al. (1996) and Dougherty et al. (1996).

4.2.2. *Detector Performance Testing*

After fabrication, electrical characteristics of the detectors were measured at LBNL to identify devices which did not meet specifications. (Many of these were subsequently reprocessed and yielded good detectors.) Measurements included: (1) depletion bias, as determined by obtaining full pulse height for alpha particles incident from the ohmic surface of the detector; (2) leakage current measured at room temperature; and (3) room temperature noise measured using pulse shaping comparable to that in the flight instrument. The leakage current and noise are generally correlated, since for the CRIS detectors the room temperature noise tends to be dominated by the fluctuations in the leakage current (shot noise). However, bad detectors were often identifiable from noise levels significantly exceeding the expected shot noise. In addition to the above tests, which were performed in air, detectors were briefly checked in vacuum to identify any devices which would become unstable under vacuum conditions. Detectors which passed this suite of tests were delivered to the CRIS instrument team for more extensive testing.

At GSFC and Caltech, the detectors were tested in vacuum at temperatures near $-25\ °C$, $20\ °C$, and $35\ °C$. These tests, which typically lasted 3 to 4 weeks, were intended to check the stability of the detector noise and leakage current under the range of conditions expected in flight. Most of this time was spent at the warmest temperature since previous experience indicated that most stability problems are encountered under warm conditions. In these thermal-vacuum (TV) tests, an automated data-acquisition system collected noise and leakage-current data for each detector every few minutes. These measurements made it possible to identify detectors subject to intermittent periods of elevated noise, even of relatively short duration. In addition, the TV tests showed that most of the CRIS detectors experienced gradual leakage current growth in vacuum (but not in air) with typical initial rates of $\sim$2% per day. Concerns over the implications of this growth for the CRIS instrument lifetime prompted more extensive TV testing of a few detectors. These tests, carried out at a temperature $\sim$8 °C, more characteristic of temperatures expected in flight, extended over more than 6 months. It was found that the leakage current growth is less (in % per day) at the lower temperature, and that the current gradually stabilizes over several months of operation.

Table V summarizes the measured characteristics of the CRIS Si(Li) detectors.

4.2.3. *Thickness and Dead Layer Mapping*

At LBNL, the thicknesses of the Si(Li) detectors were measured at 24 points distributed over the central active region using a non-contact technique in which capacitance measurements were used to determine the residual air gap that remains when a conductor (the detector with its surface contacts grounded) is placed between a pair of probes. The capacitance gauge measurements were converted to absolute thicknesses by comparing with similar measurements made on a gauge block of precisely known thickness. Average detector thicknesses were found to lie in the range 2918 to 3027 μm, with a mean of 2990 μm.

In order to obtain detailed thickness maps covering the entire central active area, measurements were made using a beam of 100 MeV nucl^{-1} ^{36}Ar nuclei from the National Superconducting Cyclotron Laboratory (NSCL) at Michigan State University (MSU). In these runs the beam penetrated the detector of interest and stopped in a second detector placed behind it. This detector pair was scanned using a small ($\sim$1 cm diameter) beam spot and the pulse heights were recorded as a function of position on the detector using particle trajectory information accurate to $\sim$1 mm obtained from a set of multiwire proportional counters. The residual energy signal from the back detector (E') provided a measure of the variation of the total thickness of the front detector (ΔE). By correlating E' values with absolute

TABLE V

Si(Li) detector characteristics

Parameter	Typical values[a]	Notes
Depletion voltage (V)	110 ± 40	a few values up to 230 V
Operating voltage (V)	400	
Noise (keV FWHM)	150 ± 50	20 °C, 2.4 μs peaking time
Leakage current (μA)		20 °C
Single-groove detectors	11 ± 4	
Double-groove detectors	17 ± 3	center and guard connected
Initial leakage current growth (% day^{-1})	2.0 ± 1.5	20 °C in vacuum
Average thickness (μm)	2970 ± 55	2990 μm average
Maximum thickness gradient (μm mm^{-1})	1.0 ± 0.25	
Average dead layer thickness (μm)	60 ± 30	lithiated surface
Central active area (cm^2)		
Single-groove detectors	68	
Double-groove detectors	57	

[a] '$\pm$' indicates typical range of values, not uncertainty.

thicknesses from capacitance gauge measurements at the same locations, full-area maps of absolute thickness were obtained.

The sum of the ΔE and E' signals measures the dead layer. The deduced dead layers are consistent with the expected thickness of heavily lithiated silicon, $\sim$55–70 μm, calculated from the detector fabrication parameters. In using the accelerator data to derive the dead layer thickness, it was assumed that no signal is collected from the ionization produced in the dead layer, while collection of ionization charge from the rest of the detector is 100% efficient. However, our measurements with alpha particles incident through the dead layer show that there is partial efficiency for collection of the ionization signal from this layer, presumably as a result of some charge diffusing out of the layer before it can be lost to recombination. This feature is likely to be significant only for nuclei that stop in or very near the dead layer.

Figure 12 shows an example of a thickness map and a dead-layer map derived from the accelerator measurements, and Table V lists the range of thicknesses and gradients that were obtained for the flight detectors.

4.2.4. *Pulse-Height Analysis of Stack Detectors*

The energy losses of particles entering the CRIS stacks are precisely measured using custom integrated-circuit pulse-height analyzers (PHAs) described in Section 5.2. Because of the large number of signals available from the CRIS Si(Li) detectors (60 centers and 44 guards), selected sets of detectors are connected in

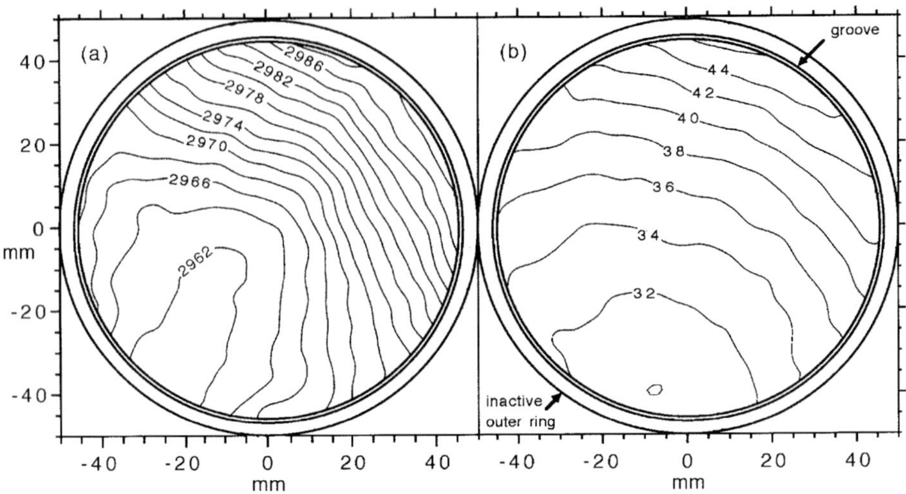

Figure 12. Contour plots of thickness (a) and dead layer (b) for a CRIS stack detector. Dead layers and relative thicknesses were derived from measurements made using beams of ^{36}Ar ions at an energy of 100 MeV nucl^{-1} from the MSU/NSCL cyclotron. Absolute thicknesses were obtained with a capacitance thickness gauge (see text). Contours are labeled in units of μm. In this example the dead layer is somewhat less than the typical value of $\sim$60 μm.

parallel for pulse-height analysis to reduce the complexity of the electronics. The detectors within a stack are labeled as indicated in Figure 11, while the four stacks are designated by the letters A, B, C, and D. Thus, for example, the first detector in stack A is called E1A. All four of the E1 detectors are individually pulse-height analyzed. However, the signals from E2A and E2B are summed into one single PHA, and the signals from E2C and E2D are summed into another. Table VI shows the combinations of detector signals that are summed in each of the 32 PHAs in CRIS. The letter 'G' denotes the guard ring of the detector.

5. Electronics and Onboard Data Processing

The CRIS instrument is functionally and electronically divided into the SOFT and 'stack' sections, allowing for parallel development and test programs. The main electronics card cage is contained in the stack section and houses all subsystems (low- and high-voltage power, logic and microprocessor, stack-detector readout circuits, analog DC control voltage generation circuitry, and housekeeping monitors) except those directly associated with the SOFT system (CCD cameras, image intensifier tubes and associated high voltage, and camera readout and data compression circuitry). The electronics design combines established and new circuit and packaging technologies with the goal of enhancing serviceability, reliability, and performance, within the confines of an aggressive development schedule. Notable design features include:

TABLE VI

Pulse-height analyzers and bias supplies

PHA label	Bias[a]	Detector elements	PHA label	Bias[a]	Detector elements
E1A	1	E1A	E1C	3	E1C
E1B	2	E1B	E1D	4	E1D
E2AB	1	E2A E2B	E2CD	3	E2C E2D
E3AB	1	E3-1A E3-2A E3-1B E3-2B	E3CD	3	E3-1C E3-2C E3-1D E3-2D
E4AB	1	E4-1A E4-2A E4-1B E4-2B	E4CD	3	E4-1C E4-2C E4-1D E4-2D
E5AB	1	E5-1A E5-2A E5-1B E5-2B	E5CD	3	E5-1C E5-2C E5-1D E5-2D
E6AB	2	E6-1A E6-2A E6-1B E6-2B	E6CD	4	E6-1C E6-2C E6-1D E6-2D
E7AB	2	E7-1A E7-2A E7-1B E7-2B	E7CD	4	E7-1C E7-2C E7-1D E7-2D
E8AB	2	E8-1A E8-2A E8-1B E8-2B	E8CD	4	E8-1C E8-2C E8-1D E8-2D
E9AB	2	E9A E9B	E9CD	4	E9C E9D
G2AB[b]		G2A G2B	G2CD[b]		G2C G2D
G3AB[b]		G3-1A G3-2A G3-1B G3-2B	G3CD[b]		G3-1C G3-2C G3-1D G3-2D
G4AB[b]		G4-1A G4-2A G4-1B G4-2B	G4CD[b]		G4-1C G4-2C G4-1D G4-2D
G5AB[b]		G5-1A G5-2A G5-1B G5-2B	G5CD[b]		G5-1C G5-2C G5-1D G5-2D
G6AB[b]		G6-1A G6-2A G6-1B G6-2B	G6CD[b]		G6-1C G6-2C G6-1D G6-2D
G7AB[b]		G7-1A G7-2A G7-1B G7-2B	G7CD[b]		G7-1C G7-2C G7-1D G7-2D

[a]Four bias supplies are shared among the 60 detectors, as indicated. Separate switches allow bias to be disconnected from the set of detectors connected to any center PHA.

[b]Detectors are biased on the ungrooved surface. Guard regions have the same bias as the corresponding central regions.

(1) The use of Actel A1020 field programmable gate arrays for essentially all logic functions.

(2) A card cage/backplane packaging concept using thermally actuated zero-insertion-force connectors.

(3) Modular packaging of the stack detectors onto the same removable printed circuit boards that carry the front-end readout electronics.

(4) A complete pulse-height analyzer hybrid circuit in which all active components reside on a custom bipolar VLSI chip.

(5) A radiation-hard microprocessor system based on the RTX2010, programmed in Forth, for low power, yet high performance real-time hardware control and data processing.

Sections 5.1–5.3 discuss individual subsystems of the CRIS electronics, while Sections 5.4 and 5.5 describe the on-board data processing and commanding functions.

                      E. C. STONE ET AL.

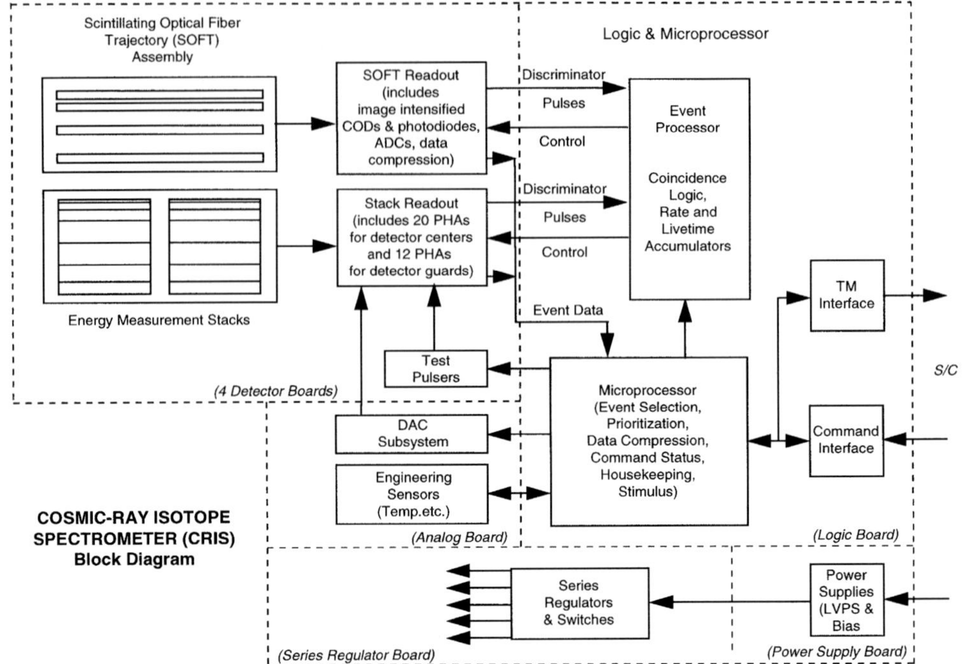

Figure 13. CRIS electronics block diagram.

5.1. LOGIC AND MICROPROCESSOR

As illustrated in Figure 13, the RTX2010 microprocessor forms the core of the system. The ease of programming and high performance level of this microprocessor influenced the hardware design from the outset, allowing simplification both of design and test. The microprocessor controls all other subsystems and also performs on-board data compression by analyzing the raw SOFT data to extract particle hit coordinates for each of the fiber layers. The main system logic and control functions were implemented as subsystems contained within Actel gate arrays which interface in a uniform way to the RTX2010 I/O bus. The logic and microprocessor system consumes less than 1 W and occupies only one single-sided printed circuit board measuring approximately 450 cm^2.

Although the capability of the RTX2010 was used to simplify hardware design, care was taken to perform the time-critical coincidence and pulse-height analysis control functions using a dedicated state machine referred to as the 'event processor'. During front-end live time, the event processor loops waiting for a trigger formed by the logical OR of discriminator outputs associated with the 32 pulse-height analyzer hybrids used to read out the stack detectors and their guard rings. Any such trigger causes the event processor to leave the 'live' state and begin the coincidence determination sequence, which has two phases. At the end of the first phase, lasting approximately 4 μs, the status (triggered or not triggered) of low level discriminators on the stack detector hybrids is latched. At the end of the second phase, lasting an additional 12 μs, the status of higher level discriminators for the stack detectors, as well as those from the SOFT trigger planes, is latched. Based on the latched data, event processing either proceeds to completion or is aborted to minimize dead time. Appendix B presents the boolean equations the CRIS logic uses to determine when a valid event has occurred.

The discriminators for the stack detector signals are set at various levels which allow an approximate sorting of events by nuclear charge. Low, medium, and high thresholds are triggered by stopping $Z \geq 1$, $Z \geq 2$, and $Z > 2$ particles, respectively, in a given detector. Thus, if only the low-level discriminators are triggered, an event receives a $Z = 1$ classification; if low and medium levels are exceeded but no high level, then $Z = 2$; and if any high level is exceeded, $Z > 2$. Processing of events classified as $Z = 1$ or $Z = 2$ may be terminated based on signals from the microprocessor which are used to throttle the rate of analysis of protons and He nuclei, ensuring adequate live time for the rarer heavy nuclei.

The CRIS electronic design provides a high degree of programmability. Discriminator thresholds and coincidence logic equations may be adjusted to compensate for noisy detectors or other anomalies. High voltage may be commanded off or on separately for various detector groups. All microprocessor code that is noncritical to communication with the spacecraft can be reprogrammed in flight.

Redundancy has been used selectively to increase reliability. Silicon detector high voltage is derived from four separate supplies. The A/B and C/D side front-end electronics are independent. The interface with the spacecraft is redundant. The SOFT system 0contains redundant camera systems, with separate interfaces to the main electronics.

5.2. STACK VLSI

0 Silicon detector wafers which form the four 'stacks' are grouped as discussed earlier to form 20 detector elements and 12 guard-ring elements. A separate pulse-height analyzer is used to read out each detector or guard element. These pulse-height analyzers and the associated stack-detector modules are mounted on four essentially identical printed circuit boards ('detector boards') housed below SOFT in the main card cage.

The pulse-height analyzers (PHAs) are hybrid circuits developed for ACE instruments during the project's design phase and manufactured at Teledyne (Cook et al., 1993a). The only active component in the PHA hybrid is a fully custom bipolar application-specific integrated circuit (ASIC) fabricated at Harris Semiconductor. This new circuit represents an evolution of discrete bipolar transistor designs which were flight proven in numerous space instruments over the past thirty years (Halpern et al., 1968; Halpern and Marshall, 1968; Harrington and Marshall, 1968, 1969; Harrington et al., 1974; Althouse et al., 1978; Cook et al.,

1993b). The PHA hybrid has improved performance relative to the prior design, while using a factor of 10 less board area and a factor of 3 less power.

Each PHA hybrid includes a preamplifier, post amplifier, amplifier/offset gate (AOG), peak detector, and Wilkinson analog-to-digital converter (ADC). Shaping time constants are 2 μs, leading to a 3.0 μs time-to-peak for the bipolar signal. Twelve-bit digitization with a maximum conversion time of 256 μs is obtained with a 16-MHz clock. The PHA hybrid was manufactured in a number of types with preamplifier feedback and bias configured for different detector capacitances and leakage currents.

PHA hybrid performance surpasses design goals (see Table III), with typical dynamic range of 2000:1, gain stability of 20 ppm/°C, offset variation of less than 0.5 channels over −20 °C to +40 °C, and deviation from linearity of less than 0.01% of full scale over the entire dynamic range. Power consumption is 40 mW. The PHA hybrids do not contain the digital counters and logic needed for the Wilkinson ADCs, which are rather implemented in nearby Actel gate arrays.

Also mounted near the PHA hybrids are test pulsers used to simulate detector signals. There are two test pulsers per detector board, one for stimulating PHA hybrids connected to detector center elements and the other for guards. The center and guard test-pulse amplitudes are controlled by separate dual gain 8-bit DACs. The test pulsers allow thorough verification of the coincidence logic and are used in routine instrument functional checks. In flight, 'stim' events are continuously generated at a low rate to facilitate the monitoring of PHA stability.

The amplifier chains in the PHA hybrids are DC-coupled from the detector through the postamplifier. This eliminates the need for a long secondary AC coupling time constant and allows the DC level at the postamplifier output to be used to measure the detector leakage current. These measurements are incorporated into the instrument housekeeping data and are useful monitors of detector health.

5.3. SOFT CAMERA ELECTRONICS

Data acquisition and control for the SOFT cameras is provided by the camera electronics unit. These electronics provide the SOFT interface to the main CRIS electronics, control the CCD operation, manage the generation of high voltage and gating signals for the image intensifier, acquire, compress, and buffer the CCD data for transfer to the CRIS microprocessor, provide trigger pulses based on the photodiode signals, and condition the calibration LED light pulses.

The CCD is a 244 × 550 pixel array frame-transfer device. Light projected onto the CCD by the fiber reducer is integrated in the device's image zone. After the image has been acquired, the entire frame of data is clocked into an adjacent memory zone which is not exposed to incident light. The CCD employs an 'antiblooming' structure which greatly reduces electron spillover when the charge on a given pixel overflows its potential well. This antiblooming structure also allows a 'quick clear' to discard old information without the necessity of reading out the entire array.

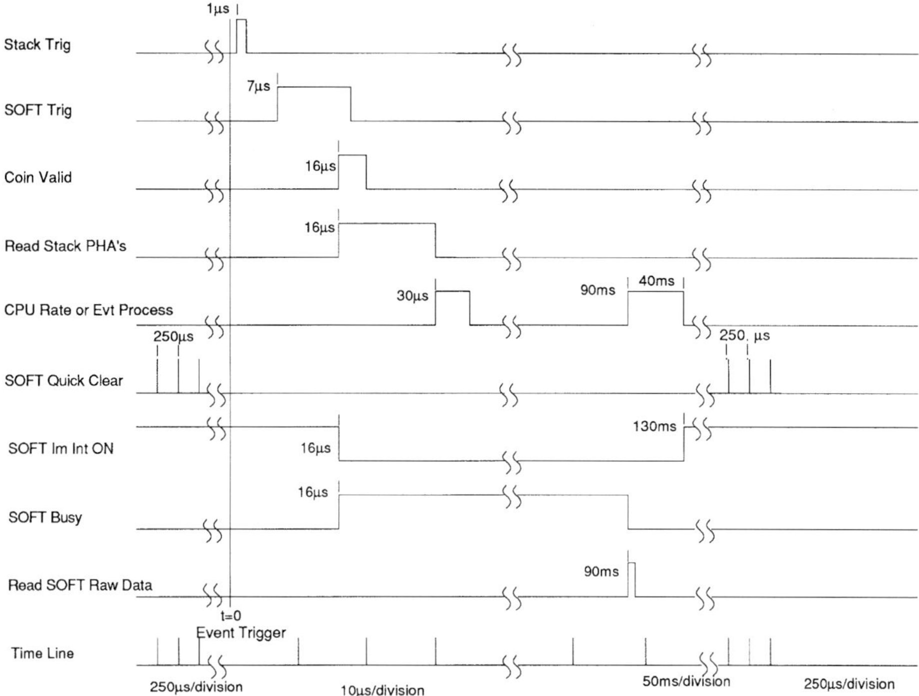

Figure 14. SOFT electronics timing diagram (see description in the text). Note that four different time scales are used, as indicated at the bottom of the figure.

Figure 14 shows the signal timing in normal operation. While waiting for an event, the camera electronics unit gates the image intensifier on and places the CCD in 'integrate' mode, i.e., data are not clocked out of the CCD. A quick clear is provided to the CCD to remove noise from the CCD image zone at a software selectable interval, typically every 192 μs. This reduces dark current build-up and minimizes background due to thermal electron emission at the cathode and to out-of-coincidence particles. When the CRIS silicon telescopes detect an event, a prompt trigger is sent to SOFT which inhibits quick clears and begins a 1.5 ms integration of the CCD (not shown). Approximately 7 μs after the prompt trigger the SOFT photodiodes provide discriminator pulses to the CRIS logic for event validation if they have also seen a signal. Events which fail the event coincidence criteria result in a camera reset and quick clears resume. If the event is valid, the 1.5 ms integration proceeds, after which the information in the image zone is transferred to the CCD memory zone in 496 μs. The pixel data from the CCD are clocked at 2.0 MHz through an analog discriminator. When a pixel is encountered which is above threshold, the clocking rate is slowed to 100 kHz and the pixel signal is passed to a 12-bit ADC. This fast/slow clocking provides a large dynamic range while minimizing the overall readout time, typically 90–300 ms.

For each CCD pixel row which has data above threshold, a row header and address are inserted into the data stream that goes into the SOFT memory. A column header and address are then inserted ahead of each contiguous segment of pixels above threshold. After an event has been read out and stored in the SOFT camera memory it is transferred to the CRIS microprocessor for compression and merging with the stack data (see Section 5.4.1).

5.4. CRIS DATA

The data returned by the CRIS instrument consist predominantly of 'events', each containing the pulse-height and trajectory measurements for a single detected nucleus. The microprocessor categorizes the events to assign priorities and optimize the mix of events that can be telemetered within the CRIS bit rate allocation. In addition, a small fraction of the data stream is used for housekeeping measurements and rates. These quantities are multiplexed over a 256-s 'instrument cycle'. Finally, when commands are sent to the instrument a command echo/command reply is inserted into the data stream. The following subsections describe each of these components of the CRIS data.

5.4.1. *Event Data*

A valid particle event in CRIS must produce a signal in the SOFT trigger plane, penetrate a certain minimum distance into the silicon-detector stack (typically at least to E2, although a greater depth requirement could be set by command), and generally not trigger the guard rings around the silicon-detector stack. For each such event, the CRIS microprocessor reads a 12-bit pulse height from all of the silicon-detector PHAs that had signals above their low-level discriminator threshold and video data from the SOFT electronics. The raw SOFT data consist of the location and intensity of every pixel above a commandable threshold intensity in whichever of the two SOFT CCD cameras is active at the time.

To maximize the number of events sent to the ground in the 464 bits-per-second telemetry rate allocated to CRIS, these event data are compressed on-board. Most importantly, the pixel data from the SOFT camera are processed to identify pixel 'clusters' (contiguous groups of pixels that meet cluster criteria set by several commandable parameters) which can be represented by a pair of centroid coordinates and an intensity. The centroids are recorded in 'half-pixel' increments for both x and y. Normal operation calls for recording the two brightest clusters in each of the fiber layers plus an additional six clusters (if they are present) for an overall maximum of 18 clusters per event. These numbers are adjustable in flight, if circumstances warrant.

In addition to data compression through the on-board extraction of SOFT centroids, further reduction is achieved by using a variable-length data format in which the event size is adjusted to match the number of sensor signals that are present. Details of the format are discussed in Appendix C.

As a result of this format, the lengths of CRIS events may vary considerably, ranging from a minimum of around 31 bytes for particles stopping in E2 in only one telescope while triggering no guards and producing six SOFT clusters, to an extreme maximum of 162 bytes for a pulser-stimulated event triggering all stack detectors and guards in both telescope pairs, with 31 SOFT clusters. A 'typical' event stopping in E4 of only one telescope, with no guards and 12 SOFT clusters is 52 bytes long.

In addition to these compressed, normal-mode events, 'diagnostic-mode' events are also sent back at a low, commandable rate (typically about one every few hours). Such an event consists of the pulse-height data from all 32 silicon stack and guard PHAs, the associated discriminator tag bits, and the position and intensity of each triggered SOFT pixel, rather than just clusters of adjacent pixels. Included with these raw data is the compressed, normal mode interpretation of the event. Such diagnostic events allow potential problems in the on-board compression algorithms to be identified so that they may be fixed by adjusting parameters in the algorithms by command. These events may be very large, up to about 6000 bytes.

5.4.2. *Priority System*

It takes the instrument about 130 ms to process a typical event (limited primarily by the SOFT system), allowing ~ 7 to 8 events per second to be processed. Given the average event lengths and the CRIS telemetry allocation of 58 bytes per second (which also includes count rate and housekeeping data), only about 1 event per second can be telemetered. Actual event rates for stopping particles with $Z > 2$ are expected to be only about 0.1 to 0.2 events per second during solar quiet times, but penetrating particles, protons, and alpha particles can increase the trigger rate to several per second. In order to maximize the return of the more interesting stopping, heavy-ion events, the CRIS microprocessor sorts the events into 61 different buffers on the basis of range, pulse height (which indicates particle species), quality of trajectory measurements, single or multiple telescope hits, and potential pulse pile-up, or 'hazard', condition. Events are read out from the buffers in a commandable order, with the highest priority going to events with long ranges, high Z, good trajectory measurements, and single telescope, non-hazard stack pulse heights. Appendix D presents the definitions of the CRIS event buffers, and the priorities that were assigned to them at launch.

Also, a commandable number of events (nominally 10) in each 256-s instrument cycle are read out using a simple polling scheme, sequentially reading from each of the buffers in turn. This ensures a more uniform sample of all event types and prevents the possibility of the lowest priority events being eliminated entirely during periods of higher particle intensity. The event buffers are 512 bytes in size, and can contain approximately 10 events.

5.4.3. *Rate and Housekeeping Data*

In addition to pulse-height data, CRIS records and transmits a number of count rates:

 – 'singles' count rate data for each of the two SOFT hodoscope trigger signals and each of the 32 silicon stack and guard PHAs,

 – 15 different coincidence rates,

 – various diagnostic rates useful for monitoring the event processing,

 – three separate live times corresponding to the $Z = 1$, $Z = 2$, and $Z > 2$ classifications (see Section 5.1),

 – rates from each of the 61 event buffers, which are needed to calculate sampling efficiencies and thus obtain absolute fluxes for each category of events.

All rates are read out during each 256-s instrument cycle. Any two of these rates may be selected as 'high-priority' rates to be read and transmitted at a higher time resolution of once every 16 s.

A number of housekeeping measurements are also recorded, including temperatures, voltage and current monitors for the low- and high-voltage power supplies, and silicon-detector leakage currents (derived from the DC outputs of the PHA hybrids). In the remaining telemetry space, the CRIS command table (844 bytes) is slowly trickled out, taking 13 instrument cycles ($\sim$56 min) to transmit the entire table.

5.4.4. *Telemetry Format*

The above data are packaged into a 58-byte-long CRIS instrument minor frame[*]. The first bit of the instrument frame is used to find the start of the 256 minor frame instrument cycle, which need not coincide with the start of a spacecraft major frame. It is set to 0 for the first 128 minor frames, and 1 for the next 128 frames of the cycle. The second bit serves as a flag to indicate that a command response precedes the event data in this minor frame. All commands are echoed back in the CRIS telemetry data, along with any output generated by the command, thus documenting the command history of the instrument within the data stream. Command responses vary in length up to 55 bytes. The first byte of a command response gives the length of the remainder of the command echo. The rest of the minor frame not used by the command response contains normal mode or diagnostic-mode data. The third bit of the instrument frame is set if the pulse-height data in this frame are from a diagnostic-mode event rather than a normal, compressed event. Diagnostic events may span any number of contiguous minor frames, and may start in the middle of a normal event (which then continues after the end of the diagnostic event). After the first three bits of the instrument frame, the remaining 21 bits in the first three bytes are allocated to rate and housekeeping data. All rates are 21

---

[*] The start of the CRIS 'instrument frame' is offset by four bytes from the beginning of the CRIS data block in the ACE spacecraft minor frame to simplify data handling by the microprocessor.

bit numbers[†] and each is assigned to a particular frame in the instrument cycle. Housekeeping data are typically 10 bit quantities, and are packaged 2 per frame, for those frames in which they appear. Those frames where the housekeeping/rate data consist of part of the command table readout contain two bytes of the command table.

The rest of the instrument frame after the first three bytes is filled out with command responses and/or event data (either normal or diagnostic), as indicated by the second and third bits of the frame. Events can span the boundary from one frame to the next, and are strung sequentially together to maximize the use of telemetry space. Only at the beginning of the 256-s instrument cycle is an event forced to start at the beginning of the event data section of the frame. The event format (see Appendix C) contains enough redundant information to allow identification of the start of an event, thereby making it possible to regain synchronization should telemetry frames be dropped within an instrument cycle.

5.5. CRIS COMMANDS

Commanding of the CRIS instrument is accomplished using strings of ASCII characters which are passed to the command interpreter of the Forth operating system running on the RTX2010 microprocessor. As such the command system has almost arbitrary flexibility to alter the operation of the instrument. In practice, most commanding is done by altering entries in a 'command table' which holds the present values of a large number of instrument parameters. Table VII summarizes the major categories of CRIS operating parameters that are set by command.

In addition, large blocks of code can be uploaded into the microprocessor's memory to alter its programming. This function is not used in normal operation, but does allow the possibility of major reprogramming should that become necessary.

6. Mechanical Design and Packaging

The mechanical design of CRIS provides the following features:
- Precise and reproducible mechanical positioning of detectors.
- Minimal use of wires and coaxial cables for interconnections.
- Close proximity of detectors to their preamplifiers with short interconnections.
- RF shielding of the detectors.
- Ease of assembly and component replacement.
- Light-tight enclosure of detectors which can vent rapidly at launch.
- Dry nitrogen purge to detectors until just prior to launch.

[†] A few of the rates which could attain large values, particularly those associated with the SOFT trigger signals, are prescaled by a factor of 32 when they exceed a value of 2^{20}. Otherwise the rates are uncompressed.

TABLE VII

CRIS command summary

Item	Comments
Coincidence logic	
Coincidence equation	Set minimum range for coincidence
Hazard event	Allow/disallow analysis of hazard events
Stim event	Give stim events highest priority
ADC2 definition	Allow/disallow trigger on any two successive detectors in stack
Stack operation	
Connect/disconnect bias	Separate switch for each detector set
Adjust ADC thresholds	Raise/lower in steps of $\sim$10%
PHA operation	Individually enable/disable any PHA
Medium/high PHA thresholds	Raise/lower in steps of 10%
H and He event timers[a]	Individually adjust from 0 to 130 s
Hodoscope operation	
Active camera	Select A or B camera
MCP quick clear	Adjust period from 32 to 512 μs
LED stim pulse	Adjust length from 0.25 to 16 μs
CCD high voltage	Adjust from 0 to 2050 V (camera A) or 2200 V (camera B)
Trigger discriminator level	Adjust from 0 to 25% of full scale
Pixel discriminator level	Adjust from 0 to 5.6% of full scale
Cluster parameters	Set cluster criteria and limits
Software parameters	
Charge buffers	Adjust boundaries
Readout priority system	Redefine event buffer readout priorities
Event format	Enable/disable transmission of guard pulse heights, 2nd telescope data
High priority rates	Select 2 rates for 16 s accumulation
Stim events	Select period and PHAs to be pulsed
Diagnostic events	Select period

[a]Timers are used to artificially create additional dead time for H and He events to throttle the rate of analysis of these lower priority events.

The SOFT trajectory system was the primary responsibility of Washington University at St. Louis, whereas the remainder of CRIS was primarily the responsibility of Caltech. This made a modular design approach particularly desirable. Thus CRIS is housed in two boxes (Figure 7) attached to one another with simple mechanical and electrical interfaces. The upper box houses the SOFT hodoscope and the lower box contains the remainder of CRIS. The lower box contains in-

Figure 15. Mounting of the CRIS detector boards in the main CRIS box. Detectors and their front-end electronics are mounted on four boards (E1/E2, E3/E4, E5/E6, and E7/E8/E9) which slide into a card cage/backplane housing. For this photograph, the detector boards were only partially inserted to illustrate the assembly approach.

dividual printed-circuit boards (PCBs) which slide into card rails and mate with zero-insertion-force (ZIF) connectors mounted on a PCB backplane at the rear of the box as shown in Figure 15. The card rails then lock the PCBs to the walls of the box using jackscrews accessible from the ends of the card rails after the boards are in place. The ZIF connectors (made by Betaphase Corporation, Inc.) on the motherboard backplane provide connections on each side of each daughter board on 0.635 mm centers (80 per inch). The Betaphase connectors are operated by applying a voltage to an internal heater which causes the jaws of the connector to open and allows the board to slide into a keyed slot. Electrical connections are made as the connector cools and clamps the board, gripping it with substantial force.

Some of the printed-circuit boards carry only electronic components, while others include module housings containing multiple detectors. The four stacks of solid-state detectors below SOFT are each formed by four vertical sets of such detector modules, one above the other. An example of a detector module is shown in Figure 16 and a sample CRIS printed-circuit board carrying four modules is shown in Figure 17. Note from Figure 17 that each module is surrounded by an aluminum 'fence' which provides an RF shield around the module.

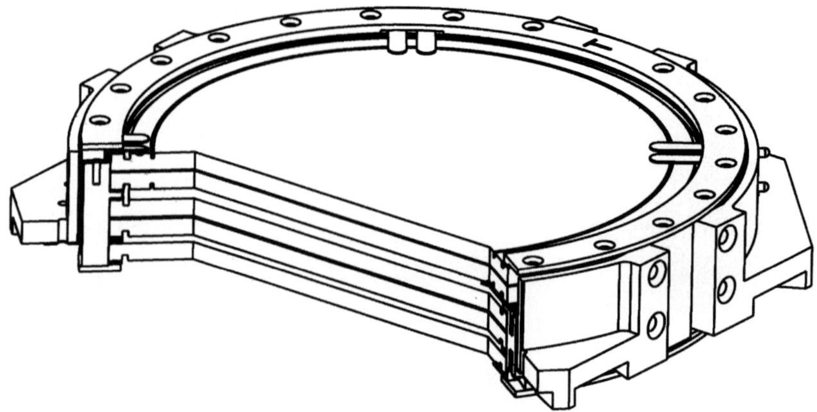

Figure 16. Cut-away drawing of a CRIS detector module (E7/E8/E9) with detectors installed. The module is machined from G10, which has a thermal coefficient of expansion well matched to that of the silicon. Electrical contacts to the center and guard ring are made with BeCu spring contacts. The detectors housed in this module are, from top to bottom, E7-1, E7-2, E8-1, E8-2, and E9. Note that for the detector pairs which are connected in parallel to a single PHA (such as E7-1 and E7-2) the ungrooved surfaces face one another, eliminating significant dead layers midway through the two-detector combination.

Connections in a module are made to each Si(Li) detector using gold-coated dimples on spring-loaded (BeCu) fingers. These fingers make pressure contact with the detector surface and then run internal to the module to fingers which make pressure contact with traces on the printed-circuit board. In the case of double-grooved detectors, separate contacts are made to the detector central areas and to the detector guard rings. Detector modules can be easily installed or removed from their respective printed-circuit boards and detectors can also be installed or removed from their module housings with relative ease.

The isotopic resolution criteria put limits on acceptable position tolerances. The allowable mechanical tolerances used in designing the CRIS instrument are listed in Table III. These levels of alignment were achieved by minimizing the stack-up of parts, which in turn keeps the tolerance accumulation to a minimum. Feet on the bottom of each RF shield extend through the printed-circuit board and mate to the top of the RF shield on the module below. This prevents build-up of position errors that otherwise might result due to the fact that the multi-layer printed-circuit board thicknesses are relatively difficult to control. After all the CRIS boards are installed (but before the card rails are tightened), five alignment bolts are inserted from the bottom of the CRIS box, extending from the bottom plate up through all the RF shields except the topmost one, and screwed into the top shield. These bolts are intended primarily to align the modules in a reproducible way. The RF shielding around each module is not perfect, since there typically remains a gap of $\sim$0.1 mm between each shield and the board above it.

Figure 17. Photograph of a CRIS detector board with all parts installed. The four detectors are surrounded by conducting 'fences' which serve as RF shielding and have precisely machined heights that permit stacking of successive detector boards with minimal build-up of tolerances. Immediately to the right of the detectors are eight hybrid pulse-height analyzers: on this board they process signals from E3AB, E4AB, E3CD, E4CD, G3AB, G4AB, G3CD, and G4CD.

CRIS has a system of purge tubes which allows dry nitrogen gas to be piped through the RF shielding system and directly into the modules where the detectors reside. Vent ports in the side walls of both SOFT and the lower CRIS box were designed to allow rapid evacuation of the boxes during launch. Baffles on the vent ports block light from entering through the instrument walls.

Apart from a front window of 254 μm (0.010″) aluminum and a thermal blanket (0.083 g cm^{-2}), the primary field of view of CRIS (45° half angle*) is completely unobstructed. The bottom of the SOFT box is a 127 μm (0.005″) aluminum sheet which isolates the two boxes while introducing only a small thickness of material into the paths of the detected particles.

CRIS is mounted on a side wall of the spacecraft with its average view direction perpendicular to the spin axis. The instrument is covered with thermal blankets on all sides except the side facing furthest away from the Sun, which is covered with

---

* The CRIS field of view is a complicated function of location on the detectors and azimuthal angle.

silver teflon and acts as a radiator to remove heat from the instrument. Each SOFT image intensifier also has an uncovered radiative cooler surface (see Figure 7).

Accelerometers are mounted between CRIS and the spacecraft on all 12 CRIS mounting feet. These provide signals to the Structural Loads and Acoustics Measurements (SLAM) instrument* which is designed to record the vibration environment during launch. These data should aid in setting more realistic vibration test specifications for future missions.

7. Resource Utilization

The CRIS instrument is composed of two rectangular boxes of slightly different dimensions which are joined together. The top box houses the SOFT system and has dimensions ($l \times w \times h$) of 53.3 cm $\times$ 43.8 cm $\times$ 10.1 cm. The bottom box contains the stack detectors and most of the electronics and has dimensions 53.3 cm $\times$ 40.6 cm $\times$ 13.4 cm. CRIS has a mass of 29.2 kg, which does not include mounting hardware, force transducers, or thermal blankets. The thermal blankets have a mass of 1.0 kg.

Under normal operating conditions, CRIS dissipates $\sim$12 W of power. However, the power consumption depends on temperature and activity level. During high rate periods at the highest temperature for which the detectors are rated (35 °C), the power is expected to rise to $\sim$16 W. Much of this increase is due to the large temperature coefficient of the leakage current of the Si(Li) detectors (factor $\sim$2 per 7 °C).

Operational heaters of 1.7 W are mounted on each image intensifier; another operational heater of 4.2 W is mounted on the backplane. Survival heaters of 1.6 W are mounted on each image intensifier; another survival heater of 15.4 W is on the outside of CRIS under the thermal blanket. The CRIS mounting bolts are thermally isolated from CRIS via Ultem bushings. The nominal internal operating temperature for CRIS is approximately -10 °C.

8. Expected Performance

In order to confirm that the CRIS instrument design would be suitable for meeting the requirements discussed in Section 3.2, a variety of calculations of the expected performance were carried out. The results of these calculations are discussed in the following subsections.

* The SLAM instrument was developed at the NASA/Goddard Space Flight Center under the direction of Scott Milne and Ken Hinkle.

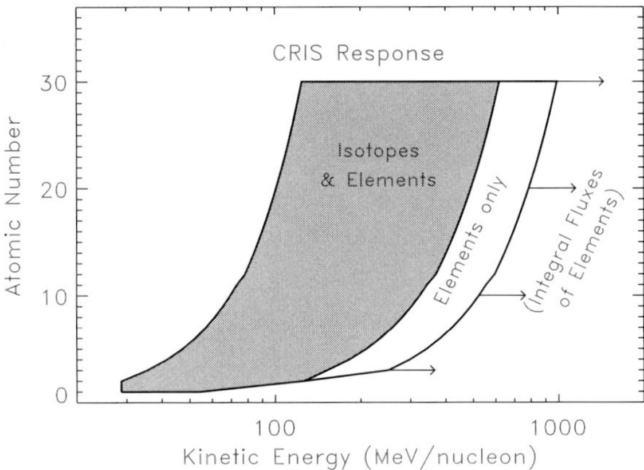

Figure 18. Energy intervals covered by the CRIS instrument as a function of the atomic number, Z, of the detected nucleus. Isotope measurements are made for particles that stop in the silicon-detector telescopes. When the particles penetrate the entire telescope but still slow significantly, charge and energy identification is still possible, but not mass determination. At the highest energies, major elements can be identified, but only lower limits can be set on their energies.

8.1. ENERGY COVERAGE VERSUS ATOMIC NUMBER

Figure 18 shows, for the range of atomic numbers observed by CRIS, the calculated energy intervals over which the instrument can identify isotopes and elements. The most precise particle identification is achieved in the shaded region, defined by the requirement that nuclei stop in one of the silicon-detector stacks between E2 and E8. In this region CRIS will make a precise measurement of each particle's total energy, as well as one or more measurements of dE/dx. From these quantities and the particle trajectories, both the charge and the mass of each detected nucleus can be derived.

In the region labeled 'Elements only' particles will penetrate the entire stack and their total energies will not be directly measured. However, these nuclei undergo significant slowing in the silicon-detector stack, resulting in measurable increases in dE/dx going from the front to the back detectors in the stack. From these patterns of dE/dx values (the Bragg curves) one can calculate the atomic number and kinetic energy of each particle. The total energy estimates are not, however, sufficiently accurate to permit reliable derivation of particle masses.

In the right-most region of the plot the particles are so energetic that their velocities do not change significantly in the stack, so it is only possible to set a lower limit on each particle's energy. However, since at high energies $dE/dx \propto Z^2$ with relatively little energy dependence, it is possible to estimate the charge. Thus for the more abundant elements it may be possible to use the measurements in this high-energy region to derive integral flux values*.

---

* The high-energy data set will include particles which pass through CRIS in the backwards direction, after having traversed a significant amount of material in the spacecraft. It remains to

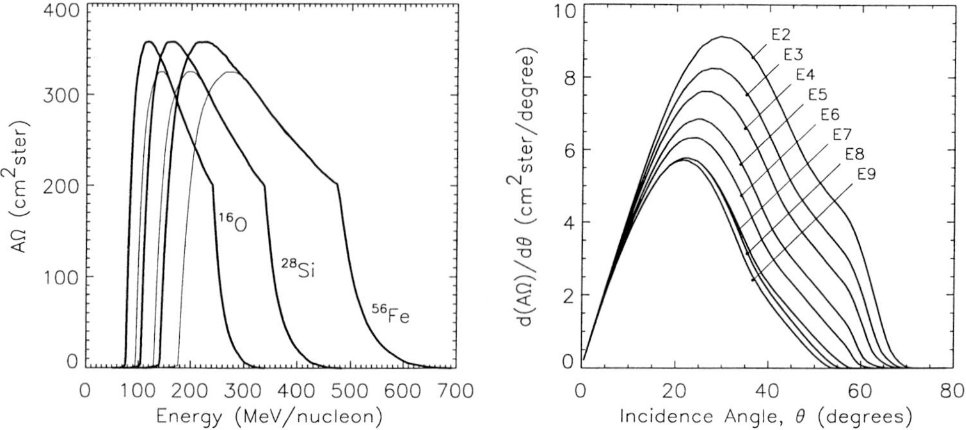

Figure 19. CRIS geometrical factor ($A\Omega$). The left panel shows the geometrical factor as a function of particle energy for ^{16}O, ^{28}Si, and ^{56}Fe nuclei. The heavy curves indicate the geometrical factors including particles which stop anywhere between E2 and E8, inclusive. When redundant dE/dx measurements are required, the particles must penetrate to at least E3. In this case the geometrical factors are reduced to the values shown with the light curves. The right panel shows the geometrical factor per unit incidence angle (θ) for particles stopping in each detector, as a function of θ. The curve labeled 'E9' also gives the geometrical factor for nuclei which penetrate the entire telescope. In both panels the ordinate values include the responses of all four CRIS telescopes.

The elements hydrogen and helium are not included among the primary objectives of CRIS: the SOFT efficiency for these species is relatively low, and at high energies they fall below threshold in the stack detectors. Nevertheless, over a limited energy interval CRIS will provide measurements of these light elements that can be used to relate the fluxes of the heavy nuclei which are the primary CRIS objectives to those of H and He, which constitute the bulk of the cosmic rays.

8.2. ANTICIPATED NUMBERS OF PARTICLES

Figure 19 shows the results of Monte Carlo calculations of the CRIS geometrical factor, $A\Omega$. In the left panel $A\Omega$ is plotted as a function of particle energy for the elements O, Si, and Fe. The right panel shows the differential geometrical factor per-unit incidence angle, d$A\Omega$/dθ, as a function of θ and of the detector in which the particle stops. The CRIS response extends to large angles, $>60°$, because of the large active area provided by the SOFT hodoscope. The curve labeled 'E9' also gives the geometrical factor distribution for penetrating particles exiting through the back of the CRIS telescopes.

    Figure 20 shows the number of good events expected from CRIS for each stable or long lived nuclide with $3 \leq Z \leq 30$. This calculation assumes data collection over a period of two years under solar minimum conditions. It takes into account

be determined whether useful integral flux measurements can be derived in the presence of this background.

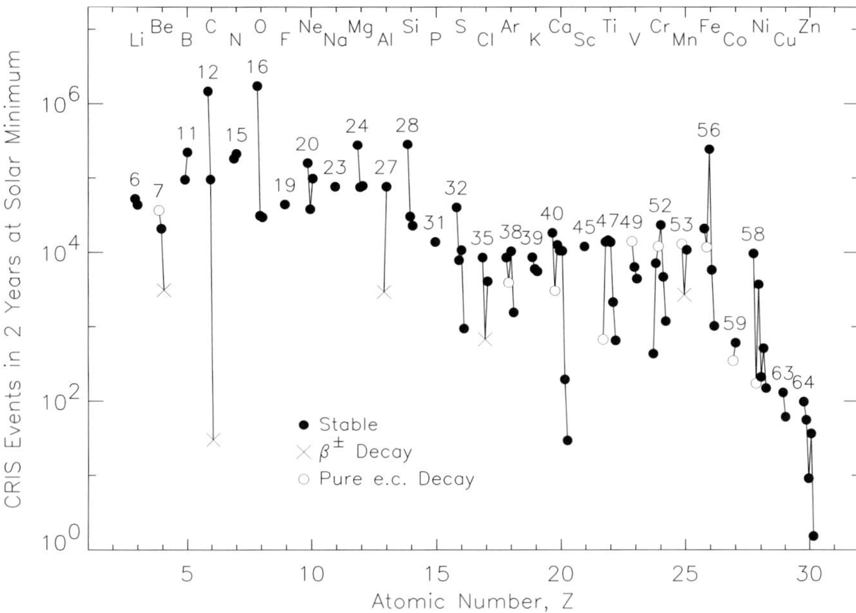

Figure 20. Number of good events expected for each isotope with two years of data collection at solar minimum. Only particles which have incidence angles <45° and which come to rest in the Si(Li) detector stacks without fragmenting are included. It is assumed that isotopes that can only decay by electron capture are effectively stable in the cosmic rays.

the geometrical factor distributions shown in Figure 19 and the energy intervals set by the thicknesses of the CRIS detectors. Particles which fragment in the instrument have been excluded, as have nuclei with angles of incidence >45° since at the largest angles the mass resolution degrades due to effects such as multiple Coulomb scattering (see Appendix A). This plot includes all particles which stop between E2 and E8, inclusive. For those which stop in E3 or beyond, one has two or more mass determinations which can be compared to remove background events. However, for particles which stop in E2, one has only a single mass determination and cannot test the energy loss data for internal consistency. Consequently, the E2-stopping events are subject to more background contamination than higher energy events. For the rarest species this extra background may degrade the quality of the abundance measurements. For such nuclides it may be desirable to restrict the analysis to those particles which stop in E3 or beyond. In this case the event yields are reduced by ∼15%, as suggested by Figure 19.

In addition to accurately measuring abundances of rare nuclei, CRIS will also collect large samples of the more abundant isotopes over short time intervals. These measurements will allow studies of spectral changes over times as short as a solar rotation, thereby enabling detailed studies of the solar modulation of galactic cosmic rays.

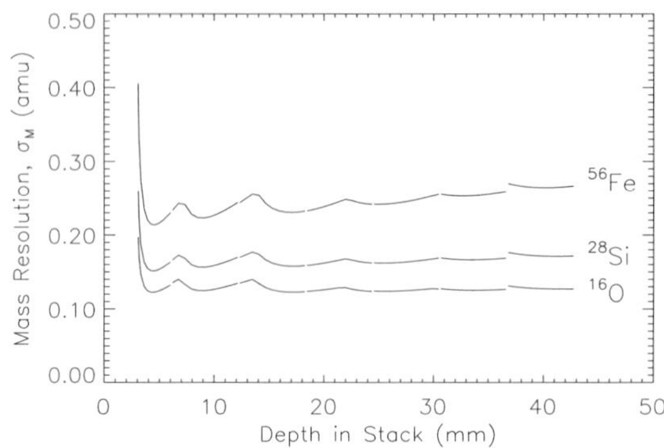

Figure 21. Calculated mass resolution for ^{16}O, ^{28}Si, and ^{56}Fe is shown as a function of the depth to which the particles penetrate in the CRIS silicon-detector stacks. The calculations were performed for an incidence angle $\theta = 20°$. The mass resolution is calculated with various combinations of detectors summed to form ΔE and E', with the best combination selected at each penetration depth. Discontinuities occur at depths where the detector combinations used for the ΔE and E' measurements are changed. The abrupt increase in the mass resolution at the front of the stack (just entering E2) is due to a large path length uncertainty due to multiple scattering for these particles which almost come to rest in E1.

8.3. EXPECTED MASS RESOLUTION

The expected mass resolution was evaluated using the formulas discussed in Appendix A and taking into account the measured characteristics of the CRIS instrument. Figure 21 shows the results of these calculations for the isotopes ^{16}O, ^{28}Si, and ^{56}Fe as a function of the depth to which the particles penetrate in the silicon-detector telescopes. For these calculations, an angle of incidence $\theta = 20°$ has been used. The overall mass resolution is dominated by a combination of Bohr/Landau fluctuations and multiple scattering in the ΔE detectors. Contributions to the mass resolution from uncertainties in the measurements of trajectories and energy losses are small compared to these fundamental contributions, as intended. The mass resolution has a significant dependence on the angle of incidence, with better resolution at smaller angles. This dependence can be used to select higher resolution data sets, when needed, at the expense of reduced statistical accuracy.

9. Accelerator Calibrations

Accelerator tests and calibrations played an important role in the development of the CRIS instrument. Prototypes of both the SOFT hodoscope and of the Si(Li) detectors were tested in a sequence of runs at the Michigan State University National Superconducting Cyclotron Laboratory (MSU/NSCL) in order to optimize the designs and to demonstrate that the performance of these sensors was sufficient to meet the CRIS requirements.

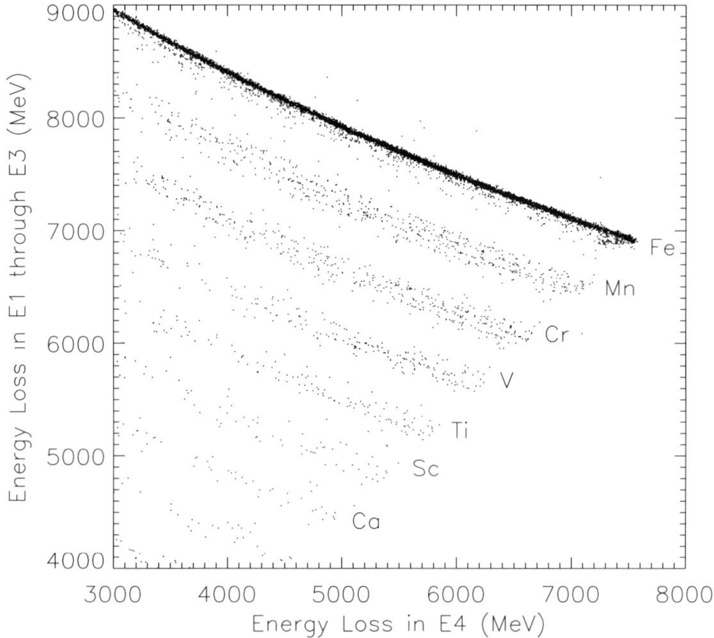

Figure 22. ΔE vs E' scatter plot from a GSI calibration run. A primary beam of ^{56}Fe nuclei was incident on a target in which a portion of the nuclei fragmented, producing lighter nuclides. 'Energy degraders' were used to distribute the stopping particles throughout the entire thickness of a CRIS stack. The nuclei had a nominal angle of incidence of 10°.

The thicknesses and dead layers of the Si(Li) detectors were mapped in detail at the MSU/NSCL, as discussed in Section 4.2.3. In other runs, the correspondence between SOFT pixel centroids and absolute positions was calibrated (Section 9.2) and the SOFT efficiencies were derived (Section 9.4).

The completed CRIS instrument was calibrated at the GSI accelerator in Darmstadt, Germany, since this facility is capable of providing beams of heavy nuclei sufficiently energetic to penetrate the entire thickness of the CRIS telescopes. These measurements provide the best pre-launch checks of the overall instrument performance.

9.1. MEASURED MASS RESOLUTION

Figure 22 shows a scatter plot from a GSI calibration run in which a beam of ^{56}Fe at 500 MeV nucl^{-1} was fragmented in a polyethylene target and was spread in energy before impinging on CRIS at an angle of 10°. This plot shows well-resolved tracks for the primary ^{56}Fe and a variety of fragment nuclei. Using these data, the charge and mass of each detected nucleus was calculated using Equation (1) (Section 3.1) and taking into account the measured gains and offsets of the pulse-height analyzers and the calibrated mean thicknesses and dead layers of the penetrated detectors

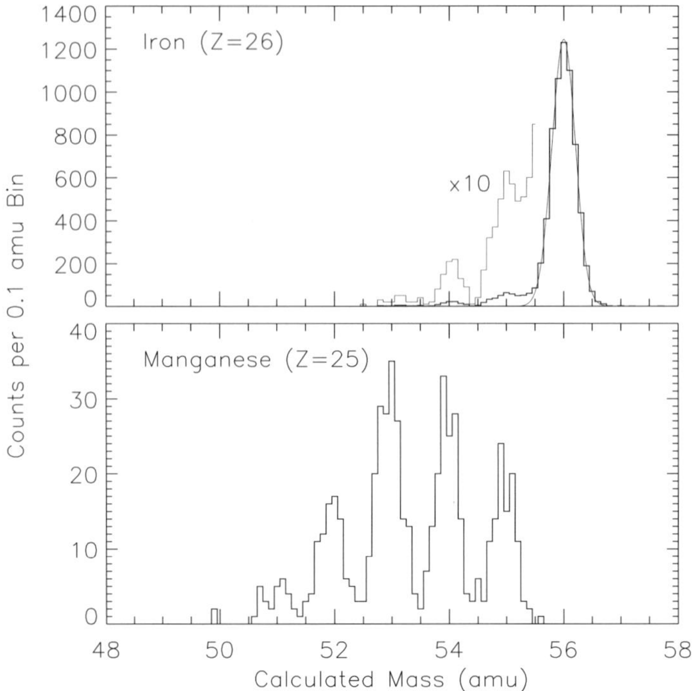

Figure 23. Mass histograms of CRIS accelerator calibration data derived from the measurements of ΔE and E' shown in Figure 22. In this analysis, no corrections were made either for the deviation of the angle from the nominal value (10°), or for the spatial variation of the detector thickness (although the beam area was small enough so that mapping corrections should be insignificant). The histograms show events stopping in E4, with masses calculated using the sum of the energy losses in E1, E2, and E3 as ΔE, and the signal from E4 as E'. The smooth curve in the iron histogram is the result of fitting the ^{56}Fe peak with a Gaussian, resulting in $\sigma_M = 0.205$ amu. The region of the measured iron distribution below 55.5 amu is shown expanded (light line) to illustrate the separation between ^{56}Fe and the lighter isotopes, which have lower abundances by a factor $\gtrsim 20$ in this run.

(but not yet using detailed thickness maps). Figure 23 shows the resulting mass histograms for Fe and Mn and indicates the instrument's capability for identifying isotopes up through the iron group.

Calibrations of this type were carried out using a variety of primary beam particles, energies, and incidence angles. Generally, it was found that the CRIS mass resolution ranges from ∼0.25 to ∼0.30 amu for iron-group elements down to ∼0.1 amu for the lightest nuclei of interest. The mass resolution increases somewhat at the lowest energies and the largest angles measured in CRIS.

9.2. SOFT COORDINATE CALIBRATION

The data required to convert pixel centroids measured in the SOFT cameras into calibrated spatial coordinates were obtained using an oxygen beam from the MSU/NSCL cyclotron. This calibration was carried out in two stages: first the correspondence between the location on the CCD pixel array and the number of the hit fiber was mapped, then the relationship between xy fiber pairs and specific fiducial points in space was measured.

Figure 24 shows a superposition in CCD pixel space of centroids obtained when the full area of SOFT was scanned with the oxygen beam. The centroids cluster around locations corresponding to the fiber centers. Fitting the mean xy pixel location for each fiber in the array and partitioning the pixel space into cells centered on each fiber yields the necessary pixel-to-fiber calibration.

The identity of the hit fibers provide initial estimates of the spatial coordinates, good to $\lesssim 1$ mm. An additional mapping step was carried out to correct for small imperfections in the straightness and spacing of the fibers. For this purpose the SOFT hodoscope was exposed to the oxygen beam through a plate having holes machined on a regular grid. As shown in Figure 25, the data from this run produced images of the fiducial holes which established the correspondence between fiber numbers and spatial coordinates. In using this map to derive coordinates for detected particles, pairs of fiber numbers (one x fiber and one y fiber) are converted to two-dimensional spatial locations, thereby taking into account any deviations of the fibers from straightness. While this adds to the complexity of the SOFT data analysis (since there are generally a number of different possible pairings of a given set of individual coordinates), it is important for fully optimizing the spatial resolution.

9.3. POSITION AND ANGULAR RESOLUTION

The SOFT detector position resolution was evaluated using the method of residuals (Crary et al., 1992; Davis et al., 1989). Cluster centroids were derived from the CCD images using an algorithm similar to that employed on board CRIS, and these were converted to spatial coordinates using the mapping procedures discussed above. Distributions were then made of the differences (ΔX2 and ΔY2) between the coordinates obtained for the H2 plane and the coordinates expected at that location based on a straight line track derived from the coordinates in the H1 and H3 planes.

Figure 26 shows one such distribution obtained from a full-area exposure of the detector using 155 MeV nucl^{-1} oxygen nuclei with an incidence angle of $0°$. The measured standard deviation, 102 μm, implies a single-coordinate spatial resolution of 83 μm (Crary et al., 1992), assuming comparable resolutions for all SOFT layers. From SOFT measurements with this resolution the angle of incidence of detected nuclei can be derived with an accuracy $\lesssim 0.1°$.

Figure 27 shows the dependence of spatial resolution on the energy deposited in SOFT. These measurements demonstrate that the SOFT hodoscope is achieving the required trajectory resolution for all elements to be studied with CRIS (see Table III).

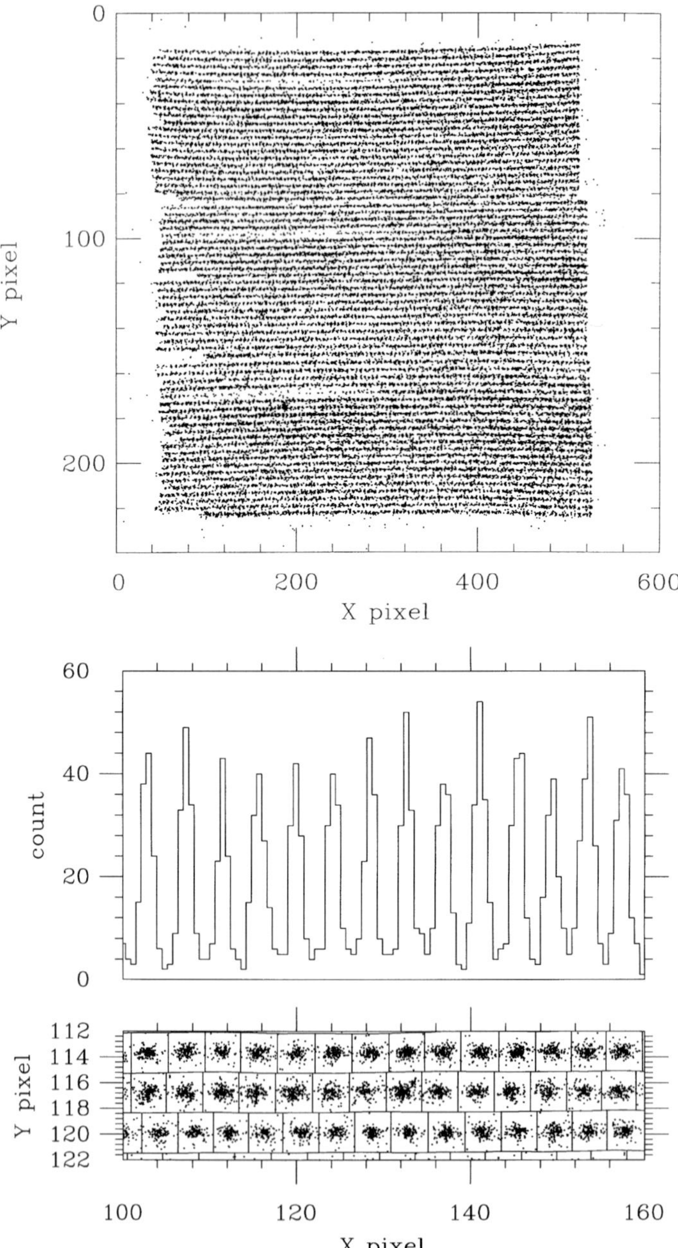

Figure 24. SOFT calibration using a heavy ion beam from the MSU/NSCL accelerator. The top panel is a cross plot of pixel cluster centroids in CCD pixel space. This figure is a superposition of many events obtained by scanning an oxygen beam over the full 26 cm × 26 cm area of the fiber hodoscope. Each row of dots in this view corresponds to a single fiber tab of width ∼2.4 cm. The bottom panel shows an expanded view of a small region in pixel space. It is seen from the cross plots that the individual fibers are clearly resolved. The histogram in the middle panel is obtained by summing the number of counts in each column for the y-pixel range from about 112 to 115. Fits to the peaks (in both the x and y dimensions) are used to derive a map of the relationship between fiber number and pixel location and establish the boundaries between fibers in the CCD image, as indicated by the grid in the bottom panel.

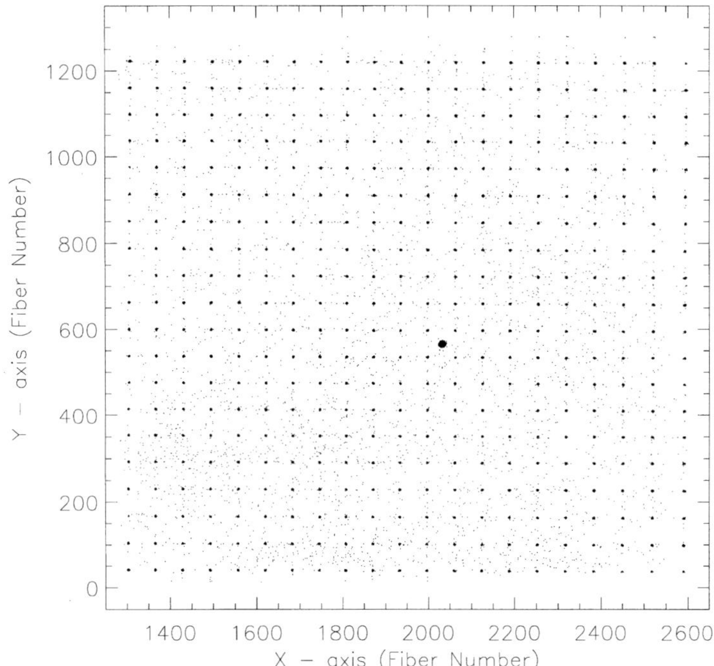

Figure 25. Absolute position calibration of SOFT. A lead mask with 1 mm diameter holes drilled on a 1.27 cm × 1.27 cm grid was used to obtain the fiber to real-space map. The regularly spaced dark spots are the superposition of events which passed through the holes. The vast majority of the events fall in these spots, which are heavily saturated in this image. The sprinkling of particles between the holes is due, in part, to nuclei which interacted in the lead plate producing long-range fragments which penetrate the plate. The large spot slightly offset from the center is the image of a 3 mm hole included for the purpose of determining the mask orientation.

In the course of calibrating the SOFT system it was observed that occasionally an otherwise-good cluster would be displaced significantly from the correct track. This has been traced to an electron 'hopping' phenomenon in which a photoelectron strikes the solid matrix of the microchannel plate and bounces into a channel up to several hundred microns away. While unimportant for heavy nuclei where the cluster centroid is derived from a large number of photoelectrons, electron hopping can affect the tracking of the lightest elements. The SOFT hodoscope was intentionally designed with the H2 plane not equidistant from the H1 and H3 planes (see Figure 6) to aid in determining which of the three layers was affected in events where hopping has altered one of the coordinate measurements. Preliminary techniques for identifying and recovering events in which hopping has occurred have been implemented.

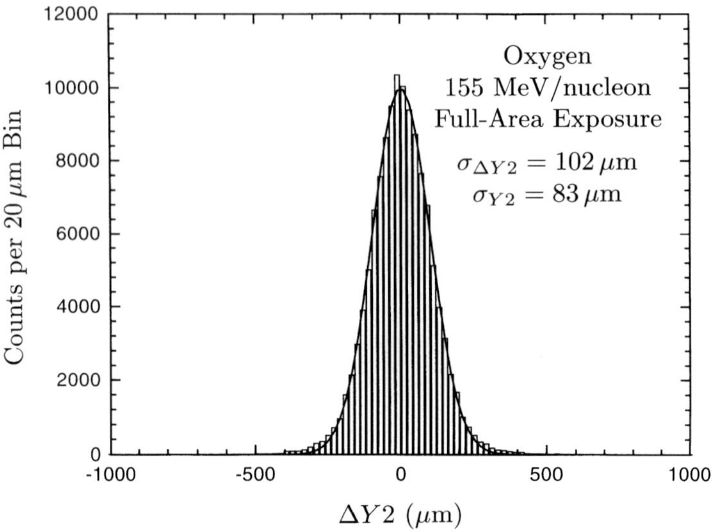

Figure 26. SOFT position resolution obtained using oxygen nuclei at $\sim$155 MeV nucl^{-1}. These events were collected over the full active area of the fiber plane. The distribution is obtained using the method of residuals (see text) and shows that a resolution $\sim$83 μm was obtained in one coordinate.

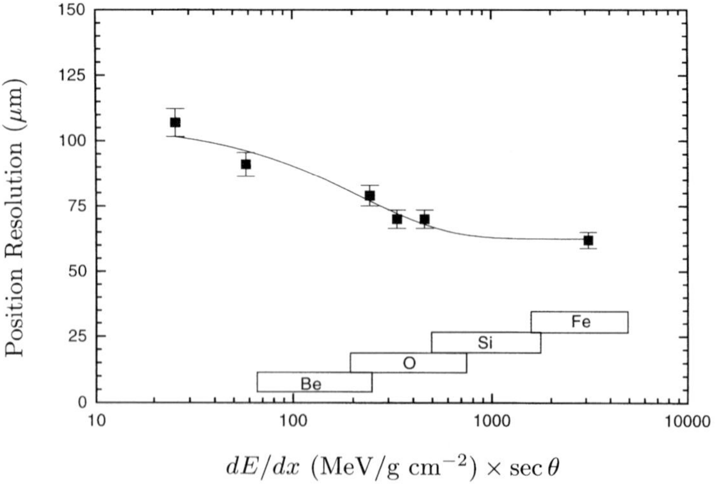

Figure 27. SOFT position resolution as a function of $dE/dx \cdot \sec\theta$, which is proportional to the energy loss in the fiber plane. The position resolution was obtained using beams of He, Li, C, N, O, and Ar nuclei (data points from left to right) with energy at the top of the instrument of 155 MeV nucl^{-1}. The particles were collected over an area $\sim$ 2 cm corresponding to the diameter of the beam spot. The bars along the abscissa indicate the range covered for several elements in flight.

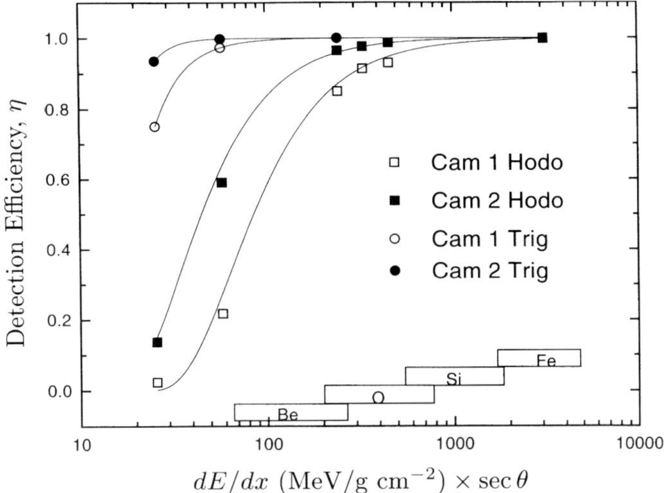

$$dE/dx \ (\mathrm{MeV/g \ cm^{-2}}) \times \sec\theta$$

Figure 28. SOFT detection efficiencies as a function of the energy loss in SOFT. The hodoscope and trigger detection efficiencies were obtained from an MSU/NSCL exposure using 155 MeV nucl^{-1} He, Li, C, N, O, and Ar nuclei (data points from left to right) incident at angles $\theta_x = \theta_y = 15°$. The hodoscope efficiencies are for '6-layer events' (i.e., events with fiber hits that lie along a straight-line trajectory in each of the six hodoscope layers). The trigger fiber efficiencies which are plotted are based on the logical OR of the Tx and Ty fiber layers. The bars along the abscissa indicate the range covered for several elements in flight.

9.4. DETECTION EFFICIENCY

The detection efficiencies of the hodoscope and trigger fibers were also measured for a range of nuclei. Coordinates were derived for the brightest cluster in each of the 6 layers, and the next 6 brightest clusters independent of layer. Events having good straight-line fits to centroids in each of the 6 layers were considered detected, and the ratio of this number to the total number of events in the data set was defined as the 6-layer detection efficiency (η). Figure 28 shows the resulting efficiencies as a function of the energy loss in SOFT.

Camera B, which has higher efficiency than Camera A, is used as the default camera, although either camera has adequate efficiency for measuring beryllium and heavier elements. The difference in efficiencies is thought to be due to differences between the fiber-optic face-plates used on the image intensifiers (Hink et al., 1996). Efficiencies for '5-layer events' are significantly greater but the event quality has not yet been studied in detail.

Figure 28 also shows the trigger-detection efficiency, which is defined as the fraction of events in which a particle was detected in at least one of the two trigger planes, since in flight the SOFT trigger is normally formed from the logical OR of the Tx and Ty signals. Trigger efficiencies using either SOFT camera are near 100% for Be and heavier nuclei.

10. Conclusions

The Cosmic-Ray Isotope Spectrometer was launched aboard the Advanced Composition Explorer spacecraft on 25 August 1997 and was turned on two days later. At the time of submission of this manuscript nearly two months of CRIS data had been returned from space. Preliminary analysis of these data indicates that CRIS has achieved essentially all of the design goals discussed above. It is expected that continued data collection over a period of several years will result in a data set with excellent mass resolution and unprecedented statistical accuracy that can be used to make major advances in understanding the astrophysics of galactic cosmic rays.

Acknowledgments

This research was supported by the National Aeronautics and Space Administration at the Space Radiation Laboratory (SRL) of the California Institute of Technology (under contract NAS5–32626 and grant NAGW-1919), the Jet Propulsion Laboratory (JPL), the Goddard Space Flight Center (GSFC), and Washington University.

We wish to thank the many individuals and organizations who made important contributions to the development of CRIS: J. H. Marshall III of Radcal Corp., J. Gill and B. Mitchell of Harris Corp., and R. McKenzie of Teledyne Corp. played major roles in the design, development, and fabrication of the custom VLSI chips and hybrid circuits, while G. Stupian and M. Leung of Aerospace Corp. provided their micro-focus x-ray facility for hybrid screening. J. Walton and his group at Lawrence Berkeley National Laboratory fabricated the silicon detectors. D. Aalami of Space Instruments was responsible for the design of the power supplies and also contributed to the design, fabrication, and testing of the other electronic assemblies. W. Blanchard of Falcon Services provided all of the board layouts except for the logic board, which was provided by J. Stelma of Design Solutions. Electronic assembly was carried out at Caltech and JPL by N. Neverida and T. Ngo-Luu of JPL with the help of T. Dea, who also assisted during conformal coating and environmental testing at JPL. V. Nguyen of Caltech was responsible for most of the subassembly testing. W. Morris of Caltech received and kitted parts and J. Valenzuela of Space Instruments assisted in PCB production and technical document control. B. Williams of JHU/APL was responsible for the thermal design and support.

At Caltech, B. Sears, R. Selesnick, J. Cummings, and L. Sollitt provided calibration and analysis support, G. Allbritton was responsible for detector testing, G. Flemming contributed to the flight software, J. Burnham provided engineering support, F. Spalding was responsible for project administration, and D. Kubly and R. Kubly performed a variety of secretarial tasks. In addition, M. Calderon, A. Davis, and T. Garrard assisted with computing issues, R. Paniagua and R. Borup provided support in the Caltech Physics shop, and B. Wong helped with a variety of laboratory tasks. J. Lopez-Tiana and E. Friese provided purchasing and contract support.

At JPL, B. Potter assisted in expediting the VLSI build at Harris Corp., D. Cipes-Cwik and R. Hill carried out the hybrid inspections at Teledyne, N. Silva contributed to the assembly of the telescope modules and instrument card cage, as well as development of specifications for magnetics parts, W. Powell served as instrument expeditor, K. Evans provided advice on parts, J. DePew provided technical support for calibrations and laboratory testing, and M. Salama and T. Scharton assisted with mechanical design issues along with P. Rentz of EER Systems Corporation. R. Pool consulted on contractual issues.

At Washington University, G. Simberger was responsible for parts procurement, electronics construction, and inspection. D. Braun was responsible for the organization and inspection of the mechanical parts, the fabrication of the fibers, and the assembly of the fiber planes. J. Cravens of Southwest Research performed the quality assurance on SOFT and provided inputs on the use of image intensifiers in space. H. Darlington of JHU/APL contributed to the design and development of the image intensified CCD cameras. S. Battel of Battel Engineering contributed to the high voltage power supply development. T. Soulanille of Altadena Instruments performed a design review for the camera electronics system. L. Geer, presently at the National Institute of Health, had the original idea to use a photodiode coupled to the image intensifiers for triggering. T. Hardt, D. Steelman, D. Huelsman, and A. Biondo fabricated substantial portions of the SOFT detector in the Washington University machine shop. D. Yates, J. Jarrell, and C. Kruger provided support with contracting, accounting, and proposal preparation. J. Howorth, D. Wilcox, and their staff at Photek developed the image intensifiers and provided helpful advice on the technical details of image intensifiers. G. M'Sadoques of K&M Electronics designed the image intensifier high-voltage power supplies, which were also fabricated by the same company.

At GSFC, B. Fridovich provided administrative support, M. Madden and B. Nahory were responsible for detector testing, and S. Hendricks of Swales and Associates, Inc. assisted with mechanical design issues.

Accelerator testing of the CRIS detectors was made possible by N. Anantaraman, R. Ronningen, and the staff of the National Superconducting Cyclotron Laboratory at Michigan State University, while H. Specht, D. Schardt, and the staff of the GSI heavy ion accelerator in Darmstadt, Germany made possible the heavy ion calibrations of the completed CRIS instrument.

The spacecraft team at JHU/APL provided assistance in many areas while CRIS was being integrated and tested on the spacecraft.

Finally, we thank A. Frandsen, G. Murphy, H. Eyerly, M. Breslof, C. Rangel, and M. McElveney of the ACE Payload Management Office for their untiring support and assistance, and D. Margolies, J. Laudadio, and the ACE Project Office at GSFC for their help in all phases of the CRIS development.

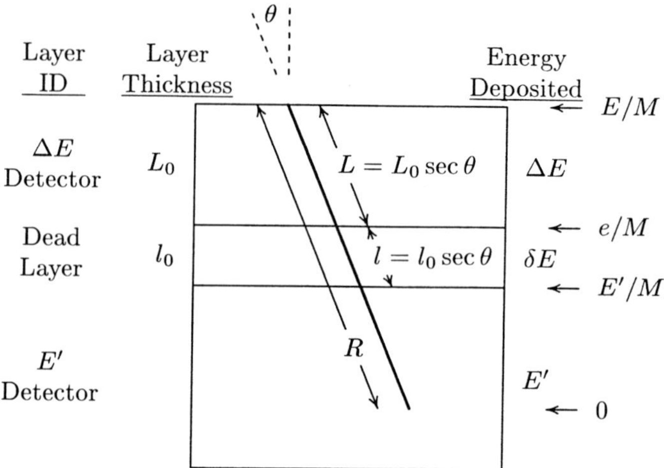

Figure 29. Schematic illustration of 2-detector energy loss vs. residual energy measurement defines the quantities used in calculating mass resolution. The heavy line represents the track of a particle in the detector layers. The far-right column shows the variables used to represent the particle's energy per nucleon at each of the detector interfaces. The layer dimensions are not drawn to scale.

Appendix A. Mass Resolution Calculations

As discussed in Section 3.1, the derivation of a particle's mass using the ΔE vs residual energy technique is based on the fact that the change in the particle's range in penetrating a detector is equal to the thickness of material penetrated. The quantities of interest are defined in Figure 29. Here we consider only the case of two energy measurements, ΔE and E'. In an instrument like CRIS where one has energy loss measurements in a larger number of layers, successive layers can be combined to obtain an equivalent two-detector configuration for which we can analyze the mass resolution using the formulas presented below.

Using $\mathcal{S}_Z(\epsilon)$ to denote the functional dependence of the specific ionization $(\mathrm{d}E/\mathrm{d}x)$ of a nucleus of atomic number Z on the particle's energy per nucleon ϵ, the range of a nucleus with charge and mass numbers Z and M_0 is $\mathcal{R}_{Z,M_0}(\epsilon) = M_0 \int_0^\epsilon d\epsilon'/\mathcal{S}(\epsilon')$. The range for another isotope (mass M) of the same element can be obtained by scaling: $\mathcal{R}_{Z,M}(\epsilon) = (M/M_0)\,\mathcal{R}_{Z,M_0}(\epsilon)$. Finally, the function $\mathcal{E}_{Z,M_0}(R)$ is defined to be the inverse of $\mathcal{R}_{Z,M_0}(\epsilon)$. These three related functions, $\mathcal{S}$, $\mathcal{R}$, and $\mathcal{E}$, are available in tabular or semi-empirical form.

Taking into account the dead layer of thickness l, one has two equations relating range differences to thickness penetrated:

$$\mathcal{R}_{Z,M_0}\left(\frac{\Delta E + \delta E + E'}{M}\right) - \mathcal{R}_{Z,M_0}\left(\frac{E'}{M}\right) = \frac{M_0}{M}(L+l), \tag{4}$$

$$\mathcal{R}_{Z,M_0}\left(\frac{\delta E + E'}{M}\right) - \mathcal{R}_{Z,M_0}\left(\frac{E'}{M}\right) = \frac{M_0}{M}l. \tag{5}$$

TABLE VIII

Partial derivatives used in mass resolution calculations

Derivative	General expression[a]
$\dfrac{\partial M}{\partial \Delta E}$	$M \dfrac{1}{\mathbf{X}\, \mathcal{S}_Z(E/M)}$
$\dfrac{\partial M}{\partial E'}$	$M \left(\dfrac{\mathcal{S}_Z(e/M)}{\mathcal{S}_Z(E/M)} - 1 \right) \dfrac{1}{\mathbf{X}\, \mathcal{S}_Z(E'/M)}$
$\dfrac{\partial M}{\partial L}$	$-M \dfrac{1}{\mathbf{X}}$
$\dfrac{\partial M}{\partial l}$	$M \left(\dfrac{\mathcal{S}_Z(e/M)}{\mathcal{S}_Z(E/M)} - 1 \right) \dfrac{1}{\mathbf{X}}$

[a]Expressions involve a common factor with dimensions of thickness:

$$\mathbf{X} \equiv \left[\frac{\Delta E}{\mathcal{S}_z(E/M)} + \left(\frac{\mathcal{S}_Z(e/M)}{\mathcal{S}_Z(E/M)} - 1 \right) \frac{E'}{\mathcal{S}_Z(E'/M)} - L + \left(\frac{\mathcal{S}_Z(e/M)}{\mathcal{S}_Z(E/M)} - 1 \right) l \right].$$

Eliminating the unknown energy loss (δE) in the dead layer between these two equations yields the basic relation needed for calculating mass resolution:

$$\frac{\Delta E}{M} = \mathcal{E}_{Z,M_0} \left(\mathcal{R}_{Z,M_0} \left(\frac{E'}{M} \right) + \frac{M_0}{M}(L + l) \right)$$
$$-\mathcal{E}_{Z,M_0} \left(\mathcal{R}_{Z,M_0} \left(\frac{E'}{M} \right) + \frac{M_0}{M} l \right). \tag{6}$$

This equation gives M implicitly in terms of the measured quantities ΔE, E', L, and l. Table VIII lists the various partial derivatives needed for calculating the effect on mass resolution due to errors in these quantities.

The following sections discuss the calculation of the contributions to the overall mass resolution caused by various sources of error.

A1. BOHR/LANDAU FLUCTUATIONS

The energy loss of a heavy nucleus in matter is the result of large numbers of Coulomb collisions with electrons in the material. As such it is a statistical process subject to fluctuations. For the energies and atomic numbers of interest for CRIS the distribution of energy losses in a thin layer of thickness x is Gaussian with r.m.s. width $\sqrt{(d\sigma^2_{\Delta E}/dx)_{\text{Landau}}\, x}$, where

$$\left(\frac{d\sigma^2_{\Delta E}}{dx} \right)_{\text{Landau}} = Z^2 \frac{(0.396 \text{ MeV})^2}{\text{g cm}^{-2}} \frac{Z_{\text{m}}}{A_{\text{m}}} \frac{\gamma^2 + 1}{2} \tag{7}$$

is the increase in the variance of the energy loss per unit path length traversed (Rossi, 1952). In this equation, γ is the Lorentz factor corresponding to the particle's energy per nucleon, Z is the particle's charge, and Z_{m} and A_{m} are the charge and mass numbers of the medium. For a particle penetrating a ΔE detector which is not thin compared to the particle's range, one must take into account the effects both of the variation of the particle's energy as it slows and of correlations between energy losses at different depths in the detector (Payne, 1969). In this case one obtains the result:

$$\sigma_{\Delta E(\text{Landau})} = \sqrt{ M \int_{e/M}^{E/M} \frac{\mathcal{S}_Z^2(e/M)}{\mathcal{S}_Z^3(\epsilon)} \left(\frac{\mathrm{d}\sigma_{\Delta E}^2}{\mathrm{d}x} \right)_{\text{Landau}} \mathrm{d}\epsilon }. \tag{8}$$

A.1.1. *Bohr/Landau Fluctuations in the ΔE Detector*

The fluctuation of a particle's energy loss in the ΔE detector results in an equal and opposite fluctuation in the residual energy E'^*. Thus the contribution of Landau fluctuations in the ΔE detector to the mass resolution is given by

$$\sigma_{M,\text{Landau in } L} = \left(\frac{\partial M}{\partial \Delta E} - \frac{\partial M}{\partial E'} \right) \sigma_{\Delta E(\text{Landau})}. \tag{9}$$

A.1.2. *Bohr/Landau Fluctuations in the Dead Layer*

The fluctuation of the energy loss in the dead layer appears only as a fluctuation in the residual energy (since the loss in the dead layer itself is not measured). This fluctuation, $\sigma_{\delta E(\text{Landau})}$, is given by an expression analogous to Equation (9) but with limits of integration E'/M and e/M. The resulting contribution to the mass resolution is

$$\sigma_{M,\text{Landau in } l} = -\frac{\partial M}{\partial E'} \sigma_{\delta E(\text{Landau})}. \tag{10}$$

A.2. MULTIPLE COULOMB SCATTERING

Coulomb collisions of the energetic nucleus with nuclei and electrons in the ΔE detector cause small deflections in the particle's trajectory. The 'multiple scattering' resulting from the cumulative effects of many such interactions alters the path length in the detector. When an energetic nucleus traverses a small thickness, x, of material the distribution of emerging trajectories has a projection on a plane containing the incoming trajectory which is approximately Gaussian, centered on the direction of the incoming trajectory, and having an r.m.s. angular width given by $\sqrt{(\mathrm{d}\sigma_{\delta\theta}^2/\mathrm{d}x)\, x}$, where

$$\frac{\mathrm{d}\sigma_{\delta\theta}^2}{\mathrm{d}x} \simeq \left(\frac{Z}{M} \frac{0.0146}{\beta^2 \gamma} \right)^2 \frac{1}{X_0} \tag{11}$$

---

* We assume that the dead layer thickness is small and the energy loss in the dead layer is not significantly affected by the fluctuation of the energy loss in the ΔE detector.

is the increase in the variance of the angular distribution (measured in radians), projected on a plane containing the incident trajectory, per unit path length traversed (Barnett et al., 1996). Here Z and M are the charge and mass numbers of the energetic nucleus, γ is its Lorentz factor, β is the corresponding velocity divided by the speed of light, and X_0 is the radiation length in the medium. For silicon, $X_0 = 21.82$ g cm^{-2}.

In the case where the thickness of the ΔE detector is sufficiently thin that the change of $\beta^2\gamma$ in it can be neglected, this angular deflection results in an r.m.s. path length error, σ_L, given by $\sigma_{L\ \text{(scat)}} = (1/\sqrt{3})\, L\, \tan\theta\, \sqrt{\left(d\sigma_{\delta\theta}^2/dx\right) x}$. When the change of the $\beta^2\gamma$ factor as the particle slows in the ΔE detector cannot be neglected, this result is replaced by an integral

$$\sigma_{L\ \text{(scat)}} = L\tan\theta\,\sqrt{M}\ \times$$
$$\times\ \sqrt{\int_{E'/M}^{E/M}\left(\frac{\mathcal{R}(E/M)-\mathcal{R}(\epsilon)}{L}\right)^2\left(\frac{d\sigma_{\delta\theta}^2}{dx}\right)_\epsilon\frac{d\epsilon}{\mathcal{S}_Z(\epsilon)}}. \tag{12}$$

The change in the particle's path length in the ΔE detector due to multiple scattering leads to a mass error given by

$$\sigma_{M,\ \text{scat}} = \frac{\partial M}{\partial L}\,\sigma_{L\ \text{(scat)}}. \tag{13}$$

A.3 TRAJECTORY MEASUREMENT ERRORS

Measured trajectories are used in two ways in the mass calculation. The slope of the particle's track is used in calculating the $\sec\theta$ factor needed to obtain a particle's path length in the ΔE detector. The extrapolation of the trajectory to the point at which it intercepts a given detector is used to look up from the calibration maps the detector thickness and dead layer at the point of penetration. Thus uncertainties in both the slope and the intercept of the trajectory contribute to the mass resolution.

To assess these contributions, it is assumed that the trajectory is derived from measurements of two or more pairs of orthogonal coordinates, x_i and y_i ($i = 1\ldots n$), made at heights z_i above the energy loss detectors. The measurements at height z_i are made with r.m.s. errors σ_{x_i} and σ_{y_i}. The projection of the trajectory on the xz plane is obtained from a fit of x_i vs z_i weighted proportional to $1/\sigma_{x_i}^2$, and similarly for the yz projection. The slope, dx/dz, of the fitted line has uncertainty

$$\sigma_{dx/dz} = \left[\sum_{i=1}^{n}\frac{1}{\sigma_{x_i}^2}(z_i - \mathbf{Z}_x)^2\right]^{-1/2}, \tag{14}$$

with the definition $\mathbf{Z}_x \equiv \sum_{i=1}^{n}(z_i/\sigma_{x_i}^2)/\sum_{i=1}^{n}(1/\sigma_{x_i}^2)$.

Furthermore the uncertainty in $x_0(z)$, the x intercept of the fitted trajectory at depth z, is given by

$$\sigma_{x_0(z)} = \sigma_{\bar{x}} \sqrt{\frac{\sum_{i=1}^{n}(z_i - z)^2/\sigma_{x_i}^2}{\sum_{i=1}^{n}(z_i - \mathbf{Z}_x)^2/\sigma_{x_i}^2}},$$

(15)

with the definition $\sigma_{\bar{x}} \equiv \left(\sum_{i=1}^{n} 1/\sigma_{x_i}^2\right)^{-1/2}$.

In the case of CRIS these expressions can be simplified by assuming that all coordinate measurement uncertainties are equal (we will denote them by σ_x) and that coordinate pairs (x_i, y_i) are measured at the same height z_i in the instrument. One then finds $\mathbf{Z}_x = \mathbf{Z}_y \equiv \mathbf{Z} = (\sum_{i=1}^{n} z_i)/n$, $\sigma_{\bar{x}} = \sigma_x/\sqrt{n}$,

$$\sigma_{dx/dz} = \sigma_{dy/dz} = \frac{\sigma_x}{\sqrt{\sum_{i=1}^{n}(z_i - \mathbf{Z})^2}},$$

(16)

and

$$\sigma_{x_0(z)} = \sigma_{y_0(z)} = \frac{\sigma_x}{\sqrt{n}} \sqrt{\frac{\sum_{i=1}^{n}(z_i - z)^2}{\sum_{i=1}^{n}(z_i - \mathbf{Z})^2}}.$$

(17)

The $\sec\theta$ factor used in calculating a particle's path length in the ΔE detector can be written in the form $\sec\theta = [1 + (dx/dz)^2 + (dy/dz)^2]^{1/2}$, and under the assumptions used for deriving Equation (17), has fractional uncertainty given by

$$\frac{\sigma_{\sec\theta}}{\sec\theta} = \frac{1}{2}\sin 2\theta\, \sigma_{dx/dz}.$$

(18)

Further simplification is realized if there are only two trajectory measurement planes or if there are three planes with one placed midway between the other two (nearly the case in CRIS). In this case $\sigma_{dx/dz} = \sqrt{2}\,\sigma_x/\Delta z$, where Δz is the spacing between the outermost two trajectory planes, and one obtains

$$\frac{\sigma_{\sec\theta}}{\sec\theta} = \frac{1}{\sqrt{2}}\sin 2\theta\, \frac{\sigma_x}{\Delta z}.$$

(19)

The error in $\sec\theta$ makes (correlated) contributions to the mass resolution due to its effects on the path lengths in both the ΔE detector and in the dead layer:

$$\sigma_{M,\sec\theta} = \frac{\partial M}{\partial L} L_0\, \sigma_{\sec\theta} + \frac{\partial M}{\partial l} l_0\, \sigma_{\sec\theta}.$$

(20)

The uncertainties in the intercepts of the derived trajectory contribute to the mass resolution through their effects on the accuracy with which thickness can be obtained from the detector calibration maps. This is discussed in the following section.

A.4 MAPPING ERRORS

Maps of the detector thickness, $L_j(x, y)$, and dead layer, $l_j(x, y)$, (with j labeling the detector number) are obtained from laboratory and accelerator measurements. They are used, in conjunction with the measured particle trajectory, to derive the path lengths for calculating particle charge and mass. Thickness errors arise due both to uncertainties in the measured maps and to uncertainties in the location at which the trajectory penetrated the detector. The overall thickness mapping error can be expressed as the sum of three uncorrelated contributions, $\sigma^2_{L(\text{map})} = \sigma^2_{L(\text{map}),L_0} + \sigma^2_{L(\text{map}),x} + \sigma^2_{L(\text{map}),y}$. When the ΔE measurement is obtained by summing signals from m successive detectors located in the stack at positions $z_1, \ldots, z_m$, these contributions can be calculated as:

$$\sigma^2_{L(\text{map}),L_0} = \sum_{j=1}^{m} \sigma^2_{L_j}, \tag{21}$$

$$\sigma^2_{L(\text{map}),x} = \left(\sum_{j=1}^{m} \frac{\partial L_j}{\partial x} \sigma_{x_0(z_j)} \right)^2, \tag{22}$$

$$\sigma^2_{L(\text{map}),y} = \left(\sum_{j=1}^{m} \frac{\partial L_j}{\partial y} \sigma_{y_0(z_j)} \right)^2. \tag{23}$$

In these expressions, each partial derivative is evaluated at the location where the particle penetrated that particular detector.

The errors related to the application of the dead-layer map are obtained from similar expressions with l substituted for L and with the locations z_j appropriate to the position of the dead layer in the stack. As noted above, only the dead layer between the ΔE and E' measurements is taken into account in this analysis; thus the sum over the individual detectors that contribute to the ΔE measurement is not required. The contributions caused by trajectory errors introduce correlations between the thickness and dead layer uncertainties.

Mapping errors (both for thickness and dead layer) make an overall contribution to mass resolution given by:

$$\sigma^2_{M(\text{map})} = \left(\frac{\partial M}{\partial L} \right)^2 \sigma^2_{L(\text{map}),L_0} + \left(\frac{\partial M}{\partial l} \right)^2 \sigma^2_{l(\text{map}),l_0} +$$

$$+ \left(\frac{\partial M}{\partial L} \sigma_{L(\text{map}),x} - \frac{\partial M}{\partial l} \sigma_{l(\text{map}),x} \right)^2 +$$

$$+ \left(\frac{\partial M}{\partial L} \sigma_{L(\text{map}),y} - \frac{\partial M}{\partial l} \sigma_{l(\text{map}),y} \right)^2 . \tag{24}$$

When designing an instrument, one normally cannot predict the directions of the thickness gradients or their correlations between different layers. Thus, in practice, the possible correlations between different contributions to the mapping error are neglected and all contributions are compounded as if uncorrelated. In addition, although correlations between the mapping errors and $\sec\theta$ errors are present since they depend on the same trajectory measurements, these correlations are neglected because their relative signs are not known.

A.5 ENERGY MEASUREMENT ERRORS

The measurement of the energy deposited in a detector or in a combination of detectors associated with a single pulse-height analyzer (PHA) is uncertain due both to the contribution of electronic noise and to the finite channel width used in digitizing the signal. For the kth PHA the energy error can be written

$$\sigma_{E_k}^2 = \sigma_{E_k(\text{noise})}^2 + \left(\frac{w_k}{\sqrt{12}} \right)^2 , \tag{25}$$

where w_k is the channel width for this PHA. The uncertainties associated with the measurements of ΔE and E' are sums in quadrature of the values of $\sigma_{E_k}^2$ from all the PHAs whose outputs are summed to form these quantities. The corresponding contributions to the mass resolution are:

$$\sigma_{M(\Delta E \text{ meas.})} = \frac{\partial M}{\partial \Delta E} \sqrt{\sum_{k \in \Delta E} \sigma_{E_k}^2} , \qquad \sigma_{M(E' \text{ meas.})} = \frac{\partial M}{\partial E'} \sqrt{\sum_{k \in E'} \sigma_{E_k}^2} . \tag{26}$$

A.6 ADDITIONAL EFFECTS

Several potential sources of mass uncertainty are omitted in our resolution calculations because they should be correctable using calibrations that can be obtained from laboratory or flight data. Effects of this type include: small rotations or displacements of the silicon detectors from their nominal placement in the stack, effects of non-linearities in the transfer functions of the pulse-height analyzers, and imperfections in the geometry of the hodoscope.

The effects of multiple Coulomb scattering in the hodoscope have also been neglected. These should be rather insignificant for the CRIS instrument due to the small amount of material above the Si(Li) detector stack.

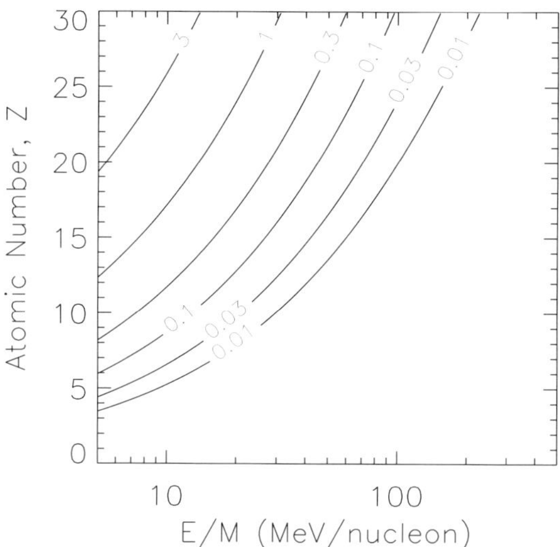

Figure 30. Contour plot of the difference between a particle's atomic number, Z, and its effective charge, q_{eff}, as a function of atomic number and particle energy per nucleon. This difference is caused by the attachment of atomic electrons. In regions of the plot where $Z - q_{\text{eff}} \gtrsim 1$ one expects that charge-state fluctuations could contribute significantly to the width of the energy loss distribution, and therefore to the mass resolution.

For heavy elements at low energy, the nuclei will have several electrons attached (Ahlen, 1980). For example, the effective charge of an iron ($Z = 26$) nucleus at 20 MeV nucl^{-1} is only +24.9. The average effect of the reduced charge on the particle's slowing is taken into account in the range energy relation that is used. However, the charge state is subject to statistical fluctuations as collisions cause attachment and stripping of electrons. These fluctuations, which should be added in quadrature to the Bohr/Landau fluctuations discussed above, have been neglected due to lack of data on the magnitude of the effects. Preliminary measurements in the regime where charge-state fluctuations should be significant suggest that calculations neglecting this effect fail to adequately account for the observed mass resolution. Figure 30 shows the amount by which the effective charge, q_{eff}, is expected to differ from fully stripped for various combinations of Z and E/M. When this difference $Z - q_{\text{eff}} \ll 1$, the calculations presented above should adequately account for the overall mass resolution. This condition is generally met for the particles of interest in CRIS, but can be more important for the lower-energy nuclei being studied with the Solar Isotope Spectrometer (see Stone et al., 1998, this issue).

A.7 MONTE CARLO CALCULATIONS

The above analytic treatment of the mass resolution is particularly useful for assessing the functional dependence of the resolution on the various parameters involved in the instrument design. As an alternative, one can carry out Monte Carlo simulations of the instrument response. The Monte Carlo approach allows one to readily average over particle parameters such as energy and incidence angle that will be encountered in flight. Both approaches were employed in designing the CRIS instrument, with generally consistent results.

Appendix B. CRIS Logic Equations

The triggering of the CRIS instrument and the determination of whether a valid event has occurred are based on boolean logic implemented using Actel gate arrays. The inputs to the logic circuits consist of discriminator pulses (some latched) from the instrument sensors, signals from the on-board stimulus and hazard circuits, and various static control bits that can be set by command. Table IX defines these inputs and Table X lists the logic equations that use these signals to determine when trigger and valid coincidence conditions are met.

The logic is designed to distinguish, approximately, among hydrogen ($Z = 1$), helium ($Z = 2$) and heavier nuclei ($Z > 2$) based on appropriately set medium level ('m') and high level ('h') discriminators. Whereas the normal coincidence requires that heavy nuclei trigger at least the E1 and E2 discriminators, an additional term ('adc2') is used to accept candidate hydrogen and helium nuclei based on triggering any two successive discriminators. This term makes it possible to include light nuclei which stop deep in the stack and do not trigger the front detectors. Pulse heights from all the detectors are, nevertheless measured (whether or not the discriminator was triggered) and are available for analysis of the event.

Appendix C. CRIS Normal Event Format

Table XI summarizes the format used for transmission of CRIS normal events. Each event starts with a fixed–length, 6–byte header, the first byte of which gives the total length of that event. As shown in Table XI, the header also indicates whether or not various optional subfields of data are included in that event, and the sizes of the variable length subfields. Among other things, the header specifies whether pulse height data from both the A/B and C/D telescopes are included, how many guard detectors were triggered and whether or not the guard pulse heights are being transmitted, the 'range' of the particle (that is, the identity of the deepest triggered detector), the number of position coordinates returned from SOFT, and the time (in instrument cycles) that the event was put into the event buffer (which is needed to match the event with the appropriate rate sample and live time). If data from both telescope pairs are included in the event, this primary header is followed by a one-byte secondary header, which lists the number of guards triggered and the range of the particle in the second telescope pair.

TABLE IX

CRIS logic inputs[a]

Signal	Description
Active inputs	
e1a, e1b, e2ab, ..., e9ab	low level discriminator signals from Si(Li) detector center PHAs
e1ma, e1mb, e2mab, ..., e8mab	medium level discriminator signals from Si(Li) detector center PHAs
e1ha, e1hb, e2hab, ..., e8hab	high level discriminator signals from Si(Li) detector center PHAs
g2ab, g3ab, ..., g7ab	low level discriminator signals from Si(Li) detector guard PHAs
g2hab, g3hab, ..., g7hab	high level discriminator signals from Si(Li) detector guard PHAs
Static inputs (commandable) [b]	
e1_mode	set to require that only e1a or e1b be present; cleared to allow both to be present (and similarly for e1c and e1d)
force_e1a	set to override requirement that e1a be present (similarly for *force_e1b, force_e2ab*)
delete_g2ab	set to override requirement that g2ab not be present (similarly for *delete_g3ab* through *delete_g7ab*)
unforce_e3ab	set to require that e3ab be present for a valid event in AB telescope pair (similarly for *unforce_e3ab* through *unforce_e8ab*)
adc2ab_disable	set to disallow as valid any events with low-level discriminator triggers from any two successive detectors
hmode0, hmode1	select one of four possible modes for trigger from SOFT hodoscope: x plane independent of y plane, y independent of x, either x or y, both x and y
force_hodo	set to override requirement that SOFT trigger be present for coincidence
hy_enable	set to allow events tentatively identified as hydrogen ($Z = 1$)
he_enable	set to allow events tentatively identified as helium ($Z = 2$)
stim_always_valid	set to treat all pulser stimulus events as valid; cleared to require that they satisfy the same coincidence requirements as particle events
haz_anti_en	set to allow hazard signal to veto events

[a]Inputs are listed only for the AB telescope pair. Corresponding signals exist from the CD telescope pair.

[b]Other command bits exist which allow any active input to be disabled and provide additional control of the logic, as described in note c to Table X.

TABLE X

CRIS logic equations

Term[a]	Definition[b]	Comments
[e1a]	$(\text{e1a} \cdot \overline{\text{e1b}} \cdot el\_mode + \text{e1a} \cdot \overline{el\_mode}) + force\_e1a$	
[e1b]	$(\text{e1b} \cdot \overline{\text{e1a}} \cdot el\_mode + \text{e1b} \cdot \overline{el\_mode}) + force\_e1b$	
glab	$\text{g2ab} \cdot \overline{delete\_g2ab} + \text{g3ab} \cdot \overline{delete\_g3ab} + \cdots + \text{g7ab} \cdot \overline{delete\_g7ab}$	any low-level ab guard discr.
[e2ab]	$\text{e2ab} + force\_e2ab$	
/e3ab/	$\text{e3ab} + \overline{unforce\_e3ab}$	similar for e4ab . . . e8ab
minrngab	$([\text{e1a}] + [\text{e1b}]) \cdot [\text{e2ab}] \cdot /\text{e3ab}/ \cdot /\text{e4ab}/ \cdot /\text{e5ab}/ \cdot /\text{e6ab}/ \cdot /\text{e7ab}/ \cdot /\text{e8ab}/$	
adc2ab	$((\text{e1a} + \text{e1b}) \cdot \text{e2ab} + \text{e2ab} \cdot \text{e3ab} + \text{e3ab} \cdot \text{e4ab} + \text{e4ab} \cdot \text{e5ab} + \text{e5ab} \cdot \text{e6ab} + \text{e6ab}$ $\text{e6ab} \cdot \text{e7ab} + \text{e7ab} \cdot \text{e8ab}) \cdot \overline{adc2ab\_disable}$	
ghor[c]	$\text{g2hab} + \text{g3hab} + \text{g4hab} + \text{g5hab} + \text{g6hab} + \text{g7hab} + \text{(corresponding terms}$ for cd)	any high-level guard discr.
mor[c]	$\text{e1ma} + \text{e1mb} + \text{e2mab} + \text{e3mab} + \text{e4mab} + \text{e5mab} + \text{e6mab} + \text{e7mab} +$ $\text{e8mab} + \text{(corresponding terms for cd)}$	any medium-level center discr.
hor[c]	$\text{e1ha} + \text{e1hb} + \text{e2hab} + \text{e3hab} + \text{e4hab} + \text{e5hab} + \text{e6hab} + \text{e7hab} + \text{e8hab} +$ (corresponding terms for cd)	any high-level center discr.
z1ab	$(\text{minrngab} + \text{adc2ab}) \cdot \overline{\text{mor}} \cdot \overline{\text{hor}} \cdot \overline{\text{glab}}$	
z2ab	$(\text{minrngab} + \text{adc2ab}) \cdot \text{mor} \cdot \overline{\text{hor}} \cdot \overline{\text{glab}}$	
z > 2ab	$\text{minrngab} \cdot \text{hor} \cdot \overline{\text{ghor}}$	
[hodo]	$\overline{hmode0} \cdot \overline{hmode1} \cdot (\text{soft0} \cdot \text{soft1}) + hmode0 \cdot \overline{hmode1} \cdot (\text{soft0} + \text{soft1}) +$ $\overline{hmode0} \cdot hmode1 \cdot \text{soft0} + hmode0 \cdot hmode1 \cdot \text{soft1} + force\_hodo$	
trigger	$\text{e1a} + \text{e1b} + \text{e2ab} + \text{e3ab} + \text{e4ab} + \text{e5ab} + \text{e6ab} + \text{e7ab} + \text{e8ab} + \text{e9ab} +$ $\text{g2ab} + \text{g3ab} + \text{g4ab} + \text{g5ab} + \text{g6ab} + \text{g7ab} + \text{(corresponding terms for cd)}$	trigger condition
valid	$[\text{hodo}] \cdot (\,(\text{z1ab} + \text{z1cd}) \cdot hy\_enable + (\text{z2ab} + \text{z2cd}) \cdot he\_enable +$ $(\text{z>2ab}) + (\text{z>2cd}) + (\text{stim} \cdot stim\_always\_valid)\,) \cdot (\overline{hazard} + \overline{haz\_anti\_en})$	coincidence requirement

[a]Inputs are defined in Table IX. Those in italics are commandable, static inputs.
[b]Most terms (other than ghor, mor, hor, [hodo], trigger, and valid) are specific to the AB telescope pair. Corresponding terms exist for the CD telescope pair.
[c]Command bits also exist which allow each term in ghor, mor, and hor to be separately deleted.

TABLE XI

CRIS normal event format

Item	Count	Bits	Sum[a]	Description
Event header				
0	1	8	8	Number of bytes in this event
1	1	1	9	Stimulated event flag
2	1	1	10	Primary telescope pair ($AB = 0$; $CD = 1$)
3	1	6	16	Event buffer ID
4	1	1	17	Guard low-level 'OR'
5	1	1	18	Hazard flag
6	1	3	21	# Low-level guards hit ($G1$) in primary telescope pair
7	1	3	24	Range ($R1 - 2$) in primary telescope pair ($R1$ values: 2–9)
8	1	1	25	Is there 2nd telescope data included in this event?
9	1	1	26	Transmit guard pulse heights in this event?
10	1	1	27	SOFT cluster centroid calculation error encountered
11	1	5	32	Number of SOFT clusters transmitted (N)[b]
12	1	8	40	SOFT # of pixels[c]
13	1	2	42	2nd telescope activity tag bits
14	1	6	48	Instrument cycle counter
2nd telescope event header (present if item 8 is true)				
15	1	2	2	Spare
16	1	3	5	# Low-level guards hit ($G2$) in secondary telescope pair
17	1	3	8	Range ($R2 - 2$) in secondary telescope pair ($R2$ values: 2–9)
SOFT hodoscope data				
18		10	10	SOFT cluster x centroids
19	N	9	19	SOFT cluster y centroids
20		5	24	SOFT cluster intensity parameter [c]
Primary telescope data				
Discriminator tag bits				
21	1	16	16	Tag bits: $G7 - G2$, $E9 - E1A$ ($E1A = $ LSB, $G7 = $ MSB)
Center pulse heights				
22	$R1 + 1$[d]	12	40 to 120	Stack-detector pulse heights $+$ 4 bits 0 pad if $R1$ is even
Guard pulse heights (if item 9 is true)				
23		12	12	Guard pulse height
24	$G1$	1	13	Spare
25		3	16	Guard address (values: 2–7)
Secondary telescope data (if item 8 is true)				
				Analagous to items 21 through 25, above
Stimulated event parameters (if item 1 is true)				
26	1	4	4	Flags indicating which guard pulsers are active
27	1	4	8	Flags indicating which center pulsers are active
28	1	8	16	guard pulser DAC level
29	1	8	24	center pulser DAC level
30	1	1	25	guard DAC high-gain mode
31	1	1	26	center DAC high-gain mode
32	1	6	32	SOFT LED pulse length

[a]Running total number of bits in each section.
[b]Maximum number of SOFT clusters read out is commandable up to $N_{max} = 31$.
[c]Encoded as a floating point value.
[d]Both E1 signals always read out.

TABLE XII

CRIS event buffers as assigned at launch

Charge	Range	Number of buffers and rates[a]	Readout priority[b]		Bad Hodo
			Good Hodo		
			HAZ $= 0$	HAZ $= 1$	HAZ $= 0$ or 1
Single-telescope-pair events (one set for A/B, one set for C/D)					
$Z \geq 3$	9 (PEN)	3	34	31	14
$Z \geq 10$	8	3	60	52	26
	5, 6, 7	3	62	54	28
	3, 4	3	61	53	27
	2	3	59	51	22
$3 \leq Z \leq 9$	8	3	56	48	23
	5, 6, 7	3	58	50	25
	3, 4	3	57	49	24
	2	3	55	47	21
$Z = 2$	9 (PEN)	3	33	30	13
	8	3	44	38	18
	5, 6, 7	3	46	40	20
	3, 4	3	45	39	19
	2	3	43	37	17
$Z = 1$	8, 9 (PEN)	3	32	29	12
	3 to 7	3	42	36	16
	2	3	41	35	15
	Total:	51			
Double-telescope-pair events					
$Z \geq 3$	9 (PEN)	1	5	5	5
	4 to 8	1	11	11	11
	2 to 3	1	10	10	10
$Z = 2$	9 (PEN)	1	4	4	4
	4 to 8	1	9	9	9
	2 to 3	1	8	8	8
$Z = 1$	9 (PEN)	1	3	3	3
	4 to 8	1	7	7	7
	2 to 3	1	6	6	6
Stim [c]	9 (typ.)	1	63	63	63
	Total:	10			

[a]Total rates: 51 (A/B single) + 51 (C/D single) + 10 (double) = 112.
 Total buffers: 51 (A/B single + C/D single) + 10 (double) = 61.
[b]Higher numbers indicate higher priorities. Buffers 0, 1, and 2 are not used.
[c]Pulser stimulus event.

Following the header(s) are the SOFT hodoscope data. Using the raw SOFT pixel locations and intensities, the on-board microprocessor determines which pixels are adjacent to each other, and calculates the total intensity and the two-dimensional centroid location of each such contiguous cluster of pixels. The resulting position and intensity information occupies three bytes for each cluster.

Silicon-detector pulse-height information follows the SOFT data, starting with the 'primary' telescope pair, that is, the one with the longest range if both telescope pairs were triggered. Tag bits indicating which of the 10 stack and six guard low level discriminators were triggered take up 2 bytes. Stack-detector pulse heights are listed after the tag bits. If guard pulse heights are to be telemetered, they follow the stack detector pulse heights. If data from the second telescope pair are included, they follow the data from the first telescope, in the same sequence (discriminator tag bits, stack pulse heights, guard pulse heights). Finally, if the event is a pulser-stimulated event, it concludes with a four-byte block listing which pulsers were active and the value of their control DAC levels.

Appendix D. CRIS Event Buffers

As discussed in Section 5.4.2, CRIS events are roughly classified onboard based on charge category, range, hazard, and trajectory quality. Events are sorted among 61 buffers according to this classification, and the corresponding rate of events falling in each buffer is counted. Each buffer is assigned a priority which is used in selecting events from the various buffers for readout. Table XII shows the priorities normally assigned to each of the event buffers. The priorities can be changed by command.

Abbreviations: ACE – Advanced Composition Explorer, ACR – anomalous cosmic ray, ADC – analog-to-digital converter, AOG – amplifier/offset gate, ASIC – application-specific integrated circuit, ASCII – American Standard Code for Information Interchange, BPBD – 2-(4-Biphenylyl)-5-(4-tert-butylphenyl)-1,3,4-oxadiazole (scintillation dye), CCD – charge coupled device, CRIS – Cosmic Ray Isotope Spectrometer, CRRES – Combined Release and Radiation Effects Satellite, DAC – digital-to-analog converter, DPOPOP – 1,4-Bis(4-methyl-5-phenyloxazol-2-yl)benzene (scintillation dye), EMA – extramural absorber, FIP – first ionization potential, GCR – galactic cosmic ray, GCRS – galactic cosmic ray source, GSFC – Goddard Space Flight Center, GSI – Gesellschaft für Schwerionenforschung mbH (accelerator laboratory in Darmstadt, Germany), ISEE – International Sun-Earth Explorer, JPL – Jet Propulsion Laboratory, LBNL – Lawrence Berkeley National Laboratory, LED – light emitting diode, LISM – local interstellar medium, MCP – microchannel plate, MLI – multilayer insulation, MSU/NSCL – Michigan State University / National Superconducting Cyclotron Laboratory, NASA – National

Aeronautics and Space Administration, PCB – printed-circuit board, PHA – pulse-height analyzer, RF – radio frequency, r.m.s. – root mean square, SAMPEX – Solar, Anomalous, and Magnetospheric Particle Explorer, S/C – spacecraft, SEP – solar energetic particle, Si(Li) – lithium-drifted silicon detector, SLAM – Structural Loads and Acoustics Measurements (instrument on ACE), SOFT – Scintillating Optical Fiber Trajectory (detector), SW – solar wind, TV – thermal-vacuum, VLSI – very large scale integrated (circuit), ZIF – zero insertion force (connector).

References

Ahlen, S. P.: 1980, 'Theoretical and Experimental Aspects of the Energy Loss of Relativistic Heavily Ionizing Particles', *Rev. Mod. Phys.* **52**, 121–173.

Allbritton, G. A., Andersen, H., Barnes, A., Christian, E. R., Cummings, A. C., Dougherty, B. L., Jensen, L., Lee, J., Leske, R. A., Madden, M. P., Mewaldt, R., Milliken, B., Nahory, B. W., O'Donnell, R., Schmidt, P., Sears, B. R., von Rosenvinge, T. T., Walton, J. T., Wiedenbeck, M. E., and Wong, Y. K.: 1996, 'Large-Diameter Lithium Compensated Silicon Detectors for the NASA Advanced Composition Explorer (ACE) Mission', *IEEE Trans. Nucl. Sci.* **43**, 1505–1509.

Althouse, W. E., Cummings, A. C., Garrard, T. L., Mewaldt, R. A., Stone, E. C., and Vogt, R. E.: 1978, 'A Cosmic Ray Isotope Spectrometer', *IEEE Trans. Geosci. Electronics* **GE-16**, 204–207.

Barnett, R. M. et al.: 1996, 'Review of Particle Properties', *Phys. Rev.* **D54**, 1.

Berezinskiĭ, V. S., Bulanov, S. V., Dogiel, V. A., Ginzburg, V. L., and Ptuskin, V. S.: 1990, *The Astrophysics of Cosmic Rays*, North-Holland, Amsterdam.

Breneman, H. H. and Stone, E. C.: 1985, 'Solar Coronal and Photospheric Abundances from Solar Energetic Particle Measurements', *Astrophys. J.* **199**, L51–L61.

Cassé, M. and Paul, J. A.: 1982, 'On The Stellar Origin of the ^{22}Ne Excess in Cosmic Rays', *Astrophys. J.* **258**, 860–863.

Connell, J. J. and Simpson, J. A.: 1995, 'The *Ulysses* Cosmic Ray Isotope Experiment: Isotopic Abundances of Fe and Ni from High Resolution Measurements', *Proc. 24th Int. Cosmic Ray Conf., Rome* **2**, 602–605.

Connell, J. J. and Simpson, J. A.: 1997a, 'High Resolution Measurements of the Isotopic Composition of Galactic Cosmic Ray C, N, O, Ne, Mg, and Si from the *Ulysses* HET', *Proc. 25th Int. Cosmic Ray Conf., Durban* **3**, 381–384.

Connell, J. J. and Simpson, J. A.: 1997b, '^{26}Al in the Galactic Cosmic Radiation: Measurements with the *Ulysses* HET Instrument', *Proc. 25th Int. Cosmic Ray Conf., Durban* **3**, 393–396.

Cook, W. R., Cummings, A., Kecman, B., Mewaldt, R. A., Aalami, D., Kleinfelder, S. A., and Marshall, J. H.: 1993a, in B. Tsuratani (ed.), 'Custom Analog VLSI for the Advanced Composition Explorer', *Small Instruments Workshop Proc.*, Pasadena, Jet Propulsion Laboratory.

Cook, W. R., Cummings, A. C., Cummings, J. R., Garrard, T. L., Kecman, B., Mewaldt, R. A., Selesnick, R. S., Stone, E. C., and von Rosenvinge, T. T.: 1993b, 'MAST: A Mass Spectrometer Telescope for Studies of the Isotopic Composition of Solar, Anomalous, and Galactic Cosmic Ray Nuclei', *IEEE. Trans. Geosci. Remote Sensing* **31**, 557–564.

Crary, D. J., Binns, W. R., Cummings, A. C., Klarmann, J., Lawrence, D. J., and Mewaldt, R. A.: 1992, 'A Bevalac Calibration of a Scintillating Optical Fiber Hodoscope', *Nucl. Inst. Meth.* **A316**, 311–317.

Davis, A. J., Hink, P. L., Binns, W. R., Epstein, J. W., Connell, J. J., Israel, M. H., Klarmann, J., Vylet, V., Kaplan, D. H., and Ruecroft, S.: 1989, 'Scintillating Optical Fiber Trajectory Detectors', *Nucl. Inst. Meth.* **A276**, 347–358.

Dougherty, B. L., Christian, E. R., Cummings, A. C., Leske, R. A., Mewaldt, R. A., Milliken, B. D., von Rosenvinge, T. T., and Wiedenbeck, M. E.: 1996, in B. A. Ramsey and T. A. Parnell (eds.), 'Characterization of Large-Area Silicon Ionization Detectors for the ACE Mission', *Gamma-Ray and Cosmic-Ray Detectors, Techniques, and Missions*, Proc. SPIE 2806, Society of Photo-Optical Instrumentation Engineers, Denver, pp. 188–198.

DuVernois, M. A. and Thayer, M. R.: 1996, 'The Elemental Composition of the Galactic Cosmic-Ray Source: *Ulysses* High-Energy Telescope Results', *Astrophys. J.* **465**, 982–984.

DuVernois, M. A., Garcia-Muñoz, M., Pyle, K. R., Simpson, J. A., and Thayer, M. R.: 1996, 'The Isotopic Composition of Galactic Cosmic-Ray Elements from Carbon to Silicon: the Combined Release and Radiation Effects Satellite Investigation', *Astrophys. J.* **466**, 457–472.

Ellison, D. C., Drury, L. O'C., and Meyer, J.-P.: 1998, 'Cosmic Rays from Supernova Remnants: A Brief Description of the Shock Acceleration of Gas and Dust', *Space Sci. Rev.* **86**, 203.

Engelmann, J. J., Ferrando, P., Soutoul, A., Goret, P., Juliusson, E., Koch-Miramond, L., Lund, N., Masse, P., Peters, B., Petrou, N., and Rasmussen, I. L.: 1990, 'Charge Composition and Energy Spectra of Cosmic-Ray Nuclei for Elements from Be to Ni. Results from HEAO-3-C2', *Astron. Astrophys.* **233**, 96–111.

Fisher, A. J., Hagen, F. A., Maehl, R. C., Ormes, J. F., and Arens, J. F.: 1976, 'The Isotopic Composition of Cosmic Rays with $5 \leq Z \leq 26$', *Astrophys. J.* **205**, 938–946.

Garcia-Muñoz, M., Mason, G. M., and Simpson, J. A.: 1977, 'The Isotopic Composition of Galactic Cosmic Ray Lithium, Beryllium, and Boron', *Proc. 15th Int. Cosmic Ray Conf., Plovdiv* **1**, 301–306.

Halpern, E. and Marshall, J. H.: 1968, 'A High Resolution Gamma-Ray Spectrometer for Use in Outer Space', *IEEE Trans. Nucl. Sci.* **NS-15**, 242–251.

Halpern, E., Marshall, J. H., and Weeks, D.: 1968, 'A Gamma-Ray Spectrometer for Space Applications', *Nucleonics in Aerospace*, Plenum Press, New York, pp. 98–106.

Harrington, T. M. and Marshall, J. H.: 1968, 'A Pulse-Height Analyzer for Charged Particle Spectroscopy on the Lunar Surface', *Rev. Sci. Instr.* **39**, 184–194.

Harrington, T. M. and Marshall, J. H.: 1969, 'An Electronics System for Gamma-Ray Spectrometry in Space Applications', *IEEE Trans. Nucl. Sci.* **NS-16**, 314–321.

Harrington, T.M., Marshall, J.H., Arnold, J.R., Peterson, L.E., Trombka, J.I., and Metzger, A.E.: 1974, 'The Apollo Gamma-Ray Spectrometer', *Nucl. Instr. Meth.* **118**, 401–411.

Hink, P. L., Binns, W. R., Klarmann, J., and Olevitch, M. A.: 1996, 'The ACE–CRIS Scintillating Optical Fiber Trajectory (SOFT) Detector: A Calibration at the NSCL', in B. A. Ramsey and T. A. Parnell (eds.), *Gamma-Ray and Cosmic-Ray Detectors, Techniques, and Missions*, Proc. SPIE 2806, Society of Photo-Optical Instrumentation Engineers, Denver, pp. 199–208.

Hubert, F., Bimbot, R., and Gauvin, H.: 1990, 'Range and Stopping-Power Tables for 2.5–500 MeV/Nucleon Heavy Ions In Solids', *Atom. Dat. Nucl. Dat. Tables* **46**, 1–213.

Leske, R. A.: 1993, 'The Elemental and Isotopic Composition of Galactic Cosmic-Ray Nuclei from Scandium through Nickel', *Astrophys. J.* **405**, 567–583.

Leske, R. A. and Wiedenbeck, M. E.: 1993, 'Composition Measurements from ISEE-3: Fluorine through Calcium', *Proc. 23rd Int. Cosmic Ray Conf., Calgary* **1**, 571–574.

Letaw, J. R., Adams, J. H., Silberberg, R., and Tsao, C. H.: 1985, 'Electron-Capture Decay of Cosmic Rays', *Astrophys. Space Sci.* **114**, 365–379.

Letaw, J. R., Silberberg, R., and Tsao, C. H.: 1993, 'Comparison of Distributed Reacceleration and Leaky-Box Models of Cosmic-Ray Abundances $(3 \leq Z \leq 28)$', *Astrophys. J.* **414**, 601–611.

Lukasiak, A., McDonald, F. B., and Webber, W. R.: 1997, 'Study of Elemental and Isotopic Composition of Cosmic Ray Nuclei Ca, Ti, V, Cr, Mn and Fe', *Proc. 25th Int. Cosmic Ray Conf., Durban* **3**, 357–360.

Lund, N.: 1989, in C. J. Waddington (ed.), 'The Abundances in the Cosmic Radiation (The Elements Lighter than Ge)', *Proceedings of the Symposium on Cosmic Abundances of Matter*, AIP Conf. Proc. 183, American Institute of Physics Press, Minneapolis, pp. 111–123.

Mewaldt, R. A.: 1989, in C. J. Waddington (ed.), 'The Abundances of Isotopes in the Cosmic Radiation', *Proceedings of the Symposium on Cosmic Abundances of Matter*, AIP Conf. Proc. 183, American Institute of Physics Press, Minneapolis, pp. 124–146.

Meyer, J.-P.: 1985, 'Solar-Stellar Outer Atmospheres and Energetic Particles, and Galactic Cosmic Rays', *Astrophys. J. Suppl.* **57**, 173–204.

Meyer, J.-P., Drury, L. O'C., and Ellison, D. C.: 1998, 'A Cosmic Ray Composition Controlled by Volatility and A/Q Ratio. SNR Shock Acceleration of Gas and Dust', *Space Sci. Rev.* **86**, 179.

Payne, M. G.: 1969, 'Energy Straggling of Heavy Charged Particles in Thick Absorbers', *Phys. Rev.* **185**, 611–623.

Ptuskin, V. and Soutoul, A.: 1998, 'Cosmic Ray Clocks', *Space Sci. Rev.* **86**, 225.

Rossi, B.: 1952, *High-Energy Particles*, Prentice-Hall, Englewood Cliffs, p. 31.

Soutoul, A., Cassé, M., and Juliusson, E.: 1978, 'Time Delay Between the Nucleosynthesis of Cosmic Rays and Their Acceleration to Relativistic Energies', *Astrophys. J.* **219**, 753–755.

Stone, E. C.: 1989, in C. J. Waddington (ed.), 'Solar Abundances as Derived from Solar Energetic Particles', *Proceedings of the Symposium on Cosmic Abundances of Matter*, AIP Conf. Proc. 183, American Institute of Physics Press, Minneapolis, pp. 72–90.

Stone, E. C. and Wiedenbeck, M. E.: 1979, 'A Secondary Tracer Approach to the Derivation of Galactic Cosmic Ray Source Isotopic Abundances', *Astrophys. J.* **231**, 606–623.

Stone, E. C., Cohen, C. M. S., Cook, W. R., Cummings, A. C., Gauld, B., Kecman, B., Leske, R. A., Mewaldt, R. A., Thayer, M. R., Dougherty, B. L., Grumm, R. L., Milliken, B. D., Radocinski, R. G., Wiedenbeck, M. E., Christian, E. R., Shuman, S., von Rosenvinge, T. T.: 1998: 'The Solar Isotope Spectrometer for the Advanced Composition Explorer', *Space Sci. Rev.* **86**, 357.

Thayer, M. R.: 1997, 'An Investigation into Sulfur Isotopes in the Galactic Cosmic Rays', *Astrophys. J.* **482**, 792–795.

Webber, W. R., Lukasiak, A., McDonald, F. B., and Ferrando, P.: 1996, 'New High-Statistical-High-Resolution Measurements of the Cosmic-Ray CNO Isotopes from a 17-Year Study Using the Voyager 1 and 2 Spacecraft', *Astrophys. J.* **457**, 435–439.

Webber, W.R., Lukasiak, A., and McDonald, F. B.: 1997, 'Voyager Measurements of the Mass Composition of Cosmic-Ray Ne, Mg, Si, and S Isotopes', *Astrophys. J.* **476**, 766–770.

Wiedenbeck, M. E. and Greiner, D. E.: 1980, 'A Cosmic-Ray Age Based on the Abundance of ^{10}Be', *Astrophys. J.* **239**, L139–L142.

Zaerpoor, K., Chan, Y. D., DiGregorio, D. E., Dragowsky, M. R., Hindi, M. M., Isaac, M. C. P., Krane, K. S., Larimer, R. M., Macciavelli, A. O., Macleod, R. W., Miocinovic, P., and Norman, E. B.: 1997, 'Galactic Confinement Time of Iron-Group Cosmic Rays Derived from the ^{54}Mn Chronometer', *Phys. Rev. Letters* **79**, 4306–4309.

THE SOLAR ISOTOPE SPECTROMETER FOR THE ADVANCED COMPOSITION EXPLORER

E. C. STONE, C. M. S. COHEN, W. R. COOK, A. C. CUMMINGS, B. GAULD,
B. KECMAN, R. A. LESKE, R. A. MEWALDT and M. R. THAYER
California Institute of Technology, Pasadena, CA 91125, U.S.A.

B. L. DOUGHERTY*, R. L. GRUMM, B. D. MILLIKEN‡, R. G. RADOCINSKI and
M. E. WIEDENBECK
Jet Propulsion Laboratory, Pasadena, CA 91109, U.S.A.

E. R. CHRISTIAN, S. SHUMAN and T. T. VON ROSENVINGE
NASA/Goddard Space Flight Center, Greenbelt, MD 20771, U.S.A.

Abstract. The Solar Isotope Spectrometer (SIS), one of nine instruments on the Advanced Composition Explorer (ACE), is designed to provide high-resolution measurements of the isotopic composition of energetic nuclei from He to Zn ($Z = 2$ to 30) over the energy range from ~ 10 to ~ 100 MeV nucl^{-1}. During large solar events SIS will measure the isotopic abundances of solar energetic particles to determine directly the composition of the solar corona and to study particle acceleration processes. During solar quiet times SIS will measure the isotopes of low-energy cosmic rays from the Galaxy and isotopes of the anomalous cosmic-ray component, which originates in the nearby interstellar medium. SIS has two telescopes composed of silicon solid-state detectors that provide measurements of the nuclear charge, mass, and kinetic energy of incident nuclei. Within each telescope, particle trajectories are measured with a pair of two-dimensional silicon-strip detectors instrumented with custom, very large-scale integrated (VLSI) electronics to provide both position and energy-loss measurements. SIS was especially designed to achieve excellent mass resolution under the extreme, high flux conditions encountered in large solar particle events. It provides a geometry factor of ~ 40 cm^2 sr, significantly greater than earlier solar particle isotope spectrometers. A microprocessor controls the instrument operation, sorts events into prioritized buffers on the basis of their charge, range, angle of incidence, and quality of trajectory determination, and formats data for readout by the spacecraft. This paper describes the design and operation of SIS and the scientific objectives that the instrument will address.

Abbreviations: ACE – Advanced Composition Explorer, ACR – anomalous cosmic ray, ADC – analog-to-digital converter, CRIS – Cosmic Ray Isotope Spectrometer, CSA – charge-sensitive amplifier, DAC – digital-to-analog converter, GCR – galactic cosmic ray, GSFC – Goddard Space Flight Center, GSI – Gesellschaft für Schwerionenforschung mbH (accelerator laboratory in Darmstadt, Germany), IDAC – current-output digital-to-analog converter, IMP – Interplanetary Monitoring Platform, ISEE – International Sun-Earth Explorer, JHU/APL – Johns Hopkins University, Applied Physics Laboratory, JPL – Jet Propulsion Laboratory, LISM – local interstellar medium, MSU – Michigan State University, NSCL – National Superconducting Cyclotron Laboratory, PCB – printed circuit board, PHA – pulse height analyzer, RTSW – real-time solar wind (data), r.m.s. – root mean square, SAMPEX – Solar, Anomalous, and Magnetospheric Particle Explorer, SEP – solar energetic particle, SIS – Solar Isotope Spectrometer, SRL – Space Radiation Laboratory, UTMC – United Technologies Microelectronics Center, VDA – vacuum deposited aluminum, VLSI – very large scale integrated (circuit).

*Present address: Mail Code 264–33, Caltech, Pasadena, CA 91125, U.S.A.
‡Present address: Lucent Technologies, Naperville, IL 60566, U.S.A.

Space Science Reviews **86**: 357–408, 1998.
© 1998 *Kluwer Academic Publishers. Printed in the Netherlands.*

1. Introduction

The Advanced Composition Explorer (ACE) will perform comprehensive studies
of the elemental, isotopic, and ionic charge state composition of energetic nuclei
from H to Zn ($1 \leq Z \leq 30$) from a vantage point located $\sim 1.5 \times 10^6$ km sunward
of Earth, near the L1 libration point (Stone et al., 1998a). These observations will
address a wide range of questions concerning the origin, acceleration, and transport
of energetic nuclei from solar, interplanetary, and galactic sources. ACE includes
six high-resolution spectrometers and three interplanetary monitoring instruments
that together will measure composition at energies ranging from < 1 keV nucl^{-1}
(characteristic of the solar wind) to $\sim$500 MeV nucl^{-1} (characteristic of galactic
cosmic radiation). The Solar Isotope Spectrometer (SIS), shown in Figure 1, is one
of these six spectrometers and is designed to measure the elemental and isotopic
composition over the energy range from $\sim$10 to $\sim$100 MeV nucl^{-1}, with excellent
mass resolution and collecting power.

The primary objective of the ACE mission will be to determine and compare the
elemental and isotopic composition of several distinct samples of matter, including
the solar corona, the interplanetary medium, the local interstellar medium, and
galactic matter. ACE will study matter from the Sun by measuring solar wind and
solar energetic particles (SEPs). Matter from the local interstellar medium (LISM)
will be observed at energies of $\sim$1 keV nucl^{-1} as interstellar neutral particles which
enter the heliosphere and are ionized to become solar wind 'pickup ions'. Some of
these ions are also observable as 'anomalous cosmic rays' (ACRs) once they have
been convected out to the solar wind termination shock and accelerated to energies
of $\sim$1 to 100 MeV nucl^{-1}. Galactic cosmic rays (GCRs) provide a sample of matter
that enters the heliosphere from other regions of the Galaxy.

Figure 2 illustrates typical energy spectra that have been measured for the GCR,
ACR, and SEP components at 1 AU. Note that the $\sim$10 to $\sim$100 MeV nucl^{-1}
energy region covered by SIS includes nuclei originating in each of the three
samples of matter to be studied by ACE, with intensities that vary on time scales
ranging from minutes to years. The GCR and ACR components, shown at their
solar minimum levels, vary by factors of $\sim$10 and $\sim$100, respectively, over the
solar cycle in this energy range, in inverse correlation with the sunspot number.
SEP fluxes, on the other hand, are transient and only rarely reach the intensity
levels shown in Figure 2. Figure 3 illustrates the time variation of GCR protons
with $>$60 MeV over the last two solar cycles. Superimposed on the gradual solar
modulation of the GCR component can be seen narrow spikes corresponding to
large solar energetic particle events. Based on an extrapolation of the onsets of the
last two solar cycles, ACE can expect to observe a marked increase in the frequency
of large SEP events just as we enter the next millennium. Up until this onset, SIS
will be able to measure the composition of ACRs and GCRs during the waning
months of the current solar minimum.

TABLE I

Summary of SIS characteristics

Characteristic	Value	Details[a]
Measurement objective	Elemental and isotopic composition of solar energetic particles, anomalous cosmic rays, and low-energy galactic cosmic rays	§2
Measurement technique	Multiple-ΔE vs residual energy plus trajectory	§3.1, Appendix A, F10
Sensor system		
Energy loss measurements	Two detector stacks, each composed of 15 ion-implanted Si detectors of varying thicknesses, all $\sim$10 cm in diameter	§3.3, §4.1, §4.2, TII
Trajectory measurements	Two 2-dimensional Si multi-strip detectors per stack; 6 cm separation; position resolution $\approx$0.29 mm	§3.4
Charge interval		§4.6, F19
Primary interval	$4 \leq Z \leq 28$	
Extended interval	$1 \leq Z \leq 30$	
Energy interval for mass analysis		§4.6, F19, F20
O	10–90 MeV nucl^{-1}	
Si	13–125 MeV nucl^{-1}	
Fe	17–170 MeV nucl^{-1}	
Field of view	95° full angle	F11
Geometrical factor	38.4 cm^2 sr	§4.6, F20
Event yields (solar minimum)		§2.2
ACR O	50 000 yr^{-1}	
ACR Ne	2500 yr^{-1}	
GCR Si	5000 yr^{-1}	
GCR Fe	5000 yr^{-1}	
Event yields (large SEP event)		F7
O	$> 10^5$	
Fe	$> 10^4$	
Mass resolution (r.m.s.)		§4.8, F10, F18
O	$\lesssim 0.15$ amu	
Fe	$\lesssim 0.35$ amu	
Resource allocation		§3.11
Dimensions ($l \times w \times h$)	$30.0 \times 41.9 \times 27.5$ cm	
Mass	21.9 kg	
Instrument power	17.8 W	
Bit Rate	1992 bits s^{-1}	

[a] Sections (§), Figures (F), and Tables (T) containing additional details.

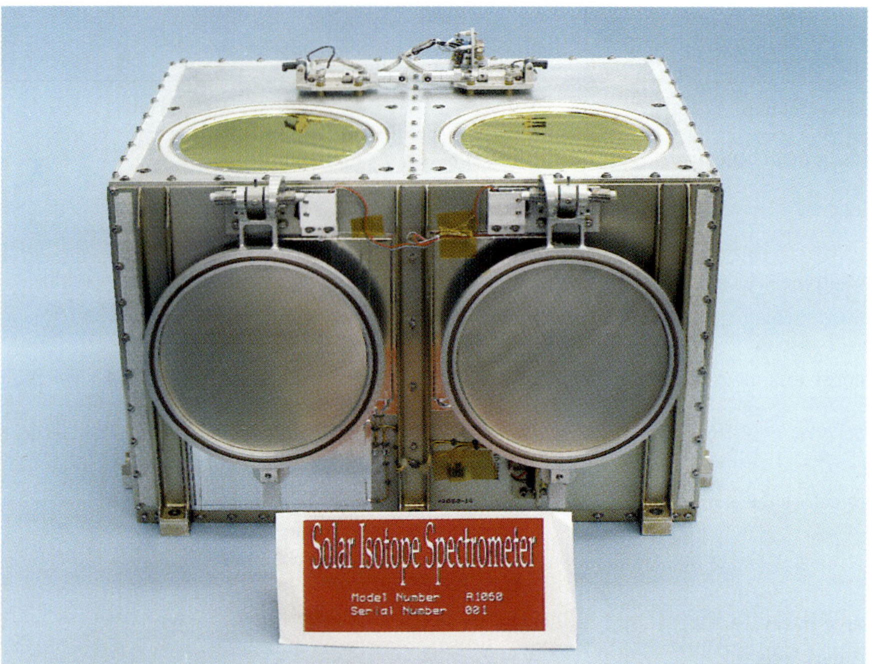

Figure 1. Photograph of the SIS instrument with the acoustic covers open. The dimensions of the housing are 30.0 × 41.9 × 27.5 cm.

The essential features of SIS are summarized in Table I. The remainder of the paper presents a more detailed discussion of the objectives, the design, and the measured and expected performance of SIS.

2. Science Objectives

2.1. THE ISOTOPIC COMPOSITION OF SOLAR ENERGETIC PARTICLES

Although the Sun contains the vast majority of solar system material, we have only limited direct knowledge of its isotopic composition. Spectroscopic observations of solar isotopes are very difficult; there are isotope observations for only a few elements and the uncertainties are large. Almost all of the isotopic abundances in the Anders and Grevesse (1989) table of 'solar system' abundances are actually based on terrestrial material, while meteoritic measurements (supplemented by spectroscopic data) serve as the standard source for elemental abundances that are used to characterize solar system material.

Solar energetic particles represent a sample of solar material that can be used to make direct measurements of the Sun's elemental and isotopic makeup and to study the most energetic acceleration processes that occur naturally in our solar system. Comprehensive surveys of SEPs have shown that there are two general classes of

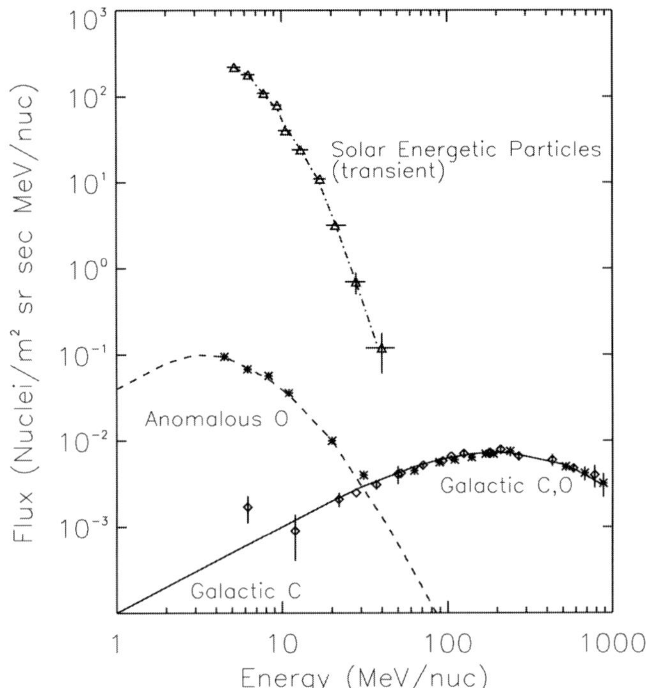

Figure 2. Solar-minimum energy spectra measured for ACR and GCR carbon and oxygen on IMP-8 are compared with the oxygen spectrum measured during the large SEP event of 23 September 1978 by ISEE-3 (Mewaldt et al., 1984a, 1984b). Curves have been drawn through the data points to guide the eye. Note that during the SEP event, solar particles dominate the observed flux below ~100 MeV nucl^{-1}.

solar particle events: 'gradual' events that are characterized by extended rise times and relatively long durations, and 'impulsive' events, generally smaller in size, which are characterized by enhanced fluxes of heavy elements such as Fe, the rare isotope ^{3}He, and energetic electrons (see, e.g., Reames, 1995). The ionic charge states in gradual events (e.g., Fe^{+15}) are characteristic of coronal temperatures of $\sim2 \times 10^{6}$ K (e.g., Leske et al., 1996b), while those in impulsive events (e.g., Fe^{+20}) require a temperature of $\sim10^{7}$ K (Luhn et al., 1987), suggesting that these particles represent the heated flare plasma. SEPs in gradual events are believed to be accelerated by shocks driven by coronal mass ejections (see, e.g., Kahler, 1992).

The compositional differences in these two classes of SEP events are illustrated in Figures 4 and 5. Note that those events with enhanced ^{3}He/^{4}He ratios all have Fe/O ratios greater than the coronal value. The Fe/O ratios in events with ^{3}He/^{4}He ratios < 0.1 are generally lower, with considerable scatter. Such composition variations are reasonably well organized by the charge to mass ratio (Q/M) of the ions, and by taking Q/M-dependent acceleration and transport effects into account it is

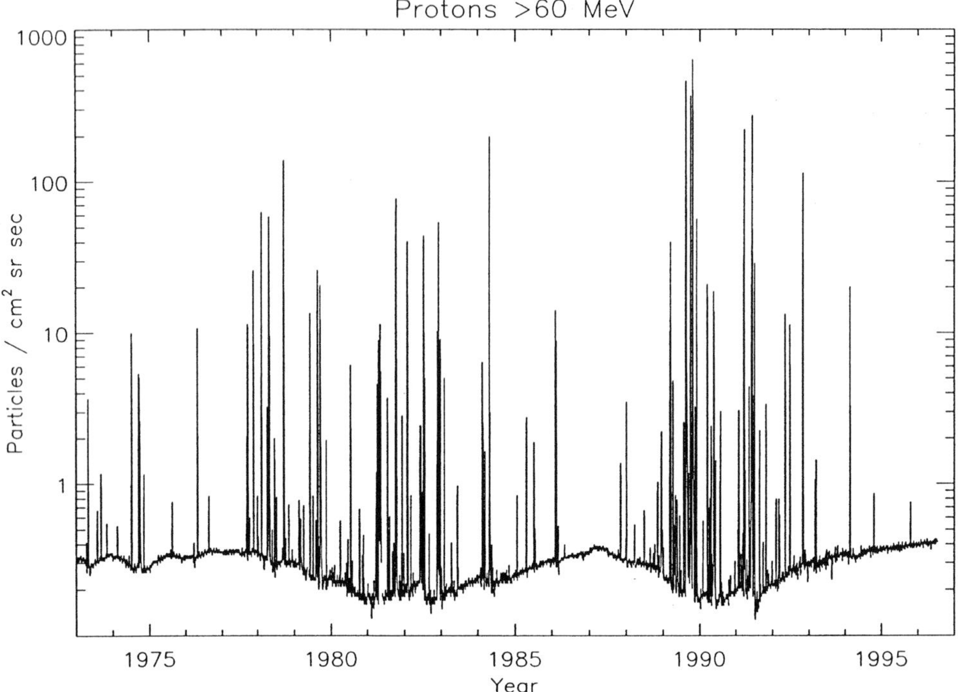

Figure 3. Measured flux of protons >60 MeV from 1973 to 1996 measured by the CPME experiment on IMP-8 (T. Armstrong, private communication). The steady flux varying from ~0.2 cm^{-2} sr^{-1} s^{-1} at solar max to ~0.4 cm^{-2} sr^{-1} s^{-1} is due primarily to galactic cosmic rays. The superimposed spikes (typically several days wide) represent solar proton events. Note that the solar cycle dependence of these two components is anticorrelated.

possible to use SEP measurements in gradual events to obtain coronal abundance measurements for a wide range of elements (Breneman and Stone, 1985).

Earlier SEP studies (see, e.g., Stone et al., 1989) have shown that the elemental composition of the solar corona differs significantly from that of the photosphere, in that the abundances of elements with first ionization potential ≥ 10 eV (including, e.g., He, N, O, and Ne) are depleted by a factor of ~4 relative to other elements. This difference apparently indicates that neutral species are less efficiently transported from the photosphere to the corona (Meyer, 1985). Although the first measurements of the SEP isotopic composition were made in the late 1970's (e.g., Mewaldt et al., 1984a), such measurements have proven to be difficult, and are presently available for only a few elements in a limited number of solar events. Figure 6 summarizes the current state of SEP isotope measurements in large gradual events. In addition, Mason et al. (1994) have presented evidence for an enrichment of heavy isotopes in impulsive solar particle events. The uncertainties in the existing measurements are relatively large as a result of statistical limitations

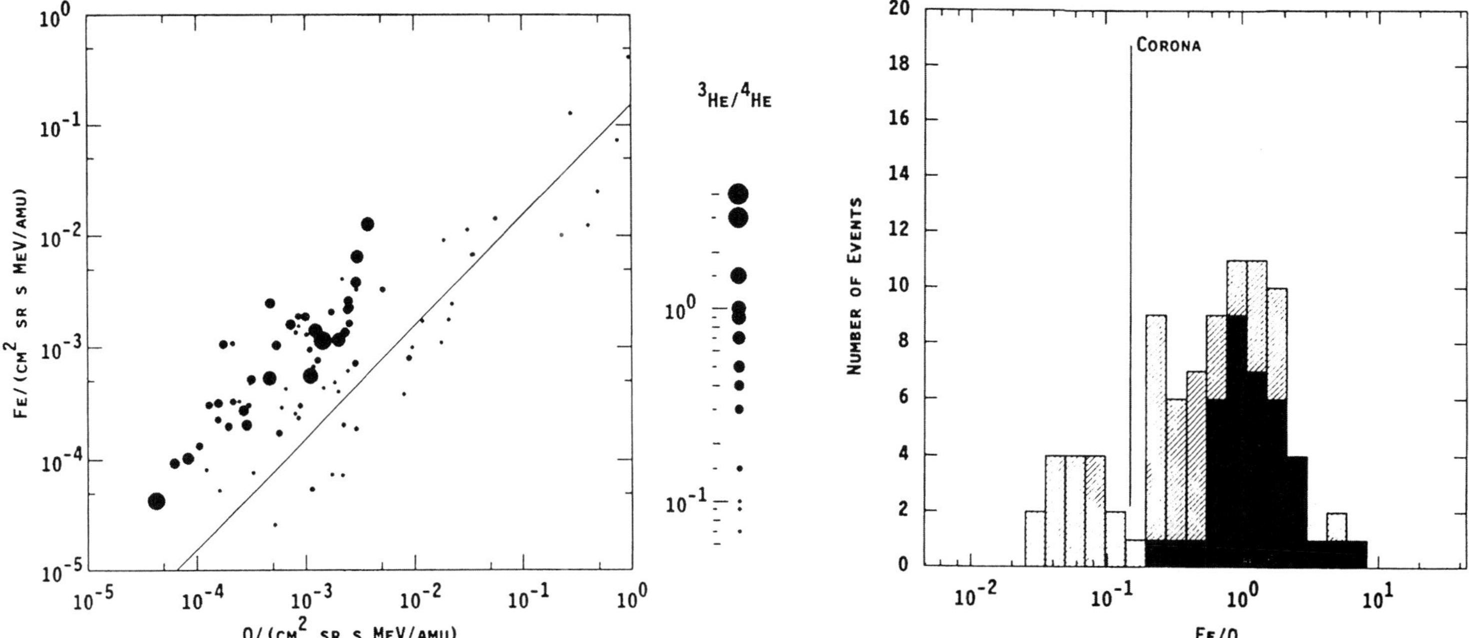

Figure 4. Characteristics of solar energetic particle events. The left-hand panel shows the intensity of 1.9–2.8 MeV amu^{-1} Fe versus the intensity of O at the same energy for 90 different electron events (reproduced from Reames et al., 1990). ^{3}He-rich events are indicated by the circle size, which increases logarithmically with ^{3}He/^{4}He as shown. The right-hand panel shows the histogram of Fe/O for the same electron events, with the ^{3}He-rich subset blackened. The diagonal line in the left-hand panel is drawn at the 'mass-unbiased' coronal abundance of Meyer (1985) and the same value is indicated in the right-hand panel. Note that the largest ^{3}He-rich events are generally 1 to 3 orders of magnitude smaller in size than the largest gradual events.

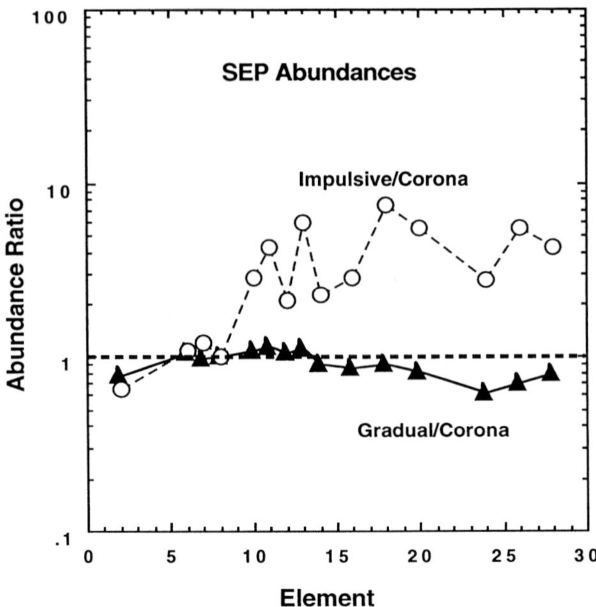

Figure 5. Average composition of gradual and impulsive SEP events (from Reames, 1993) normalized to coronal abundances tabulated by Garrard and Stone (1993).

and, in some cases, as a result of background associated with the high count rates in the largest solar events.

With its greatly improved collecting power over earlier instruments, it is hoped that SIS can make a major advance in our knowledge of SEP isotopic composition. Figure 7 shows an estimate of the number of events that SIS would have observed in the gradual SEP event of 30 October 1992 measured by SAMPEX (see Selesnick et al., 1993).

2.2. THE ISOTOPIC COMPOSITION OF ANOMALOUS COSMIC RAYS AND THE LOCAL INTERSTELLAR MEDIUM

During solar minimum conditions there are seven elements (H, He, C, N, O, Ne, and Ar) whose energy spectra have shown anomalous increases in flux above the quiet time galactic cosmic-ray spectrum (see, e.g., the reviews in Klecker, 1995, and Simpson, 1995). This anomalous cosmic ray (ACR) component is now thought to represent neutral interstellar particles that have drifted into the heliosphere, become ionized by the solar wind or ultra-violet radiation, and then been accelerated to energies >10 MeV nucl^{-1} (Fisk et al., 1974), most likely at the solar wind termination shock (Pesses et al., 1981). A unique prediction of this model is that anomalous cosmic rays should be singly ionized, unlike galactic cosmic rays which are essentially fully stripped, and unlike SEPs, which generally have charge states representative of coronal temperatures. Recent measurements from SAMPEX have

SOLAR CORONAL ISOTOPE ABUNDANCES

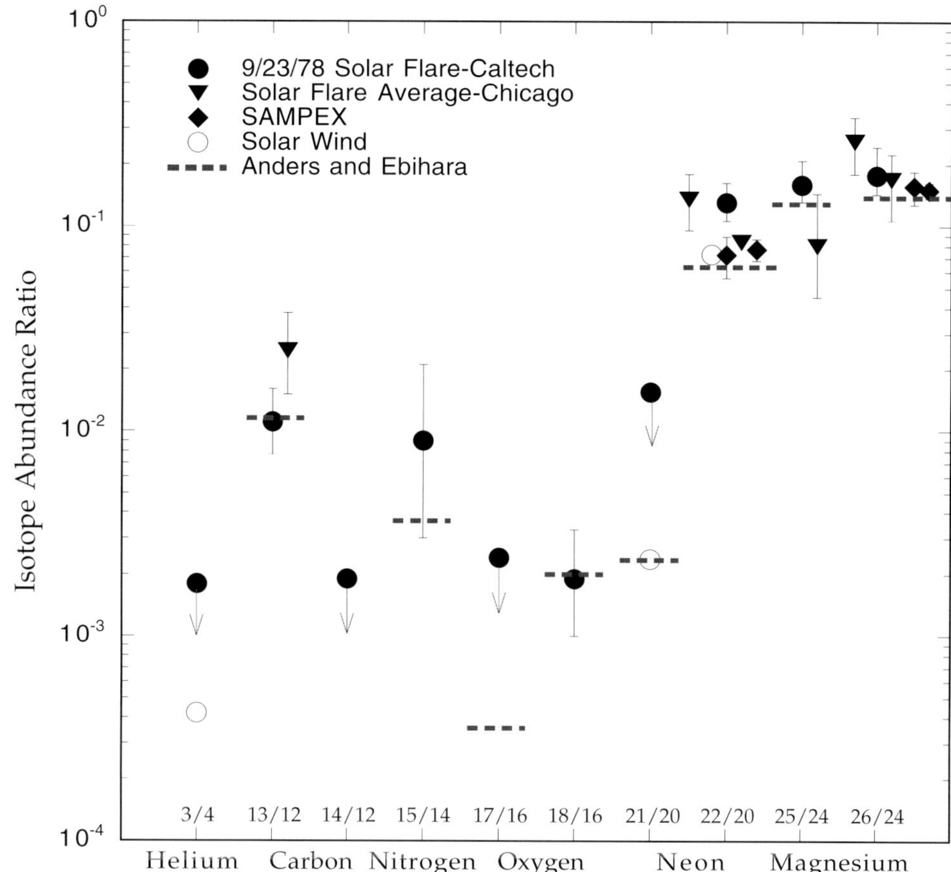

Figure 6. Summary of published measurements of isotope ratios in gradual SEP events for the elements He, C, N, O, Ne, and Mg, based on a similar figure from Selesnick et al. (1993). Shown are measurements from SAMPEX (Selesnick et al., 1993), ISEE-3 (Mewaldt et al., 1984a; Mewaldt and Stone, 1989), Phoenix-I/S81-1 (Simpson et al., 1983), and IMP-8 (Dietrich and Simpson, 1979). Also shown are solar wind measurements from Geiss et al. (1972) and solar system abundances from Anders and Ebihara (1982).

shown that the bulk of ACR N, O, and Ne with energies <20 MeV nucl^{-1} are indeed singly charged (Klecker et al., 1995), confirming the neutral-gas-origin model. However, some fraction of the ACR ions are ionized further, probably during the acceleration process, and consequently at higher energies most ACRs are multiply charged (Mewaldt et al., 1996). The observed charge states imply that ACRs have passed through less than 1 μg cm^{-2} of material since the time of their acceleration.

Anomalous cosmic-ray observations offer a unique opportunity to study a sample of matter from local interstellar space. Because the ACR component apparently

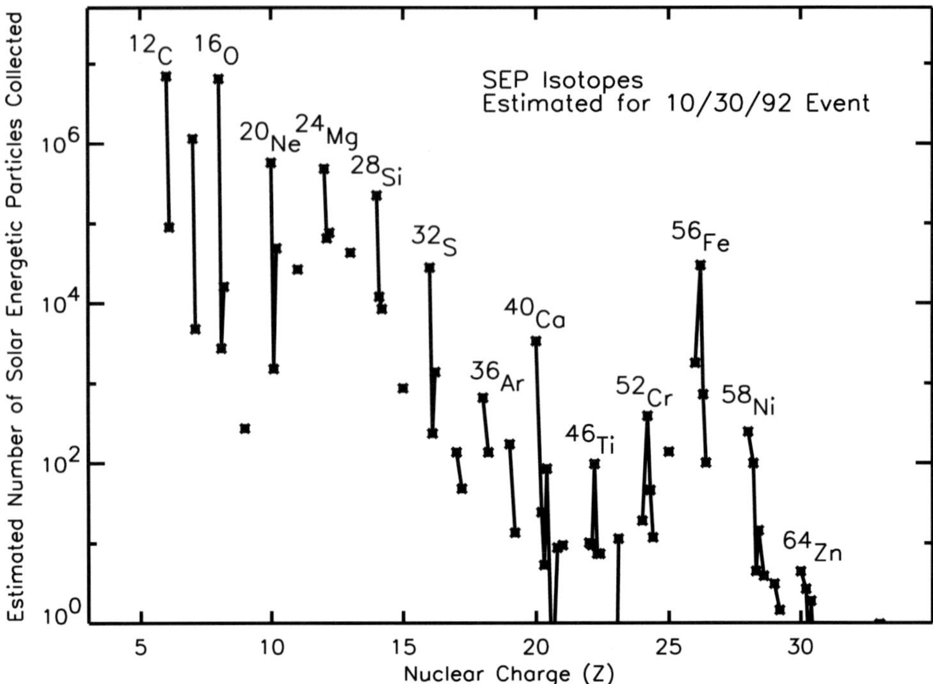

Figure 7. Estimated number of events of each isotope that SIS would have observed in the large, gradual SEP event of 30 October 1992, based on composition measurements from SAMPEX (D. Williams, private communication). The abundances of the isotopes within each element (connected by a solid line) are assumed to be the same as in Anders and Grevesse (1989).

represents a direct sample of the local interstellar medium, it carries important information about galactic evolution in the solar neighborhood over the time interval since the formation of the solar nebula some 4.6 billion years ago. This information can be obtained by comparing the isotopic composition of ACR nuclei with that of solar system abundances, including those measured by SIS in solar energetic particles. Radio and optical spectroscopy studies of isotope ratios in interstellar molecules exhibit considerable scatter, but they indicate that isotope abundance ratios such as $^{13}C/^{12}C$, $^{15}N/^{14}N$, and $^{18}O/^{16}O$ are dependent on distance from the Galactic center, including variations of as much as a factor of ~ 2 from their solar system values (see Figure 8). Variations of this magnitude can be obtained by some galactic evolution models (see, e.g., Audouze, 1983; and Tosi, 1982). Under solar minimum conditions, SIS will observe $\sim 50\,000$ ACR oxygen, ~ 8000 ACR nitrogen, and ~ 2500 ACR neon per year. Isotope studies of these elements can lead to improved estimates of the composition of interstellar matter in the vicinity of the Sun.

Recently, there have been new measurements from Wind, Geotail, and Voyager that provide possible evidence for ACR contributions to additional elements, including S, Si, and Fe (see Reames et al., 1997; Takashima et al., 1997; Stone and

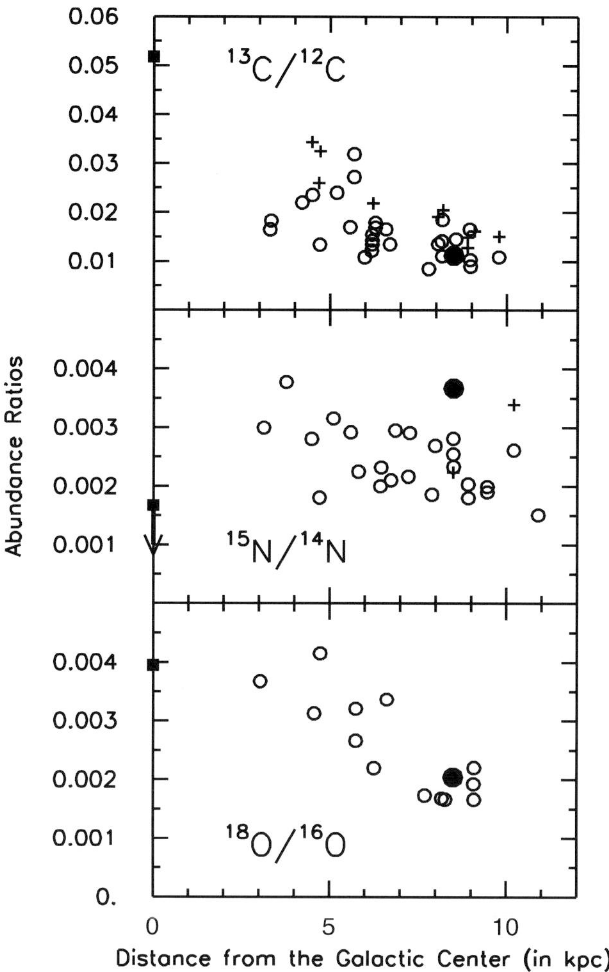

Figure 8. Survey of isotope measurements as a function of distance from the Galactic center (based on data in Figure 2 of Wilson and Rood, 1994). Data for the $^{13}C/^{12}C$ ratio are based on radio astronomy measurements of CO (open circle) and H_2CO (plus). The $^{15}N/^{14}N$ data are based on measurements of HCN (open circle) and NH_3 (plus), and the $^{18}O/^{16}O$ data are based on H_2CO. The Galactic center data are shown as filled squares, and the solar system values from Anders and Grevesse (1989) are shown as filled circles at 8.5 kpc.

Cummings, 1997), with relative abundances $\sim 0.1\%$ that of oxygen. While these elements are expected to be generally ionized in the interstellar medium, because their first ionization potentials are considerably lower than that of H (13.6 eV), it is possible that the observed fluxes represent the neutral abundances of these species in the LISM. It is also possible that they originate from other sources of pickup ions (see, e.g., Geiss et al., 1996), since there is as yet no evidence that these enhanced fluxes have ionic charge states of $Q \leq 4$, as would be expected for ACR nuclei (see Klecker et al., 1997). With its large geometry factor, SIS will be able to search for ACR contributions down to a level of $\sim 10^{-4}$ the intensity of oxygen.

2.3. STUDIES OF LOW ENERGY GALACTIC COSMIC-RAY ISOTOPES

Galactic cosmic rays represent an accessible sample of matter that originates outside the solar system. The isotopic composition of this sample of matter contains a record of the nuclear history of cosmic-ray material, including its synthesis in stars, and subsequent nuclear interactions with the interstellar gas. The primary instrument for studying galactic cosmic-ray isotopes on ACE is the Cosmic Ray Isotope Spectrometer (CRIS) which will cover the energy interval from ~ 100 to 500 MeV nucl^{-1} with a geometry factor of >200 cm^2 sr (see Stone et al., 1998b). SIS will extend isotope studies to lower energies with a collecting power that is still several times greater than that of earlier satellite instruments. These measurements will be particularly useful for studying the energy dependence of the isotopic composition.

Neon isotope studies are of particular interest, because there are a variety of different compositions of neon isotopes observed in the solar system (see, e.g., Podosek, 1978). The right-hand panel of Figure 9 summarizes reported measurements of the ^{22}Ne/^{20}Ne ratio in cosmic rays as a function of energy/nucleon. It is well known that GCR neon is enriched in ^{22}Ne, even when corrections are made for contributions from the fragmentation of heavier comic rays during their transport through the interstellar medium. Note that below ~ 50 MeV nucl^{-1} there is a sudden decrease in the ^{22}Ne/^{20}Ne ratio, as the ACR component begins to dominate. The magnitude of this decrease will, of course, depend on the ^{22}Ne/^{20}Ne in the LISM, and the curves in Figure 9 illustrate several possibilities.

Figure 9 also summarizes ^{22}Ne/^{20}Ne measurements from the solar system, galactic cosmic rays, and from the LISM as derived from previous studies of ACRs. The meteoritic and lunar component known as 'Neon-B' apparently represents directly implanted solar wind, while 'Neon-A' is apparently a primordial component (see, e.g., Podosek, 1978). Galactic cosmic-ray source material has an overabundance of ^{22}Ne/^{20}Ne by a factor of 2.3 to 7 compared to Neon-A and Neon-B. Studies of ACR isotopes to date are limited by statistical uncertainties, but clearly show a ratio much lower than in GCRs (Leske et al., 1997). This implies that GCRs are not simply an accelerated sample of interstellar matter; they must also include contributions from sources that are rich in ^{22}Ne, such as Wolf-Rayet stars (Cassé and Paul, 1982; Prantzos et al., 1987). With two years of solar minimum measurements from SIS it should be possible to distinguish whether the ACR ^{22}Ne/^{20}Ne ratio agrees with either that of Neon-A or Neon-B.

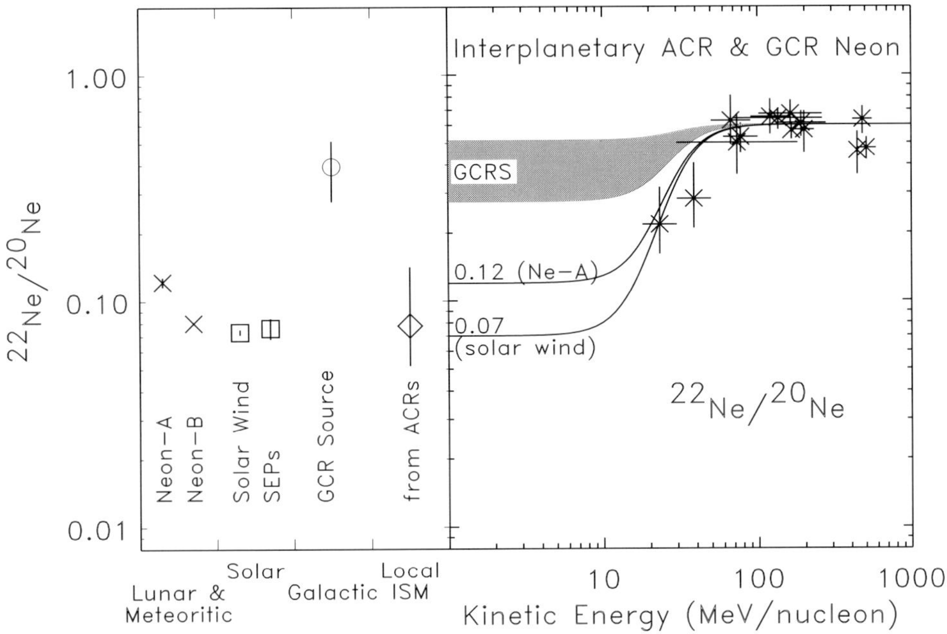

Figure 9. Summary of measurements of the ^{22}Ne/^{20}Ne ratio (adapted from Leske et al., 1996a). On the right are shown measurements of cosmic rays. When the GCR ratio of $\sim$0.6 (assumed to be valid $\geq$50 MeV nucl^{-1}) is corrected for contributions from cosmic-ray secondaries, it results in a cosmic-ray source ratio of $\sim$0.27 to $\sim$0.5 when cross section and other uncertainties are included. Below $\sim$40 MeV nucl^{-1} ACRs dominate and the measured ratio decreases; the curves show the ratio expected for various assumed abundances in the local interstellar medium. On the left are summarized lunar and meteoritic components (Neon-A and Neon-B; see, e.g., Podosek, 1978); the solar wind value from Apollo (Geiss et al., 1972), SEP measurements from SAMPEX (Selesnick et al., 1993), the GCR source value spanning the measurements from *Ulysses* (Connell and Simpson, 1993) and Voyager (Lukasiak et al., 1994), and the average of the LISM value based on ACR isotope measurements from Voyager and SAMPEX (Cummings et al., 1991; Leske et al., 1996a).

3. Instrument Description

3.1. APPROACH

Identification of the charge and mass of energetic nuclei in SIS relies on a refined version of the standard $dE/dx - E$ technique. A particle of charge Z, mass M, and velocity v has kinetic energy E proportional to Mv^2 and loses energy as it penetrates matter at a rate dE/dx proportional to Z^2/v^2. Taking the product of these two quantities yields

$$(dE/dx)E \propto Z^2M \tag{1}$$

independent of the particle's velocity. A plot of dE/dx vs. E for a particular nuclide over a range of velocities forms a hyperbola, and the hyperbolas for different nuclides are separated according to their values of Z^2M.

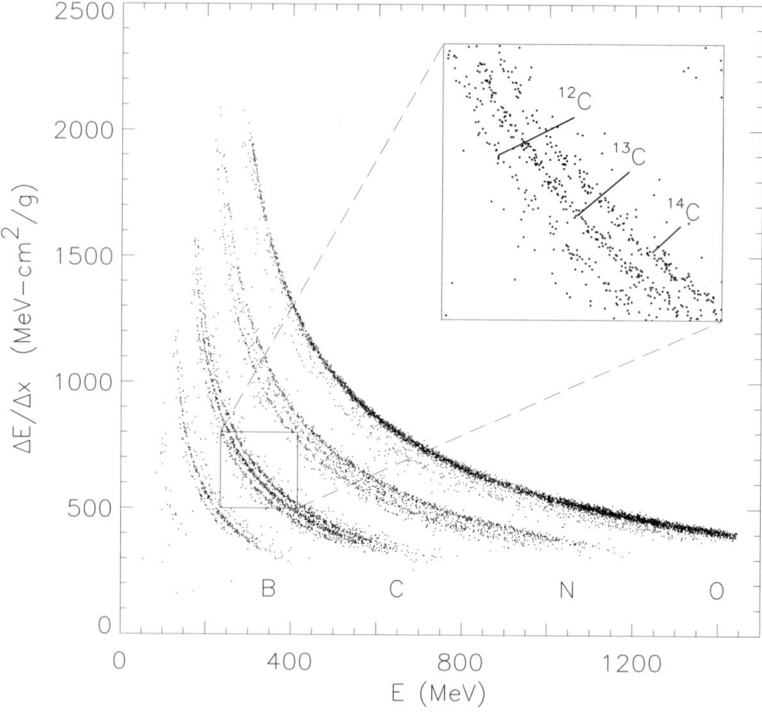

Figure 10. Illustration of the dE/dx $-$ E technique using data acquired during an accelerator calibration of SIS at GSI in Darmstadt, Germany.

This approach is illustrated in Figure 10 in which we approximate dE/dx by the ratio $\Delta E/\Delta x$ where ΔE is the energy deposited by the particle in penetrating a thickness Δx $\sim$0.10 g cm^{-2} of silicon. The total energy is obtained by adding to ΔE the energy deposited in a following stack of silicon detectors in which the particle comes to rest. Each point on the plot corresponds to a measured heavy nucleus. The heavy hyperbolic 'tracks' corresponding to the elements boron through oxygen are well separated. Each of these bands is made up of several more closely spaced tracks corresponding to different isotopes of the element (see inset in Figure 10). The wide spacing of adjacent elements and the closer spacing of successive isotopes results from the quadratic dependence on Z and linear dependence on M in Equation (1). Typically, the spacing between isotopes is $\sim\frac{1}{8}$ the spacing between elements, so there is no ambiguity in the particle identification as long as the elements of interest have fewer than 8 isotopes present. This is the case for all of the elements to be measured with SIS.

The resolution of the particle identification can be significantly improved by measuring ΔE in a detector having a thickness, L, that is a significant fraction of the particle's range. An equation for the charge and mass of a detected nucleus can be obtained by equating the change in the particle's range in passing through this detector to the thickness of material penetrated. Letting $R_{Z,M}(E/M)$ be the range of a particle of charge Z, mass M, and kinetic energy E we can write:

$$R_{Z,M}(E/M) - R_{Z,M}(E'/M) = L \sec(\theta) , \tag{2}$$

where θ is the angle at which the particle penetrates the ΔE detector and E' is its residual energy when it emerges. Tabulations of the function $R_{Z,M}(E/M)$ are available (e.g., Hubert et al., 1990).

Equation (2) is an implicit expression giving Z and M in terms of the measured quantities E', $\Delta E = E - E'$, L, and θ. Solving it typically begins by assuming a mass-to-charge ratio and solving for Z. In the resulting charge histogram, different elements are well separated, allowing one to then assign an integer value of Z to each event. With Z known, we then use Equation (2) to solve for M. Practical techniques for carrying out these solutions are described in Appendix A.

Using Equation (2), the fundamental equation on which the $\Delta E - E'$ technique is based, one can calculate the contributions to the mass resolution from various physical effects such as energy loss fluctuations and multiple scattering and from uncertainties in the measured values of ΔE, E', L, and θ. Additional insight can be obtained using a power-law approximation to the function $R_{Z,M}(E/M)$, which permits an explicit solution for M. These techniques are discussed in Stone et al. (1998b).

3.2. DESIGN REQUIREMENTS

If an instrument is to achieve the objectives in Section 2 and make a significant contribution beyond previous instruments that have measured solar energetic particles and anomalous cosmic rays, it must satisfy several design requirements.

To resolve adjacent isotopes that differ in abundance by factors of $\sim$100 requires a mass resolution of $\sim$0.25 amu or better (see, e.g., Stone, 1973). For an instrument that uses the multiple $dE/dx - E$ approach, the mass resolution is a sum of several separate contributions (see Appendix A of Stone et al., 1998b) that can be evaluated by taking appropriate partial derivatives of Equation (2) above. Some of these depend on the physics of the interaction of charged particles with matter (e.g., energy-loss fluctuations, multiple scattering, and electron pickup and loss), while others depend on instrument design (e.g., trajectory resolution, uniformity of detector response, and electronic noise levels and stability).

It is also necessary that SIS be capable of returning accurate composition measurements in the presence of high fluxes of low-energy protons and helium nuclei. Measurements of solar energetic particles are usually made in a very hostile environment in which the flux of protons >1 MeV may exceed 10^5 cm^{-2} sr^{-1} s^{-1}. Chance coincidences between these low-energy protons and the heavier nuclei that are of primary interest to SIS can potentially lead to ambiguous trajectories or distorted energy loss measurements. To ensure that as few as possible of the nuclei with $Z \geq 10$ are missed requires that the instrument be capable of selecting the most interesting nuclei for analysis, and that the bit rate be sufficient to transmit several events per second.

Measurements of anomalous cosmic rays, free from contamination of solar and interplanetary particles at lower energy, and free from galactic cosmic rays with higher energies, are best made in the energy interval from $\sim$5 to 25 MeV nucl^{-1}, where the flux is a decreasing function of energy. Similarly, solar energetic particle spectra typically decrease rather steeply with increasing energy (see Figure 2). It follows that to maximize the number of detected particles for both of these species requires the use of thin detectors with as low a threshold for penetration as possible, combined with large geometry factors. For this reason SIS has two telescopes composed of the largest area Si devices available ($\sim$65 cm^2). It is also of interest to extend measurements of solar energetic particles to as high an energy as possible in order to understand the acceleration process in these events. The SIS detector stack is composed of devices of graduated thicknesses in order to cover a broad energy range.

The scientific goals described above lead to a number of design requirements on the various subsystems in SIS. For example, the custom VLSI electronics designed for SIS and CRIS are required to have a dynamic range of >1000 to accommodate the desired element range, and they need to be both linear and stable over the expected temperature range. Because SIS is designed mainly for heavy ions, the requirements on noise are not as severe as they would be for singly-charged particles. The hodoscope for SIS must identify multiple particles with a range of energy losses varying by a factor of $\sim$1000 and achieve a typical angular resolution of a few tenths of a degree. The thickness of each detector has to be known to an accuracy of $\sim$0.1%. The following discussion will document how these design requirements were achieved.

3.3. THE TELESCOPES

There are two identical telescopes in SIS, referred to as 'A' and 'B', respectively. Each is composed of 17 high-purity silicon detectors as shown in Figure 11 and summarized in Table II. The top two detectors, M1 and M2, are position sensitive ('matrix') devices that form the hodoscope and measure both the trajectory and energy loss of incident nuclei. Below M1 and M2 is an energy-loss stack composed of thicker, single-electrode silicon devices. Stack detectors have 65 cm^2 active areas and would be completely circular in shape (with 46 mm radii) but for a 42 mm long orientation flat. Thicknesses range from 100 μm to 3.75 mm and progress with depth in each telescope. Note that T6 and T7 are composite devices made by summing the outputs of three and six wafers, respectively. The crystal axes of all detectors are aligned, which allows particle channeling effects to be more easily recognized.

The apertures of the SIS telescopes are covered by three Kapton windows that provide protection from sunlight and micro-meteorites (see Section 3.10). The

TABLE II

SIS detector and ADC characteristics

Range	Detector Name	Nominal detector thickness (mm)	Active area (cm^2)	Geometry factor[c] $(cm^2\ sr)$	Nominal[d] ADC threshold (MeV)	Nominal ADC full scale (MeV)
–	M1	0.075	34	–	4, 0.5[e]	700
R0	M2	0.075	34	38.4	4, 0.5[e]	700
R1	T1	0.1	65	38.4	1.1	900
R2	T2	0.1	65	37.2	1.2	900
R3	T3	0.25	65	36.6	1.0	1800
R4	T4	0.5	65	34.4	2.2	2800
R5	T5	0.75	65	33.0	2	4200
R6	T6	2.65[a]	65	27.2	5	8000
R7	T7	3.75[b]	65	19.4	7.5	12000
R8	T8	1	65	19.4	1.7	540

[a]T6 is composed of three wafers with thicknesses of 1.0, 0.9, and 0.75 mm.
[b]Composed of six wafers with thicknesses 1.0, 0.75, 0.5, 0.5, 0.5, and 0.5 mm.
[c]Sum of two telescopes; limited to trajectories intersecting the top of the following detector.
[d]Thresholds are commandable.
[e]Initial trigger requires 4 MeV; strips detect $\geq$ 0.5 MeV (see Section 3.7.1).

combined thickness of these windows has a stopping power that is equivalent to $\sim$31 μm of silicon.

3.4. THE MATRIX DETECTOR TRAJECTORY SYSTEM

The matrix detectors are octagonal in shape, $\sim$75 μm in thickness, and have 34 cm^2 active areas that are divided into 64 strips (see Figure 12). The strips are 0.96 mm wide, separated by 0.040 mm gaps, and oriented in perpendicular directions on opposite sides to provide both X and Y axis readout. These ion-implanted detectors were produced for SIS by Micron Semiconductor Ltd. using n-type $\langle 111 \rangle$ float-zone-refined silicon. The thin silicon wafers are bonded into G-10 mounts with a silicone resin (Shin-etsu KJR-9022E). Redundant sets of aluminum wirebonds are used to connect the aluminum strip contacts on each detector to the corresponding gold-plated copper pads on its mount. Signal lines follow copper traces on a Kapton 'flex-strip' to a rigid G-10 connector designed to mate with a high-density connector on the board containing the matrix detector pulse-height analyzer (PHA) electronics. A detailed discussion of the design and testing of these devices has been given by Wiedenbeck et al. (1996).

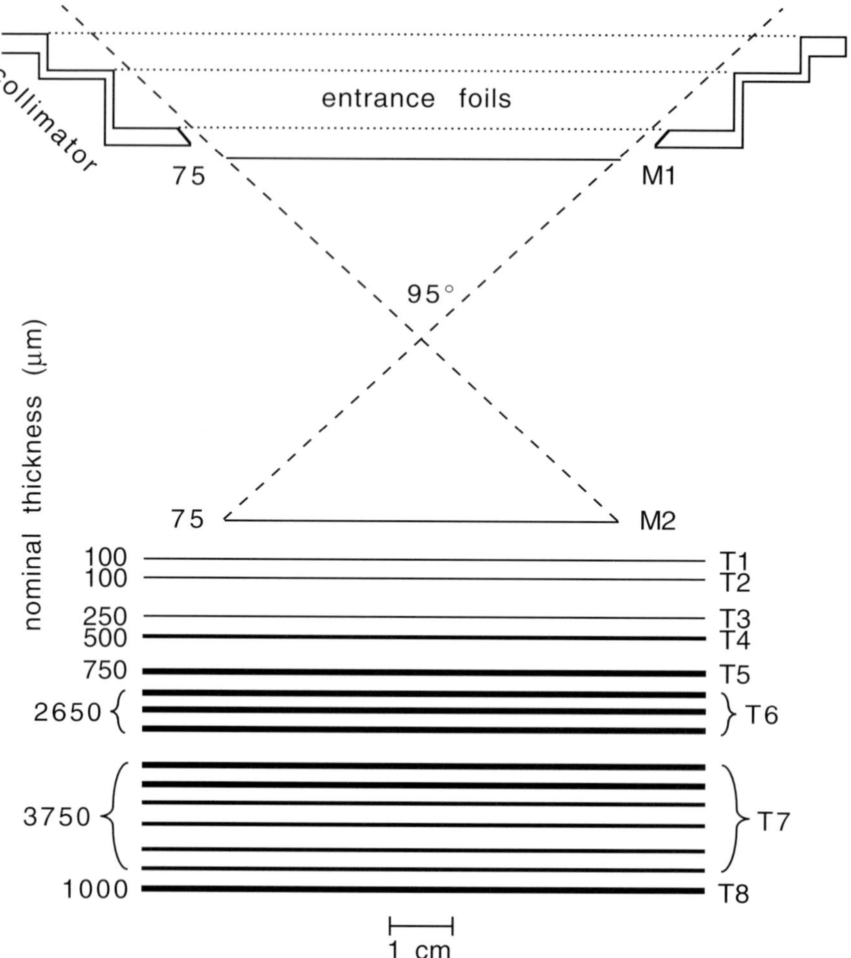

Figure 11. Scale drawing of one of the two SIS telescopes, showing only the active regions of the detectors. Detector designations are listed at the right, while the total silicon thicknesses are listed at the left. The metallized Kapton entrance foils total $\sim$7.2 mg cm^{-2} thick. The opening angle of 95° is an average which accounts for the octagonal shape of the matrix detectors.

Valid events in SIS typically require a coincidence between M1A and M2A or between M1B and M2B (where A and B refer to the A and B telescopes, respectively). Each of the strips on M1 and M2 is then individually pulse-height analyzed with its own 12-bit analog-to-digital converter (ADC) so that the trajectory of heavy ions traversing the system can be separated from the tracks of low-energy H or He that might happen to hit one of these detectors at the same time. The custom VLSI circuitry that accomplishes this is described in detail in Section 3.5.1 and Appendix B. Detectors M1 and M2 are separated by 6 cm; the resulting r.m.s. angular resolution of the system is $\sim$0.25°, averaged over all angles.

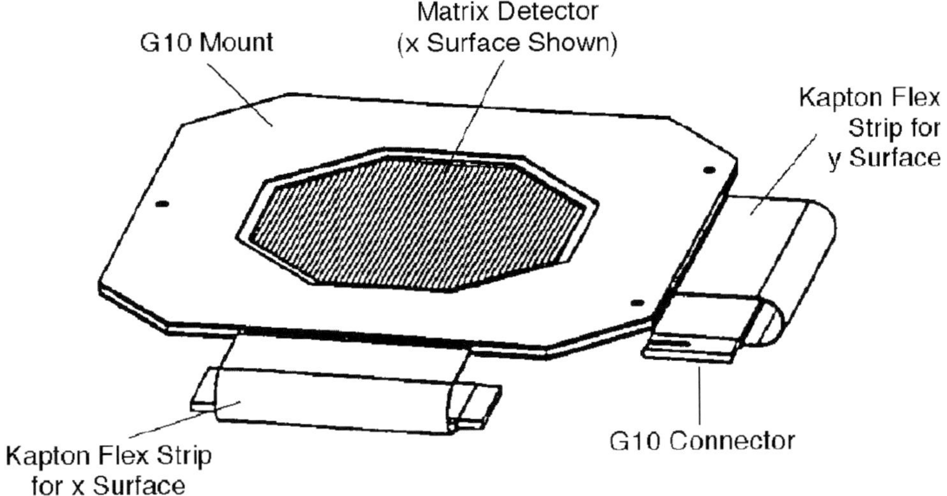

Figure 12. Schematic of a SIS matrix detector in its mount. Each surface of the detector has 64 metallized strips. The X-surface strips are orthogonal to the Y-surface strips and all strips are individually pulse-height analyzed.

3.5. ENERGY-LOSS STACK DETECTORS

The stack detectors were also made by Micron Semiconductor Ltd. using n-type $\langle 111 \rangle$ float-zone-refined silicon with ion-implanted electrodes. All detectors were bonded into custom-designed G10 mounts, with electrical connections provided by redundant wires bonded to pads on these mounts. The detectors were screened under bias in a thermally cycled vacuum chamber for several weeks.

Each detector in SIS is biased at least 20 V beyond its depletion voltage, which in turn was established by scanning alpha particle sources across the ohmic surface while varying the bias and noting when complete charge collection efficiency was obtained. Electronics noise for stack channels T1 to T7 ranges between 100 and 400 keV r.m.s.. Further fabrication, testing, and selection details can be found in Dougherty et al. (1996).

3.6. ELECTRONICS

The SIS electronics system is organized as shown in the block diagram in Figure 13. The electronics and packaging design follows closely that of CRIS (see Stone et al., 1998b). Key new features include programmable gate arrays, the card-cage/backplane style packaging using thermally actuated zero insertion force connectors, the modular packaging of silicon detectors onto the same PC boards that carry the front-end electronics, the RTX2010 microprocessor as central controller, and a newly developed pulse-height-analysis hybrid for stack detector read-

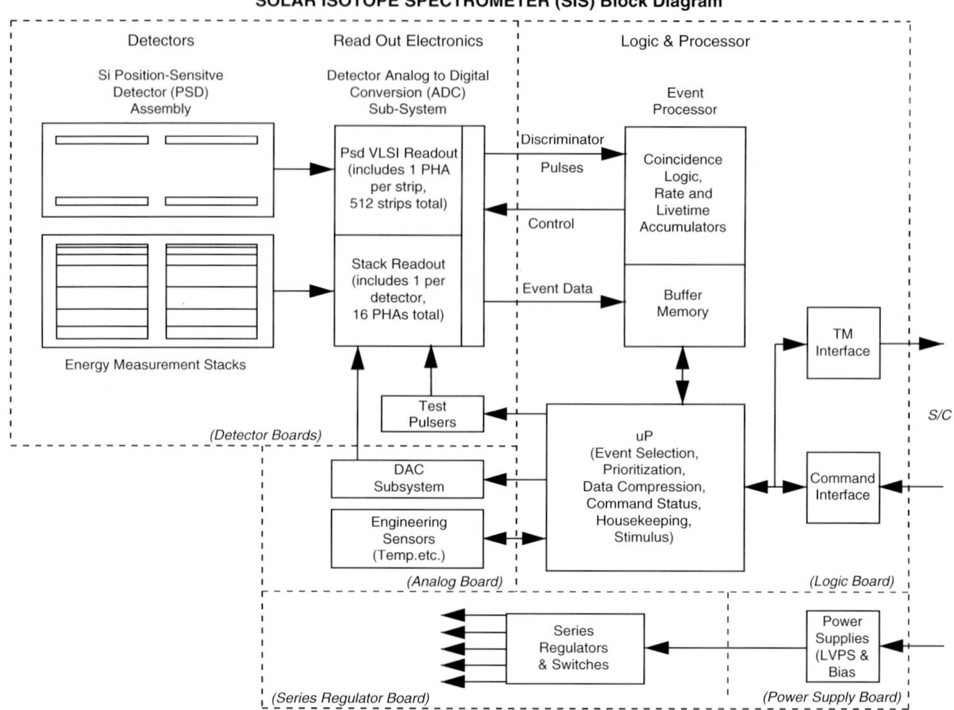

Figure 13. Functional block diagram for the SIS electronics.

out. In addition, the SIS matrix detectors are read out using a new CMOS VLSI circuit containing 16 complete pulse-height analyzers on a single chip.

The electronic design provides a high degree of programmability and selective redundancy to increase reliability. Discriminator thresholds and coincidence logic equations may be adjusted to compensate for noisy detectors or other anomalies. High voltage may be commanded off or on separately for each detector element. The A and B telescope front-end electronics are independent and the interface with the spacecraft is redundant. All microprocessor code that is noncritical to communication with the spacecraft can be reprogrammed in flight. The microprocessor can be booted from any one of four identical copies of the system code.

3.6.1. *Matrix Detector Electronics*

Matrix detector signals obtained from each of 512 strip electrodes are individually pulse-height analyzed in order to avoid degradation of resolution at the high count rates expected in large solar energetic particle events. The matrix detector pulse-height analyzers were implemented using a custom CMOS VLSI circuit especially developed for SIS. Table III lists the characteristics of this device, and Appendix B discusses the internal design and operation of the circuit. A total of 32 of these 16-channel PHA chips are used in SIS.

TABLE III

Matrix VLSI characteristics

Parameter	Value
PHAs per chip	16
Power per channel	13 mW
Full scale signal	31 pC (700 MeV Si)
Dynamic range (full scale:threshold)	1400:1[a]
Gain variation	$\leq 10\%$[b]
Threshold variation	$\sim 10\%$ r.m.s. at 3 MeV
ADC type	12-bit Wilkinson
Dead time per event	$(\sim 6 + 0.125\ N)\ \mu s$[c]
Integral nonlinearity ($+$ input)	$\sim 1\%$ of full scale
Integral nonlinearity ($-$ input)	$\sim 0.1\%$ of full scale
Max. leakage current cancellation	$1\ \mu A$
Package size	$(13.5\ \text{mm})^2$, 84 pins
Temperature coeff. of gain	$< \pm 20$ ppm/°C
Offset temp. variation (-20 °C to 40 °C)	<1 chan (after correction)
Cross talk for full-scale pulse	2 chan
Radiation tolerance	~ 100 kRads

[a]Noise ≤ 100 keV r.m.s., excluding quantization error.
[b]Among channels on one chip. Chip-to-chip variations can be larger.
[c]N = pulse-height channel number, including pedestal.

3.6.2. *Stack Electronics*

Silicon detector wafers which form the two 'stacks' are grouped as discussed earlier to form a total of 16 detector elements, each instrumented with a separate PHA. These analyzers and the associated stack detector modules are mounted on eight nearly identical printed circuit boards ('detector boards'), four in each stack (see Figure 14).

The PHAs are hybrid circuits (see Figure 15) manufactured at Teledyne (Cook et al., 1993b; Stone et al., 1998b). The only active component in the PHA hybrid is a fully custom bipolar application-specific integrated circuit fabricated at Harris Semiconductor. This new circuit represents an evolution of discrete bipolar transistor designs which were flight proven in numerous space instruments over the past thirty years (see references in Cook et al., 1993a). The PHA hybrid has improved performance relative to the prior design, while using a factor of 10 less board area and a factor of 3 less power.

Each PHA hybrid includes a preamplifier, postamplifier, normally open amplifier/offset gate, and peak detector/Wilkinson ADC. Shaping time constants are $2\ \mu s$ with a $3\ \mu s$ peaking time of the bipolar signal. Twelve-bit quantization with a maximum conversion time of 256 μs is obtained with a 16 MHz clock. The PHA

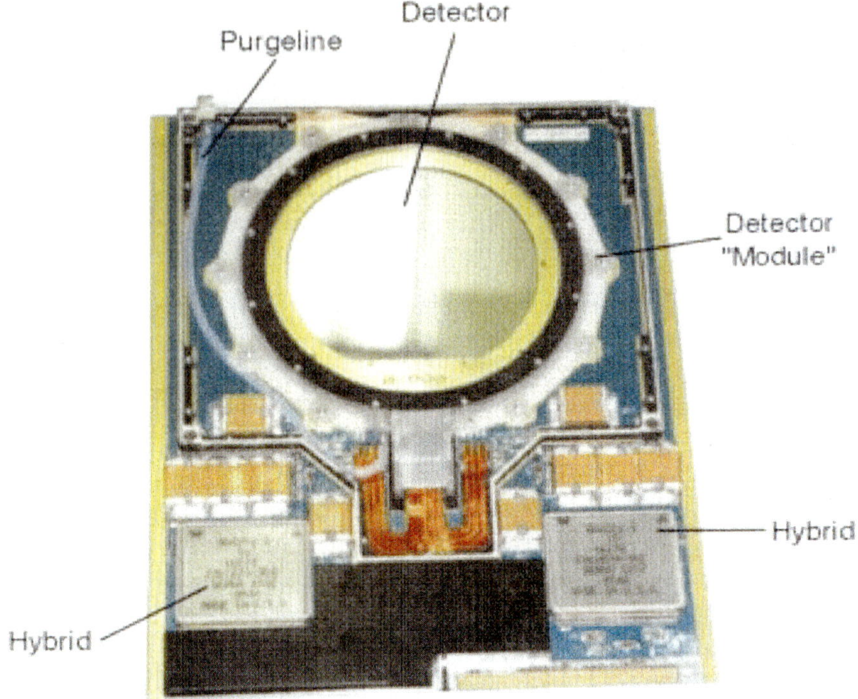

Figure 14. Photograph of a SIS stack detector module mounted in its board. There are two or more detectors mounted in the module on each stack board (e.g., T1 and T2).

hybrid was manufactured in a number of types with preamp feedback and bias configured for different detector capacitances and leakage currents.

PHA hybrid performance exceeds design goals, with typical dynamic range of 2000, gain stability of 20 ppm/°C, offset variation of less than 0.5 channels over −20 °C to +40 °C, and deviation from linearity of less than 0.01% of full scale over the entire dynamic range. Power consumption is 40 mW. The PHA hybrids do not contain the digital counters and logic needed for the Wilkinson ADCs, which are rather implemented in a nearby ACTEL gate array on the detector board.

Also mounted near the PHA hybrids are test pulsers used to simulate detector signals. There is one test pulser per detector board. The test pulse amplitudes are controlled independently for the A and B stacks by two dual-gain 8-bit digital-to-analog converters (DACs). The test pulsers allow thorough verification of the coincidence logic and are used in routine instrument functional checks.

The amplifier chains in the PHA hybrids are DC coupled from the detector through the postamplifier. This eliminates the need for a long secondary AC coupling time constant and allows the DC level at the postamp output to be used to measure the detector leakage current. Such measurements are routinely incorpo-

Figure 15. Photograph of a SIS stack hybrid showing the custom bipolar VLSI chip developed for ACE surrounded by passive feedback and filtering components.

rated into the instrument housekeeping data and are useful monitors of detector health.

3.7. ON-BOARD PROCESSING

The central controller of the SIS instrument is an RTX2010 microprocessor system. The ease of programming and high performance level of this microprocessor allowed simplification of hardware design and testing. The microprocessor controls all instrument subsystems, performs on-board data compression and prioritization, executes commands from the spacecraft, collects engineering data, and formats and transfers the telemetry stream. A unique task of the processor is the management of the matrix detectors. The activities of balancing the individual strip leakage currents and measuring the pedestals (discussed in Appendix B) are carried out by sophisticated routines that have been verified to operate successfully at the high particle rates anticipated during major solar events.

3.7.1. *Coincidence Logic*

Whenever any detector in the SIS instrument is triggered, the coincidence logic decides whether a valid event has occurred. The normal requirement for an event is a coincidence between either M1A and M2A, or between M1B and M2B. In the normal mode, the thresholds of M1 and M2 are set at $\sim$4 MeV in order to be insen-

TABLE IV

SIS coincidence requirements

Species	Equation	Hy-enable	He-enable	Matrix threshold state
$Z \geq 3$	M1M $\cdot$ M2M $\cdot$ Hor	–	–	Low
$Z \geq 3$	M1H $\cdot$ M2H $\cdot$ Hor	False	False	High
$Z \geq 3$	M1L $\cdot$ M2L $\cdot$ Hor $\cdot$ ADC3	False	False	High
$Z = 2$	M1M $\cdot$ M2M $\cdot$ $\overline{\text{Hor}}$	–	True	–
$Z = 2$	M1L $\cdot$ M2L $\cdot$ Mor $\cdot$ $\overline{\text{Hor}}$ $\cdot$ ADC2	–	True	–
$Z = 1$	M1L $\cdot$ M2L $\cdot$ $\overline{\text{Mor}}$ $\cdot$ ADC2	True	–	–

Matrix threshold state = commandable two-state (High/Low) flag.
M1H = commandable discriminator threshold set nominally at $\sim$16 MeV in M1.
M2H = commandable discriminator threshold set nominally at $\sim$16 MeV in M2.
M1M = commandable discriminator threshold set nominally at $\sim$4 MeV in M1.
M2M = commandable discriminator threshold set nominally at $\sim$4 MeV in M2.
M1L = energy loss $\geq$0.5 MeV in M1, software threshold.
M2L = energy loss $\geq$0.5 MeV in M2, software threshold.
Mor = logical 'or' of all M ($Z \geq 2$) stack thresholds.
Hor = logical 'or' of all H ($Z \geq 3$) thresholds.
ADC2 = coincidence between two consecutive stack ADCs.
ADC3 = coincidence between three consecutive stack ADCs.
Hy-Enable = true only after expiration of Hy-timer.
He-Enable = true only after expiration of He-timer.
– = not applicable

sitive to low-energy protons. However, because He and other light ions stopping near the back of the detector stacks do not typically deposit more than 4 MeV in M1 and M2, it was necessary to also incorporate additional trigger conditions, as discussed below and summarized in Table IV.

In order to maximize the analysis of the heavy ($Z > 3$) nuclei of interest, it is desirable to rapidly characterize the charge of an incident nucleus. To accomplish this, each of the stack ADCs (T1 to T8) has two programmable discriminator levels. The 'medium' levels in each ADC are set just above the maximum energy loss that can be deposited by a stopping proton at the maximum acceptable incidence angle (e.g., $\sim$4 MeV in the 100 μm thick T1 detectors) and the 'high' levels are set just above the maximum energy loss of a stopping ^{4}He. The logic conditions 'Mor' (corresponding to the logical 'or' of all medium discriminators from T1 to T7) and 'Hor' (corresponding to the logical 'or' of all high discriminators from M1 through T7) can then be used to achieve the desired separation, as illustrated in Table IV.

The large flux of H and He during large SEP events would cause considerable dead time if all of these particles were analyzed. On the other hand, it is still of scientific interest to monitor the abundance and energy spectra of these species. To accomplish this, the SIS trigger logic incorporates two programmable timers that can be adjusted from 0 to 130 s. The 'Hy' timer starts whenever an H event has been fully analyzed; until this timer expires all events recognized to have $Z = 1$ (on the basis of the Mor and Hor discriminator conditions) will be rapidly rejected. This process takes $\sim$20 μs, much less than the typical time of 10 ms necessary to fully process an H event. Once the Hy timer expires, H analysis is enabled and the logic will again accept the next valid H event. There is also a 'He' timer that serves to throttle the analysis of He events in the same manner. It is expected that these timers will normally be set at $\sim$10 s; long enough that H and He will each account for at most $\sim$1% of analyzed events under high rate conditions.

The 'ADC2' trigger mode is used to detect H and He nuclei that do not trigger the thresholds of M1 and M2 (called M1M and M2M in Table IV). If the Hy and/or He timers have expired and two consecutive stack ADCs (T1 through T8) are triggered in coincidence, but not M1M·M2M, the logic recognizes an ADC2 condition and initiates ADC rundown of all stack and matrix ADCs. The microprocessor then polls the resulting energy losses in the matrix detectors and declares a valid event if there are detectable energy losses $\geq$0.5 MeV in at least one X and one Y strip in both M1A and M2A or at least one X and one Y strip in both M1B and M2B. If neither of these conditions is met, the logic is reset to wait for the next event. Events recognized by the ADC2 condition are sorted into event buffers in the same manner as M1M·M2M events (see Section 3.7.2). The ADC2 mode can be disabled by command if desired.

There is also an optional analysis mode (ADC3) in which the matrix thresholds are raised to $\sim$16 MeV, just beyond the maximum energy loss of an alpha particle, in order to further reduce the dead time resulting from triggering of M1 and M2 by low-energy He. In this mode, light nuclei such as C, N, and O that stop near the back of the detector stack do not all trigger M1 and M2; these events are recognized (in a manner analogous to the ADC2 mode) if they trigger three consecutive stack ADCs (ADC3 condition) and also trigger Hor. In this mode the matrix thresholds are lowered to their nominal levels and the coincidence logic reverts back to its nominal state if either of the Hy or He timers has expired.

3.7.2. *Priority System*

During the largest solar particle events there may be $\gtrsim$100 $Z \geq 6$ ions per second triggering SIS. Although SIS is capable of analyzing most of these, the allocated telemetry rate of 1992 bps allows only 10 to 15 events per second to be stored in the spacecraft solid-state recorder and transmitted to the ground. In order to ensure that the most interesting events are not missed, SIS employs a series of discriminators to identify H, He, and $Z \geq 3$ nuclei, and then stores events in a series of 95 prioritized buffers (see Appendix C). The buffers distinguish four element categories (H, He, $3 \leq Z \leq 9$, and $Z \geq 10$); range in the instrument (as identified by the last detector triggered); zenith angle ($\leq 15°$, $15°$ to $25°$, and $\geq 25°$; quality of the hodoscope data (number of non-adjacent strips per detector surface); and the 'Hazard' flag, which identifies events which occur within $\sim$20 μsec of the most recent previous ADC trigger. Each buffer can typically accommodate several events, depending on their length, and includes events from both telescopes.

Events are typically read out of these buffers in order of their priority; however, the first N events (where N is a commandable integer) of each 256-second major frame are read out by cycling sequentially through the buffers, in order to ensure that all buffers are periodically sampled during high rate periods. Further details of the priority system are discussed in Appendix C.

3.8. SIS DATA

Data produced by the SIS instrument consists primarily of 'event' data, which record the pulse height and trajectory measurements of each detected nucleus, count-rate data, and housekeeping parameters. For each event, the SIS microprocessor reads a 12-bit pulse height from each of the 8 stack-detector PHAs and 256 matrix-detector PHAs in the telescope that was hit. If all this information were to be telemetered to the ground, about 1.5 minor frames (1.5 s) would be needed to send each event within the 1992 bits per second telemetry rate allocated to SIS. Since many of the pulse heights are typically 0, especially for the matrix strips, it is possible to compress the data to a much smaller volume. Essentially only the pulse heights in the detectors and strips that were triggered are telemetered, along with their identification. Details of the SIS variable-length, compressed event format are given in Appendix D.

With this event compression scheme, event length varies considerably, ranging from a minimum of 15 bytes for particles stopping in M2 triggering only 4 matrix strips to an extreme maximum of 109 bytes for a pulser-stimulated event ('Stim' event) triggering all stack detectors and 31 matrix strips, with the optional extended header (see Appendix D). A 'typical' event stopping in T4 with 6 strips (allowing for chance coincidences or noisy strips) is 20 bytes long. After allowing telemetry space for rate and housekeeping data, an average of $\sim$10 events per second may be transmitted, which is better than an order of magnitude improvement over that possible without using a compressed event format. Along with the 'class' selection and buffer prioritization systems described in Appendix C, this innovation allows SIS to send a large fraction of the most interesting pulse-height events even during intense solar particle events.

In addition to pulse heights, SIS records and transmits a variety of count rates, some primarily to assess the health of the instrument, others crucial for determining event detection efficiencies in order to properly calculate absolute particle fluxes. These rates are summed over the entire 256 s instrument cycle (SIS does not measure 'sector' rates over fractions of the $\sim$12 s spacecraft spin period and is not designed to look for particle anisotropies). Up to 18 of the 213 total measured rates may be selected by command to be read and transmitted at a higher time resolution of once every 32 s.

TABLE V

SIS command summary

Item	Comments
Coincidence logic	
Coincidence equation	Include any/all terms of M1·M2·T1·T2
Hazard event	Allow/disallow analysis of hazard events
Stim events	Give Stim events highest priority
ADC2 definition	Enable/disable front/back halves of telescope
ADC3 definition	Enable/disable front/back halves of telescope
Instrument operation	
Adjust ADC thresholds	Raise/lower in $\sim$10% steps
ADC operation	Enable/disable each individually
Matrix strip operation	Enable/disable individual strips
Matrix strip thresholds	Raise/lower in 10% steps
Medium and high ADC thresholds	Raise/lower in 10% steps
Hy and He event timers	Adjust from 0 to 130 s
Stack detector bias voltages	64 steps; T1–T2 (10–35 V), T3 (10–60 V), T4 (30–120 V), T5–T6 and guard (120–250 V)
Matrix detector bias voltages	64 steps; 0–30 V
Software parameters	
Angle and charge Buffers	Adjust boundaries
Readout priority system	Redefine event buffer readout priorities
Number of strips selected per matrix layer	Adjust max and select values
Event format	Include extended header
High priority rates	Select which rates are included
Stim events	Change frequency and masks
Diagnostic events	Change frequency

A number of housekeeping parameters such as temperatures and voltages, as well as indications of the leakage current in each of the stack detectors and matrix strips, are also included in the data. Further information about the rate and housekeeping data and the packaging of all this information into the telemetry format is given in Appendix D.

3.9. COMMAND SYSTEM

There are a wide range of parameters and functions of the operation of the SIS instrument that can be modified by command in order to optimize its performance and to preserve as many as possible of the instrument functions in the event of inflight component failures or degradation. Table V summarizes these commandable functions. There are a total of sixteen hardware command bits that control the instrument logic for the two telescopes. In addition, there are a wide variety of commandable parameters that include ADC thresholds, detector bias voltages, and in-flight calibration. It is also possible to modify the priority system, the event format, and various software parameters that define the classification and priority of events. If a detector should become noisy, or fail, it is possible to turn off its bias, and/or its ADC.

3.10. MECHANICAL DESIGN

The mechanical design of SIS provides the following features:
 – Precise and reproducible mechanical positioning of detectors.
 – Elimination of coaxial cables.
 – Close proximity of detectors to their preamplifiers with a minimal number of connection points.
 – A continuous RF shield around the detectors.
 – A card-cage design for ease of assembly and reassembly of individual components.
 – Dry nitrogen purge to detectors until just prior to launch.

These goals have been achieved by utilizing a box design in which individual printed circuit boards (PCBs) slide into card rails and mate with zero-insertion-force connectors mounted on a PCB backplane at the rear of the box (Figure 16). The card rails then lock the PCBs to the walls of the box using jackscrews accessible from the ends of the card rails after the boards are in place.

Some of the printed circuit boards carry only electronic components, while others include module housings containing multiple detectors (see Figure 14). Each SIS telescope is formed by 6 sets of such detector modules (4 stack detector modules and 2 matrix detector modules), one above the other. The SIS detectors were designed to have their signals and bias voltages carried by flex circuits (printed-circuit traces inside flexible Kapton ribbons; see Figure 12) which is an integral part of each detector and its mount. This approach greatly reduced the number of parts needed for assembly and significantly reduced the cost and time involved in the assembly process. The matrix detector flex circuits plug directly into 2 inch (5 cm) connectors, two per detector, thereby greatly reducing the amount of room needed to interconnect a detector that has 128 independent connections to preamplifier circuitry. The remaining SIS detectors (the 'stack' detectors) use flex circuits which terminate more conventionally through Miraco connectors.

The use of flex circuitry in SIS has generally eliminated the need for coaxial cables between detectors and preamplifiers. Previous telescope designs utilized coaxial cables, but experience has shown that the center conductors are fragile and often break during assembly. Even when coaxial cables are assembled successfully, it has been found that electronic pickup can be highly dependent on how they are

Figure 16. Cut-away view of SIS, showing one of the two SIS telescopes. At the top of the cut-away section, one can see three entrance foils supported by the entrance collimator. Below are two matrix detectors followed by the four modules which hold T1+T2, T3+T4, T5+T6, and T7+T8. The modules are cut away to show the detectors. The power supply appears below the telescope. The acoustic doors, shown in Figure 1, are not included here.

routed. By contrast, the routing of flex circuitry and printed-circuit-board traces are under the direct control of the designer. The connector reliability has been found to be generally acceptable.

The isotopic resolution criteria discussed above put limits on acceptable detector-position tolerances. These tolerances are specified in Table VI.

The stackup of parts was minimized in order to keep the tolerance accumulation to a minimum. Each detector module is surrounded by an iridited aluminum fence which is attached to a ground plane on the upper surface of the corresponding printed circuit board. Feet on the bottom of each such RF shield extend through the printed circuit board and mate to the top of the RF shield on the next module down. This prevents build-up of position errors that might arise from the fact that the multi-layer printed-circuit-board thicknesses are relatively difficult to control. After the SIS boards are all installed, but before the card rails are tightened, alignment bolts are inserted from the top of the SIS box, extending from the top plate down through the collimator, through all RF shields except the bottommost one, and screwed into the bottom shield. These bolts are intended primarily to align

TABLE VI

SIS mechanical tolerance requirements

Lateral position tolerance	Matrix:	0.1 mm
	Stack:	1.0 mm
Detector co-planarity	Matrix:	0.1°
	Stack:	0.1°
Detector azimuthal orientation	Matrix:	0.1°
	Stack:	1.0°

the modules in a reproducible way. The RF shielding around each module is not perfect, since in general there remains a gap of 0.1 mm between each shield and the board above it.

SIS uses a compact purging system integrated into the center wall of the instrument which allows purge gas to be piped through the RF shielding system and directly into the modules where the detectors reside. Evacuation vents were also designed in for launch. Baffles on the vent ports block light from entering through the instrument walls.

Tests of the matrix detectors subsequent to the initial design showed that the severe acoustic environment expected during launch had a potential to break the bond wires which connect the matrix detector strips to individual preamplifiers. As a consequence, acoustic doors which can be opened by firing explosive cable cutters were added to the design in order to minimize penetration of acoustic energy into the SIS box via the telescope apertures. At the same time, an additional layer of aluminum, 1.6 mm thick, was bonded to the exterior walls of the box using a 50 μm thick layer of acoustic dampening film. These measures reduced the acoustic levels inside the instrument by about 13 dB relative to those outside.

The acoustic doors are mounted on the top surface of the SIS box in such a way that they are thermally isolated from the SIS box. Behind each door is a mechanical collimator within which are mounted three Kapton foils. The outermost foil is 25 μm thick and has a vacuum deposited aluminum coating (VDA) on the inner surface only. The two inner foils are each 8 μm Kapton with VDA on both sides. The purpose of the foils is to block sunlight and provide protection from micro-meteorites. Apart from these foils, the field of view of each SIS telescope is completely unobstructed. The outermost foil surface is bare Kapton in order to provide an effective radiator of whatever sunlight is not reflected from the inner VDA surface.

The SIS instrument is mounted on the top deck of the spacecraft on a tilted bracket which points the two telescopes at 25° to the $+Z$-axis of the spacecraft.

The exterior of SIS, on the sides of the box which face away from the Sun, is covered with silver-coated Teflon to create radiator surfaces. The sides which face the Sun are covered with a thermal blanket. An operational heater is mounted internal to SIS, while a survival heater is located under the thermal blanket. The SIS mounting bolts are thermally isolated from SIS via Ultem bushings. The nominal internal operating temperature for SIS is approximately 20 °C.

3.11. RESOURCES

The SIS instrument is housed in a single box measuring $30.0 \times 41.9 \times 27.5$ cm. It has a mass of 21.9 kg, which includes the hardware to attach it to the tilted bracket. The tilted bracket has a mass of 2.0 kg, which includes the hardware which attaches the bracket to the spacecraft. There are two thermal blankets which together have a mass of 0.40 kg.

SIS has a nominal power dissipation of 17.8 W, but the power is activity dependent. During high rate periods the power is expected to rise to 18.3 W. In addition, there are 4.2 W of operational heaters and 18.2 W of survival heaters in SIS.

Table I summarizes these and other resource allocations for SIS.

3.12. ELECTRICAL GROUND SUPPORT EQUIPMENT

Electrical ground support equipment was developed to support the testing and calibration of SIS prior to integration on the ACE spacecraft. Hardware interfaces to the instrument (commands, telemetry, power, and housekeeping) were provided by a portable spacecraft simulator unit developed at Johns Hopkins University, Applied Physics Laboratory (JHU/APL). The simulator collects data from the instrument for each minor frame (one per second), formats them, attaches time and other status information, and transmits the resulting data block over an RS232 serial line. It also receives commands over the RS232 connection. Two separate classes of commands to the spacecraft simulator are provided: those to be passed to the instrument and those to be used for configuring the simulator itself.

A Unix workstation is used to collect the data from the simulator and to send commands to it. A set of software applications was developed to parse the data stream and extract useful quantities from the telemetry packet, to make basic displays for monitoring instrument health and function, and to enable generation of commands and display of command replies. Also, during periods when the instrument was on the spacecraft, socket connections were provided for passing telemetry to the Unix work station for display. These connections were available both on the area net and from remote sites via the Internet. This capability allowed more scientists and engineers to watch instrument performance during critical tests without having to travel to the test site.

Additional software handles the generation of higher-level template-driven displays for on-line monitoring of instrument performance. This software also allows accumulation and live display of histograms and scatter plots of user-selected variables. These displays are particularly useful during particle calibrations of the instrument.

4. Calibrations and Performance

4.1. CALIBRATIONS OF INDIVIDUAL STACK DETECTORS

Uncertainties in the detector thickness at a given point introduce corresponding uncertainties in the derived mass. To ensure that this latter error is less than 0.1 amu, even for Fe-group nuclei, requires that total and dead-layer thicknesses be known to within $\sim$0.1% of the total thickness over the entire area of each detector.

Absolute thickness measurements of the SIS detectors were made using mono-energetic beams of ^{36}Ar nuclei at the Michigan State University National Superconducting Cyclotron Laboratory (MSU/NSCL). For the thinner detectors, these beams were sent through nine selected spots on each wafer, and stopped in well-characterized, thick, residual-energy (E') detectors placed immediately behind. Thicker SIS detectors were raster scanned, producing complete maps of E'. Care was taken to avoid channeling, and resulting energy-loss uncertainties, by orienting the beam away from crystal axes and planes. To calibrate depth versus E', the beam was also sent through 'slivers' of silicon left over after orientation flats were cut from other detectors. The thicknesses of these slivers were measured to within 0.12 μm using a high-precision mechanical micrometer.

A check on radiation hardness was made when the flux of 870 MeV ^{36}Ar was gradually incremented by factors of 3 until $\sim$2 $\times$ 10^8 particles cm^{-2} had been accumulated over an area of $\sim$1 cm^2 at approximately the 390 μm depth of one 490 μm thick detector. Immediately after the exposure, the leakage current had risen by a factor of $\sim$4 to $\sim$7 μA. Subsequently, no radiation-damage effects were visible in alpha-particle maps of charge collection efficiency.

Maps made using radioactive alpha-particle sources revealed that the dead layers were uniform to within $\pm$0.05 μm on both faces of each detector, with a range from 0.1 to 0.6 μm in thickness. The charge collection response near the edge of each detector was also mapped, as were depletion and breakdown characteristics. Devices which performed poorly in any one of these tests were avoided in selecting detectors for the flight telescopes.

4.2. INTERFEROMETER MAPPING

The 30 detectors making up the two SIS stacks were selected from a total of more than 60 large area devices with thicknesses ranging from 100 μm to 1000 μm, each of which had to be mapped to determine its thickness profile. Because mapping with accelerator beams of heavy ions requires large amounts of beam time and entails a very extensive data analysis effort, a dual-laser interferometer system was developed that is capable of making automated, high-precision measurements of detector thickness variations (see description in Milliken et al., 1995).

A comparison of the particle and interferometer approaches to detector mapping is shown in Figure 17. The interferometer data were the primary source of stack detector maps.

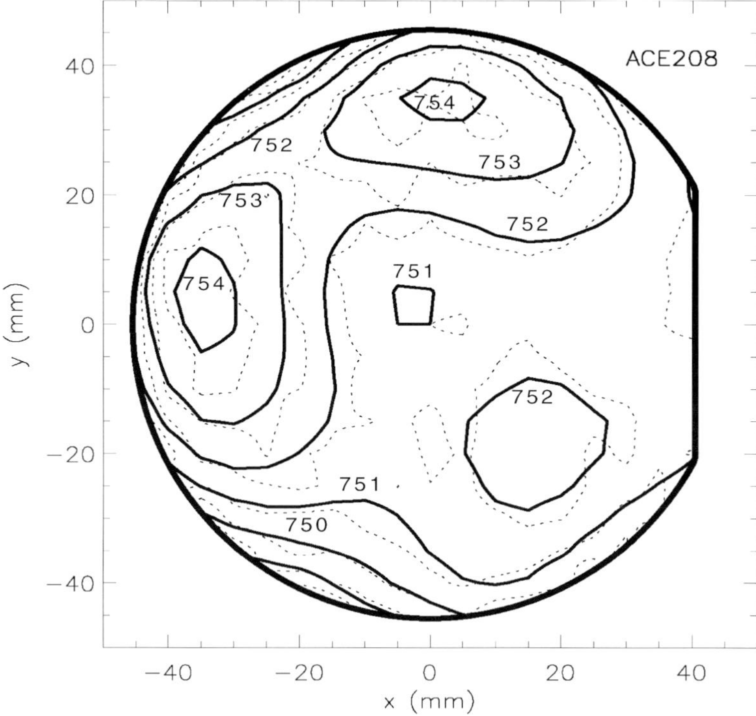

Figure 17. Comparison of particle (dotted) and interferometer (solid) maps of one 750 μm thick stack detector. The contours are labeled for absolute thickness in units of microns. The particle map was produced using residual-energy measurements from a beam of penetrating [36]Ar nuclei that was raster-scanned across the detector. Statistical fluctuations are indicated by the raggedness of this map. The interferometry used two aligned laser beams simultaneously reflecting off the two mirrored surfaces of the detector.

4.3. ELECTRONIC CALIBRATIONS

The electronic stimulus system in SIS has one test pulser associated with each matrix detector strip (512 total) and one test pulser for each stack detector board (four in each telescope). All test pulsers may be pulsed independently, or in coincidence, in any combination. There are two stack detector pulser reference voltages (one for each telescope), and three matrix detector pulser reference voltages (one for the detectors' high-voltage side in each telescope, and one for the detectors' low-voltage side, common to both telescopes). Each reference voltage is generated by an 8-bit DAC with two gain states. The system has the following capabilities: (1) periodically pulse all PHAs with a grid of input levels that will provide relative comparisons of PHA stability and (2) pulse a known pattern of inputs to check whether stimulation events are correctly processed, stored, and read out of the correct buffers. These capabilities were used throughout the pre-flight testing period to monitor the functionality and stability of the instrument electronics, and they provide the basis for in-flight electronic calibrations.

Prior to their installation in the instrument, all of the stack hybrids were individually calibrated with a laboratory test system that included a computer-controlled DAC and calibrated test capacitors. These calibrations were carried out at three temperatures (-20 °C, $+20$ °C, and $+40$ °C).

4.4. INSTRUMENT CALIBRATIONS

In addition to the accelerator calibrations of individual detectors that were described in Section 4.1, an engineering model of SIS was calibrated at the MSU/NSCL in February, 1996 with 100 MeV nucl^{-1} beams of ^{20}Ne, ^{40}Ar, and ^{60}Ni. In addition, a second, essentially identically instrumented calibration unit, composed of spare detectors and flight electronics, was calibrated at MSU in April, 1997. The SIS flight instrument was calibrated at the Gesellschaft für Schwerionenforschung mbH (GSI) accelerator in Darmstadt, Germany in June, 1996, with beams of 300 MeV nucl^{-1} ^{18}O and 300, 500, and 700 MeV nucl^{-1} ^{56}Fe.

These data were used to calibrate and check the performance of a variety of instrument functions, including: (1) the response and uniformity of the matrix detector trajectory system, (2) detector thresholds, (3) the coincidence logic, (4) ADC response, (5) counting rates, (6) the priority and readout systems, (7) the variation of detector and telescope response with energy and angle, (8) the accuracy of heavy-ion range-energy relations, and (9) the balancing of the matrix strip leakage currents in a high-rate environment. In these calibrations the instrument was mounted on a movable stage with two translational and two rotational degrees of freedom so that particles with a variety of angles of incidence could be studied. A system of calibrated absorbers was used to degrade the energy of the beam and cause it to stop at desired ranges within SIS. In addition, a system of polyethylene targets was used to cause the primary beam to fragment into lighter species. Figure 18 shows an example of data obtained during one of the fragmentation runs.

4.5. IN-FLIGHT CALIBRATIONS

Using the electronic stimulation system described in Section 4.3, it is possible to continually monitor the stability of each of the 528 PHAs in SIS. Approximately once per minute the 10 detectors in each of the two telescopes are pulsed at input levels that are gradually incremented through the available dynamic range. These events are flagged and may be used to check PHA performance (pedestals, noise levels, and gain stability) and the instrument logic over the course of time.

SIS will also be calibrated using in-flight particles. The stability of the detectors and electronics will be monitored by observing whether the charge and mass peaks of prominent isotopes such as ^{16}O, ^{28}Si, and ^{56}Fe vary with time or temperature. It is also planned to check the co-planarity and alignment of individual detectors (relative to the ensemble) by comparing their response to abundant species as a function of zenith and azimuth angles. Such checks might, for example, detect any

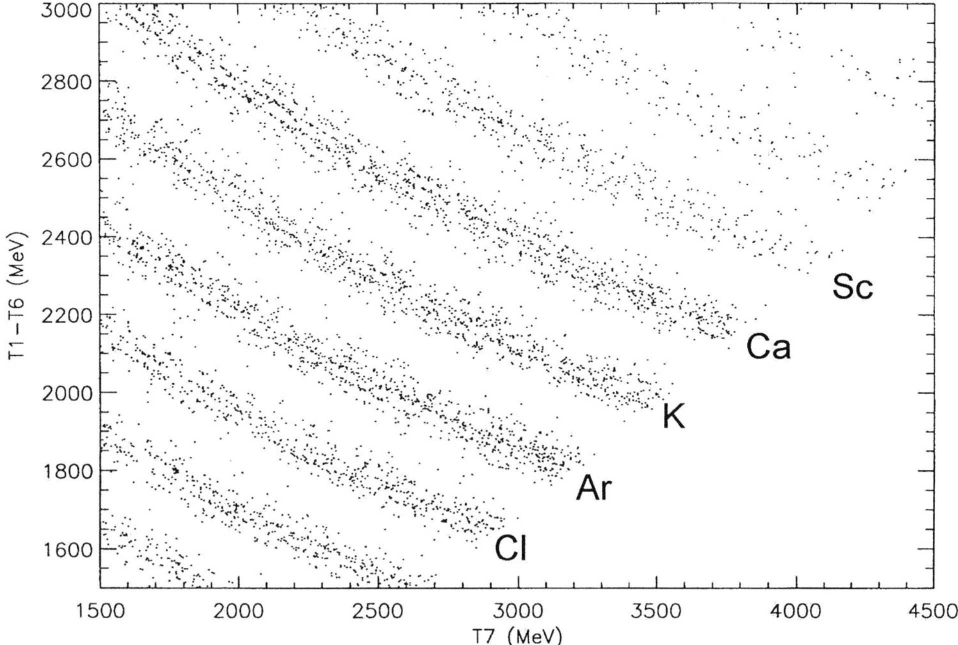

Figure 18. Scatter plot of ΔE vs E' for events collected in a SIS accelerator calibration run at the GSI accelerator in Darmstadt, Germany. The data that are plotted were collected for heavy nuclei that stopped in T7 (used as E') and the sum of the signals from T1 through T6 is used as ΔE. In this run a primary beam of ^{56}Fe at 500 MeV nucl^{-1} incident a 5° was allowed to fragment in a target. Data are shown for elements near $Z = 20$. Tracks corresponding to a number of isotopes of each element are clearly visible in these data.

relative motion of the detectors that occurred as a result of the launch environment. Large solar particle events provide the best in-flight calibration data because they provide excellent statistical accuracy over a short time period.

4.6. ENERGY RANGE AND COLLECTING POWER

The energy range of nuclei that can be measured by SIS (Figure 19) is determined by their range-energy relationship: the minimum energy is for nuclei stopping in M2 and corresponds to the thickness of M1 plus the three windows (75 + 31 μm of equivalent thickness of silicon), while the maximum energy for nuclei stopping in T7 (at normal incidence) corresponds to a nominal thickness of ~8.25 mm of silicon.

At wider angles somewhat higher energy particles stop within the telescope. For Ranges 1 to 7 there are redundant estimates of the charge and mass which provide improved mass resolution. The geometry factor for stopping particles is shown in Figure 20 as a function of energy/nucleon for several abundant species (summed over the two telescopes).

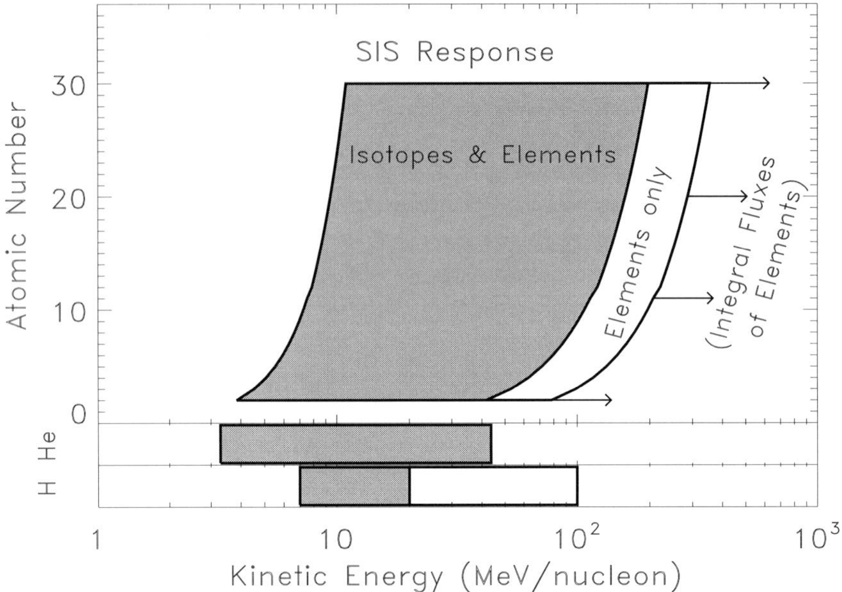

Figure 19. SIS energy range for nuclei with $1 \leq Z \leq 30$. In addition, SIS will make exploratory measurements of SEP nuclei with $31 \leq Z \leq 40$.

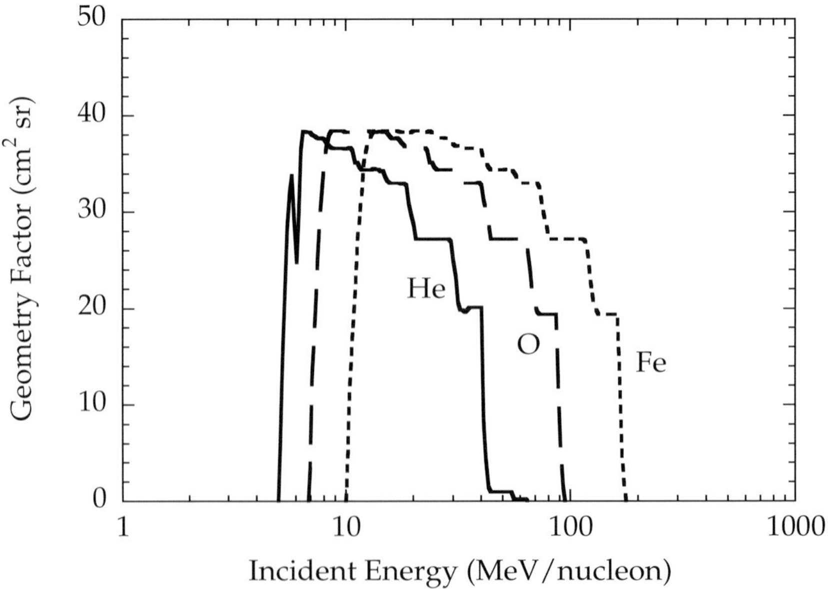

Figure 20. Geometry factor versus energy for the two SIS telescopes combined, where events passing out the side of the telescopes are eliminated.

It is also possible to identify the charge of nuclei that penetrate the entire telescope by comparing their energy losses in the ten devices, M1 to T8. The maximum energy to which these penetrating particles can be identified is expected to be at least twice that of T7 events and this range is reflected in the 'Elements only' band in Figure 19.

The energy range over which SIS can measure H and He depends on the exact settings of the matrix detector thresholds (M1M and M2M). For a nominal threshold setting of $\sim$4 MeV, ^{3}He and ^{4}He that stop in M2 and T1 will trigger M1M and M2M. For deeper ranges the energy loss in M1 and then M2 drops below 4 MeV, but H and He can still be identified using the 'ADC2' trigger mode (see Section 3.7 and Table IV) if the energy loss in M1 and M2 is above the software thresholds M1L and M2L. For M1L $\approx$ M2L $\approx$ 0.5 MeV, this mode allows ^{4}He to be identified to the end of T7 ($\sim$40 MeV nucl^{-1}). If either of these thresholds is raised above their nominal settings, the energy range for He will be decreased.

SIS has much more limited capability to measure protons. For the nominal thresholds discussed above, the proton response extends only from $\sim$6 to $\sim$11 MeV. The flux of somewhat higher energy protons can be monitored using count rates from detectors deeper in the telescope (see Section 4.7). SIS does not respond at all to electrons.

4.7. SIS CONTRIBUTIONS TO THE REAL TIME SOLAR WIND SYSTEM

In June of 1996, just months before the scheduled delivery of the SIS instrument to the spacecraft contractor, NASA Headquarters personnel inquired about the possibility of using ACE instruments to provide real-time measurements of $\sim$10 to $\sim$100 MeV protons in order to help assess the radiation risk to astronauts on the Space Station during large SEP events. A survey of various instruments on ACE indicated that none had appropriate coincidence count rates for protons >10 MeV (see Stone et al., 1998a). Although SIS has very limited response to protons that are required to trigger M1 and M2 (see Section 4.6), it was realized that the count rates of detectors deeper in the SIS stack provide a means of monitoring intensity variations of higher-energy particles. To implement this approach, the SIS microprocessor was reprogrammed to provide two such rates and the spacecraft real-time data system was reprogrammed to read out these rates every 32 s and include them in the real-time solar wind (RTSW) data format (see Zwickl et al., 1998).

The two selected rates are T4 singles and a coincidence between T6 and T7. For the nominal thresholds in Table II and for particles incident through the telescope aperture, T4 responds to $\sim$10 to $\sim$30 MeV protons, while T6$\cdot$T7 responds to $\sim$30 to $\sim$80 MeV. There will, of course, be a significant background due to higher energy cosmic rays that pass through the sides of the SIS box, but the minimum required energy is considerably higher in this case, and this background should vary slowly over the solar cycle. Most of the response for typical solar particle energy spectra will be through the front of the telescope, and not through the sides. The nominal geometry factor for T4 singles is $\sim$55 cm^2 sr (summed over the 2 telescopes), while that for T6·T7 coincidences is $\sim$35 cm^2 sr. Data from these two rates are now part of the RTSW data set that also includes data from SWEPAM, MAG, and EPAM (Zwickl et al., 1998). These data are available to NOAA in real time for forecasts of space weather and are also available to the rest of the community through the RTSW website.

4.8. IN-FLIGHT PERFORMANCE

On August 27, two days after the launch of ACE, SIS was turned on and found to be operating as designed. During the next few months, SIS acquired quiet-time measurements of nuclei with $2 \leq Z \leq 28$ and was able to identify low-energy enhancements in the energy spectra of the ACR elements C, N, O, Ne, and Ar, as well as reporting the first energy spectra for ACR ^{22}Ne and ^{18}O. These quiet-time data, obtained during the passage of ACE to L1, provided the opportunity to test the various instrument operation modes and to adjust some of the thresholds.

In early November of 1997, there were two large SEP events that provided the first test of the instrument under high-rate conditions. The 6 November 1997 event, in particular, provided statistically accurate composition and energy spectra measurements for most of the elements from C to Zn ($6 \leq Z \leq 30$) up to energies of $\sim$100 MeV nucl^{-1}. Preliminary analysis also indicates that there are a number of elements (including C, O, Ne, Mg, Si, S, Ar, Ca, Fe, and Ni) where more than one isotope is resolved and where it will be possible to make measurements of SEP abundance ratios such as ^{13}C/^{12}C, ^{34}S/^{32}S, and ^{54}Fe/^{56}Fe.

Although considerable work remains to optimize the mass resolution, preliminary measurements in ranges 4–7 include $\sigma \approx 0.17$ amu r.m.s. at O and $\sigma \approx 0.40$ amu r.m.s. at Fe during both quiet time and the large solar events of November, 1997. These preliminary values will undoubtedly improve when additional corrections are applied, but there is presently no indication that the mass resolution degrades during times of peak intensity. In addition, the matrix detector readout approach did prove to be successful in identifying heavy-ion trajectories in those events where a chance coincidence occurred between a heavy ion and a low-energy proton. During the first few months of the mission, SIS was also able to identify several impulsive ^{3}He-rich events that were also observed by lower-energy instruments on ACE.

In summary, an ideal ACE was launched at time to allow SIS to sample both solar minimum and solar maximum conditions and is fully expected that the science objectives outlined in Section 2 will be achieved.

Acknowledgments

This research was supported by the National Aeronautics and Space Administration at the Space Radiation Laboratory (SRL) of the California Institute of Technology (under contract NAS5-32626 and grant NAGW-1919), the Jet Propulsion Laboratory (JPL), and the Goddard Space Flight Center (GSFC).

We wish to thank the many individuals and organizations who contributed to the development of SIS: J. H. Marshall III of Radcal Corp., J. Gill and B. Mitchell of Harris Corp., and R. McKenzie of Teledyne Corp. played major roles in the design, development, and fabrication of the custom stack VLSI chips and hybrid circuits, while G. Stupian and M. Leung of Aerospace Corp. provided their microfocus X-ray facility for hybrid screening. S. Kleinfelder helped design the custom matrix VLSI chip which was fabricated by UTMC. C. Wilburn and his staff at Micron Semiconductor Ltd. fabricated the silicon detectors. D. Aalami of Space Instruments was responsible for the design of the power supplies and also contributed to the design, fabrication, and testing of the other electronic assemblies. W. Blanchard of Falcon Services provided all of the board layouts except for the logic and matrix boards, which were provided by J. Stelma of Design Solutions. Electronic assembly was carried out at Caltech and JPL by N. Neverida and T. Ngo-Luu of JPL with the help of T. Dea, who also assisted during conformal coating and environmental testing at JPL. V. Nguyen of Caltech was responsible for most of the subassembly testing. W. Morris of Caltech received and kitted parts and J. Valenzuela of Space Instruments assisted in PCB production and technical document control. B. Williams and J. Krein of JHU/APL were responsible for thermal design and support. L. Stillman of the same institution was responsible for the mechanical design of the door mechanism.

At Caltech, B. Sears, R. Selesnick, J. Cummings, and L. Sollitt provided calibration and analysis support, G. Allbritton was responsible for detector testing, G. Flemming contributed to the flight software, J. Burnham provided engineering support, F. Spalding was responsible for project administration, and D. Kubly and R. Kubly performed a variety of secretarial tasks. In addition, M. Calderon, A. Davis, and T. Garrard assisted with computing issues, R. Paniagua and R. Borup provided support in the Caltech Physics shop, and B. Wong helped with a variety of laboratory tasks. We thank H. Issaian and A. Jefferson for their help with neutron calibrations and R. Kavanaugh for use of the Am-Be neutron source. J. Lopez-Tiana and E. Friese provided purchasing and contract support.

At JPL, J. Rice and C. Cruzan provided assistance with wire bonding of the detectors, B. Potter assisted in expediting the VLSI build at Harris Corp., D. Cipes-Cwik and R. Hill carried out the hybrid inspections at Teledyne, N. Silva contributed to the assembly of the telescope modules and instrument card cage, as well as development of specifications for magnetics parts, W. Powell served as instrument expediter, K. Evans provided advice on parts, J. DePew provided technical support for calibrations and laboratory testing, and M. Salama and T. Scharton assisted with mechanical design issues along with P. Rentz of EER Systems Corporation. R. Pool consulted on contractual issues.

At GSFC, B. Fridovich provided administrative support, M. Madden and B. Nahory were responsible for detector testing, and S. Hendricks of Swales and Associates, Inc. assisted with mechanical design issues.

Some accelerator testing of the SIS detectors was made possible by N. Anantaraman, R. Ronningen, and the staff of the MSU/NSCL, while H. Specht, D. Schardt, and the staff of the GSI heavy ion accelerator in Darmstadt, Germany made possible the heavy ion calibrations of the SIS instrument. R. Koga provided access to the 'Aerospace' chamber at the Lawrence Berkeley Laboratory's 88" Cyclotron where additional detector calibrations and IC latch-up testing took place. The spacecraft team at JHU/APL provided assistance in many areas while SIS was being integrated and tested on the spacecraft.

Finally, we thank A. Frandsen, G. Murphy, H. Eyerly, M. Breslof, C. Rangel, and M. McElveney of the ACE Payload Management Office for their untiring support and assistance, and D. Margolies, J. Laudadio, and the ACE Project Office at GSFC for their help in all phases of the SIS development.

Appendix A. Iterative Calculation of Particle Masses

As discussed in Section 3.1 the fundamental relationship used for deriving a particle's mass from measurements of ΔE, the energy that it loses in traversing a thickness of material L, and E', the residual energy with which it emerges, is $\mathcal{R}_{Z,M}\left(\left(\Delta E + E'\right)/M\right) - \mathcal{R}_{Z,M}\left(E'/M\right) = L$. Using the fact that the particle range is proportional to its mass, this can be written as

$$\mathcal{R}_{Z,M_0}\left(\left(\Delta E + E'\right)/M\right) - \mathcal{R}_{Z,M_0}\left(E'/M\right) = (M_0/M)\,L\,, \tag{A1}$$

where M_0 is the mass of any selected reference isotope of element Z.

The solution of Equation (A1) for M is complicated by the fact that M appears in the argument of the range function. In fact, this dependence, which is approximately proportional to $M^{-1.7}$, is stronger than the explicit M^{-1} dependence on the right-hand side.

It is useful to develop a formula which isolates most of the M dependence in Equation (A1) on one side of an equation so that one can do an iterative calculation of mass which converges in a relatively small number of steps. To obtain such a formula we note that the range–energy relation is reasonably well represented by a power law, $\mathcal{R}_{Z,M}\left(E/M\right) \simeq kM/Z^2\,(E/M)^a$. Then to a good approximation we can write $\mathcal{R}_{Z,M_0}\left(E/M\right) \simeq (M_0/M)^a\,\mathcal{R}_{Z,M_0}\left(E/M_0\right)$. With this substitution Equation (A1) can be solved for M yielding

$$M \simeq M_0\left[\frac{\mathcal{R}_{Z,M_0}\left(\left(\Delta E + E'\right)/M_0\right) - \mathcal{R}_{Z,M_0}\left(E'/M_0\right)}{L}\right]^{1/(a-1)}. \tag{A2}$$

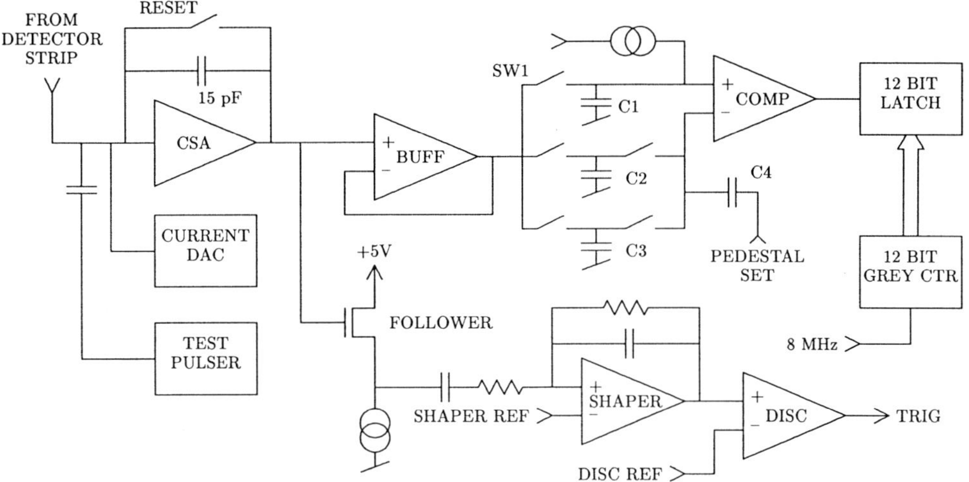

Figure 21. Block diagram of one matrix PHA.

This equation is used as the basis for an iterative calculation of the particle's mass: replacing M_0 with an estimate of the mass, the formula yields an improved estimate. Although the value of a has a weak dependence on energy, reasonable convergence is obtained by simply using a constant $a = 1.7$.

An additional complication is introduced by the presence of a dead layer between the layers of material in which ΔE and E' are measured. A particle traversing a dead layer of thickness l suffers an energy loss δE which goes unmeasured. However, we can calculate the expected value of δE based on an estimate of the particle's mass. Then if we replace ΔE in Equation (A2) by $\Delta E + \delta E$ and replace L by $L + l$ we obtain a formula which can be applied even when there is a significant dead layer.

Appendix B. Details of the Matrix Detector Electronics

As discussed in Section 3.6.1, the signals from the four SIS matrix detectors comprise a total of 512 strips, each of which is individually pulse-height analyzed using custom VLSI circuitry developed especially for ACE. A block diagram of one PHA is shown in Figure 21. Each matrix detector strip is DC coupled to the input of a charge-sensitive amplifier (CSA). The output voltage of a precision, unity gain buffer which follows the CSA is used to charge one of three hold capacitors (C1 through C3) selected by externally controlled switches. Under quiescent conditions, when no detector pulse is present, the CSA signal is switched between C2 and C3 every 3 μs. Thus one of these capacitors is charging while the other holds a steady sample of the baseline (which changes slowly over time due to unbalanced DC current integrated by the CSA, see below).

When a heavy nucleus passes through the detector, the ionization charge it produces is integrated by the CSA causing its output to step up (time constant $\sim$100 ns) from the present baseline by an amount proportional to the collected charge. Under the control of the external logic, switch SW1 is closed causing capacitor C1 to charge to the final voltage reached by the CSA output. At the same time the alternate sampling of the baseline into C2 and C3 is halted and the capacitor with the most recent good sample is connected to the reference input of a comparator. A DC current source causes the voltage on C1 to linearly ramp down until it reaches the stored baseline voltage. The output of the Wilkinson comparator is then a logic pulse with width proportional to the detected ionization charge. To ensure that the 'rundown' time is non-zero even for very small input signals, a constant pedestal voltage is applied via C4 before beginning the ramp. At the end of the rundown interval the state of a counter running at 8 MHz is latched to give a 12-bit digitized representation of the pulse height.

A second branch of the circuit is used to produce discriminator signals which are used as inputs to external control logic. The CSA output is processed by a source follower and a shaping amplifier with a peaking time of $\sim$500 ns. Two discriminators (one of which is shown in Figure 21) compare the shaped pulse with externally provided reference voltages and produce logic pulses for signals above these thresholds. In the operation mode in which SIS is optimized for large SEP events these reference voltages can be set to correspond to the maximum energy loss of a proton in M1 or M2 ($\sim$4 MeV) and the maximum energy loss of an alpha-particle ($\sim$16 MeV), and the coincidence logic can then be operated to respond only to $Z > 1$ and/or $Z > 2$ events.

After an event has been processed, an externally-controlled reset switch is closed to discharge the integration capacitor. Because the CSA input is DC coupled to the detector strip, it integrates the strip's DC current as well as the signal charge. The CSA reset switch must be periodically closed to clear this charge from the capacitor. In order to keep this reset interval relatively long (milliseconds) a current-output digital-to-analog converter (IDAC) is provided to add a programmable current to cancel the leakage current to within $\sim$2 nA. Under external control the CSA can be allowed to integrate for an extended time ($\sim$500 ms) to obtain a precise measurement of the residual unbalanced current. This value is then used as a basis for updating the IDAC setting. In SIS this leakage current balancing is performed every 512 s. The IDAC settings are read out as part of the housekeeping data, making it possible to track the leakage current on each strip and monitor the health of the detector. The pedestal values are also monitored by periodically triggering a conversion when no input signal is present.

A CMOS VLSI chip (see Figure 22) was developed which contains 16 of the matrix detector PHA circuits. The processing of signals from the 128 strips on each detector therefore requires a total of eight of these chips. The circuit can be configured to handle either positive or negative charge inputs, so the same types of

Figure 22. Photograph of a matrix VLSI chip (packaged). Each chip contains 16 complete ADC channels and dissipates 13 mW/channel.

devices can be used to process signals from both detector surfaces. The flight chips were fabricated using a radiation-hard 1.2 micron process by UTMC.

For chips containing 16 PHAs a number of signals have been made common to all channels to reduce the complexity of the interfaces to the external circuitry. These include the switch controls for the sample and hold and the CSA reset, the start of the rundown interval, the injection of the pedestal signal, and the triggering of the test pulser. A logical OR is formed of the 16 discriminator outputs to produce a single trigger signal which is brought out for use by external control logic. The OR of the 16 Wilkinson comparator signals is also brought off-chip to indicate when all conversions are complete and to provide a relatively prompt indication of the size of the largest detected signal.

Each PHA has an on-chip test pulser. The analog DC reference for this pulser and the logic pulse used to trigger a pulse are common to all 16 channels. Similarly, all channels share common discriminator reference voltages.

The IDACs used for leakage current compensation are separate for each of the channels, but the digital control inputs for all 16 IDACs are serially loaded through a single input. This serial DC command loop is also used to load other configuration bits, including masks which can be used to remove selected PHAs from the ORs of the trigger and signal comparator signals. This feature is useful for disabling strips which may be generating excessive noise. Another serial connection is used for reading out the digitized pulse heights from all sixteen PHAs.

In order to obtain pulse-height information from these detectors with precision sufficient for isotope identification (few tenths of a percent), it is necessary to correct for strip-to-strip differences of PHA gains, pedestals, and nonlinearities. Furthermore, in large SEP events, it is important to be able to distinguish H, He, and heavier elements on board so that the limited bit rate available to SIS can be devoted primarily to high-Z, high-energy events, which are of greatest interest. This identification requires on-board pedestal corrections. The pedestals, which are measured every 512 seconds and included in the SIS housekeeping data, are subtracted from each event before it is classified according to charge group. The pedestal-subtracted strip pulse heights are included in event telemetry, but the original pulse heights can be reconstructed if necessary.

Deviations of the matrix PHAs from linearity were measured using a precision pulser prior to installation in the SIS instrument. The linearity is typically excellent (few channels) when the VLSI chips are configured for negative inputs (corresponding to the junction surface of the detector). When used for positive input, the integral nonlinearity is typically a factor of 10 larger, but still correctable.

The relative gains of the 128 PHAs for a given detector were calibrated using heavy ion data. Since a typical particle produces signals corresponding to the same energy deposition for a single strip on each surface of the detector, the ratio of these signals is just the ratio of the gains between the two strips (offset corrections having previously been made). A data set from either an accelerator calibration or collected in flight will contain events corresponding to all possible pairs of front/back strips. An iterative procedure can be used to derive all of the strip PHA gains relative to the gain of a selected strip.

Appendix C. Details of the SIS Priority System

As discussed in Section 3.7.2, SIS employs a priority system to ensure that the most interesting events are not missed during large solar particle events, when the analyzed event rate will often exceed the telemetry capacity of 10 to 15 events per second. SIS uses a system of discriminators on the ADC outputs of each detector to identify H, He, and $Z \geq 3$ nuclei. This information, supplemented by pulse-height and trajectory information from the analyzed events, is used to store events in a series of 95 prioritized buffers listed in Table VII. The buffers are distinguished by range in the instrument, zenith angle, quality of the hodoscope data, and by the 'Hazard flag' (see Section 3.7.2). Events are read out of these buffers in order of their priority, with the exception that the first N events (where N is a commandable integer) of each 256-s major frame are read out by cycling sequentially through the buffers. The priority of all buffers can be re-assigned by command.

In addition to the priority system discussed above, which is intended to optimize the use of telemetry, an additional system, referred to as the 'class' system, is used to prevent the saturation of the on-board computing capability used to select and format events. Before complete readout and processing of the matrix detector data associated with each event (the most computationally-intensive part of event processing), the on-board microprocessor places each event into one of four classes (0, 1, 2 or 3), based on the approximate charge and range of the event. Events of class 0 include the most desirable high-Z events and are always passed along for

TABLE VII

SIS event buffers and priorities

| Charge | Range | mh=0 hz=0 | | | mh=0 hz=1 | | | mh=1 hz=0 | mh=1 hz=1 |
		0–15°	15–25°	≥ 25°	0–15°	15–25°	≥ 25°	0–50°	0–50°
$Z \geq 10$	8	62	62	62	61	61	61	60	60
	5, 6, 7	126	125	116	124	123	115	118	117
	3, 4	122	121	112	120	119	111	114	113
	1, 2	94	93	74	92	91	72	86	84
	0	90	88	73	89	87	71	85	83
$3 \leq Z \leq 9$	8	62	62	62	61	61	61	60	60
	5, 6, 7	110	109	106	104	103	102	98	97
	3, 4	108	107	105	101	100	99	96	95
	1, 2	82	81	70	79	78	69	66	64
	0	80	76	68	77	75	67	65	63
$Z = 2$	8	45	45	45	44	44	44	34	34
	5, 6, 7	59	59	59	54	54	54	41	41
	3, 4	58	58	58	53	53	53	40	40
	1, 2	57	57	57	52	52	52	39	39
	0	49	49	49	48	48	48	38	38
$Z = 1$	8	43	43	43	42	42	42	33	33
	5, 6, 7	56	56	56	51	51	51	36	36
	3, 4	55	55	55	50	50	50	35	35
	2	47	47	47	46	46	46	37	37
Stim		127	127	127	127	127	127	127	127

Notes:

Buffer numbers correspond to readout priorities; highest number = highest priority (buffer numbers 0–32 are unassigned spares).

'Range' refers to the last detector in which an energy loss is measured, i.e., Range 3 means no signal is recorded from beyond T3; range 0 means particle stopped in M2.

'mh' (multi-hodoscope) events have two non-adjacent matrix strips in one or more layers.

'hz' (hazard) events occur within 20 μs of a previous PHA trigger.

complete processing. However, events of class 1, 2, or 3 are discarded if more than 20 events of that class are already awaiting telemetry readout. The class count limit (nominally 20) is commandable separately for each of the three lower priority classes to allow the system to be optimized for SEP count rate environments. In order to allow the proper measurement of event counting rates, the microprocessor maintains counts of the number of events of each class which are examined and of the number in each class which are accepted for complete processing. These counts are included in telemetry every 256 s, along with other count rate data.

Appendix D. SIS Data Formats

D.1. EVENT DATA

For each valid particle event, the SIS microprocessor reads a 12-bit pulse height from each of the 8 stack-detector PHAs and 256 matrix-detector PHAs in the telescope that was hit. To maximize the number of events sent to the ground in the 1992 bits-per-second telemetry rate allocated to SIS, these event data are compressed into a flexible length format illustrated in Table VIII and described below.

Each normal SIS event starts with a fixed-length, 5-byte header, the first byte of which gives the total length of that event. Also included in the header are tag bits indicating whether or not various optional subfields of data are included in that event, and how many stack pulse heights and matrix strip pulse heights and coordinates are present. Additional information about the event is found in the header, such as whether it was a pulser-produced stimulated event, which event buffer it was assigned to, whether it was from the A or B telescope, and whether any triggers at all were seen in the other telescope, how many matrix strips had non-zero pulse heights, and how long ago (in instrument cycles) the event was put into the event buffer (which is needed to match the event with the appropriate rate sample time). Engineering information may follow in an optional 4-byte extended header, giving the average of the strip PHA offsets from each of the 4 surfaces of the matrix detectors in the hit telescope, as well as the time to peak for the matrix and T1 pulse heights.

Following the header are the pulse heights from the stack detectors, starting from T1 and continuing through the deepest detector triggered. This consists of a string of 12-bit pulse heights, with an additional 4 bits of zeroes to fill out the last byte if needed.

Matrix detector information is next. The surface number, strip number, and pulse height uses 20 bits for each strip with a non-zero pulse height. By default, information from a maximum of 10 strips on the top of M1, 9 strips on the bottom of M1, 6 strips on the top of M2, and 6 on the bottom of M2 may be telemetered; if more strips are triggered, only those with the largest pulse heights are sent. More strips are allocated by default to M1 than M2 since it is more exposed and should see a higher count rate. These strip allocations may be changed by command to any pattern in which the maximum total is 31 strips.

If the event is pulser-generated, the normal event format concludes with a 10-byte block listing which pulsers were active and the value of their control DAC levels.

In addition to these compressed, normal-mode events, the instrument can be commanded to send 'diagnostic' mode events. Such an event consists of the compressed event data and raw pulse-height data from all 8 stack PHAs and the 256 matrix strips of the triggered telescope, along with tags indicating which discriminators fired and a variety of tag bits revealing the coincidence signals processed by

TABLE VIII
SIS normal event format

Item #	Count	Bits	Sum	Description
Event header (fixed length):				
0	1	1	1	Matrix detector not triggered
1	1	7	8	Number of bytes in this event
2	1	1	9	Stimulated event flag
3	1	7	16	Event buffer ID
4	1	1	17	Primary telescope (A=0; B=1)
5	1	1	18	Other telescope also hit?
6	1	1	19	Include time to peak and average matrix offsets data block?
7	1	1	20	Hazard flag
8	1	4	24	Range ($R1$) in primary telescope
9	1	3	27	# Other matrix groups not transmitted (encoded)
10	1	5	32	# Matrix strips ($N1$) in primary telescope
11	1	3	35	'Time tag': 3 LSBs of major frame counter (256 s)
12	1	5	40	'Phase'
Time to peak and matrix offset data (if item #6 true; fixed length):				
13	1	4	4	Time to peak for T1 pulse height, in units of 0.25 μs
14	1	4	8	Time to peak for M2 pulse height, in units of 0.25 μs
15	1	4	12	Time to peak for M1 pulse height, in units of 0.25 μs
16	1	5	17	Average of M1 ground side strip offsets
17	1	5	22	Average of M1 high voltage side offsets
18	1	5	27	Average of M2 ground side strip offsets
19	1	5	32	Average of M2 high voltage side strip offsets
Stack (variable length):				
20	$R1$	12	0 to 96	Stack detector pulse heights + 4 bits 0 pad if $R1$ is odd
Matrix (variable length):				
21	$N1$	2	2	Plane number
		6	8	Strip number
		12	20	Strip pulse height
		.	.	.
		.	.	.
		.	.	.
			≤ 620	+ 4 bits of 0 pad if $N1$ is odd
				(Defaults of 10, 9, 6, and 6 strips/plane respectively; commandable to max total of 31 strips)

TABLE VIII

Continued

Item #	Count	Bits	Sum	Description
Stimulated event parameters (if item #2 true; fixed length):				
22	1	8	8	Stim mask – flags for which pulsers are active, MSB
23	1	8	16	Stim mask – flags for which pulsers are active, LSB
24	1	8	24	VREF1 – tel. A stack DAC level
25	1	8	32	VREF2 – tel. B stack DAC level
26	1	8	40	Matrix ground side VREF
27	1	8	48	Matrix HV side VREF, tel. A
28	1	8	56	Matrix HV side VREF, tel. B
29	1	2	58	Spare
30	1	6	64	Gain bits
31	1	16	80	Spare

the on-board logic. Such diagnostic events allow potential problems in the onboard compression algorithms to be identified so that they may be fixed by making the appropriate changes to the flight firmware. The raw data part of these events are a fixed 540 bytes in length.

D.2. RATE AND HOUSEKEEPING DATA

In addition to pulse-height data, SIS records and transmits 'singles' count rate data from each of the 16 stack and 4 matrix detectors (useful for monitoring detector health), a variety of different coincidence rates and event processing monitor rates, 3 separate live times, and rates from each of the 96 event buffers, which are needed to calculate efficiencies and thus obtain absolute fluxes for each category of events, for a total of 213 different rates. All these rates are accumulated over one 256-s instrument cycle. Any 18 of these 213 rates may be selected by command to be read and transmitted at a higher time resolution of once every 32 s. These so-called 'high-priority' rates have their appropriate live time corrections done on board by the microprocessor. Various housekeeping measurements are also recorded, including voltage monitors for the low-voltage power supplies, current monitors for the high-voltage supplies, silicon-detector leakage currents, and temperatures. Every 512 s, the fine and coarse current DAC settings and matrix detector PHA offsets for all 512 matrix strips are sent to the ground. In the remaining telemetry space, the SIS command table is slowly trickled out, taking 15 instrument cycles (64 min) to transmit the entire table.

D.3. TELEMETRY FORMAT

The event, rate, and housekeeping data are packaged into a 249 byte long SIS instrument minor frame. The first bit of this minor frame is used to find the start of the 256-s instrument cycle. It is set to 0 for the first 128 frames, and 1 for the next 128 frames of the cycle. The second bit serves as a flag to indicate that a command response appears in this minor frame, beginning in the seventh byte of the frame. The first byte of a command response gives the length of the remainder of that command echo. The maximum length of a command echo is programmable, but always less than a full minor frame; the rest of the frame after the command response contains normal or diagnostic mode data. All commands are echoed back in the SIS telemetry data, along with any output generated by the command, thus documenting the command history of the instrument within the data stream. The third bit is set if the pulse-height data in this frame represent a diagnostic mode event rather than a normal compressed event. Diagnostic events may span any number of contiguous minor frames, and may start in the middle of a normal event (which then continues after the end of the diagnostic event). After these first 3 bits, the remaining 45 bits in the first 6 bytes are allocated to rate and housekeeping data.

Each item of rate or housekeeping data is assigned to one of the 256 minor frames per instrument cycle. These data are packaged to maximize the use of the 45 bits per frame allocated to them. All rates are compressed into 15-bit numbers, and are packaged three per frame for those frames in which they appear. House-keeping data are typically 10-bit quantities packed four per frame, while other frames contain five 8-bit matrix detector PHA offsets or nine 5-bit matrix detector leakage current DAC values. Frames containing part of the command table readout each include five bytes. Normal mode pulse-height data, diagnostic event data, or command responses (whichever is indicated by the second and third bits of the frame) fill out the rest of the minor frame after the first 6 bytes.

Events can span the boundary from one frame to the next and are strung sequentially together to maximize the use of telemetry space. Only at the beginning of the 256-s instrument cycle is an event forced to start at the beginning of the event data section of the frame. If one or more frames in the middle of an instrument cycle should be lost due to telemetry problems, it is important to be able to find the starting point of the next event to avoid losing the rest of the pulse-height data from that cycle. This is readily done as follows. Both the total length of the event and the length of each subsection (from which the total length may be independently calculated) are included in fixed positions in the event header (see Table VIII). By assuming that any given byte is the start of an event and comparing its value of the apparent event length with that calculated from the supposed event header, one can test whether the start of an event may indeed have been found. If so, the next byte after its end should also satisfy the criteria to be the start of an event, and so on until the end of the instrument cycle.

References

Anders, E. and Ebihara, M.: 1982, 'Solar-System Abundances of the Elements', *Geochim. Cosmochim. Acta* **46**, 2363–2380.

Anders, E. and Grevesse, N.: 1989, 'Abundances of the Elements: Meteoritic and Solar', *Geochim. Cosmochim. Acta* **53**, 197–214.

Audouze, J.: 1983, in A. Maeder and A. Renzini (eds), 'Observational Tests of Stellar Evolution Theory', *IAU Symp.* **105**, 541.

Breneman, H. H. and Stone, E. C.: 1985, 'Solar Photospheric and Coronal Abundances from Solar Energetic Particle Measurements', *Astrophys. J.* **199**, L57–L61.

Cassé, M. and Paul, J. A.: 1982, 'On the Stellar Origin of the Ne-22 Excess in Cosmic-Rays', *Astrophys. J.* **258**, 860–863.

Connell, J. J. and Simpson, J. A.: 1993, 'The Ulysses Cosmic Ray Isotope Experiment II: Source Abundances of Ne, Mg and Si Derived from High Resolution Measurements', *Proc. 23rd Int. Cosmic Ray Conf., Calgary* **1**, 559–562.

Cook, W. R., Cummings, A. C., Cummings, J. R., Garrard, T. L., Kecman, B., Mewaldt, R. A., Selesnick, R. S., Stone, E. C., and von Rosenvinge, T. T.: 1993a, 'MAST: A Mass Spectrometer Telescope for Studies of the Isotopic Composition of Solar, Anomalous, and Galactic Cosmic Ray Nuclei', *IEEE Trans. Geosci. Remote Sensing* **31**, 557–564.

Cook, W. R., Cummings, A. C., Kecman, B., Mewaldt, R. A., Aalami, D., Kleinfelder, S. A., and Marshall, J. H.: 1993b, 'Custom Analog VLSI for the Advanced Composition Explorer', *Small Instruments Workshop Proc.*, Pasadena, CA.

Cummings, A. C., Stone, E. C., and Webber, W. R.: 1991, 'The Isotopic Composition of Anomalous Cosmic-Ray Neon', *Proc. 22nd Int. Cosmic Ray Conf., Dublin* **3**, 362–365.

Dietrich, W. F. and Simpson, J. A.: 1979, 'The Isotopic and Elemental Abundances of Neon Nuclei Accelerated in Solar Flares', *Astrophys. J. Lett.* **231**, L91–L94.

Dougherty, B. L., Christian, E. R., Cummings, A. C., Leske, R. A., Mewaldt, R. A., Milliken, B. D., von Rosenvinge, T. T., and Wiedenbeck, M. E.: 1996, 'Characterization of Large-Area Silicon Ionization Detectors for the ACE Mission', *SPIE Conf. Proc.* **2806**, 188–198.

Fisk, L. A., Kozlovsky, B. and Ramaty, R.: 1974, 'An Interpretation of the Observed Oxygen and Nitrogen Enhancements in Low-Energy Cosmic Rays', *Astrophys. J. Lett.* **190**, L35–L38.

Garrard, T. L. and Stone, E. C.: 1993, 'New SEP-Based Solar Abundances', *Proc. 23rd Int. Cosmic Ray Conf., Calgary* **3**, 384–387.

Geiss, J., Gloeckler, G., and von Steiger, R.: 1996, 'Origin of C+ Ions in the Heliosphere', *Space Sci. Rev.* **78**, 43–52.

Geiss, J. F., Buehler, H., Cerutti, H., Eberhardt, P., and Filleux, Ch.: 1972, 'Solar Wind Composition Experiment', *Apollo-16 Prelim. Sci. Report, NASA SP-315* **231**, 14-1.

Hubert, F., Bimbot, R., and Gauvin, H.: 1990, 'Range and Stopping-Power Tables for 2.5–500 MeV/Nucleon Heavy Ions In Solids', *Atom. Dat. Nucl. Dat. Tables* **46**, 1-213.

Kahler, S. W.: 1992, 'Solar Flares and Coronal Mass Ejections', *Ann. Rev. Astron. Astrophys.* **30**, 113–141.

Klecker, B.: 1995, 'The Anomalous Component of Cosmic Rays in the 3-D Heliosphere', *Space Sci. Rev.* **72**, 419–430.

Klecker, B., McNab, M. C., Blake, J. B., Hamilton, D. C., Hovestadt, D., Kästle, H., Looper, M. D., Mason, G. M., Mazur, J. E., and Scholer, M.: 1995, 'Charge State of Anomalous Cosmic-Ray Nitrogen, Oxygen, and Neon: SAMPEX Observations', *Astrophys. J.* **442**, L69–L72.

Klecker, B., Oetliker, M., Blake, J. B., Hovestadt, D., Mason, G. M., Mazur, J. E. and McNab, M. C.: 1997, 'Multiply Charged Anomalous Cosmic Ray N, O, and Ne: Observations With HILT/SAMPEX', *Proc. 25th Int. Cosmic Ray Conf., Durban* **2**, 273–276.

Leske, R. A., Mewaldt, R. A., Cummings, A. C., Cummings, J. R., Stone, E. C., and von Rosenvinge, T. T.: 1996a, 'The Isotopic Composition of Anomalous Cosmic Rays from SAMPEX', *Space Sci. Rev.* **78**, 149–154.

Leske, R. A., Cummings, J. R., Mewaldt, R. A., Stone, E. C., and von Rosenvinge, T. T.: 1996b, 'Measurements of the Ionic Charge States of Solar Energetic Particles at 15-70 MeV/nucleon Using the Geomagnetic Field', *AIP Conf. Proc.* **374**, 86–95.

Leske, R. A., Mewaldt, R. A., Cummings, A. C., Stone, E. C., and von Rosenvinge, T. T.: 1997, 'Updated Measurements of the Isotopic Composition of Interplanetary and Geomagnetically Trapped Anomalous Cosmic Rays', *Proc. 25th Int. Cosmic Ray Conf., Durban* **2**, 321–324.

Luhn, A., Klecker, B., Hovestadt, D. and Möbius, E.: 1987, 'The Mean Ionic Charge State of Silicon in ^{3}He-rich Flares', *Astrophys. J.* **317**, 951–955.

Lukasiak, A., Ferrando, P., McDonald, F. B., and Webber, W. R.: 1994, 'Cosmic-Ray Isotopic Composition of C, N, O, Ne, Mg, Si Nuclei in the Energy Range 50-200 MeV per Nucleon Measured by the Voyager Spacecraft During the Solar Minimum Period', *Astrophys. J.* **426**, 366–372.

Mason, G. M., Mazur, J. E. and Hamilton, D. C.: 1994, 'Heavy-Ion Isotopic Anomalies in ^{3}He-Rich Solar Particle Events', *Astrophys. J.* **425**, 843–848.

Mewaldt, R. A. and Stone, E. C.: 1989, 'Isotope Abundances of Solar Coronal Material Derived from Solar Energetic Particle Measurements', *Astrophys. J.* **337**, 959–963.

Mewaldt, R. A., Selesnick, R. S., Cummings, J. R., Stone, E. C., and von Rosenvinge, T. T.: 1996, 'Evidence for Multiply-Charged Anomalous Cosmic Rays', *Astrophys. J.* **466**, L43–L46.

Mewaldt, R. A., Spalding, J. D., and Stone, E. C.: 1984a, 'A High-Resolution Study of the Isotopes of Solar Flare Nuclei', *Astrophys. J.* **280**, 892–901.

Mewaldt, R. A., Spalding, J. D., and Stone, E. C.: 1984b, 'The Isotopic Composition of the Anomalous Low-Energy Cosmic Rays', *Astrophys. J.* **283**, 450–456.

Meyer, J. P.: 1985, 'Solar-Stellar Outer Atmospheres and Energetic Particles, and Galactic Cosmic Rays', *Astrophys. J. Suppl.* **57**, 173-204.

Milliken, B. Leske, R. A., and Wiedenbeck, M. E.: 1995, 'Silicon Detector Studies with an Interferometric Thickness Mapper', *Proc. 24th Int. Cosmic Ray Conf., Rome* **4**, 1283–1286.

Pesses, M. E., Jokipii, J. R., and Eichler, D.: 1981, 'Cosmic Ray Drift, Shock Wave Acceleration, and the Anomalous Component of Cosmic Rays', *Astrophys. J.* **246**, L85–L89.

Podosek, F.: 1978, 'Isotopic Structures in Solar System Materials', *Ann. Rev. Astron. Astrophys.* **16**, 293–334.

Prantzos, N., Arnould, M., and Arcoragi, J. P.: 1987, 'Neutron-Capture Nucleosynthesis During Core Helium Burning in Massive Stars', *Astrophys. J.* **315**, 209–228.

Reames, D. V.: 1993, 'Mean Element Abundances in Energetic Particles from Impulsive Flares', *Proc. 23rd Int. Cosmic Ray Conf., Calgary* **3**, 388–391.

Reames, D. V.: 1995, 'Solar Energetic Particles – A Paradigm Shift', *Rev. Geophys. Suppl.* **33**, 585–589.

Reames, D. V., Barbier, L. M., and von Rosenvinge, T. T.: 1997, 'Wind/EPACT Observations of Anomalous Cosmic Rays', *Adv. Space Res.* **19**, 809–812.

Reames, D. V., Cane, H. V., and von Rosenvinge, T. T.: 1990, 'Energetic Particle Abundances in Solar Electron Events', *Astrophys. J.* **357**, 259–270.

Selesnick, R. S., Cummings, A. C., Cummings, J. R., Leske, R. A., Mewaldt, R. A., Stone, E. C., and von Rosenvinge, T. T.: 1993, 'Coronal Abundances of Neon and Magnesium Isotopes from Solar Energetic Particles', *Astrophys. J.* **418**, L45–L48.

Simpson, J. A.: 1995, 'The Anomalous Nuclear Component in the Three-Dimensional Heliosphere', *Adv. Space Res.* **16** (9), 135–149.

Simpson, J. A., Wefel, J. P., and Zamow, R.: 1983, 'Isotopic and Elemental Composition of Solar Energetic Particles', *Proc. 18th Int. Cosmic Ray Conf., Bangalore* **10**, 322–325.

Stone, E. C.: 1973, 'Cosmic Ray Isotopes', *Proc. 13th Int. Cosmic Ray Conf., Denver* **5**, 3615–3626.

Stone, E. C. and Cummings, A. C.: 1997, 'Evidence for Anomalous Cosmic Ray S, Si, and Fe in the Outer Heliosphere and for a Non-ACR Source of S at 1 AU', *Proc. 25th Int. Cosmic Ray Conf., Durban* **2**, 289–292.

Stone, E. C., Burlaga, L. F., Cummings, A. C., Feldman, W. C., Frain, W. E., Geiss, J., Gloeckler, G., Gold, R., Hovestadt, D., Krimigis, S. M., Mason, G. M., McComas, D., Mewaldt, R. A., Simpson, J. A., von Rosenvinge, T. T., and Wiedenbeck, M. E.: 1989, 'The Advanced Composition Explorer', *AIP Conf. Proc.* **203**, 48–58.

Stone, E. C., Frandsen, A. M., Mewaldt, R. A., Christian, E. R., Margolies, D., Ormes, J. F., and Snow, F.: 1998a, 'The Advanced Composition Explorer', *Space Sci. Rev.* **86**, 1.

Stone, E. C., Cohen, C. M. S., Cook, W. R., Cummings, A. C., Gauld, B., Kecman, B., Leske, R. A., Mewaldt, R. A., Thayer, M. R., Dougherty, B. L., Grumm, R. L., Milliken, B. D., Radocinski, R. G., Wiedenbeck, M. E., Christian, E. R., Shuman, S., Trexel, H., von Rosenvinge, T. T., Binns, W. R., Crary, D. J., W. R., Dowkontt, P., Epstein, J., Hink, P. L., Klarmann, J., Lijowski, M., and Olevitch, M. A.: 1998b, 'The Cosmic Ray Isotope Spectrometer for the Advanced Composition Explorer', *Space Sci. Rev.* **86**, 285.

Takashima, T., Doke, T., Hayashi, T., Kobayashi, M., Shirai, H., Takehana, N., Ehara, M., Yamada, Y., Yanagita, S., Hasebe, N., Kashiwagi, T., Kato, C., Munakata, K., Kohno, T., Kondoh, K., Murakami, H., Nakamoto, A., Yanagimachi, T., Reames, D. V., and von Rosenvinge, T. T.: 1997, 'The First Observation of Sulfur in Anomalous Cosmic Rays by the Geotail and the Wind Spacecrafts', *Astrophys. J.* **477**, L111–L113.

Tosi, M.: 1982, 'CNO Isotopes and Galactic Chemical Evolution', *Astrophys. J.* **254**, 699–707.

Wiedenbeck, M. E., Christian, E. R., Cook, W. R., Cummings, A. C., Dougherty, B. L., Leske, R. A., Mewaldt, R. A., Stone, E. C., and von Rosenvinge, T. T.: 1996, 'Two-Dimensional Position-Sensitive Silicon Detectors for the ACE Solar Isotope Spectrometer', *SPIE Conf. Proc.* **2806**, 176–187.

Wilson, T. L. and Rood, R. T.: 1994, 'Abundances in the Interstellar Medium', *Ann. Rev. Astron. Astrophys.* **32**, 191–226.

Zwickl, R. D., Sahm, S., Barrett, B., Grubb, R., Detman, T., Raben, V., Smith, C. W., Riley, P., Gold, R., Mewaldt, R. A., and Maruyama, T.: 1998, 'The NOAA Real-Time-Solar-Wind (RTSW) System Using ACE Data', *Space Sci. Rev.* **86**, 635.

THE ULTRA-LOW-ENERGY ISOTOPE SPECTROMETER (ULEIS) FOR THE ACE SPACECRAFT

G. M. MASON[1,2], R. E. GOLD[3], S. M. KRIMIGIS[3], J. E. MAZUR[1,*],
G. B. ANDREWS[3], K. A. DALEY[1], J. R. DWYER[1], K. F. HEUERMAN[1],
T. L. JAMES[1], M. J. KENNEDY[3], T. LEFEVERE[3], H. MALCOLM[3], B. TOSSMAN[3]
and P. H. WALPOLE[1]

[1]*Department of Physics, University of Maryland, College Park, MD 20742, U.S.A.*
[2]*Institute for Physical Science and Technology, University of Maryland, U.S.A.*
[3]*Johns Hopkins University/Applied Physics Laboratory, Laurel, MD 20723, U.S.A.*

Abstract. The Ultra Low Energy Isotope Spectrometer (ULEIS) on the ACE spacecraft is an ultra high resolution mass spectrometer designed to measure particle composition and energy spectra of elements He–Ni with energies from $\sim$45 keV nucl^{-1} to a few MeV nucl^{-1}. ULEIS will investigate particles accelerated in solar energetic particle events, interplanetary shocks, and at the solar wind termination shock. By determining energy spectra, mass composition, and their temporal variations in conjunction with other ACE instruments, ULEIS will greatly improve our knowledge of solar abundances, as well as other reservoirs such as the local interstellar medium. ULEIS is designed to combine the high sensitivity required to measure low particle fluxes, along with the capability to operate in the largest solar particle or interplanetary shock events. In addition to detailed information for individual ions, ULEIS features a wide range of count rates for different ions and energies that will allow accurate determination of particle fluxes and anisotropies over short ($\sim$few minutes) time scales.

1. Scientific Goals

The origin and evolution of the solar system and galaxy can be studied by measuring the composition of their condensed and gaseous materials. These compositional data contain information on processes that created the material in the big bang, in later nucleosynthetic processing, and, more recently, in fractionation processes associated with the condensation of the material from the interstellar medium into stars, planets, and smaller objects. Since astrophysical objects routinely accelerate ions to high energies, multiple streams of particles from a wide variety of sites fill the interplanetary space near Earth. These particle streams may be sampled, studied, and probed for information about their parent sites and the processes that accelerated and brought them here.

In the pioneering studies at the start of the space program, energetic particles detected in the interplanetary medium near Earth were thought to have two primary sources: the sun and galaxy. Solar energetic particles were energized in association

---

* Now at Aerospace Corp., Los Angeles, CA 90009, U.S.A.

Space Science Reviews **86**: 409–448, 1998.
© 1998 *Kluwer Academic Publishers. Printed in the Netherlands.*

with explosions (flares) visible on the solar surface. Galactic cosmic rays arrived from outside the solar system from an unknown source of such power that only processes involving supernovae seemed likely. Later, an additional particle source was identified wherein neutral atoms from the local interstellar gas had a single electron removed by (UV) sunlight, and then were energized by a shock in the outer solar system. Most recently, evidence has mounted for acceleration of ions from a new particle source: interstellar and interplanetary dust grains.

These samples of matter contain a rich lode of information about their parent sites, about astrophysical plasma processes that accelerate ions, and the conditions under which they can be transported over enormous distances to be found near Earth. Studies carried out to date have revealed that while each of these material samples contains a grossly similar composition, variations routinely occur, and these contain important new clues about the particle sites and the processes taking place there. For example, material flowing out of the sun (the solar wind) in the equatorial regions often shows enhancements of materials such as Iron and Magnesium compared to the abundances in the solar photosphere. Another case is the accelerated interstellar neutral material, which is rich in Nitrogen and Oxygen, and has very little Carbon in contrast to other sources in nature.

Recent advances in knowledge of the solar wind composition, and the composition of galactic cosmic rays, have highlighted the need for advanced instrumentation with the ability to detect differences between the various matter samples and other material reservoirs. For example, compared to meteoritic abundances the isotope ^{22}Ne had been found to be enriched in some samples of solar material, and not in others. The solar wind elemental composition over the solar poles shows striking differences with that observed near the ecliptic plane. By sampling these energetic particle compositions, both elemental and isotopic, with greatly increased precision, we can systematically investigate and interpret these abundance differences, and identify others that have to date remained beyond the reach of experimental study.

This paper describes the Ultra Low Energy Isotope Spectrometer (ULEIS), one of the 4 advanced mass spectrometers for the ACE mission that, as a group, cover the range from solar wind (1 keV nucl^{-1}) energies to galactic cosmic rays (100–1000 MeV nucl^{-1}), thereby allowing a global investigation of these different energetic material samples (Stone et al., 1989). ULEIS operates in the energy range from ~45 keV nucl^{-1} to a few MeV nucl^{-1}, thereby sampling particles from the sun, from interplanetary shock accelerated material, and from the interstellar neutrals accelerated at the termination shock. Since the scientific investigations to be undertaken with ULEIS are coordinated with the other ACE instruments, other papers in this issue contain discussions that deal with the issues discussed below (Gloeckler et al., 1998; McComas et al., 1998; Möbius et al., 1998; Stone et al., 1998b) .

1.1. Elemental and isotopic composition of matter

The bulk of our knowledge of solar system abundances derives primarily from studies of terrestrial and meteoritic material (Anders and Ebihara, 1982). Compilations of these abundances serve as a baseline for a broad range of astrophysical, solar, and planetary studies. Improving our knowledge of the elemental and isotopic abundances solar system reservoirs is therefore of fundamental importance.

1.1.1. *Solar Isotopic Abundances*

Spectroscopy of solar emission lines has long been used to study the *elemental* composition of the sun, but this technique is not effective in determining *isotopic* abundances. However, shock waves in the solar corona routinely accelerate material to high energies; these particles often escape into the inner solar system where they can be studied in detail. These isotopic measurements are sufficiently difficult that only a handful of such particle events have been studied, and these with limited statistical accuracy (Mewaldt and Stone, 1989; Selesnick et al., 1993). For C, N, O, and Mg, isotopic abundance measurements to date are all consistent with solar wind (where available) and meteoritic abundances (Mewaldt and Stone, 1989). On the other hand, the first measurements of ^{22}Ne/^{20}Ne showed a significant (factor of 2) excess abundance of ^{22}Ne, while more recent results from SAMPEX show ^{22}Ne/^{20}Ne similar to solar wind values as might be expected (Selesnick et al., 1993). These differences might be due to differences between individual solar particle events. By measuring solar isotopic abundances in a large number of particle events, it will be possible to characterize the event-to-event variations, and correct for systematic trends.

1.1.2. *Coronal Composition*

Material in the solar corona originates from the photosphere of the sun, moving outwards by poorly understood processes. Recent evidence from the Ulysses solar wind measurements (Galvin et al., 1984, 1995) shows that the long-known first ionization potential (FIP) enhancements observed in the ecliptic plane are largely reduced or absent altogether in the fast solar wind observed over the solar polar regions. This gives evidence for the importance of magnetic fields in the operation of an ion-neutral separation process that causes the FIP bias (von Steiger and Geiss, 1989; Fisk et al., 1998). Large ('gradual') solar particle events appear to accelerate material over large regions of the corona itself or regions close to the sun (Mason et al., 1984; Reames et al., 1996). Thus, the elemental composition of solar energetic particle nuclei in gradual events also carries information about the coronal composition. At higher energies (10 s of MeV nucl^{-1}) there are systematic abundance variations from photospheric values that correlate well with particle charge-to-mass ratio (Breneman and Stone, 1985). As shown in Figure 1, for particles of energies near 1 MeV nucl^{-1} and below, these deviations are smaller (Mazur et al., 1993). By studying the abundances over a broad energy range, it is possible to construct a table of coronal abundances from the energetic particle population, for comparison with abundances obtained from solar wind measurements and photospheric spectroscopy.

1.1.3. *Corotating Interaction Regions and Pickup Ions*

High speed solar wind streams emanate from magnetically open portions of the corona, and in the solar ecliptic plane these regions typically alternate with regions of closed coronal magnetic field lines that are associated with the source of 'slow' solar wind. Under solar minimum conditions, when other activity on the sun is relatively low, the high and low speed streams can continue for months, setting up nearly steady state in the inner heliosphere. In these situations, the high speed solar wind streams overtake the slow wind, and interact so as to form forward and backward moving shock waves (Barnes and Simpson, 1976; McDonald et al., 1976). This interaction is strongest at radial distances of a few AU, and since the regions appear to co-rotate with the coronal features that are the origin of the fast and slow wind, they are called 'co-rotating interaction regions' (CIRs). Ions energized in these regions flow into the inner heliosphere, and can be observed near Earth (Mewaldt et al., 1978; Gloeckler et al., 1979; Richardson and Hynds, 1990). Comparison with recent solar wind abundance determinations shows that the CIR abundances at 1 AU are close to the *average* between fast vs slow solar wind, as shown in Figure 2 (Mason et al., 1997). In addition, some CIR elemental abundances such as C show a dependence on solar wind speed which is not expected from models developed to date (Fisk and Lee, 1980). It may be that in addition to a solar wind source, CIRs also accelerate other low energy components in the interplanetary medium, such as pickup ions. Pickup ion sources include interstellar neutrals, and products from the sputtering and evaporation of interplanetary dust, and so these studies may yield new insights into other material in the interplanetary space (Gloeckler et al., 1994).

1.1.4. *Anomalous Cosmic-Ray Composition*

Anomalous cosmic rays (ACR) arise from interstellar neutrals of high ionization potential that enter the heliosphere, are ionized by solar UV or charge exchange, and then transported outwards again to the termination shock region where they are energized to $\sim$10 MeV nucl^{-1} (Fisk et al., 1974; Pesses et al., 1981; Klecker, 1995; Cummings and Stone, 1996). Recent observations from SAMPEX of a transition from singly ionized Oxygen to multiply charged Oxygen near 20 MeV nucl^{-1} (Mewaldt et al., 1996; Klecker et al., 1997) have revealed the overall energy limits of the anomalous cosmic-ray acceleration mechanism (to $\sim$240 MV/Q), and have put estimates of the time scale for acceleration ($\sim$1 year) on a firmer basis (Jokipii, 1996). The ACR thus provide a sample of the neutral gas in the local interstellar medium. Although the elemental abundance of the ACR shows extreme fractionation (hence the name), the isotopic ratios should be relatively little affected by the acceleration process. Thus studies of the ACR isotopic composition can be

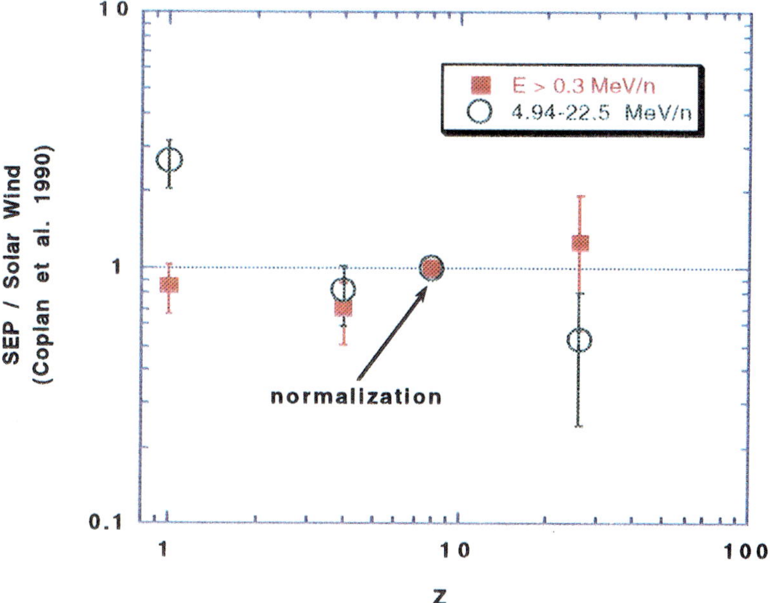

Figure 1. Abundances measured in 10 large solar particle events compared to solar wind abundances. Note that for abundances averaged over E > 0.3 MeV nucl^{-1}, there is little difference from solar wind abundances, while for the same particle events, the abundances for 4.9–22 MeV nucl^{-1} show significant differences from the solar wind (Mazur et al., 1993).

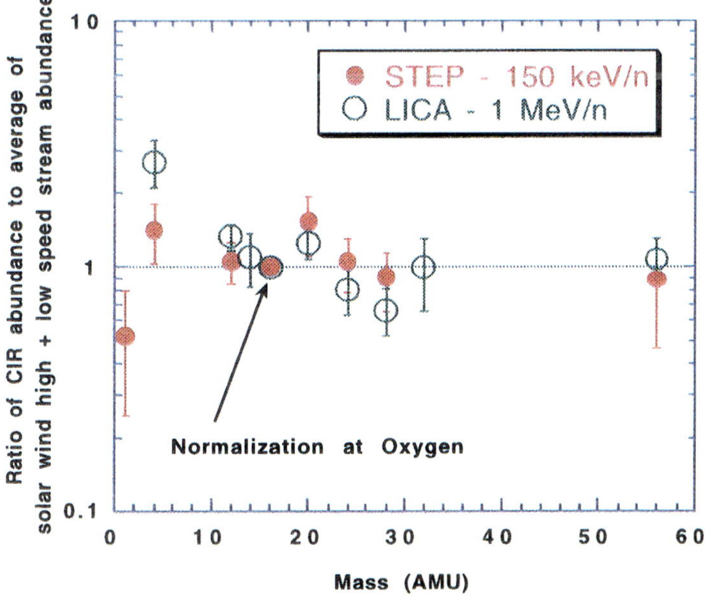

Figure 2. Comparison of CIR element abundances at 150 keV nucl^{-1} and 1 MeV nucl^{-1} with the average of *Ulysses* fast and slow speed solar wind composition (Mason et al., 1997).

used as a new tool to determine local galactic isotopic composition. Since ACR fluxes are very low, instrumental studies to date have not had sufficient precision to determine the isotopic ratios with sufficient detail to reveal anomalies (if any) with solar system abundances (Mewaldt et al., 1976).

1.2. PARTICLE ACCELERATION ON THE SUN AND IN THE HELIOSPHERE: MODELS

In addition to measurements of the abundances of particles in large solar particle events and in impulsive solar flares, the distribution of particles with energy is an important diagnostic of the acceleration and transport processes. An acceleration model not only must account for the observed elemental and isotopic abundances, but it also must describe how the energy is distributed via electric and magnetic fields among ions with differing mass to charge ratios. The energy spectra are key to our understanding of how these samples of solar material become energized and how they propagate to the observer.

1.2.1. *Energy Spectra of Large Solar Particle Events*

Large solar particle events have been observed in interplanetary space simultaneously with several spacecraft; these observations, together with the association of the events with interplanetary Type II radio bursts indicate that a traveling interplanetary shock accelerates the particles, possibly out of the ambient solar wind. Included in this seed population might also be a contribution from interstellar pickup ions (Gloeckler et al., 1994). Figure 3(a) shows the energy spectra of several particle species measured in a single solar energetic particle event (Mazur et al., 1992). The solid curves are fits to the data using a model of stochastic particle acceleration that might occur at such a shock. Figure 3(b) shows the abundance ratios as functions of energy for the same event; the trends with energy indicate that the acceleration process is more efficient for particles with lower energies and lower mass to charge ratios.

The spectra in Figure 3 represent an average of the particle intensity over a several day-long time period. An alternative method shown in Figure 4 (Reames et al., 1997) measures the low energy particle spectra in the event's decay phase after the shock has traveled past the observer. In the region downstream of the shock, the spectra continue to rise as power laws at the lower energies, in agreement with the predictions of shock acceleration. These two examples of large solar particle event spectra illustrate the importance of measuring the spectra over a wide energy range for many particle species. They not only show how the acceleration process might favor one species over another, but they also shed light on the most appropriate way to use the spectra as a diagnostic of an acceleration process that evolves with time as the shock propagates out from the sun.

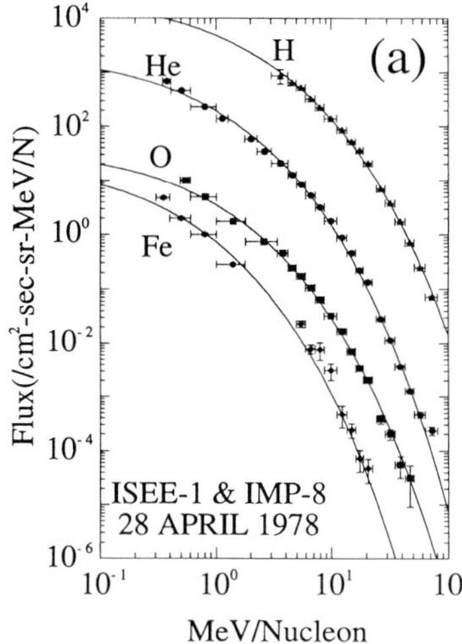
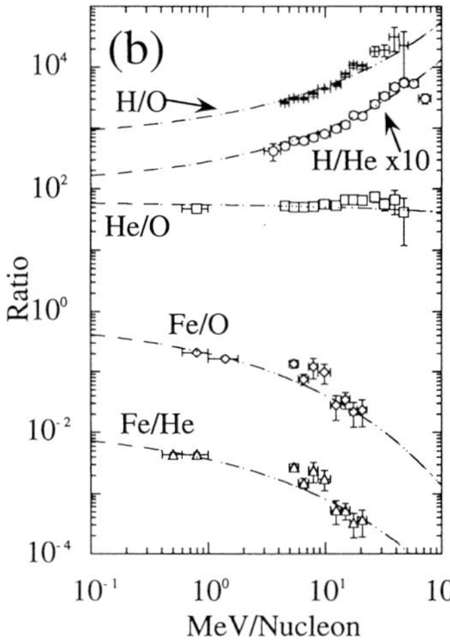

Figure 3. (a) Energy spectra of H, He, O, and Fe ions in a large solar particle event (Mazur et al., 1992). Solid curves are fits to a model of stochastic particle acceleration. (b) Abundance ratios as functions of energy from the same event. The systematic trends with energy suggest that the acceleration mechanism is more efficient for particles with lower energy and smaller mass to charge ratios.

1.2.2. *Energy Spectra of Impulsive Solar Flares*

Impulsive solar flares produce bursts of $\sim$100 keV electrons that stream along the magnetic field lines from the sun, as well as a charged particle population that is distinct from any other sample of matter in the heliosphere. These particle events, which typically last a few hours to a day, are enhanced in the ^{3}He isotope by a factor of 10^3–10^4 when compared to the composition of the solar corona. Compared to Oxygen, the elements Neon through Sulfur are also enhanced, but by a comparatively small factor of 3–5; Iron is more abundant by a factor of $\sim$10. Isotopic anomalies in ^{3}He-rich flares have been observed only once, during a single series of events in July 1992; these events showed modest enhancements in neutron-rich isotopes of Neon and Magnesium (Mason et al., 1994). Impulsive flares particles also exhibit high ionization states (Luhn et al., 1985) and are also associated with non-relativistic electrons (Reames et al., 1985), indicating that these particles are likely a direct sample of the material that is accelerated at the flare site in the corona.

In the initial work on this subject, the most successful mechanism that accounted for the high abundance of ^{3}He used a plasma resonance method to preferentially heat the ^{3}He (Fisk, 1978), which has a unique charge to mass ratio. After the preheating, a second stage acceleration by stochastic processes (Möbius et al., 1982) predicted rounded spectra at the lowest energies; this is in contrast to the observed spectra shown in Figure 5. The trend for the energy spectra to continue to rise even at energies near 20 keV nucl^{-1} suggests that the low energy particles contribute significantly to the total energy released in the flare. These are energies

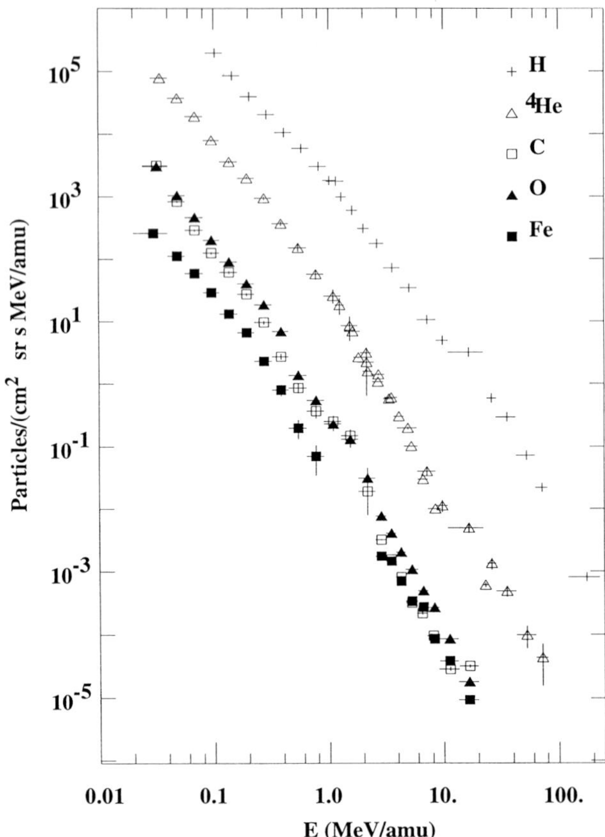

Figure 4. Gradual solar particle event spectra measured by the EPACT instrument on the WIND spacecraft down to $\sim$20 keV nucl^{-1} (Reames et al., 1997) during 21–25 October 1995. The low energy portions of the spectra were accumulated after an interplanetary shock passed the spacecraft on 22 October.

well below those probed by gamma-ray observations in flares (Ramaty et al., 1995). More recent theories using cascading plasma waves (Miller and Reames, 1995; Miller, 1998) are still being developed, but promise to account for the abundance anomalies in both elements and isotopes as well as the shape of the energy spectra.

2. Design Requirements

As the preceding discussion has shown, critical new information about the origin and evolution of the solar system and local galactic region can be obtained from high resolution elemental and isotopic composition of energetic particles from matter reservoirs on the solar corona, the solar wind, the local interstellar gas, and, perhaps, interplanetary dust. Elements over the range He to the Fe group are well suited for this study since they comprise the bulk of the material heavier than

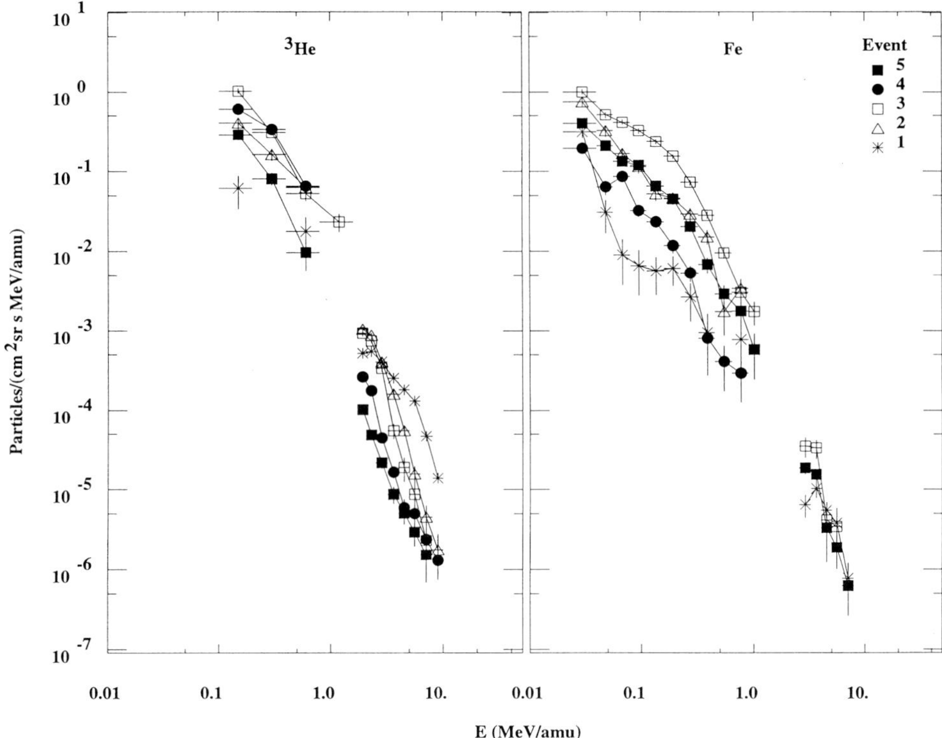

Figure 5. Impulsive flare energy spectra measured in five events by the EPACT instrument on the WIND spacecraft (Reames et al., 1997). Ions are accelerated beyond 10 MeV nucl^{-1}, and the spectra continue to rise at the lowest measured energies. Both the upper and lower observed energies, as well as the spectral forms, place tight constraints on the particle acceleration mechanism.

protons, and are the tracers of the primary nucleosynthetic processes that formed the solar system.

2.1. TELESCOPE

The Ultra Low Energy Isotope Spectrometer (ULEIS) for the ACE mission is one of a suite of high resolution spectrometers designed to give coordinated observations over the range of energies beginning at the solar wind ($\sim$1 keV nucl^{-1}) and extending up to moderate energy galactic cosmic rays (several hundred MeV nucl^{-1}). This broad energy range encompasses a huge dynamic range of flux levels (about 19 orders of magnitude) and time variations, and requires multiple techniques to span the entire range. ULEIS covers a range intermediate between the solar wind (SWIMS and SWICS) instruments (Gloeckler et al., 1998), and the solar isotope spectrometer (SIS) instrument (Stone et al., 1998b) – i.e., from about 45 keV nucl^{-1} to a few MeV nucl^{-1}. Exploratory measurements of ultra-heavy

ULEIS Telescope Cross Section

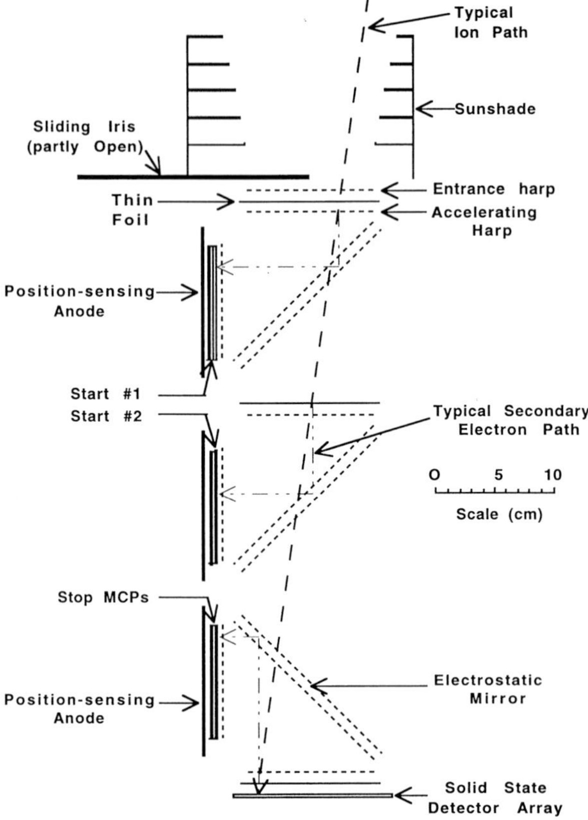

Figure 6. A cross sectional view of the ULEIS telescope.

species (mass range above Ni) will also be performed in a more limited energy range near 0.5 MeV nucl^{-1}.

In the ULEIS energy range the particle sources available for study include: solar energetic particles, particles accelerated by interplanetary shocks including Coronal Mass Ejection (CME) events, CIR ions, and anomalous cosmic rays. Fluxes in this range cover about 9 orders of magnitude from a large interplanetary shock to quiet periods when anomalous cosmic rays dominate. The isotopic abundances in these particle sources are expected to be close to solar or meteoritic values, which is to say that the main isotope peaks can be 50–100 times the abundance of the adjacent rarer isotopes. Making statistically significant determinations of the rare isotope abundances requires both high mass resolution ($\sigma_M \sim 0.2$ amu) and low background. In addition, substantial data rates are required since resolution of the rare isotope peaks is sufficiently difficult that it must be done on the ground. This means that for each rare isotope event detected, large numbers of a nearby common isotope must be telemetered, since the on board instrument data processing unit

TABLE I
ULEIS design goals

	Goal	Science driver
Geometric factor	1 cm^2 sr	adequate counting statistics for – small particle events – anomalous cosmic rays
Particle species measured	$2 \leq Z \leq 28$	– He–Ni covers most abundant material – main nucleosynthetic sequence
Elemental energy range	0.3–2.0 MeV/n	– large solar particle events – impulsive solar flares; – interplanetary shocks – corotating interaction regions
Mass resolution	$\sigma_m < 0.15$ amu ($Z = 6$) $\sigma_m < 0.5$ amu ($Z = 26$)	– resolve adjacent isotopes for C-Si – even isotopes for heavier nuclei
Event rate range, R	1/week $< R < 10^5$ s^{-1}	low background to cover – anomalous cosmic rays – high event rate in large solar particle events

is unable to separate adjacent isotopes for elements heavier than helium. Table I summarizes key design goals for ULEIS, and lists the primary scientific goals that lead to those requirements.

2.2. DATA PROCESSING UNIT

The ULEIS telemetry bit rate is limited to approximately 1 kb s^{-1}, which is insufficient to transmit detailed data on all particles triggering the telescope. In order to follow, e.g., anisotropies and the evolution of particle spectra during intense periods, particles need to be identified in broad mass and energy bins and counted in so-called 'matrix-rates'. The low triggering efficiency of the telescope for protons and helium is helpful in this regard; in addition, when counts become too high, there is a closable cover that is activated to shield the telescope aperture and thereby lower the triggering rate. In order to count all particles for binning the matrix rate processing must be able to handle up to $\sim$3000 counts s^{-1}.

Figure 7. The ULEIS instrument shown with the telescope on its mounting bracket, and connected to the analog electronics and DPU boxes.

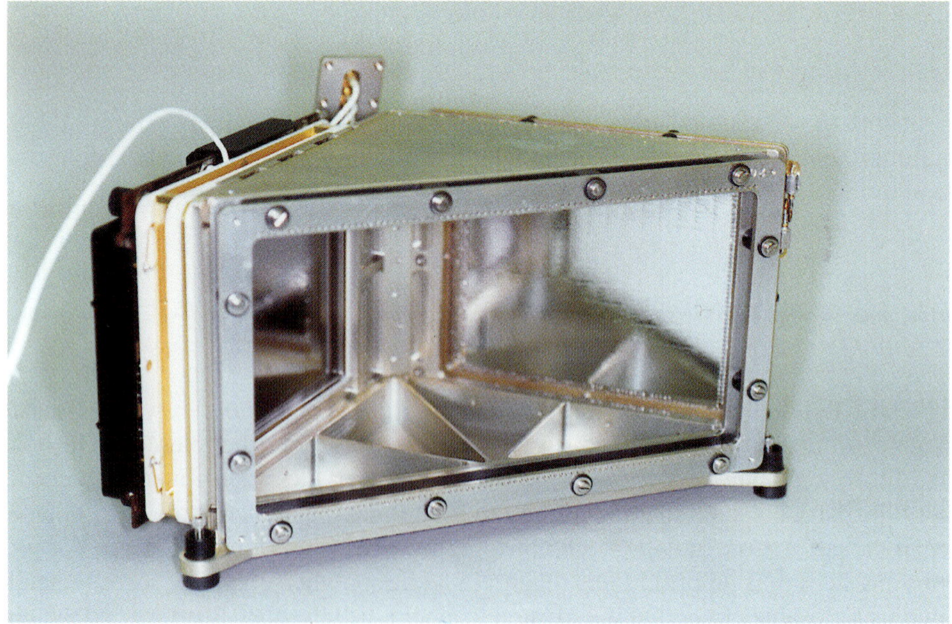

Figure 8. A single 'wedge' assembly consisting of thin foil, electrostatic mirrors, and microchannel plate assembly. The view is through the electrostatic mirror.

TABLE II

ULEIS telescope typical energy ranges

| | Energy ranges (keV nucl^{-1}) | | | | | |
| | $\sigma_m < 0.3$ amu | | $\sigma_m < 0.6$ amu | | $\sigma_m < 2$ amu | |
Species	Lo[a]	Hi[b]	Lo[a]	Hi[b]	Lo[a]	Hi[b]
^{4}He	90–2000	197–8100	65–2000	197–8100	55–2000	197–8100
^{12}C	120–770	154–13300	78–770	105–13300	42–770	105–13300
^{16}O	153–620	180–7250	90–620	105–10000	45–620	87–10000
^{28}Si	280–405	290–2300	146–405	153–5700	50–405	58–5700
^{56}Fe	–	–	–	270–2360	80–220	82–3015

[a] Small solid-state detectors (D1-D4).
[b] Large solid-state detectors (D5-D7).

3. Instrument Description

3.1. Operating Principles

A cross section of the ULEIS telescope is shown in Figure 6, and typical energy ranges are summarized in Table II. The telescope box is mounted on a bracket (see Figure 7) so that its view cone points in the sunward hemisphere at an angle of $60°$ to the ACE spin axis. The instrument is a time-of-flight mass spectrometer that identifies incident ion mass and energy by simultaneously measuring the time-of-flight, τ, and residual kinetic energy, E, of particles that enter the telescope acceptance cone and stop in one of the array of seven silicon solid-state detectors (SSDs) in the back of the telescope. The time-of-flight is redundantly determined by two start pulses, START-1 and START-2, and a STOP pulse from 'Z-stack' microchannel plate (MCP) assemblies that detect secondary electrons that are emitted from the thin foils when the ion passes through them. These secondary electrons are accelerated to ~ 1 kV and deflected onto the MCPs by electrostatic mirrors. The design of the secondary electron optics yields isochronous flight paths for all secondary electrons emitted normally from the foil or detector surface. The measured energy, $E = \frac{1}{2}mv^2$, and the velocity, $v = L/\tau$ (where L is the path length in the telescope), are combined to yield the mass of the ion,

$$m = 2E \left(\frac{\tau}{L}\right)^2 , \tag{1}$$

and the energy per nucleon, E/m, inside the telescope. The ion incident energy is obtained after correcting for the energy loss in the foils and the front contact of the solid-state detector. ULEIS measures only the mass of the incident particles, and so cannot distinguish isomers such as ^{40}Ca and ^{40}Ar. However, for isotopes lighter than Ni, there are only a few such isomers, and in all cases for solar-system abundances, one isotope has the dominant abundance (Anders and Ebihara, 1982), so this is not an important source of ambiguity in the data. The design of ULEIS evolves from consideration of a number of important scientific goals for particle measurements on the ACE mission. First is the need to measure a large number of isotopes over a broad energy range for solar and interplanetary particle populations. Second is the requirement to perform accurate measurements over a wide dynamic range of particle intensities, e.g., from intense solar particle events and strong shock events to small, ^{3}He-rich flares that are near the threshold of sensitivity for previous instruments. Third is the requirement to operate reliably over a multi-year lifetime. These considerations require an ion detector whose design achieves a low energy threshold, high mass resolution, and large geometrical factor, and that employs technologies of proven long-term performance on spacecraft. The low energy threshold is achieved by using the time-of-flight technique with a thin entrance foil on the telescope. The high mass resolution is attained by combining time-of-flight measurements of high accuracy (<300 ps FWHM) with a long ($\sim$50 cm) flight path whose length is determined accurately by measuring the (x, y) positions where the ion penetrates each of the three thin foils. Finally, the large geometrical factor is obtained by using large-area microchannel plates (8 cm $\times$ 10 cm) and an array of solid-state detectors with total area of $\sim$73 cm^2. ULEIS has a full instrument geometrical factor of $\sim$1.3 cm^2 sr with 100% duty cycle for all species.

3.2. TELESCOPE SCHEMATIC

This section describes details of the telescope components, starting at the top of Figure 6 and moving towards the bottom.

3.2.1. *Closable Cover*

Solar ultraviolet radiation may cause secondary electron emission from the telescope foil, thus increasing the background in the instrument. The ULEIS design avoids this by pointing the telescope at 60° to the spin axis and including a sunshade that prevents sunlight from striking the foil. Since the ACE spin axis is maintained pointing directly at the sun (Chiu et al., 1998), the ULEIS field of view scans a band in the sky perpendicular to the ecliptic. The ULEIS sunshade serves a secondary role by preventing particles incident from outside the instrument view cone from striking the front foil, thereby significantly lowering the peak START-1 count rate in intense solar flare or interplanetary shock events. However, in some events, the ULEIS front foil count rates could exceed $\sim 5 \times 10^6$ s^{-1}, thereby causing accidental triggering in the time-of-flight system and pulse pile up in the solid-state detectors.

To prevent this, a mechanical shutter ('sliding iris' in Figure 6) slides closed under on-board control depending on the START-1 MCP singles counting rate. The iris is controlled by a Clifton size 11 stepper motor, and has four discrete settings corresponding to 100%, 25%, 6%, and 1% of full opening, thereby making it possible for ULEIS to operate in the most intense events with excellent mass resolution. The four iris settings are sensed by Optek optical switches. If the iris is positioned at some other intermediate location, the optical switches provide no information on the location. However, in this case the cover can be reset by ground command, and its location determined by counting motor steps. The 6% and 1% settings of the iris are achieved using holes in the iris cover itself, and so are not dependent on precise mechanical location of the cover. The iris is tightly closed at launch to provide an acoustic seal; release post-launch is done with redundant miniature wire cutters and springs.

Even when the geometrical factor is reduced by the iris, in intense events ULEIS will record sufficient events to completely fill its allocated telemetry, and so statistical accuracy will not suffer. Changing of the iris position takes place only on boundaries of the instrument's internal collection cycle, in order to prevent data being taken in an intermediate state. An encoded set of light-emitting diodes on the iris provides positive feedback to the control circuitry that the desired setting has been reached. Two radioactive sources are located on the inside of the iris cover in order to permit in-flight calibration when the cover is closed: 1 μCi of ^{244}Cm (18 year half life; 5.80 and 5.76 MeV alphas), and 0.3 μCi of ^{148}Gd ($\sim$35 y half life; 3.2 MeV alpha).

Behind the sliding iris but in front of the entrance thin foil, there is a single entrance 'harp' assembly (see Section 3.2.2) held at ground potential, in order to electrostatically shield the entrance foil that is at a negative HV potential.

3.2.2. *Wedges*

The telescope's three timing signals (START-1, START-2, and STOP) each originate from modular assemblies called 'wedges' due to their triangular shape. Figure 8 shows a single START wedge assembly. The three 'active' surfaces of each wedge consist of a thin foil, an electrostatic mirror, and a microchannel plate assembly (see also Figure 6). In order to have the signal output of the microchannel plate assemblies at ground potential, the wedge is biased with the foil at negative HV (typically -4000 V). Electrons emitted from the inner surface of the foil are accelerated towards the accelerating harp, which is biased at about -3000 V. They then enter a uniform potential region bounded by the accelerating harp, inner electrostatic mirror harp, and a harp in front of the microchannel plates. The outer electrostatic mirror harp is at ground potential. Secondary electrons coast in the uniform potential region, entering the electrostatic mirror where they are reflected $90°$, and re-emerge moving towards the microchannel plates. After passing through another harp in front of the MCPs, they strike the front surface of the MCP. The MCP front surface is biased at about 50 V more positive than the uniform potential region, in order to reduce background.

Each harp is strung with $0.001''$ stainless steel wire with 1.0 mm spacing, for a net transparency of 97.5% for harps perpendicular to the telescope boresight, and 96.4% for the mirror harps at $45°$ to the telescope axis. The transparency of the 3 harps in each wedge assembly is thus $0.975 \times 0.964 \times 0.964 = 0.906$. The harp transparency is combined with foil transparency in cumulative fashion in Table III.

TABLE III

ULEIS telescope transparency budget

Element	Transparency	Cumulative transparency
Entrance harp	0.975	0.975
START-1 foil mesh	0.937	0.914
START wedge harps	0.906	0.828
START-2 foil mesh	0.937	0.776
START wedge harps	0.906	0.703
STOP wedge harps	0.906	0.637
STOP foil mesh	0.937	0.596
Total		0.596

Since multiple secondary electrons are emitted by particles heavier than He over the ULEIS energy range, the harps (including the one in front of the MCPs) have very low probability of stopping all the electrons emitted by passage of a single ion.

The design of the mirrors and mounting pieces was optimized to prevent fringe fields from distorting the flight paths of the secondary electrons. The spacing of acceleration regions between the thin foil and the accelerating harp, and in between the 2 harps in the electrostatic mirror is 1.0 cm, which is 10 times the harp wire spacing. The mirror assemblies used in ULEIS are based on smaller designs developed for use in low energy nuclear accelerators (Starzecki et al., 1982). These smaller systems have estimated intrinsic time dispersions of $\sim$115 ps for fission fragments (Starzecki et al., 1982). Timing performance data for the ULEIS wedges is presented in Section 4.2.

3.2.3. *Foils*

The ULEIS START-1 foil is a 2000 Å thick polyimide foil, coated on both sides with 300 Å of Al, supplied by Luxel Co., of Friday Harbor, Washington (Powell et al., 1997). The double coating prevents pinholes in the Al on one side of the foil from letting scattered light into the telescope chamber. The sunshade, along with the telescope bracket angle of $\sim$60° to the solar direction, ensure that direct sunlight never strikes the START-1 foil. Omnidirectional Lα backscattered UV is a potential concern with thin foils exposed to a large solid angle of sky. However, for typical interplanetary UV levels as measured on the Pioneer–Venus probe (Ajello, 1990) the ULEIS start foil is sufficiently opaque to limit counting rates due to this source to less than a few hundred per second (Postlaunch note: the actual START-1 count rates observed shortly after full bias was applied were about 2000 counts s^{-1}

TABLE IV

Material in front of ULEIS solid-state detectors

Element	Material	Thickness (Å)[a]	Thickness (μg cm^{-2})
START-1 foil outer Al layer	Al	325 ± 32	8.8 ± 0.9
START-1 foil substrate	polyimide	2075 ± 100	22.8 ± 1.1
START-1 foil inner Al layer	Al	300 ± 30	8.1 ± 0.8
START-2 foil substrate	polyimide	2175 ± 100	23.9 ± 1.1
START-2 foil Al layer	Al	300 ± 30	8.1 ± 0.8
STOP foil Al layer	polyimide	287 ± 28	7.7 ± 0.8
STOP foil substrate	Al	2185 ± 100	24.0 ± 1.1
Solid-State Detector front contact (average)	Al	2000 ± 1000	54.0 ± 27.0
Total			157.5 ± 33.5[b]

[a]Manufacturer's specifications.
[b]Comprised of polyimide and Al absorbers.

for the START-1 time-of-flight discriminator, and 5000 counts s^{-1} for the START-1 WSA electronics). The START-2 and STOP foils are 2000 Å thick polyimide, with a single 300 Å Al coating on the side of the foil facing into the wedge. A complete list of absorbing material in front of the solid-state detectors is given in Table IV.

The foil thickness is chosen to be the minimum consistent with sufficient mechanical strength to survive launch acoustics and vibration. All the foils are mounted on a double mesh combination consisting of a custom-etched Au plated Be–Cu mesh with wire thickness of 0.005″, and line spacings of 4.0 lines per inch (transparency 96%). On this mesh, a second, fine mesh (0.0015″ wires, 16 lines per inch) is mounted along with the foil (transparency 97.6%). The transparency of the two meshes is 0.937 (see Table III). The Be–Cu mesh is nearly mechanically decoupled from its mounting frame by means of thin attachment prongs that are highly compliant. This mesh then 'floats' within a restraining frame so that flexing of the wedge assembly is not transmitted directly to the foil mesh. The wedges themselves are mechanically decoupled from the rest of the telescope via attachment points solely at the base of the telescope box. To provide protection against severe acoustic levels at launch, the 'sliding iris' cover is closed securely during launch, and opened with redundant squibs at a later time. The sliding iris cover itself, as well as the walls of the telescope box, were thickened to provide additional acoustic shielding. To provide for rapid venting of the telescope chamber during launch ascent, without at the same time allowing significant acoustic energy into the telescope chamber, the telescope chamber is vented through a 'muffler' tuned to dampen at frequencies near the peak of the acoustic spectrum.

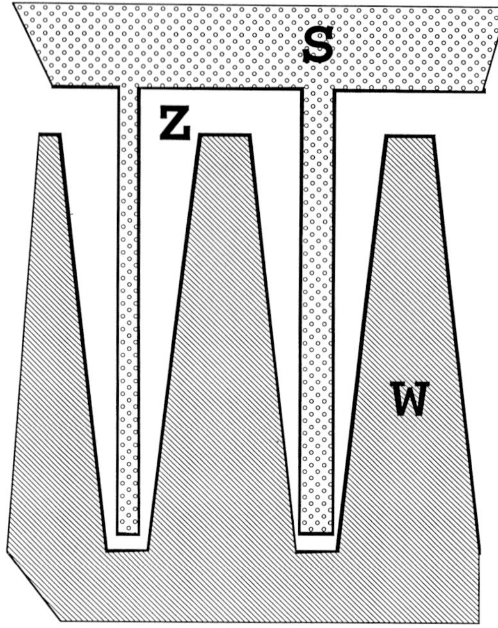

Figure 9. 'Wedge and strip' anode schematic.

3.2.4. *MCP Assemblies*

Each MCP assembly consists of three microchannel plates in a 'Z-stack' configuration, which was chosen to optimize signal FWHM and timing (Siegmund et al., 1986b). The plate size is 80×100 mm, with 1 mm thickness, 25 μm pore size, and 19° bias angle. The nominal gain is 5×10^6 at 3000 V/set, with a leakage current range of 10–25 μA/plate at 1000 V bias. Each Z-stack was procured (from Galileo Electro-Optical, Inc.) as a matched set, and the three plates were mounted directly in contact with each other, without interplate spacers or contacts. The electron cloud output from the last plate in each stack is collected by the wedge-and-strip anode, which was placed 1.0 cm in back of the plate in order to allow the electron cloud to spread.

Achieving uniformity of response over the entire area of these large MCPs is important for high performance operation of the system. The combination of large area and thin (1 mm) plate diameter can lead to large differences in response due to deviations in the plates from strict flatness. Since the plates are held together at the edges, deviations from flatness tend to occur towards the center of the plates. Where the plates are no longer touching each other, the gain is larger since the electron cloud can spread out between adjacent plates, thus triggering more channels deeper in the stack than would be the case if the plates were rigidly held together. This leads to an apparent decrease of gain (by a factor of $\sim$5) in the outer portion of each MCP stack. For the ULEIS flight instrument, the orientations of the individual MCPs in each Z-stack were selected in order to achieve a nearly uniform response to a UV source (by Siegmund Scientific, Berkeley, CA). After launch, the plates may change their 'flatness' over time due to outgassing of water, and in response to stresses within the stacks themselves. This may degrade the timing performance of the system, although the constant-fraction discriminators in the time-of-flight electronics should help minimize these effects.

3.2.5. *Wedge and Strip Anode*

Electrons from the MCPs fall on three collector wedge-and-strip anodes (WSA) (Anger, 1966) supplied by Siegmund Scientific, Berkeley, CA. Figure 9 shows a sample three collector WSA (magnified) pattern where the regions marked W, S, and Z ('wedge, strip, and zigzag') are separate conductive regions (Lapington and Schwarz, 1986). For the ULEIS instrument, the anodes are applied with a thick film technique on a 1 mm thick alumina substrate with a pattern consisting of 30 cells covering an area 8.5×10.4 cm, with gaps between the anode elements of $0.002''$. For electron cloud sizes larger than the individual scale sizes shown, the anodes collect a charge Q_W, Q_S, and Q_Z that can be analyzed to give the location of the centroid of the charge cloud. The pulse sizes on all three electrodes are analyzed and telemetered with each event. A rough, unscaled (x', y') location is then obtained from the equations:

$$x' = \frac{Q_S - X_{\text{talk}} Q_Z}{Q_W + Q_S + Q_Z} \tag{2}$$

and,

$$y' = \frac{Q_W - X_{\text{talk}} Q_Z}{Q_W + Q_S + Q_Z} , \tag{3}$$

where X_{talk} is a factor that corrects for some of the cross-talk distortions that arise due to capacitive coupling (few hundred pF typical) between the elements of the anode (Siegmund et al., 1986a). Calibrated locations on the anode are then obtained from Equations (2) and (3) by applying an offset from 0, a scaling factor, and then a final re-mapping that removes the residual distortions in the image. The (x, y) coordinate of the charge cloud center on the MCP has a one-to-one correspondence with the location where the ion penetrated the foils due to the simple reflective property of the electrostatic mirrors (see Figure 6). The overall accuracy of this technique locates the centroid of impact of the charge cloud on the anode to within a few mm, which exceeds the 1 cm accuracy needed to make path length corrections required for isotopic resolution (see Section 4.3.1.1).

3.2.6. *Solid-State Detectors*

The 8×10 cm active area at the back of the telescope is too large to be covered by a single, low noise solid-state detector. In order to avoid dead areas between the active elements, a custom layout was used, as shown in Figure 10. In order to cover the lower energy range of the instrument with very low noise detectors,

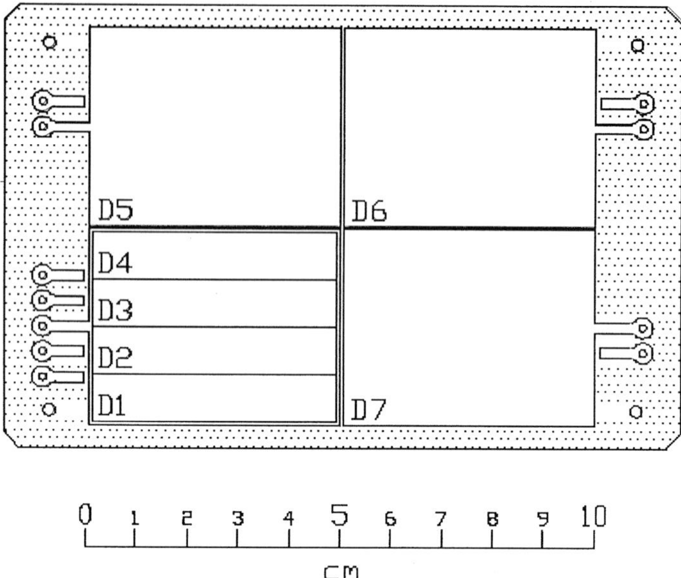

Figure 10. ULEIS solid-state detector array.

4 detector elements (D1, D2, D3, D4) in the array have 9.4 mm × 48 mm active areas, and three large elements (D5, D6, D7) have active areas of 38 × 48 mm. This arrangement covers over 90% of the nominal 80 cm² active area as determined by the MCP size. The silicon thickness for all elements is 500 ± 30 μm, and the metalizing on the active surface is 2000 ± 1000 Å Al. At nominal operating bias of 50 V, the alpha-peak FWHM is 25 keV for the small elements, and 45 keV for the large elements (typical). Leakage current at 50 V bias is typically 20 nA for the small elements, and 50 nA for the large elements. The detectors are mounted on a 0.080″ ceramic substrate. Signal leads were kept a minimum length by routing each detector signal through a pin that passes through the substrate and plugs directly into a circuit board that contains the 7 pre-amplifiers for the system. The detector assembly was provided by Micron Semiconductor, Inc., of Longwood, Florida.

3.3. FUNCTIONAL BLOCK DIAGRAMS

3.3.1. *Telescope Electronics*

Figure 11 is a block diagram of the ULEIS telescope box electronics showing the primary measurements and data flow. The selection of electronics located in the ULEIS telescope box is a compromise between the requirement for low-noise connections to the sensor, vs a requirement for low-temperature ($\sim$0 °C) operation of the solid-state detectors which limits the amount of power dissipation that can be tolerated in the telescope box itself. Relatively high power electronics such as the fast time-of-flight circuitry, are therefore located in a separate box (see below).

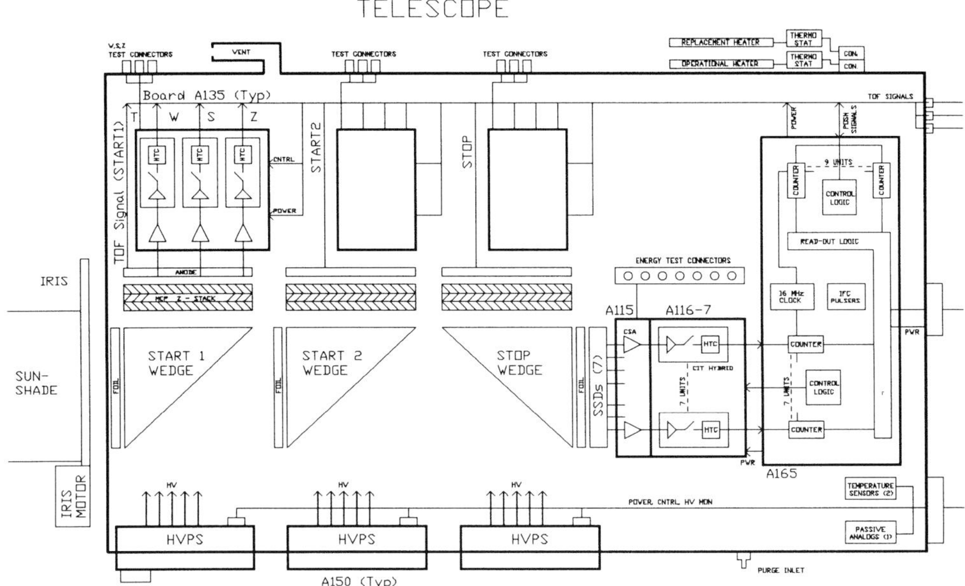

Figure 11. ULEIS telescope box block diagram.

Time-of-flight signals are routed directly from the telescope to the analog electronics box, (see below) where signals are examined to see if a valid stop has occurred (START-1 followed by a STOP within about 300 ns). If a valid stop occurs, then a signal is sent back to the telescope box to determine if one of the solid-state detectors has triggered in coincidence with the valid stop. If a coincidence occurs between the valid stop and a solid-state detector trigger, an event is recorded, consisting of the following information: energy deposit in the solid-state detector that was triggered; time-of-flight between START-1 and STOP ($\equiv$TOF-1), time-of-flight between START-2 and STOP ($\equiv$TOF-2), wedge ($\equiv$W), strip ($\equiv$S), and zigzag ($\equiv$Z) pulse sizes from the START-1, START-2, and STOP wedges, and status bits indicating which detector was triggered. These data are relayed to the data processing unit (DPU) for further processing (see below). Each event whose detailed pulse heights are telemetered is formatted into a 14-word event (1 word = 12 bits) whose contents are given in Table V. Four such events are telemetered each second.

3.3.1.1. *Wedge and strip anodes.* The three signals from the wedge, strip, and zigzag anodes are directed to a circuit board mounted directly in back of the anode. On this board, the three signals go through a single stage of wide dynamic range pre-amplification, and are then sent to hybrid circuits where they are digitized. The digital outputs are then sent to the analog electronics box (see below) for further processing. The hybrids used in ULEIS are a modification of the hybrids used in the ACE SIS and CRIS instruments, and were developed by the California

TABLE V

ULEIS telemetered PHA event contents

Word	Contents (12 bit words)
0	START-1 Wedge PHA
1	START-1 Strip PHA
2	START-1 Zigzag PHA
3	START-2 Wedge PHA
4	START-2 Strip PHA
5	START-2 Zigzag PHA
6	STOP Wedge PHA
7	STOP Strip PHA
8	STOP Zigzag PHA
9	SSD Energy
10	Time-of-flight 1 (START-1 - STOP)
11	Time-of-flight 2 (START-2 - STOP)
12	See below
13	See below

Normal Mode:

Word 12:	HAZ	LA1	LA0	SA1	SA0	D7	D6	D5	D4	D3	D2	D1
Word 13:	zero	zero	Event Box #						CO	ES	TOF2	TOF1
	MSB											LSB

Codes:

HAZ = Hazard flag.

LAn = Large solid-state detector address: 00 = D5, 01 = D6, 10 = D7, 11 = error.

SAn = Small solid-state detector address: 00 = D1, 01 = D2, 10 = D3, 11 = D4.

Dn = Discriminator number that fired.

Event box # = matrix rate box number assigned to this PHA event (6 bits).

CO = Calibrate/normal mode flag (1 = calibrate; 0 = normal).

ES = Energy system ID: 0 = large; 1 = small.

TOFn = which TOF system fired.

Calibrate Mode:

Word 12:	CE11	CE10	CE9	CE8	CE7	CE6	CE5	CE4	CE3	CE2	CE1	CE0
Word 13:	SSD ID			ES	CAL STEP		CM	CO	zero	TOF2	TOF1
	MSB											LSB

CE = Energy calibration step number.

SSD ID = Solid-state detector ID: 0 = D1, 1 = D2,... 6 = D7 (3bits).

CAL STEP = Time-of-flight calibration step number (3 bits).

CM = Calibrate mode speed (1 = slow).

TABLE VI
ULEIS energy and timing nominal specifications

System	Nominal specifications
Low threshold solid-state detector system	Detectors: 4×4.5 cm^2 500 μm thick detectors detector alpha peak = 25 keV FWHM Electronics: discriminator threshold 55 keV amplifier total energy range 55 keV–8 MeV analyzer: 12 bit height-to-time converter, 1 per detector
High threshold solid-state detector system	Detectors: 3×18.2 cm^2 500-μ thick detectors detector alpha peak = 45 keV FWHM Electronics: discriminator threshold: 500 keV amplifier total energy range 0.5–160 MeV analyzer: 12 bit height-to-time converter, 1 per detector
Time-of-flight system	Ranges: TOF-1: 5–335 ns; TOF-2: 5–210 ns discriminator: constant fraction analyzer: 12 bit successive approximation ADC, 1 per channel

Institute of Technology ACE group (Stone et al., 1998a, 1998b) Control circuitry for the hybrids resides in the back of the telescope box with the solid-state detector electronics.

3.3.1.2. *Solid-state detectors.*   Solid-state detector signals are processed in a stack of 4 boards mounted directly in back of the detector array. The board closest to the array has preamplifiers for each of the 7 detectors. These signals are then processed in hybrid circuits mounted on the next two boards in the stack. The fourth board contains ACTEL CMOS field programmable gate array (FPGA) control circuitry for the solid-state detector hybrids, as well as for the hybrids on each of the wedge and strip anode circuit boards. The nominal thresholds and gains for the solid-state detectors are summarized in Table VI. The hybrids are Caltech hybrids similar to those used for the wedge and strip anodes.

3.3.1.3. *High voltage bias and miscellaneous.*   The telescope box also contains three high voltage power supplies that are used to bias the foils, harps, and MCP assemblies (see Figure 11). The supplies are separately commandable in order to provide capability to bias microchannel Z-stacks of different initial gains, and to

TABLE VII

Singles event rates

Rate #[a,b]	Name	Logic[c]
0	D1	D1
1	D2	D2
2	D3	D3
3	D4	D4
4	D5	D5
5	D6	D6
6	D7	D7
7	START 1	START 1
8	START 2	START 2
9	STOP	STOP
10	Valid Stop 1[d]	VS1
11	Valid Stop 2[d]	VS2
12	Event	$VS * E * / BUSY$[e]
13	START 1 Wedge	START-1 $(W + S + Z)$
14	START 2 Wedge	START-2 $(W + S + Z)$
15	STOP Wedge	STOP $(W + S + Z)$

[a]Each rate is divided into 8 sectors per spin.
[b]Accumulation time: 24 s (2 spins).
[c]Discriminator output name shown; $*$ = logical AND; $+$ = logical OR; / = logical NOT.
[d]Valid stop: START-n followed by a STOP signal within 300 ns.
[e]E = Solid-state detector trigger (any of D1 … D7).

allow for recovery of gain late in the mission in the event that prolonged radiation exposure decreases the output of the plates. To provide information on operating temperatures, four temperature transducers are placed at critical locations near the solid-state detector array, near the front of the telescope, in the solid-state detector electronics, and in the WSZ circuitry.

3.3.2. *Analog and DPU Electronics*

The remainder of the instrument electronics are housed in two separate boxes (see Figure 7). The analog electronics box contains the time-of-flight circuits, some control circuits for the telescope logic, and the low voltage power supply. Output from the analog electronics box is sent to the data processing unit (DPU), which formats the data for transfer to the spacecraft telemetry system. The DPU also processes commands and processes telescope pulse-height data for the matrix rates.

TABLE VIII

Example of ULEIS matrix rate bins

Energy nucl^{-1}	^{1}H	^{3}He	^{4}He	^{12}C	^{16}O	^{28}Si	^{56}Fe
20							20
28							
40				40	40	40	40
57		57	57				
80	75		80	80	80	80	80
113			113				
160	160	160	160	160	160	160	160
226			226				
320			320	320	320	320	320
453	453	453	453				
640			640	640	640	640	640
905			905				
1280	1280	1280	1280	1280	1280	1280	1280
1810			1810				
2560			2560	2560	2560	2560	2560
3620	3620	3620	3620				
5120			5120	5120	5120	5120	
7241			7241				
10240		10240	10240		10240		
13000	13000			13000			
Readout frequency (s)	12	12	12	24	24	24	24
Priority	2	3	3	4	4	4	5

Priorities rotate, so each species has highest priority depending on recent telemetry history; 'background' box is priority 1.

3.4. COUNT RATES

3.4.1. *Singles*

Counting rates of individual detector elements and various logic combinations are collected in 24-bit accumulators, and sectored into 8 sectors per spin. The rates are read out once for every two spins of the spacecraft (1 spin = 12 s). Table VII lists the singles rate definitions. All rates are compressed in the DPU before being telemetered.

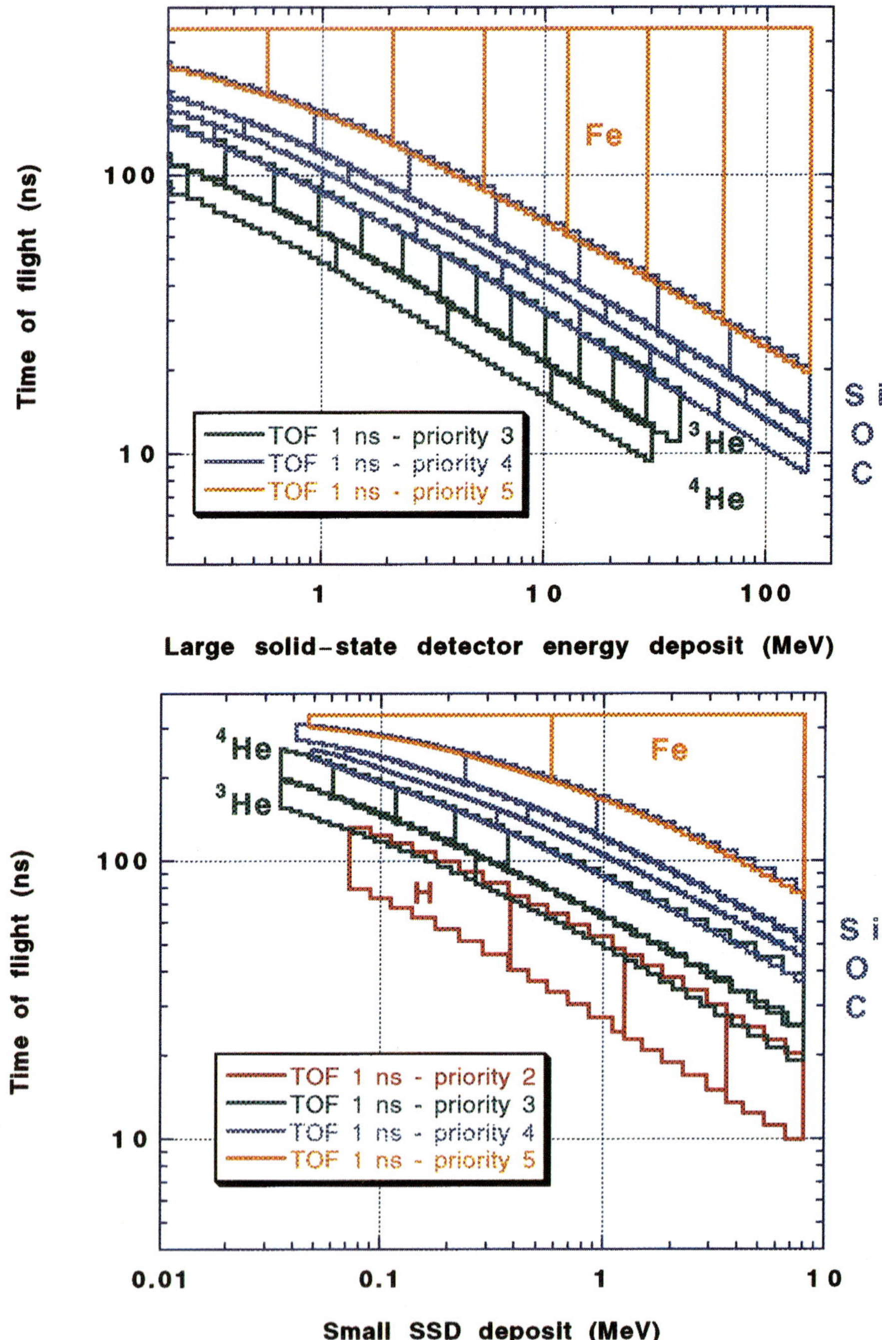

Figure 12. Example of ULEIS 'matrix rate' boxes in the time-of-flight vs energy matrices for both small and large solid-state detector systems (see also Table 7.)

3.4.2. *Matrix Rates*

Only a small fraction of the events triggering the telescope can be telemetered with complete pulse-height information. Therefore a second channel is used, consisting of matrix rates that assign individual events to species and energy boxes based on on-board processing of their pulse heights in the time-of-flight vs energy matrices. This processing takes place in the DPU. Table VIII shows an example of the types of coverage possible with this system, and Figure 12 shows locations of boxes in the time-of-flight versus energy matrices. The major species H, He, C, O, Si, and Fe each has assigned boxes, along with ^{3}He, which is important to monitor in small impulsive solar particle events. In the table, the entries show the boundaries of the boxes (start or stop energy per nucleon). The number of points in each spectrum varies from species to species. There are 34 matrix rates read out every spin (12 s), and 42 rates read out every 2 spins (24 s). In the example shown here, H, ^{3}He, and ^{4}He are read out every spin, and heavier species are read out every two spins.

The matrix rate boxes are defined with up-loadable tables, and thus can be re-configured during the mission. Because the trimming and offsets for each of the 4 small solid-state detectors and 3 large solid-state detectors may differ from each other, the box definitions (in analyzer channel numbers) are given separately for each detector. This makes it possible also to recover from any changes in detector response during the mission. The matrix box definitions also include priority assignments for use in the algorithm that allocates telemetry to PHA events.

3.5. RESOURCE SUMMARY

Overall resources used by ULEIS are summarized in Table IX.

4. Performance

4.1. GEOMETRY FACTOR – TIME-OF-FLIGHT PATH LENGTH

Table X summarizes the geometry factors for ULEIS, taking account of the transmissions of all high voltage harps and foil meshes, as well as the closable iris settings. Detector areas for the 7 solid-state detector elements are in Table VI. The flight paths for particles entering the telescope parallel to the axis are:
 – START-1 foil to STOP foil: 50.0 ± 0.1 cm.
 – START-2 foil to STOP foil: 32.6 ± 0.1 cm.

The uncertainties in the path length are due to the compliant mounting of the foil substrates (see Section 3.2.3).

TABLE IX

ULEIS resources

Mass[a,b]		
telescope and sunshade	11.5 kg	
analog electronics box	4.0	
data processing unit box	1.9	
interbox harnesses	1.0	
Total mass	18.4 kg	
Power	**Nominal**	**Peak**
telescope	2.2W	2.3
analog electronics box	10.1	10.9
DPU	2.5	3.4
operational heater	2.6	2.6
survival heater		4.6[c]
iris motor	0.0	2.0
Total power	17.4 W	21.2 W
Data rate	1000 bps	

[a]Telescope mounting bracket (2.7 kg) not included in mass.
[b]Thermal blankets not included.
[c]Survival heater is not included in peak power total since is it not operated when the instrument is powered.

4.2. TIMING ACCURACY

One key contribution to the isotopic resolution of ULEIS is the time-of-flight system. ULEIS uses a constant-fraction discrimination design supplied by Siegmund Scientific, Berkeley, CA, in order to achieve a timing dispersion of <300 ps FWHM. The constant-fraction discrimination minimizes the effects of time-of-flight walk, wherein start and stop timing signals of different amplitude produce amplitude-dependent times-of-flight. Walk is the major potential source of dispersion in the time-of-flight measurement, since the statistical nature of the secondary electron amplification in the MCPs produces broad ranges of START-1, START-2, and STOP timing signals, even for identical secondary electron yields from each of the foils. Section 4.3.1.2 discusses examples of the ULEIS time-of-flight system response with accelerator measurements.

TABLE X

ULEIS geometry factors

Detector	Individual geometry factor (cm^2 sr)			
Iris setting:	100%	25%	6%	1%
D1	0.078	0.019	0.0058	0.0010
D2	0.079	0.019	0.0058	0.0010
D3	0.079	0.019	0.0059	0.0010
D4	0.079	0.019	0.0059	0.0010
D5	0.318	0.077	0.0236	0.0040
D6	0.318	0.077	0.0236	0.0040
D7	0.318	0.077	0.0236	0.0040
Small SSD total	0.314	0.076	0.0234	0.0040
Large SSD total	0.954	0.231	0.0709	0.0121
Total	1.268	0.307	0.0942	0.0161

Note: Figures include loss of sensitivity due to transmission of harps and foil meshes (see Table III).

4.3. MASS RESOLUTION: ENERGY AND MASS DEPENDENCIES

The mass resolution of ULEIS is dependent on particle species and energy. Using Equation (1) to obtain the mass of an ion, we can determine the uncertainty in the measurement of its mass by

$$\left(\frac{\sigma_m}{m}\right)^2 = \left(\frac{\sigma_E}{E}\right)^2 + \left(\frac{2\sigma_\tau}{\tau}\right)^2 + \left(\frac{2\sigma_L}{L}\right)^2 , \tag{4}$$

where the terms on the right hand side of Equation (4) are the contributions from the uncertainties in the measured energy, time-of-flight, and path length, respectively. Figure 13 shows the dependence of each term on the right hand side of Equation (4) on the particle's incident energy using a calculation of the energy losses in the 3 ULEIS foils and nominal dispersions of the energy, time-of-flight, and position systems, and assuming $m = 16$ (Oxygen). At the instrument threshold near $\sim$45 keV nucl^{-1}, the uncertainty in the energy measurement dominates the mass dispersion since the noise in the energy system is a considerable fraction of the deposited energy (Ipavich et al., 1978; Zabel et al., 1980; Galvin, 1982). As the energy of the ions increases, the energy dispersion of the solid-state detectors remains about the same, and so the σ_E/E ratio and σ_m/m ratio decrease as shown in the figure. In contrast, the timing dispersion term $2\sigma_\tau/\tau$ increases with energy, since the time-of-flight becomes smaller. The energy and timing contributions to the dispersion σ_m becomes comparable near 0.6 MeV nucl^{-1}. As the ion energy

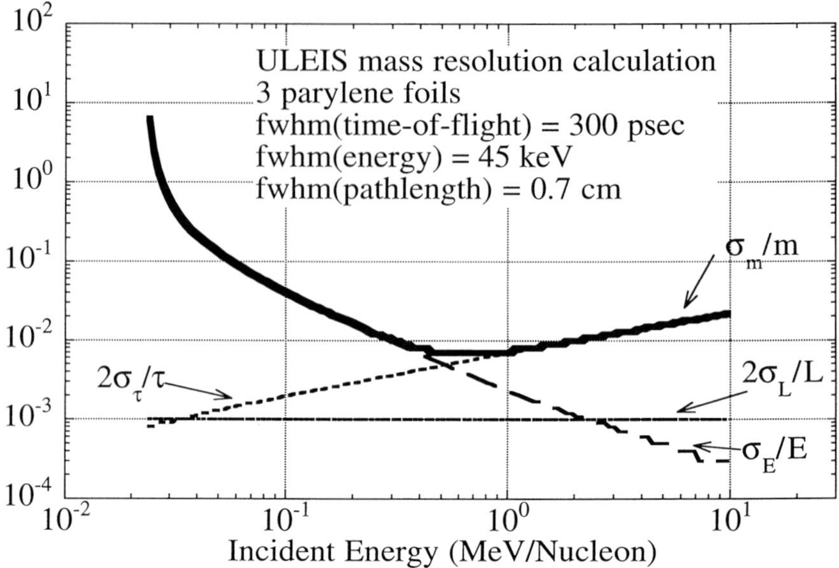

Figure 13. Calculated relative dispersions for ^{16}O in position, energy, and time-of-flight measurements for ULEIS versus incident particle energy.

increases further, the timing dispersion term continues to rise (dominating the mass dispersion) since for faster particles the measured time-of-flight τ is smaller. The path length dispersion term σ_L/L is much smaller than the other two (see Section 4.3.1.1). Thus, the three terms in Equation (4) combine to yield the σ_m/m curve with a characteristic minimum somewhat below 1 MeV nucl^{-1}. Examples of these dispersions using accelerator data are discussed below in Section 4.3.1.

The energy dependence of the energy and time-of-flight dispersions yields a minimum in the mass resolution near 1 MeV nucl^{-1} (see Figure 13). A given mass resolution therefore applies over a finite energy range. Table II lists sample energy ranges and reflects the dependence of σ_m on particle mass.

4.3.1. *Instrument Resolution: Examples*

We have measured the response of ULEIS to heavy ions at Brookhaven National Laboratory's Tandem van de Graaff and Lawrence Berkeley Laboratory's 88-inch cyclotron. Routine tests of ULEIS in vacuum with a $\sim$1 MeV nucl^{-1} ^{4}He source verify the instrument's basic functions, but the larger amplitude timing signals from heavy ions at large accelerators are a more sensitive measure of the instrument's mass resolution.

4.3.1.1. *Position system.*

Figure 14 shows the measured size of a 2 mm diameter beam of 1.85 MeV nucl^{-1} ^{40}Ar at the START-1, START-2, and STOP foils. The x-direction is along the 10-cm dimension of the MCPs. The particle beam was at normal incidence to the foils and collimated with a 2 mm diameter aperture

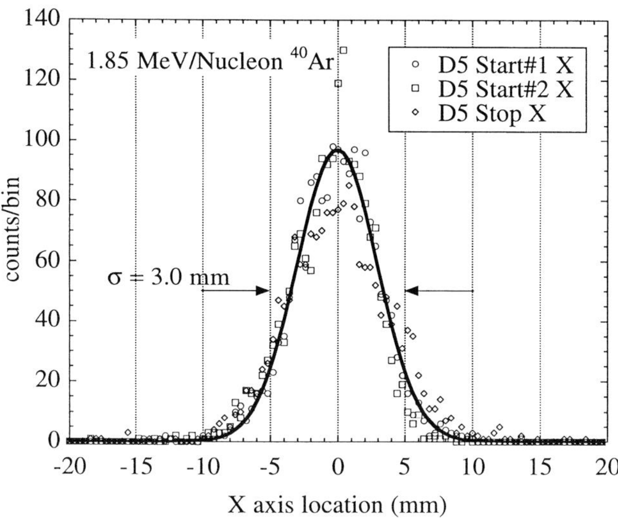

Figure 14. Position resolution for a normal incidence beam of 1.85 MeV nucl^{-1} ^{40}Ar measured at the 3 ULEIS foils.

upstream of ULEIS. The three position measurements have standard deviations of ~3 mm, due to the cosine distribution of the secondary electrons emitted from the foils (Meckbach, 1976). Figure 14 also shows that, for this beam, there is little coulomb scattering in the ULEIS foils, since the beam diameter is virtually unchanged for the three foils. Propagated through the instrument from START-1 to STOP, an uncertainty of 3 mm in the calculated positions yields a typical uncertainty in the path length of $\sigma_L/L \sim 0.0006$. As shown in Figure 13, this is much less than other sources of dispersion in the instrument discussed below.

4.3.1.2. *Time-of-flight system.* Figure 15 is a histogram of the measured times-of-flight of 1.84 MeV nucl^{-1} ^{15}N measured between START-1 and STOP. The standard deviation is $\sigma_\tau = 114$ ps, ~12% better than the design goal of 130 ps. Ions at this energy have small losses and straggling in the 3 foils compared to the total ion energy, so the measured 114 ps uncertainty is a result of the dispersions in the mirror and wedge assemblies and any residual walk in the time-of-flight electronics. Measured with ULEIS, heavy nuclei near 1 MeV nucl^{-1} typically have $\sigma_\tau < 130$ ps. The time-of-flight for 1.84 MeV nucl^{-1} ^{15}N is 26.6 ns between START-1 and STOP, so that $\sigma_\tau/\tau = 0.0045$.

The overall timing goal of <130 ps is very challenging, and requires excellent stability from the electronics and the MCP assemblies themselves. Since ULEIS particle measurements involve, e.g., interplanetary shocks or solar particle events that last only a few days, small drifts in the system over very long periods do not cause a problem with isotope identification during an event. Since ACE is spinning, and the instrument deck is continuously in sunlight, temperature variations

                                    G. M. MASON ET AL.

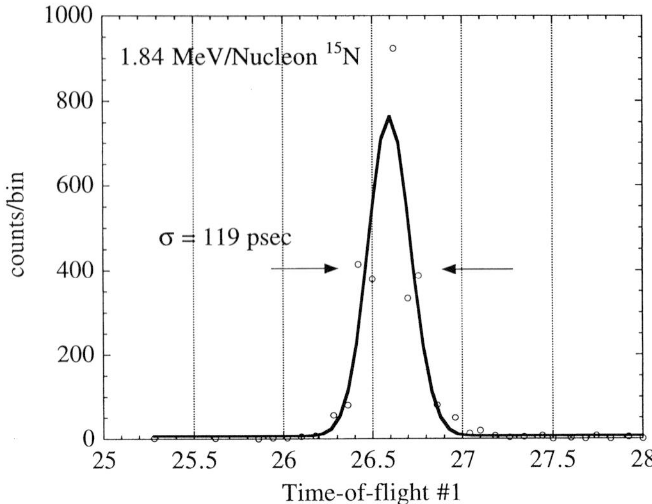

Figure 15. Time-of-flight distribution for mono-energetic ^{15}N measured between the START-1 and STOP foils.

over periods of a week should be very small. The most critical aspect of meeting the timing accuracy is due to the nonuniform response of the MCP stacks (see Section 3.2.4). Since the instrument pre-launch calibration only covered a few ion species and energies such as shown in Figure 15, it remains to be seen whether this timing resolution will be achieved over the entire sensitive area of the solid-state detectors and over the whole range of species and energies. If the timing dispersion is > 130 ps, the mass resolution degrades incrementally, with the effect being most important at the high energy end of the instrument response window (see Figure 13 and Table II) and for higher mass isotopes.

4.3.1.3. *Energy system.* To determine if the ULEIS solid-state detector system is adequate for the mass resolution shown in Figure 13, we can compute the standard deviation of the energy measurements of the solid-state detector system for various 1–2 MeV nucl^{-1} heavy ion beams. The measured widths include effects of detector resolution and electronics noise; losses and straggling in the foils and pulse-height defect are small at these energies. For example, 2 MeV nucl^{-1} ^{16}O deposits 32.0 MeV in the solid-state detector with a σ_E of 0.160 MeV. Thus, $\sigma_E/E = 0.005$ and if all the uncertainty in mass was due to the energy system dispersion, then $\sigma_m/m = 0.005$ and $\sigma_m = 0.08$ amu at mass 16. This is a factor of 2 less than the $\sigma_m = 0.16$ amu design goal at mass 16.

4.3.1.4. *Mass resolution.* Figure 16 shows the mass resolution of ULEIS for several monoenergetic species near 1 MeV nucl^{-1}; masses were calculated directly from Equation (1). In applying Equation (1) we neglected the path length dispersion since the beam was parallel to the telescope axis and scattering is minimal

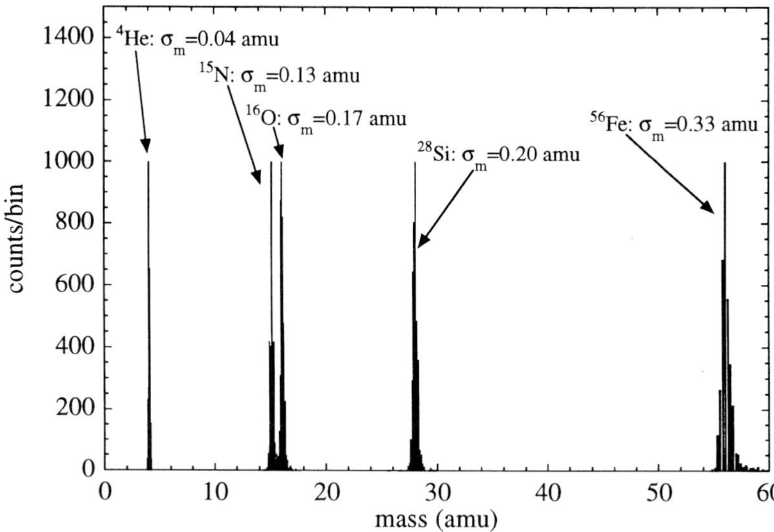

Figure 16. ULEIS mass distributions for ^{4}He and several heavy ion beams. Incident particle energies are all between 1 and 2 MeV nucl^{-1}.

at these energies. The mass histograms in Figure 16 also include the effects of dispersion in the energy system since they include all the measured energies for a given incident beam. Figure 17 compares the standard deviations of these mass distributions with the calculated response of ULEIS. In each case, the measured resolution is within $\sim$20% of that expected from Equation (4). Although there is only one calibration point per species shown in Figure 17 they are close to the energies where the instrument's mass dispersion is at its minimum value, and is dominated by the timing resolution.

4.4. BACKGROUND REJECTION

ULEIS has a naturally low background due to the coincidence requirements between (1) a start and a stop timing signal, producing a valid stop, and (2) between a valid stop and the trigger of a solid-state detector. ULEIS employs two times-of-flight, START-1 to STOP (TOF-1) and START-2 to STOP (TOF-2) in order to further reduce the contribution of possible background sources such as time-of-flight ringing and accidental coincidences. The two times-of-flight allow the rejection of events in the ground-based processing that do not have consistent times-of-flight.

Figure 18(a) shows the TOF-1/TOF-2 ratio as a function of TOF-1 for 1.84 MeV nucl^{-1} ^{15}N. Each point is an individual ion that triggered both times-of-flight systems and a solid-state detector; the solid horizontal lines show $\pm$5% limits on the ratio of times-of-flight calculated from the distances between the foils. Figure 18(b) shows the effect of the time-of-flight constraint on the TOF-1 distribution. The bulk

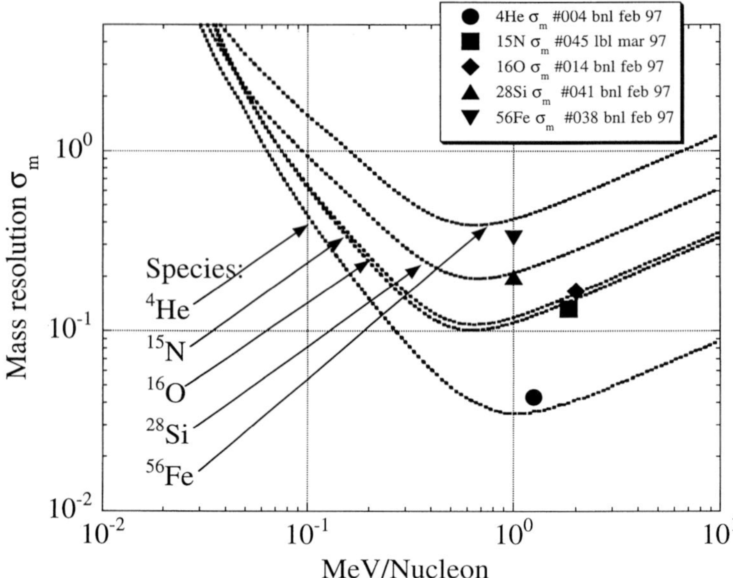

Figure 17. Comparison of measurements of the ULEIS mass resolution with calculations based on Equation (4).

of the events satisfy the constraint, while single-event outliers do not; these outliers would subsequently be rejected in the ground-based analysis. The two-parameter measurement is important in studies of rare isotopes such as ^{18}O whose abundance is only ~0.2% of its neighbor ^{16}O.

In Figure 18(a) the points outside the valid limits form a 'Y' shaped pattern. The three branches of the 'Y' correspond to three different types of events. First, the upper right branch of the 'Y' is due to events whose TOF-1 is about 4 nanoseconds too long; the relatively long time offset, and the clustering of points around a single value indicates that these events are associated with ringing somewhere in the electronics that causes a time-of-flight that is too long. The vertical branch of the 'Y' is caused by the same effect except for the TOF-2 signal: this makes the TOF-1/TOF-2 ratio too small, and is uncorrelated with the TOF-1 value as seen. Thirdly, there is the upper left branch of the 'Y', which has values of TOF-1 that are too low, and these occur in correlation with values of TOF-2 that are too low, but that are not in the expected ratio of TOF-1/TOF-2. The mechanisms leading to these types of background events are not understood. Note that the background events in Figure 18(a) are less than 1% of the main peak (see Figure 18(b)).

4.5. EFFICIENCY

Secondary electron yields for ions passing through thin foils roughly follow the dE/dx of the particles, and are therefore both species and energy dependent. For H and He, the forward secondary electron yield (appropriate for the two START

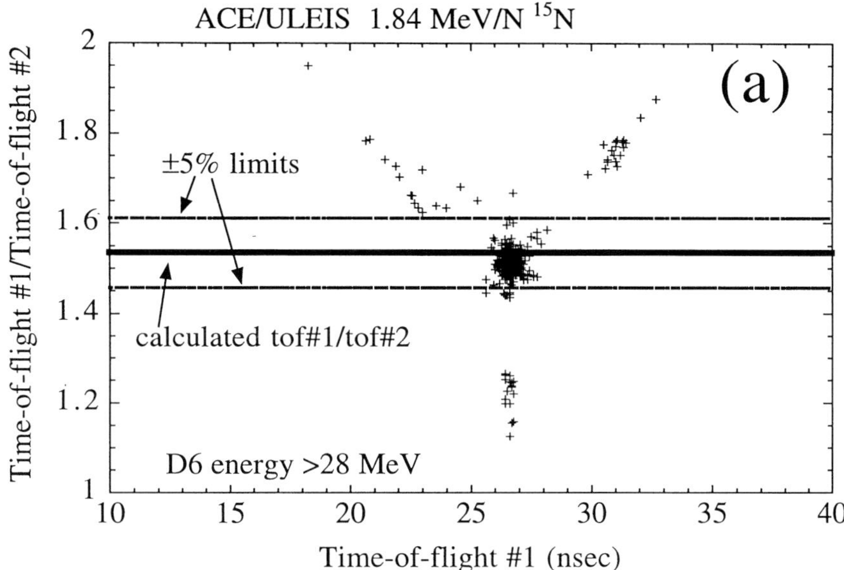

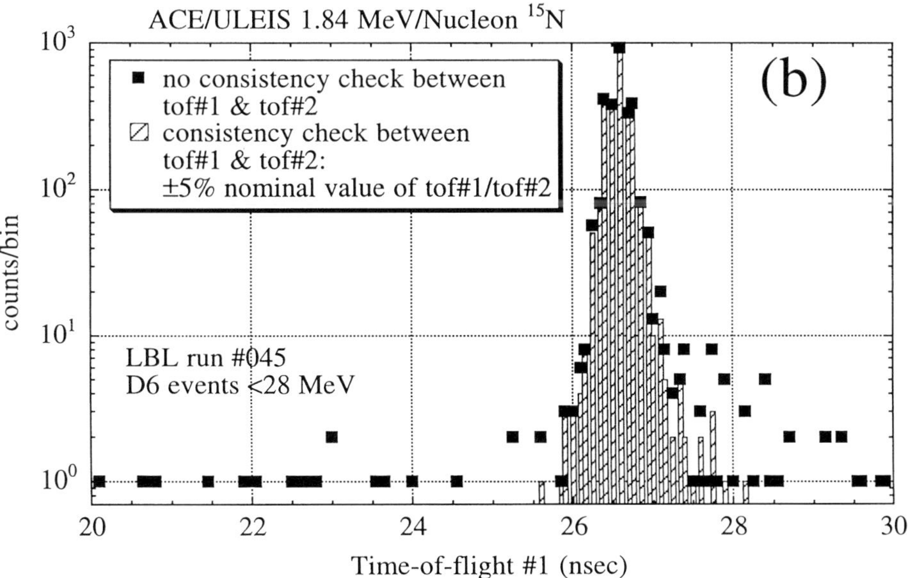

Figure 18. (a) Comparison of expected ratio of times-of-flight with measurements of 1.84 MeV nucl^{-1} ^{15}N. Most events satisfy the ±5% constraint on TOF-1/TOF-2. (b) Effect of time-of-flight consistency check on the measured TOF-1 distribution.

TABLE XI

Triggering efficiencies for EPACT/STEP telescope (estimated) at 500 keV nucl^{-1}

Element	Efficiency
H	0.03
He	0.22
C	0.97
O	1.00
Fe	1.00

foils) peaks at roughly 3–10 electrons around a few hundred keV nucl^{-1}, while the backward emission (appropriate for the STOP foil) is a factor of 2 lower (Clerc et al., 1973; Meckbach, 1976; Pferdekämper and Clerc, 1977). An additional important effect is the dead area of the MCP front surfaces, which is approximately 50%. The triggering efficiency of the plates for single electrons hitting a channel may also be less than unity since the plate operating bias is usually selected to minimize ion feedback. Since both a start and a stop MCP trigger are required to analyze an event, the overall telescope efficiency is proportional to the square of the efficiency for a single MCP stack trigger. These effects combine to produce overall triggering efficiencies that are much less than unity for H and He, while for heavier ions the efficiency is close to 1. At the current writing, these efficiencies have not been fully determined for the ULEIS telescope, but as a guide, Table XI lists 0.5 MeV nucl^{-1} efficiencies for the EPACT/STEP telescope on the WIND spacecraft (von Rosenvinge et al., 1995), which covers a similar energy range to ULEIS. Because of the dependence on $\mathrm{d}E/\mathrm{d}x$ which peaks in the range 0.1–1 MeV nucl^{-1}, the efficiencies are lower outside that energy range.

5. Flight Operations

ULEIS operates in its normal mode continuously except for about one hour/month when an on-board calibrator is activated by ground command. Dual small calibration alpha-particle sources are mounted inside the iris cover for testing instrument particle response in flight.

The multiparameter measurements returned by ULEIS for each ion allow for self-calibration of the instrument in space. Any shifts in detector response will result in a shift of particle tracks in the time-vs-energy data, that can be easily detected and corrected for. If necessary, there is a provision for uploading new look-up tables for the matrix rates to take account of possible drifts. The ability to adjust detector and MCP bias and multiple redundancy in the solid-state detector array helps protect against a wide range of possible drifts and malfunctions.

Acknowledgements

We would like to thank the many dedicated individuals in the Space Physics Group of the Department of Physics at the University of Maryland, and in the Space Department of the Johns Hopkins University Applied Physics Laboratory for the design, construction, checkout, calibration, and flight qualification of the ULEIS instrument. We thank the staffs of the Brookhaven National Laboratory tandem Van de Graaff, and the Lawrence Berkeley Laboratory 88-inch cyclotron, including the Aerospace Corporation Space Sciences Department, for their support in accelerator runs for the development and calibration of the ULEIS instrument. Finally, we thank the ACE project office at the Goddard Space Flight Center, and the ACE Payload Management Office at Caltech for their enthusiastic and continuing support of the ULEIS effort. This work was supported in part by NASA under cooperative agreement PC 38597 with the California Institute of Technology.

References

Ajello, J. M.: 1990, 'Solar Minimum $L\alpha$ Background Observations from Pioneer Venus Orbiter Ultraviolet Spectrometer: Solar Wind Latitude Variation', *J. Geophys. Res.* **95**(A9), 14855

Anders, E. and Ebihara, M.: 1982, 'Solar-System Abundances of the Elements', *Geochim. Cosmochim. Acta.* **46**, 2363.

Anger, H. O.: 1966, *Trans. Instr. Soc. Am.* **5**, 311.

Barnes, C. W. and Simpson, J. A.: 1976, 'Evidence for Interplanetary Acceleration of Nucleons in Corotating Interaction Regions', *Astrophys. J.* **210**, L91.

Breneman, H. H. and Stone, E. C.: 1985, 'Solar Coronal and Photospheric Abundances from Solar Energetic Particle Measurements', *Astrophys. J.* **299**, L57.

Chiu, M. C., et al.: 1998, 'ACE Spacecraft', *Space Sci. Rev.* **86**, 257.

Clerc, H.-G., Gehrhardt, H. J., Richter, L., and Schmidt, K. H.: 1973, 'Heavy-Ion Induced Secondary Electron Emission – a Possible Method for Z-Identification', *Nucl. Instr. Methods* **113**, 325.

Coplan, M. A., Ogilvie, K. W., Bochsler, P., and Geiss, J.: 1990, 'Space-Based Measurements of Elemental Abundances and Their Relation to Solar Abundances', *Solar Phys.* **128**, 195.

Cummings, A. C. and Stone, E. C.: 1996, 'Composition of Anomalous Cosmic Rays and Implications for the Heliosphere', *Adv. Space Res.* **78**, 117.

Fisk, L. A.: 1978, '^{3}He-Rich Flares: a Possible Explanation', *Astrophys. J.* **224**, 1048.

Fisk, L. A., Kozlovsky, B., and Ramaty, R.: 1974, 'An Interpretation of the Observed Oxygen and Nitrogen Enhancements in Low-Energy Cosmic Rays', *Astrophys. J.* **190**, L35.

Fisk, L. A. and Lee, M. A.: 1980, 'Shock Acceleration of Energetic Particles in Corotating Interaction Regions in the Solar Wind', *Astrophys. J.* **237**, 620.

Fisk, L. A., Schwadron, N. A., and Zurbuchen, T. H.: 1998, 'On the Slow Solar Wind', *Space Sci. Rev.* **86**, 51.

Galvin, A. B.: 1982, *Charge States of Heavy Ions in the Energy Range ~30–130 keV/Q Observed in Upstream Events Associated with the Earth's Bow Shock*, Dept. of Physics, University of Maryland.

Galvin, A. B., Ipavich, F. M., Cohen, C. M. S., Gloeckler, G., and von Steiger, R.: 1995, 'Solar Wind Charge States Measured by Ulysses/SWICS in the Solar Polar Hole', *Space Sci. Rev.* **72**, 65.

Galvin, A. B., Ipavich, F. M., Gloeckler, G., Hovestadt, D., Klecker, B., and Scholer, M.: 1984, 'Solar Wind Ionization Temperatures Inferred from the Charge State Composition of Diffuse Events', *J. Geophys. Res.* **89**, 2655.

Gloeckler, G., Bedini, P., Bochsler, P., Fisk, L. A., Geiss, J., Ipavich, F. M., Cani, J., Fischer, J., Kallenbach, R., Miller, J., Tums, E. V., and Wimmer, R.: 1998 'Investigation of the Composition of Solar and Interstellar Matter Using Solar Wind and Pickup Ion Measurements with SWICS and SWIMS on the ACE Spacecraft', *Space Sci. Rev.* **86**, 497.

Gloeckler, G., Geiss, J., Roelof, E. C., Fisk, L. A., Ipavich, F. M., Ogilvie, K. W., Lanzerotti, L. J., von Steiger, R., and Wilken, B.: 1994, 'Acceleration of Interstellar Pickup Ions in the Disturbed Solar Wind Observed on Ulysses', *J. Geophys. Res.* **99**, 17637.

Gloeckler, G., Hovestadt, D., and Fisk, L. A.: 1979, 'Observed Distribution Functions of H, He, C, O, and Fe in Corotating Energetic Particle Streams: Implications for Interplanetary Acceleration and Propagation', *Astrophys. J.* **230**, L191.

Ipavich, F. M., Lundgren, R. A., Lambird, B. A., and Gloeckler, G.: 1978, 'Measurement of Pulse-Height Defect in Au-Si Detector for H, He, C, N, O, Ne, Ar, Kr from $\sim$2 to $\sim$400 keV nucl^{-1}', *Nuc. Instr. Methods* **154**, 291.

Jokipii, J. R.: 1996, 'Theory of Multiply-Charged Anomalous Cosmic Rays', *Astrophys. J.* **466**, 47.

Klecker, B.: 1995, 'The Anomalous Component of Cosmic Rays in the 3-D Heliosphere', *Space Sci. Rev.* **72**, 419.

Klecker, B., Oetliker, M., Blake, J. B., Hovestadt, D., Mason, G. M., Mazur, J. E., and McNab, M. C.: 1997, 'Multiply Charged Anomalous Cosmic Ray N, O, and Ne: Observations with HILT/SAMPEX', *Proc. 25th Int. Cosmic Ray Conf., Durban* **1**, 273.

Lapington, J. S. and Schwarz, H. E.: 1986, 'The Design and Manufacture of Wedge and Strip Anodes', *IEEE Trans. Nucl. Sci.* **NS-33**, 288.

Luhn, A., Klecker, B., Hovestadt, D., and Möbius, E.: 1985, 'The Mean Ionic Charge State of Silicon in ^{3}He-Rich Solar Flares', *Proc. 19th Int. Cosmic Ray Conf., La Jolla* **4**, 285.

Mason, G. M., Gloeckler, G., and Hovestadt, D.: 1984, 'Temporal Variations of Nucleonic Abundances in Solar Flare Energetic Particle Events. II. Evidence for Large Scale Shock Acceleration', *Astrophys. J.* **280**, 902.

Mason, G. M., Mazur, J. E., Dwyer, J. R., Reames, D. V., and von Rosenvinge, T. T.: 1997, 'New Spectral and Abundance Features of Interplanetary Heavy Ions in Corotating Interaction Regions', *Astrophys. J.* **486**, L149.

Mason, G. M., Mazur, J. E., and Hamilton, D. C.: 1994, 'Heavy Ion Isotopic Anomalies in ^{3}He-Rich Solar Particle Events', *Astrophys. J.* **425**, 843.

Mazur, J. E., Mason, G. M., Klecker, B., and McGuire, R. E.: 1992, 'The Energy Spectra of Solar Flare Hydrogen, Helium, Oxygen, and Iron: Evidence for Stochastic Acceleration', *Astrophys. J.* **401**, 398.

Mazur, J. E., Mason, G. M., Klecker, B., and McGuire, R. E.: 1993, 'The Abundances of Hydrogen, Helium, Oxygen, and Iron Accelerated in Large Solar Particle Events', *Astrophys. J.* **404**, 810.

McComas, D. J., Bame, S. J., Barker, P., Feldman, W. C., Phillips, J. L., Riley, P., and Griffee, J. W.: 1998, 'Solar Wind Electron Proton Alpha Monitor (SWEPAM) for the Advanced Composition Explorer', *Space Sci. Rev.* **86**, 563.

McDonald, F. B., Teegarden, B. J., Trainor, J. H., von Rosenvinge, T. T., and Webber, W. R.: 1976, 'The Interplanetary Acceleration of Energetic Nucleons', 1976, *Astrophys. J.* **203**, L149.

Meckbach, W.: 1976, 'Secondary Electron Emission from Foils Traversed by Ion Beams', in I. A. Sellin and D. J. Pegg (eds.), *Beam-Foil Spectroscopy*, Plenum Press, New York, p. 577.

Mewaldt, R. A., Selesnick, R. S., Cummings, J. R., Stone, E. C., and von Rosenvinge, T. T.: 1996, 'Evidence for Multiply Charged Anomalous Cosmic Rays', *Astrophys. J.* **466**, L43.

Mewaldt, R. A. and Stone, E. C.: 1989, 'Isotope Abundances of Solar Coronal Material Derived from Solar Energetic Particle Measurements', *Astrophys. J.* **337**, 959.

Mewaldt, R. A., Stone, E. C., Vidor, B., and Vogt, R. E.: 1976, 'Isotopic and Elemental Composition of the Anomalous Low-Energy Cosmic Ray Fluxes', *Astrophys. J.* **205**, 931.

Mewaldt, R. A., Stone, E. C., and Vogt, R. E.: 1978, 'The Radial Diffusion Coefficient of 1.3–2.3 MeV Protons in Recurrent Proton Streams', *Geophys. Res. Letters* **5**, 965.

Miller, J. A.: 1998, 'Particle Acceleration in Impulsive Solar Flares', *Space Sci. Rev.* **86**, 79.

Miller, J. A. and Reames, D. V.: 1995, 'Heavy Ion Acceleration by Cascading Alfvén Waves in Impulsive Solar Flares', *High Energy Solar Physics*, AIP Conf. Proc. 374, Am. Inst. Physics, New York, p. 450.

Möbius, E., et al.: 1998, 'The Solar Energetic Particle Ionic Charge Analyzer (SEPICA) and the Data Processing Unit (S3DPU) for SWICS, SWIMS, and SEPICA', *Space Sci. Rev* **86**, 449.

Möbius, E., Scholer, M., Hovestadt, D., Klecker, B., and Gloeckler, G.: 1982, 'Comparison of Helium and Heavy Ion Spectra in ^{3}He-Rich Solar Flares with Model Calculations Based on Stochastic Fermi Acceleration in Alfvén Turbulence', *Astrophys. J.* **259**, 397.

Pesses, M. E., Jokipii, J. R., and Eichler, D.: 1981, 'Cosmic Ray Drift, Shock Wave Acceleration, and the Anomalous Component of Cosmic Rays', *Astrophys. J.* **246**, L85.

Pferdekämper, K. E. and Clerc, H.-G.: 1977, 'Energy Spectra of Secondary Electrons Ejected by Ions from Foils', *Zeitschrift Physik A* **280** 155.

Powell, F. R., Keski-Kuha, R. A. M., Zombeck, M. V., Goddard, R. E., Chartas, G., Townsley, L. K., Möbius, E., Davis, J. M., and Mason, G. M.: 1997, 'Metalized Polyimide Filters for X-ray Astronomy and Other Applications', Invited Review Paper, *Proc. SPIE (The International Society for Optical Engineering)*.

Ramaty, R., Mandzhavidze, N., and Kozlovsky, B.: 1995, 'Solar Atmospheric Abundances from Gamma Ray Spectroscopy', *High Energy Solar Physics* p. 172.

Reames, D. V., Barbier, L. M., and Ng, C. K.: 1996, 'The Spatial Distribution of Particles Accelerated by Coronal Mass Ejection-Driven Shocks', *Astrophys. J.* **466**, 473.

Reames, D. V., Barbier, L. M., von Rosenvinge, T. T., Mason, G. M., Mazur, J. E., and Dwyer, J. R.: 1997, 'Energy Spectra of Ions Accelerated in Impulsive and Gradual Solar Events', *Astrophys. J.* **483**, 515.

Reames, D. V., von Rosenvinge, T. T., and Lin, R. P.: 1985, 'Solar ^{3}He-Rich Events and Nonrelativistic Electron Events: a New Association', *Astrophys. J.* **292**, 716.

Richardson, I. G. and Hynds, R. J.: 1990, 'Spectra of >35 keV ions in Corotating Ion Enhancements at 1 AU', *Proc. 21st Int. Cosmic Ray Conf., Dublin* **5**, 337.

Selesnick, R. S., Cummings, A. C., Cummings, J. R., Leske, R. A., Mewaldt, R. A., Stone, E. C., and von Rosenvinge, T. T.: 1993, 'Coronal Abundances of Neon and Magnesium Isotopes from Solar Energetic Particles', *Astrophys. J.* **418**, L45.

Siegmund, O. H. W., Lampton, M., Bixler, J., Bowyer, S., and Malina, R. F.: 1986a, 'Operation Characteristics of Wedge and Strip Readout Systems', *IEEE Trans. Nucl. Sci.* **33**(1), 724.

Siegmund, O. H. W., Lampton, M., Bixler, J., Chakrabarti, S., Vallerga, J., Bowyer, S., and Malina, R. F.: 1986b, *J. Opt. Soc. Amer. A* **3**, 2139.

Starzecki, W., Stefanini, A. M., Lunardi, S., and Signorini, C.: 1982, 'A Compact Time-Zero Detector for Mass Identification of Heavy Ions', *Nucl. Instr. Method.* **193**, 499.

Stone, E. C., et al.: 1989, '*Phase A Study of an Advanced Composition Explorer*', California Institute of Technology.

Stone, E. C., et al.: 1998a, 'The Cosmic Ray Isotope Spectrometer for the Advanced Composition Explorer', *Space Sci. Rev.* **86**, 285.

Stone, E. C., et al.: 1998b, 'The Solar Isotope Spectrometer for the Advanced Composition Explorer', *Space Sci. Rev.* **86**, 355.

von Rosenvinge, T. T., et al.: 1995, 'The Energetic Particles: Acceleration, Composition, and Transport (EPACT) Investigation on the Wind Spacecraft', *Space. Sci. Rev.* **71**, 155.

von Steiger, R. and Geiss, J.: 1989, 'Supply of Fractionated Gases to the Corona', *Astron. Astrophys.* **225**, 222.

Zabel, T. H., Fewell, M. P., Kean, D. C., and Spear, R. H.: 1980, 'The Response of Silicon Surface-Barrier Detectors to Light and Heavy Ions', *Nucl. Instr. Methods* **174**, 459.

THE SOLAR ENERGETIC PARTICLE IONIC CHARGE ANALYZER (SEPICA) AND THE DATA PROCESSING UNIT (S3DPU) FOR SWICS, SWIMS AND SEPICA

E. MÖBIUS, L. M. KISTLER, M. A. POPECKI, K. N. CROCKER, M. GRANOFF,
S. TURCO, A. ANDERSON, P. DEMAIN, J. DISTELBRINK, I. DORS, P. DUNPHY,
S. ELLIS , J. GAIDOS, J. GOOGINS, R. HAYES, G. HUMPHREY, H. KÄSTLE,
J. LAVASSEUR, E. J. LUND, R. MILLER, E. SARTORI, M. SHAPPIRIO, S. TAYLOR,
P. VACHON, M. VOSBURY and V. YE

*Institute for the Study of Earth, Oceans and Space, University of New Hampshire, Durham,
NH 03824, U.S.A.*

D. HOVESTADT, B. KLECKER, H. ARBINGER, E. KÜNNETH, E. PFEFFERMANN
and E. SEIDENSCHWANG

Max-Planck-Institut für Extraterrestrische Physik, Postfach 1603, D-85740 Garching, Germany

F. GLIEM, K.-U. REICHE, K. STÖCKNER and W. WIEWESIEK

*Institut für Datenverarbeitung, Technische Universität Braunschweig, Postfach 3229,
D-38106 Braunschweig, Germany*

A. HARASIM and J. SCHIMPFLE

*Fachbereich Elektrotechnik, Fachhochschule Landshut, Am Lurzenhof 4, D-84036 Landshut,
Germany*

S. BATTELL

Battel Engineering, 10020 North 58th Street, Scottsdale, AZ 85253, U.S.A.

J. CRAVENS

Cravens Engineering, 1309 Bristol Drive, Iowa City, IA 52240, U.S.A.

G. MURPHY

Design Net Engineering, 1968 Mountain Maple Ave., Highlands Ranch, CO 80126, U.S.A.

Abstract. The Solar Energetic Particle Ionic Charge Analyzer (SEPICA) is the main instrument on the Advanced Composition Explorer (ACE) to determine the ionic charge states of solar and interplanetary energetic particles in the energy range from ≈ 0.2 MeV nucl^{-1} to ≈ 5 MeV charge^{-1}. The charge state of energetic ions contains key information to unravel source temperatures, acceleration, fractionation and transport processes for these particle populations. SEPICA will have the ability to resolve individual charge states and have a substantially larger geometric factor than its predecessor ULEZEQ on ISEE-1 and -3, on which SEPICA is based. To achieve these two requirements at the same time, SEPICA is composed of one high-charge resolution sensor section and two low-charge resolution, but large geometric factor sections. The charge resolution is achieved by the focusing of the incoming ions, through a multi-slit mechanical collimator, deflection in an electrostatic analyzer with a voltage up to 30 kV, and measurement of the impact position in the detector system. To determine the nuclear charge (element) and energy of the incoming ions, the combination of thin-window flow-through proportional counters with isobutane as counter gas and ion-implanted solid state detectors provide for 3 independent ΔE (energy loss) versus E (residual energy) telescopes. The multi-wire proportional counter simultaneously determines the energy loss ΔE and the impact

Space Science Reviews **86**: 449–495, 1998.

position of the ions. Suppression of background from penetrating cosmic radiation is provided by an anti-coincidence system with a CsI scintillator and Si-photodiodes. The data are compressed and formatted in a data processing unit (S3DPU) that also handles the commanding and various automated functions of the instrument. The S3DPU is shared with the Solar Wind Ion Charge Spectrometer (SWICS) and the Solar Wind Ion Mass Spectrometer (SWIMS) and thus provides the same services for three of the ACE instruments. It has evolved out of a long family of data processing units for particle spectrometers.

1. Introduction and Scientific Objectives

The ionic charge state Q is an important parameter for deciphering the local conditions at the origin of energetic particle populations as well as the processes involved in their selection, acceleration and transport. First direct measurements of the charge state of solar energetic particles in the energy range of 0.3–3 MeV nucl^{-1} were carried out using the Ultra Low Energy Z E Q (ULEZEQ) sensor on the International Sun-Earth Explorers (ISEE-1 and -3) (Hovestadt et al., 1978). More recently the measurements have been extended to higher energies by utilizing the Earth's magnetic field as a spectrometer with several instruments on the Solar Anomalous and Magnetospheric Explorer (SAMPEX) mission (Mason et al., 1991; Baker et al., 1993; Klecker et al., 1993). These approaches have allowed the determination of average charge states in solar cosmic rays (Hovestadt et al., 1981), interplanetary accelerated particles (Hovestadt et al., 1982) and the confirmation that anomalous cosmic rays (ACR) are mostly singly charged (Klecker et al., 1995). However, questions concerning the detailed charge state and/or individual physical processes involved have been left open because of either the lack of individual charge-state resolution or poor counting statistics.

1.1. SOLAR ENERGETIC PARTICLES

Over the last decade it has been recognized that solar energetic particle events (SEPs) can be traced back to basically two classes of solar flares. Impulsive flares are characterized by short time scale (several minutes) electromagnetic (radio and X-ray) emission, generally low fluxes of energetic particles in interplanetary space, a high electron to ion ratio, and substantial enhancements in the abundances of heavy ions and ^{3}He over ^{4}He (Mason et al., 1986; Reames, 1990). Gradual solar flares are accompanied by long duration radio emissions that are generally associated with shocks in the corona and emit high fluxes of energetic particles with a low electron to ion ratio and abundances that reflect normal solar corona conditions (e.g., Reames, 1992). It is assumed that the ionic charge state of energetic ions from these flares carries the information on the coronal temperature within the flare site, because the charge state is quickly frozen in during the acceleration process and the column density of matter between the sun and the observing spacecraft is very low.

It is common to deduce a coronal temperature from the measured average charge state assuming ionization equilibrium (e.g., Arnaud and Rothenflug, 1985).

The average charge state of heavy ions during impulsive ^{3}He-rich flares has been found to reflect temperatures of the order of 10^7 K (Klecker et al., 1984; Luhn et al., 1987). However, because of the low energetic ion fluxes during these events, only average values over the full complement of events during the entire observation period could be obtained. At this point it is not known whether the relatively wide charge distribution that has been seen is due to substantial variations from event to event or is a real feature during individual events. Clearly the collecting power of a charge sensitive instrument has to be substantially increased in order to allow a detailed study of individual impulsive events in order to make progress on local heating and selective acceleration processes prevalent in these flares.

Contrary to the impulsive flares, gradual events exhibit charge states that reflect substantially lower temperatures in the neighborhood of $1\text{--}2 \times 10^6$ K (Hovestadt et al., 1981). In addition, different species seem to indicate different temperatures. Attempts to explain these variations in terms of nonthermal conditions in the flare site or through interaction of the accelerated particles with coronal material have been unsuccessful (Luhn et al., 1984). In the most promising interpretation Mullan and Waldron (1986) suggested that heating and ionization in flares through local X-ray generation plays a role. In order to study the significance of such processes on the ionic charge states it is necessary to narrow the measurements down to individual charge states, since this will allow a much more precise deconvolution of the energy distribution of the electrons and/or X-rays that have produced the observed charge states. The goal requires a substantial improvement in the charge-state resolution over previous instruments.

1.2. INTERPLANETARY ACCELERATED PARTICLES

Two types of interplanetary accelerated particles have been observed, ion populations associated with co-rotating interaction regions (CIR), where high and low speed solar wind streams meet, and coronal mass ejection (CME) related particle events, where acceleration occurs at the corresponding shock waves. During the operational phase of the ISEE instrument (around solar maximum) no good signatures of CIRs were observed. In CME related events charge states were found that are basically compatible with the charge states in gradual solar flare particle populations (Hovestadt et al., 1982). It came as a surprise that a class of apparently interplanetary events, i.e., events that could not be correlated with solar flare activity, was detected that had an unusually high abundance of He$^+$ (Hovestadt et al., 1984). No accompanying low-charge-state heavy ions were found beyond the detection threshold of the instrument. More recently Gloeckler et al. (1994) have pointed out that the high He$^+$/He^{2+} ratio observed by ISEE may be produced by the acceleration of interstellar pickup ions at interplanetary shocks. With their wide velocity distribution these ions may present a source population for very efficient further acceleration in interplanetary space (Möbius et al., 1985). To delineate the source distributions of CIRs and CME related energetic particle events again a larger collecting power of the instrument is needed.

1.3. ANOMALOUS COSMIC RAYS

The anomalous cosmic rays are thought to be a final product after pickup of interstellar ions and their acceleration at the heliospheric termination shock (e.g., Fisk et al., 1974; Jokipii, 1986). With the SAMPEX mission it has been well established now that these ions are indeed mostly singly charged (Klecker et al., 1995). However, the measurements using the cut-off rigidity in the Earth's magnetic field have been restricted to energies >8 MeV nucl^{-1}. A low-background large-geometric-factor instrument for ionic charge-state determination will be able to extend these results to the lower end of the ACR spectrum.

1.4. INSTRUMENT PERFORMANCE REQUIREMENTS

Within a complement of high-resolution spectrometers to measure the composition of solar, local interstellar, and galactic cosmic-ray material on the Advanced Composition Explorer (ACE) (Stone et al., 1998a) the SEPICA instrument is the prime sensor for the determination of the charge-state distribution of energetic particles. Its basic design requirements are to measure the ionic charge state, Q, the kinetic energy, E, and the nuclear charge, Z, of energetic ions in the energy range 0.2–3 MeV nucl^{-1} for elements H through Fe. As a consequence of the scientific objectives listed above several substantial improvements over previous instrumentation are required. The need to resolve individual charge states at least in gradual flares for species such as oxygen and carbon requires a charge-state resolution $\Delta Q/Q \leq 0.1$ over a substantial energy range. For practical reasons charge resolution is limited towards higher energies. Therefore, we adopt a goal for single charge-state resolution up to ≈ 1 MeV charge^{-1}. Since ^{3}He-rich flares are to be studied, SEPICA also needs to separate isotopes for low mass number species, in particular ^{3}He and ^{4}He. On the other hand these impulsive flares as well as CIRs and ACR are known for low particle fluxes. In order to accumulate data with sufficient counting statistics a substantially larger geometrical factor than SEPICA's predecessor, the ULEZEQ sensor on ISEE-1 and -3, is necessary. To improve on previous measurements by at least one order of magnitude, we adopt a geometric factor of ≥ 0.2 cm^2 sr. For the very low fluxes in CIRs and ACR the large geometric factor must be accompanied by a low-background susceptibility.

It should be pointed out here that generally the two requirements of (a) higher charge-state resolution and (b) larger geometric factor are mutually exclusive, if an instrument of the same size and with the same resources is to be built. To achieve higher resolution with any deflection technique, a collimation to a smaller acceptance angle range has to be chosen, which in turn leads to a reduced collecting power of the sensor. Therefore, the SEPICA sensor is significantly larger than the

ULEZEQ sensor, and we have chosen a modular approach with three individual fans that allows a separate optimization of these competing goals in individual sensor sections.

We start this paper with the principles of operation in Section 2, followed by the description of the SEPICA sensor subsystems in Section 3. The sensor electronics and the internal data selection and processing are described in Section 4 followed by the sensor calibration and performance in Section 5. The supporting subsystems, such as high-voltage generation and gas regulation are presented in Section 6. The mechanical packaging and thermal aspects are covered in Sections 7 and 8. The description of the common data processing unit (S3DPU) for SWICS, SWIMS and SEPICA that has been designed and built by the Technische Universität Braunschweig (TUB) under guidance of the University of New Hampshire (UNH) is included in this paper in the concluding Section 9. It should be noted that several subsystems provided challenges which were met with novel designs that have pushed technical limits. These technical aspects are covered in the corresponding chapters. For example, the design of the high-precision collimator system and the integration of many electrical feedthroughs into the ceramic frames of the multi-wire proportional counter are described in Section 3. A novel arrangement for a 30 kV supply and the first use of micro-machined bi-metallic valves in the isobuthane gas flow system are presented in Section 6. Thus the paper combines the functional description of the SEPICA instrument with a reflection on how some of its biggest technical challenges were met.

2. Principles of Operation and Basic Design

To simultaneously determine the energy E, nuclear charge Z and ionic charge Q of incoming particles the SEPICA sensor combines several different measurement methods. The analysis of each ion starts with the determination of the electrostatic deflection of incoming ions in a collimator-analyzer assembly. Then the energy loss and the residual energy are measured in a $dE/dx - E$ telescope. Potential background from penetrating radiation is suppressed by the use of an anti-coincidence detector. The instrument is based on the general design of the ULEZEQ sensor flown on the ISEE spacecraft (Hovestadt et al., 1978).

2.1. MEASUREMENT METHODS

SEPICA consists of three independent sensor units, called 'fans'. Each of the three fans is symmetric about the plane with the high-voltage deflection plate. A schematic view of one individual sensor unit is shown in Figure 2.1 together with the basic measuring principles. Shown is a single side of one SEPICA fan. Energetic particles enter a multi-slit collimator, which selects those incoming particles that target a narrow line in the detector plane (indicated by F, the 'focal line'). They

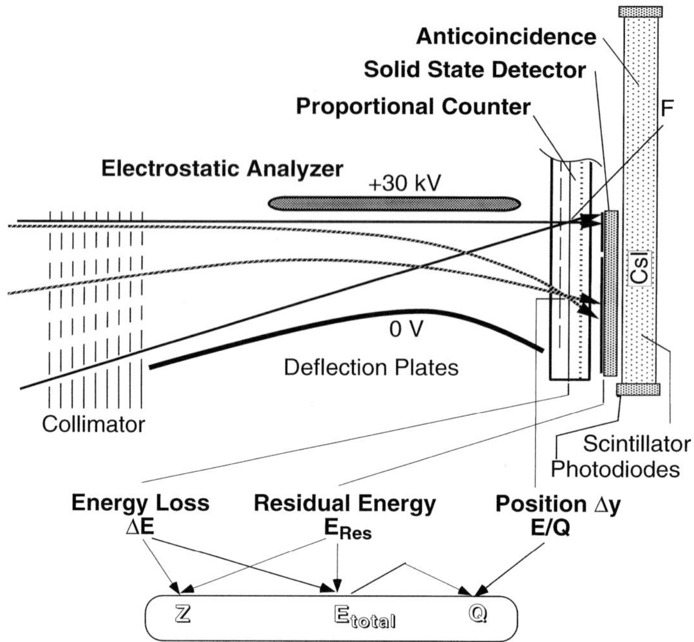

Figure 2.1. Schematic view and principles of operation of the SEPICA instrument.

are electrostatically deflected between a set of electrode plates. The curved plate is on ground potential, while the flat center plate is supplied with a positive high voltage up to 30 kV (to be set by telecommand). In Figure 2.1 a pair of undeflected ion trajectories is shown along with a pair of deflected trajectories. The deflection, which is inversely proportional to energy per charge, E/Q, is determined in a multi-wire thin-window proportional counter. The proportional counter is also used to measure the specific energy loss dE/dx of the ions, which depends on the energy E and the nuclear charge Z of the particle. Finally, the residual energy of the particle, E_{Res}, is directly determined in a silicon solid-state detector (SSD). An anti-coincidence system covers the rear side of each detector assembly to suppress background signals from penetrating high energy particles. This is of particular importance for the study of low fluxes in weak solar events and during quiet times. It consists of a CsI scintillator which is viewed by silicon photodiodes.

The identity of each incoming ion is determined in two consecutive steps that are illustrated in Figure 2.2. For energies above approximately 1 MeV nucl^{-1} the specific energy loss dE/dx of ions can be approximated by the relation

$$dE/dx = \beta \times Z^2 (E/A)^\alpha , \tag{2.1}$$

i.e., it is only a function of the nuclear charge Z and the energy/mass E/A, where A denotes the atomic mass number. α is negative and of the order of 1, i.e., the energy loss falls with increasing energy. β is a constant that summarizes the effects of the target gas in the proportional counter. Below 1 MeV nucl^{-1} partial recombination

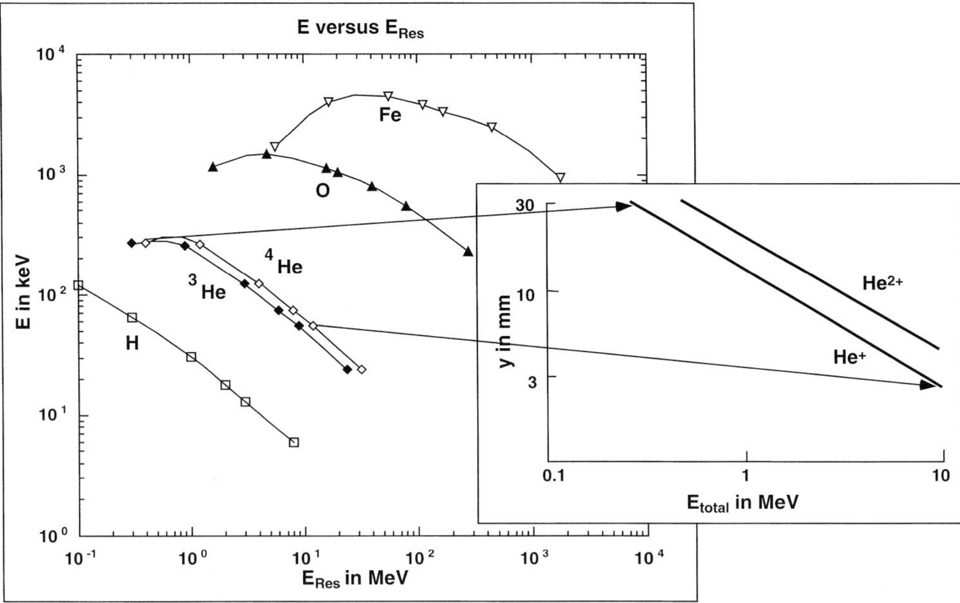

Figure 2.2. Energy loss (ΔE) versus residual energy (E_{Res}) curves for H, ^{3}He, ^{4}He, O, and Fe (background panel). After selection of one element or isotope curve (here ^{4}He) the charge states can be separated from their deflection Δy in the electrostatic analyzer (foreground panel).

in the interaction with the target material reduces the effective charge of the ion to values $Z^* < Z$ so that the specific energy loss decreases with decreasing energy. Therefore, the typical $\mathrm{d}E/\mathrm{d}x$ versus E_{Res} curves in the left panel of Figure 2.2 have a maximum at ≈ 1 MeV nucl^{-1}. Because the energy loss is a function of energy/mass (E/A) and not of total energy, the curves for different isotopes of the same element are separated by their total energy according to the isotope mass A as indicated for He.

$$E_{\mathrm{total}} = (E/A)A \, . \tag{2.2}$$

This leads to a differentiation in the measured residual energy E_{Res}, and thus isotopes can be identified with SEPICA for low Z ions, such as He. The final parameters for the sensor are obtained in calibration. The same method for the determination of elements and isotopes at higher energies (>10 MeV nucl^{-1}) is used in the Solar Isotope Spectrometer (SIS) (Stone et al., 1998b) and the Cosmic Ray Isotope Spectrometer (CRIS) (Stone et al., 1998c) on ACE.

With the knowledge of the species the measured residual energy can now be corrected for the accumulated energy loss effects in the sensor according to

$$E_{\mathrm{total}} = E_{\mathrm{Res}} + \Delta E_{\mathrm{Window}} + \Delta E_{\mathrm{PC}} + \Delta E_{\mathrm{SSD}} \, , \tag{2.3}$$

where $\Delta E_{\mathrm{Window}}$ is the energy loss in the front and rear windows of the proportional counter, ΔE_{PC} is the loss in the counter gas which includes contributions from additional dead layers in the counter, and ΔE_{SSD} is the loss in the entrance layers of the solid state detector. All these contributions are determined in sensor calibrations with radioactive sources and at particle accelerators. According to

$$Q = (Q/E)E_{\text{total}} \tag{2.4}$$

the ionic charge state can be computed by combining the total energy (E) with the energy/charge (E/Q) as derived from the deflection Y of the ions in the electrostatic analyzer voltage U_{defl}

$$Y = \delta U_{\text{defl}} Q/E \,, \tag{2.5}$$

where δ is a sensor constant that depends on the size and geometry of the analyzer. SEPICA has been designed so that a typical deflection of 10 mm is achieved for 1 MeV charge^{-1} ions with a deflection voltage of 30 kV. The spread of charge states for a selected ion species is sketched in the right panel of Figure 2.2. The charge resolution is reduced with increasing energy. The determination of average ionic charge states can still be achieved up to $\approx$5 MeV charge^{-1}. The ion species and energy can be derived for all ions that are stopped by the SSD.

2.2. Design Goals and Capabilities

The instrument capabilities of SEPICA are compiled in Table 2.1. The capability of SEPICA to resolve details of the ionic charge states is restricted to the energy range between 0.5 and 3 MeV charge^{-1}, with determination of the average charge state up to $\approx$5 MeV charge^{-1}. Over a wider energy range (e.g., up to about 6 MeV nucl^{-1} for Fe and O) element identification with energy determination will be possible. For SEPICA two specific goals were set: For energies $\leq$1 MeV charge^{-1} a resolution $\Delta Q/Q \leq 0.1$ (FWHM) should be achieved. On the other hand the instrument should have a total geometric factor of at least $\approx$0.2 cm^2 sr.

In order to achieve the two seemingly mutually exclusive goals, i.e., to provide high sensitivity for the study of small solar energetic particle (SEP) events and to reach high charge-state resolution to resolve individual charge states, SEPICA has been subdivided into three nearly independent sensor fans. The detector systems, their respective electronics, and the electrostatic analyzer are identical for all three fans. However, there are two different versions of the entrance collimator. Two fans are identical in the geometry of the slit collimator and have a combined geometric factor of about 0.2 cm^2 sr, which is sufficient in the expected counting rate for even small SEP events such as ^{3}He-rich solar flares. The third fan has a slit system, whose open width is narrower by a factor of three, in order to allow for a charge resolution which is higher by a factor of three. As a result, the geometrical factor of this fan is reduced to $\approx$0.03 cm^2 sr. Therefore, the sensitivity of this portion will be sufficient only during relatively large SEP, CME, and CIR events, when the particle fluxes are high or when a long accumulation time is possible. With these capabilities SEPICA is greatly improved (in both sensitivity and charge resolution)

TABLE 2.1

SEPICA performance

Energy range	Resolution	Elements
H: 0.2–3 MeV		
He: 0.3–6 MeV nucl^{-1}	Isotopic, $\Delta A < 1$	H to He
O: 0.2–15 MeV nucl^{-1}	Individual elements, $\Delta Z < 1$	H to O
Fe: 0.1–5.4 MeV nucl^{-1}	Groups, $\Delta Z \approx 2$–3	O to Fe
	Low res. fans	High res. fan
Geometric factor	0.2 cm^2 sr	0.03 cm^2 sr
$\Delta Q/Q$, $E \leq 1$ MeV/Q	0.3	0.11
$\Delta Q/Q$, $E \leq 3$ MeV/Q	1	0.32

TABLE 2.2

SEPICA resources

Resource	Measured
Mass	37.4 kg
Power*	18.5 W nominal, 19.5 W peak
Average data rate**	608 bps

*Includes 2 W of power for internal thermal control.
**Peak data rate of SEPICA must be coordinated with the
data requirements of the SWICS and SWIMS instruments
such that the total rate available to the S3DPU (1624 bps)
is never exceeded.

over its predecessor, the ULEZEQ sensor on ISEE-1 and -3. The required resources
are compiled in Table 2.2.

3. Sensor Description

The SEPICA sensor consists of six separate sections that are functionally identical
to the schematics in Figure 2.1. Two sections that are the mirror image of each
other are combined into sensor fans. Each sensor fan consists of a collimator, electrostatic deflection plates, a proportional counter, two solid-state detectors (SSD)
and the anti-coincidence system. The back portion of each fan is composed of one
dE/dx thin-window proportional counter, serving two instrument sections. A set
of six solid-state detectors, one for each section, is placed behind the counters.
Three anti-coincidence assemblies are placed behind the solid-state detector plane;
each serving two sections. The fans have been built as separate units that could

    E. MÖBIUS ET AL.

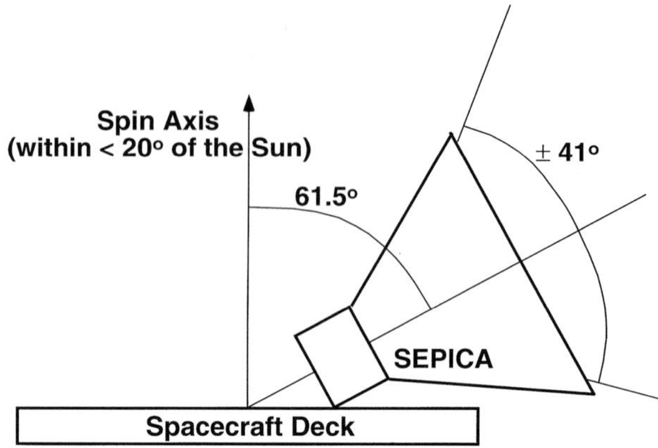

Figure 3.1. Schematic of the orientation and the field-of-view of the SEPICA sensor on the ACE spacecraft.

be integrated and tested independently of each other. All three fans have the same field-of-view and are mounted such that their center viewing direction points at an angle of 61.5° with respect to the spin axis as shown in Figure 3.1. SEPICA sweeps out an angle band of ±41° by 360° centered at this direction. The instantaneous field-of-view is 82° by 17°. Because the spin axis remains within <20° of the sun, no sunlight will enter the SEPICA sensor.

3.1. ELECTROSTATIC ANALYZER SECTIONS

One geometrically flat, positive high-voltage plate (nominally at 30 kV) is employed for two neighboring instrument sections. When the high voltage is applied, the particles, which are originally passively collimated by the entrance collimator towards the focus 'F' (in Figure 2.1), will be deflected towards the curved ground-plates in each section. The high-voltage plate has been designed as a hollow Ultem structure which is sandwiched between two 0.5 mm Al plates. It has been plated with a sequence of Cu and Au plating. At the ends the plate is thickened to minimize edge effects and to allow the Ultem insulator mounting and thus the very sensitive high-voltage triple conjunctions (between conductor, insulator and vacuum) to be recessed into a low electric-field region. The 30 kV HV cable is inserted through the insulating mounting into the center of the plate, so that the physical connection is made in the interior which is a perfect equipotential region. This connection also contains a 40 MΩ resistor that filters the remaining ripple of the HV supply together with the intrinsic capacitance of the analyzer of 40 pF.

The collimator consists of a flat stack of plates which produce a well-defined focal line in the detector plane. The deflector plates are trapezoid shaped. The geometry is shown in Figure 3.2 together with sample ion trajectories. As a consequence of the flat collimator-analyzer geometry, incoming particles with the same velocity v_0 perpendicular to the detector plane are deflected by the same amount.

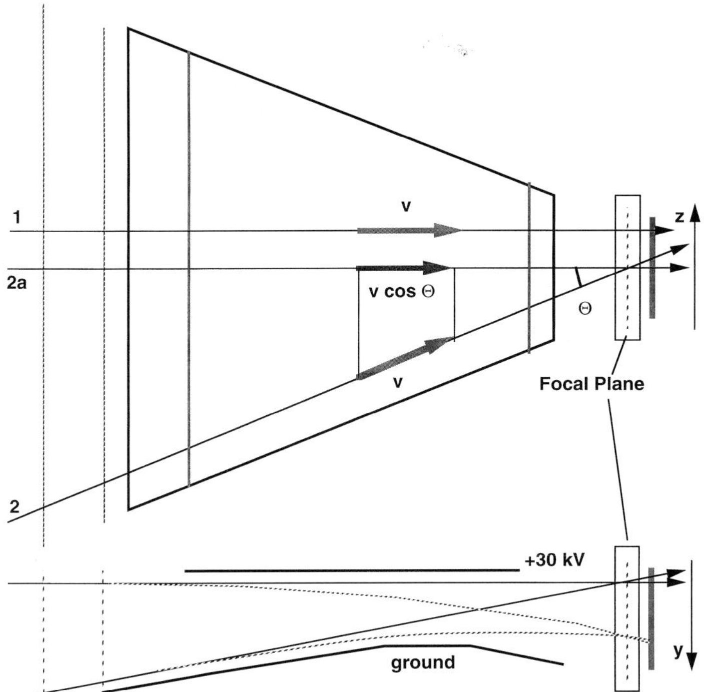

Figure 3.2. Top view and cross-sectional view of electrostatic analyzer and sample trajectories. The trajectory of the projection 2a of trajectory 2 is the same as that of 1, if only the projection of the ion velocity of 2 is considered.

Therefore, a correction has to be applied to ions that enter the sensor at oblique angles, but with the same total energy. This necessitates the determination of the incoming angle Θ in the plane parallel to the deflection plate. For the computation of the ionic charge Q, the energy component perpendicular to the detector plane

$$E_\perp = E \, \cos^2 \Theta \qquad\qquad (3.1)$$

is used, where E represents the total energy of the incoming ion that is actually measured in the sensor. The angle Θ is determined with two position measurements in the z-direction (parallel to the deflection plate) of the instrument.

While this approach requires additional resources in the sensor to provide the information on the angle Θ, it has several advantages over the implementation in the ULEZEQ sensor (Hovestadt et al., 1978) where the trajectories through the electrostatic analyzer have been optimized using a curved collimator plate geometry even for oblique angles. Firstly, such an optimization cannot be used simultaneously for a wide acceptance angle range and a large area detector system, a combination that is required to achieve a large geometric factor. Secondly, the mechanical precision of the collimator stack is greatly improved for a flat geometry

Figure 3.3. Front view of SEPICA with all three apertures.

along with a substantial simplification of the mechanical integration. The entrance apertures of the SEPICA sensor are shown in Figure 3.3.

3.2. COLLIMATOR DESIGN

The collimators are comprised each of two half stacks of precision etched, 0.10 mm thick BeCu collimator plates. The half stacks are a mirror image in elevation, with one half stack covering the upper 41° and the other stack covering the lower 41° in elevation. Each plate has 32 rows of 5 mm long slits that extend ±260 mm in elevation grouped into two 40 mm wide columns on either side of the HV deflection plate. For the high-resolution collimator, the slit width is 0.15 mm, with a 0.15 mm separation between slits at the outlet plate. The two low-resolution collimators have outlet plates with 0.45 mm slit widths and 0.155 mm slit spacing. Each stack contains 13 etched plates that are separated by precision thickness (±0.025 mm tolerance) aluminum and magnesium spacer plates for a total stack thickness of 40 mm. The collimators are assembled by placing alternating layers of BeCu and spacer plates onto three precision $\frac{3}{16}''$ stainless steel dowel pins protruding from an 8 mm thick aluminum mounting plate. Alignment is accomplished by gently pushing (with silicone rubber faced right angle blocks) the edges of the plates so that the slots become tangent with the three pins. An aluminum clamping plate is placed over the inlet etched plate to uniformly clamp the stack, and to couple the upper and lower half stacks together.

The collimator assemblies achieve nearly zero alignment tolerance by placing the majority of the burden on the etching and machining precision, in lieu of some elaborate alignment scheme. By direct CAD file transfer, errors in graphics interpretation were avoided. Further, by holding the etching vendor (Tecomat) to ± 0.01 mm locational and size tolerances for the slits, and ± 0.005 mm for the alignment slots, the plates established a new benchmark in precision wet etching. Combining these breakthroughs in etching tolerance with CNC drilling of the mounting plate pin holes, and careful (perpendicular) installation of the screened (size and straightness) dowel pins, assembly was possible with only 0.013 mm clearance about the alignment pins. This clearance represented less than 10% of the critical spacing between adjacent slits for both the high- and low-resolution outlet plates. Clearly the bulk of the alignment was designed and fabricated in the parts.

3.3. MULTI-WIRE PROPORTIONAL COUNTERS

Three geometrically flat, multi-wire, dE/dx proportional counters are used, each serving two instrument sections in a single sensor fan. The proportional counters serve several purposes in the SEPICA sensor:

(1) They are used to determine the specific energy loss dE/dx of the incoming ions in the counter gas. This signal is taken from the central anode wire plane of the counter, whose wires are all interconnected and to which a positive bias voltage is applied. The dE/dx signal is also used as the master trigger signal for the electronic readout and event identification circuits of the instrument.

(2) In the two cathode layers between which the anode plane is sandwiched, the position of the penetrating ion is determined in two dimensions. The cathode wires are grouped in bands of seven wires each that are read out by individual signal channels. In the front cathode the position in the y-direction, i.e., the direction of electrostatic deflection, is determined. In the rear cathode the position in z is determined. Together with another z-measurement in the solid-state detectors the incoming angle Θ of the ions in elevation is derived.

Such a multi-wire proportional counter has been successfully flown as an imaging detector for X-ray astronomy in the focal plane of the Wolter telescope on the German X-ray satellite ROSAT (Pfeffermann et al., 1987). In order to achieve optimum position resolution the center anode plane of the counter is the focal plane of the collimator.

Energy-Loss Determination
The energy loss of the ions in the counter depends on the gas density and the thickness of the gas layer. The ions lose energy mainly due to ionizing collisions with electrons. The resulting charge of the ion-electron pair production is proportional

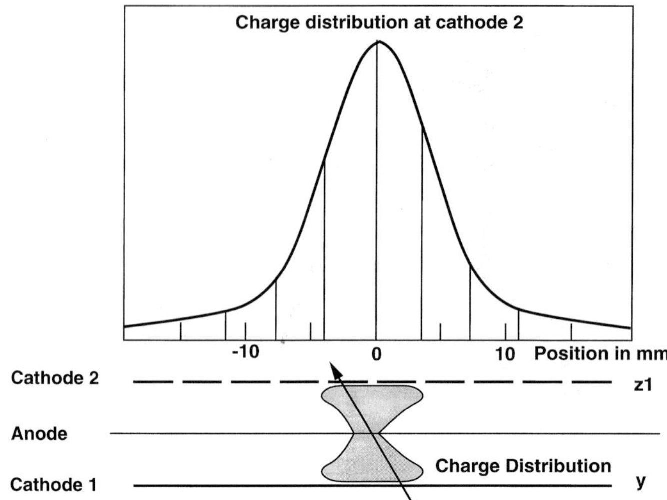

Figure 3.4. Distribution of charged particles generated by an ion that penetrates the proportional counter. The position is determined from the center of gravity of the distribution over five adjacent cathode wire strips (3.5 mm width and 0.5 mm separation) on both cathode 1 and 2.

to the total energy loss. The counter is operated in its proportional regime, i.e., a constant gas amplification of this charge signal is achieved by electron multiplication in the increased electric field close to the anode wires. The anode wires are 10 μm diameter tungsten wires, separated from each other by 2 mm. With an isobutane filling of nominally 20 Torr and an operational voltage of 1000–1100 V an amplification by ≈ 100 is achieved.

Position Measurement

The position of an incoming ion is determined by the localization of the charges at the anode. Therefore, the anode has been chosen as the focal plane. The charges are concentrated on only a few anode wires. However, the image charges of the electron signal produced on both cathodes and the ion clouds that migrate from the anode to the cathodes are distributed over several adjacent wire strips in both cathode planes, as is depicted in Figure 3.4. As a result a charge distribution is recorded whose width is comparable to the anode cathode separation distance. The cathodes consist of 50 μm stainless steel wires with a pitch of 0.5 mm, grouped into signal strips of eight wires each (3.5 mm wide and 0.5 mm apart). This configuration produces an optimum position resolution for the counter geometry. The charge clouds are typically distributed over five such strips. The position is derived from the center of gravity of the distribution. In the front cathode the position in the direction of the electrostatic deflection (y-direction) is determined, while in the rear cathode the position along the focal line (z-direction) is measured.

Counter Assembly

The proportional counters are sealed volumes through which isobutane is flowing at a constant rate such that an operational pressure of $\approx$20 Torr is maintained. In order to minimize the build-up of carbon compounds from the isobutane on the anode wires during operation, the gas is completely exchanged every 4 hours. In addition, care has been taken in the choice of materials in the counter volume and plumbing. Further an admixture of 2% methanol reduces the destruction of isobutane molecules under radiation (Va'vra, 1986).

Because the SSDs are mechanically separated from the counters, two windows (front and rear) are needed to close the volume. In order to minimize the energy loss of the incoming particles in the windows, they had to be made as thin as possible while still providing a sufficient margin for the differential pressure applied. In addition, a conducting surface was necessary to provide equipotential surfaces inside the counter volume. Polyimide windows with a thickness of 0.5 μm and a 500 Å Al layer on an Al frame with a stainless steel support grid, supplied by Luxel Corp., Friday Harbor, WA, were used for this purpose. This allowed a significant reduction of the energy loss in the windows compared with the 450 μg/cm^2 Ni window of the ISEE ULEZEQ sensor, although SEPICA needs two windows. As an illustration for this improvement, take the energy loss of ^{4}He at 0.34 MeV nucl^{-1}, the low energy threshold for He with ULEZEQ: the loss in the combination of both polyimide windows is 0.24 MeV compared with 0.34 MeV in the single Ni window. Together with a 24 mm column of isobutane at 20 Torr in the counter, a reduction of the total energy loss by 40% over that of ULEZEQ has been achieved which allows a significant reduction of the low-energy threshold for SEPICA.

The windows have withstood a pressure of up to 60 Torr without a sign of permanent stretching. Much higher overpressures could be applied without rupture. The danger of destruction due to micrometeorites is very low. Due to the presence of the slit collimator in front of the proportional counter the probability of damage to the windows from micrometeorite impact during a 5-year mission lifetime is estimated to be <0.5%.

The counter design requires one anode plane and two cathode planes with 10 μm and 50 μm wires exactly spaced by 2 mm and 0.5 mm, respectively. The wires are placed on structural Macor frames on a custom-made precision winding machine, then bonded in place with adhesive and soldered to metal traces. The challenge in this configuration is to carry the signal lines from the cathode plates out to the support electronics while maintaining the pressure volume integrity. In the ROSAT counter the anode and cathode plates were placed within the gas volume thereby requiring a containment vessel and additional signal feedthroughs. With SEPICA, to allow three counter systems close to each other and to minimize the instrument size and mass, the novel approach of integrating the containment and the signal feedthroughs into the counter plates was undertaken. Therefore, the assembly of the counter is broken into six primary component layers with O-ring seals between each of them: the inlet and outlet thin film windows on Al frames, a gas interface plate made from Ultem, the front and rear cathode plates, and the center anode plate, all three of them made from Macor.

The biggest challenge was introduced by the need to hermetically seal 21 electric feedthroughs in each cathode plate. In the SEPICA counter design this was achieved by developing 'Surface Trace Feedthroughs' in cooperation with Ceramic-to-Metal Seals, Melrose MA. Signal line feedthroughs were fabricated into the surface of the Macor cathode plates by first milling shallow pockets into the Macor, filling the pockets with Dupont Silver #1773, firing the metallization, and then grinding the surface until the traces and Macor are flush. The bond between the Macor and metallization is virtually leak tight, and the resulting surface finish after grinding is O-ring seal quality. The surface traces are solderable in their fired state (2% Ag solder required), allowing the cathode wires to be soldered inside the O-ring envelope, and the electronics interface connector to be soldered outside the envelope. The remaining single 2 kV feedthrough for the anode plate is a more conventional single O-ring seal with a Ni plated Ti conductor compressing a small O-ring in a reamed bore in the anode plate.

The assembly is integrated from the anode plate outward, i.e., each cathode plate is fastened separately with approximately 20 vented cap screws to the anode plate. A custom 1 mm cross section Buna-N O-ring is placed between each plate to form the seal. The Ultem 1000 gas plate is fastened to the front cathode. The inlet and outlet windows are attached to the gas plate and the back of the second cathode, respectively. A proportional counter assembly is shown in Figure 3.5.

3.4. SOLID-STATE DETECTORS

Solid-state detectors (SSD) are used for the determination of the total energy E of the incoming ions and of a second position along the focal line. Together with the corresponding position in the proportional counter the incoming angle of the ion can be determined in order to apply the necessary correction to the deflection in the electrostatic analyzer. Because of their excellent signal-to-noise ratio and to allow position determination ion-implanted silicon pixel detectors have been chosen. To cover the energy under consideration the thickness of the detectors is 500 μm. To facilitate the additional z-position measurement the SSDs are segmented into strips oriented along the y-direction. The layout of the detector is shown in Figure 3.6.

Another challenge for the SSDs is the required dynamic range of total ion energies from 0.1 to 300 MeV. However, by making use of the electrostatic deflection, ions with energies ≥ 1 MeV charge^{-1} are contained to within ≈ 10 mm from the undeflected focal line. Therefore, the SSDs have also been segmented into a high- (1–300 MeV) and a low- (0.1–30 MeV) energy section. The high-energy section contains eight strips, while the low-energy section has twelve strips. This arrangement has been optimized to the need of better angular correction in the low-energy regime where a charge resolution $\Delta Q / Q = 10\%$ can be achieved. To avoid cross talk on the SSD from the high- to the low-energy section a guard structure has

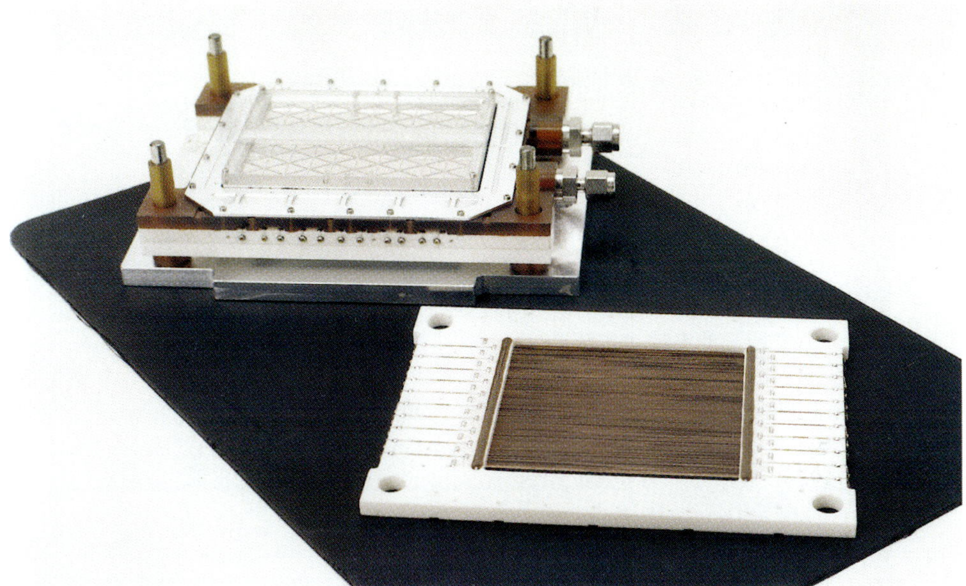

Figure 3.5. SEPICA proportional counter assembly, as seen from the front side. The gas in- and outlet are on the left. In the Fan assembly the center bar of the window support is located exactly behind the HV deflection plate of the electrostatic analyzer.

been implemented in the layout. The stop ring avoids noise from imperfections at the edge of the SSD wafer. The two SSDs for each fan are mounted side by side directly on the printed circuit board with the SSD electronics. In this way noise pickup and electronic cross talk between channels could be minimized.

3.5. ANTICOINCIDENCE DETECTORS

The anti-coincidence counters completely cover the back of the solid-state detectors and extend beyond the sensitive area of the SSDs sufficiently to detect ions which penetrate the detector volume at oblique angles. CsI scintillators are used with Si photodiodes (by SILICON Sensors GmbH, Berlin) as light detecting devices. A similar arrangement has been successfully implemented in the SAMPEX HILT experiment (Klecker et al., 1993). Each of the three anti-coincidence units consists of a $\approx 100 \times 60$ mm CsI crystal with a thickness of 5 mm and a 3 mm acrylic backing. The scintillation signals are picked off by 20 Si photodiodes of 10×8 mm, 4 on each short side and 6 on each long side of the CsI crystal. The anti-coincidence assembly is mounted to the rear side of the SSD circuit board and covered with a ground shield. To avoid cross talk between anti-coincidence and SSD channels the front side of the CsI crystal has been covered with a vapor deposited 1000 Å layer of Al. Together with the reflective paint that is applied to the backing this maximizes the light collection of the detector system.

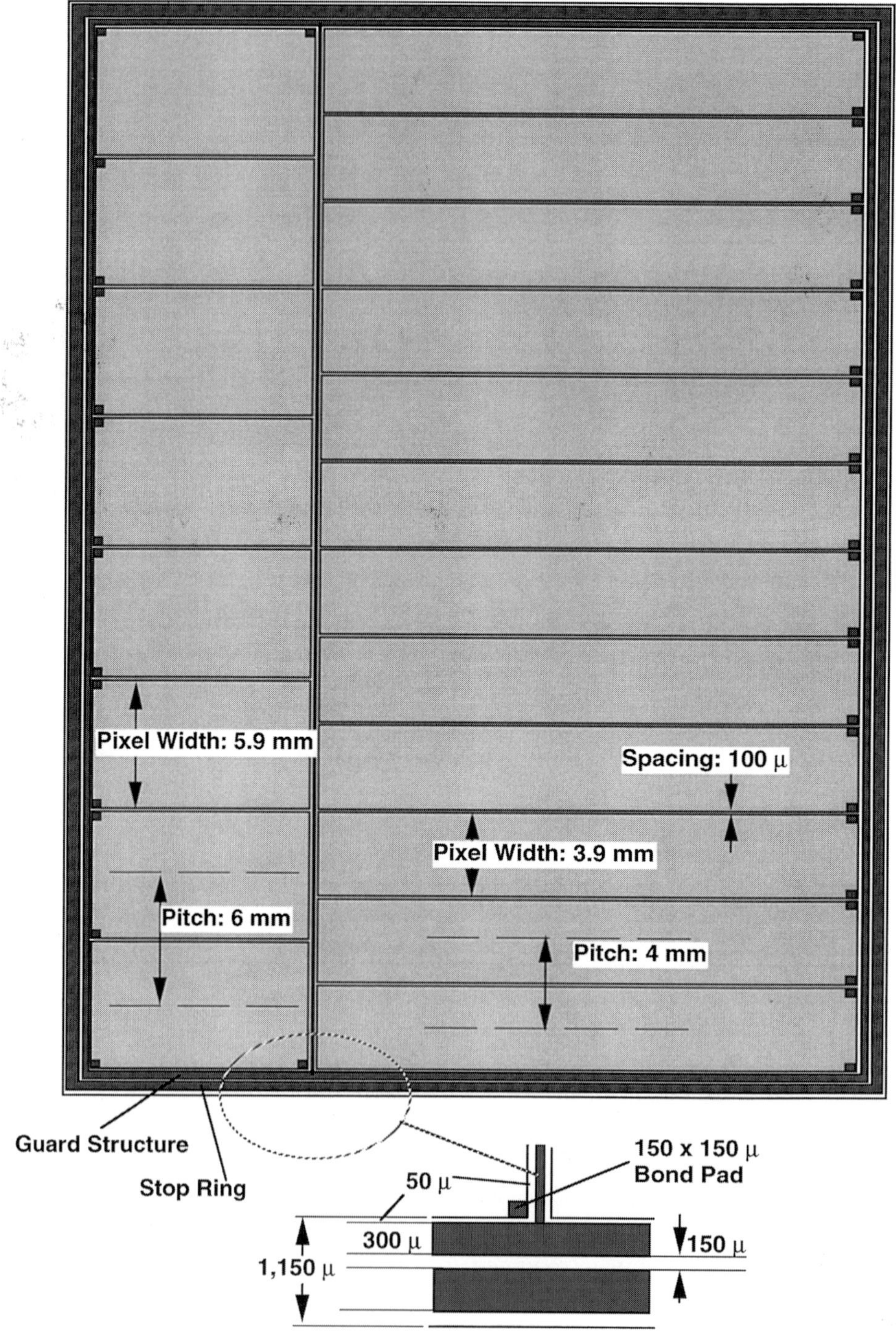

Figure 3.6. Layout of the rear side of the ion-implanted silicon pixel detectors by Canberra, Belgium. The front side consists of a <1000 Å thick dead layer with vacuum deposited Al.

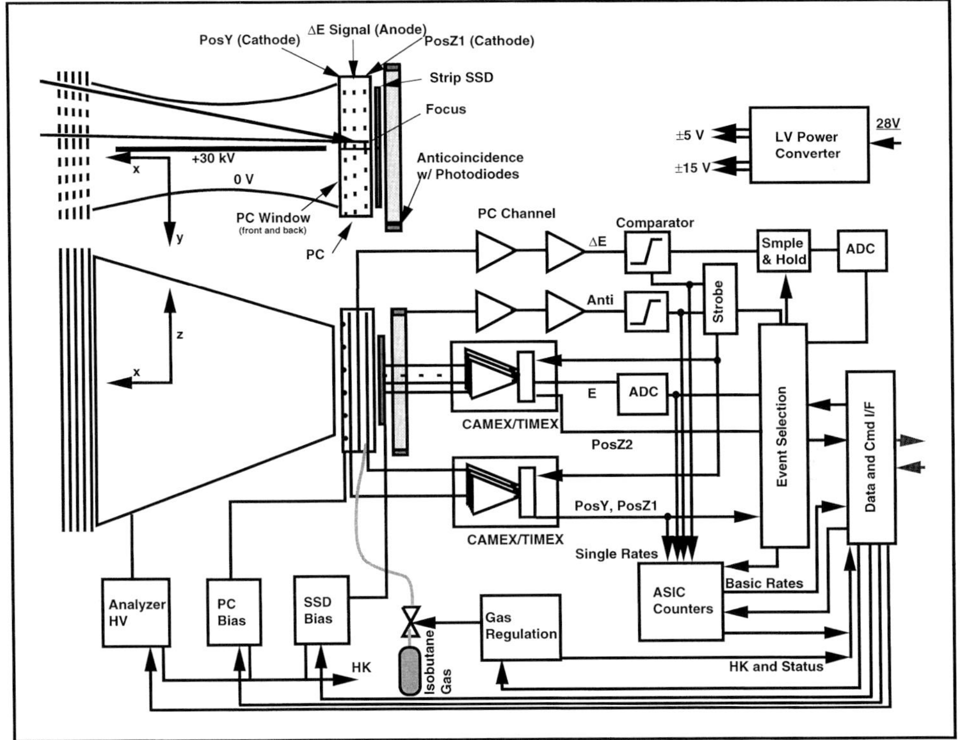

Figure 4.1. SEPICA block diagram.

The assembled proportional counter is mounted together with the SSD assembly
and anti-coincidence detector as well as the two remaining boards of the sensor
electronics over the four titanium alignment rods on a mounting frame. This mounting frame is used for integration with the electrostatic analyzer section of the sensor
fan. Clamping nuts secure the PC and electronics, and a ground enclosure that
contains the signal and power connectors shields the complete detector assembly.

4. Sensor Electronics

All sensor analog electronics is contained in the sensor fans, while the conversion
to digital signals and the event selection is carried out in the electronics box. This
also houses most of the supporting electronics. Figure 4.1 shows the main functions
in an electronics block diagram.

4.1. ANALOG ELECTRONICS

The analog electronics consists of individual amplifier chains for the dE/dx and
anti-coincidence channels as well as CMOS VLSI circuits (CAMEX and TIMEX)

which serve the position determination in the proportional counters and the energy measurement in the solid-state pixel detectors. The individual amplifier chains for the PC anode and the anti-coincidence consist of an AMPTEK A250 hybrid circuit with a FET preamplifier stage, followed by pulse shaping. Each of the three fans contains one dE/dx and one anti-coincidence channel.

Each CAMEX (CMOS Multichannel Analog Multiplexer) contains 32 charge-sensitive amplifier channels whose outputs are sampled by fast switches into storage capacitors. After completion of the measurement cycle the charge difference accumulated during the event is read from all channels sequentially into the following electronics (Lutz et al., 1987). The CAMEX was originally designed for high-energy ion beam experiments, in which event timing is known *a priori*; i.e., the detection cycle can be treated synchronously with the experiment. In space applications the automatic sampling cycle of the CAMEX has to be interrupted at the detection of an incoming particle to allow a readout sequence. This readout is triggered by a signal in the corresponding dE/dx channel for the proportional counter/solid-state detector unit, provided that no high-energy particle is signaled by the anti-coincidence detector. The necessary control electronics for the timing of the sampling and the readout is implemented in a custom VLSI called TIMEX. In order to miniaturize the electronics and to reduce cross talk the CAMEX/TIMEX combination, including some passive circuitry, has been implemented on a custom hybrid, developed at the Fachhochschule Landshut, Germany. A similar hybrid was previously flown in the CELIAS instrument on SOHO (Hovestadt et al., 1995). On each CAMEX the user can select the gain for a group of sixteen channels from 0.12 and 2.2 V/fC.

Four CAMEX/TIMEX amplifier hybrids are used in each fan: two serve the position sensing in the PCs and the other two, energy determination with the solid-state detectors. Each proportional counter CAMEX is connected to 21 cathode wire groups. The connections are interleaved so that position information in both the rear (deflection direction) and front (focal line direction) planes is available on each CAMEX, thus providing partial redundancy. The high-gain state of the CAMEX is used for the PC. Each SSD CAMEX serves one half of each of the two detectors with twenty individual pixels. The CAMEX channels that serve the eight pixels next to the deflection plate in the center of each fan, where the high-energy ions hit, are switched to the low-gain, while the remaining twelve pixels are set to the high-gain state. The ratio of the two SSD pixel gains is 20, thus providing a nominal energy range of 0.1–300 MeV.

4.2. SEPICA DIGITAL ELECTRONICS

The SEPICA digital electronics consists of three 'ADC/peak boards' (one per fan) on which the analog signals (dE/dx, PC positions, and SSD) are digitized and the peak positions are determined, one 'event board' which selects valid events from the three fans and accumulates monitor rates from the event signals, and a

TABLE 4.1

Trigger conditions and related count rates for each of the three fans

Rate	Description
PPC_Trig	Trigger from the proportional counter
PPC_Trig∗Anti-coincidence Trigger	Anti-coincidence signal in coincidence with PPC_Trig
PPC_Trig∗$\overline{\text{Anti-coincidence Trigger}}$	PPC_Trig NOT in coincidence with Anti-coincidence
Valid Event	Event satisfying the commandable Valid Event Criteria (given below)

'preprocessor board', which carries out initial processing of the data to reduce the number of bits per event into the data processing unit (S3DPU). The preprocessor was also necessary so that the same interface to the S3DPU could be used for SEPICA, SWICS and SWIMS.

Two basic types of data are generated in the SEPICA electronics: rate data and pulse-height analysis (PHA) event data. Event data contains detailed information about each particle which enters the instrument. Rate data contains count rates of individual signals and coincidences between signals, as shown in Table 4.1.

Event processing begins when a signal above threshold is detected by the PC anode and the selected conditions are met. In this case the signals are read from the ΔE channel as well as from the cathode and SSD CAMEXes. They are converted in 11 bit ADCs. The single signal from the anode of the proportional counter goes through a high-gain amplifier and a low-gain amplifier. The difference between the two gains is a factor of 15. If the high-gain signal exceeds a commandable threshold, the low-gain is used. The outputs from the CAMEXes, the SSD signals and the position signals from the PC go through further processing on the ADC/peak board. The baseline level for each CAMEX output, which is stored in a lookup table, is subtracted from each measurement. The SSD pixel as well as the Y and Z strips in the PC with the largest signals are determined from the resulting signals. The energy signal is read from the selected SSD pixel. To determine the positions in the proportional counter in Y and Z, the peak signal plus the signals from the two adjacent strips on each side are selected. Prior to passing the data to the preprocesser board, it is determined whether the event is valid according to criteria in Table 4.2.

Any combination of these PC Y, PC Z1, and SSD criteria, including disabling all checks, are allowed and can be set by telecommand. The default valid event condition is

$$\text{PC Y-position} \bullet \text{PC Z1-Position} \bullet \text{(SSD-1)} \tag{4.1}$$

TABLE 4.2

Criteria for the validation of events in SEPICA

Criteria	Description
PC Y-Position	At least one wire strip had a signal over threshold
PC Z1-Position	At least one wire strip had a signal over threshold
SSD-1	One and only one SSD strip had a signal over threshold.
SSD-2	At least one SSD strip had a signal over threshold

TABLE 4.3

Pulse height signals transmitted in engineering mode

Signal	Description	Quantity	Bits/item
Δ+ gain bit	PC anode signal (both gains)	2	12
E	Energy (8 low, 12 high-gain pixels, both SSDs)	40	11
Z2	Strip number of the peak SSD pixel	1	5
YC	Pulse height of all strips in Y	21	11
YP	Strip number of peak Y strip -2	1	5
Z1C	Pulse height of all strips in Z1	21	11
Z1P	Strip number of peak Z strip -2	1	5
Fan ID	Fan ID	1	2
SSD ID	Identifies top or bottom half of fan	1	1

There are two sensor modes, 'engineering mode' and 'science mode'. In engineering mode, the full set of pulse-height signals as shown in Table 4.3 is transmitted unchanged for a limited number of events.

In science mode the information for each valid event is already limited to the most important signals upon transfer into the preprocessor board according to Table 4.4.

In science mode, the preprocessor board reduces the information to be transmitted even further. It calculates the total deflection Y of the incoming ion in the deflection field according to

$$Y = \Sigma_i (YP_i * YC_i)/\Sigma_i (YC_i) \, , \tag{4.2}$$

where YP_i is the wire position and YC_i is the pulse-height amplitude for each cathode strip. The Z position $Z1$ is calculated from the amplitude distribution at the rear cathode in a similar way, whereas the Z position $Z2$ taken in the SSDs is the center position of the SSD pixel with the maximum signal. Finally, the two Z positions $Z1$ and $Z2$ are combined into the difference $|Z1 - Z2|$ which

TABLE 4.4

Pulse-height signals passed into the preprocessor in science mode

Signal	Description	Quantity	Bits/item
$\Delta E+$ gain bit	The proportional counter signal	1	12
E	Peak energy from the SSD's	1	11
Z2	Strip number of the peak SSD	1	5
YC	Pulse height of strips surrounding Y peak	5	11
YP	Strip number of corresponding Y strips	5	5
Z1C	Pulse height of strips surrounding Z1 peak	5	11
Z1P	Strip number of corresponding Z strips	5	5
Fan ID	Fan ID	1	2
SSD ID	Identifies top or bottom half of fan	1	1

TABLE 4.5

PHA event data as transferred into the S3DPU

Signal	Description	Quantity	Bits/item
$\Delta E' +$ gain bit	PC anode signal	1	11
$E +$ gain bit	Peak Energy from the SSDs	1	12
Y	Computed position in deflection direction	1	10
$|Z1 - Z2|$	Computed angle in fan direction	1	8
Fan	Fan ID	1	2
SSD ID	Identifies top or bottom half of fan	1	1
Mode	0 = Science, 1 = Engineering	1	1

is proportional to the tangent of the incoming angle Θ parallel to the fan plane. The locations of the center of each wire strip and of each SSD pixel are stored in a look-up table so that the SSD position, $Z2$, and the PC position, $Z1$, will be in the same units, with the origins of $Z1$ and $Z2$ aligned such that $Z1 = 0$ and $Z2 = 0$ corresponds to $\theta = 0$. Figure 4.2 shows the $Z1$ (PC) and $Z2$ (SSD) planes and indicates for a sample trajectory how $|Z1 - Z2|$ is used to determine the incident fan angle Θ. The PC cathode strip numbers and the high gain pixels of the SSD are shown. The preprocessor also drops the LSB from the ΔE signal thus returning $\Delta E'$, and it adds a gain bit to the $\Delta E'$ and the SSD energy signal.

The processed PHA event data that is transferred into the S3DPU is compiled in Table 4.5.

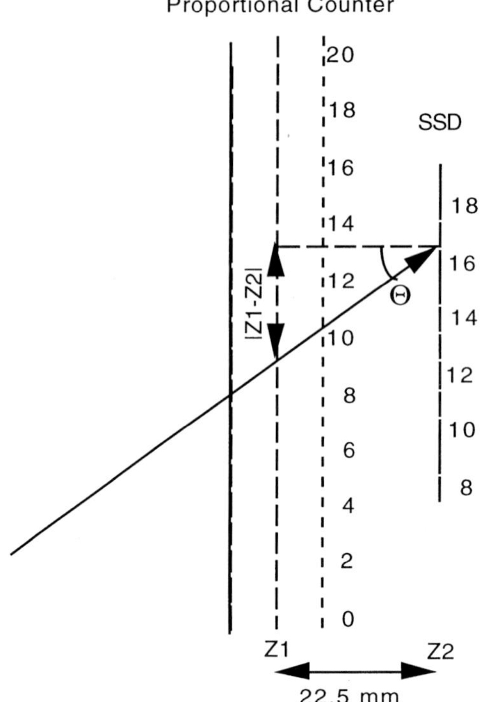

Figure 4.2. Angle θ, and the $Z1$ and $Z2$ directions. Schematic of fan.

4.3. EVENT PROCESSING IN THE S3DPU

The SEPICA instrument is served by a common Data Processing Unit (S3DPU) together with the SWICS and SWIMS experiments. In this section the requirements for the S3DPU from the side of SEPICA and the data stream through the unit are compiled. A brief description of the S3DPU architecture is given in Section 9.

Basic and Matrix Rate Data
In the S3DPU, the event data is accumulated on board into bins representing specific species, energies and charge states. Prior to the accumulation, two corrections are applied to the data. (i) The ΔE signal is corrected for the path length through the PC:

$$\Delta E' = \Delta E \, \cos \theta \, . \tag{4.5}$$

(ii) The deflection Y is corrected for traversal of the electrostatic analyzer at oblique angles:

$$Y' = Y \, \cos^2 \theta \, . \tag{4.6}$$

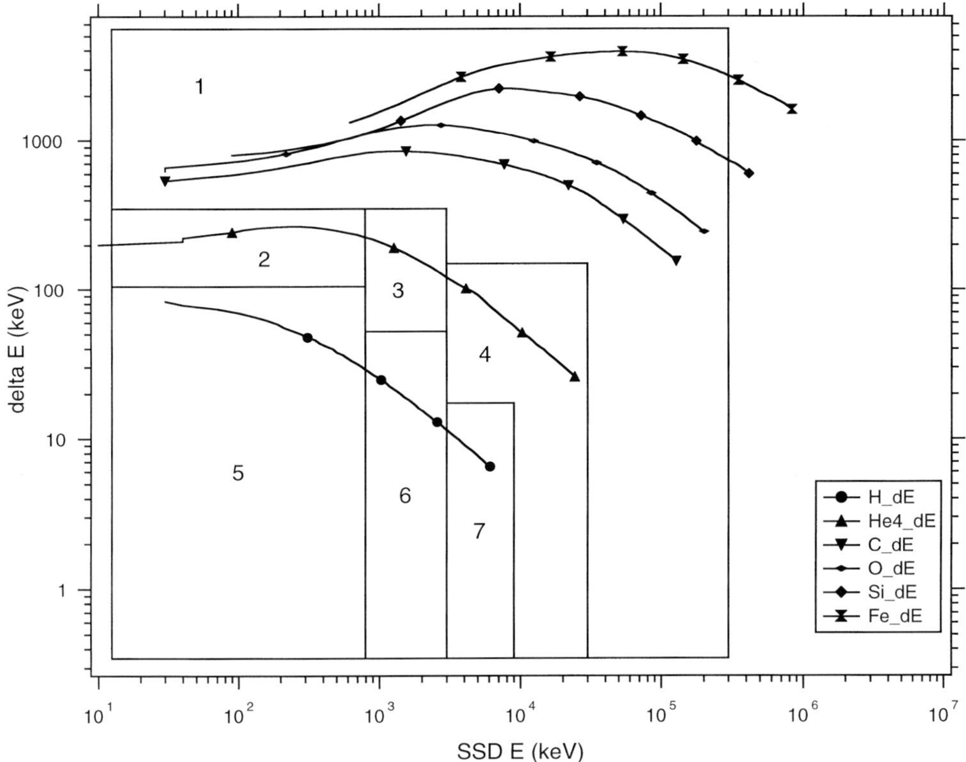

Figure 4.3. Basic Rate Boxes in $E - \Delta E$ space. Hydrogen (both neutral and ionized) is contained in the BP rates and priorities (4, 5, 6) and He (charged and neutral) in the BA rates and priorities (1, 2, 3). All other ions and neutrals are contained in the BH rate and priority (0).

These corrections are made using a look-up table for $\cos \Theta$ and $\cos^2 \Theta$, given the value $|Z1 - Z2|$. The charge state is then calculated according to

$$Q^* = C1Y'E_{\text{Res}} , \tag{4.7}$$

where $C1$ is a constant determined by calibration.

Using the values of E_{Res}, $\Delta E'$, and $Q*$, each event is accumulated into two types of 'Classification Rates': Basic Rates and Matrix Rates. Both sets of rates cover the full valid event range in E_{Res}, $\Delta E'$, and $Q*$. The Basic Rates are larger bins, designed to grossly divide up the data into protons, helium, and heavy ions. The basic rates are not further divided into charge-state ranges. The boundaries for the basic rates are also the boundaries that determine the priorities for the PHA Data. The Matrix Rates divide the data more finely into individual species and into charge-state ranges. The boundaries for the Basic Rates and Matrix Rates as defined for the start of the mission are shown in Figures 4.3 and 4.4. These boundaries are adjustable by telecommand.

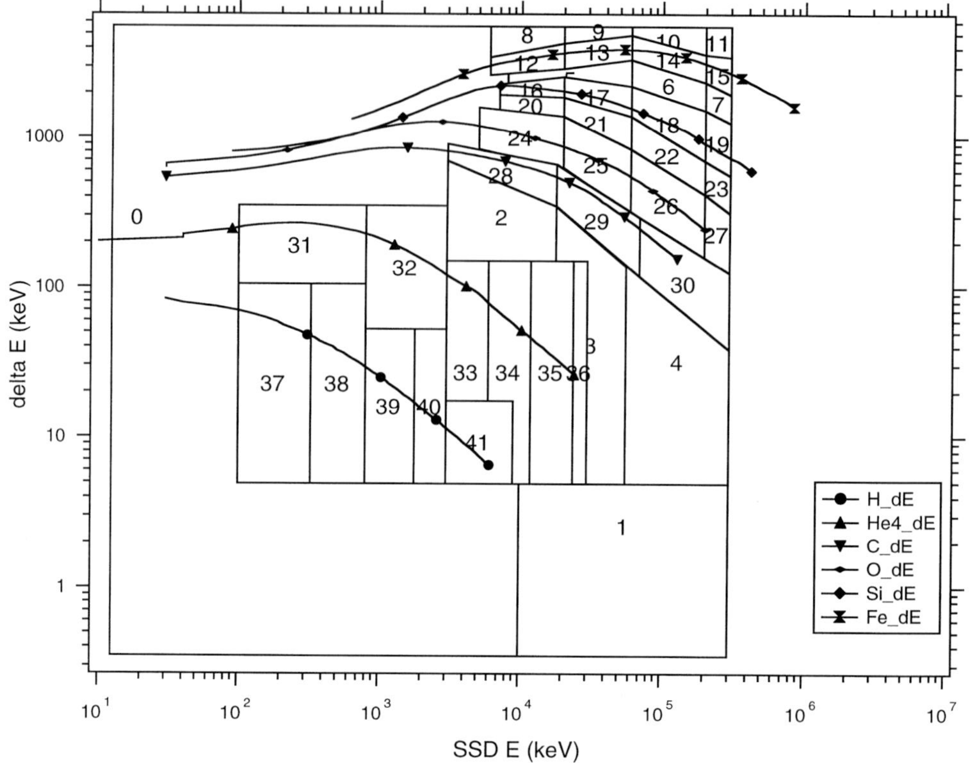

Figure 4.4. Matrix boxes for the SEPICA sensor.

The charge-state represents the third dimension in Figure 4.4. The shape of the boxes in E and ΔE remains the same throughout the entire Q range. The limits for the charge ranges are defined individually for each element and energy range. The ranges in Q are contiguous, and there is no overlap. The maximum of the last box is the maximum allowed Q, 31 (or 5 bit for Q). The Q ranges for each of the matrix elements are shown in Figure 4.5. The physically reasonable range for each element is shaded.

PHA (Pulse-Height Analysis) Data

A small sample of the raw events from the preprocessor board are added to the telemetry. The S3DPU adds two quantities to these events: the priority code and the spin sector in which the event occurred. In addition, it log compresses the energy and energy loss signals. The priority code indicates which range the event occurred in. The priority range boundaries correspond to the basic rate boundaries. In order to emphasize heavy ions in the transmitted data, the fraction of events from each range to be put into the telemetry can be specified. If there are not enough events of a high-priority range to fill the available telemetry, it is filled by events of lower priority.

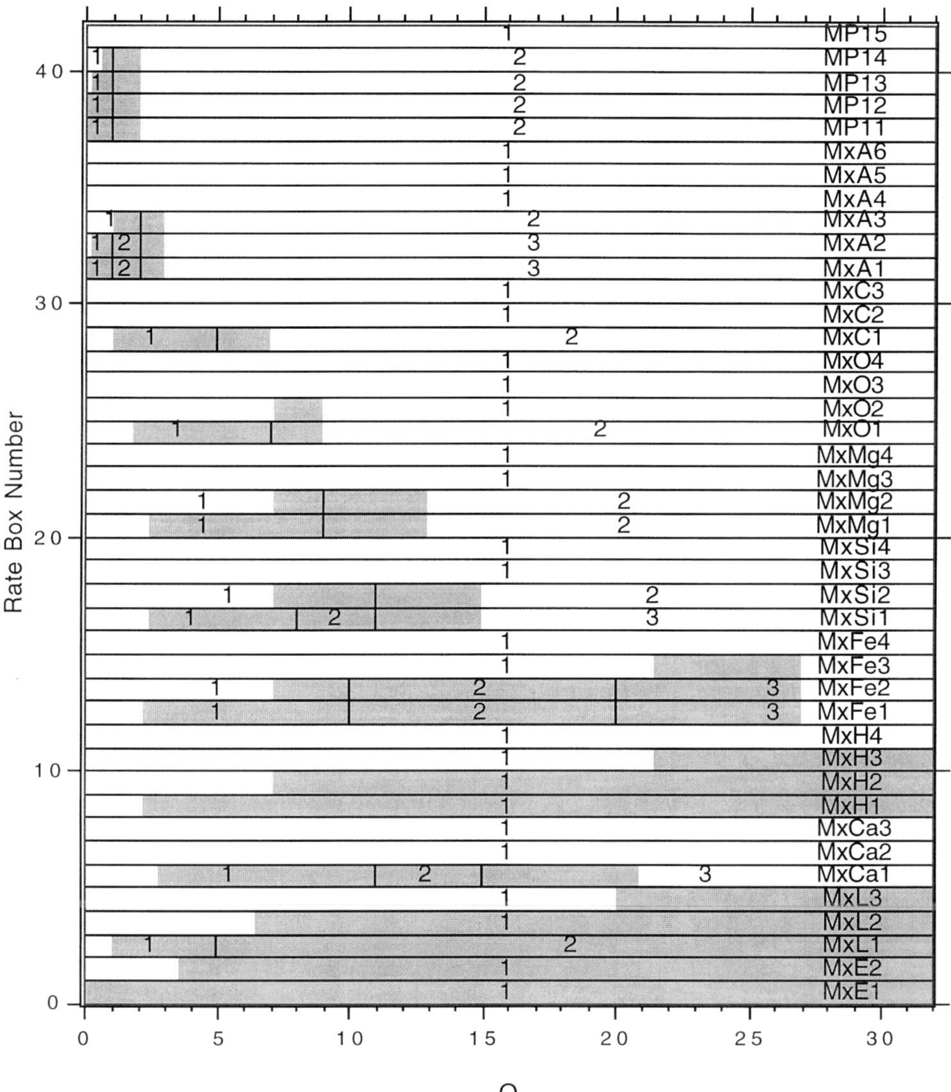

Figure 4.5. Charge-state bins for each matrix rate. The shaded region shows the physically meaning-ful regime. The minimum of the shaded region is the charge which will give 2 mm deflection for the minimum energy in the energy range, and the maximum is the fully stripped charge for the ion. For energy range in the boxes with no shading the deflection is <2 mm.

In order to emphasize data from the high-resolution fan (fan 1) when it is available, the S3DPU checks the BH rate for fan 1. If it exceeds a programmable threshold, only PHA events from fan 1 are included in the PHA data. Otherwise, the PHA data comes from all fans. The final PHA events that are transmitted in science mode contain 48 bits as shown in Table 4.6.

TABLE 4.6

PHA events as transmitted into telemetry in science mode

Signal	Description	Quantity	Bits/item
$\Delta E'$ + gain bit	PC anode signal, log compressed	1	10
E + gain bit	Peak energy from SSDs, log compressed	1	11
$Z1 - Z2$	Angle in fan direction	1	8
Y	Computed position in deflection direction	1	10
Fan ID	Fan ID	1	2
SSD ID	Identifies top or bottom half of fan	1	1
Priority	Priority code	1	3
Azimuth	Sector in which event was measured	1	3
Total			48

5. Calibration and Performance

The calibration of the SEPICA can be separated into the determination of the geometric factor, the evaluation of the charge resolution (separately for each fan), and the identification of elements and charge states to define the Matrix boxes in Figures 4.3 and 4.4. The geometric factor of each fan is determined by the effective aperture area and the acceptance angle of the respective collimator and the transparency of any obstructing structure in the detector assemblies, such as grids and the proportional counter wires. Also the charge-state resolution depends mainly on the geometry of the collimator. Therefore, these sensor attributes can be determined from a geometric characterization of the collimators, which has been performed before the fan integration activities. However, the element and charge-state identification are properties of the entire sensor fan. Therefore, the latter activity was performed with complete fans or even the total sensor using α-sources and accelerator beams.

5.1. GEOMETRIC CHARACTERIZATION OF THE COLLIMATORS

Focal-line width and throughput of the SEPICA collimators were verified in a series of optical tests. The focal-line widths were measured with the full collimator illuminated. The focal-line FWHMs for the three fans are 0.982 ± 0.007 mm, 3.23 ± 0.04 mm, and 3.47 ± 0.03 mm. Ideal per design values are 0.984 mm, 2.95 mm, and 2.95 mm, respectively. The slightly increased values for the focal line of fans 2 and 3 are probably due to taking the measurement below the focal line. Evaluation of data with partial illumination of the collimator showed a distinct separation of the respective focal line images. Because the collimators were already integrated at this time, the measurement could not be repeated. By comparing the centers of the focal line separately for the upper and lower half of the collimators it was verified that the two halves of the collimator are aligned to within 2.5 μm.

The collimator throughput was determined for the engineering model by a comparison of the light transmission through a single slit for a two-plate collimator (entrance and exit only) with that through the full collimator assembly. Any misalignment due to tolerances reduces the transmission for the full collimator over the two-plate assembly, which represents ideal transmission for the design geometry. Because this more precise method requires excessive handling of the collimator plates, the relative transmission of the flight collimators was verified by determining the reduction of the focal-line width over the design geometry for several 5 mm $\times$ 5 mm sample areas across the collimator. The square of the focal line width reduction is equivalent to the reduction in transmission, because both entrance aperture and acceptance angle are affected. For fan 1 an average transmission across the collimator area of $67 \pm 4\%$, close to the goal of 70% was found. This translates into a net geometric factor of 0.031 cm^2 sr. For the other two fans, the transmission is consistent with a combined geometric factor of 0.20 cm^2 sr. This value includes all reductions of the instrument detection efficiency due to collimator transmission and the transparency of the detector assemblies. The detection efficiency computed from the individual grid transparencies and the actual efficiency as derived from the comparison between the proportional counter trigger rate and the coincidence rate agree within the instrument accuracy.

The deflection in the electrostatic analyzer and position resolution of the collimator-analyzer system were tested with a low-energy (2 keV) ion beam. A microchannel plate imaging system was used for detection in the focal plane, and the deflection voltage was scaled to the lower energies. The deflection test was repeated in the full fan configuration with α-particles from an ^{241}Am source. These measurements have verified that SEPICA deflects 1 MeV/Q ions by 10 mm with the full deflection voltage of 30 kV.

5.2. CALIBRATION OF THE ELEMENT AND CHARGE IDENTIFICATION OF SEPICA

The element identification in SEPICA is achieved in a ΔE versus residual energy E parameter analysis of the sensor. While the energy response of SSDs is fixed by their charge conversion of ≈ 3 eV per e-ion pair, the ΔE signal as measured by a PC depends on its geometry, the gas composition and density, as well as the electric field strength around the anode wires. In isobuthane the creation of an e-ion pair requires 23 eV. For nominal operation of the SEPICA fans with 20 Torr PC pressure at 25 °C and an anode voltage of 1000 V, a gas gain of 115 is achieved with a doubling of the signal for each additional 90 V. The electronic threshold for the ΔE signal is set to 6 fC which is equivalent to an energy loss of 8 keV in the PC. This corresponds to the signal of a 3 MeV proton. The high-gain section of the amplifier chain tops at 195 fC (equivalent to the maximum energy loss of He ions),

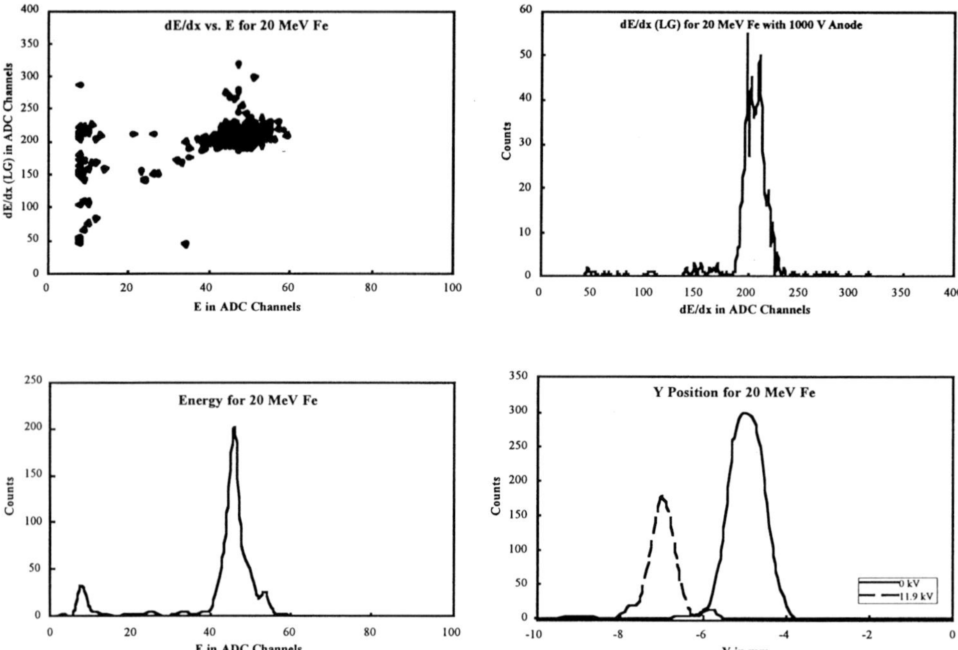

Figure 5.1. Sample calibration measurement as taken with the SEPICA spare fan at BNL with a 20 MeV Fe beam at 11.5 kV deflection voltage. A dE/dx versus E chart is shown in the upper left along with the corresponding dE/dx and E spectra to its right and bottom. The fourth panel on the lower right shows the deflection of the ions in comparison with the deflection voltage turned off.

while the low-gain section extends to 4.27 pC thus including also 60 MeV Fe with the maximum energy loss in the operational regime.

Figure 5.1 shows a sample of calibration data taken at the Brookhaven National Lab (BNL) with a 20 MeV Fe beam. Both element identification capability from dE/dx versus E and charge resolution from ion deflection are illustrated. The FWHM of the dE/dx distribution is ≈10% and for residual energy ≈8%. Some events are found along a line towards lower dE/dx and E and a line at constant dE/dx. The first group probably has lost energy at the support grid of the entrance window and thus falls on the Fe track at lower energies, while the latter group must have lost energy after passing the PC. The event groups with $E = 0$ mark ions that were stopped in one of the grids or wire planes of the PC and thus were degraded in E and/or dE/dx. The relatively wide foot in energy may be attributed to the large energy loss and straggling experienced by 0.36 MeV nucl^{-1} Fe ions. The lower right panel of Figure 5.1 demonstrates the position resolution and ion deflection in a comparison of two measurements with 11.5 kV and 0 kV deflection voltage applied. The undeflected focal line is at the nominal Y position of 5 mm, while an additional deflection of 2 mm is achieved in the 11.5 kV run. The result is compatible with the beam charge state of 10 to 11 and a slightly less than 1 mm wide focal line in the PC.

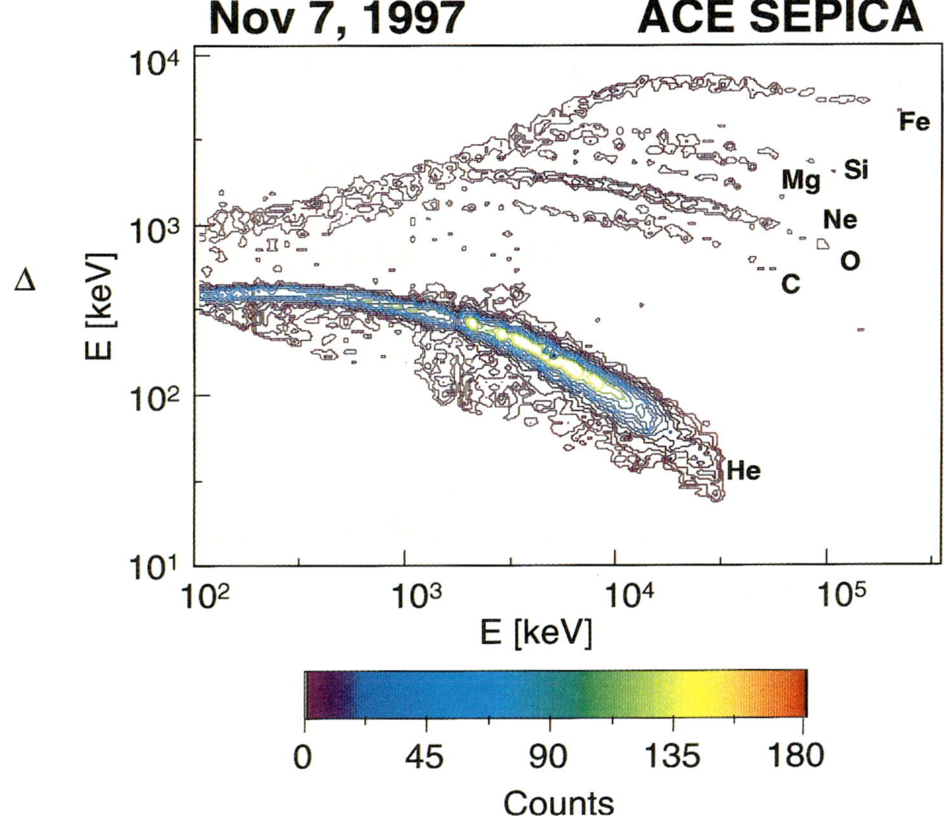

Figure 5.2. Color-coded pulse-height event matrix in energy loss (ΔE) versus residual energy (E_{res}) for one of data from fan 1 of SEPICA during a solar flare on November 7, 1997. The tracks for the different elements are clearly separated. Hydrogen events have been taken out during the data processing to reduce the dynamic range in total counts for the display.

The element resolution of SEPICA and the actual position of the element tracks for the calibration of the Matrix Rate boxes as shown in Figure 4.3 are demonstrated for in-flight operation from the strong solar flare on November 5, 1997, in Figure 5.2. SEPICA can indeed resolve individual elements up to Ne and provides a good resolution of the major ion groups up to Fe. The long-term behavior of the inflight calibration can be monitored through three ^{241}Am α-sources with 50 μCi activity, one in each sensor fan. The calibration α-particles are directed into the low-energy part of the detector in order to avoid interference with the natural He population. The stability of the electronics is monitored separately through a self-stimulation sequence that can be initiated by telecommand.

TABLE 6.1

Power as used by subsystem

Subsystem	Average power (W)	Peak power (W)
Detectors	6.93	6.93
E-box	7.71	8.71
Gas regulation	0.36	0.36
30 kV supply	1.50	1.50
Totals	16.5	17.5

6. Supporting Subsystems

The functionality of the SEPICA instrument is not only based on the sensor performance and the processing electronics, it also requires the trouble-free operation of sophisticated support systems, such as high-voltage supplies for the deflection unit, PCs and SSDs, as well as a gas-flow control system for the three PCs. These subsystems, which required a high design effort, are described below. The power distribution into subsystems is compiled in Table 6.1.

6.1. DEFLECTION HV SUPPLY

The deflection HV supply will provide a positive HV up to 30 kV to each of the positive deflection plates in the electrostatic analyzers. The 30 kV power supply consists of an amplitude controlled oscillator and a Cockroft–Walton multiplier. The voltage available to the bottom of the voltage multiplier can be varied from zero to 2000 V peak by comparing an input analog command with feedback from the high voltage and using the error signal to control the current in the oscillator. A hard limit is set at 2000 V peak by the DC power supply to the oscillator. The multiplier consists of 35 stages which allow the output voltage to reach about 36 kV.

The power supply is divided into two cavities, one with the low-voltage circuits up to the output transformer, the other with the voltage multiplier and the 5000 MΩ feedback resistors. The low-voltage circuitry is packaged in a standard printed wire board, the transformer is mounted on the metallic housing, and the high-voltage output wire is routed very carefully to minimize the danger of discharges. The voltage multiplier is subdivided into seven circular printed wire boards, each with five stages of multiplication and a metal shield ring around the circumference that provides a low electric field environment on the board and shapes the field between the electronics and the housing. The boards have a center hole and are mounted on an Ultem tube that contains the high-voltage feedback resistors. Short tubes of silicon nitride are used to set the spacing between the boards and to provide a low

Figure 6.1. 30 kV HV supply, interior of multiplier assembly and insulation insert (top), driver board and individual multiplier stage (bottom).

thermal resistance to the metal housing at the bottom of the structure. There is an insulating sleeve of fiberglass inside the external aluminum cylindrical housing. The high-voltage end is capped by a field-shaping torus and supported through an Ultem end piece. The output is through high-voltage cables, with their metal shield terminated on the metal housing and the center wire and insulation extending through the Ultem support to the center of the torus. Here the connections are made by means of pin and socket in a low field region. A view of the open HV supply is shown in Figure 6.1.

The electric-field distribution in the high-voltage section was studied and optimized throughout the design phase using the ELECTRO simulation program by Integrated Engineering Software. The engineering model supply was tested up to 40 kV by raising the input voltage supply beyond the specified value. It was operated for 400 hours at 36 kV for flight qualification.

6.2. PROPORTIONAL COUNTER BIAS SUPPLY

The PC bias supply provides a positive high voltage for each of the three PCs, which can be commanded separately with 8-bit resolution up to 2.5 kV. The effective source resistance of each of the three outputs is 500 MΩ and provides sufficient safeguards for each of the PCs. The high-voltage of the Cockroft–Walton multiplier stack is determined and loop-controlled by the highest set value. The individual output voltages for the three PCs are independently controlled and regulated by separate feedback loops with high-voltage opto couplers. In case of a failure in one of the PCs, the corresponding bias voltage can be switched off by setting its command value to zero. This approach saves mass, power and space.

The Cockroft–Walton multiplier stack consists of five stages. The opto couplers are customized assemblies of a transparent high-voltage diode and four IR-LEDs completely enclosed with an optically transparent potting material.

6.3. SOLID-STATE DETECTOR BIAS SUPPLY

The SSDs are biased with a negative voltage up to -75 V, which is provided by a single transformer. The primary side of the transformer is fed with a feedback loop controlled AC voltage. The secondary transformer output is rectified and separately supplied over a 100 kΩ filtering network to each of the six SSDs. The actual bias voltage can be set with 8-bit resolution up to 75 V.

6.4. GAS REGULATION SYSTEM

The SEPICA gas regulation system controls the flow of isobutane gas through each of the 210 cm^3 PC volumes for all three fans to maintain a fixed gas density. The PCs are operated in a flow-through mode in order to keep them supplied with fresh gas. This minimizes the build-up of cracked hydrocarbons on the counter wires. Therefore, the counter gas pressure is balanced against a calibrated precision orifice, which provides a gas flow of 0.75 cm^3/min at a PC pressure of 20 Torr. In essence the control system is referenced to a pressure set value that is adjusted internally for constant gas density according to the sensor temperature as measured by a thermistor. With the combined outflow rates of all three fans the SEPICA gas supply will be sufficient for full operation over more than 6 years.

Description of the Gas Regulation
A schematic overview of the gas system is shown in the block diagram in Figure 6.2. The regulation of the gas flow is accomplished individually for each PC by using a membrane pressure transducer and a micro-machined bi-metallic silicone valve by IC Sensors, Milpitas, CA. The voltage output from the pressure transducer is compared with a value set by the S3DPU based on telecommand and sensor temperature. The control loop of the system uses pulse-width modulation for the on-time of the valve to adjust the pressure to the preset value. Each proportional counter is serviced by the system in sequence, i.e., the maximum on-time for each PC can be one third of the time. After a maximum on-time of 9 s the gas control will time out and automatically shut off the gas, unless the set value has been reached before that time or the S3DPU issues an overriding 'off' command. The maximum on-time is used during the initial filling of the PCs. During nominal operation the fill intervals typically last 100–400 ms.

The regulation band for the gas pressure is $\pm 1\%$ and can be set by command through the S3DPU to any value between 10 and 30 Torr. Nominal operation is anticipated at 20 Torr. Gas regulation of 1% precision is assured with a 10-bit ADC. The control algorithm and additional safety functions are implemented in an Actel gate array.

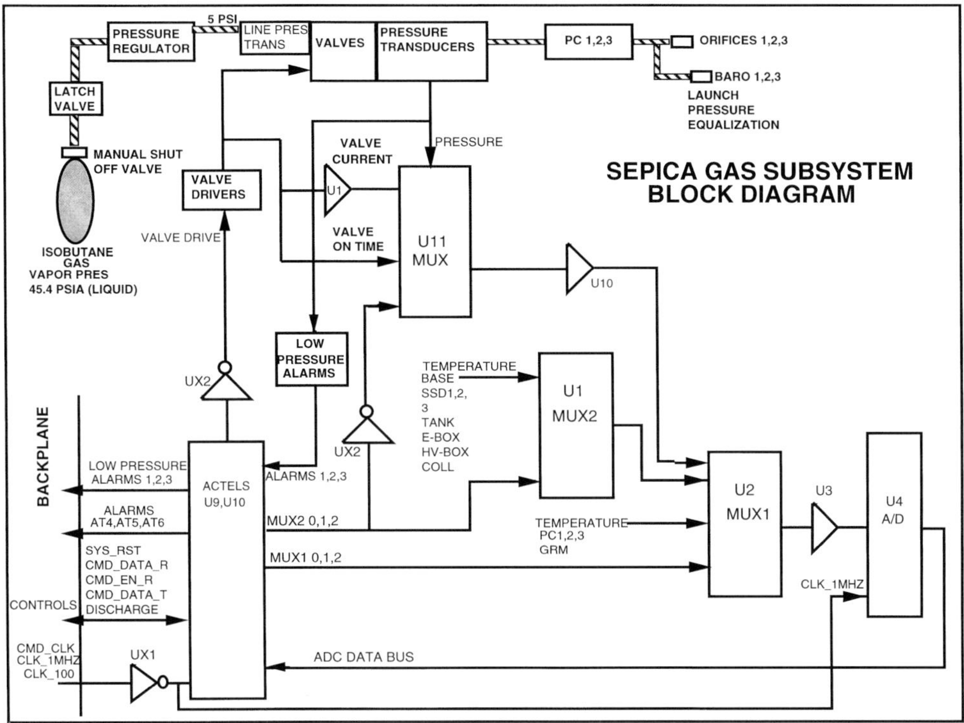

Figure 6.2. Block diagram of SEPICA gas flow control.

The gas regulation works off a supply pressure of 250 Torr. Preregulation of the gas pressure from the tank, which is at the isobutane vapor pressure, is achieved with a mechanical pressure reducing valve. A fourth pressure transducer monitors the preregulated gas at 250 Torr which feeds the three gas regulation valves. During launch the tank volume is sealed by a magnetic latch valve that is controlled by spacecraft command.

System Safeguards

During the ascent of the spacecraft through the atmosphere all three counter systems are open to the outside through a baro switch that closes at an outside pressure of ≈60 Torr. In this way the integrity of the PC windows is assured. The remainder of the air trapped in the counter volume will then bleed slowly through the precision orifice.

During instrument operation the PC gas pressure and the valve on-times will be continuously monitored by the S3DPU. If the gas pressure is too high or too low, or if the gas valve on-time is too long, the S3DPU will automatically take corrective action without human intervention. In case of an overpressure, the corresponding regulation valve will be power-cycled with a pulse of full power automatically, which is a known cure in case a valve stays open after an excessive temperature swing. In case of an underpressure the corresponding regulation valve will be shut off and the PC HV for this unit will be disabled to avoid damage of the PC. In all cases an over- or underpressure alarm will be set in the telemetry.

7. Structural Design and Packaging

SEPICA is a modular instrument with 3 distinct sensor fans, an isobutane gas supply and regulation system, a 30 kV high-voltage supply, an electronics box and a passive thermal radiator all tied to a common base-plate. This modular design allows independent assembly, testing and integration of each fan and supporting subsystem with minimal impact on the balance of the instrument. All subsystems are through bolted to the base plate. SEPICA weighs 37.4 kg ready for launch with basic outline dimensions of 71 cm × 46 cm and 53 cm height. The mass distribution according to subsystems is compiled in Table 7.1.

The bulk of the SEPICA structure is aluminum, with alloy 6061 used in low-stress areas, and alloys 7075 and 2219 used in higher stress parts such as the base-plate, or in elaborate parts requiring precise flatness and tolerance control such as the collimator mounting and spacer plates as well as the radiator. A limited supply of magnesium AZ-31B plate is used in the collimator spacer plates and some of the structure for mass reduction purposes. As a high electrical performance, low outgassing material, Ultem 1000 polyetherimide is used for electrical and HV insulation in the detector assembly and the 30 kV HV supply. Macor machinable ceramic is utilized for the proportional counter cathode and anode plates. The isobutane tank is a carbon fiber overwrapped aluminum vessel, custom fabricated for this application.

The ACE spacecraft mechanical interface is through 10 mounting lugs with titanium $\frac{1}{4}''$–28 aerospace quality socket head cap screws, and upper/lower Ultem 1000 thermal isolation bushings. These high-strength fasteners are torqued to 11 Nm, thus providing adequate compression to avoid slippage between the deck and bushings. All electrical interface harnesses connect with SEPICA at the instrument electronics box via standard D-subminiature connectors. In addition, two sets of $\frac{3}{16}''$ Teflon tubing lines run bundled into a spacecraft harness. One set is utilized to purge the SEPICA detectors with dry nitrogen during integration and testing and the other one to carry the isobutane gas, that is vented from the proportional counters, away from the spacecraft.

8. Thermal Control

SEPICA is thermally controlled via radiative dissipation to space. The instrument is conductively isolated from the spacecraft deck by Ultem in order to minimize spacecraft dissipation requirements. The resulting thermal resistance to the deck is

TABLE 7.1

Mass distributed by subsystem

Subsystem	Mass (kg.)
HV supply & dist. cables	1.7
Gas supply system	1.9
Gas dist. & regulation	3.7
Fan assemblies(3)	18.7
Base plate & brackets	4.9
Thermal items/radiators	1.6
E-box	4.9
Total	37.4

TABLE 8.1

Temperature limits for each subsystem

Subsystem	In spec. operating	Design/test limits	No-op. survival
Detectors	-15 to $+35$ °C	-20 to $+45$ °C	-20 to $+60$ °C
Electronics	-15 to $+60$ °C	-25 to $+70$ °C	-25 to $+75$ °C
30 kV supply	-15 to $+55$ °C	-25 to $+65$ °C	-25 to $+70$ °C
Gas supply tank	-15 to $+40$ °C	-20 to $+50$ °C	-25 to $+60$ °C
Gas regulation	-10 to $+40$ °C	-20 to $+50$ °C	-20 to $+60$ °C

SEPICA requires 2 W of operational heater power to maintain the gas regulation subsystem within its in-spec. operating limits. Additionally, SEPICA requires 8 W of survival heater power to maintain the nonoperating limits.

15 °C/W. A 1000 cm^2 passive radiator mounted on the starboard side of SEPICA, covered with silver Teflon film, provides the necessary radiative shunt. Other than the instrument aperture and the radiator, the balance of SEPICA is covered with multilayer insulation blankets. The preferred operational thermal range for the bulk of SEPICA is -10 °C to $+40$ °C, which brackets the demonstrated cold and hot case limits for the detectors on the spacecraft of $+5.8$ °C and $+39.5$ °C. The thermal constraints of all essential subsystems are compiled in Table 8.1

The SEPICA instrument electronics and power supplies dissipate approximately 16.5 W average during normal operation. All instrument subsystems are surface treated with high ϵ, black anodized hard coat for optimal emittance into the cavity under the blanket, and eventually to the back of the radiator. The aperture (collimator) plates are coated with Goddard Composite Coating (GCC), a unique high ϵ, low α vacuum-deposited film, intended to minimize solar heating effects. Thermal baffles between the three detector apertures and forward edge of the radiator are also covered with silver Teflon.

A thermostatically actuated operational heater circuit of 2 W is provided on the isobutane gas regulation module to protect the regulation transducers from low temperature extremes (below 0 °C). An 8 W survival heater circuit is provided on the instrument electronics box in the event of an extreme cold temperature turn-on or other casualty. The survival heater has been demonstrated to maintain the SEPICA cavity temperature at an average of +17 °C.

9. Common Data Processing Unit (S3DPU) for SWICS, SWIMS and SEPICA

Three of the ACE-sensors, SWICS, SWIMS and SEPICA, are served by one common data processing unit, whose heritage goes back to the DPUs for AMPTE-CHEM, Ulysses-SWICS, PHOBOS-SOWICOMS, Geotail-EPIC and SOHO-CELIAS, which have been designed and implemented by the Technical University Braunschweig, Germany. The S3DPU is a direct derivative of the SOHO-CELIAS DPU (Hovestadt et al., 1995). A common characteristic of this DPU family is an event driven, table-oriented fast preprocessor for classification and priority identification, that is very flexible and can serve a variety of different sensors with identical hardware. The general hardware and software functions as well as the design philosophy of the S3DPU are described in the following section, while the data structure and the scientific contents are discussed in conjunction with the individual sensors. For SEPICA the data flow from the sensor to the spacecraft has been described in Section 4. of this paper. The corresponding description for SWICS and SWIMS may be found in the paper by Gloeckler et al. (1998) in this volume.

9.1. Tasks and general structure

The S3DPU has been designed to perform the following tasks:

(1) Path-length correction for passage through the electrostatic analyzer and the proportional counter (PC) according to the impact positions in the PC and the solid state detectors for SEPICA;

(2) Classification of pulse-height analysis (PHA) events into:

— a two-dimensional M versus M/Q matrix for SWICS

— a one-dimensional M vector for SWIMS. It accumulates them separately into Matrix Rates (with low resolution in matrix space, energy separation and high time resolution) and into Matrix Elements (with high resolution in matrix space, energy integrated and low time resolution)

— two-dimensional Z versus Q Matrix Rates for SEPICA;

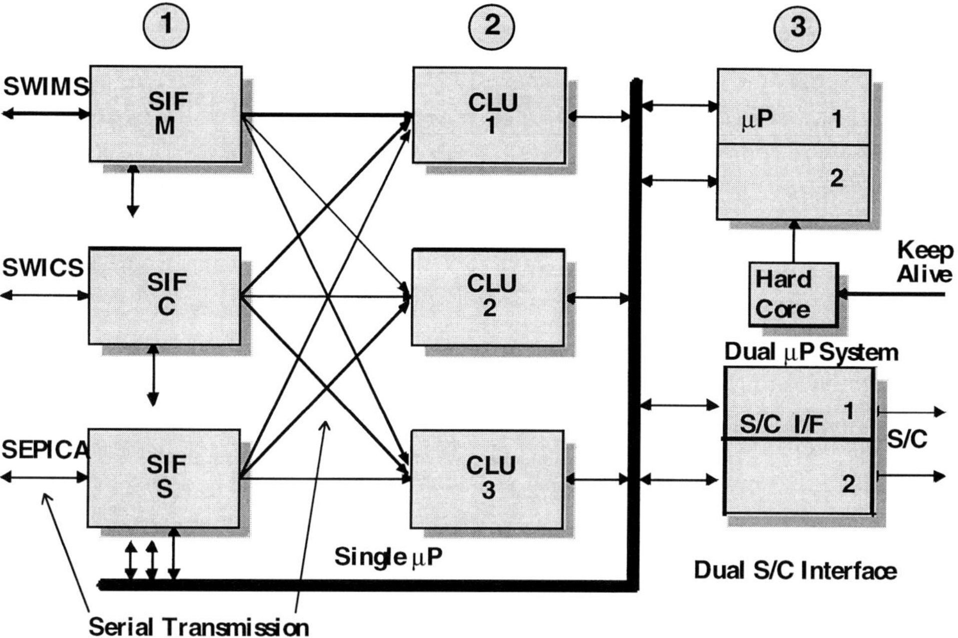

Figure 9.1. Schematic block diagram of the S3DPU.

(3) Prioritization of PHA events from each sensor according to a predefined scheme and insertion of these events into the PHA Section of the Experiment Data Blocks;

(4) Formatting and transfer of the Experiment Data Block into the spacecraft telemetry;

(5) Receiving, decoding and execution of memory load commands;

(6) Control of the deflection voltage stepping according to a predetermined scheme for SWICS and SWIMS;

(7) Automated control of the sensor status and the reconfigurable S3DPU parts that can also be altered or initiated by telecommand;

(8) Initialization of the S3DPU program after a latch-up-induced power-down period;

(9) Monitoring of the housekeeping (HK) values for the S3DPU and the sensors with autonomous reaction on out-of-spec values for critical items, such as HV and gas pressure;

(10) Execution of preprogrammed inflight calibration cycles for each sensor on telemetry request (sequences with different Inflight Calibration commands, various settings of the stimulation DACs for all sensor elements are stored in the S3DPU);

The S3DPU architecture is structured into three levels as shown in Figure 9.1:

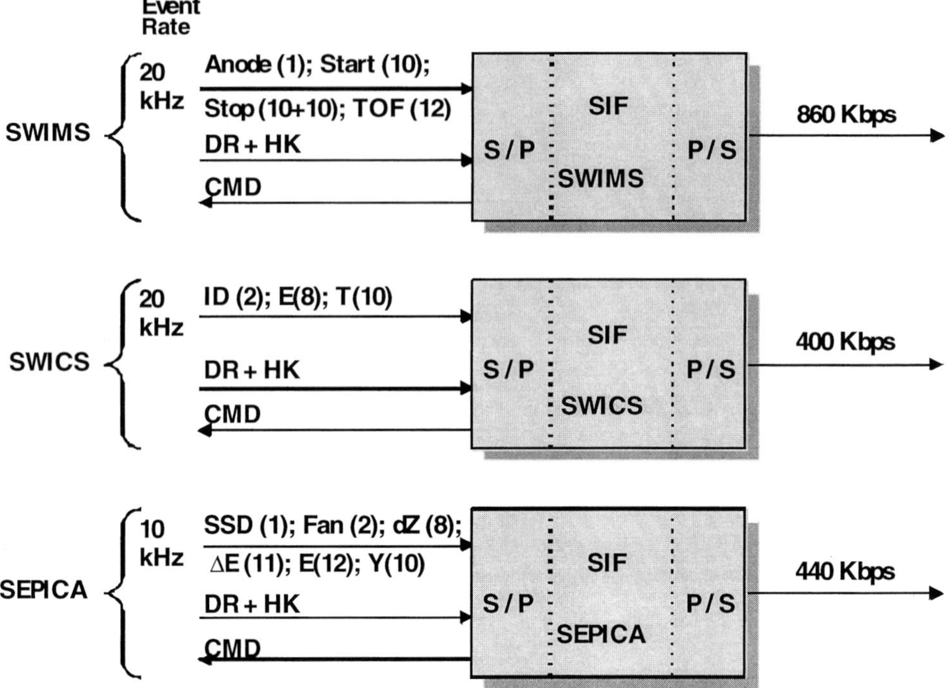

Figure 9.2. Block diagram, data structure and rates for the three sensor interfaces of the S3DPU.

(1) The heart of the S3DPU is a dual redundant microprocessor system (80C86) with a dual redundant S/C interface, including redundant power supply and watch-dog circuitry.

(2) To connect to the three sensors there are three hardware-identical Sensor Interfaces (SIF).

(3) Finally, there are three hardware-identical Classification Units (CLU). Both, SIFs and CLUs, can be configured by the microprocessor system for any sensor.

The level (1) processing (core software) is sequenced by a tailored Real Time Operating System, which responds to interrupts (from the spacecraft interface, the CLUs, the watch-dog circuitry and the latch-up protection circuitry) according to a priority scheme. The sequencing of level (2) and (3) processing is event driven. Within level (3) each of the three CLUs can be allocated to each SIF. This switching is supported by serial data transmission from the SIFs to the CLUs. If one or even two of the CLUs should fail, the remaining ones will be shared by all three sensors according to a preprogrammed or commanded stepping scheme.

9.2. SENSOR INTERFACES

Figure 9.2 shows the SIFs in more detail. The electronics of each individual sensor and the SIFs communicate through serial transmission via balanced twisted pair lines. Accidental high-voltage surges, that may enter the SIFs as a common mode signal, are rejected by balloon ring coils. The SIFs comprise all the preprocessing of sensor data, which cannot be handled by the identical CLUs. The amount of these dedicated preprocessing tasks increases in the order of SWICS, SEPICA, SWIMS. In order to conform with the heritage SIF from SOHO a special pre-processor board has been included in the SEPICA sensor electronics as described in Section 4. The maximum event rates of 20 kHz (SWIMS), 20 kHz (SWICS) and 10 kHz (SEPICA) result in maximum output rates of approximately 1 Mbps (SIF-M) and approximately 0.5 Mbps, (SIF-C, SIF-S), respectively.

9.3. CLASSIFICATION UNITS

In the classification units, the incoming particles are sorted according to their relevant parameters which then allows separate onboard accumulation. With SEPICA the energy loss (ΔE) in the PC, the residual energy (E_{Res}) after passing the PC, and the deflection (y) in the known electric field of the electrostatic analyzer are measured to determine the element (Z), energy (E), and ionic charge (Q) of each particle, as described above. SWIMS measures the time-of-flight (τ) in a retarding harmonic potential which determines the mass/charge (M/Q) of the particle with high precision. In addition, the energy/charge (E/Q) is recorded. In SWICS the original energy/charge (E/Q) of the incoming particle as well as its time-of-flight (τ) and residual energy (E_{Res}) after post-acceleration by the voltage U_{acc} are measured. From these quantities the mass (M) and mass/ionic charge (M/Q) are derived and recorded together with the original E/Q of each particle. For all three instruments the derived quantities form the basis for their classification.

To illustrate the classification scheme we take SWICS as an example. The actual relations (Gloeckler et al., 1998)

$$M = \beta(E,\ M)\tau^2 E_{\mathrm{Res}} \,, \tag{9.1}$$

$$M/Q = \alpha(E,\ M)\tau^2 (E/Q + U_{\mathrm{acc}}) \,, \tag{9.2}$$

relate the measured quantities E/Q, τ, and E_{Res} to the mass (M) and mass/charge (M/Q) of the incoming ions. The species and energy dependent parameters α and β describe the energy loss in the entrance foil of the time-of-flight sensor and the energy defect in the solid-state detectors, respectively. Relations (9.1) and (9.2) are modeled by power series expressions

$$\ln M = b_0 + b_1 * v + b_2 * u b_3 * u * v + b_4 * v^2 + b_5 * u^3 \,, \tag{9.3}$$

$$\begin{aligned}
\ln M/Q &= a_0 + a_1 * u + a_2 u^4 \,, \\
a_0 &= A_{00} + A_{01} * w + A_{02} w^2 \\
a_1 &= A_{10} + A_{11} * w + A_{12} w^2 \\
a_2 &= A_{20} + A_{21} * w + A_{22} w^2
\end{aligned} \tag{9.4}$$

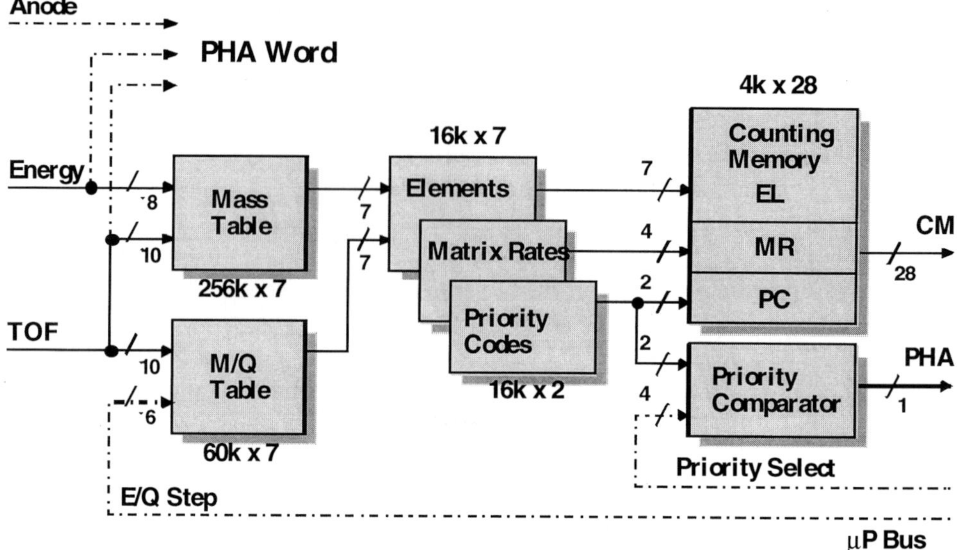

Figure 9.3. Flow diagram of the CLU coding for SWICS as an example.

where $u = \ln(\tau/ns)$; $v = \ln(E_{\mathrm{Res}}/\mathrm{keV})$; $w = \ln((E/Q + U_{\mathrm{acc}})/kV)$. The coefficients represent sensor parameters that are determined by the sensor calibrations. With the appropriate modifications, similar relations can be applied to other sensors with a three-parameter measurement. In this way the example described above is generic for all sensors that are supported by the S3DPU.

In case of low-event rates up to ≈ 1 kHz, a numerical calculation of these relations can be performed by a standard microprocessor. Such a software-based approach has been applied, e.g., to the energetic particle and ion composition instrument on Geotail (Williams et al., 1994). Higher event rates call for a table-based evaluation within a dedicated preprocessor. For illustration purposes Figure 9.3 shows a straightforward implementation as a hard-wired two-stage pipeline of fixed-sized tables. The two-dimensional table of 28 entries in energy and 210 entries in time-of-flight delivers a seven-bit value for $M = f(E, \tau)$. Correspondingly, the two-dimensional table of 210 entries in time-of-flight and 27 entries in energy per charge delivers a seven-bit value for M/Q. The combination of both table outputs defines one out of 214 fine resolution bins within the $M - M/Q$-plane. The three channel $(M, M/Q, E/Q)$ address tables in the second pipeline stage compact a set of adjacent fine resolution bins into an accumulation channel into either a Matrix Rate, a Matrix Element, or a Priority Class for PHA events. The E/Q entry of the M/Q table is constant over the period of each deflection voltage step, generally one spacecraft spin. Therefore, the size of the M/Q-table can be reduced to 1 K $\times$ 7 by loading only the valid table section for each E/Q step, provided the entire table is stored within additional random access memory (RAM).

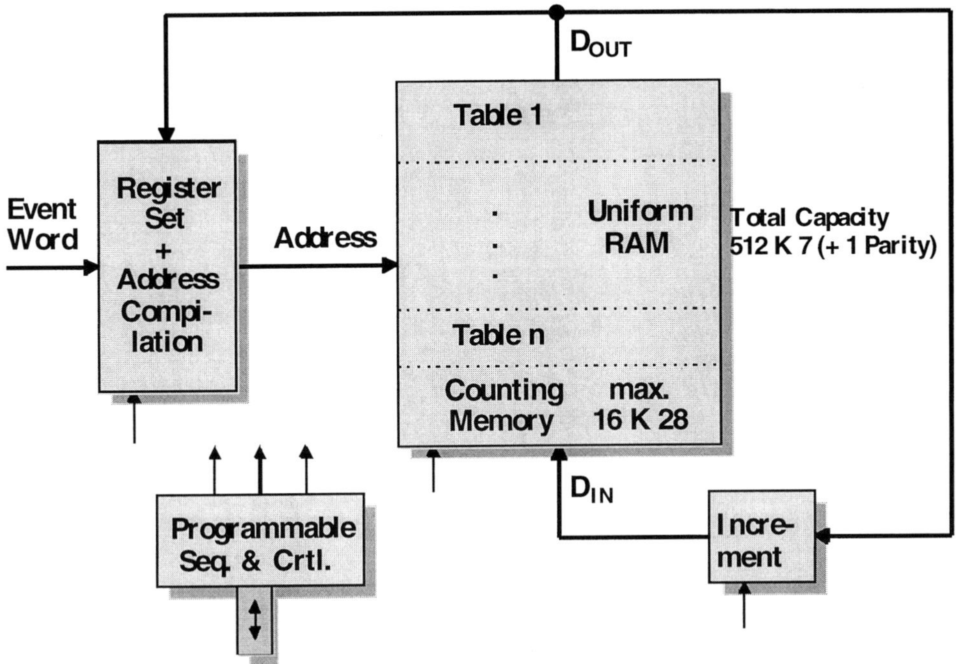

Figure 9.4. Flow diagram of a typical classification software architecture along with the allocation of memory space to the relevant classification tables.

At the extreme end of the implementation, a task specific hard-wired pipeline structure can support very high event rates (up to several 10^6 events s^{-1}), but at the expense of flexibility. For medium event rates (10^4–10^5 events s^{-1}) the tables may be accessed in sequence. A compromise of five subsequent steps has been chosen for the S3DPU. The transition from parallel to sequential table access leads to a very flexible approach that is characterized by a large uniform address space and by mapping of the data flow diagram into a RAM-controlled state machine, which executes the various table operations step by step. Then the number of tables, their size and the interconnection scheme can be configured in software. In this way the table contents are also configurable. Thus the processor can individualize the CLU for each sensor.

The RAM address space of the S3DPU CLUs can be configured into nine tables plus counting channels as seen in Figure 9.4. Typically, an individual table lookup returns another address, which after appropriate modification, e.g., by adding a base address, represents the pointer for the subsequent table access. The final result is the pointer to the counting channels to be incremented (up to three in our example). The contents of the tables are routinely calculated and transferred into the table locations by the active processor. Computation algorithms and coefficients are stored in the micro-processor software. These EPROM-stored coefficients can be modified via telecommand. In this way adjustments or even substantial adaptations to the algorithms can be implemented during the mission.

In addition to the two-dimensional matrix spectra, PHA words of selected events are inserted into the experiment data frame. A priority code that biases against the most abundant species, such as H and He, is assigned to each event. This priority scheme is based on the same classification algorithm as the matrix elements, but with a much coarser resolution. For each PHA event the resulting priority code is compared with a priority select mask that is computed by the processor based on previous events. If the priority of the current PHA event is accepted, the event is read and added to the data block. The detailed requirements for the matrices and the PHA event processing are described for SEPICA in Section 4. The corresponding information for SWICS and SWIMS can be found in the paper by Gloeckler et al. (1998).

9.4. DATA VALIDATION AND REDUNDANCY CONCEPTS

The S3DPU presents a potential single-point failure for three instruments and therefore has to be inherently reliable. Therefore, great care has been taken to avoid single-point failures in the S3DPU and to allow sharing of identical multiple hardware as much as possible. In particular, the S3DPU contains a dual redundant processor system. All core hardware that supports the basic operating system is dual redundant with the exception of the bus system. This concept includes a dual redundant spacecraft interface and a dual redundant power converter. Selection of the interface is controlled by a flip-flop, which toggles automatically with each power cycle of the S3DPU. The processor can be selected by toggling a selection flip-flop via a specific pulse command issued by the spacecraft. The flip-flops belong to the minimum core hardware and are powered by a keep-alive line from the spacecraft. A high degree of redundancy is achieved for the three identical CLUs by cross-strapping. If one or even two CLUs should fail, the remaining CLUs can be shared by all three sensors according to a cyclic automatic reconfiguration scheme. In case of the failure of one of the sensors, the CLUs can be configured such that the data stream will be completely filled with data from the remaining active sensors.

The S3DPU is equipped with protection switches against latch-up through penetrating radiation. The S3DPU circuitry is subdivided into eight independent supply partitions. If a partition-individual threshold for the supply current is exceeded, the S3DPU is shut down completely, with the exception of the status memory and some status flip-flops. This core hardware is then powered by the keep-alive voltage. After a short delay time of ≈ 5 s the central switch will be closed again, and the S3DPU is reinitialized. Since the current of the partitions varies substantially with the activity, a single threshold would have to be set so high that even a sub-threshold latch-up current could overheat a device after some time. Therefore, two thresholds are monitored separately, a high threshold during regular S3DPU operation and a low threshold every 156 ms within a short time window of ≈ 1 ms, when the S3DPU is driven into a sleeping state.

Great care has been taken with the software to run the unit inherently reliably and to ensure data integrity. The software is protected by hamming single error correction (SEC) for both the processor-RAM and -EPROM. For example, single event upset (SEU) induced bit flips are corrected automatically by a SEC coder/decoder hardware in the S3DPU. The parity bit protection is extended to the CLU table RAM in order to avoid SEU induced misclassifications. In addition, the table contents are routinely refreshed by recalculation from the stored parameter set.

9.5. PACKAGING

The S3DPU is built in classic card-cage technique. A stack of nine multilayer daughter boards is interconnected by a multilayer motherboard with flexible extensions to all box connectors. The box walls have been designed to a thickness of 3 mm Al in order to provide sufficient shielding for the high-density standard RAM/EPROM devices within the CLUs and the microprocessor systems, whose tolerance dose is approximately 15 krad. The S3DPU weighs 3.9 kg, uses 3.0 W on average (4.0 W peak) and can operate within $-55\,^\circ$C and $+85\,^\circ$C.

Acknowledgements

The investigators gratefully acknowledge the enthusiastic support with contracts, rush purchasing and accounting by the staff of the Space Science Center: Jeanne Bartlett, Susan Roy, and Pat Stevenson as well as the dedicated day-to-day assistance by Phyllis Kelly and Robin Pendexter. They wish to include in their thanks the many unnamed individuals at the hardware contributing institutions. Four test and calibration campaigns at Brookhaven National Lab were made a success due to the help by Charles W. Carlson and his team. We also would like to say a special thanks to personnel of the Goddard Space Flight Center test facilities who went to great lengths to help SEPICA in the end game. In addition, we would like to extend special thanks to the ACE Payload Management Office under Al Frandsen and the Explorer Project Office under Don Margolies for their continued assistance to keep the instrument implementation effort on track.

The University of New Hampshire would like to extend special thanks to the following people and their corporations. The expertise and dedication they brought to the project helped make SEPICA become a true technological achievement for all involved. In no special order we thank: Mark Amrich, Thermo Electron (Tecomet), for the fabrication of the photo-etched collimator plates; Forbes Powell, Luxel Corp., for the fabrication of the proportional counter polyimide windows; Paul Burger, Canberra Semiconductor, for the fabrication of the solid state detectors;

Hal Jerman, IC Sensors, for the fabrication of the micro-machined silicon regulation valves; Dave Broderick, Ceramic-to-Metal Seals, for the development and fabrication of the proportional counter anode and cathode plates; Dave Ferguson, MOOG Space Products Division, for the fabrication of the main isobutane gas latch valve and regulator; Aleck Papanicolopoulos, Structural Composite Industries, for the fabrication of the isobutane tank; Jennifer Molin, Northeast EDM, for the fabrication of the collimator spacer plates; Steve Dunham, Dunham Engineering, for the design effort with the collimator assembly. Other notable companies with major contributions to SEPICA are Stone Machine, American Electroplating, Seacoast Machine, Bicron, AOTCO Metal Finishing, Janco P/C, Eastern Manufacturing Corp., Sovereign Circuits, Lockheed Sanders, Assurance Technology Corp., and Plating for Electronics.

Without the continued help and dedication of all these institutions, companies and the many individuals, this instrument would not have been possible. The design, fabrication and testing effort for SEPICA was supported under NASA Contract #NAS5–32626 to the California Institute of Technology through subcontract # PC112778 to the University of New Hampshire.

References

Arnaud, M. and Rothenflug, R.: 1985, 'An Updated Evaluation of Recombination and Ionization Rates', *Astron. Astrophys. Suppl. Ser.* **60**, 425.

Baker, D. N., Mason, G. M., Figueroa, O., Colon, G., Watzin, J. G., and Aleman, R. M.: 1993, 'An Overview of the Solar, Anomalous, and Magnetospheric Particle Explorer Mission', *IEEE Trans. Geosci. Remote Sens.* **31**, 531.

Fisk, L. A., Kozlovsky, B., and Ramaty, R.: 1974, 'An Interpretation of the Observed Oxygen and Nitrogen Enhancements in Low-Energy Cosmic Rays', *Astrophys. J.* **190**, L39.

Gloeckler, G., Geiss, J., Roelof, E. C., Fisk, L. A., Ipavich, F. M., Ogilvie, K., Lanzarotti, L. J., von Steiger, R., and Wilken, B.: 1994, 'Acceleraton of Interstellar Pickup Ions in the Disturbed Solar Wind Observed on *Ulysses*', *J. Geophys. Res.* **99**, 17637.

Gloeckler, G., Bedini, P., Bochsler, P., Fisk, L. A., Geiss, J., Ipavich, F. M., Cain, J., Fischer, J., Kallenbach, R., Miller, J., Tums, O., and Winner, R.: 1998, 'Investigation of the Composition of Solar Wind and Interstellar Matter Using Solar Wind and Pickup Ion Measurements with SWICS and SWIMS on the ACE Spacecraft', *Space Sci. Rev.* **86**, 497.

Hovestadt, D. et al.: 1978, 'The Nuclear and Ionic Charge Distribution Particle Experiments on the ISEE-1 and ISEE-C Spacecraft', *IEEE Trans. Geosci. El.* **GE-16**, 166.

Hovestadt, D., Gloeckler, G., Höfner, H., Klecker, B., Fan, C. Y., Fisk, L. A., Ipavich, F. M., O'Gallagher, J. J., and Scholer, M.: 1981, 'Direct Observation of Charge State Abundances of Energetic He, C, O, and Fe Emitted in Solar Flares', *Adv. Space Res.* **1**, 61.

Hovestadt, D., Klecker, B., Höfner, H., Scholer, M., Gloeckler, G., and Ipavich, F. M.: 1982, 'Ionic Charge State Distribution of Helium, Carbon, Oxygen, and Iron in an Energetic Storm Particle Enhancement', *Astrophys. J.* **258**, L57.

Hovestadt, D., Gloeckler, G., Klecker, B., and Scholer, M.: 1984, 'Ionic Charge State Measurements During He$^+$-Rich Solar Energetic Particle Events', *Astrophys. J.* **281**, 463.

Hovestadt, D. et al.: 1995, CELIAS - Charge, Element and Isotope Analysis System for SOHO, in: The SOHO Mission, *Solar Phys.* **162**, 441.

Jokipii, J.R.: 1986, 'Particle Acceleration at the Termination Shock 1. Applications to the Solar Wind and the Anomalous Component', *J. Geophys. Res.* **91**, 2929.

Klecker, B., Hovestadt, D., Gloeckler, G., Ipavich, F. M., Scholer, M., Fan, C. Y., and Fisk, L. A.: 1984, 'Direct Determination of the Ionic Charge Distribution of Helium and Iron in ^{3}He-Rich Solar Energetic Particle Events', *Astrophys. J.* **281**, 458.

Klecker, B. et al.: 1993, 'HILT: A Heavy Ion Large Area Proportional Counter Telescope for Solar and Anomalous Cosmic Rays', *IEEE Trans. Geosci. Remote Sens.* **31**, 542.

Klecker, B. et al.: 1995, 'Charge States of Anomalous Cosmic-Ray Nitrogen, Oxygen, and Neon: SAMPEX Observations', *Astrophys. J.* **442**, L69.

Luhn, A., Klecker, B., Hovestadt, D., Gloeckler, G., Ipavich, F. M., Scholer, M., Fan, C. Y., and Fisk, L. A.: 1984, 'Ionic Charge States of N, Ne, Mg, Si, and S in Solar Energetic Particle Events', *Adv. Space Res.* **4**, 161.

Luhn, A., Klecker, B., Hovestadt, D., and Möbius, E.: 1987, 'The Mean Ionic Charge of Silicon in ^{3}He-Rich Solar Flares', *Astrophys. J.* **317**, 951.

Lutz, G., Buttler, W., Bergmann, H., Holl, P., Hosticka, B. J., Manfredi, P. F., and Zimmer, G.: 1987, 'Low Noise Monolithic Front End Electronics', *MPI-PAE Report, Exp. El.* **170**.

Mason, G. M. et al.: 1991, 'The Solar, Anomalous, and Magnetospheric Particle Explorer (SAMPEX)', *Adv. Space Sci.*

Mason, G. M., Reames, D. V., Klecker, B., Hovestadt, D., and von Rosenvinge, T. T.: 1986, 'The Heavy Compositional Signature in ^{3}He Solar Particle Events', *Astrophys. J.* **303**, 849.

Möbius, E., Hovestadt, D., Klecker, B., Scholer, M., Gloeckler, G., and Ipavich, F. M.: 1985, 'Direct Observation of He$^+$ Pick-up Ions of Interstellar Origin in the Solar Wind', *Nature* **318**, 426.

Mullan, D. J. and Waldron, W. L.: 1986, 'Ionic Charge States of Solar Energetic Particles: Effects of Flare X-Rays', *Astrophys. J.* **308**, L21.

Pfeffermann, E. et al.: 1987, 'The Focal Plane Instrumentation of the ROSAT Telescope', *SPIE* **733**, 519.

Reames, D. V.: 1990, 'Energetic Particles from Impulsive Solar Flares', *Astrophys. J.* **73**, 235.

Reames, D. V.: 1992, 'Particle Acceleration in Solar Flares: Observations', in G. P. Zank and T. K. Gaisser (eds.), *Particle Acceleration in Cosmic Plasmas, AIP Conf. Proc.* **264**, 213.

Stone, E. C., Frandsen, A. M., Mewaldt, R. A., Christian, E. R., Margolies, D., Ormes, J. F., and Snow, F.: 1998a, 'The Advanced Composition Explorer', *Space Sci. Rev.* **86**, 1.

Stone, E. C., Cohen, C. M. S., Cook, W. R., Cummings, A. C., Gauld, B., Kecman, B., Leske, R. A., Mewaldt, R. A., Thayer, M. R., Dougherty, R. L., Grumm, R. L., Milliken, B. D., Radocinski, R. G., Wiedenbeck, M. E., Christian, E. R., Shurman, S., and von Rosenvinge, T. T.: 1998b, 'The Solar Isotope Spectrometer (SIS)', *Space Sci. Rev.* **86**, 357.

Stone, E. C., Cohen, C. M. S., Cook, W. R., Cummings, A. C., Gauld, Kecman, B., Leske, R. A., Mewaldt, R. A., Thayer, M. R., Dougherty, R. L., Grumm, R. L., Milliken, B. D., Radocinski, R. G., Wiedenbeck, M. E., Christian, E. R., Shurman, S., Trexel, S., von Rosenvinge, T. T., Binns, W. R., Dowkonn, P., Epstein, J., Hink, P. L., Klarmann, J., Eijowski, M., and Oletich, M. A.: 1998c, 'The Cosmic Ray Isotope Spectrometer (CRIS)', *Space Sci. Rev.* **86**, 283.

Va'vra, J.: 1986, Review of wire chamber aging, Proc.of Wire Chamber Conf., Vienna.

Williams, D. R., McEntire, R. W., Schlemm II, C., Lui, A. T. Y., Gloeckler, G., Christon, S. P., and Gliem, F.: 1994, 'Geotail Energetic Particles and Ion Composition Instrument', *J. Geomag. Geoelectr.* **46**, 39.

INVESTIGATION OF THE COMPOSITION OF SOLAR AND INTERSTELLAR MATTER USING SOLAR WIND AND PICKUP ION MEASUREMENTS WITH SWICS AND SWIMS ON THE ACE SPACECRAFT

G. GLOECKLER*, J. CAIN, F. M. IPAVICH and E. O. TUMS
Department of Physics, University of Maryland, College Park, U.S.A.

P. BEDINI, L. A. FISK and T. H. ZURBUCHEN
Department of Atmospheric, Oceanic, and Space Sciences, University of Michigan, Ann Arbor, U.S.A.

P. BOCHSLER, J. FISCHER and R. F. WIMMER-SCHWEINGRUBER
Physikalisches Institut, University of Bern, Switzerland

J. GEISS and R. KALLENBACH
International Space Science Institute, Bern, Switzerland

Abstract. The Solar Wind Ion Composition Spectrometer (SWICS) and the Solar Wind Ions Mass Spectrometer (SWIMS) on ACE are instruments optimized for measurements of the chemical and isotopic composition of solar and interstellar matter. SWICS determines uniquely the chemical and ionic-charge composition of the solar wind, the thermal and mean speeds of all major solar wind ions from H through Fe at all solar wind speeds above 300 km s^{-1} (protons) and 170 km s^{-1} (Fe^{+16}), and resolves H and He isotopes of both solar and interstellar sources. SWICS will measure the distribution functions of both the interstellar cloud and dust cloud pickup ions up to energies of 100 keV e^{-1}. SWIMS will measure the chemical, isotopic and charge state composition of the solar wind for every element between He and Ni. Each of the two instruments uses electrostatic analysis followed by a time-of-flight and, as required, an energy measurement. The observations made with SWICS and SWIMS will make valuable contributions to the ISTP objectives by providing information regarding the composition and energy distribution of matter entering the magnetosphere. In addition, SWICS and SWIMS results will have an impact on many areas of solar and heliospheric physics, in particular providing important and unique information on: (i) conditions and processes in the region of the corona where the solar wind is accelerated; (ii) the location of the source regions of the solar wind in the corona; (iii) coronal heating processes; (iv) the extent and causes of variations in the composition of the solar atmosphere; (v) plasma processes in the solar wind; (vi) the acceleration of particles in the solar wind; (vii) the physics of the pickup process of interstellar He in the solar wind; and (viii) the spatial distribution and characteristics of sources of neutral matter in the inner heliosphere.

*Also at: Department of Atmospheric, Oceanic, and Space Sciences, University of Michigan, Ann Arbor, U.S.A.

Space Science Reviews **86:** 497–539, 1998.
© 1998 *Kluwer Academic Publishers. Printed in the Netherlands.*

1. Introduction

The solar wind has now been observed for many decades and, while our knowledge of its properties and dynamics has increased tremendously over the years, a proper description of its origin and acceleration is still ahead of us. Measurements of the detailed composition of the solar wind are still relatively new but offer the best hope for solving this long standing problem, especially when combined with correlative remote sensing observations of the Sun's interior, surface, and atmosphere.

The discovery by SWICS on *Ulysses* of vast sources of both interstellar and interplanetary matter in the inner heliosphere, followed up by the initial discovery of He$^+$ by SULEICA on AMPTE, opened up a new avenue of galactic research and has produced knowledge far beyond expectations. At the 1 AU location of ACE, characteristics of the neutral interplanetary sources in particular can be studied to a degree not possible before and, in combination with observations with a nearly identical SWICS instrument on *Ulysses*, we have the unique opportunity to map the three-dimensional spatial distribution and the long term time variability of these sources.

The Solar Wind Ion Composition Spectrometer (SWICS) and the Solar Wind Ions Mass Spectrometer (SWIMS) on ACE will measure with unprecedented detail the chemical and isotopic composition of the solar wind as well as interstellar and interplanetary neutral matter under all conceivable solar wind flow conditions. We will derive routinely the bulk speeds, densities and kinetic temperatures* of the dominant solar wind ion species, and the charge state measurements will allow us to infer coronal temperatures. These experiments will also provide in-ecliptic, 1 AU baseline measurements of solar wind composition and conditions as well as the density of pickup helium (and probably oxygen and neon), important for interpretation of results from solar wind instruments on the *Ulysses* solar polar mission, as well as from space and ground based remote sensing observations of the Sun. SWICS and SWIMS will also characterize the matter entering the magnetosphere by measuring the elemental, isotopic, and charge state composition of the solar wind. The two instruments compliment each other, with SWICS providing elemental and charge state composition of the most abundant heavy solar wind ions under all circumstances. SWIMS, on the other hand, will determine with high mass resolution the elemental and isotopic composition of the solar wind over longer time scales and will measure the charge states of nearly every element between C and Fe in cold solar wind flows.

SWICS, the improved flight spare of SWICS on *Ulysses*, makes use of energy per charge analysis followed by the time-of-flight vs energy technique to determine both the mass per charge and mass of ions. Post-acceleration is used in order to measure the mass of ions of solar wind energies which otherwise would fall

---

*With the term 'kinetic temperature' we denote the second moment of the energy distribution function measured by the sensor. It should be roughly equivalent to the second order moment of the 3D-velocity distribution in the Sun–Earth direction.

below the threshold energy of the solid-state detectors. SWIMS, incorporating the best features of the MASS instrument of the WIND-SMS experiment and the MTOF instrument of the SOHO-CELIAS experiment, has high-mass resolution ($M/\Delta M > 100$) capabilities for measurements of the solar wind elemental and isotopic composition. SWIMS uses energy per charge analysis and acceleration or deceleration, followed by a time-of-flight measurement in a retarding, quadratically changing electric potential.

The design, fabrication, and testing of the SWIMS instrument was a collaborative effort involving the University of Maryland and the University of Bern. The SWICS experiment was modified at the University of Maryland using the *Ulysses* flight spare unit developed jointly by the University of Maryland, University of Bern, and the Max-Planck-Institut für Aeronomie. The combined DPU unit for SWICS, SWIMS, and SEPICA was developed by the Technical University of Braunschweig.

2. Major Scientific Objectives

Representative examples of major scientific problems the SWICS and SWIMS investigation on ACE will address are outlined below.

(1) Determine solar abundances by measuring with SWIMS the average elemental and isotopic composition of the solar wind in different solar wind flows found in the rise towards maximum portion of the solar cycle (e.g., in-ecliptic fast versus slow wind, CMEs, post-shock flows, etc.).

(2) Study solar wind acceleration, especially of heavy ions, using solar wind composition and charge state measurements made with SWICS with moderate time resolution.

(3) Study physical processes in the solar atmosphere, such as atom-ion separation in the upper chromosphere, by measuring with SWIMS the abundances of all elements below Ni, thus spanning the full range of first ionization potentials.

(4) Characterize the physical properties of the acceleration regions (in the lower corona) by measuring with SWICS and SWIMS the solar wind charge state distributions of several ion species (e.g., C, O, Mg, Si, and Fe).

(5) Study plasma processes affecting the solar wind kinetic properties and the velocity distributions of pickup and other suprathermal ions by measuring with SWICS the distribution functions of major ion species from $\sim$0.5 to $\sim$100 keV e^{-1}.

(6) Study interplanetary acceleration mechanisms producing shock-accelerated energetic storm particle events, CIR particles events, and upstream ions using measurements made with SWICS of the characteristics (composition and charge states) of the source particle populations.

(7) Provide stringent constraints on galactic chemical evolution over the past 4.6 Gy and deduce the primordial baryon density by determining with good preci-

sion the ^{3}He/^{4}He ratio both in the interstellar medium and in the fast and slow solar wind.

(8) Characterize the physical and chemical properties of the local interstellar cloud by determining with good precision the Neon and Oxygen abundance, the ^{22}Ne/^{20}Ne ratio, and the kinetic temperature of He and O atoms in this region.

(9) Determine the size, distribution, and strength of sources of neutral particles in the inner heliosphere by analyzing the velocity distribution functions of pickup C^+, O^+, and heavier singly charged ions for ion speed below the solar wind speed.

(10) In addition, SWICS and SWIMS measurements on ACE will provide the instantaneous characteristics of matter entering the Earth's magnetosphere through measurements of the mass, charge state, and energy distributions of solar wind and supra-thermal ions. The unique features of the ACE mission, including its complement of science instruments, make it possible to address a number of science topics and perform correlative studies with other ACE investigations. These include the study of dynamic processes in the foreshock region producing reflected, diffuse, and upstream ions, and acceleration of solar energetic particles.

The major scientific objectives of the SWICS/SWIMS investigation are discussed in some detail below.

2.1. STUDIES OF THE SOLAR CORONA

The acceleration of the solar wind and of solar energetic particles are topics of great interest but remain as yet to be fully understood. Solar wind acceleration models require a knowledge of boundary conditions (primarily temperature and density profiles) in the acceleration region and must make predictions that are consistent with all the measured properties of the solar wind at 1 AU. SWICS and SWIMS will provide information about some of the characteristics in the acceleration region, but of even greater importance will be the detailed characterization of the 1 AU solar wind in terms of composition and charge states, as well as bulk speeds and kinetic temperatures of some 40 ion types. Such measurements of the solar wind properties will play a crucial role in testing and constraining solar wind acceleration models, and should lead to a better understanding of where and how the solar wind is formed. Our current knowledge regarding solar wind composition measurements and models of solar wind origin is summarized by Fisk (1998) in this issue.

2.1.1. *Physical Processes in the Solar Corona*

The SWICS/SWIMS measurements will provide essential information on the conditions characterizing, and the physical processes operating in, that region of the corona where most of the solar wind and solar energetic particle acceleration occurs. The solar wind acceleration region is supposed to lie relatively high in the corona (typically $\sim$1–3 solar radii in altitude) where the density is very low, especially in coronal holes. Before SOHO, spectroscopic measurements of this region were difficult to obtain and many of the main questions are still an issue of much debate (see, e.g., Habbal et al., 1995). As we discuss below, however, an abundance of information regarding the acceleration region is available from *in situ* measurements of solar wind and suprathermal ions. Indeed, it is quite convenient that the very particles whose acceleration and flow are being studied carry much of the necessary information within their composition and behavior.

When the ionization and recombination times characteristic of the dominant ion states of a particular ion species in the solar wind (e.g., the +6, +7, and +8 charge states of oxygen, the +4, +5, and +6 charge states of carbon, etc.) become long compared with the solar wind expansion time, the ionization state of that species becomes fixed (i.e., *frozen in*) and remains essentially unchanged as the ions flow through the heliosphere (Hundhausen et al., 1968). The relative abundances of the several ionization states of a particular ion species in the solar wind depend strongly on the electron temperature in the coronal region where the *freezing in* occurs. The measured ionization state of solar wind ions from SWICS and SWIMS will provide a direct measure of this electron temperature. The altitude at which the ionization state of solar wind ions is frozen-in is different for different ion species, as was inferred by SWICS *Ulysses* measurements (Geiss et al., 1995). For example, under typical coronal conditions, the ionization state of oxygen freezes in near 1.5 solar radii, whereas that of iron is not fixed until $\sim$3 solar radii. Because the ionization state depends so strongly on the electron temperature where freezing-in occurs, a comparison of the solar wind ionization states of oxygen with those of iron leads to a description of the radial profile of the coronal electron temperature. SWICS will measure not only the ionization state (or charge fraction) of O and Fe, but also of C, Mg, and Si, which will provide important information on the coronal temperature and the temperature gradient over a temperature range (from less than 10^6 K to much greater than 2×10^6 K). With SWIMS, charge states of elements such as Na, Al, Ca, and Cr will be observed.

2.1.2. *Acceleration of the Solar Wind*

The acceleration of heavy ions in the solar wind (helium and $Z > 2$) is expected to be different from the acceleration of the basic proton-electron plasma. Frictional coupling can be important in the heavy ion expansion (Bürgi and Geiss, 1986), and wave-particle interactions can have a different role in the acceleration of heavy ions than in that of protons (Hollweg, 1978). The composition, temperatures, and mean speeds of the solar wind minor ions measured by SWICS will reveal the effects of the several physical processes important in their expansion. These measurements will provide information concerning the overall proton-electron expansion, for they reveal the nature of the coronal environment in which both the heavy ions and protons are accelerated.

Frictional effects are likely to play an important role in the acceleration of heavy ions, as they serve to couple the heavy-ion expansion to the basic proton-electron expansion. The occurrence of frictional acceleration should be apparent in measurements of the mean bulk speeds of different ion species. If Coulomb interactions are important, ions with large charge-squared-to-mass ratios should show evidence of preferential acceleration or deceleration. Other effects can preferentially alter particle speeds. For instance, a polarization electric field arising from electron-proton charge separation accelerates minor ions in proportion to their charge/mass ratio.

Wave-particle interactions and other processes in the corona may selectively increase the temperatures of heavy ions (relative to those of protons and electrons), thus facilitating their escape in the solar wind (e.g., Isenberg and Hollweg, 1983). Such heating processes may be reflected in the kinetic temperatures and mean flow speeds of heavy ions observed in the solar wind. Indeed, solar wind measurements have revealed that ions, at least in selected periods, tend toward equal thermal speeds, leading to an increasing temperature with increasing particle mass (Neuge-bauer, 1981; Schmidt et al., 1980; Ogilvie et al., 1980, Collier et al., 1996). With SWICS we will measure the kinetic temperatures and mean speeds of all major ion species including elements up to iron.

Heavy ions (as well as protons) may be directly accelerated through interaction with a wave field. In fact, this is almost certainly the case for the polar coronal hole wind (Geiss et al., 1995). The resonant acceleration of ions by Alfvén waves has been suggested as a potentially important process (Hollweg, 1978). Such processes may have various dependencies on the ion charge and mass, or no dependence whatsoever. Again, measurements of the mean flow speeds and temperatures of minor ions in the solar wind should reveal the occurrence of these processes.

2.1.3. *Atom-Ion Separation Processes*

It is well established that solar energetic particles (SEP) and solar wind elements that have first ionization potentials (FIP) less than $\sim$10 eV are over abundant relative to the corresponding photospheric abundances (e.g., Breneman and Stone, 1985; Gloeckler and Geiss, 1989). Therefore, a process seems to be operating that preferentially supplies the corona with elements that are easily ionized. An additional (and possibly related) mechanism appears to separate ions from neutrals on a time scale corresponding to their ionization rate and to preferentially feed the ionized fraction to the corona. Various models attempting to explain this so-called FIP effect have been proposed (e.g., von Steiger and Geiss, 1989; Geiss and Bürgi, 1986; Vauclair and Meyer, 1985), but none has been completely successful. The breakpoint that separates the *high* (normal abundance) and *low* (over abundant) FIP elements appears to be not a sharp step function, but rather a transition region that typically includes the elements carbon, sulfur, and phosphorus. Sometimes this region shifts in FIP to include elements such as oxygen. The magnitude of the relative abundance enhancement of the low FIP elements appears to vary, possibly as a function of solar wind type. For example, the solar wind from coronal holes appears to have only a small FIP effect, having a composition that most closely resembles that of the photosphere (Gloeckler et al., 1989; von Steiger et al., 1992). This could be a consequence of the different solar magnetic field conditions. With

SWICS and SWIMS we will study the temporal variation of the abundance of heavy ions in the solar wind, which will help in distinguishing among different modeling efforts. In particular, SWIMS will provide the continuous measurements of rarer elements and isotopes in the solar wind, including such elements as Ni, Ca, Al, and Na with low FIP ($<$ 8 eV) and S and possibly P with a FIP in the transition region.

2.1.4. *Isotopic Composition of the Solar Wind*

Current measurements of solar wind isotopes heavier than helium are restricted, with a majority coming from lunar sample analysis and the foil collection technique (Geiss et al., 1972). SWICS will provide isotopic measurements for helium (^{3}He and ^{4}He) in the solar wind and the local interstellar cloud. With the SWIMS sensor we will obtain a continuous record of isotopic abundances of solar wind heavy ions (e.g., Ne, Mg, and Si). Solar wind isotopic abundance measurements with SWIMS will also allow for a direct correlation with simultaneous SEP isotopic measurements (by the energetic particle experiments ULEIS and SIS on ACE), which may resolve apparent disagreements in, e.g., the ^{20}Ne/^{22}Ne ratio (Geiss et al., 1972; Gloeckler and Geiss, 1989).

2.1.5. *Coronal Transients*

Coronal mass ejection (CME) events appear to originate (initially) in magnetically closed or quasi-closed coronal regions (e.g., Burlaga, 1984). The mass ejections from such regions move outward from the Sun to form a small but not insignificant part of the solar wind. (Fast CME-associated interplanetary shocks are the most dramatic solar wind manifestation of coronal transients in the low-latitude inner heliosphere.) As noted in the preceding sections, solar wind heavy ion measurements provide important information regarding the properties of the coronal region in which the wind originates. Hence, measurements by SWICS and SWIMS of heavy ions in solar wind plasma ejected as a coronal transient will give information regarding the properties of magnetically closed coronal regions, perhaps yielding important clues to the mechanism(s) whereby mass-ejection coronal transients are driven (e.g., Galvin et al., 1987). The interplanetary manifestations of these magnetically (quasi) closed structures sometimes include the bi-directional streaming of suprathermal ions (Marsden et al., 1987), which would be measured by SWICS.

2.2. STUDIES OF HELIOSPHERIC PHENOMENA

2.2.1. *Particle Acceleration*

Particles are accelerated in the solar wind up to energies $\sim$1 MeV nucl^{-1} by propagating shock waves that are generated by CMEs. They are also accelerated, particularly in solar-minimum conditions, in association with stream-stream interaction regions in the solar wind (Tan et al., 1989; Reames et al., 1991). During solar minimum conditions, the Sun produces relatively steady high-speed solar wind streams, which, upon interacting with lower-speed wind at several AU from the Sun, accelerate particles that propagate back toward Earth in steady co-rotating particle streams. Each of these examples of acceleration is important to study since they can provide, by analogy, information on particle acceleration in large-scale astrophysical plasmas in less accessible regions of the universe.

Knowledge of the mass and the ionic charge, and thus of the rigidity, of the accelerated particles, and of their spectra over a wide energy range, can give answers to presently unsolved problems associated with the acceleration, viz., the source of the accelerated particles and the nature of the acceleration.

Particles accelerated in the solar wind could come directly from the solar wind plasma, or they could originate as more energetic particles of solar origin (such as high velocity tails of the solar wind), or as pickup ions. *Ulysses* observations (Gloeckler and Geiss, 1998) have shown that in the inner heliosphere, especially inside the orbit of Earth, the dominant sources of pickup ions are not interstellar atoms but could be neutrals desorbed from interplanetary dust as well as neutrals produced by evaporation of interstellar grains. If the former occurs, the requirements on the acceleration process are of course more severe. With the capability of SWICS to measure uniquely the composition of the solar wind and pickup ions, we can constrain the likely source of the accelerated particles.

As with solar flare and/or CME associated particles, the nature of acceleration processes in the solar wind is revealed in the rigidity and/or energy dependence of the spectra that they produce. SWICS, with its capability to measure the mass, the ionic charge, and the spectrum of ion species from H to Fe, will provide detailed constraints on these dependencies up to ~ 100 keV charge^{-1}; i.e., the energy range from where the particles are injected into the acceleration process up to some of the higher energies obtained in typical events. Data from the SEPICA instrument, which can measure the charge states of more energetic particles, will allow us to examine these processes up to energies of several MeV nucl^{-1} to which ions are typically accelerated.

2.2.2. *Plasma Processes in the Solar Wind*

The SWICS and SWIMS experiments on ACE also provide the opportunity to investigate a variety of fundamental plasma processes in the solar wind. For example, there can be local heating of the solar wind at propagating shock waves or in stream-stream interaction regions; there can be instabilities driven by heat flux and other means. SWICS, with its capability of measuring the kinetic temperatures of all major solar wind ions, will provide an important probe of these processes. The dominant instabilities occurring can be revealed in the response of the kinetic temperatures of solar wind ions with different charge/mass ratios. As a simple example, when the ions cyclotron-damp the instabilities, the cyclotron frequency of the ions that absorb the most energy reveals the dominant frequency excited. Spectral measurements by SWICS and SWIMS will also reveal when and how plasma processes generate extended suprathermal tails on the solar wind ion distributions.

2.3. DETAILED STUDIES OF PICKUP IONS

Heliospheric pickup ions are produced wherever there is a source of neutral molecules or atoms (see Gloeckler and Geiss, 1998; this issue). Ionized by solar UV and/or the solar wind, these newly formed ions are immediately accelerated by the $\mathbf{V}_{sw} \times \mathbf{B}$ electric field, gyrating about the ambient B field, thus forming a *ring*-type distribution in velocity space at $V = V_{sw}$ in the solar wind reference frame. Because pitch-angle scattering is weak (Gloeckler et al., 1995a; Fisk et al., 1997; Möbius et al., 1997), the spread into a *shell* distribution is far less rapid than previously assumed, leading to anisotropic distributions. Adiabatic deceleration in the expanding solar wind fills in the velocity space for $V < V_{sw}$ to a degree that depends on the spatial distribution of neutrals between the Sun and the observation point. Energy diffusion and acceleration to speeds beyond V_{sw} have been predicted (e.g., Isenberg, 1991), but is observed to be weak (Gloeckler et al., 1995a).

Pickup ions that have been observed so far include : (a) cometary pickup ions (e.g., Ipavich et al., 1986; Gloeckler et al., 1986), (b) interstellar pickup ions (Möbius et al., 1985; Gloeckler et al., 1993), (c) lunar pickup ions (Hilchenbach et al., 1991), and more recently (d) pickup ions from distributed inner sources of neutrals located inside several AU (Geiss et al., 1995). At the 1 AU orbit of ACE these inner sources will dominate, producing the largest fluxes of heavy pickup ions. Of the interstellar species only He, Ne and possibly O will be observable.

Thus, one of the principal objectives of SWICS on ACE will be the detailed mapping of the spatial distribution of these inner sources. This will be accomplished by measurements of the distribution functions of C^+, N^+, O^+, and perhaps some heavier ions over ion speeds from ~ 0.6 to ~ 2 times the solar wind speed. Because of adiabatic cooling, the speed of the pickup ions, V/V_{sw}, is uniquely related to the distance from the Sun, $R/R_\odot (R_\odot = 1$ AU) where they were picked up. Thus the density of neutrals at a distance $R(< 1$ AU) is determined from the pickup ion phase space density measured at the corresponding V/V_{sw}. In one orbit of ACE around the Sun we will obtain the spatial distribution of the neutral density of C, N, and O both in longitude and radial distance from the Sun. Subsequent orbits will allow us to examine solar cycle variability (if any) of these regions. Combining these ACE observations with similar measurements from SWICS on *Ulysses* at high latitudes will give us a truly three dimensional view of inner sources of pickup ions and will allow us to speculate on the origin of these sources.

The ACE orbit is optimum for detailed studies of interstellar pickup He. The neutral helium density is not reduced very much even at 1 AU from its interstellar value and gravitational focusing increases the flux of pickup He many fold in the *down wind* direction (Möbius et al., 1985, 1995). These statements also apply to Ne, although to a lesser degree. Among the prime objectives of the SWICS/

SWIMS investigation on ACE will be the more precise (than was possible with SWICS on *Ulysses*) measurements of the ^{3}He/^{4}He ratio in the local interstellar gas, a value of fundamental importance for models of Big Bang cosmology and galactic chemical evolution. In addition, the longitudinal distribution of interstellar neutral helium will be mapped using measurements of the fluxes of pickup He$^+$ and solar EUV flux provided by SOHO. This information, along with measurements of pickup He$^+$ made with SWICS on *Ulysses*, will allow us to infer the three-dimensional distribution of interstellar helium in the inner heliosphere, and to determine the ionization (loss) rate of neutral helium.

2.4. MULTI-MISSION STUDIES

2.4.1. *Long Baseline Measurements*
A SWICS sensor (Gloeckler et al., 1983, 1992) is successfully flying on the *Ulysses* spacecraft and is currently making detailed measurements of the elemental, ionic-charge composition and flow properties of solar wind and interstellar pickup ions at all heliographic latitudes (Gloeckler et al., 1993; Gloeckler, 1996). *Ulysses* was launched in October 1990 and began the out-of-the-ecliptic phase of the mission in February 1992. The first polar pass was completed in 1994. A comparison of solar wind elemental and ionic charge compositions at two different latitudes, and of the coronal conditions they imply, will be made using data from the two similar SWICS sensors on the two spacecraft.

We will use these simultaneous measurements from ACE (in the ecliptic) and *Ulysses* (out of the ecliptic) to help separate spatial and temporal variations and thus to determine how coronal conditions and processes vary with heliographic latitude and distance as discussed above.

2.4.2. *Short Baseline Measurements*
There are two instrument packages on currently active space missions which are very complementary to the SWICS and SWIMS sensors on ACE. On the satellite WIND, the SMS sensors measure the composition of the major components in the solar wind and some isotopes (for details, see Gloeckler et al., 1995b). Due to the fact that the distance between ACE and WIND is changing constantly due to differences in their respective orbits, the actual 3D structure of compositional boundaries can be mapped. For comparisons of isotopic and elemental abundances, data from the CELIAS sensors on SOHO will also be very interesting. Since SOHO is a three axis stabilized spacecraft, the geometrical factor of the instruments should be large (for details, see, Hovestadt et al., 1995). However, SOHO does not carry a magnetometer nor a 3D-plasma instrument and therefore the data analysis will be likely to rely on ACE data as well.

We will use these simultaneous measurements from ACE, WIND and SOHO to analyze spatial and temporal variations of the solar wind composition. We will therefore be able to determine the small-scale structure of coronal conditions rele-

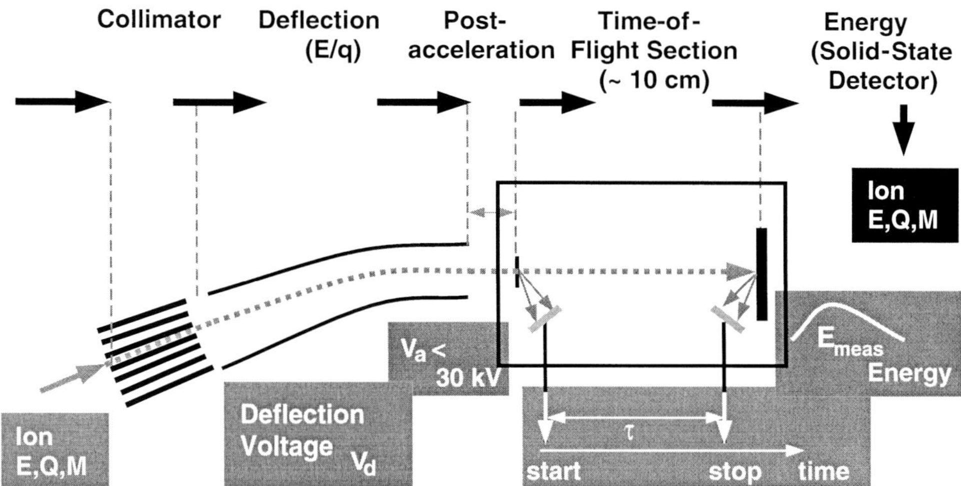

Figure 1. Schematic of the measurement technique used in SWICS, showing the functions of each of the five basic elements.

vant for the origin of the slow and fast solar wind and also of transient phenomena such as CMEs.

3. Instrument Descriptions

3.1. THE SOLAR WIND ION COMPOSITION SPECTROMETER (SWICS)

3.1.1. *SWICS Principle of Operation*

The operation of the SWICS sensor is based on techniques of particle identification using a combination of electrostatic deflection, post-acceleration, time-of-flight, and energy measurement (Gloeckler, 1977; Gloeckler and Hsieh, 1979; Gloeckler et al., 1992). Figure 1 shows schematically the principle of operation of SWICS and illustrates the function of the five basic sensor elements used:

(1) Ions of kinetic energy E, mass M, and charge (ionization) state Q enter the sensor through a large-area, multi-slit collimator that selects proper entrance trajectories of the particles.

(2) The electrostatic deflection analyzer serves both as a UV trap and an energy per charge (E/Q) filter, allowing only ions within a given energy per charge interval (determined by a stepped deflection voltage) to enter the time-of-flight vs energy system.

(3) Ions are post-accelerated by up to a 30 kV potential just before entering the time-of-flight vs energy system. The energy they gain is sufficient to be adequately measured by the solid-state detectors which typically have a 25 to 35 keV energy threshold. An energy measurement is required for determining the mass

composition of an ion population, and ions with energies below $\sim$30 keV must be accelerated if their mass is to be identified.

(4) In the time-of-flight (TOF) system the speed of each ion is determined by measuring the travel time τ of the particle between the start and stop detectors separated by a distance of 10 cm.

(5) The particle identification is completed by measuring the residual energy of the ions in a conventional low-noise solid-state detector.

From simultaneous measurements of the time-of-flight, τ, and residual energy, E_{meas}, and a knowledge of E/Q and the post-acceleration voltage, U_a, we can determine the mass (M), charge state (Q), incident energy (E), or incident speed (V_{ion}) of each ion as follows:

$$M \quad = 2(\tau/d)^2(E_{\mathrm{meas}}/\alpha),$$

$$Q \quad = \frac{E_{\mathrm{meas}}/\alpha}{(U_a+E/Q)\cdot\beta} \approx (E_{\mathrm{meas}}/\alpha)/U_a,$$

$$M/Q = 2(\tau/d)^2(U_a + E/Q)\beta \approx 2(\tau/d)^2 U_a, \tag{1}$$

$$E_{\mathrm{ion}} \quad = Q \cdot (E/Q),$$

$$V_{\mathrm{ion}} \quad = 438 \cdot [(E/Q)/(M/Q)]^{1/2},$$

where d is the flight path, β takes account of the small energy loss of ions in the thin foil of the start-time detector and α is the nuclear defect in solid state detectors (Ipavich et al., 1978). The units of V_{ion} are km s^{-1} when E/Q is in keV e^{-1} and M/Q is in amu e^{-1}. The approximate expressions for Q and M/Q hold for typical solar wind ions.

3.1.2. *Description of the SWICS Instrument*

The SWICS experiment consists of three separately mounted units which are electronically interconnected: the Sensor, the -30 kV Post Acceleration Power Supply (PAPS), and the Command and Data Translator (CDT). These units in turn contain various subsystems that will be described more fully below.

Sensor

A simplified cross-section of the SWICS sensor consisting of the deflection analyzer and the high-voltage bubble is shown in Figure 2.

The cylindrically shaped high-voltage bubble to which a post-acceleration voltage of up to -30 kV may be applied contains the TOF telescope and a proton/helium detector, the analog electronics, and the sensor power supplies. Each of these subsystems is supported by a G-11 insulator bulkhead and enclosed by a vapor-deposited, parylene-coated, machined-aluminum container. This container's outer surface is separated from the parylene-coated inner surface of the outer housing by a 6 mm gap.

The ultra-clean TOF compartment is physically isolated from the bubble electronics, with venting provided through the entrance slits and the collimator-deflection

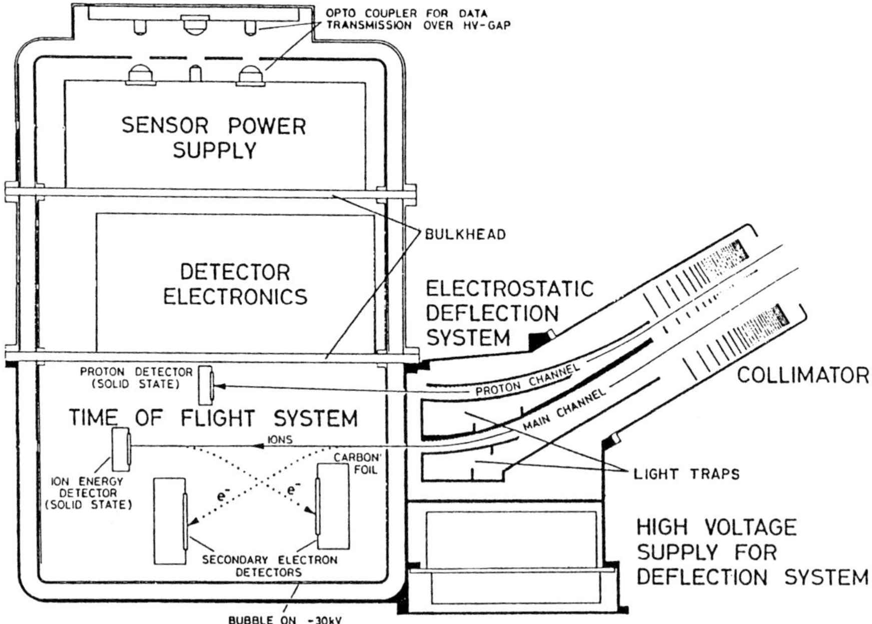

Figure 2. Cross-section of the SWICS sensor showing the collimator, the two-channel deflection system and its deflection power supply, the time-of-flight system and proton/helium detector, analog electronics, sensor bias and power supply, and opto-couplers for digital data transmission. The three inner compartments are supported by two bulkheads and are maintained at the post-acceleration voltage (-15 kV to -30 kV). The outer diameter of the cylindrically shaped outer housing is 15 cm.

system. Digital signals are transferred to the CDT across the 6 mm gap by six opto-couplers. The opto-coupler openings in the housings also serve as venting ports for the sensor electronics and power supplies. Power is supplied to the high-voltage bubble from the PAPS by means of an isolation transformer through a six-pin high-voltage feed-through (not shown) connected to the upper compartment. The photograph of the sensor (Figure 3) shows the outer configuration of the cylindrical bubble housing, the opto-coupler box, and the deflection system with the collima-tor opening covered by a dust/acoustic protective cover which swings open after launch. The gold-plated, cylindrically shaped container houses the -30 kV supply. The sensor is mounted on the top deck of the spacecraft, in the same orientation as shown in the photograph.

Deflection analyzer. The three-dimensional configuration of the deflection ana-lyzer may be visualized by revolving the cross-sectional view shown in Figure 2 by 69° about the symmetry axis of the co-axial cylindrical sensor containers (HV bubble and outer shell). A single conical collimator (see Table 1 for details) ser-vices the two separate deflection regions of the analyzer. The multi-slit collimator is similar in construction to the collimator on our AMPTE instrument (Gloeckler et al., 1985) and allows us to extend the upper energy limit of our analyzer system to

TABLE I

Collimator and deflection analyzer characteristics for the SWICS sensor*

Sensor subsystem	Main channel	H/He channel
Collimator		
Type	conical	conical
Size		
Thickness (cm)	35	35
Slit area (cm^2)	87	31
No. of plates	18	18
Channel Configuration		
No. of channels	2960	1040
Cross-sectional area (cm^2)	3×10^{-3}	3×10^{-3}
Geometrical factor/channel(cm^{2}sr)	7×10^{-7}	7×10^{-7}
Sensor geometrical factor		
Isotropic (cm^{2}sr)	2×10^{-3}	7×10^{-4}
Directional (cm^2)	2×10^{-2}	9×10^{-3}
Deflection analyzer		
Type	small angle, conical with light trap	
Ion-optical properties		
Energy/charge range (keV/q)	0.49–100.0	0.11–15.05
Analyzer constant	12.46	15.93
Analyzer resolution	6.4 %	5.2 %
Analyzer dispersion	< 0.5 %	< 0.5 %
Deflection plates		
Configuration	serrated, black coated	
Shape	pie-shaped 70°	
Area (cm^2)	125	125
Gap (mm)	5 to 12	5 to 8
No. of voltage steps/cycle	60	60
Step size	1.0744/1.0365	1.0744/1.0365
Deflection voltage(V)	46.09–6300	6.91–945

*Note that the overall sensitivity of the integrated system depends not only on the geometrical factor of the subsystems, as given here, but is also dependent on the efficiencies of detecting various ions in the time-of-flight section.

Figure 3. Photograph of the SWICS instrument.

$\sim$100 keV charge^{-1} while maintaining a reasonably large geometrical factor. The widths of the individual channels in the collimator are such as to limit dispersions in the analyzer and flight-path differences in the TOF system to <0.5%. The two inner deflection plates are connected to separate outputs of a variable voltage supply which is housed immediately below the deflection system and which typically increments the deflection voltage of both plates simultaneously in logarithmic steps. The maximum voltages on the upper and lower deflection plates are +1 kV and +10 kV, respectively. Serration, black-coating, and light traps are used to eliminate reflection of visible and UV radiation into the TOF system. Figure 4 shows a top view of the sensor with the electrostatic deflection plates and collimator plates exposed.

The smaller (upper) of the two deflection analyzer regions (proton/helium channel) normally covers an energy range from 0.16 to 15.05 keV charge^{-1}, has a resolution of 5.2%, and will routinely analyze solar wind protons, He, and heavier ions. These ions will be post-accelerated and will be counted by a single rectangular solid-state detector at two threshold levels 20–45 keV, and above 45 keV, corresponding to the energies of post-accelerated solar wind protons ($\sim$30 keV) and post-accelerated He and heavier ions (>$\sim$60 keV). As the voltage is stepped over the full range, this system provides separate E/Q spectra for solar wind

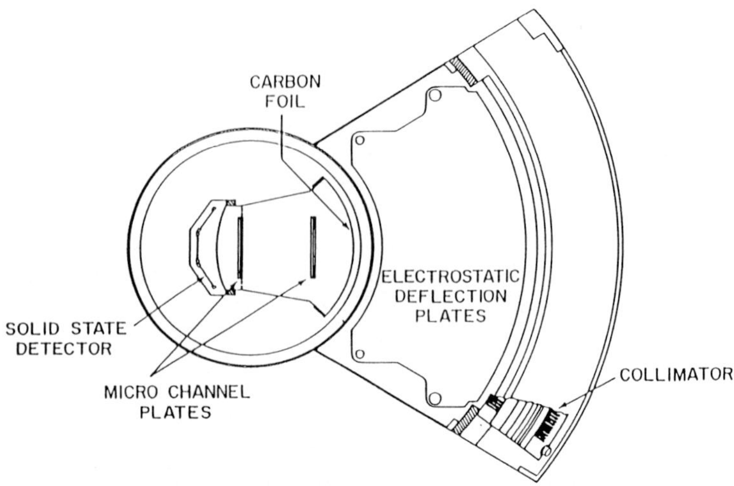

Figure 4. Top view of the SWICS instrument, showing the cross section of the time-of-flight telescope, and the positions of the three solid-state detectors, two microchannel plate assemblies, and
the curved, grid supported carbon foil. The shapes of the deflection plates and the individual plates
of the collimator are also shown.

protons and for solar wind He plus heavier elements, allowing us to determine
in a simple manner, and at all solar wind temperatures, the bulk speed, density, and
temperature of H^+ and He^{++} in the solar wind.

The larger (lower) deflection analyzer (main channel) has a 6.4% energy/charge
resolution and is used for the full M vs M/Q analysis of solar wind He and heavier
ions and of all suprathermal ions in the range 0.49–100.0 keV/charge. The TOF vs
E system is placed behind the exit slit of the analyzer and inside the high-voltage
bubble (Figure 2). At any given voltage step, the analyzer passes ions that have
equal (to within the 6.4%) energy per charge. These ions are then post-accelerated
and their mass, charge state, and energy measured by the TOF vs E system as
described below.

Time-of-flight vs energy system.   An important advantage of the TOF technique
which measures the velocities of ions over a system which selects a narrow range
of velocities (using, for example, crossed electric and magnetic fields) is that stepping over a velocity range is not required in our instruments. The TOF system
accepts a wide range of velocities (or M/Q ratios) simultaneously, resulting in
factors of 10 to 20 increases in both the time resolution and sensitivity. A second
major advantage is that coincidence measurements used in TOF systems reduce
background levels many orders of magnitude below the typical 10^{-2} to 1 count s^{-1}
background measured in, for example, the singles rates of solid-state detectors on
a typical spacecraft.

Figure 5 shows the cross section of the SWICS TOF vs E assembly consisting
of a 'start' and 'stop' detector 10 cm apart. The start and stop signals are derived

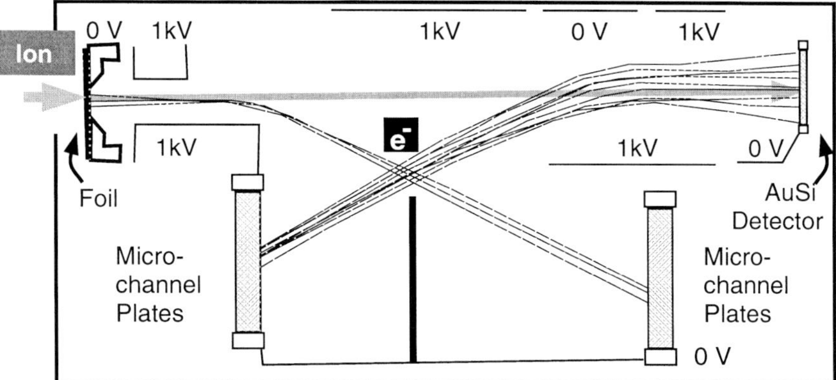

Figure 5. Cross section of the SWICS time-of-flight vs energy telescope showing computer-generated trajectories of secondary electrons emitted from the carbon foil and solid-state detector. The front surface of each of the two chevron-assembly microchannel plates (MCP) is biased slightly negatively with respect to the housing to repel low-energy (<100 eV) secondary electrons. A physical partition between the two MCPs prevents secondary electrons from one MCP triggering the other.

from secondary electrons (Gloeckler and Hsieh, 1979) which are released with a mean energy of a few eV when an ion enters or leaves a solid surface. The surface material used for the start detector is a thin foil (typically 1.5 μg cm^{-2} carbon foil supported on an 86% transmission nickel grid), and for the stop detector the gold front surface of the ion implant solid-state detector. The secondary electrons from the start and stop detectors are accelerated to $\sim$1 kV and then deflected by a system of acceleration gaps and deflection surfaces and strike the respective microchannel plate (MCP) assemblies, each of which consists of two rectangular MCPs in a chevron arrangement. A common supply (1 kV) is used to both accelerate and deflect the electrons. The output signals from the start and stop MCP assemblies are impedance matched and capacitively coupled across 4 kV in order to keep the foil and solid-state detector at local ground potential. The MCPs are normally biased at $\sim$3 kV to operate at a gain of about 2×10^6. This common bias voltage is adjustable by ground command. However, because both MCPs share the same bias supply, their gains cannot be separately adjusted.

The top view of the TOF vs energy telescope (Figure 4) shows the positions of the start and stop MCPs, the three solid-state detectors, and the curved carbon foil. The wide field-of-view and the three detectors are necessary to provide look-directions towards the Sun for all orientations of the ACE spacecraft. This also enables measurements of the suprathermal and pickup ion distribution functions over a large portion of phase space. The slight difference in the flight path of the secondary electrons introduces a timing uncertainty of $<\sim$0.2 ns FWHM, which is smaller than the $<\sim$ 0.5 ns FWHM resolution of the analog electronics. Ions are practically unaffected by the electric fields of the TOF assembly because of their

TABLE II

Time-of-flight and energy-system characteristics for the SWICS instrument

Subassembly	Characteristics
Time-of-flight assembly (TOF)	
Flight path of	10 cm
start element	carbon foil (1.5 μg cm^{-2}, grid supported)(0.4 $\times$ 8 cm^2)
Stop element of	3 rectangular solid-state detectors
main channel	(11.9 mm $\times$ 13.9 cm $\times$ 309 μm each)
Stop element of	1 rectangular solid-state detector
auxiliary channel	(6.0 mm $\times$ 6.6 mm $\times$ 309 μm)
Microchannel plates (MCP)	
Type	chevron stack
Size	2 rectangular (1.5 $\times$ 4.3 cm^2 each)
TOF telescope dispersion	
Path length, Δd/d	< 0.005
Secondary electrons	0.3 ns
Energy measurement circuitry	
Range (keV)	35–600
Electronic noise (FWHM)	8 keV
E-ADC range	256 channels
E-ADC resolution	2.34 keV/channel
Time-of-flight measurement circuitry	
Range (ns)	10–180
Electronic noise (FWHM)	0.5 ns*
T-ADC range	1024 channels
T-ADC resolution	0.176 ns/channel

*Includes rms variations and drift of the analog electronics.

higher energy. The path-length dispersion $\Delta d/d$ for ions is negligible in the plane of Figure 5 and $<\sim$0.005 FWHM in the other direction.

The residual energy is measured by one of the three rectangular solid-state detectors. Table II gives additional details for the SWICS TOF vs E assembly.

Electronics

The SWICS electronics (see Figure 6) consists of an analog subsystem which is built into the central compartment of the high-voltage bubble (Figure 2), a variable deflection voltage supply and a detector/MCP bias supply both contained in the sensor, and the -30 kV post-acceleration power supply (PAPS) in its own cylindri-

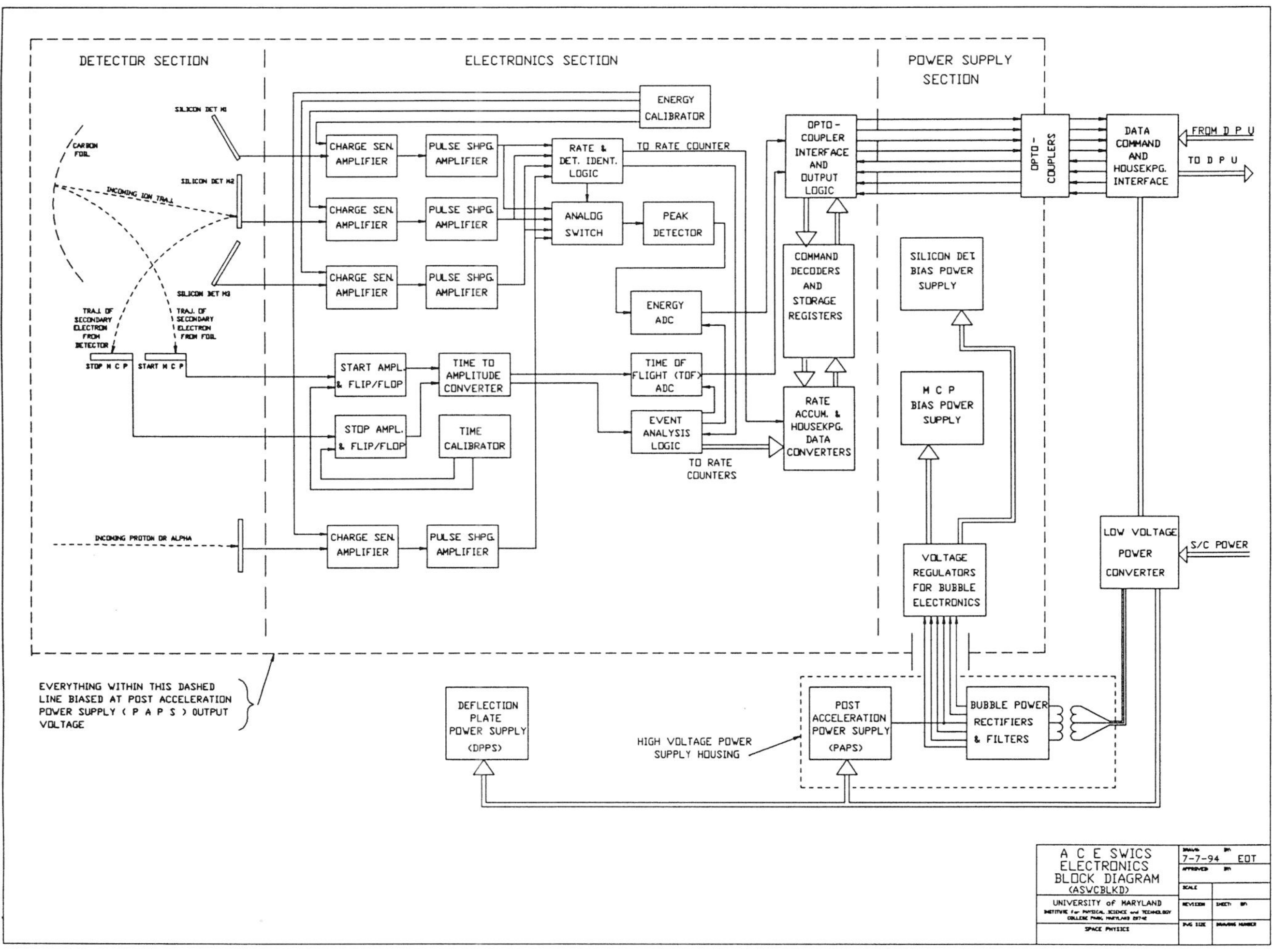

Figure 6. Electronics block diagram for the SWICS instrument.

cal housing. The Command and Data Translator (CDT) and the low-voltage power converter (LVPC) are mounted together in a housing beneath the sensor (not shown in Figure 2).

Analog electronics.   The main function of the SWICS analog electronics is to measure the time-of-flight τ and energy E of ions triggering the TOF vs E system. In addition, solar wind protons and He are quickly identified and counted, and a number of coincidence conditions are established and their occurrence counted.

Time-of-flight measurement.   Each MCP assembly output is capacitively coupled to a fast preamplifier whose function is to accept the 0.9 ns rise-time MCP output signals, shape, amplify, and feed them into a fast timing discriminator using tunnel diodes. The output signals from the start and stop timing discriminator are used as inputs to the Time-to-Amplitude Converter (TAC) which will produce: (a) an output pulse whose amplitude is proportional to the time interval between the trigger of the start and stop discriminators (T signal), and (b) a logic pulse (valid τ) provided the stop signal follows the start signal in <200 ns. The T signal is stretched and pulse-height-analyzed by a 10-bit 'Time' Amplitude-to-Digital Converter (T-ADC) with a 40 ns conversion time. The valid τ logic pulse is used to establish logic conditions and increment counting rates. We have measured the overall timing resolution of the analog electronics to be <0.2 ns when both the start and stop MCP pulses exceed 100 mV.

Energy measurement.   The output of each of the three solid-state detectors of the TOF telescope is coupled to a low-noise (5 keV FWHM) preamplifier and shaping amplifier (unipolar pulses of 1 μs duration at 10%). All amplifiers have been hybridized to minimize weight and reduce cross talk. The output signals of each amplifier chain (E-signals) are pulse-height-analyzed by a common 8-bit 'Energy' Amplitude-to-Digital Converter (E-ADC) and are fed to respective threshold discriminators, whose output is used to identify the triggered detector and to establish logic conditions and increment counting rates. Two of the three main solid-state detectors (SSD) are ion implant SSDs, whereas the one farthest off the spin axis is a surface barrier SSD.

Solar wind proton/helium channel.   The output of the low-noise H/He solid-state detector also goes through a preamplifier/shaping-amplifier chain which then feeds two threshold discriminators (20 keV and 45 keV) whose outputs increment the 'proton' and 'He' rate channels, respectively. It is possible to command the instrument into a mode such that the output of the H/He detector is analyzed instead of that of the three main telescope detectors. We also note that it is possible to select a deflection voltage stepping sequence that prevents solar wind protons from entering the time-of-flight system, but not the proton/helium channel.

TABLE III

SWICS telemetry allocation for data items generated in the analog electronics and S^3DPU

Data item	# per spin	Origin	Corresponding physical parameter	Bits per item per spin*	Bits per item per second
FSR	1×8	AE**	start detector rate	8×8	5.33
DCR	1×8	AE	start-stop coincidence rate (valid T)	8×8	5.33
TCR	1×8	AE	valid T - solid-state detector (SSD) coincidence rate	8×8	5.33
MSS	1×8	AE	combined count rate of three main channel SSDs	8×8	5.33
PROT	1×8	AE	solar wind proton rate	8×8	5.33
ALFA	1×8	AE	solar wind helium and heavy-ion rate	8×8	5.33
MEi	2	DPU	15 matrix elements, defined by up to 30 m vs m/q boxes	2×8	1.33
MRi	8×8	DPU***	counting rate of 8 selected ion species, defined by up to 20 m vs m/q boxes	64×8	42.67
BRi	4×8	DPU	4 sectored basic rates (incl. error rate), defined by up to 10 m vs m/q boxes	32×8	21.33
PHA	194	AE	194 24-bit E and T pulse-height events	194×24	388.00
HK	25	DPU	housekeeping data	25×8	16.67
Status	5	DPU	status bytes	5×8	40.00
Total					505.33 bit s^{-1}

*At the nominal spacecraft spin rate of 5.0 rpm the spin period is 12 s.

**AE = analog electronics.

***One set of the 15 sectored matrix elements (190 bytes) is collected over one complete voltage stepping cycle of 60 spins (12 min) and read out during the following cycle, 2 bytes per spin.

Monitor rates. The Monitor Rates are those rates produced in the analog electronics. The Front SEDA Rate (FSR) is incremented every time the Start detector is fired, the Double Coincidence Rate (DCR) indicates a valid τ, meaning that a Stop signal occurred within 200 ns of a Start signal, the Main Solid State (MSS) rate increments every time one of the three main solid-state detectors fires, and the Triple Coincidence Rate (TCR) counts when there is a valid τ signal in coincidence with an MSS count. The PROT and ALFA rates are, respectively, counts from 20–45 kV and greater than 45 kV discrimination of the proton/helium channel of SWICS.

Logic conditions and in-flight calibration. Table III lists the data items (rates and pulse heights) generated by the analog electronics. Pulse-height analysis of the

T signal is normally started by a valid τ signal whether or not an energy signal is present. Pulse-height analysis of the E signal normally requires a triple coincidence condition (start-stop and energy). It is possible, however, to change the T-ADC analysis logic by ground command to require also a triple coincidence for its analysis. An in-flight calibration provides on command a sequence of timing and amplitude pulses which are fed, respectively, to the fast amplifiers or preamplifiers of each MCP or solid-state detector. When the instrument is being calibrated, the trigger logic prevents all except the calibrator pulses from being analyzed.

3.2. THE SOLAR WIND ION MASS SPECTROMETER (SWIMS)

The Solar Wind Ion Mass Spectrometer (SWIMS) will determine the elemental and isotopic composition of solar wind ions over a wide range of solar wind conditions using a novel application of the time-of-flight measurement technique that results in exceptionally high mass resolution ($M/\Delta M > 100$). The high mass resolution SWIMS sensor will determine the abundance of nearly all of the chemical elements and most of their isotopes in the mass range 3 to 60 amu. SWIMS will measure the charge states of elements not resolved by SWICS, especially during periods of low kinetic temperature solar wind flow. The velocity range of the sensor is mass dependent, extending from $\sim$200 to $\sim$1000 km s^{-1} for C and from $\sim$200 to $\sim$500 km s^{-1} for Fe. This allows measurement of the abundances of the rarer solar wind species, including some elements heavier than iron and most isotopic ratios. The SWIMS sensor consists of an electrostatic deflection system, a time-of-flight High Mass Resolution Spectrometer (HMRS) (Hamilton et al., 1990; Gloeckler, 1990), associated high voltage supplies and low voltage power converter, and analog and digital electronics. A simplified cross section and block diagram of the SWIMS sensor is given in Figure 7. A photograph of the SWIMS instrument is shown in Figure 8.

3.2.1. *SWIMS Principle of Operation*

Solar wind ions of kinetic energy E, mass M, and ionization state Q enter the electrostatic deflection system which acts as an ultraviolet (UV) trap and an energy per charge (E/Q) passband filter. Ions which exit the deflection system have retained their original charge state and are confined to an energy per charge interval determined by the stepped deflection voltage.

Solar wind ions with an energy per charge E/Q corresponding to the value set by the deflection system enter the time-of-flight High Mass Resolution Spectrometer (HMRS) after passing through a thin carbon foil. In passing through the carbon foil, the ions reach charge equilibrium and emerge predominantly as either neutrals or singly charged ions, with a smaller fraction having charges $Q^* \geq 2$ or $Q^* = -1$.

The carbon foil (located at $z = 0$) also acts as the entrance to the TOF high mass resolution spectrometer. As the ion emerges from the foil, a small number (few to few tens) of secondary electrons are ejected from the foil. These electrons

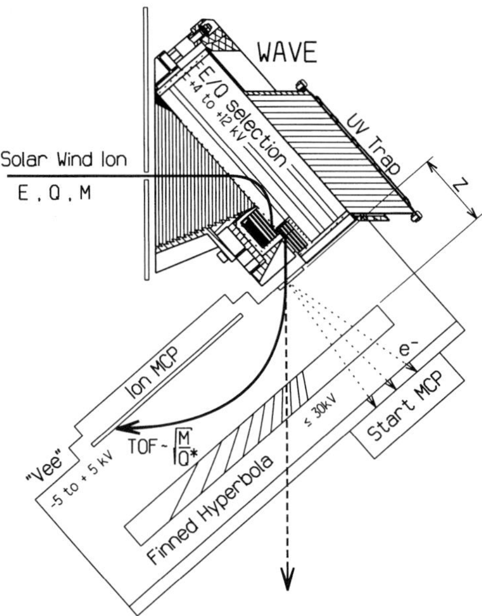

Figure 7. Cross-sectional schematic of the SWIMS instrument showing the WAVE entrance system, start MCP location, rail hyperbolic deflection surface, and position sensing stop MCP. A typical ion trajectory is shown.

Figure 8. Photograph of the SWIMS instrument.

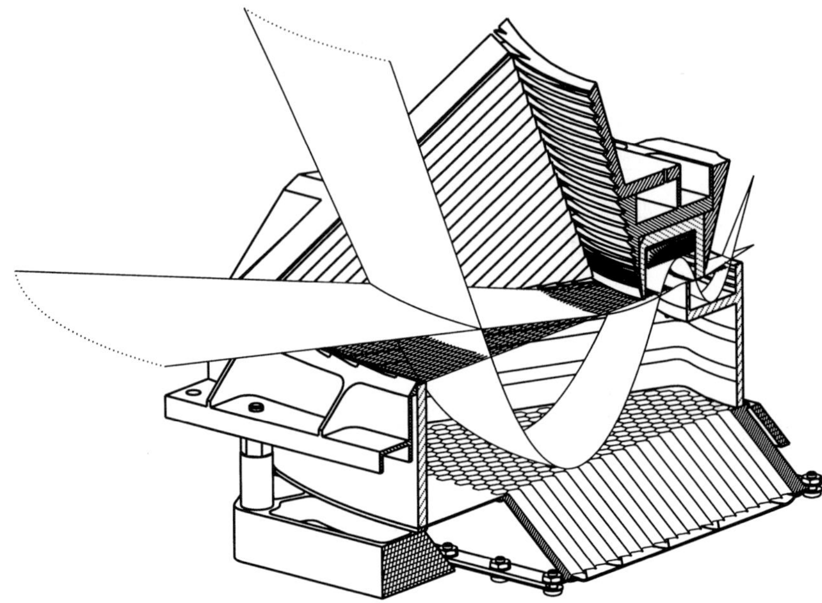

Figure 9. Cross section of the SWIMS deflection system. The two bands represent the effective field of view.

are accelerated to a microchannel plate assembly (MCP) and generate the start signal for the time-of-flight analysis. Positively charged ions are deflected back to a second, large area MCP assembly (also located at $z = 0$), thereby generating the stop signal for the time-of-flight analysis. The principle of operation of the mass analyzer is to measure the time-of-flight (τ) of the predominantly singly ionized particles in a static retarding harmonic electric potential ($V \propto z^2$, where z is the vertical distance from the carbon foil). In such a potential τ is independent of the magnitude and angle of the velocity of the emerging ion and depends only on its M/Q^* value:

$$\tau \propto (M/Q^*)^{1/2}, \tag{2}$$

Hence, a measurement of τ in this harmonic potential yields an unambiguous value of M/Q^* and, for $Q^* = 1$, of M. Typical times-of-flight range from 60 to 460 ns. Since τ can easily be measured with the precision of a fraction of a nanosecond, the ion mass is determined to a high degree of accuracy ($M/\Delta M > 100$). The required electric field is produced by a combination of a hyperbolic plate set at a large positive voltage V_H (typically 25 to 30 kV) and a V-shaped plate at ground potential (see Figure 10). The value of V_H determines the maximum ion energy that can be deflected by the electric field but does not affect the mass resolution.

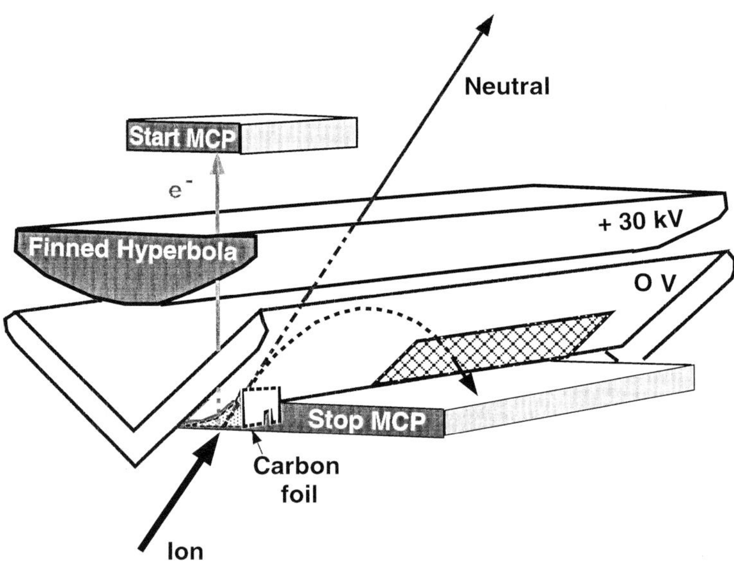

Figure 10. Schematic of the high mass resolution section of the SWIMS instrument. The high transparency of the hyperbola prevents neutrals from scattering onto the Stop MCP.

3.2.2. *Description of the SWIMS Instrument*

Unlike SWICS, the SWIMS instrument consists of one singly mounted box as shown in Figure 8. The assemblies contained in that unit are described below.

SWIMS deflection analyzer

Ions enter the SWIMS sensor through an entrance slit placed in front of the Wide-Angle, Variable Energy/charge (WAVE) passband deflection system. The WAVE deflection system is a wide angle electrostatic, multiple-chamber ion reflection system (see Figures 7 and 9) with the deflection plates connected to a stepped voltage supply. A black coated trap at the rear of the first (and largest) chamber prevents ultraviolet (UV) radiation from entering the time-of-flight analyzer. The entrance slit converts the WAVE system (which has a wide energy/charge passband of order one) to a narrow passband stepped deflection analyzer. The width and position of the entrance slit define the analyzer constant and energy passband $(\Delta(E/Q)/(E/Q) = 0.05)$. The high voltage power supply steps the voltage on the reflection plates once per spin while the entrance grid is held at ground. The voltage steps are logarithmically spaced from 0.6 kV (0.5 keV e^{-1}) to 12.2 kV (9.5 keV e^{-1}) in 60 steps. The system has $\sim$5 % energy/charge resolution.

SWIMS Time-of-flight High Mass Resolution Spectrometer (HMRS)

Ions which exit the WAVE system have retained their original charge state and are confined to a 5% energy/charge band. Ions pass through a 4 mm×15 mm grid-supported carbon foil ($<$2 μg cm^{-2}) cause emission of secondary electrons which are rapidly accelerated through a hole in the hyperbola (see Figure 10) and reach the start chevron microchannel plate (MCP) assembly where their arrival time is recorded. Ions emerging from the start foil are most likely to have a 0 or +1 charge state with a much smaller fraction being +2 or (for some species) −1. Neutrals and negatively charged ions will strike the high voltage plate (hyperbola) of the HMRS. A finned hyperbola is used in order to suppress background caused by sputtering of those neutrals striking it.

A static harmonic potential region is created in the mass analyzer system between the finned, hyperbolically shaped positive potential plate and a V-shaped plate at relative ground potential. Positively charged ions emerging from the start foil will be deflected in this region and will hit the large area (100 mm × 15 mm) stop chevron MCP assembly. Because of the harmonic potential inside the mass analyzer, the times of flight of positively charged ions depend only on their mass/charge and, since most have charge state +1, only on their mass. The maximum voltage, V_H, of the hyperbolic deflection plate is +30 kV (but adjustable downward). The value of V_H determines the maximum ion energy that can be deflected by the electric field, but does not affect the mass resolution.

The anodes for both MCPs provide not only the required timing signals, but also amplitude signals that are pulse-height-analyzed by the electronics. These amplitude signals contain some information about particle mass and energy and will be used to further reduce residual background. In addition, the Stop MCP anode provides one-dimensional position information. This allows us to determine the approximate incident energy of an ion, from which its initial ionization state or charge Q may then be determined using the known E/Q setting of the deflection analyzer. The position information can also be used for additional rejection of background events.

SWIMS acceleration/deceleration system

The HMRS and its associated electronics are floated at an adjustable acceleration/deceleration potential V_F. This potential is adjusted once per spin and may be set negative to accelerate low-energy solar wind ions through the foil and thus increase the yield of +1 ions after the foil, or set positive to decelerate higher energy solar wind ions so that they are contained within the time-of-flight section. V_F has 256 linearly-spaced values from −5.0 to +5.0 kV. V_F is adjusted at each of the 60 steps of the WAVE deflection system to a preselected set of values related to the solar wind speed measured in the previous cycle.

SWIMS modes of operation

In its normal mode of operation the SWIMS deflection and acceleration/ deceleration voltages will be stepped once per spin. The deflection analyzer has 60 settings (0–59) with 5.1% separation between steps. The nominal sequence will comprise 60 steps starting with step 59 (9.5 keV e^{-1}) and ending with step 0 (0.5 keV e^{-1}). The acceleration/deceleration voltage will depend on the current step number and the speed of solar wind alpha particles measured during the previous cycle.

There will be optional modes in which any stepping sequence (of 60 steps) is selected. Such modes can be used to avoid (e.g., step over) the solar wind protons for most of the time and thus reduce fluence on the Start MCP. Another option is to set a given deflection voltage for each of the 60 steps in the sequence and specify the value of V_F. This option was used during calibration, for example.

Electronic and data system

The electronics for the SWIMS instrument (see Figure 11) consist of two digital and four analog boards that are packaged in an isolated box attached to the rear of the sensor. All circuitry dedicated to timing discrimination, internal calibrator operation, and pulse-height analysis of the Start and Stop microchannel plate signals are referenced to V_f, whereas the circuitry for the command/data interface to the S^3DPU are referenced to ground. The voltage difference between these two types of circuits is bridged by the use of opto-couplers.

Time-of-flight measurement. Fast preamplifiers accept the output signals from each of the two microchannel plate assemblies (Start and Stop), shape them, amplify them, and feed them to fast timing discriminators. The output from these discriminators are used as input to the Time-to-Amplitude Converter (TAC) which produces an output pulse whose amplitude is proportional to the time interval between the triggering of the Start and Stop discriminators, and a logic pulse if the Stop signal follows the Start signal within about 200 ns. The output pulse is shaped and pulse-height-analyzed by an Analog to Digital Converter (ADC), producing a 12-bit digital value. The logic pulse is used to increment counting rates.

Position and pulse-height information. Both Start and Stop micro-channel plate assemblies provide not only the required timing signals, but also amplitude signals that are pulse-height-analyzed by the electronics. These amplitude signals provide some information about particle mass and energy. In addition, the Stop MCP anodes provide one-dimensional position information, which allows us to determine the energy of an ion, and is also useful for rejecting background events. The position sensing is accomplished with a simple strip-strip anode composed of two interwoven electrodes of increasing and decreasing widths, respectively. The ratio of the signal from one electrode to the sum of the two signals indicates where along the anode an event occurs.

Power system

There are a total of eight power supplies in the SWIMS instrument, all located in the main sensor compartment. A Low-Voltage Power Converter (LVPC) accepts the spacecraft main bus 28V line and provides a variety of regulated and unregulated low voltages for the electronic circuitry. The Energy/Charge Power Supply (EQPS) outputs voltage up to 12 kV and is used to select ions of particular energy/charge. The V_f Power Supply (V$_f$PS) provides -5 kV to $+5$ kV for the acceleration or

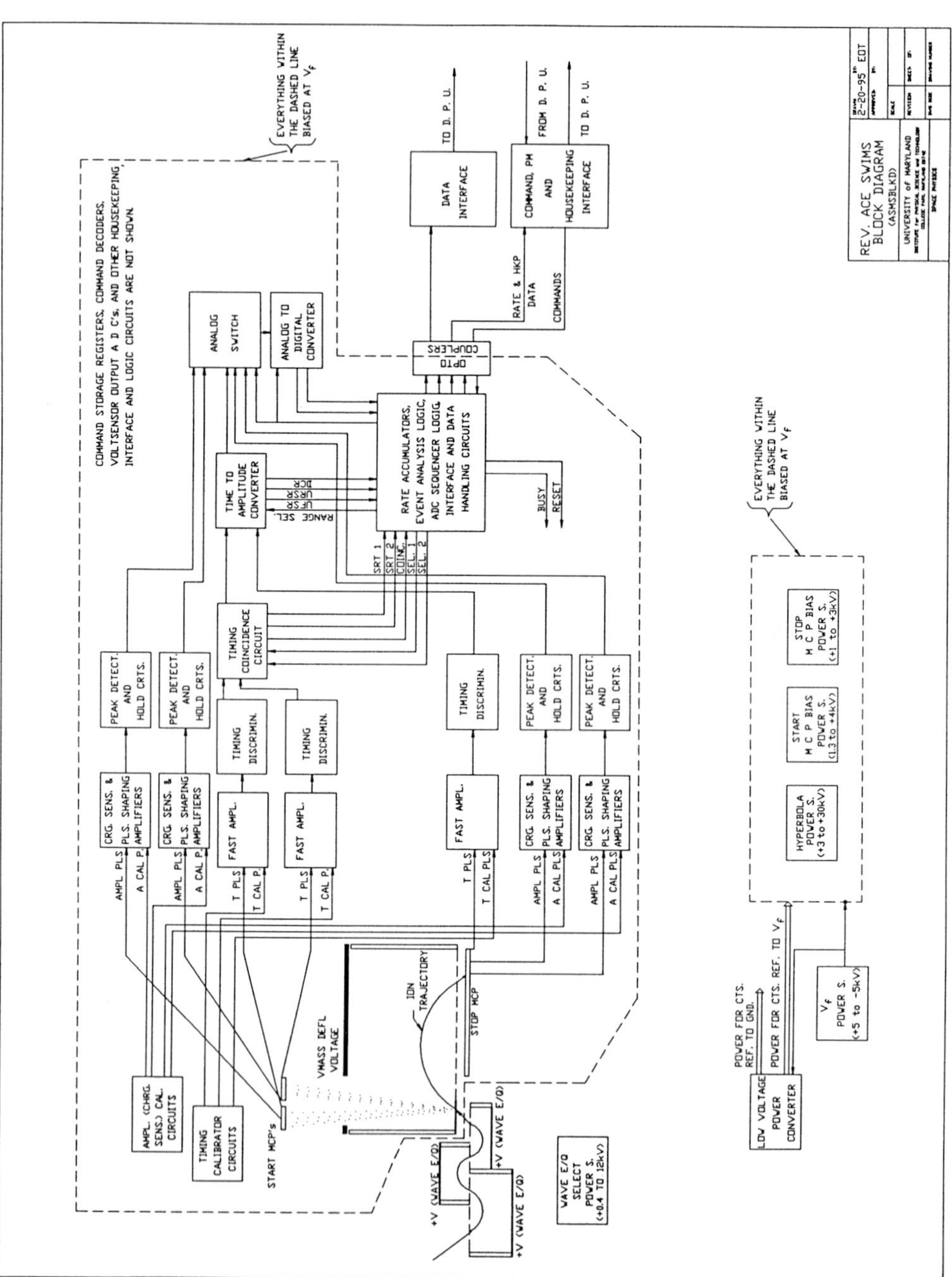

Figure 11. Electronics block diagram for the SWIMS instrument.

deceleration of these ions, respectively, and the Hyperbola Power Supply (HPS) provides up to 30 kV to the hyperbolic electrode in the HMRS to bend the ion trajectories down onto the Stop microchannel plate (MCP) assembly. For each of the two MCP assemblies there are two redundant power supplies so that at any one time either one set or another is operating. This is switchable by ground command. Unlike SWICS, the gain of each MCP assembly can be set independently by ground command to the power supplies.

3.3. FUNCTIONS OF THE DATA PROCESSING UNIT (S^3DPU)

Since a full description of the operation of the S^3DPU is given elsewhere (Möbius, et al., this issue), only aspects related to the SWICS/SWIMS operation will be discussed here. For each spin period of the spacecraft the S^3DPU produces an experiment data block containing housekeeping, rate, and pulse-height data for the SEPICA instrument, SWICS and SWIMS. For summary of the SWICS and SWIMS data items refer to Tables III and IV.

The basic pulse-height data (Time and Energy for SWICS; Time alone for SWIMS) provided by the sensors lend themselves to straightforward on-board processing which simplifies ground-based data reduction. The S^3DPU performs this function and provides the interface between the spacecraft and the SWICS and SWIMS instruments. The other principal SWICS- and SWIMS-related functions of the S^3DPU are to: (a) execute fast classification of ions according to the ion mass (M) and mass per charge (M/Q); (b) collect and store count-rate and pulse-height data, determine event priority, and execute appropriate event sequencing, compress the contents of each counting-rate register into an 8-bit floating-point representation, format all data into 8-bit words, and transfer this information to the spacecraft; and (c) perform all necessary control functions for the experiments, accept and execute ground commands, monitor the experiments status, and execute on-board calibration sequences of the analog electronics initiated by ground command.

3.3.1. *Fast On-board Classification of Ions and Rates*

For every ion for which pulse heights (T and E for SWICS; T for SWIMS) are available the S^3DPU executes fast classification according to ion mass and mass-per-charge (SWICS only) for collection and storage of ion species rates. Fast look-up table techniques are used to establish a correspondence between energy (E) and time-of-flight channel (T) pulse-height data contained in the SWICS event words and the positions of the mass (M) and mass-per-charge (M/Q) surfaces in the T vs E parameter space. For SWIMS only T and the hyperbola voltage are used. The classification algorithms used to generate the look-up tables is derived from the fundamental relations presented in Section 3.1 (SWICS) and Section 3.2 (SWIMS) and take into consideration several instrumental parameters such as the energy-per-charge value of the deflection voltage step, the post-acceleration voltage, the energy

TABLE IV

SWIMS telemetry allocation for data items generated in the analog electronics and S^{3}DPU

Data item	# per spin	Origin	Corresponding physical parameter	Bits per item per spin*	Bits per item per second
FSR	1×8	DPU**	start detector rate	8×8	5.33
FSRA	1×8	AE***	start detector rate – A anode (not used)	8×8	5.33
FSRB	1×8	AE	start detector rate – B anode	8×8	5.33
FSRAB	1×8	DPU**	start detector rate – A anode .AND. B anode (not used)	8×8	5.33
RSR	1×8	AE	stop detector rate	8×8	5.33
DCR	1×8	AE	valid start – stop coincidence	8×8	5.33
MFSR	1×8	AE	multiple FSRs rate	8×8	5.33
MDCR	1×8	AE	multiple DCRs rate	8×8	5.33
BRi	4×8	DPU	four basic rates, sectored	32×8	21.33
TOFi	35	DPU	two 1024-byte TOF rates are accumulated over one 60-spin cycle; 35 bytes are readout each spin of the following cycle	35×8	23.33
PHA	99	AE	99 48-bit pulse-height events	99×48	396.00
HK	35	DPU	housekeeping data	35×8	23.33
Status	5	DPU	status information	5×8	3.33
Total					510.0 bit s^{-1}

*At the nominal spacecraft spin rate of 5.0 rpm the spin period is 12 s.
**The S^{3}DPU determines if FSR=FSRA, FSRB, or FSRAB. This is changeable by ground command.
***AE = analog electronics.

loss in the carbon foil, and (for SWICS) the pulse-height defect in the solid-state detector. The M and M/Q values for SWICS returned by the classifier are used to increment appropriate storage registers corresponding to Basic Rates, Matrix Rates, and Matrix Elements. For SWIMS only Basic Rates are collected.

Directional information

The S^{3}DPU uses the spacecraft spin clock and Sun pulse to apportion the azimuthal (spacecraft) plane into eight equally sized sectors, nominally 45° each. For each type of rate (Monitor, Basic, Matrix, and Element) the S^{3}DPU collects eight rates per spin period – one for each of the azimuthal sectors.

Basic rates

For SWICS, Basic Rates correspond to counting rates of ion species groups that fall into specified regions of $E - \tau$ (or τ only) space defined by coarse resolution rectangular 'boxes' in M vs M/Q or upper and lower limits in M/Q space when no E is measured. For SWIMS these rates are defined by upper and lower limits in each of the four pulse-height data items: Time-of-flight τ, Start Amplitude A, Stop Amplitude K, and Stop Position X. The boundaries of all boxes are defined in S^3DPU internal tables that can be uploaded by ground command. A Basic Rate is incremented by the S^3DPU when the M and M/Q calculated for an event word fall within the appropriate pulse-height range. These rates are used for the normalization of the direct pulse-height words (see section below on direct pulse-height analysis). For SWICS a maximum of four Basic Rates can be defined, using a maximum of 10 box definitions. For SWIMS there are also four Basic Rates.

Matrix rates (SWICS only)
Matrix rates correspond to the counting rates of selected ion species defined by medium resolution rectangular 'boxes' in M vs M/Q space or upper and lower limits in M/Q space when no E is measured (so called 'mass zero' events). The boundaries of these boxes are defined in a S^3DPU internal table which can be uploaded by ground command. A Matrix Rate is incremented by the S^3DPU when the M and M/Q (or only M/Q) calculated for an event word fall within the appropriate pulse-height range. A maximum of 8 Matrix Rates can be defined, using a maximum of 20 box definitions. Positions of the Matrix Rates in M vs M/Q space are shown in Figure 12 overlaid on raw flight data discussed below.

Matrix elements (SWICS) and time-of-flight spectra (SWIMS)
For SWICS, Matrix Elements (and for SWIMS, T Spectra) are fine resolution Matrix Rates that are collected over an entire deflection voltage stepping cycle (60 spins) and read out during the following cycle. As a result, information indicating the voltage step (hence E/Q) over which a particular event occurred is not retained. The boundaries of these boxes (upper and lower limits in M and M/Q) are defined in a S^3DPU internal table which can be uploaded by ground command. For SWICS a maximum of 15 Matrix Elements can be defined, using a maximum of 30 box definitions. For SWIMS there are two 1024-byte Time-of-flight Spectra that are defined by upper and lower limits in Start Amplitude A, Stop Amplitude K, and Stop Position X. The Stop Amplitude K is the sum of the amplitudes for the two electrodes ($k1$ and $k2$) of the Stop position-sensing anode. The Stop Position X is defined as $k1/(k1 + k2)$.

3.3.2. *Direct Pulse-Height Analysis Data and Priority Selection*
The most detailed information about the composition, arrival direction, and energy of ions is contained in the pulse-height analysis (PHA) words. For SWICS each word is 24 bits long: 8 bits for energy, 10 bits for time, 3 bits to indicate in which of eight sectors the event occurred, and 3 bits to identify both the priority and the associated detector. The SWIMS PHA word is 48 bits long: 12 bits time, 10 bits each for Start Amplitude, Stop Amplitude, and Stop Position, and 3 bits each for Range (priority category) and Sector information. Nominally the S^3DPU records

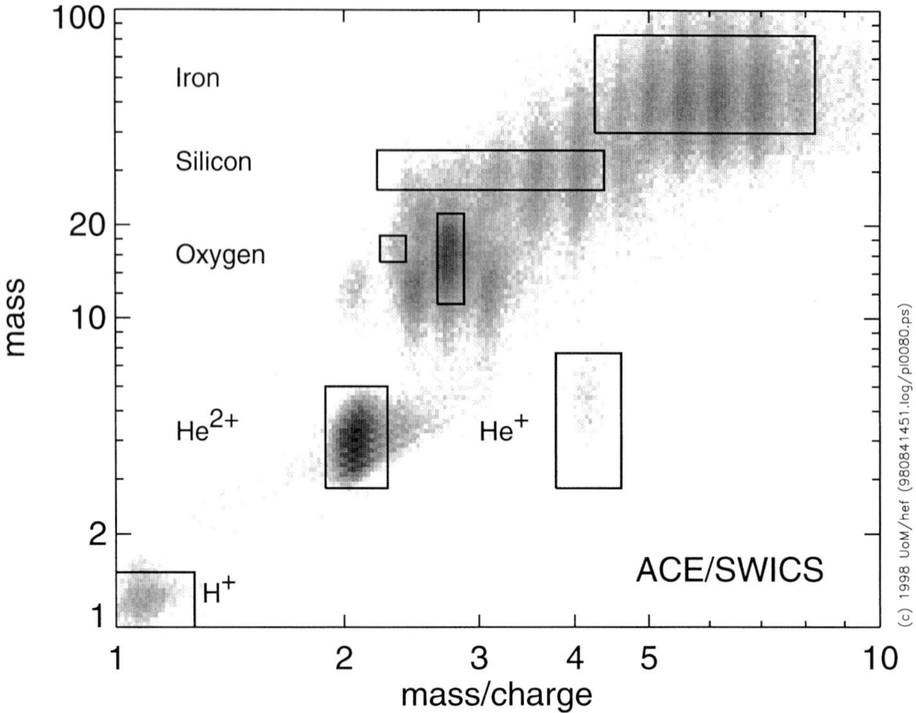

Figure 12. Matrix rate box definitions overlaid on flight data. These definitions are used by the S^{3}DPU to calculate rates which are also used for the calculation of browse parameters.

194 SWICS pulse-height events each spin period, and 99 SWIMS events. The priority categories are defined by the Basic Rates (see discussion of Basic Rates above).

4. Instrument Requirements

4.1. TELEMETRY AND COMMAND

On-board processing reduces considerably the telemetry that would otherwise be required to send back the information acquired by the sensors. The telemetry required to transmit the basic rates, matrix rates, matrix elements, and direct pulse heights is given in Tables III and IV. The total bit rate for SWICS is 505 bit s^{-1}, and for SWIMS is 510 bit s^{-1}. The instruments are controlled by a total of four spacecraft relay commands each: instrument power, operational heater power, survival heater power, and pyro power (for opening of the acoustic cover). Sensor configuration and deflection voltage sequencing is accomplished by the S^{3}DPU, which sends the current set of sensor commands to the sensors every 1.5 s. (That set of commands is changed by ground command to the S^{3}DPU.)

4.2. CLEANLINESS AND THERMAL

Because of the susceptibility of the MCPs and solid-state detectors to contaminants and because thin foils are employed in both sensors, dust/acoustic covers are utilized. These covers were commanded to swing open after the spacecraft was injected into its orbit. To avoid damage to the MCPs prior to launch, the instruments were purged continuously with dry nitrogen through ports provided in the sensors.

Thermal-design requirements for SWICS and SWIMS are an orbital temperature of $-25°$ C to $0°$ C, preferably $-20°$ C to $-10°$ C. For SWICS the thermal requirements are driven by solid-state-detector operating and survival temperature limits. For SWIMS the requirements are governed by the operating limits of the telescope assembly since high temperatures lead to a high background due to the ionization of outgassed water. The requirements are achieved through use of thermal reflecting coatings, blankets and radiators, and operational heaters.

5. Instrument Capabilities and Performance

The SWICS and SWIMS instruments are capable of measuring the solar wind elemental, isotopic and charge-state composition under all conceivable solar wind conditions. The sensitivity and dynamic range of the instrument are such that the mean speeds, temperatures, and densities of the major elements in the solar wind may be determined with a time resolution of 12 min to $\sim$1 hr.

5.1. ENERGY RANGE

For SWICS the combined energy range of the main channel and the H/He channel extends from 110 eV/q (145 km s^{-1} for protons) to 100 keV/q (4380 km s^{-1} for protons; 1660 km s^{-1} for Fe^{8+}). This large dynamic range of 1000 in energy per charge will allow us to carry out solar wind composition measurements under all possible flow conditions, as well as to study pickup ions and the suprathermal tails of the distribution functions of, for example H, He, and O. Combining SWICS and SWIMS data allows the determination of charge state distributions of isotopes that can not be resolved by SWICS alone.

5.2. SENSITIVITY

The multi-slit collimator used in SWICS makes it possible to obtain a large geometrical factor and a wide field-of-view without sacrificing the energy resolution of the deflection analyzer. Since SWICS will predominantly measure solar wind ions, the relevant measure of sensitivity is related to the effective area presented to an incoming parallel beam (see Table I). The entrance system of SWIMS allows particles to enter from a wide range of angles.

The counting efficiencies in both TOF systems depend on the degree of scattering of ions in the front foil (most pronounced for low-energy heavy ions) and the number of secondary electrons produced by ions at the surface of the foil. The SWICS efficiency also depends on the secondary electron production by the solid-state detector (lowest number for higher energy protons). For SWIMS the efficiency depends on the charge state of the ions after passing through the carbon foil. The majority of the incident ions become neutrals as they pass through the carbon foil. In order to minimize the background caused by reflected neutrals, a finned hyperbola was used to allow the neutrals to be trapped. As a result SWIMS operates with a very low background.

We have calibrated SWICS and SWIMS using beams of selected ions available prior to launch from H through Kr. Some of our results of efficiencies are presented in Section 5.5 below.

5.3. Intensity dynamic range

With the SWICS and SWIMS instruments we are able to achieve an intensity dynamic range of $\sim 10^9$ because: (a) the most intense solar wind fluxes (protons) are generally measured only in the SWICS H/He channel; (b) the rare elements are analyzed in the TOF systems with high priority; and (c) the high immunity to background makes it possible to detect and identify rare elements and isotopes. The maximum proton flux of $\sim 10^9$ cm^{-2} s^{-1} that we can measure is determined by the peak counting rate of our H/He detector ($\sim 5 \times 10^5$ count s^{-1}), the proton temperature, and the energy bandwidth and geometrical factor of the deflection analyzer.

5.4. Mass and mass-per-charge resolution

We have determined the mass and mass/charge resolution of SWICS for every major ion species using all known aspects of the instrument. These include effects due to: (a) deflection-analyzer system resolution and dispersion; (b) electronic noise in the TOF measurement and path-length dispersion of secondary electrons; (c) path-length differences of ions in the TOF system; (d) energy dispersion associated with nuclear defects in solid-state detectors and electronic noise in the energy measurement; and (e) timing dispersions caused by energy straggling of ions in the carbon foil. These effects which have been measured or are determined by the geometry of the system are combined in quadrature to give the FWHM resolutions in mass, (M), and mass/charge, (M/Q). Table VI summarizes the mass and mass/charge resolution in SWICS for common solar wind ions.

For SWIMS the mass resolution Δ_M is roughly proportional to $\sqrt{M}$ and also depends somewhat on the scattering of ions in the carbon foil. In general, $\Delta_M/M < 0.02$).

TABLE V

Mass and power for SWICS and SWIMS

Subsystem	Mass (g)	Power (mW)
SWICS		
−30 kV supply (PAPS)	**566**	**550**
Sensor	**3426**	**1900**
Deflection system and deflection supply	1483	400
Analog electronics and bias supply	908	1200
Sensor elements and structure	1035	300
CDT Assembly	**1698**	**2500**
CDT	800	1000
Low-voltage power converter	800	1500
Thermal blankets	**280**	
SWICS total	**5970 g**	**4950* mW**

* Average power; peak power 6.11 W

Subsystem	Mass (g)	Power (mW)
SWIMS		
WAVE entrance system	**586**	
Sensor	**7014**	**6800**
Power supplies	1450	2850
Analog/Digital electronics	980	3950
HMRS telescope	740	
Structure, etc.	4044	
Thermal items	**450**	
SWIMS total	**8050 g**	**6800** mW**

**Average power; peak power 7.2 W

Examples of the mass (SWICS and SWIMS) and mass/charge (SWICS only) resolution are shown in Figure 13 for SWICS and Figure 14 for SWIMS which show raw flight data.

5.5. INSTRUMENT CALIBRATION

The SWICS instrument on ACE is the improved flight spare unit of SWICS on *Ulysses* (Gloeckler et al., 1992), launched in 1990, and is similar to SWICS on WIND (Gloeckler et al., 1995b) launched in 1994. SWIMS is an improved version of the SOHO CELIAS-MTOF instrument, (Hovestadt et al., 1995) launched in 1995, incorporating energy per charge analysis used in the MASS instrument on WIND. There is preflight calibration and flight data available for these instruments allowing for cross checking of instrument responses. Pre-flight calibration

TABLE VI

SWICS resolution characteristics for typical solar wind ions*

Element	Mass (amu)	Charge (e)	Energy** (keV)	Time-of-flight (ns)	$\Delta_{m/q}/(m/q)$ (FWHM)	Δ_m/m (FWHM)
H	1	1	26	38.5	0.052	0.472
He	4	2	52	53.1	0.043	0.253
C	12	6	143	52.6	0.035	0.163
N	14	7	164	52.6	0.035	0.167
O	16	6	131	60.2	0.035	0.221
Ne	20	8	166	58.1	0.028	0.220
Si	28	9	174	64.1	0.028	0.246
S	32	10	189	64.9	0.028	0.254
Fe	56	11	157	79.3	0.026	0.302

*For 440 km s^{-1} solar wind speed and 30 kV post-acceleration.
**Measured by solid-state detector.

for SWICS was performed in November–December 1995, June 1996, and June 1997 at the University of Bern Accelerator Facility. SWIMS pre-flight calibration was performed in June–July 1996 and June 1997 at the University of Bern. Both SWICS and SWIMS were additionally calibrated at the Justus-Liebig University in Giessen, Germany in April–June 1996. We measured the energy and MCP bias dependence of efficiencies for various ion species, geometrical factors, and the angular and energy responses of the electrostatic deflection systems for both sensors. SWIMS was also checked for temperature effects on HV-noise in the Thermal Vacuum chamber at Bern. The Bern accelerator facility is well suited for the calibration of solar-wind-type instruments. It is possible to obtain uniform, stable beams of gas-derived ion species (H^+, He^+, He^{++}, Ne^{+4}, Ar^{+4}, Kr^{+5}) in the range of 0.5 to 60 keV e^{-1}. The Giessen facility, using an ECR (Electron Cyclotron Resonance) ion source, was used to deliver heavy solar wind ions (Si, S, Na, Fe) as well as high charge-state ions (O^{+5}) to SWICS and a wide range of metals (Ti, Ni, Co, Zn, Cr) to SWIMS. The SWIMS flight spare unit is available for extensive calibrations with heavy metal ion beams after the launch of ACE.

5.5.1. *SWICS Efficiencies*

Knowledge of the efficiencies of the MCPs as a function of energy of the various elements is required to obtain fluxes and relative abundances of solar wind ions. The stop efficiencies of SWICS for He, C, O, Ar, and Kr as a function of energy from 0.5 to 20 keV amu^{-1} are shown in Figure 15. Start efficiencies have a similar energy and species dependence.

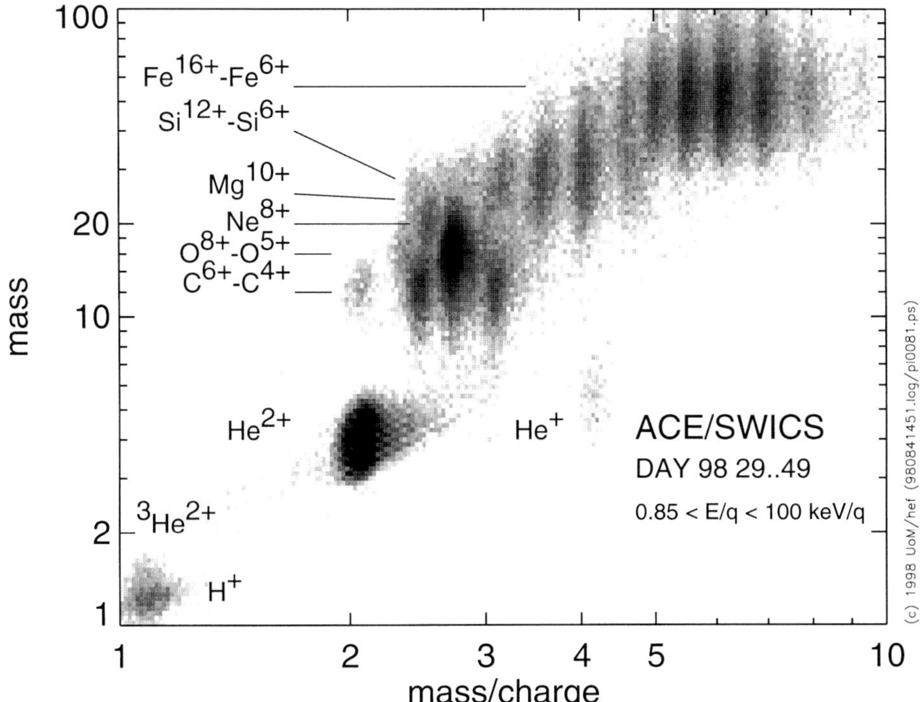

Figure 13. Display of the mass vs mass-per-charge distributions of solar wind ions derived from the raw energy and time-of-flight pulse-height data collected by SWICS during days 29–49 of 1998. The mass and mass/charge values were computed on the ground using algorithms identical to those employed by the S^{3}DPU for the on-board classification of solar wind ions. The density profiles show well-resolved peaks of the major solar wind heavy elements and their dominant charge states. The relative abundances of the various charge states of elements that can be derived from data such as shown here may be used to infer the temperatures and temperature profiles of the low solar corona.

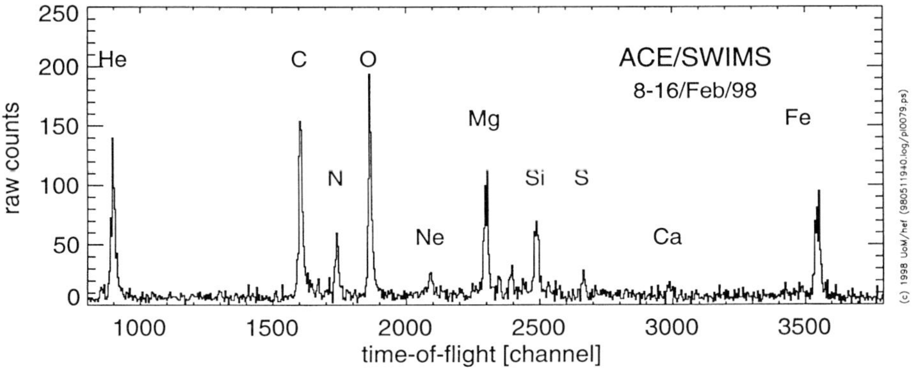

Figure 14. Display of the mass distribution of solar wind ions derived from the raw time-of-flight pulse-height data collected by SWIMS.

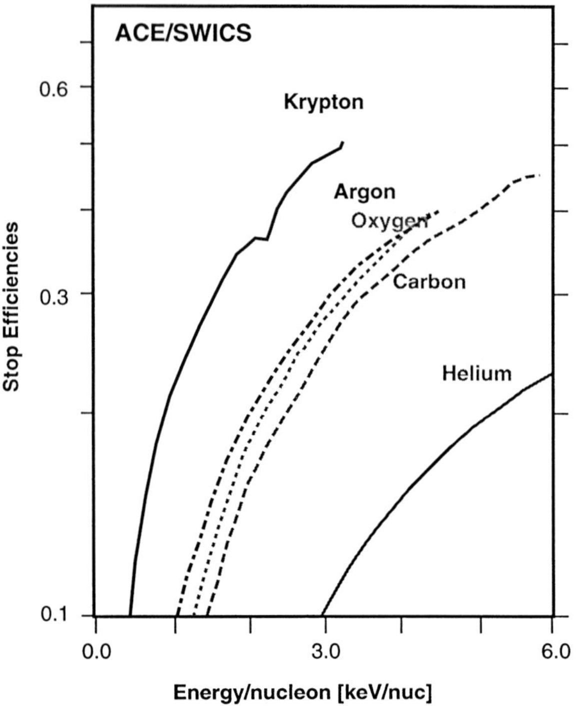

Figure 15. Measured dependence of the SWICS stop efficiency on energy per nucleon for the indicated elements derived from calibration data. The stop efficiency is the ratio of the Double Coincidence Rate to the Start Rate and reflects the effects of ion scattering caused by the carbon foil, the efficiency of generating back-scattered secondary electrons from the solid-state detectors and adjacent surfaces, the efficiency of focusing these electrons onto the stop microchannel plate (MCP), and the efficiency of detecting these electrons by the stop MCP.

5.5.2. *SWIMS Elemental and Isotopic Resolution*

SWIMS has excellent mass resolution, which allows it to resolve all the solar wind elements and isotopes up to and including Ni. Figure 16 shows the mass resolution of SWIMS, which resolves peaks for C (from the thin carbon foil), Ne (the primary gas in the plasma), Mg, Ti, Ni, Zn, and Rb. There are no significant isotope peaks due to the mass-per-charge selection of the bending magnets in the calibration facility. Figure 17 shows raw calibration data measured with the SWIMS spare unit that shows clearly resolved isotopes of Ne. Notice that a fraction of the Ne ions become doubly charged as they pass though the carbon foil.

6. Browse Parameter Data

The SWICS/SWIMS investigation will provide selected solar wind parameters for several heavy ion species averaged over 12 min (or multiples of 12 min). The following data will be included:

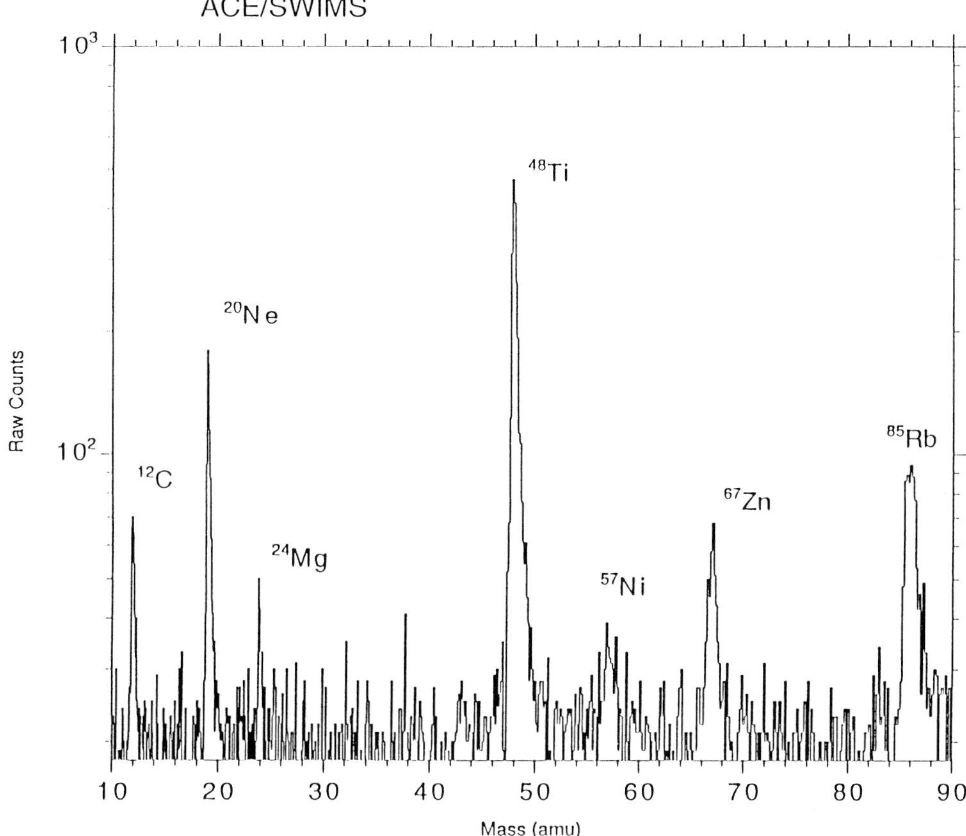

Figure 16. Mass spectrum obtained by SWIMS during calibration using an alloy rod as the primary ion source. Several peaks from the sample (Mg, Ti, Ni, Zn, Rb) are clearly resolved, as well as a Neon peak from the ionizing gas. A Carbon peak is also present from ions knocked out of the thin carbon foil at the entrance of the instrument.

(1) Bulk speed and kinetic temperature of solar wind He^{++} and O^{6+}.
(2) Coronal electron freezing-in temperature derived from O^{6+}/O^{7+}.
(3) Fe/O (Low FIP/ High FIP) abundance ratio.
(4) He/O abundance ratio.

7. Summary

The SWICS/SWIMS Investigation on the ACE spacecraft will obtain detailed information on the thermal and suprathermal ion populations in the solar wind upstream of the Earth's magnetosphere, with an emphasis on compositions of these plasmas. This information is required for studies of a variety of solar, heliospheric and interstellar phenomena, and is important for interpreting results from other

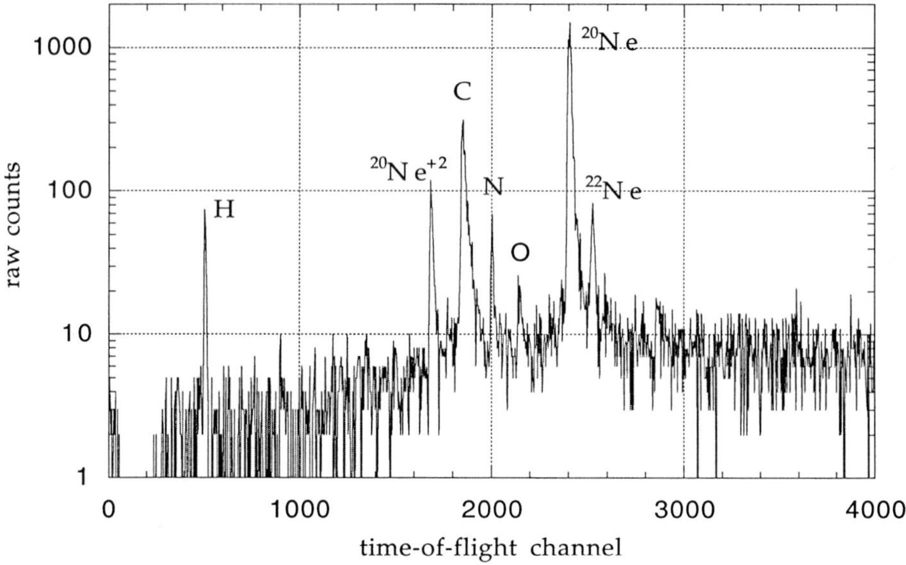

Figure 17. Display of the isotopic resolution of Neon derived from calibration data using Ne^{4+} as a primary beam.

heliospheric missions such as *Ulysses*, ISTP spacecraft such as WIND and SOHO, and various space and ground-based observations of the solar photosphere and lower corona. These data, especially in combination with measurements from other instruments on ACE, other GGS and ISTP spacecraft and *Ulysses*, as well as ground and space observations of the Sun, will make possible a number of new studies of solar, heliospheric and magnetospheric phenomena. These scientific objectives require measurements provided by the two complementary instruments of this investigation. The Solar Wind Ion Composition Spectrometer (SWICS) will determine routinely and with good time resolution the charge states and distribution functions of the most abundant solar wind heavy ions (H, He, C, O, Ne, Mg, Si, and Fe) from 0.5 to 30 keV e^{-1}, and will measure pickup ions of interplanetary and interstellar origin. The Solar Wind Ion Mass Spectrometer (SWIMS) will measure the elemental and isotopic composition and charge state of the solar wind plasma with far greater precision than is possible with SWICS (i.e., elements Na, Al, S, Ar, Ca, Cr, and Ni, and isotopes of Ne, Mg, Si, S, Ar, Ca, and Fe).

Acknowledgements

The development of the SWICS and SWIMS experiments was an international effort involving the Space Physics group in the Department of Physics at the University of Maryland, the Physikalisches Institut of the Universität Bern in Bern,

Switzerland, and the Institut für Datenverarbeitungsanlagen of the Universität Braunschweig in Braunschweig, Germany. Many individuals at these institutions contributed greatly in the design, development, fabrication, testing, and calibration of the two instruments and the S^{3}DPU. We are particularly grateful to the following individuals: At the University of Maryland to Fred Wire for the design and testing of digital subsystems for both instruments; to Richard Pappalardo for the developing and testing the power systems for SWIMS; to Bob Cates and Charles Tomasevich for the mechanical design of SWICS and SWIMS; to Maurice Pairel and Pota Floros who did an excellent job laying out and assembling the wiring harnesses and the many electronics boards for the instruments; to Scott Lasley for his laboratory testing of sensor detectors and for his invaluable support during the integration, testing and calibrations of the instruments; to Wayne Shanks and Jeff Miller for their help during instrument testing and calibration; to Pat Ipavich and Jo Ann Simms for the financial management and reporting for the experiments. At the University of Bern to Kurt Bratschi and Paul Liechti, who developed, fabricated and tested the deflection systems for SWIMS and SWICS, including the SWICS entrance collimator and deployable cover; to Raoul Pärli, Rosmarie Neukomm, Olivier Kern, and Peter Wurz for their assistance during the calibrations of SWIMS and SWICS both in Bern and in Giessen; and to Urs Schwab, Martin Steinacher, and Adrian Etter for the operation of the Bern Calibration Facility during the many 24-hour days of calibrations of the instruments. At the Technical University of Braunschweig to Wolfgang Wiewesiek for the electrical design and testing of the DPU hardware; to Bjoern Fiethe for the design and testing of the EGSE hardware; and to Kay-Uwe Reiche and Kai Stöckner who wrote and debugged the extensive DPU software. At the Justus-Liebig University in Giessen, Germany to Professor Dr Erhard Salzborn, Roland Trassl, and Frank Broetz for their valuable and generous support during the testing there. We would like to thank Bruce Williams and Julie Krein of the Applied Physics Laboratory (APL) for their thermal design of both instruments. APL's spacecraft team, led by Mary Chiu and Judi von Mehlem, was also of great assistance from the beginning. In particular, we recognize the extraordinary efforts of James Roberts, Joseph Staiger, and Elliot Rodberg and their integration and test teams, without whose support and flexibility, the quality of SWICS and SWIMS experiments would surely have suffered. We appreciate the support of the ACE Project Office at the Goddard Space Flight Center, headed by Mr Don Margolics, and of the ACE management team at the California Institute of Technology, led by Mr Al Frandsen, throughout the long years of development and preparation of the SWICS and SWIMS experiments. The SWICS and SWIMS investigations were funded by NASA through contract NAS5-32626, by the Swiss National Science Foundation, and by the Bundesminister für Forschung und Technologie in Germany.

References

Breneman, H. H. and Stone, E. C.: 1985, *Astrophys. J.* **299**, L57.

Bürgi, A. and Geiss, J.: 1986, *Solar Phys.* **103**, 347.

Burlaga, L. F.: 1984, *Space Sci. Rev.* **39**, 255.

Collier, M. R., Hamilton, Gloeckler, G., Bochsler, P., and Sheldon, R. B.: 1996, *Geophys. Res. Lett.* **23**, 1191.

Fisk, L. A., Schwadron, N. A., and Gloeckler, G.: 1997, *Geophys. Res. Lett.* **24**, 93.

Fisk, L. A., Schwadron, N. A., and Zurbuchen, T. H.: 1998, *Space Sci. Rev.* **86**, 51.

Galvin, A. B., Ipavich, F. M., Gloeckler, G., Hovestadt, D., Bame, S. J., Klecker, B., Scholer, M., and Tsurutani, B. T.: 1987, *J. Geophys. Res.* **92**, 12069.

Geiss, J. and Bürgi, A.: 1986, *Astron. Astrophys.* **159** 1.

Geiss, J., Bühler, F., Cerutti, H., Eberhardt, P., and Filleux, C.: 1972, *Sec. 14 of the Apollo 16 Preliminary Science Report* **NASA SP-315**.

Geiss, J., Gloeckler, G., von Steiger, R., Balsiger, H., Fisk, L. A., Galvin, A. B., Ipavich, F. M., Livi, S., McKenzie, J. F., Ogilvie, K. W., Wilken, B.: 1995, *Science* **268**, 1033.

Geiss, J., Gloeckler, G., and von Steiger, R.: 1996, in R. von Steiger, R. Lallement, and M. A. Lee (eds), 'The Heliosphere in the Local Interstellar Medium, Proceedings of the First ISSI Workshop', *Space Sci. Rev.* **78**, 43.

Gloeckler, G.: 1977, University of Maryland Technical Report TR 77-043 (unpublished).

Gloeckler, G.: 1990, *Rev. Sci. Instrum.* **61** , 3613.

Gloeckler, G.: 1996, in R. von Steiger, R. Lallement, and M. A. Lee (eds), 'The Heliosphere in the Local Interstellar Medium, Proceedings of the First ISSI Workshop', *Space Sci. Rev.* **78**, 335.

Gloeckler, G. and Geiss, J.: 1989, in J. C. Waddington (ed.), *AIP Conf. Proceedings* **183**, 49.

Gloeckler, G. and Geiss, J.: 1998, *Space Sci. Rev.* **86**, 127.

Gloeckler, G. and Hsieh, K. C.: 1979, *Nucl. Inst. Meth.* **165**, 537.

Gloeckler, G., Geiss, J., Balsiger, H., Fisk, L. A., Gliem, F., Ipavich, F. M., Ogilvie, K. W., Stüdemann, W., and Wilken, B.: 1983, *ESA Spec. Publ.* **SP-1050**, 77.

Gloeckler, G., Ipavich, F. M., Stüdemann, W., Wilken, B., Hamilton, D. C., Kremser, G., Hovestadt, D., Gliem, F., Lundgren, R. A., Rieck, W., Tums, E. D., Cain, J. C., Masung, L. S., Weiss, W., and Winterhof, P.: 1985, *IEEE Trans. on Geosci. and Remote Sensing* **GE-233**, 234.

Gloeckler, G., Hovestadt, D., Ipavich, F. M., Scholar, M., Clicker, B., and Galvin, A. B.: 1986, *Geophys. Res. Letters* **13**, 251.

Gloeckler, G., Ipavich, F. M., Hamilton, D. C., Wilken, B., and Kremser, G.: 1989, *EOS Trans.* **70**, 424.

Gloeckler, G., Geiss, J., Balsiger, H., Bedini, P., Cain, J. C., Fischer, J., Fisk, L. A., Galvin, A. B., Gliem, F., Hamilton, D. C., Hollweg, J. V., Ipavich, F. M., Joos, R., Livi, S., Lundgren, R. A., Mall, U., McKenzie, J. F., Ogilvie, K. W., Ottens, F., Rieck, W., Tums, E. O., von Steiger, R., Weiss, W., and Wilken, B.: 1992, *Astron. Astrophys. Suppl. Ser.* **92**, 267.

Gloeckler, G., Geiss, J., Balsiger, H., Fisk, L. A., Galvin, A. B., Ipavich, F. M., Ogilvie, K. W., von Steiger, R., and Wilken, B.: 1993, *Science* **261**, 70.

Gloeckler, G., Schwadron, N. A., Fisk, L. A., and Geiss, J.: 1995a, *Geophys. Res. Lett.* **22**, 2665.

Gloeckler, G., Balsiger, H., Bürgi, A., Bochsler, P., Fisk, L. A., Galvin, A. B., Geiss, J., Gliem, F., Hamilton, D. C., Holzer, T. E., Hovestadt, D., Ipavich, F. M., Kirsch, E., Lundgren, R. A., Ogilvie, K. W., Sheldon, R. B., and Wilken, B.: 1995b, *Space Sci. Rev.* **71**, 79.

Habbal, S. R., Esser, R., Guhathakurta, M., and Fisher, R. P.: 1995, *Geophys. Res. Lett.* **22**, 1465.

Hamilton, D. C., Gloeckler, G., Ipavich, F. M., Lundgren, R. A., Sheldon, R. B., and Hovestadt, D.: 1990, *Rev. Sci. Instrum.* **61**, 3104.

Hilchenbach, M., Hovestadt, D., Klecker, B., Möbius, E.: 1991, *Proceedings Solar Wind Seven (E. Marsch, G. Schwenn ,ed.), Goslar, Germany, Pergamon Press* **349**.

Hollweg, J. V.: 1978, *Rev. Geophys. Space Phys.* **16**, 689.

Hovestadt, D., Hilchenbach, M., Bürgi, A., Klecker, B., Laeverenz, P., Scholer, M., Grünwaldt, H., Axford, W. I., Livi, S., Marsch, E., Wilken, B., Winterhoff, H. P., Ipavich, F. M., Bedini, P., Coplan, M. A., Galvin, A. B., Gloeckler, G., Bochsler, P., Balsiger, H., Fischer, J., Geiss, J., Kallenbach, R., Wurz, P., Reiche, K. U., Gliem, F., Judge, D. L., Ogawa, H. S., Hsieh, K. C., Möbius, E., Lee, M. A., Managadze, G. G., Verigin, M. I., and Neugebauer, M.: 1995, *Solar Phys.* **162**, 441.

Hundhausen, A., Gilbert, H., and Bame, S.: 1968, *J. Geophys. Res.* **73**, 5485.

Ipavich, F. M., Lundgren, R. A., Lambird, B. A., and Gloeckler, G.: 1978, *Nucl. Inst. and Meth.* **154**, 291.

Ipavich, F. M., Galvin, A. B., Gloeckler, G., Scholer, M., and Hovestadt, D.: 1981, *J. Geophys. Res.* **86**, 4337.

Ipavich, F. M, Galvin, A. B., Gloeckler, G., Hovestadt, D., Klecker, B., and Scholer, M.: 1986, *Science* **232**, 366.

Isenberg, P. A., and Hollweg, J. V.: 1983, *J. Geophys. Res.* **88**, 3923.

Isenberg, P. A.: 1991, *J. Geophys. Res.* **96**, 155.

Marsden, R. G., Sanderson, T. R., Tranquille, C., Wenzel, K.-P., and Smith, E. J.: 1987, *J. Geophys. Res.* **92**, 11009.

Möbius, E., Hovestadt, D., Klecker, B., Scholer, M., Gloeckler, G., and Ipavich, F. M.: 1985, *Nature* **318**, 426.

Möbius, E., Hovestadt, D., Klecker, B., Kistler, L. M., Popecki, M. A., Crocker, K. N., Gliem, F., Granoff, M., Turco, S., Anderson, A., Arbinger, H., Battell, S., Cravens, J., Demain, P., Distelbrink, J., Dors, I., Dunphy, P., Ellis, S., Gaidos, J., Googins, J., Harasim, A., Hayes, R., Humphrey, G., Kastle, H., Kunneth, E., Lavasseur, J., Lund, E. J., Miller, R., Murphy, G., Pfeffermann, K., Reiche, K.-U., Sartori, E., Schimpfle, J., Seidenschwang, E., Shappirio, M., Stockner, K., Taylor, S. C., Vachon, P., Vosbury, M., Wiewesiek, and Ye, V. 1998, 'The Solar Energetic Particle Ionic Charge Analyzer (SEPICA) and the Data Processing Unit (S3DPU) for SWICS, SWIMS and SEPICA', *Space Sci. Rev.* **86**, 449.

Möbius, E., Rucinski, D., Hovestadt, D., and Klecker, B.: 1995, *Astron. Astrophys.* **304**, 505.

Möbius, E., Rucinski, D., Lee, M. A., and Isenberg, P. A.: 1997, *J. Geophys. Res.* (submitted).

Neugebauer, M.: 1981, *Fundamentals of Cosmic Physics* **7**, 131.

Ogilvie, K. W., Bochsler, P., Geiss, J, and Coplan, M. A.: 1980, *J. Geophys. Res.* **85**, 6069.

Reames, D. V., Richardson, I. G., and Barbier, L. M.: 1991, *Astrophys. J.* **382**, L43.

Schmidt, W. K. H., Rosenbauer, H., Shelley, E. G., and Geiss, J.: 1980, *Geophys. Res. Lett.* **7**, 697.

Tan, L. C., Mason, G. M., Klecker, B., and Hovestadt, D.: 1989, *Astrophys. J.* **345**, 572.

von Steiger, R., and Geiss, J.: 1989, *Astron. Astrophys.* **225**, 222.

von Steiger, R., Christon, S. P., Gloeckler, G., and Ipavich, F. M.: 1992, *Astrophys. J.* **389**, 791.

Vauclair, S., and Meyer, J.-P.: 1985, *Proc. 19th Int. Cosmic Ray Conf.* **4**, 233.

ELECTRON, PROTON, AND ALPHA MONITOR ON THE ADVANCED COMPOSITION EXPLORER SPACECRAFT

R. E. GOLD, S. M. KRIMIGIS, S. E. HAWKINS, III, D. K. HAGGERTY, D. A. LOHR
and E. FIORE

The Johns Hopkins University Applied Physics Laboratory, Laurel, MD 20732–6099, U.S.A.

T. P. ARMSTRONG and G. HOLLAND

University of Kansas, U.S.A.

L. J. LANZEROTTI

Bell Labs, Lucent Technology, U.S.A.

Abstract. The Electron, Proton, and Alpha Monitor (EPAM) is designed to make measurements of ions and electrons over a broad range of energy and intensity. Through five separate solid-state detector telescopes oriented so as to provide nearly full coverage of the unit-sphere, EPAM can uniquely distinguish ions ($E_i \gtrsim 50$ keV) and electrons ($E_e \gtrsim 40$ keV) providing the context for the measurements of the high sensitivity instruments on ACE. Using a $\Delta E \times E$ telescope, the instrument can determine ion elemental abundances ($E \gtrsim 0.5$ MeV nucl^{-1}). The large angular coverage and high time resolution will serve to alert the other instruments on ACE of interesting anisotropic events. The experiment is controlled by a microprocessor-based data system, and the entire instrument has been reconfigured from the HI-SCALE instrument on the *Ulysses* spacecraft. Inflight calibration is achieved using a variety of radioactive sources mounted on the reclosable telescope covers. Besides the coarse (8 channel) ion and (4 channel) electron energy spectra, the instrument is also capable of providing energy spectra with 32 logarithmically spaced channels using a pulse-height-analyzer. The instrument, along with its mounting bracket and radiators weighs 11.8 kg and uses about 4.0 W of power. To demonstrate some of the capabilities of the instrument, some initial performance data are included from a solar energetic particle event in November 1997.

1. Science Objectives

The Electron, Proton, and Alpha Monitor (EPAM) instrument on the ACE spacecraft is designed to measure a broad range of energetic particles over nearly the full unit-sphere at high time resolution. Such measurements of ions and electrons in the range of a few tens of keV to several MeV are essential to understand the dynamics of solar flares, co-rotating interaction regions (CIRs), interplanetary shock acceleration, and upstream terrestrial events.

The primary objective of EPAM is to support the high sensitivity instruments on ACE. These are the Cosmic Ray Isotope Spectrometer (CRIS), the Solar Energetic Particle Ionic Charge Analyzer (SEPICA), the Solar Isotope Spectrometer (SIS), the Solar Wind Ion Mass Spectrometer (SWIMS), the Solar Wind Ion Composition Spectrometer (SWICS), and the Ultra Low Energy Isotope Spectrometer

Space Science Reviews **86**: 541–562, 1998.
© 1998 *Kluwer Academic Publishers. Printed in the Netherlands.*

(ULEIS). As a monitor instrument EPAM will provide the context for the measurements made by the high sensitivity instruments. The large dynamic range of EPAM extends from about 50 keV to 5 MeV for ions, and 40 keV to about 350 keV for electrons. To complement its electron and ion measurements, EPAM is also equipped with a Composition Aperture (CA) which unambiguously identifies ion species reported as species group rates and/or individual pulse-height events. The instrument achieves its large spatial coverage through five telescopes oriented at various angles to the spacecraft spin axis. The low-energy particle measurements, obtained at time resolutions between 1.5 and 24 s, and the ability of the instrument to observe particle anisotropies in three dimensions make EPAM an excellent resource to provide the interplanetary context for studies using other instruments on ACE.

EPAM is part of the Real-Time Solar Wind (RTSW) system developed by NASA and NOAA. The instrument provides 24-hr coverage of the space weather environment at ACE. The data products include measurements of the real-time flux levels, energy spectra, and anisotropy of energetic ions.

EPAM has some unique capabilities that enable it to provide significant insights in the understanding of energetic particle acceleration and propagation in the heliosphere. Low-energy solar particles will act as probes of the morphology of coronal and interplanetary field structures as they evolve during the rise in solar activity from the near solar minimum conditions at the launch of ACE, through solar maximum. EPAM is the only instrument on the ACE spacecraft that observes the full unit-sphere distribution of particle fluxes; it is the only instrument that measures electrons; it has the highest time resolution of the energetic particle instruments and it covers a wide range of energies with sufficient dynamic range to handle everything from very small events to the most intense solar flares and interplanetary shock spikes. These capabilities will enable EPAM to contribute to the study of a large variety of ACE mission science objectives.

EPAM will examine the acceleration and propagation of ions and electrons in solar energetic particle events. It covers relativistic energies for electrons and non-relativistic ion energies. EPAM covers more than three decades of ion energy including the crucial energy regime spanning the transition from prompt ion propagation at and above 5 MeV nucl^{-1} down to the slower probes of interplanetary field structures at 50 keV. A 32-channel energy spectrum accumulator in EPAM can follow the velocity dispersion of the outflowing ions.

Interplanetary shock acceleration of low-energy ions and electrons is a prominent feature of particle fluxes in the energy range from 50 keV to a few MeV. With its three-dimensional coverage, EPAM can measure the true flow patterns of particles accelerated near shocks. CIRs are also important accelerators of ions, primarily at radial distances beyond 1 AU. The structure and dynamics of particle acceleration at CIRs can be examined in three dimensions by EPAM as these particles flow back into the inner heliosphere. The nearly full unit-sphere coverage of EPAM is important for separating the apparently similar anisotropies observed resulting from particle flows or cross-field gradients in particle fluxes when they are observed in only two dimensions.

En route from the Earth to the L_1 libration point, ACE observed many bursts of outflowing magnetospheric particles. These bursts, which have been examined both at Jupiter and at Earth, contain many details of particle acceleration and the propagation of the particles upstream to ACE. EPAM will examine the rapid time variations, strong three-dimensional anisotropies and the composition of upstream bursts.

2. Instrument

EPAM is designed to make *in situ* measurements of ions and electrons in interplanetary space. The instrument was adapted from the flight-spare unit of the HI-SCALE instrument aboard the *Ulysses* spacecraft. HI-SCALE was built by the Johns Hopkins University Applied Physics Laboratory with Dr Louis J. Lanzerotti of Lucent Technologies as Principal Investigator; detailed descriptions of HI-SCALE may be found elsewhere (Lanzerotti et al., 1983, 1992, 1993). Although HI-SCALE and EPAM are identical in design and functionality, the mechanical layout of the detectors making up the HI-SCALE instrument was designed for an unimpeded field-of-view when mounted on the corner of the box-like structure of *Ulysses*. In order to provide EPAM with an unimpeded field-of-view on the irregular octagon structure of ACE, the two telescope assemblies were moved to a mounting fixture, elevated above the spacecraft top deck and located between two solar arrays (see Chiu et al., this issue). An interface box was added to the EPAM electronics to convert the control signals sent to and received from the *ACE* spacecraft, in order to use the *Ulysses* spacecraft interface.

EPAM measures energetic particles in the energy range from about 50 keV to 5 MeV. Its configuration consists of five solid-state detector systems on two stub telescopes, referred to as the 2A and 2B assemblies. The telescopes are mounted on a support bracket, which also mounts and encloses the instrument electronics box. The EPAM instrument alone weighs approximately 5.64 kg and the instrument along with its mounting bracket weighs 11.8 kg. Although survival heaters are present, the nominal instrument temperature is well above the thermostatic set point of these heaters, and so they will not cycle on. EPAM consumes about 4.0 W of power during normal operations. Table I summarizes the characteristics of EPAM.

Each stub telescope supports a Low-Energy Magnetic Spectrometer (LEMS) to measure ions, and a Low-Energy Foil Spectrometer (LEFS) that measures both ions and electrons. These two detector systems together form a LEMS/LEFS pair which is identical in design on each stub-arm telescope. The two LEMS sensors are oriented 30° and 120° from the spin axis and are identified as LEMS30 and LEMS120. The LEFS are pointed 60° and 150° from the spacecraft spin axis and are known as the LEFS60 and LEFS150 telescopes. The last detector system

544

TABLE I

ACE/EPAM instrument characteristics

Mass	instrument	5.64 kg
	bracket	6.13 kg
Power		4.0 W
EPAM data rate		168 bits/s , continuous
Thermal control		passive
Spatial coverage		5 apertures, $\approx$full unit sphere
Time resolution		1.5–6 s
Energetic particles	electrons	40–$\approx$350 keV
	ions	46–4800 keV
	species groups	H, He, CNO, Fe
Geometric factors	LEMS	0.428 cm^2 sr
	LEFS	0.397 cm^2 sr
	CA	0.103 cm^2 sr

measures ion composition, and is referred to as the Composition Aperture (CA). Its look-direction is oriented 60° from the spacecraft spin-axis and so is known as the CA60 telescope.

The two LEMS telescopes and the CA60 telescope have covers that may be commanded closed by heating bi-metallic springs attached to each cover. Radioactive sources mounted in the covers provide inflight calibration of the LEMS and CA detector systems. A photograph of the instrument is shown in Figure 1, with labels indicating the five detector assemblies.

ACE is a spin-stabilized spacecraft which rotates about an axis directed within 20° of the Sun at a rate of about 5 revolutions per minute and hence a spin period of 12 s. As the spacecraft spins, these five telescopes sweep out swaths of space, providing nearly full three-dimensional coverage for measuring ion anisotropies and approximately 40% coverage for electron anisotropies. As the spacecraft spins, the instrument electronics sample the detectors such that the swath of space swept out by each telescope is divided into nearly equally spaced sectors.

Figure 2 shows the EPAM look directions and the approximate sector definitions, projected onto the unit sphere, as well as a cylindrical projection of these look directions for the five sensor heads. The abscissa denotes the clock angle in degrees from the spacecraft x-axis. The ordinate gives the polar angle from ACE's spin vector. In the case of the detector pair, LEMS30/LEFS150, four sectors are used, each 90° wide. These sectors are referred to by the upper case letters A to D for LEMS30 and the numbers 1 to 4 for the LEFS150. The LEMS120/LEFS60 detector pair is divided into eight sectors, each 45°. These sectors are denoted in the figure by the numbers 1 through 8 for LEMS120, and the letters a to h for the

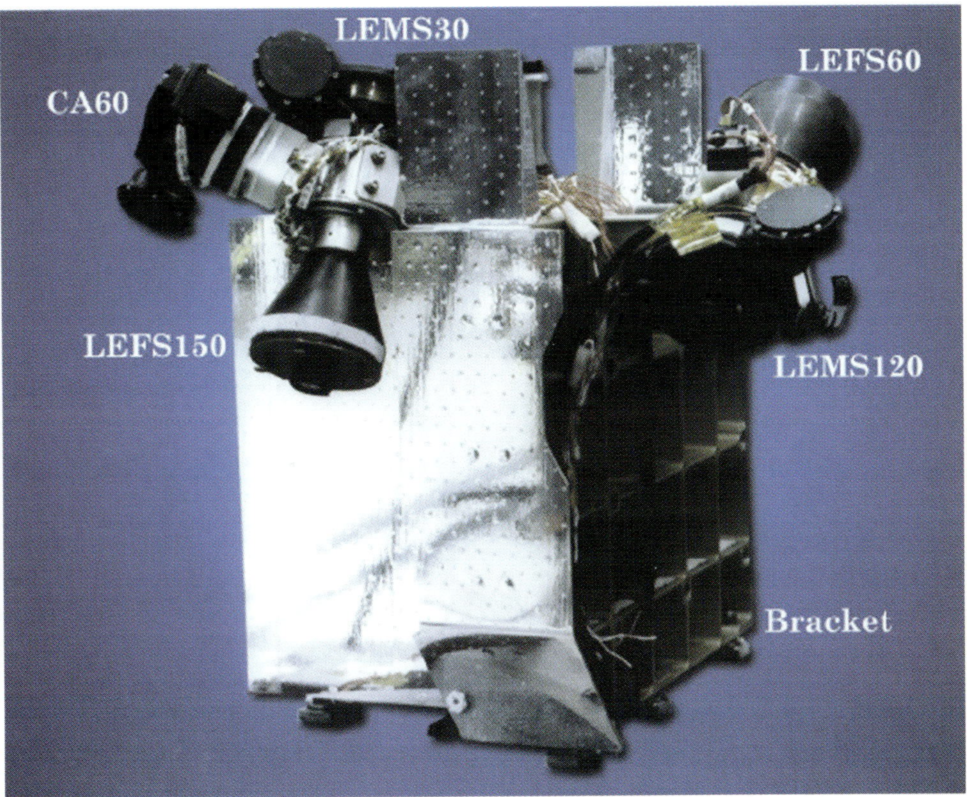

Figure 1. The EPAM instrument on ACE. The five detector orientations on the two stub-arm telescopes are designed to provide nearly full coverage of the unit sphere as the spacecraft spins. The LEMS and the CA detectors have reclosable covers and are used for inflight calibrations.

LEFS60 system. The CA60 telescope also is sampled in 8 sectors. From this figure, it is clear that EPAM covers most of the unit-sphere, with considerable overlap in spatial coverage.

2.1. DETECTOR SYSTEMS

A schematic view of the EPAM 2A and 2B stub-arm telescope configurations is shown in Figure 3. The EPAM 2A configuration shows the LEMS120/LEFS60 telescope pair. The 2B configuration shows the LEMS30/LEFS150 pair, as well as the CA60 telescope and its detectors (discussed below). The two LEFS detectors have a full-cone angle of 53° and a geometric factor of 0.397 cm^2 sr. The two LEMS detectors have a full-cone angle of 51° and a geometric factor of 0.428 cm^2 sr. The CA has a reduced geometric factor of 0.103 cm^2 sr and a full-cone angle of 45°.

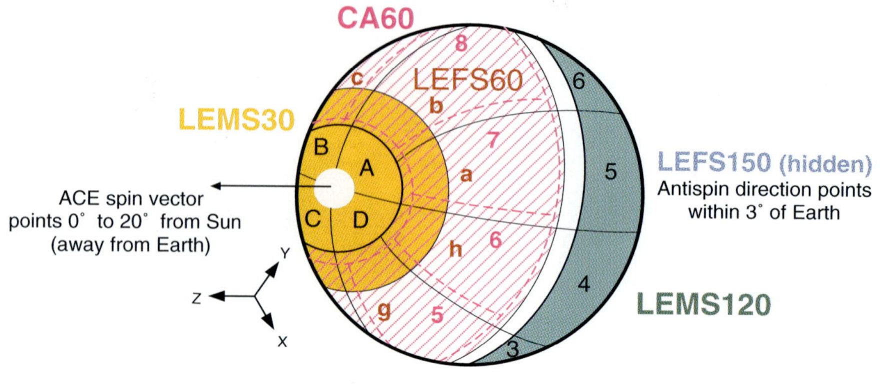

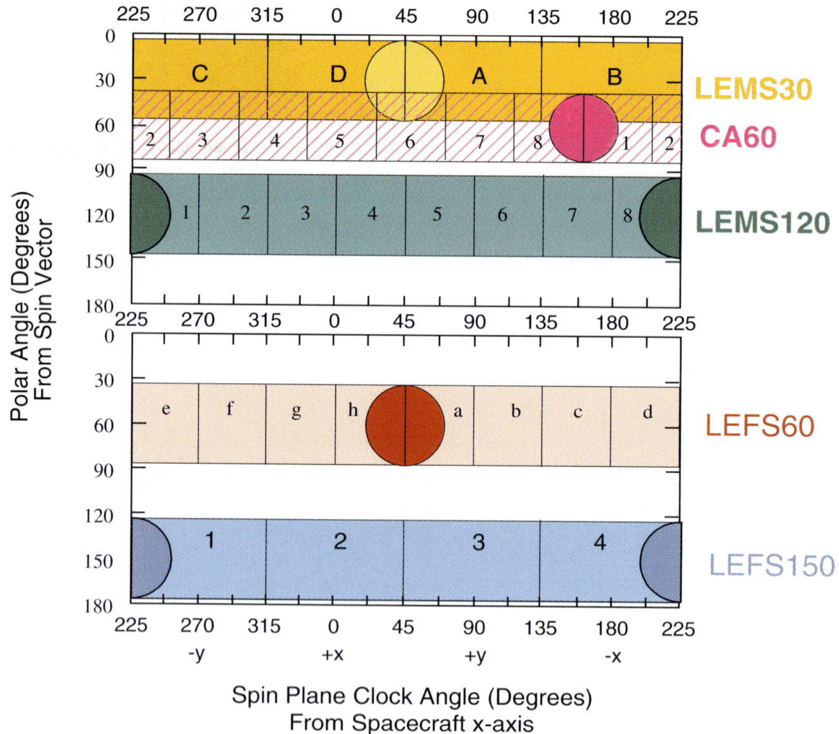

Figure 2. The five EPAM detector assemblies are shown in a cylindrical projection. The energy channels for each of the telescopes are abbreviated for the electrons as E (LEFS150) and E' (LEFS60). The ion channels are denoted by P (LEMS30) and P' (LEMS120), and the composition channels by W (CA60). The abscissa shows the clock angle measured from the spacecraft's x-axis. Depending on the detector, either 4 or 8 sectors divide a spin into approximately equally spaced regions. The ordinate shows the polar angle measured from the spin vector of ACE.

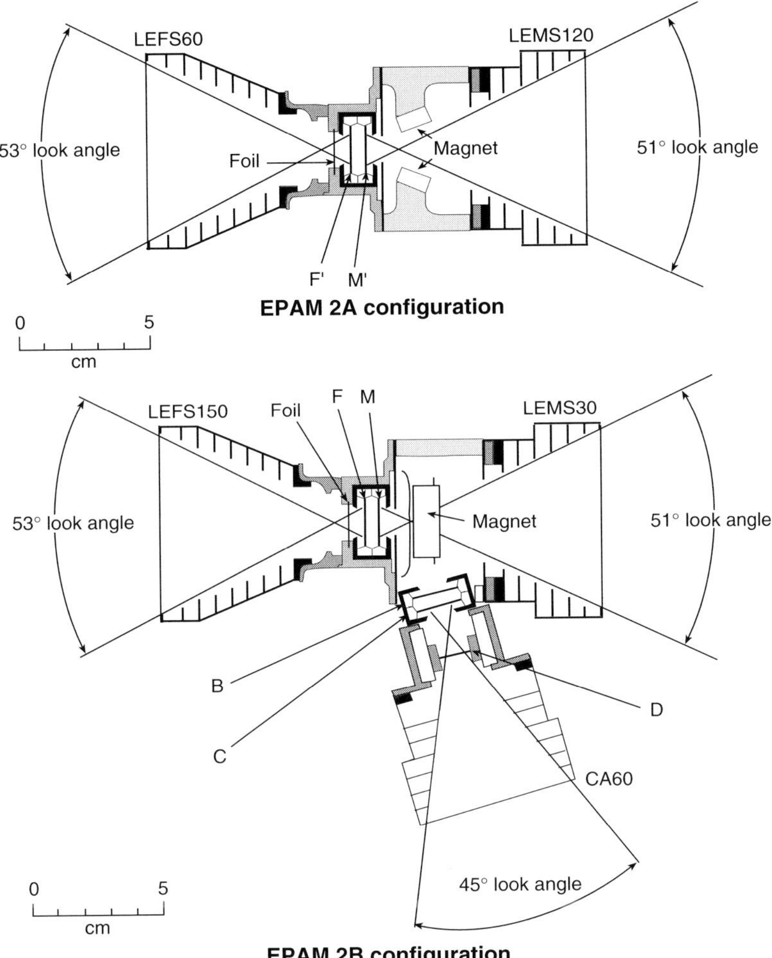

Figure 3. Schematic outline of the EPAM instrument's detector assemblies. The detectors are identified by the letters M, F, M', F', B, C and are totally depleted surface barrier Si detectors, each approximately 200 μm thick. The D detector is 4.8 μm thick.

2.1.1. *LEMS/LEFS*

The LEFS60 and LEFS150 telescopes measure electrons with energies less than about 350 keV. An aluminized Parylene foil, nominally 0.35 mg cm^{-2} thick, is used to absorb ions with energies below approximately 350 keV, while allowing electrons with energies above about 35 keV to pass through to the solid-state detector. The detector assemblies are referred to as F and F' for the LEFS150 and LEFS60 telescopes, respectively.

The two LEMS telescopes measure ions and these detectors are referred to as M for the LEMS30 telescope, and M' for the LEMS120 telescope. A rare-earth magnet in front of each of the two LEMS detectors, M and M', sweeps out any electrons

TABLE II

Low-energy magnetic spectrometer detector systems

LEMS30			LEMS120		
Energy channel	Logic	Passband (MeV)	Energy Channel	Logic	Passband (MeV)
$P1$	$M1\overline{M2F}$	0.046–0.067	$P'1$	$M'1\overline{M'2F'}$	0.047–0.068
$P2$	$M2\overline{M3F}$	0.067–0.115	$P'2$	$M'2\overline{M'3F'}$	0.068–0.115
$P3$	$M3\overline{M4F}$	0.115–0.193	$P'3$	$M'3\overline{M'4F'}$	0.115–0.195
$P4$	$M4\overline{M5F}$	0.193–0.315	$P'4$	$M'4\overline{M'5F'}$	0.195–0.321
$P5$	$M5\overline{M6F}$	0.315–0.580	$P'5$	$M'5\overline{M'6F'}$	0.310–0.580
$P6$	$M6\overline{M7F}$	0.580–1.060	$P'6$	$M'6\overline{M'7F'}$	0.587–1.060
$P7$	$M7\overline{M8F}$	1.060–1.880	$P'7$	$M'7\overline{M'8F'}$	1.060–1.900
$P8$	$M8\overline{F}$	1.880–4.700	$P'8$	$M'8\overline{F'}$	1.900–4.800

with energy below about 500 keV. In the LEMS30 telescope, these electrons are measured in the B detector which is located at the back of the CA60 telescope assembly (described in Section 2.1.3). The B detector is also the anti-coincidence detector of the CA. Because the B detector measures deflected electrons, it is a pure electron detector and is not generally susceptible to ion contamination.

Each of the four detectors (M, F, M', F') used in the LEMS/LEFS telescopes is a totally depleted, solid-state, silicon surface barrier detector with a total thickness of approximately 200 μm. The penetrating particle creates electron-hole pairs in the detector, the number of which is related to the incident kinetic energy of the particle. Provided the detector stops the particle, its total energy can be inferred. Completely analogous to LEMS30, the LEMS120 telescope uses the M' detector, and the differential flux is measured in eight rate channels, $P'1$ through $P'8$. Table II gives the energy channel name, the logic equation implemented by the instrument electronics, and the energy passbands for the two LEMS detector systems. The overall EPAM block diagram (adapted from Lanzerotti et al., 1992) is shown in Figure 4.

The LEFS150 telescope uses detector F to measure rate and energy and detector M serves as its anti-coincidence detector. The differential flux of the particles is determined from the total electron energy in three energy channels, referred to as $E1$ to $E3$. Four additional energy channels are used to measure ions above about 400 keV which penetrate the foil and stop in the F detector. These channels are referred to as $FP4$ to $FP7$ where the notation 'FP' is used to emphasize that these particles are foil protons (ions). The energy of these ions may be inferred by adding the energy loss through the foil to the electronic energy channel passbands. Pure electrons are measured in the B detector of the CA60 telescope, described below. Table III gives the energy channel name, the logic equation implemented

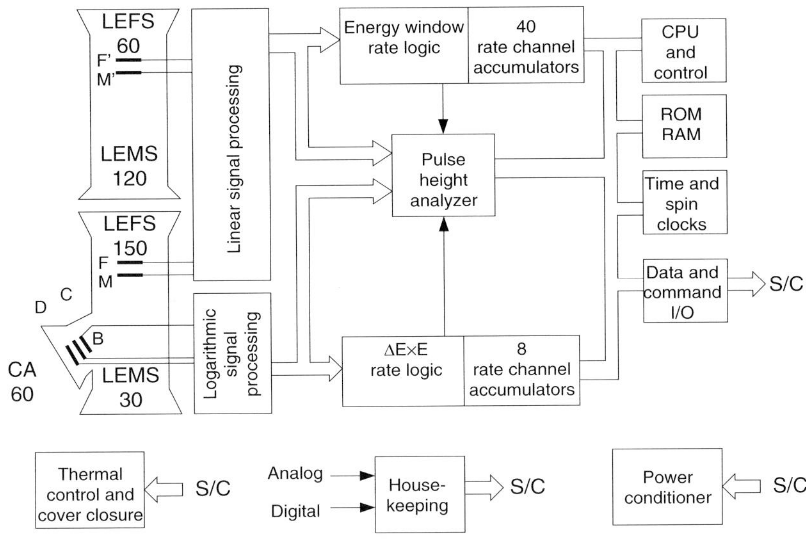

Figure 4. Block diagram of the EPAM instrument electronics.

TABLE III

Low Energy Foil Spectrometer detector systems

LEFS150			LEFS60		
Energy channel	Logic	Passband (MeV)	Energy Channel	Logic	Passband (MeV)
$E1$	$F1\overline{F2M}$	0.044–0.058	$E'1$	$F'1\overline{F'2M'}$	0.045–0.062
$E2$	$F2\overline{F3M}$	0.058–0.104	$E'2$	$F'2\overline{F'3M'}$	0.062–0.102
$E3$	$F3\overline{F4M}$	0.104–0.180	$E'3$	$F'3\overline{FP'4M'}$	0.102–0.175
$FP4$	$FP4\overline{F5M}$	0.463–0.568	$FP'4$	$F'4\overline{F'5M'}$	0.463–0.568
$FP5$	$F5\overline{F6M}$	0.568–0.796	$FP'5$	$F'5\overline{F'6M'}$	0.568–0.795
$FP6$	$F6\overline{F7M}$	0.796–1.114	$FP'6$	$F'6\overline{F'7M'}$	0.795–1.193
$FP7$	$F7\overline{M}$	1.114–4.924	$FP'7$	$F'7\overline{M'}$	1.193–4.924

by the instrument electronics, and the energy passbands for the two LEFS detector systems. The LEFS60 telescope uses the same detection scheme with the F' detector to measure rate and energy with M' used in anti-coincidence. Energetic electrons above about 350 keV will pass through the foil and the 200 μm thick detector $F(F')$, and will be detected by the anti-coincidence circuit when they are detected in the $M(M')$ detector.

2.1.2. *MF Spectrum Accumulator*

The MF Spectrum Accumulator, or MFSA, provides much higher energy resolution for the LEMS and LEFS detector systems than is available from the rate channels. This feature of the EPAM electronics produces a 32-point logarithmically spaced energy spectrum, accumulated from each of the LEMS/LEFS detectors. These data are processed according to a schedule which rotates through each sector of the detectors M, F, M', F'.

The MFSA samples the same analog signal as the rate channels and converts it to a logarithmic signal, then digitizes it to 8 bits rather than processing the analog signal with discriminators and accumulators, as is done with the rate channels. The MFSA digitization is identified in Figure 4 as the central block labeled 'Pulse-height analyzer.' At the time the instrument was designed, space qualified 8-bit analog-to-digital converters required about 30 μs to complete one conversion. This slow rate could not keep up with the analog portion of the instrument, hence only a subset of events could be processed.

The MFSA accumulation scheme is summarized in Table IV. Each row corresponds to a schedule in the MFSA accumulation. The MFSA data system consists of a bank of four memories which are multiplexed to store the 32-bin, 8-bit spectra, on a per sector basis. These four memory banks are denoted by the Roman numerals I–IV in Table IV. The LEMS30 and LEFS150 telescopes each have four sectors, and therefore accumulate a spectrum from detectors M, F for each sector in a single schedule of 128 s. The eight sectored detectors (M', F') require two schedules to accumulate a 32-bin spectrum for each sector. The sectors in Table IV are denoted by the letters A–D for the four sectored telescopes, and the numbers 1–8 for the eight sectored ones. An entire collection cycle requires 1024 s.

2.1.3. *Composition Aperture*

The CA telescope is capable of determining ion composition using a $\Delta E \times E$ detection scheme. Although the principal responsibility of EPAM is to monitor electrons, protons, and alphas, the CA provides an unambiguous determination of ion composition, unlike the LEMS detectors. The CA60 telescope is comprised of three solid-state detectors (cf., Figure 3): a thin, 4.8 μm epitaxial silicon detector referred to as the D detector, and two thick ($\sim$200 μm) totally depleted surface barrier silicon detectors known as C and B. The B detector, as mentioned previously, measures deflected electrons from the LEMS30 head, but also acts as the anti-coincidence detector for the CA.

The CA system uses log amplifiers to extend the dynamic range of the detector. These amplifiers are extremely temperature sensitive, and therefore are thermally regulated with heaters to maintain calibration. The logic used in the CA depends on slanted discriminators to define each species group. The eight CA rate channels, denoted by the symbols W1 to W8, count all particles in a given energy nucl^{-1} range. Multiple species may therefore be associated with a single CA rate channel. As a result, a species group is identified by the dominant species in that group.

TABLE IV

Schedule for the accumulation of the 32 point spectra produced by the MFSA for each detector and sector of the LEMS and LEFS systems. Each schedule (row) requires 128 s to process and each sector's spectrum is stored in one of four 32 byte memory banks (I–IV).

Schedule	Detector	Sector in memory bank			
		I	II	III	IV
0	M	A	B	C	D
1	F'	a	c	e	g
2	M'	1	3	5	7
3	F'	b	d	f	h
4	M'	2	4	6	8
5	F'	a	c	e	g
6	F	1	2	3	4
7	F'	b	d	f	h

Table V lists the detailed information for each group. Defined for each energy channel W is the count rate logic, the energy passbands of the dominant species in each species group, the identifier of the group, and the atomic number response. An example of the multiplicity of species for a given species group is given by the O-species group. This group is defined by the rate channels denoted W5 and W6 and is dominated by oxygen; however, there is also a significant contribution from carbon and nitrogen. This group is therefore also identified as the CNO group. Similarly, the Fe-group is made up of all species with $10 \leq Z \leq 28$, but here iron is the dominant species. The energy passbands for the magnetically deflected electrons from the LEMS30 are also shown in Table V. Note that the lowest energy channel in detector B has a selectable low threshold level.

Like the MFSA system described above, timing and downlink constraints make it impossible to process every event from the two detectors D and C using an 8-bit analog-to-digital converter. Therefore, an adaptive priority scheme was designed into the instrument to selectively process PHA events. In this scheme, the less abundant species such as Fe and O are preferentially selected over the dominant constituents of H and He. The method makes use of the fast but coarse characterization of the rate channels W1 to W8. The logarithmic analog signal from the detector is digitized to 8 bits provided that it falls within the appropriate rate channel.

The adaptive priority scheme, shown in Table VI, determines what species group (cf., Table V) to analyze, based on the last species group analyzed. This

TABLE V

Composition aperture detector system

Energy channel	Logic	Passband (MeV nucl^{-1})	Species group*	Z
W1	$C1 D1 \overline{C2 D2 B}$	0.521–1.048	H	1
W2	$C2 D1 \overline{D4 D2 S1 B}$	1.048–1.734	H	1
W3	$C1(D2 + S1) \overline{C3 D3 B}$	0.389–1.278	He	2
W4	$C3 D1 S1 \overline{C4 D3 B}$	1.278–6.984	He	2
W5	$C2 D3 \overline{C4 D4 B}$	0.546–1.831	O	6–9
W6	$C4 D2 S2 \overline{D4 S3 B}$	1.831–19.107	O	6–9
W7	$C2 D4 \overline{C4 B}$	0.298–0.955	Fe	10–28
W8	$C4 D3 S3 \overline{B}$	0.955–92.663	Fe	10–28
		(keV)		
DE1	$B1a \overline{B2 C}$	38–53	electrons	
	$B1b \overline{B2 C}$	48–53	electrons	
DE2	$B2 \overline{B3 C}$	53–103	electrons	
DE3	$B3 \overline{B4 C}$	103–175	electrons	
DE4	$B4 \overline{B5 C}$	175–315	electrons	

*These species and passbands correspond to the dominant species in each group. The reader is referred to Table III. of Lanzerotti et al. (1993) for the passbands of additional species.

TABLE VI

Adaptive priority scheme used in the CA system. Note that each species in the table refers to a species group defined in Table V.

		Species group of previous event			
		H	He	O	Fe
Priority for current event	Highest	Fe	H	He	O
		O	Fe	Fe	Fe
		He	O	O	He
	Lowest	H	He	H	H

is done for each of the eight sectors of the CA. For example, if in sector 3, the last species group analyzed was Fe, then the priority for the current event would be taken from the last column of Table VI. Thus, if an oxygen event were available, it would get processed, otherwise, the instrument would process Fe, He, or H, in that order, based on the availability of each. In this way, the instrument should analyze a larger number of the rare species.

Two PHA events are downlinked for each $\sim$1.5 s sector for an average rate of about 1.3 events per second. Of these two events, it is the lower priority event that is used to drive the adaptive priority scheme for the same sector on the next spacecraft rotation.

3. Calibration

A ground-based calibration was performed on the EPAM instrument prior to launch. Such a characterization of the instrument is essential in order to understanding the inflight measurements made by EPAM. Such calibration data include determination of the electronic thresholds and verification of these thresholds using calibrated pulsers. The ground-based calibration also included characterization of the instrument response to accelerated beams of ions and electrons in order to verify the energy channel thresholds. EPAM also has the capability to perform inflight calibrations.

3.1. PREFLIGHT CALIBRATION

The basic instrument functions and electronic thresholds of EPAM have been exercised and measured both as the flight spare instrument of HI-SCALE, and after the instrument was reconfigured as EPAM. The LEMS and LEFS telescopes of the EPAM instrument were characterized using the two particle accelerators at the NASA Goddard Space Flight Center. These accelerators were able to stimulate a limited number of EPAM's energy channels and verify the channel thresholds.

An electrostatic accelerator was used to calibrate the lower energy channel thresholds of the LEMS and LEFS telescopes using protons and electrons, respectively. These accelerated particle beams ranged in energy from 32 to 114 keV for protons and 40 to 117 keV for the electrons. The higher energy channels of the LEMS telescopes were stimulated by protons and helium ions that were accelerated by a Van de Graaff generator. The proton beam was varied in energy from 0.391 to 1.492 MeV and the helium ions ranged from 1.500 to 1.608 MeV.

3.2. INFLIGHT CALIBRATION

EPAM was designed with inflight calibration capabilities that provide discrete high-energy helium and electron emissions to verify proper instrument operation. Calibration sources are mounted inside the covers of the CA and LEMS telescopes.

TABLE VII

EPAM inflight calibration sources

Telescope	Source	Strength (μCi)	Particle Energy, Type	X-ray (keV)	Half-life (yrs)
CA60	^{244}Cm	1.0	5.81 MeV, α		18.11
	^{148}Gd	0.16	3.18 MeV, α		75
LEMS120	^{241}Am	1.0	5.49 MeV, α	59.5	432.2
LEMS30	^{241}Am	1.0	5.49 MeV, α	59.5	432.2
	^{133}Ba	1.0	45 keV, β	31, 80, 302, 356	10.5

By commanding heaters to turn on, the bi-metallic springs attached to each cover of the LEMS and CA telescopes will slowly close, permitting an inflight calibration to be performed and the instrument's response measured.

Each LEMS cover contains a 1.0 μCi ^{241}Am source which emits α particles at 5.486 MeV which will be observed in the highest LEMS energy channel. The source also emits $\sim$60 keV X-rays which are observable in the lowest LEMS energy channel. The LEMS30 telescope cover also contains a 1.0 μCi ^{133}Ba calibration source. The barium source provides a range of energetic electrons that enter the LEMS30 telescope and are magnetically deflected into the B detector of the CA. The CA60 telescope cover contains a dual 1.0 μCi ^{244}Cm and 0.16 μCi ^{148}Gd calibration source that provides 5.8 MeV and 3.183 MeV alpha particles. These peaks are observed in the W3 and W4 channels. Table VII summarizes the radioactive sources for each telescope, the particle type and energy, and selected X-ray emissions.

4. Inflight Performance

The ACE spacecraft was launched from Cape Canaveral on 25 August 1997 and two days later EPAM was turned on. An initial 75-min inflight calibration test was performed prior to commanding the pyrotechnic devices to fire, releasing the covers on the LEMS and CA telescopes, containing the radioactive sources. Since that time, EPAM has performed well with immediate observations of upstream magnetospheric events, solar events, and other interplanetary phenomena.

4.1. IONS AND ELECTRONS

An example of the wide range of observations possible with EPAM is shown in Figure 5. Six selected channels from the instrument are plotted for the twelve-day

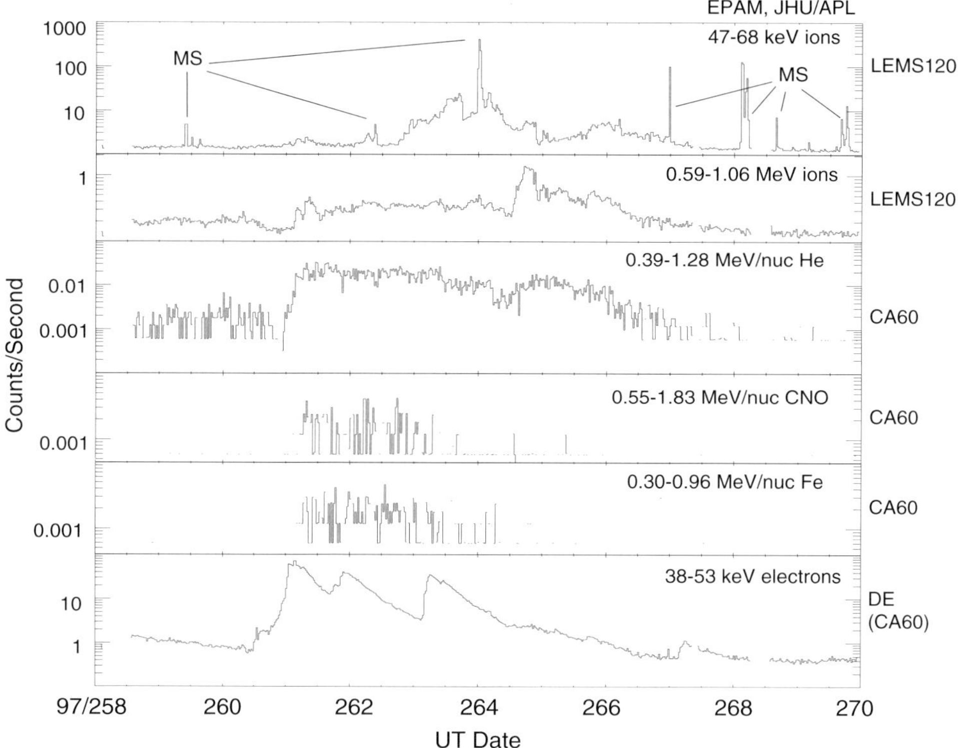

Figure 5. Selected channels from EPAM during a 12 day interval about a month after the instrument was turned on.

period from 15–27 September 1997. The top curve shows LEMS120 response to the 47–68 keV ions. The LEMS120 telescope, with its slightly oblique view of the terrestrial magnetosphere, resolves numerous upstream magnetospheric events, labeled MS, during this 12 day period. Nearly one hundred upstream, magnetospheric events were observed in the first 30 days of the mission. The second panel shows LEMS120 response to ions in the 0.587–1.060 MeV energy range. The magnetospheric events shown for this time interval have relatively soft spectra and do not extend beyond 500 keV, thus a very different profile is seen in the higher energy ions. The third, fourth, and fifth panels in Figure 5 show the instrument's capabilities at measuring and differentiating the composition of the observed ions at higher energies.

The CA60 telescope samples ions at $60°$ from the spin axis, and since it does not point back to the magnetosphere, it observed a very different ion profile than was observed at $120°$ from the spin axis. The third panel shows a steep rise on day 261 in the 0.389–1.278 MeV nucl^{-1} helium flux. The intensity of the helium ions remains enhanced for a number of days before decaying away after day 266. The fourth and fifth panels show responses to 0.546-1.831 MeV nucl^{-1} CNO group ions and 0.298–0.955 MeV nucl^{-1} Fe group ions, respectively. The ion composition during the first portion of this small solar event is very rich in He and heavy nuclei with the Fe group ions about as plentiful as the CNO group. The last panel in Figure 5 shows the magnetically deflected electrons, 38–53 keV, from the LEMS30 telescope. The onset of the triple solar electron event is seen in phase with the onset of the helium ions, and the smaller enhancement of the 0.587–1.060 MeV channel of the LEMS120 ions.

Figure 6 shows about four hours of data during a series of upstream magnetospheric events with higher temporal and angular resolution than was shown in Figure 5. The eight curves are the responses to 47–68 keV ions of the eight angular sectors in the LEMS120 cone. The event onset began shortly after 11:00 UT, labeled A, where the count rates rose to about 100 c s^{-1} in all sectors and remained at that intensity for about 10 min. The intensities then jumped to about 1000 c s^{-1}, however, the varying profiles from one sector to the next are indicative of the highly structured nature of some upstream events.

A second event, labeled B, shows a period where count rates in sectors 2, 3, and 4 were very intense, but the opposite looking sectors 7 and 8, showed no significant enhancement. This event illustrates the effectiveness of the angular resolution of the EPAM instrument, and the strong anisotropic behavior characteristic of many upstream events. The third event during this time, labeled C, shows a small but clear temporal difference in the event onset between sector 2, arriving just after the event, and sector 6, arriving just before the event. The time difference between the event onset in these two sectors is about 7 min.

In Figure 7, two anisotropy pie-plots are shown during the events labeled (B) and (D) in Figure 6. The large anisotropy is clearly seen in Figure 7(a) where the ion count rates in sectors 3 and 4 reach nearly 1000 c s^{-1} while virtually no counts above background are observed in sectors 7 and 8. Figure 7(b) shows the nearly isotropic behavior about 1.5 hour later.

A large solar event was observed by EPAM on 6 November 1997 and the instrument's electron, ion, and species composition responses are shown in Figure 8. The top group of curves shows the 38 to 315 keV magnetically deflected electrons (cf., Table V), averaged over a spacecraft spin and a 1-hr time interval. The middle panel shows selected responses to 47 to 1900 keV ions, and the final group shows the CA responses to 0.521–1.048 MeV nucl^{-1} protons, 0.389–1.278 MeV nucl^{-1} He, 0.546–1.831 MeV nucl^{-1} CNO group ions, and 0.298–0.955 MeV nucl^{-1} Fe group ions. The onset of the event is clearly seen from the energetic electrons, while a strong shock spike is seen in the ion responses later in the day.

Figure 9 shows four selected energy spectra from the MF Spectrum Accumulator which has 32 logarithmically spaced channels. Figure 9(a) shows the first spectrum which is a 128-s average of ions observed at 120° from the spin axis, prior to the onset of the first solar event. The second spectrum (b) shows the higher energy ions, $E > 1$ MeV, arriving at the spacecraft while the lower energy ions are still at pre-event intensities. The third spectrum (c) shows the enhancement

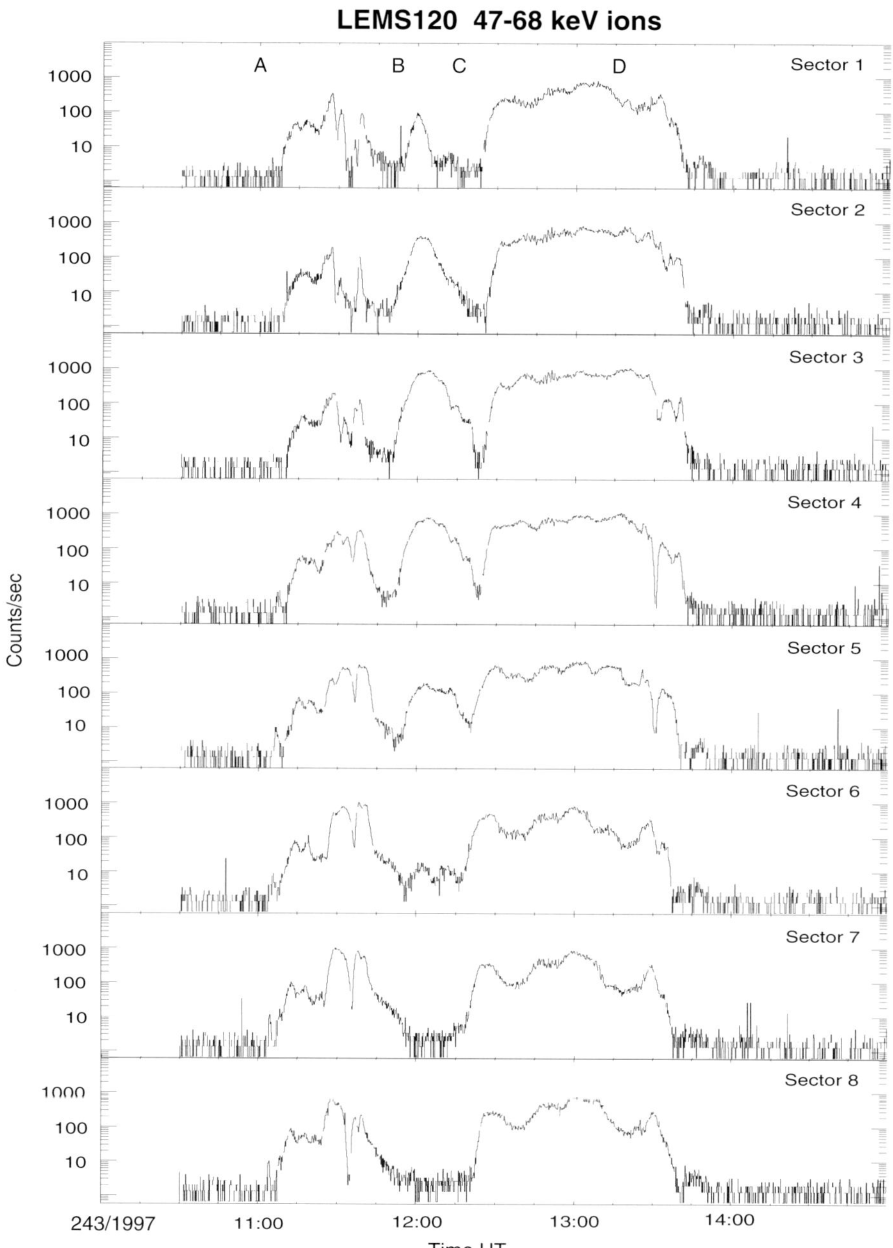

Figure 6. High time resolution rate channel data from day 243 (1 September 1997). All eight sectors of the LEMS120 telescope are shown and depict the high anisotropy of this event.

558

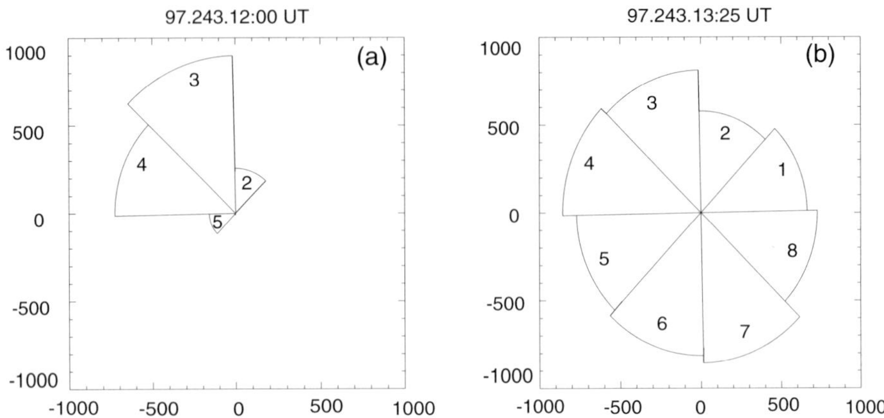

Figure 7. Anisotropy plots from EPAM showing very different distributions for two times on day 243 (1 September 1997). In (a) the strong anisotropy is seen in sectors 3 and 4, measured at 12:00 UT (labeled B in Figure 6), whereas in (b) the nearly isotropic distribution was measured at 13:25 UT (label D in Figure 6).

extending down to low energies. The fourth spectrum (d) shows the ion responses right at the shock spike. The energy dependence of the shock accelerated ions is clearly evident in the EPAM spectrum.

All events in the CA telescope are pulse height analyzed and the two highest priority events in each sector are telemetered to the ground. Figure 10 is a 12-hr accumulation of the pulse height matrix plotted by the energy deposited in detectors C and D for each event. Individual atomic species form clearly defined tracks which show the species resolution of EPAM.

The boundaries labeled W1–W8 show the mapping of the CA species-group rate channels into pulse-height space using the logic equations given in Table V. Within these limits, all particles are counted, independent of the adaptive priority scheme, and these rates are used to normalize the tracks shown in the PHA matrix.

5. Summary

The EPAM instrument provides comprehensive energy, angular, and species coverage with good resolution over a key parameter space in studies of energetic particles in the near Earth interplanetary medium. As such, it serves to highlight the context in which the detailed compositional, isotopic, and charge-state measurements of the principal ACE instruments can be properly assessed and interpreted. Further, several key channels are used to provide real-time information on space weather status (see Zwickl et al., 1998) that is utilized by NOAA and broadcast worldwide to all users. Finally, EPAM serves as an excellent baseline at 1 AU for studies of radial and latitudinal gradients in conjunction with an instrument of

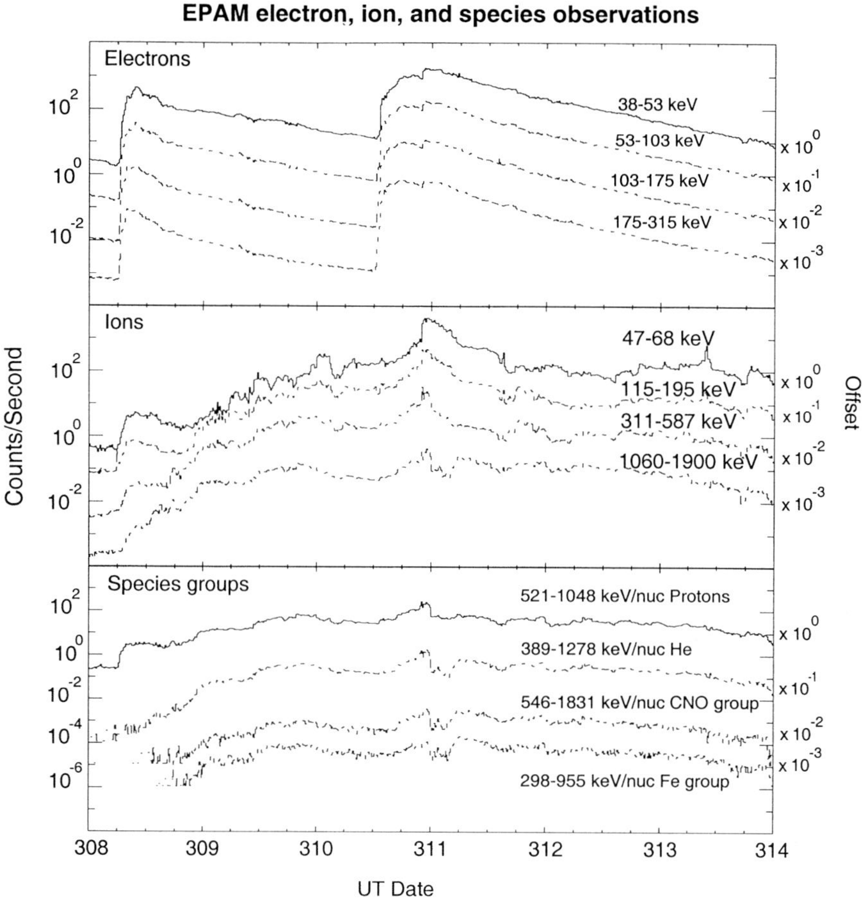

Figure 8. Rate channel data from the EPAM magnetically deflected electrons, LEMS120 ions, and species groups from the CA showing the time progression of the particles resulting from a large solar energetic particle event in November 1997.

identical design on the *Ulysses* spacecraft. The quality of the initial data shown in this paper is such that there is every expectation for a long-term, continuous data stream documenting the onset of the new solar cycle.

Acknowledgements

Although EPAM started from a working *Ulysses* spare instrument, a great deal of effort was required to convert the HI-SCALE spare unit into the final EPAM instrument. We cannot acknowledge all of those who have helped in this effort, however, a few people who worked very long and hard must be recognized. B. E. Tossman served as the project manager, G. B. Andrews was the system engineer during the first half of the EPAM project, T. G. Sholar was the mechanical engineer,

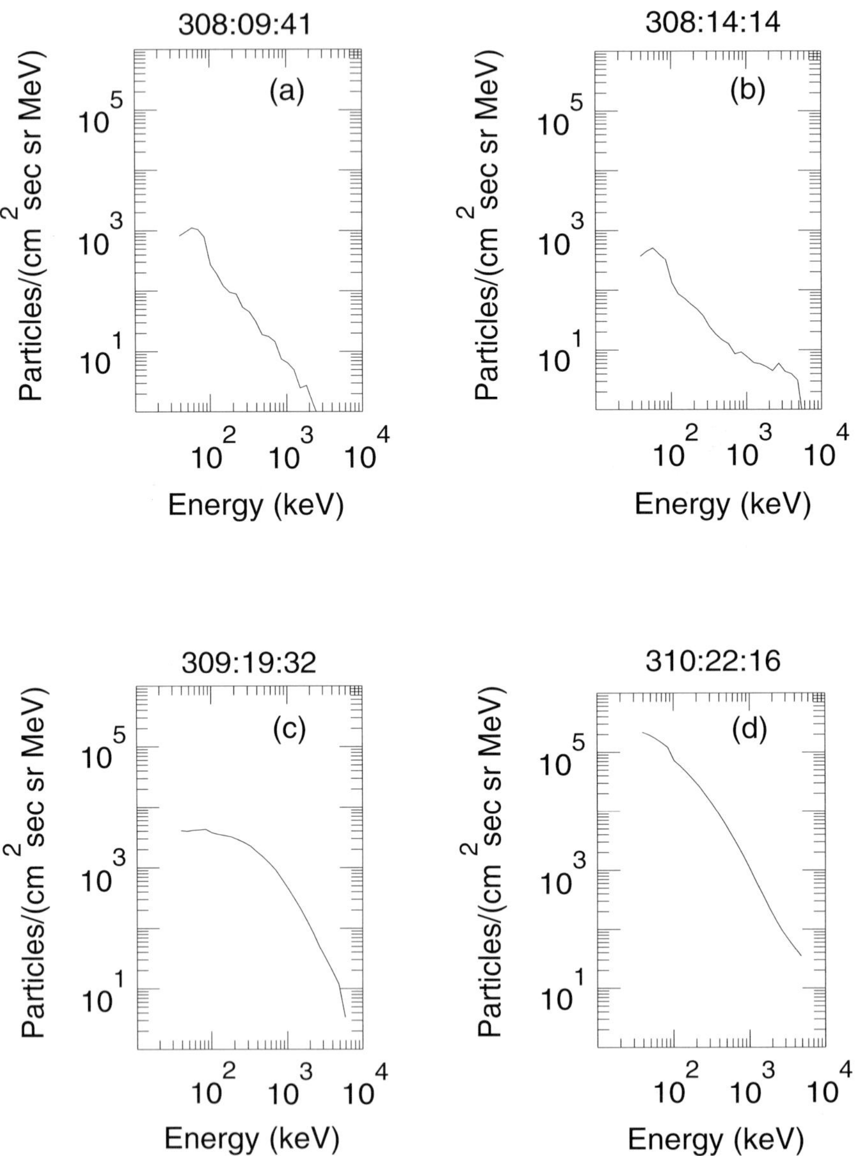

Figure 9. The 32 channel MFSA energy spectrum from the M detector (LEMS30 telescope) for four times following a large solar flare event in November 1997. In panel (d) note the large increase in the low-energy particles.

EPAM PHA Matrix
Start time = 1997 Day 311 06:00, End time = 1997 Day 311 18:00

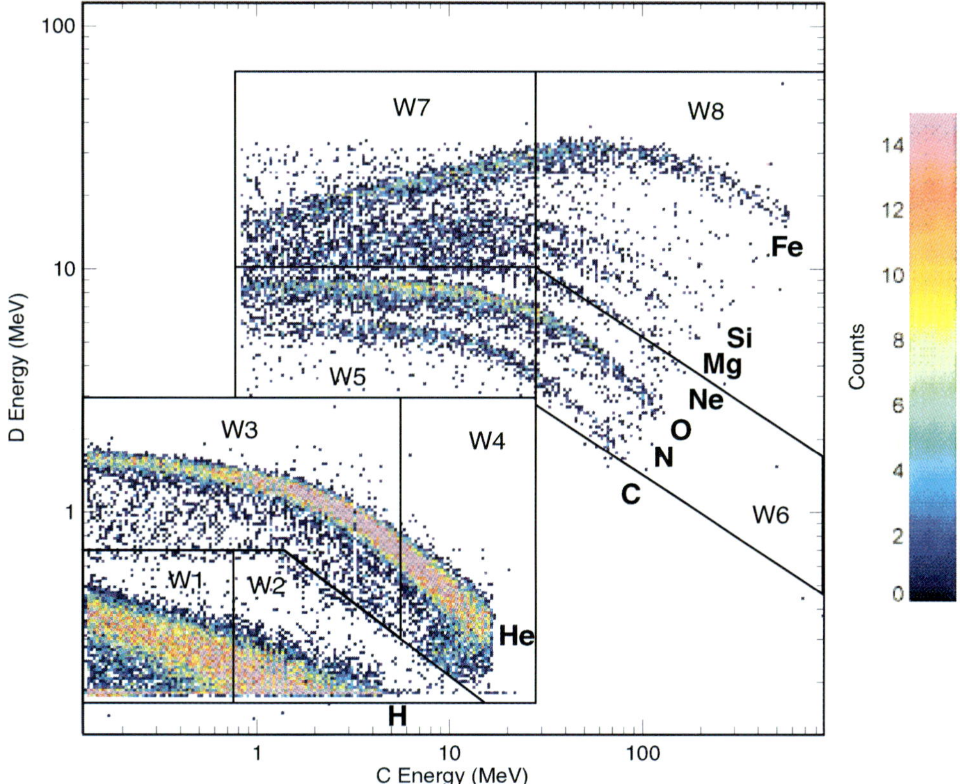

Figure 10. EPAM PHA matrix during the peak fluxes of the November 1997 solar flare event.

B. D. Williams and J. A. Krein provided thermal engineering. R. E. Thompson and J. A. Buttermore did all of the disassembly of HI-SCALE and assembly of the reconfigured instrument, EPAM.

References

Chiu, M. C., von-Mehlem, U. I., Willey, C. E., Betenbaugh, T. M., Maynard, J. J., Krein, J. A., Conde, R. F., Gray, W. T., Hunt, Jr., J. W., Mosher, L. E., McCullough, M. G., Panneton, P. E., Staiger, J. P., and Rodberg, E. H.: 1998, 'ACE, Spacecraft', *Space Sci. Rev.* **86**, 257.

Lanzerotti, L. J., Armstrong, T. P., Maclennan, C. G., Simnett, G. M., Cheng, A. F., Gold, R. E., Thomson, D. J., Krimigis, S. M., Anderson, K. A., Hawkins, S. E., III, Pick, M., Roelof, E. C., Sarris, E. T., and Tappin, S. J.: 1993, 'Measurements of Hot Plasma in the Magnetosphere of Jupiter', *Planetary Space Sci.* **41**, 893.

Lanzerotti, L. J., Gold, R. E., Anderson, K. A., Armstrong, T. P., Lin, R. P., Krimigis, S. M., Pick, M., Roelof, E. C., Sarris, E. T., Simnett, G., and Frain, W. E.: 1983, in K.-P. Wenzel, R. G. Marsden, and B. Battrick (eds.), 'The ISPM Experiment for Spectra, Composition, and Anisotropy Measurements of Charged Particles at Low Energies', *The International Solar Polar Mission — Its Scientific Investigations*, Noordwijk, The Netherlands, 141–154.

Lanzerotti, L. J., Gold, R. E., Anderson, K. A., Armstrong, T. P., Lin, R. P., Krimigis, S. M., Pick, M., Roelof, E. C., Sarris, E. T., Simnett, G., and Frain, W. E.: 1992, 'Heliosphere Instrument for Spectra, Composition, and Anisotropy at Low Energies', *Astron. Astrophys.* **92**, 349.

Zwickl, R. D., Doggett, K. A., Sahm, S., Barrett, W. P., Grubb, R. N., Detman, T. R., Raben, V. J., Smith, C. W., Riley, P., Gold, R. E., Mewaldt, R. A., and Maruyama, T.: 1998, 'The NOAA Real-Time Solar-Wind (RTSW) System Using ACE Data', *Space Sci. Rev.* **86**, 633.

SOLAR WIND ELECTRON PROTON ALPHA MONITOR (SWEPAM) FOR THE ADVANCED COMPOSITION EXPLORER

D. J. McCOMAS, S. J. BAME, P. BARKER, W. C. FELDMAN, J. L. PHILLIPS and P. RILEY

Los Alamos National Laboratory, Los Alamos, NM 87545, U.S.A.

J. W. GRIFFEE

Sandia National Laboratory, Albuquerque, NM 87185, U.S.A.

Abstract. The Solar Wind Electron Proton Alpha Monitor (SWEPAM) experiment provides the bulk solar wind observations for the Advanced Composition Explorer (ACE). These observations provide the context for elemental and isotopic composition measurements made on ACE as well as allowing the direct examination of numerous solar wind phenomena such as coronal mass ejections, interplanetary shocks, and solar wind fine structure, with advanced, 3-D plasma instrumentation. They also provide an ideal data set for both heliospheric and magnetospheric multi-spacecraft studies where they can be used in conjunction with other, simultaneous observations from spacecraft such as *Ulysses*. The SWEPAM observations are made simultaneously with independent electron and ion instruments. In order to save costs for the ACE project, we recycled the flight spares from the joint NASA/ESA *Ulysses* mission. Both instruments have undergone selective refurbishment as well as modernization and modifications required to meet the ACE mission and spacecraft accommodation requirements. Both incorporate electrostatic analyzers whose fan-shaped fields of view sweep out all pertinent look directions as the spacecraft spins. Enhancements in the SWEPAM instruments from their original forms as *Ulysses* spare instruments include (1) a factor of 16 increase in the accumulation interval (and hence sensitivity) for high energy, halo electrons; (2) halving of the effective ion-detecting CEM spacing from $\sim$5° on *Ulysses* to $\sim$2.5° for ACE; and (3) the inclusion of a 20° conical swath of enhanced sensitivity coverage in order to measure suprathermal ions outside of the solar wind beam. New control electronics and programming provide for 64-s resolution of the full electron and ion distribution functions and cull out a subset of these observations for continuous real-time telemetry for space weather purposes.

1. Introduction

The Advanced Composition Explorer (ACE) mission was developed to examine both the isotopic and elemental composition of our Galaxy from the four reservoirs of matter that are accessible to a direct measurement from near the Earth. The first two of these reservoirs connect back to the Sun through the solar wind and through solar energetic particles, the third traces to the local interstellar medium (which is also the source population for anomalous cosmic rays) by way of locally ionized interstellar neutrals which form singly charged ions that are then picked up by the solar wind, and the fourth traces to our Galaxy at large by way of galactic cosmic rays. These observations are carried out with a suite of six experiments,

Space Science Reviews **86**: 563–612, 1998.
© 1998 *Kluwer Academic Publishers. Printed in the Netherlands.*

which jointly measure ion composition over the energy range from ~100 eV to several hundred MeV per nucleon, with sensitivities that are one to two decades larger than previously achieved. In addition, ACE carries three experiments that monitor the interplanetary medium, thereby providing the context for the advanced ion-composition experiments aboard ACE. These experiments are the Solar Wind Electron Proton Alpha Monitor (SWEPAM), the energetic Electron, Proton and Alpha particle Monitor (EPAM), and two magnetometers (MAG).

SWEPAM plays a critical role in ACE by providing the solar wind observations that enable the understanding of the ACE composition measurements in the context of previous knowledge about the solar wind. This knowledge has been developed over more than three decades of solar wind research. Specifically, SWEPAM will place the ACE composition measurements in the context of low-speed streamer belt flows, the high-speed solar wind from coronal holes, coronal mass ejections (CMEs), the various types and strengths of interplanetary shocks, magnetic connection to the Earth's bow shock, and other solar wind structures. Table I, from the ACE project's Level 1 requirements definition, lists the scientific objectives of the ACE mission. SWEPAM is considered a primary measurement (P) for six of these objectives and an important contributing measurement (C) for another five. Thus, SWEPAM measurements of the solar wind provide one of the critical cornerstones for the ACE mission.

The SWEPAM experiment provides state-of-the-art measurements of electron and ion distribution functions in three dimensions over all of the velocity space needed to characterize the bulk flow and kinetic properties of the solar wind. SWEPAM utilizes the refurbished flight spare instruments from the Solar Wind Over the Poles of the Sun (SWOOPS) experiment (Bame et al., 1992) onboard the joint NASA/ESA *Ulysses* mission. *Ulysses* was launched on October 6, 1990. It was deflected out of the ecliptic plane into its present high-inclination polar orbit about the Sun, using a gravitational assist at Jupiter, in late February 1992. SWOOPS has operated flawlessly since launch, thereby allowing exploration of the solar wind as a function of heliolatitude for the first time. Figure 1 summarizes the SWOOPS solar wind speed observations as a function of latitude in a polar plot format, similar to that shown by McComas et al. (1998). Note that while the high latitude solar wind is a comparatively steady flow of high-speed wind from both polar coronal holes, the low-latitude wind where ACE will reside is a complicated mixture of slow flows from the streamer belt and higher speed coronal hole flows.

The SWOOPS experiment comprises two separate instrument units, one for electrons and one for ions. These instruments, although different in size and detail, are based on the same basic principles. Both are electrostatic energy per charge (E/q) analyzers followed by sets of channel electron multiplier (CEM) sensors, which allow individual particles transmitted through the analyzers to be counted. In both cases, full complements of low and high voltage power supplies, memory, data processing capabilities, and spacecraft interfaces are provided as a part of the package such that each of the two instrument units is operated completely

TABLE I

ACE level 1 science objectives

	CRIS	SIS	ULEIS	SEPICA	SWIMS	SWICS	EPAM	SWEPAM	MAG
Composition of matter									
Generate table of solar isotropic abundance		P	P	P	P	C	C	C	
Determine coronal composition		P	P	P	P	P	C	P	
Compare cosmic-ray and solar isotope pattern	P	P	P	P	P	C	C	C	
Measure interstellar/interplanetary pickup ions				C	C	P		C	C
Determine anomalous cosmic-ray composition	P	P	P	C			C		
Origin/evolution of elements									
Identify solar/meteoritic composition differences		P	P	P	P	P	C	C	
Solar particle contributions to Moon/planets/meteorites		P	P	C	P	P	C	C	
Identify cosmic-ray nucleosynthesis process	P	P							
Determine age of cosmic-ray source material	P	P							
Search for evidence of galactic evolution	P	P	P	C	P	C			
Corona formation/solar wind acceleration									
Isolate coronal formation processes		P	P	P	P	P	C	P	
Study solar plasma conditions				P	P	P		P	C
Study solar wind acceleration/fractionation					P	P		P	P
Particle acceleration/transport									
Fractionation in solar flare/interplanetary acceleration		P	P	P	P	P	C	P	C
Constrain particle acceleration models	P	P	P	P	P	P	P	P	P
Test 3 He-rich and gamma-ray flare models	C	P	P	P			P	P	C
Measure cosmic-ray acceleration/transport time scale	P	P							
Test anomalous cosmic-ray origin	P	P	P	P					

P = primary measurements.
C = contributing measurements.

TABLE I

Continued

Acronym	Name	Lead institution
CRIS	Cosmic-Ray Isotope Spectrometer	CIT
EPAM	Electron, Proton, and Alpha-Particle Monitor	APL
MAG	Magnetic Field Monitor	Bartol
SEPICA	Solar Energetic Particle Ionic Charge State Analyzer	UNH
SIS	Solar Isotope Spectrometer	CIT
SWEPAM	Solar Wind Electron, Proton, and Alpha-Particle Monitor	Los Alamos
SWICS	Solar Wind Ion Composition Spectrometer	UMD
SWIMS	Solar Wind Ion Mass Spectrometer	UMD
ULEIS	Ultra Low Energy Ion Spectrometer	UMD

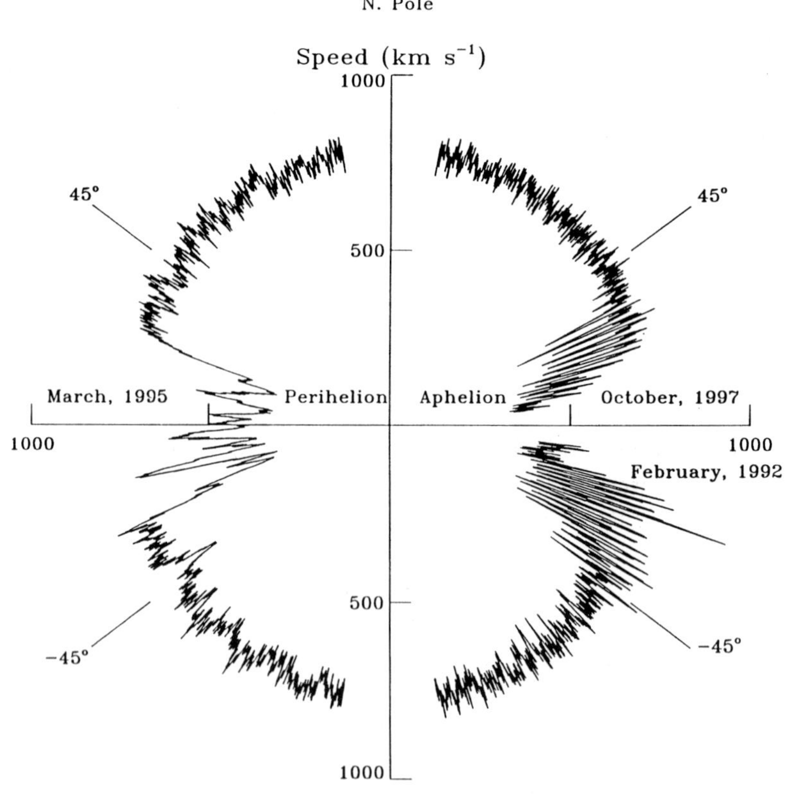

Figure 1. Polar plot of solar wind speed measured by the SWOOPS experiment on *Ulysses* over nearly one complete polar orbit about the Sun. Cyclic variations between high and low speeds, observed on the right side of the plot at mid-latitudes, are due to rotation of the Sun.

independently of the other. For the SWEPAM experiment on the ACE mission we have selectively modified, refurbished, and improved the SWOOPS flight spare instruments. In particular, SWEPAM provides enhanced sensitivity to suprathermal electrons ($E > 100$ eV) to allow a greater sensitivity for detecting CMEs and for establishing their magnetic topology and evolution in interplanetary space. SWEPAM also provides higher angular resolution measurements of solar wind ions as well as increased sensitivity to suprathermal ions coming from directions adjacent to the bulk solar wind in order to provide a bridge between the energetic ions detected using the suite of ACE composition experiments, and the solar wind thermal population.

This paper provides the primary documentation source for the ACE/SWEPAM experiment. In Section 2 we briefly review solar wind science and place the anticipated SWEPAM observations and scientific results in the context of previous observations. Sections 3 and 4 provide detailed descriptions of the SWEPAM ion and electron instruments, respectively. In each we: (1) summarize the electro-optical properties and general specifications of the sensor, (2) show the mechanical design of the sensor, (3) discuss the electronics associated with each instrument, and (4) show the results of sensor calibrations. While sufficient detail is given for the interested reader to understand the implemented mechanical and electronics designs, these two sections are arranged so that the detailed subsections can be skipped without losing an understanding of the top-level design. Section 5 describes the commanding, data handling, modes of operation, and the real-time solar wind capability of SWEPAM.

2. Scientific Objectives

The primary purpose served by SWEPAM is to provide the solar wind context for the other experiments on ACE. Connection to global heliospheric structure will be facilitated by correlation of SWEPAM data with simultaneous measurements made using the SWOOPS experiment on *Ulysses*. During the early portion of the ACE mission, *Ulysses* will be positioned near 5.4 AU, close to the ecliptic plane. The heliographic locations of ACE relative to *Ulysses* between 1997 and 2002 are shown in Figure 2. Times of longitudinal lineup are indicated by the intersections between the dotted and solid traces in the top panel. These lineups occur in early 1998, 1999, and 2000.

It will also be critical to place the new ACE observations in the context of synoptic changes in solar wind structure that are associated with phases of the solar cycle. The time period of the ACE mission will extend from near solar minimum toward the coming solar maximum. Over this portion of a solar cycle the photospheric and coronal magnetic fields evolve from simple, dipolar-like configurations, with axes nearly perpendicular to the ecliptic plane, to configurations of considerable complexity. This complexity is characterized by a rapid spatial evolution of the

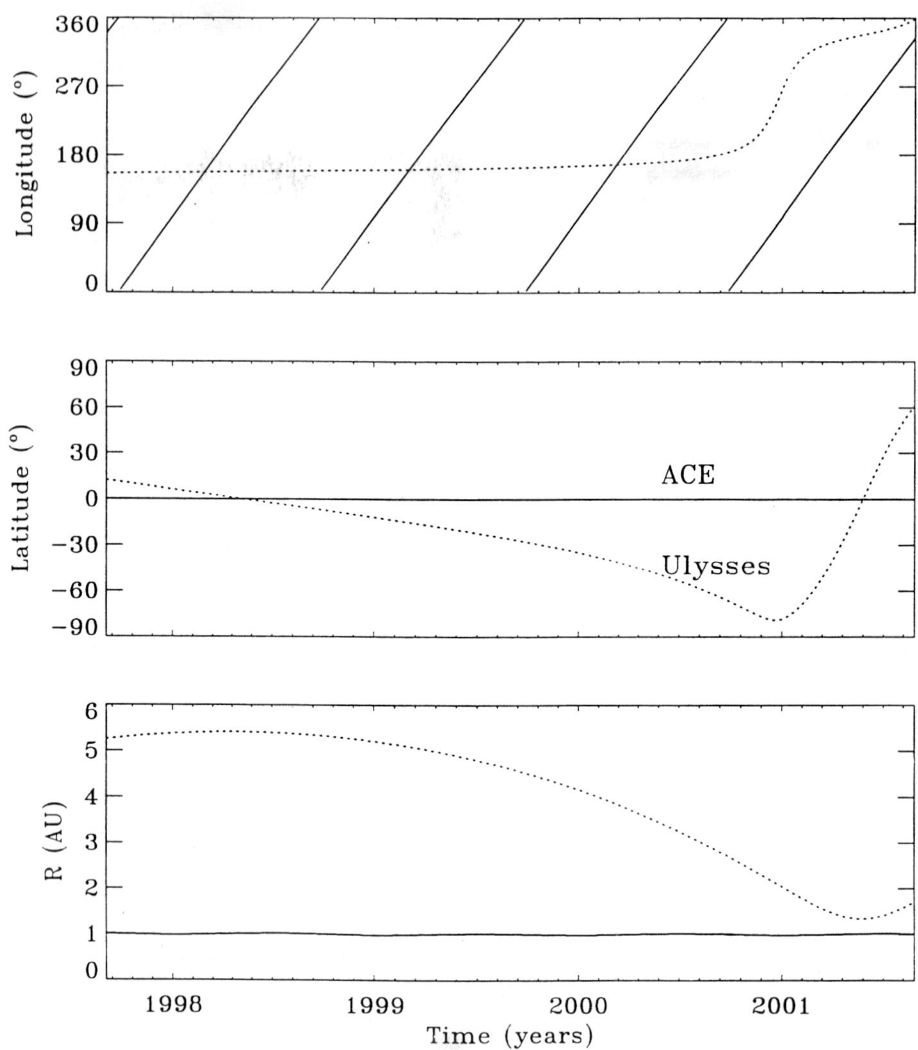

Figure 2. Heliographic longitudes, latitudes, and radial distances of ACE and *Ulysses*, spanning the period from the launch of ACE in late 1997 through 2002.

photospheric magnetic field and accompanying time variability, first associated with the appearance of magnetic active regions at relatively high heliographic latitudes, and later to their appearance at progressively more equatorial latitudes. Understanding ACE observations as a function of the solar activity cycle will be facilitated by comparison of these observations with measurements made from the particle and fields experiments onboard ISEE-3 in the late 1970s and early 1980s. These observations were made from the same Sun–Earth–L1 neighborhood to be occupied by ACE, but in a previous solar cycle. Specific contributions by

SWEPAM in support of a number of the major scientific thrusts of the ACE mission are presented below.

2.1. FORMATION OF THE SOLAR CORONA AND ACCELERATION OF THE SOLAR WIND

Synoptic studies of solar and solar wind structure during the 1970s revealed an evolving global configuration that followed the pattern of changes in photospheric and coronal magnetism mentioned above. Near solar minimum, the global solar wind consists of high-speed flows emanating from both solar poles separated by a roughly planar, low-speed streamer belt at equatorial latitudes that is nearly parallel to the ecliptic plane (see Figure 20(d) of Hundhausen, 1977). A similar global structure was observed by *Ulysses* during its fast-latitude scan from $-80.2°$ to $+80.2°$ heliographic latitude between September 1994 and August 1995, as shown on the left side of Figure 1. This interval occurred just prior to the minimum of solar activity during the present activity cycle. An expanded version of this traversal, including the observed proton density scaled to 1 AU, is shown in Figure 3 (assembled by B. E. Goldstein using SWOOPS plasma data and a *Yohkoh* X-ray image of the corona, kindly provided by the *Yohkoh* mission of ISAS, Japan). *Ulysses* observed nearly structure-free, high-speed solar wind ($V > 700$ km s^{-1}) for all heliographic latitudes poleward of about 22°, and variable, but generally low-speed solar wind equatorward of 22°. This high-speed–variable-speed transition was recently reconfirmed by *Ulysses* as it approached aphelion within about 29° of the ecliptic plane (upper right quadrant of Figure 1) (McComas et al., 1998).

If these conditions prevail throughout the early portion of the ACE mission, as expected, then phenomena associated with the streamer belt and its embedded heliospheric current sheet will dominate observations made by all experiments onboard ACE (see the yellow stripe in the bottom panel of Figure 3). Observations made by ISEE-3 during the same phase of the solar cycle between 1978 and 1980, and by *Ulysses* just after its launch early in the current solar cycle, suggest that ACE will also likely spend a fraction of its time in those portions of the coronal-hole-associated high-speed streams that are adjacent to streamer-belt flows. Transitions between streamer-belt and coronal-hole-associated flows correspond to the flow-speed variability encountered by *Ulysses* at intermediate heliographic latitudes shown in Figures 1 and 3. The correspondence between the solar wind flow state observed by *Ulysses* and the structure of the corona as observed in X-rays by *Yohkoh*, is shown by comparing the panels in Figure 3. The high-speed wind is clearly associated with the large polar coronal holes. While most, if not all, of the solar wind must expand along open field lines, the low-speed wind seems to be associated with the bright, largely closed magnetic structures. The intermediate and variable solar wind appears to come from the margins between these two regions.

During the approach to solar maximum, an ever increasing number of magnetically active regions rise through the photosphere. These then interact both with

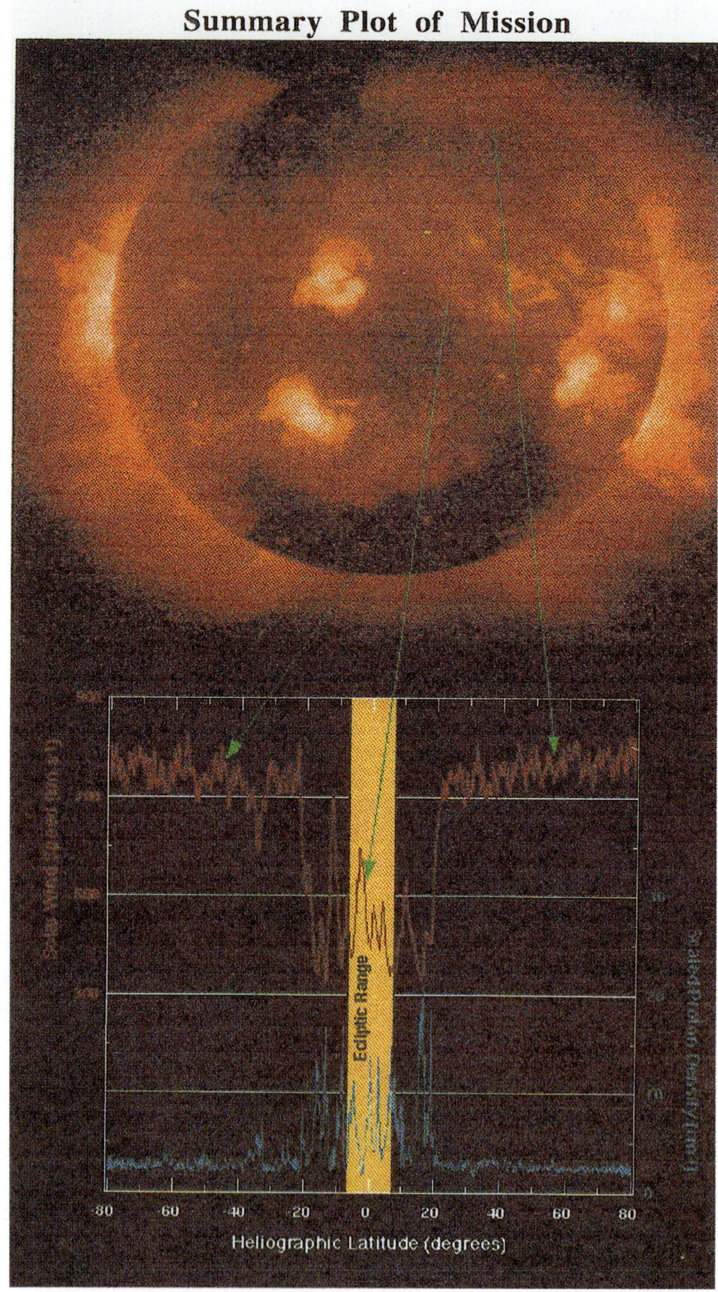

Figure 3. Identification of the solar wind flow state, as given by the proton bulk speed and density (projected to 1 AU), measured by *Ulysses* during its fast latitude scan, compared with the configuration of the solar corona, as measured in soft X-rays by *Yohkoh*. *Yohkoh* data was kindly provided by the *Yohkoh* mission of ISAS, Japan.

the preexisting dipolar magnetic field and with each other, producing a series of rapid reconfigurations that contribute to a pattern of solar magnetism characterized by a regular (and at times, extreme) time variability. These reconfigurations lead to an increasingly complex magnetic configuration. The global structure of the corona and resultant solar wind change accordingly in response to these underlying changes. In particular, an ever greater fraction of the solar wind encountered far from the Sun is associated with coronal mass ejections (CMEs) and other transient flows. In addition, some of the apparently time-stationary high-speed flows encountered outside of the streamer belt originate from relatively small unipolar photospheric regions that evolve on timescales of months, in contrast to the timescale of years over which the much larger area coronal holes evolve near solar minimum.

Although much is already known about the evolution of the large-scale pattern of solar wind structure in response to the solar activity cycle, very little is known about the mechanisms that connect the flow state of the slow wind with plasma conditions in the corona. Even more fundamental, we do not yet know the mechanisms that heat the solar corona and accelerate its plasma outward to form the solar wind. One of the major scientific goals of SWEPAM is to provide the link between the data from the composition experiments on ACE and the sources of the various types of solar wind back in the solar corona.

2.1.1. *Streamer Belt*

Bright, high-density streamers are clearly seen during solar eclipses and are prominent in coronagraph images of the Sun. These streamers, which are also associated with the equatorial bright regions seen in the soft X-ray images of the Sun (e.g., Figure 3), give rise to highly structured but slow solar wind flows. These flows are characterized by large variations in speed, density, temperature, heat flux, helium abundance, and heavy ion charge-states (e.g., Feldman et al., 1977, 1981; Borrini et al., 1981). Some of these variations probably reflect fine spatial-scale structure in the streamer belt owing, at least in part, to the meandering of the neutral line in the photosphere. The interplanetary extension of this neutral line is a current sheet that is embedded in streamer belt associated flows. In addition, variability in the low-speed wind may be caused by quasi-continuous, fine-scale magnetic restructuring at low altitudes above the photosphere (Axford, 1985). Some of this restructuring may be caused by the different rate of rotation of fine-scale photospheric magnetic fields relative to that of the deep-seated currents that are responsible for large-scale patterns of solar magnetism (Sheeley et al., 1975), and some may be due to the emergence and decay of active regions within the streamer belt.

Attempts have been made to associate peculiarities in the internal state of solar wind ions and electrons with each of the foregoing possibilities, without definitive success. The problem in interpretation stems from essential ambiguities in relating the plasma at 1 AU with production mechanisms in the corona. A partial list of solar wind observables includes: (1) double ion streams as shown in Figure 4,

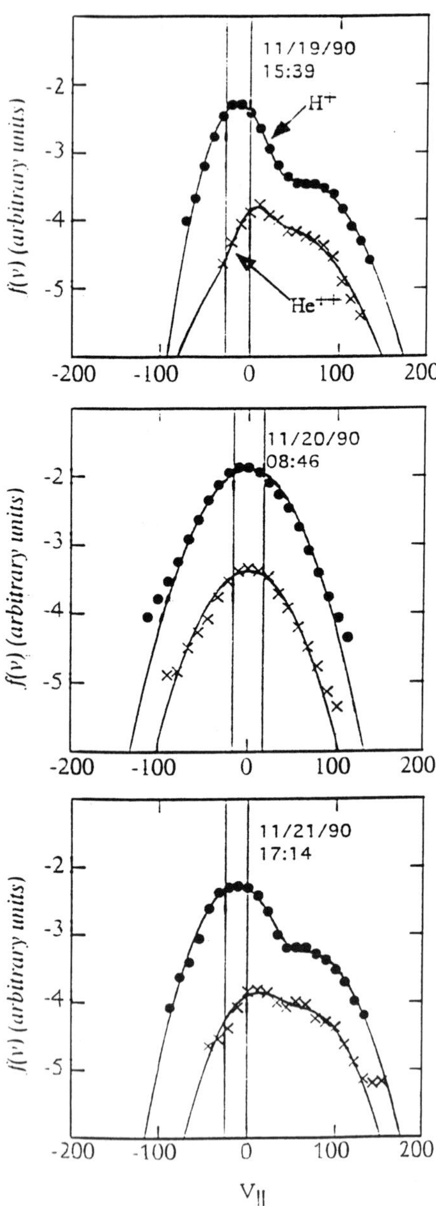

Figure 4. Evidence for magnetic reconnection within the streamer belt as shown by SWOOPS observations of double proton and alpha-particle streams on either side of (top and bottom panels), but not at the center of (center panel) the heliospheric current sheet.

possibly driven by magnetic reconnection (Hammond et al., 1995; Feldman et al., 1996); (2) heat flux dropouts, possibly signifying magnetic disconnection from the corona (McComas et al., 1989); (3) heat flux enhancements, possibly caused by transient-driven shocks that are magnetically connected to the spacecraft, (Feldman and Marsch, 1997); (4) non-compressive density enhancements (NCDEs) (Gosling et al., 1977); (5) the intensity and shape of suprathermal electron velocity distributions, possibly signifying major magnetic restructuring of the inner heliosphere; (Gosling et al., 1987; Feldman and Marsch, 1997); and (6) the He to H relative abundance variations, which may be caused by changes in the structure of the chromosphere-corona transition zone (e.g., von Steiger and Geiss, 1989). A major scientific objective of the SWEPAM experiment is to relate such observables to peculiarities in isotopic and elemental abundances measured by the composition experiments on ACE.

2.1.2. *Coronal Mass Ejections*

The interplanetary manifestations of CMEs in the solar wind are varied. Perhaps the single best identifier is the occurrence of counter-streaming suprathermal electrons (e.g., Gosling, 1996, and references therein). The appearance of CMEs in coronagraph images shows radially evolving sets of structured loops of enhanced density; these structures are interpreted as tracing sets of nested magnetic loops that are closed at both ends to the Sun. Connection to the hot corona at both ends of these loops should, in turn, drive an outward-directed electron heat flux along both coronal footpoints. The consequent field-aligned streams of suprathermal electrons that carry this heat flux are thought to be the counter-streaming electron signatures observed in interplanetary space. An example of such a signature is shown in C and D of Figure 5 (from McComas et al., 1994). By comparison, regions A and E show more typical halo electron distributions where direct connection to the hot corona is made in only one direction along the magnetic field.

Although the presence of counter-streaming suprathermal electrons is highly likely, it is neither a necessary nor a sufficient interplanetary signature of a CME. Many other plasma and magnetic field identifiers have also been associated with CMEs in the solar wind. These identifiers include: (1) a generally enhanced helium abundance that is often characterized by large temporal and/or spatial variations (Hirshberg et al., 1972); (2) unusually low ion and electron temperatures (Gosling et al., 1973; Montgomery et al., 1974); (3) macroscopic rotations of the IMF interpreted as a magnetic cloud (Klein and Burlaga, 1982); (4) charge states of heavy ions that indicate a coronal temperature larger than about 2×10^6 K (Bame et al., 1979); and (5) counter-streaming suprathermal ions (Marsden et al., 1987).

The solar wind identified with CMEs is often highly structured. Not only do the various observables used to identify a CME often vary considerably during any given event, they often do not coexist in any given parcel of CME plasma. Indeed, sometimes they do not even overlap spatially. These observations have led to a picture of CMEs in the solar wind as consisting of a raisin-pudding conglomerate

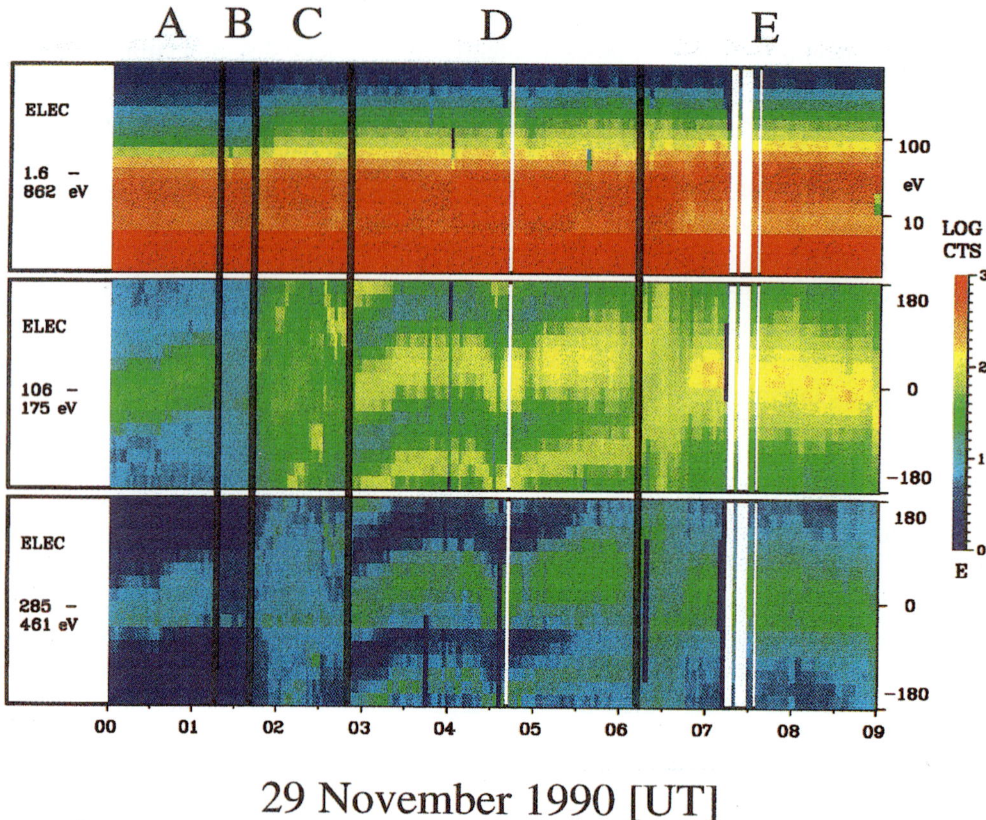

29 November 1990 [UT]

Figure 5. Color-coded plots of suprathermal electron counts for a CME (C, D) observed by *Ulysses* on 29 November 1990. Region B displays a heat flux dropout, which may signify a region that is magnetically disconnected from the corona at either end while regions A and E show typical mono-directional electron heat fluxes. Figure taken from McComas et al. (1994).

of small-scale structures convecting together in the ambient interplanetary medium. Figure 6 shows a schematic sketch of a CME in the solar wind that combines features presented by Bame et al. (1979) and Gosling and McComas (1987). This basic picture has been reinforced by recent observations of a CME observed simultaneously at high and low heliographic latitudes by *Ulysses* and IMP-8 (Gosling et al., 1995). It is curious that while the He abundance in this CME was >10% at times at low latitude, in the same CME at high latitude, it remained constant at the high-speed wind value of ~4.4% (Barraclough et al., 1996).

2.1.3. *Outstanding Questions*

A common thread that connects streamer belt flows with CMEs is that they both originate from regions that overlie complex magnetic topologies in the photosphere that, at times, undergo large reconfigurations. The streamer belt generally overlies most zones of high magnetic activity that consist of complex, strong magnetic

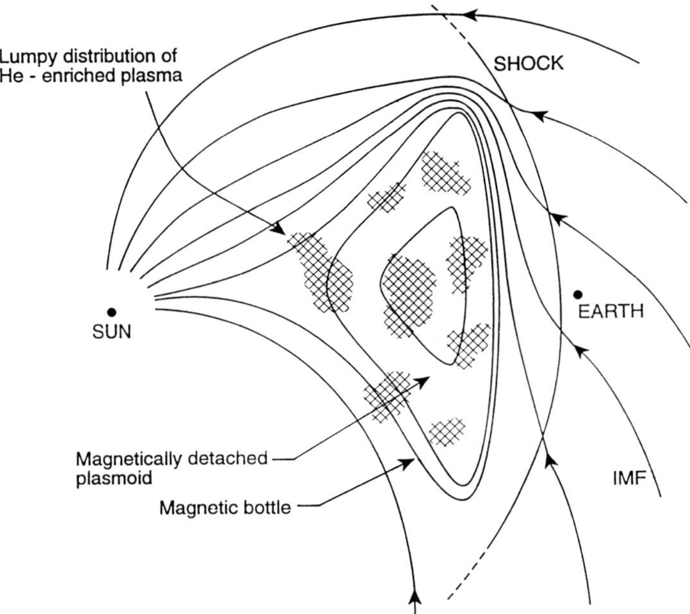

Figure 6. Schematic diagram of a CME that drives a shock wave in the ambient upstream solar wind. The hypothetical disturbance that is sketched contains both a plasmoid (magnetically disconnected from the Sun) and a bottle (magnetically connected to the Sun at two footpoints). The entire disturbance is filled with isolated patches of He-enhanced plasma that cross boundaries between the three different magnetic topologies present in interplanetary space.

field configurations having a high multipolarity and a relatively short lifetime (e.g., Hundhasuen, 1977). CMEs are thought to result from a large-scale restructuring of longer-lived magnetic field structures (Low, 1996). Time sequence observations of the corona in white light (see, e.g., Hundhausen, 1988), EUV (Dere, 1994), and in X-rays (Tsuneta, 1996) have graphically illustrated the consequences of these reconfigurations on the structure of the transition region and overlying corona. These range from the ponderous uplift of large volumes of coronal gas (a CME) to the injection of fine-scale, high-speed jets of relatively cold (Hα surges and EUV explosive events, 10^4 to 10^5 K) and very hot (X-ray jets, 10^6 to 10^7 K) gas into the pre-existing, ambient solar wind.

All the foregoing phcnomcna and thcir conscqucnt solar wind extensions to interplanetary space are intrinsically time dependent. Although the relatively gross aspects of their origin (free energy contained in the photospheric magnetic field), and their effects on the low corona (convected enthalpy and kinetic energy of plasma bulk motion), are known from remote-sensing observations, not very much is known about the mechanisms that drive them. It seems reasonable to presume that magnetic reconnection in some form is the prime mechanism that effects the transfer of energy from the preexisting magnetic field to the plasma and energetic particle population. However, we do not know the altitude at which reconnection

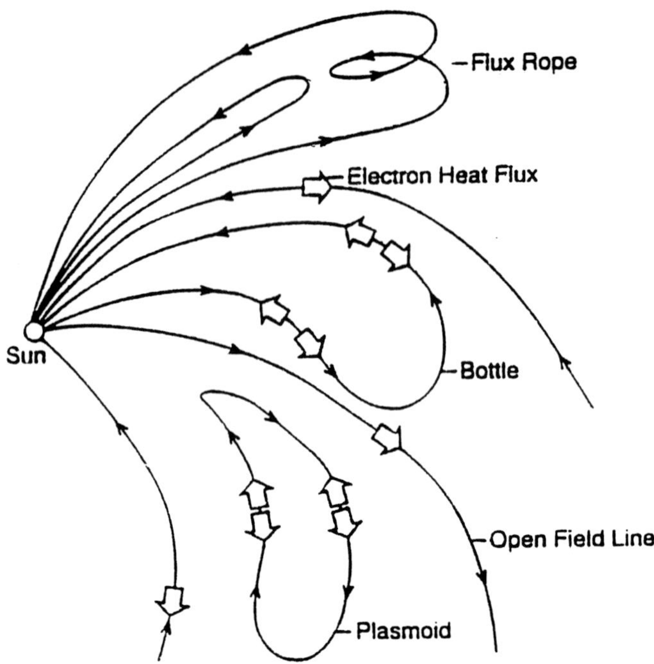

Figure 7. Schematic illustrations of four generic types of interplanetary magnetic topologies. At bottom is a simple open topology, connected at one end to the Sun and at the other end to the interstellar medium. Heat flux is expected to be unidirectional away from the Sun as shown by the arrow. Just above is a completely disconnected plasmoid. Heat flux is expected to be either absent or bidirectional. Next above is a magnetic bottle or tongue, connected at both ends to the Sun. The heat flux here is also expected to be bidirectional. At the top is pictured the simplest flux-rope topology. It is also connected at both ends to the Sun but contains one or more loops. Figure from Feldman and Marsch (1997).

is initiated. Nor do we know the plasma properties in regions that undergo reconnection, or the relative fraction of the magnetic energy that goes into plasma bulk motion, ion heating, electron heating, and the energetic particle population. Are magnetically-closed structures ever produced by this reconnection, or do they always produce complex flux-rope topologies containing a mixture of open, closed, and disconnected field lines (i.e., Gosling et al., 1995)? Schematic examples of several possible magnetic topologies are shown in Figure 7. What is the time scale of the initial reconnection event? Are slow-mode shocks generated by these events as suggested by Petschek (1964), and if so, what is the contribution to the plasma structure of the streamer belt? At what altitude are fast-mode shocks initiated? Do the flux ropes continue to evolve as they expand outward into the solar wind? Is there a component of the streamer belt flow at 1 AU that is not associated with solar activity (i.e., that is driven by a quasi-static pressure gradient as first suggested by Parker (1958))? If so, what is the heating mechanism that supports a quasi-static, million-degree corona and where does this heating occur?

Answers to many of these questions have been offered based on analyses of previous particle and fields observations. None of these, however, is definitive because of essential ambiguities in interpretations of most of the data that was available. Specifically, many characteristics of these data (such as plasma ion and electron temperature and particle energy spectra) are affected by so many processes that occur between the Sun and Earth that they contain highly ambiguous information about conditions near the coronal base where the solar wind is launched. However, the ambiguity in interpretation of the isotopic, elemental, and charge-state composition of plasma and energetic particles from the Sun is considerably reduced because these parameters do not change greatly above the coronal base during their transport to 1 AU. We therefore expect that it should be possible to make considerable progress in finding solid answers to many of the foregoing questions through a joint analysis of the solar wind data returned by SWEPAM and the composition measurements made by the comprehensive composition payload on ACE.

2.2. PARTICLE ACCELERATION AND TRANSPORT

Another major thrust of the ACE mission is to increase our understanding of the origin and evolution of the energetic particle populations that are found almost ubiquitously throughout the inner heliosphere. An important component of this population is accelerated in interplanetary space, and most of it must be transported across large distances in the solar wind before being detected at 1 AU. Numerous analyses have shown that the forward and reverse shocks that bound co-rotating interaction regions (CIRs) beyond 1 AU are copious accelerators of particles. An example of one such interaction region observed by *Ulysses* at 5 AU is shown in Figure 8 (Gosling et al., 1993). Note the large enhancements in the electron suprathermal population extending to considerable distances upstream from both the forward shock (day 20 ~05:00) and reverse shock (day 22 ~02:00) that bound the CIR. Other important sites of particle acceleration occur near forward shocks that are driven through the ambient plasma by fast CMEs.

The seed population for particle acceleration can originate in the thermal and suprathermal components of the solar wind, or in the population of pickup ions that are injected into the solar wind from the neutral component of the interstellar medium, which interpenetrates the inner heliosphere. This latter injection results from (1) charge-exchange collisions with solar wind ions; (2) photoionization by solar EUV photons; and (3) impact ionization by solar wind electrons. Although charge-exchange collisions and photoionization generally dominate impact ionization in the injection of interstellar neutrals into the solar wind, shock-heated electrons upstream of some CMEs can sometimes make important (and for helium, dominant) contributions, as illustrated by the impact ionization rate curves (marked with circles and squares) in Figure 9 (Isenberg and Feldman, 1995; Feldman et al., 1996).

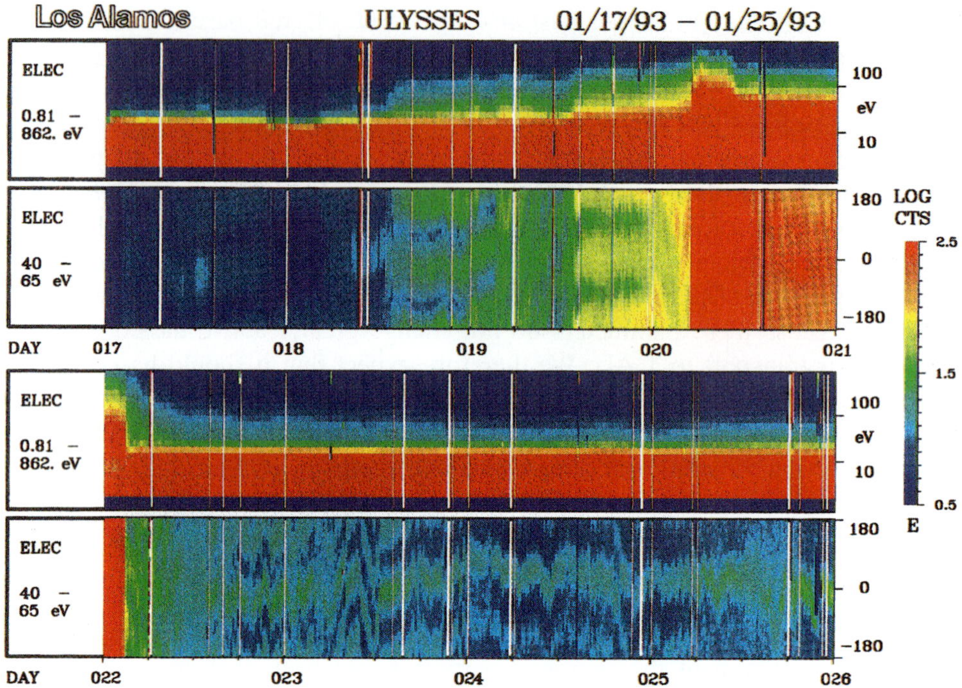

Figure 8. Color-coded plot of solar wind electron data for the co-rotating interaction region encountered by *Ulysses* at ∼5 AU in January 1993. Figure taken from Gosling et al. (1993).

Although the shapes of particle energy distributions largely result from processes that occur in the neighborhood of shocks, they evolve in important and characteristic ways while streaming from these shocks to their point of detection. This evolution is controlled both by the topology of the IMF that is carried by the solar wind and by the nature of the waves that fill interplanetary space. The effect of waves on energetic particles is enhanced if the magnetic topology is that of a complex flux rope or if the imbedded magnetic field intersects a shock front at more than one location. In such cases, the time available for scattering a seed particle in transit through a given interplanetary distance is longer and there is more time for interactions.

2.2.1. *Co-rotating Interaction Regions*

Shocks driven by co-rotating interaction regions do not generally form at radial distances less than about 2 to 3 AU, and so will only occasionally be directly detected by ACE near 1 AU. SWEPAM data can, however, be used to predict CIR onsets and their locations relative to Earth using appropriate MHD models. Predictions of these models can be checked using the solar wind observations made by the SWOOPS instrument onboard *Ulysses*, which will be located just beyond 5 AU near the ecliptic plane throughout the prime portion of the ACE mission (Figure 2). This combination of dual-point observations and MHD model predictions

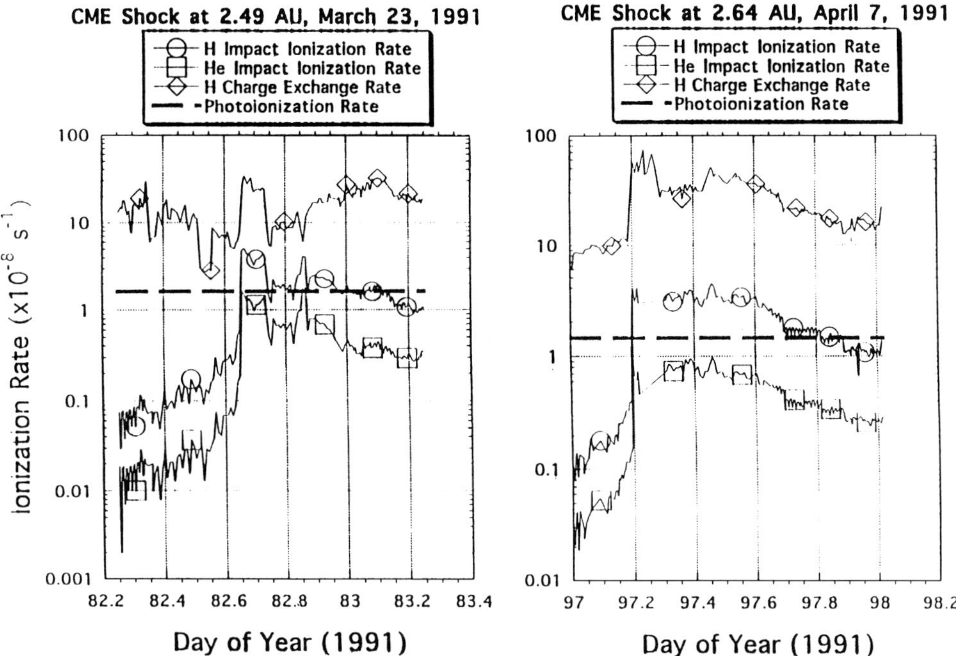

Figure 9. The various ionization rates of interstellar hydrogen and helium atoms calculated from data measured by *Ulysses* across two CME-driven shock wave disturbances. Identification of the ionization processes for the different curves is given in the legend. Figure taken from Feldman et al. (1995).

should provide the needed context for composition observations of backstreaming populations of energetic ions accelerated by the shock pairs that bound CIRs at large heliocentric distances.

2.2.2. *Forward Shocks*

Forward shocks, driven by fast CMEs, accelerate significant numbers of thermal and suprathermal ions and electrons to high energies (see, e.g., Figure 8). These shocks, and the associated enhanced populations of suprathermal ions and electrons, will be detected and fully characterized by SWEPAM. Special attention was given in upgrading the physical configuration of the SWEPAM ion analyzer, and in implementing data accumulation cycles for both plasma ion and electron experiments, to return fully three-dimensional energy-angle spectra having sufficient velocity-space resolution and counting statistics to provide a bridge between the solar wind thermal population and the energy spectra of energetic particles. The shapes of these distributions should be sufficiently detailed to enable definition of the magnetic topology of the ambient plasma through which the shocks propagate. This topology will be of fundamental importance in assessing the type and efficiency of acceleration processes at work at these shocks.

2.2.3. *Pickup Ions of Interstellar Origin*

SWEPAM data will also be used to support studies of the solar wind pickup of newly ionized interstellar neutral atoms. Most important for the pickup of interstellar hydrogen will be SWEPAM's measurement of the solar wind mass flux because the ionization rate is dominated by charge-exchange collisions with solar wind protons.

Past studies of ISEE-3 and *Ulysses* electron data show that just upstream and downstream of strong interplanetary shocks, electron impact ionization can enhance ionization and hence pickup of interstellar hydrogen atoms by roughly 10% (Isenberg and Feldman, 1995; Feldman et al., 1996). These same measurements also indicate that electron impact ionization can dominate photoionization for interstellar helium atoms (see, e.g., Figure 9). These effects are doubly significant because the subsequent acceleration of these newly injected ions to higher energy is enhanced in the neighborhood of shocks. SWEPAM data can also be used to help in studying particle transport through the solar wind by providing knowledge of the magnetic topology in the neighborhood of these shocks. This knowledge will result from study of the shapes and intensity of suprathermal electron and ion velocity distributions.

3. Solar Wind Ion Instrument (SWEPAM-I)

The solar wind ion instrument, SWEPAM-I, contains a spherical section electrostatic analyzer (ESA). Ions enter the sensor through its aperture and are immediately introduced into the gap between the electrostatic analyzer plates. A negative high voltage on the inner plate of the electrostatic analyzer biases the analyzer such that only ions within a narrow energy per charge range ($\sim$5%) and azimuthal angular range (3–4.5°, depending on polar angle) are permitted to pass through the analyzer. The transmitted ions are detected in the channel electron multipliers (CEMs) located immediately behind the analyzers. The aperture of the sensor is aligned such that its fan-shaped field-of-view rotates about the spacecraft spin axis. Figure 10 schematically displays the SWEPAM-I pointing and electro-optics. We define two angles through the normal to the sensor aperture: the polar angle, θ, in the plane of this fan (plane of the right hand sketch), where 0° is parallel to the Sun-pointing spin-axis direction of the ACE spacecraft. The azimuthal angle, ϕ, is in the plane perpendicular to the fan (plane of the left hand sketch).

The normal to SWEPAM-I's aperture is pointed 18.75° in polar angle away from the sunward pointing spin axis direction on the ACE spacecraft. This allows SWEPAM-I to measure ions arriving at polar angles from 0° (along the Sun-pointing spin axis) out to $\sim$65°. Ions in the main solar wind beam, which usually arrive from polar angles of <25°, are measured by CEMs on both sides of the fan. The offset angle of the sensor's pointing will interleave observations from one half of the rotation with those from the other half with a consequent improvement

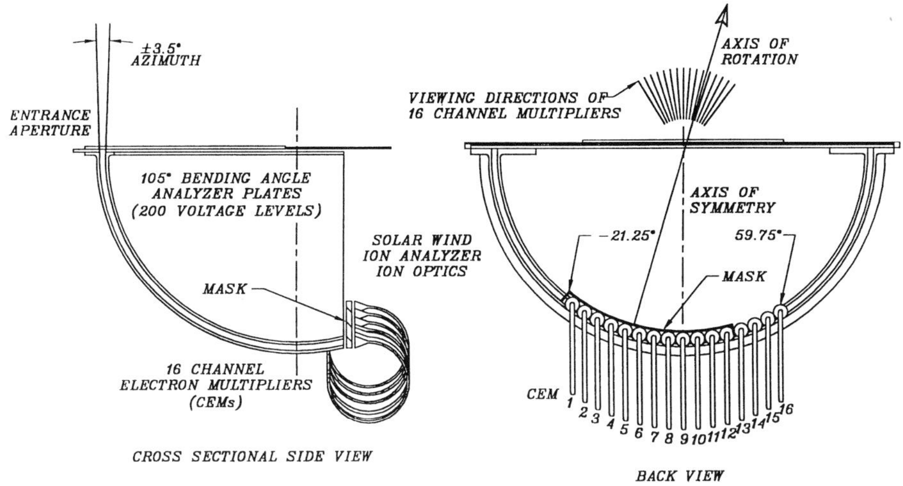

Figure 10. Schematic diagram of the SWEPAM-I electro-optics. The ion sensor is tilted with respect to the spin axis such that observations from CEMs 1-5 should interleave with those from CEMs 6-10. The exit aperture mask that has been added for SWEPAM is indicated in both images (see text).

in polar angle resolution from 5° down to 2.5°, if the spacecraft spin axis is perpendicular to the Sun-facing deck of the ACE spacecraft. While the ACE design goal is for the spin axis to be perpendicular to this deck to within 0.5°, the exact orientation will not be known until after deployment of the solar panels on orbit.

Ions entering the ESA at different polar angles with respect to the aperture normal end up being counted in different CEMs, thus identifying the incident polar angle of the ion in the plane of the acceptance fan of the instrument. As ACE spins, the CEMs sweep out small conical segments of the sky, centered on the spacecraft spin axis direction. The combination of CEM identification and spin phase of the spacecraft provide unique directional information about all ions measured in the instrument. Further, the high voltage step of the ESA selects the E/q of transmitted ions. This combination of CEM number, spin phase, and ESA step level, which is equivalent to polar angle, azimuthal angle, and energy of the incident ions, respectively, provides the basic matrix of measurements in which SWEPAM-I counts are collected.

SWEPAM-I instrument capabilities and hardware parameters are summarized in Tables II and III, respectively. This instrument makes full 3-D plasma measurements of protons and alpha particles every 64 s. In its search mode (typically once every 32 minutes) SWEPAM-I covers the entire 260 eV q^{-1}–36 keV q^{-1} energy range. Between search modes, SWEPAM-I tracks the solar wind and measures every other of 40 energy levels around the peak; track mode provides 5% energy resolution measurements every 64 s. In addition, observations from two adjacent

TABLE II

SWEPAM instrument capabilities

	SWEPAM-I	SWEPAM-E
Particle species measured	3-D protons and alphas	3-D electrons
Energy range	260 eV/q–36 keV/q	1.6 eV–1350 eV
Energy resolution $\Delta E/E$ (FWHM)	5% or 2.5% (1)	12%
Analyzer K (ion energy/ESA voltage)	17.1	4.3
Polar angle FOV	$(-)25°$ to 65°	10°–170°
Polar angle resolution (FWHM)	5° or 2.5° (2)	21°
Azimuthal angle resolution (FWHM)	3° to 4.5° (3)	9° to 28° (3)
G-factor/pixel (cm^2 sr eV/eV)	$1–20 \times 10^{-6}$	$2–7 \times 10^{-4}$
Time resolution	64 s	64 s

(1) 2.5% for energy interleaved data (128 s resolution).

(2) 2.5° if orientation of S/C spin axis allows interleaving data from both sides of fan.

(3) Smaller angles are for normal incidence; larger angles are at the maximum polar ranges.

track modes can be interleaved to give 2.5% energy resolution with 128 s cadence. Section 5 describes the modes of operation in more detail.

Figure 11 displays a photograph of the SWEPAM-I instrument which mounts on the sunward-facing instrument platform of the ACE spacecraft. The instrument is comprised of two major subassemblies, a large cylindrical 'sensor' assembly, and a rectangular 'electronics box'. The two are joined by angled mounting brackets, which set the pointing angle of the sensor head and, hence, instrument aperture with respect to the spacecraft spin axis.

3.1. SWEPAM-I SENSOR

Components of the ion sensor are housed in the ribbed, thin-walled, drum-shaped container visible in Figure 11. All of the parts are attached to the end covers of the drum. The curved-plate electrostatic analyzer is mounted on the inside of the sunward drum face, and views space through a single, oversized entrance aperture. The ESA is composed of nested spherical section plates having a 105° bending angle. Conically-shaped stiffening ribs were machined as integral parts of the curved aluminum alloy plates in order to maintain their dimensional stability. Both of the plates were coated with copper and blackened using an Ebanol-C process in order to reduce UV scattering into the sensor. The electrostatic analyzer has a gap width of 2.84 mm between plates and an average radius of 100 mm.

Sixteen channel electron multiplier detectors, or CEMs, are arrayed such that their 7 mm funnels have 5° polar angle separations around the edge of the gap where the particles exit the analyzer plates. Each is individually mounted on a ceramic card which also carries glass-coated thin film resistors and bare ceramic

TABLE III

SWEPAM hardware parameters

	SWEPAM-I	SWEPAM-E
Box size (L × W × H) (cm)	36 × 24 × 30	25 × 18 × 19
Mass (kg)	3.7	2.5
Power, average (W)	3.1	2.7
Power, peak (W)	3.3	2.9
Telemetry rate (b/s)	540	460
Number of CEMs	16	7
Temperature limits (°C)		
Preferred operating	0 to +20	0 to +20
In calibration	−20 to +45	−20 to +45
Operating survival	−25 to +50	−25 to +50
Non-operating survival	−30 to +60	−30 to +60
EMC interference		
DC magnetic	0.01 nT @ 10′	0.03 nT @ 10′
AC magnetic	BDL	BDL
AC electrical	BDL	BDL
Ordnance	2 dimple motors	2 dimple motors
Red tags (HV safe/arm)	2	2

BDL: Below Detectable Levels.

coupling capacitors needed to bias the anode slightly positive with respect to the exit end of the CEM, and to couple out the signal. This arrangement maintains cleanliness in order to guarantee long CEM lifetimes (e.g., McComas and Bame, 1984 and references therein), permits replacement of individual CEMs as needed, and enables the precise spacing that is required for the fine angular resolution of this instrument.

In the twin instrument on the continuing *Ulysses* mission, a motor-actuated aperture wheel provides the variable sensitivity required to make measurements across the wide range of heliocentric distances covered by that mission. For ACE, which remains at a fixed distance of 0.99 AU, this mechanism was replaced with a single, oversized entrance aperture wide enough to expose the entire width of the analyzer gap, and extending 7.8 mm along the gap. The aperture is covered with a simple, spring-loaded dust cover. Redundant pyro-activated dimple motors actuate the release mechanism by command, on orbit.

The ACE mission limits the Sun-spacecraft spin axis angle to between 4° and 20°, far less than the *Ulysses* mission. Therefore, we were able to build new instru-

Figure 11. Photograph of the SWEPAM-I instrument. The large cylindrical housing is the sensor head; the rectangular electronics box behind it houses the high and low voltage power supplies and control electronics.

ment mounting brackets, visible in Figure 11, that are configured so that, when the instrument is mounted on the spacecraft, rotation of CEMs 1-12 about the spin axis covers the total angular range of solar wind ions expected throughout the mission. The central axis of the measurement array, parallel to the spacecraft spin axis, is located one quarter of the way between the centroids of CEMs 5 and 6 ($\theta = 18.75°$).

The new mounting scheme also makes it possible to devote four of the SWEPAM-I polar angle detectors to high-sensitivity measurements of suprathermal ions. We therefore added a thin aluminum exit aperture mask covered with 70 lines per inch, 92% transmissive nickel mesh interposed between the exit of the analyzer plates and the CEMs. This mesh, biased at the same voltage as the funnels, precludes electric fields from the ESA from reaching into the funnels and affecting the CEM counting efficiencies. This mask also provides two different sizes of exit apertures. Analyzed ions enter funnels 1–12 through apertures tailored for solar wind ion fluxes near 1 AU. These apertures are 0.4 mm wide, selecting ions exiting only from the very center of the gap. CEM funnels 13–16, which will be exposed only to suprathermal ions and not to the solar wind ion beam because of their angular positions, have circular apertures in the exit mask large enough to transmit ions from across the ESA gap into the CEM funnels.

Each of the sixteen CEMs is connected to its own dedicated amplifier-discriminator circuit on its own miniature printed circuit board, located on the outside of the sensor box. Signal lines are fully shielded, only a few centimeters long, and pass through the sensor box wall on individual miniature feedthroughs, in order to avoid any cross-talk or noise pickup at the inputs to these sensitive circuits. These circuits amplify and discriminate pulses from the CEMs; each discriminator has two, selectable level settings of 1×10^6 and 2×10^7 electrons per pulse. The lower, more sensitive setting is used for all science observations while a combination of alternating between these levels and stepping the CEM high voltage is used for periodic CEM gain calibrations. Normalized pulses from the 16 circuits are cabled to counters in the electronics box.

Since the pumping speed through the aperture is small, SWEPAM-I is fitted with an additional pump-out baffle box to enhance the evacuation of the interior of the instrument. The blackened baffle box requires a minimum of seven surface reflections between its two open ends. This design, fully proven on *Ulysses*, does not allow for any appreciable background from photons or charged particles entering through the pump-out channel. To preserve the cleanliness of the sensor interior, the aperture and baffle openings were hermetically sealed during the long storage period after *Ulysses* development was completed. For the bulk of the ACE spacecraft testing the instrument has been continuously purged.

3.2. SWEPAM-I ELECTRONICS

With the exception of the sensor-assembly amplifier-discriminator circuits described above, the ion instrument's electronics are housed in the attached electronics box (Figure 11). The purpose of the support electronics is to (1) furnish programmed high voltages for the analyzer which sets the ion energy per charge passband, (2) bias the CEMs at commandable high voltages appropriate to achieve high and stable counting efficiency, (3) condition and provide low voltage power for the various electronic circuits and high voltage power supplies, and (4) process and accumulate pulses to form a digitized data string for input into the spacecraft command and data handling system. Figure 12 provides a block diagram of the SWEPAM-I electronics.

In order to configure the *Ulysses* SWOOPS instruments for use on ACE, two fundamental changes were required. First, the instrument's *Ulysses* interfaces needed to be adapted to the ACE spacecraft. Second, the flight processing capabilities needed to be increased in order to meet the ACE science objectives. In order to achieve these objectives, we took advantage of the instrument's modular design and replaced a select number of logic modules and the back plane (motherboard), leaving the heritage high voltage and low voltage supplies untouched.

The electronics box has four cavities: the low voltage power supply, high voltage power supply, motherboard and main electronics cavities. The motherboard cavity is the primary wiring cavity, interconnecting the spacecraft interface con-

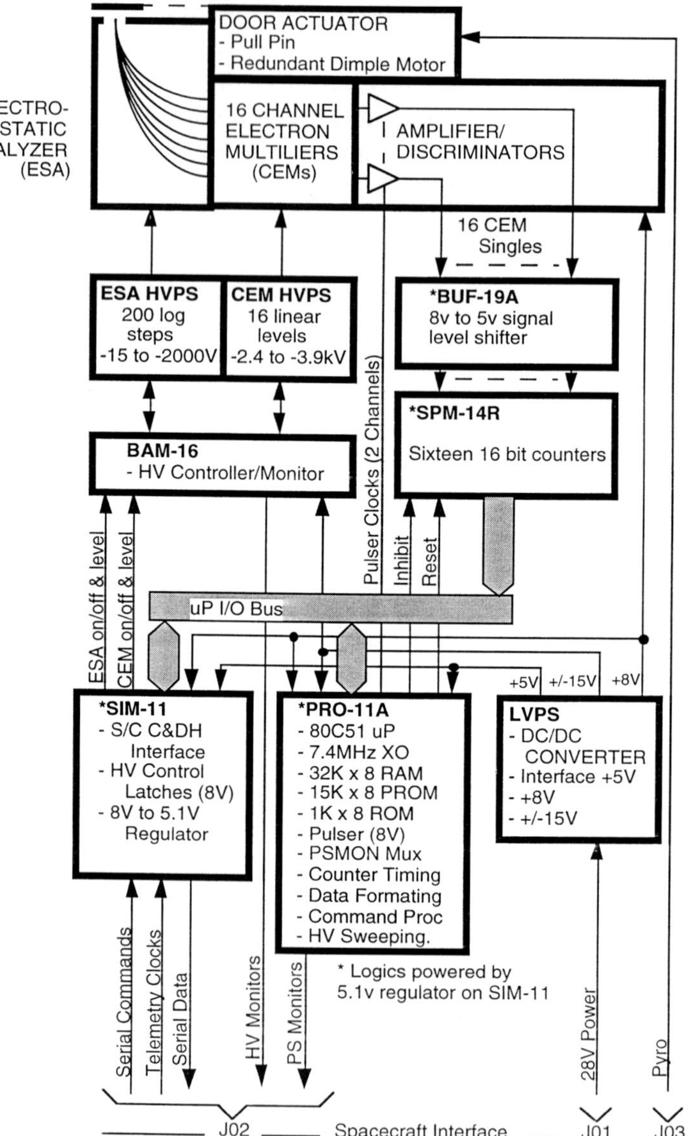

Figure 12. Block diagram of the SWEPAM-I electronics.

nectors, power supplies, processing electronics and sensor electronics. All signals enter and exit this cavity on connectors. The main electronics cavity (tall, central portion of the electronics box) contains five boards or modules: the processor (PRO-11), counters (SPM-14), signal buffer/level-shifter module (BUF-19), spacecraft interface (SIM-11), and high voltage controller (BAM-16) module.

The combined ion instrument has four external connectors: spacecraft command and data handling, spacecraft power, pyro for cover actuation, and a test connector.

The first three are mated into the spacecraft wire harness while the latter is used for test purposes only. SWEPAM-I has two grounds. Signal ground is isolated from the electronics box housing but connected to the electrically isolated sensor head. The signal ground is carried back to the spacecraft single point ground on a wire in the spacecraft harness. SWEPAM-I dissipates 3.1 W when it is fully functional at expected count rates. There are no survival heaters but the instrument is thermally bonded to the thermally controlled spacecraft structure. The instrument temperature is monitored by a sensor mounted on the bulkhead between the motherboard and the low voltage power supply.

3.2.1. *Power Supplies and Controller*

The SWEPAM-I electronics box connects directly to the spacecraft 28 V power and contains both low and high voltage power supplies to produce and condition voltages used in the instrument. Two boards make up the low voltage power supply. The first is a DC/DC converter which isolates the spacecraft primary power ground from the instrument's secondary power ground. This heritage supply 'chopper' was designed to be synchronized to *Ulysses'* power system. The ACE spacecraft does not have power system synchronization so this supply free runs at 55 ± 0.5 kHz. The second board of the low voltage supply rectifies, regulates and filters the outputs from the multi-tap isolation transformer. Spacecraft interface signal line drivers and receivers are powered by $+5$ V; ±15 V is provided for the high voltage control module and used to condition analog telemetry signals output to the spacecraft; $+8$ V is provided for the sensor amplifier-discriminator circuits; the balance of the processing electronics are powered by an 8 V to 5 V linear power regulator on the spacecraft interface module.

Two encapsulated high voltage supplies are provided: a -2.4 kV to -3.9 kV CEM bias supply and a -15 V to -2 kV electrostatic analyzer field supply. Both supplies are driven by the BAM-16 module. The high voltage supply lines are shielded, coaxial cables; the supply return current is carried on the shield. The CEM bias supply assembly includes the feedback controlled drive circuit, drive current monitor, step-up transformer, five-stage diode/capacitor voltage multiplier and the high voltage monitor/feedback resistor divider network which supplies the reference level for the feedback controller. The ESA field supply assembly includes the feedback controlled drive circuit, drive current monitor, step-up transformer, five stage diode/capacitor voltage multiplier, feedback controlled high voltage shunt regulator and a regulator performance monitor. In order to improve efficiency, this supply is designed to vary both the feedback controlled multiplier and regulator references such that the multiplier output is 10% above the regulator output. Most of this 10% head room is required to allow for the multipliers' slow recovery rate in the 'flyback' to start the next sweep. It turned out that the CEM and ESA high voltage supplies did not function well at the 55 kHz free running power supply chopper frequency; consequently, it was decided to allow these clocks to free run at their natural frequency (57 kHz) so none of the power supplies are synchronized to the spacecraft power system or to each other.

The BAM-16 high voltage control module provides the supplies with switched input power, drive clocks, reference levels, monitor conditioning and over-current protection. Each supply is activated by its own ON/OFF control signal; an over-current condition will turn this switch off requiring the ON/OFF control signals to be cycled to reactivate the supplies. The processor selects the desired CEM bias level with a four bit control word. The BAM-16 translates this into 16 equally spaced levels from -2.4 kV to -3.9 kV (100 V spacing). The processor also selects the desired ESA field level with an eight bit control word. The BAM-16 translates this into 256 logarithmically spaced levels of which only the first 200 (-15 V to -2 kV) are used.

3.2.2. *Counters, Processor, and Interface Modules*

The SWOOPS digital electronics were all powered with 8 V. It was not uncommon in the late 1970s and early 1980s when the SWOOPS instruments were developed to find CMOS processors, memories and peripherals that would operate at these or even higher levels. By the 1990s when the retrofit of the SWOOPS to the SWEPAM was made, essentially all CMOS had been standardized to run on 5 V. In order to reuse the heritage counter module (SPM-14), a new module (BUF-19) was added to level shift the 8 V amplifier-discriminator signals to 5 V level before routing to the counter module. The SPM-14 module contains 16 counters, one per CEM, and the bus interface circuits. The TA585 counters used are custom 16-bit counters which interface directly to an eight bit processor bus. While these devices do not have a spill flag, the SWEPAM sample intervals and electronic count rates preclude an overflow.

The *Ulysses* 1802 microprocessor-based processor module and two memory modules were replaced with a single new processor module, PRO-11. PRO-11 includes an 80C51 microcontroller, 7.4 MHz crystal controlled clock, 2K Bootstrap PROM, 28K EEPROM , 48K RAM, watchdog timer, inflight calibration pulser and an 8 channel power supply monitor multiplexer. The timer, pulser, multiplexer controller, memory controller and I/O decoding are all implemented in a single Field Programmable Gate Array (FPGA).

Using lookup tables, the spacecraft telemetry clock (not the spin clock) and its own chip counter/timer circuit, the microcontroller generates a timing function which sequentially disables the counter inputs, commands a new ESA level, moves the data from the counters into RAM, clears the counters, allows high voltage to settle then re-enables the counter inputs to collect data at the new E/q level. Each instrument data mode has a lookup table which defines the mode in terms of the E/q level and count interval, high voltage settling times, number of E/q levels per spin sector, number of spin sectors per spectral segment (roughly equivalent to a spin) and the number of spectral segments per mode.

Bootstrap PROMs contain the firmware which initialize the instrument when the processor is first activated and provide the facilities to load all flight software into EEPROM or RAM. These devices are read-only and contain two copies of the bootstrap codes to ensure there is always a way to correct or tune the flight software. The EEPROMs contain the bulk of the flight software. The instrument will be launched with the most current version of the flight software resident in the EEPROM. If any software uploads are required after launch, they will most likely be limited to revisions of the data mode lookup tables that are moved to RAM by the startup codes.

To accommodate these instruments on the ACE spacecraft, a new spacecraft interface module was designed to replace the *Ulysses* heritage interface. This module, the SIM-11, includes the command receiver, framing pulse receiver, data transmitter, high voltage control latches (CEM and ESA ON/OFF, CEM Level and ESA Level) and an 8 V to 5 V linear power regulator to supply the processor electronics. In order to keep the number of processor interrupts to a minimum, the data transmitter uses a first-in/first-out (FIFO) memory device permitting the processor to buffer an entire minor frame's worth of data between readouts. Except for this FIFO and the linear regulator, all functions on this module are implemented in a single FPGA.

3.3. SWEPAM-I CALIBRATION

Calibration of SWEPAM-I was performed at the Los Alamos plasma analyzer calibration facility. The angular and energy response functions for all sixteen CEMs were measured using a known, unidirectional, 5 kV proton beam. In order to maintain absolute cleanliness for the CEMs during the calibration, we chose to mount only the cylindrical sensor head, and not the electronics box, within the vacuum chamber. The sensor head was mounted on a computer-controlled three-axis table (two orthogonal rotation stages and one translation stage) which allowed ions to be directed at the aperture over all possible input angles for the instrument. A 3-dimensional array of sensor transmission/counting efficiencies was obtained as a function of polar angle, azimuth angle, and ESA voltage (giving E/q) for each CEM. The response function of the instrument is completely defined by these sixteen 3-dimensional arrays.

An example of a 2-dimensional cut through the 3-dimensional response function is shown for CEM 4 in Figure 13. Levels of transmitted ions from an ion beam arriving at a fixed 20° polar angle are color coded as a function of azimuthal angle and ESA voltage. The azimuthal and energy responses are coupled as expected for a spherical section electrostatic analyzer (Gosling et al., 1978); the polar response is essentially decoupled from these two. Figure 14 shows three 1-dimensional cuts through the response function for channel 4 at the centroids of each of the other two parameters. In each, the measured number of counts (solid line) is fit with a theoretical curve (dotted line). In Figure 14(a) the distribution of counts for CEM

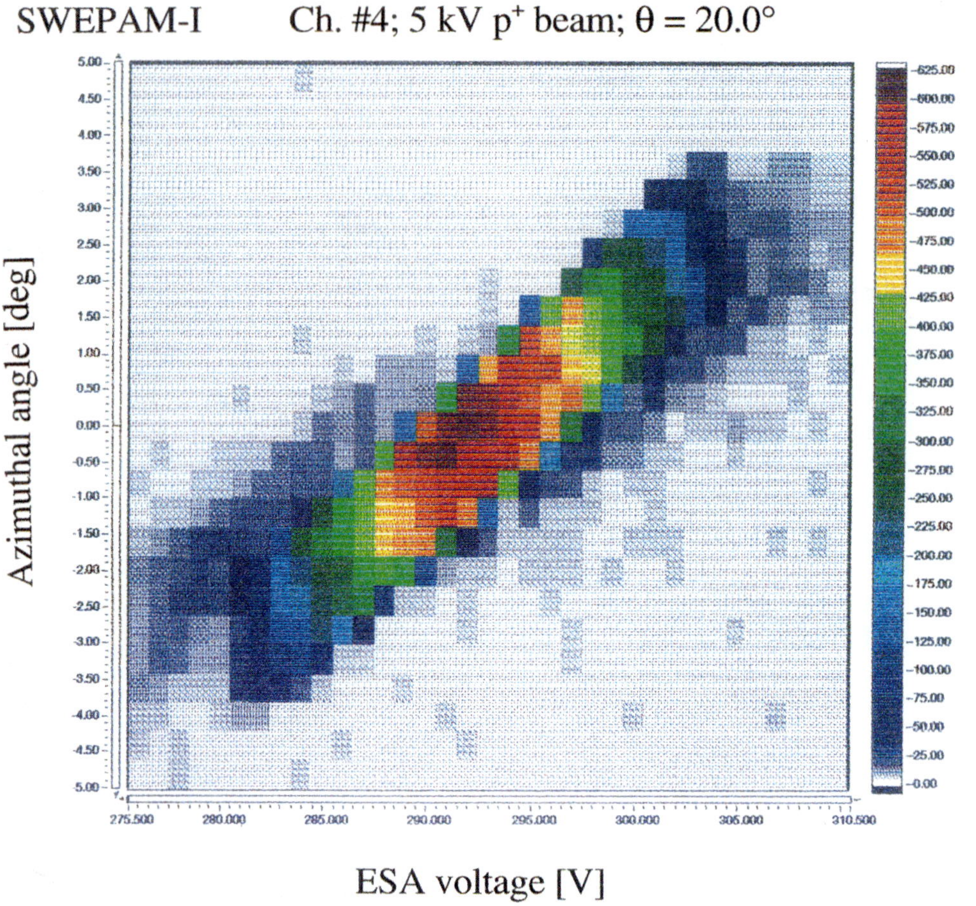

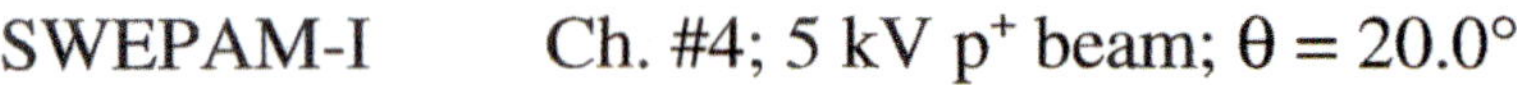

Figure 13. Color-coded plot of transmitted counts through channel 4 of SWEPAM-I as a function of azimuthal angle and ESA voltage (equivalent to E/q) for a 5 kV proton beam incident at a polar angle of 20°.

4 is shown as a function of analyzer plate voltage for the 5 keV proton beam in a look direction given by $\phi = 0°$ and $\theta = 25.8°$. The response is approximated by a gaussian profile. The azimuthal profile of the CEM at an analyzer plate voltage of 293 V and polar angle of $\theta = 25.8°$ is also approximated by a gaussian profile and is shown in Figure 14(b). While the responses as a function of both analyzer plate voltage and azimuth are approximated by a gaussian, the response in polar angle is best approximated by a trapezoid. Comparison of the measured counts with the best-fit trapezoid is shown in Figure 14(c) for an azimuthal angle of 0° and analyzer plate voltage of 293 V. The measured analyzer constant for SWEPAM-I, given by the ratio of incident particle energy to plate voltage, is $\sim$17.1.

Calibration data similar to those shown for CEM 4 were taken for each of the 16 CEMs in SWEPAM-I. Figure 15 shows a polar cut through all 16 channels at

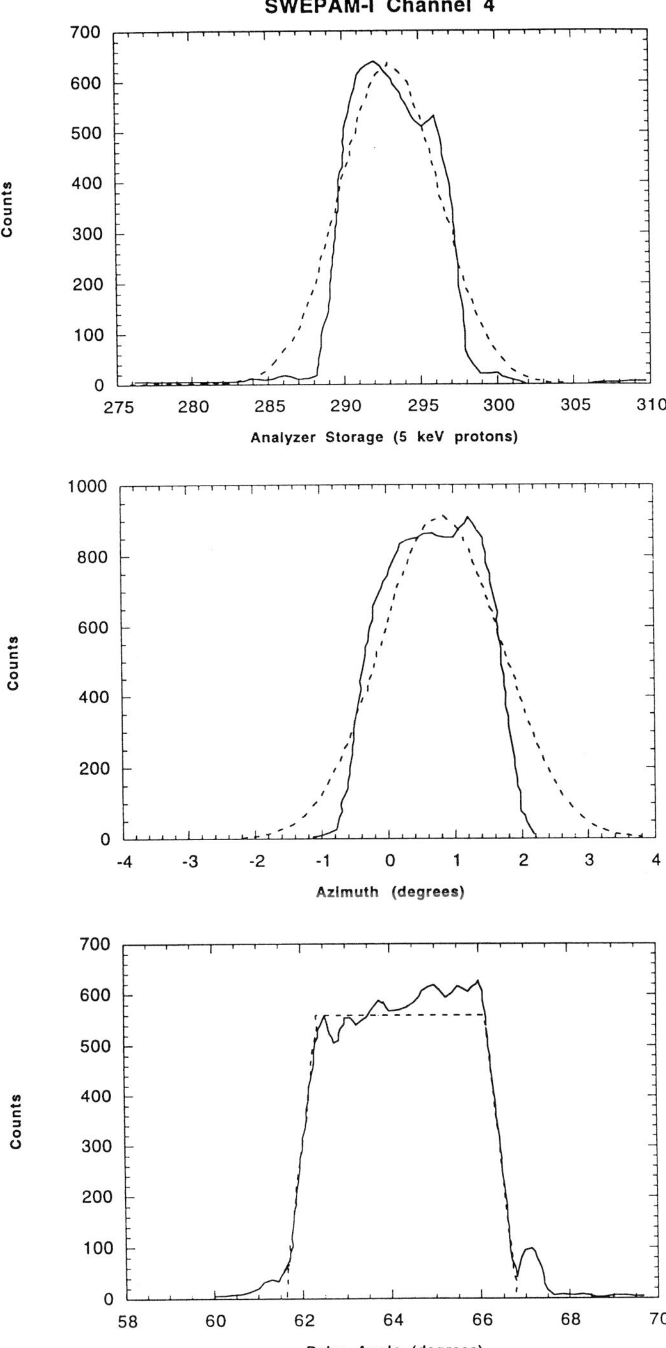

Figure 14. Central energy (a), azimuthal angle (b), and polar angular (c) cuts through the channel 4 response function of SWEPAM-I. The solid lines show the calibration data while the dotted curves display theoretical fits.

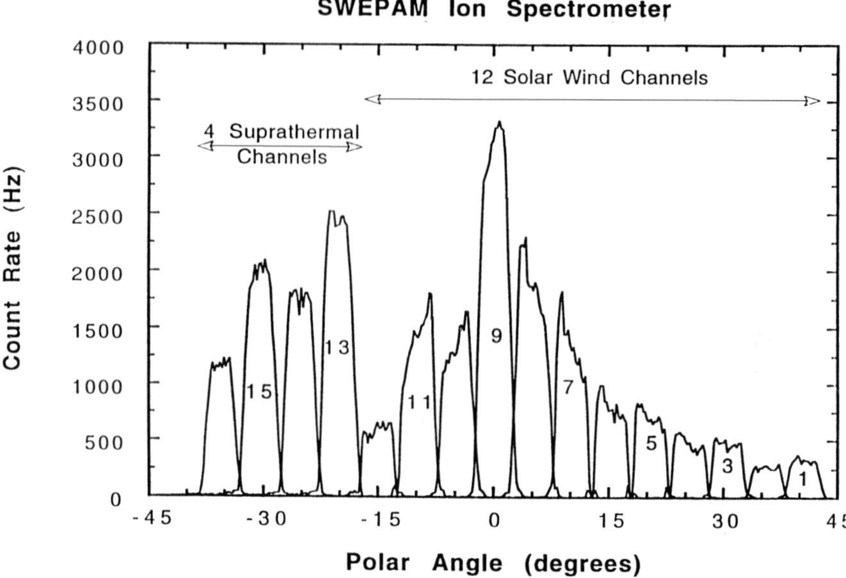

Figure 15. Polar angular cut through all 16 channels of SWEPAM-I at $0°$ azimuth and the peak of the E/q transmission curve. The four unmasked, suprathermal channels have appreciably larger transmission than they otherwise would for such large polar angles.

$0°$ azimuth. This cut appears to be more variable than one might have expected, however, multiple runs confirmed the variations shown. In addition, the roughness of these curves is at least in part attributable to the fact that the central azimuth was slightly different ($<1°$) for each channel and that the very narrow angular response functions were cut in slightly different places. In any case, the full response function information is used in characterizing each channel's calibration curves.

Once these 16 arrays were fully measured, the polar, azimuthal, and energy responses were characterized and the geometric factors for each CEM were calculated. Geometric factors were calculated by numerical integration of the three response functions (energy, θ-angle, and ϕ-angle) over all energies, and all angles. These geometric factors provide the necessary conversion from the measured counts to more physically meaningful distribution functions, from which one can calculate the moments of the distribution such as density, flow velocity vector, temperature matrix, and the heat flux. Table IV summarizes the calibrated geometric factors for SWEPAM-I.

4. Solar Wind Electron Instrument (SWEPAM-E)

Like SWEPAM-I, the electro-optical design of the solar wind electron instrument, SWEPAM-E, is based around a spherical section electrostatic analyzer. For SWEPAM-E, the bending angle of the analyzer is $120°$. Figure 16 schematically

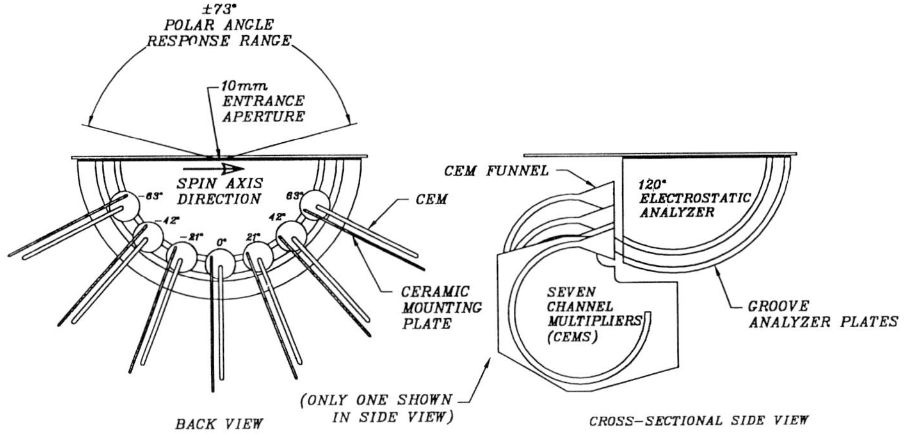

SWEPAM ELECTRON INSTRUMENT

Figure 16. Schematic diagram of the SWEPAM-E electro-optics, similar to Figure 10.

TABLE IV

Geometric factors for SWEPAM (10^{-6} cm^2 sr eV eV^{-1})

CEM	G	CEM	G	CEM	G	CEM	G
1	7.62	5	8.50	9	0.92	13*	16.67
2	7.42	6	6.01	10	1.59	14*	15.20
3	9.23	7	5.07	11	2.86	15*	20.77
4	8.67	8	1.63	12	4.69	16*	14.88

* Used for measuring suprathermal instead of solar wind ions.

displays the SWEPAM-E electro-optics. Electrons enter the instrument through its aperture which is pointed normal to the spacecraft spin axis. A positive high voltage on the inner plate of the ESA biases the analyzer such that only electrons within a range of energies (12%) and azimuthal angles (9° at $\theta = 0°$ to 28° at $\theta = \pm70°$) are permitted to pass through the analyzer and are detected in the seven channel electron multipliers (CEMs) located immediately behind the analyzer.

Polar and azimuthal angles are defined just as they were for SWEPAM-I. Electrons entering the ESA at different polar angles with respect to the aperture normal end up being counted in one of the seven CEMs which gives 21° polar angular resolution for electrons entering in the wide azimuthal angle, fan-shaped field-of-view (FOV). The plane of symmetry of the fan is parallel to the spacecraft spin axis and the middle CEM (#4) viewing angle is centered on a plane parallel to the spacecraft equator ($\theta = 90°$). As ACE spins, the SWEPAM-E fan-shaped field-of-view sweeps out >95% of 4π sr, missing only small ($\sim10°$ half angle) conical

Figure 17. Photograph of SWEPAM-E; similar to Figure 11.

holes centered along directions parallel and anti-parallel to the spin axis (roughly along the radial direction to and from the Sun). Electrons with energies from 1.6 eV–1.35 keV are measured in 20 contiguous energy bins determined by the instrument's high voltage stepping. Again, the combination of CEM number, spin phase, and ESA step level, which is equivalent to polar angle, azimuthal angle, and energy of the incident electrons, respectively, provides the basic SWEPAM-E measurement matrix. The general SWEPAM-E instrument and hardware parameters are summarized in Tables II and III, respectively.

Figure 17 displays a photograph of the SWEPAM-E instrument which also mounts on the sunward-facing instrument platform of the ACE spacecraft. Like SWEPAM-I, this instrument is comprised of two major subassemblies, a ribbed, thin-walled, cylindrical sensor assembly, and a roughly rectangular electronics box. The sensor cylinder mounts directly onto an extension of the electronics box such that the instrument aperture points at right angles to the spacecraft spin axis.

4.1. SWEPAM-E SENSOR

The curved-plate electrostatic analyzer is mounted on the inside of the outward pointing drum face, and views space through a single, rectangular entrance aperture. The analyzer is composed of nested 120° spherical section plates of aluminum alloy that are grooved in addition to being blackened in order to minimize backgrounds caused by UV scattering through the plates and photoelectron and secondary electron production in the plates themselves. The electrostatic analyzer plate system has a gap width of 3.5 mm and an average radius of 41.9 mm. Seven large-funnel (11 mm diameter) CEMs are arrayed with 21° separations along the analyzer plate's exit. A thin aluminum exit aperture mask covered with 70 lines per inch, 92% transmissive nickel mesh (not shown in the simplified drawing) is again interposed between the analyzer exit and CEM funnels to preclude electric fields from affecting the CEM counting efficiencies. Each CEM is mounted on its own ceramic card which also carries the resistors and coupling capacitors needed to bias the anodes slightly positive with respect to the exit of the CEMs and to couple out the signals. Just as for SWEPAM-I, this arrangement maintains cleanliness, permits replacement of individual CEMs as needed, and enables the precise spacing that is required for the fine angular resolution of this instrument. Each of the seven CEMs is connected to its own dedicated amplifier-discriminator circuit; these circuit boards are identical to those used for SWEPAM-I.

Electrons enter the analyzer through an entrance aperture which is wider than the analyzer gap and runs 10 mm along it. At launch, this aperture is also covered with a spring-loaded dust cover to protect the CEMs from contamination. Redundant pyro-activated dimple motors release a mechanism by command on orbit, and allow the cover to be retracted from over the aperture. A blackened pump-out baffle box, similar to the one used on SWEPAM-I, is provided to enhance the evacuation of the interior of the instrument. Throughout most of the observatory level testing SWEPAM-E was also under continuous purge.

Because of the sensitivity of low-energy electrons to spacecraft surface charges, the SWEPAM-E requires a special spacecraft thermal blanket. This blanket maintains the thermal characteristics of the spacecraft, is conductive to carry off any surface charge build-up to the spacecraft bus, and also produces no magnetic field. The side of the SWEPAM-E cover containing the aperture is made with an outer surface of indium/tin oxide (ITO) coated Kapton. This material has the same thermal properties as the regular spacecraft blanket (which is made of beta cloth – a fiberglass material with a Teflon coating). A short (∼25 cm) vertical section of the blanket covering the spacecraft panel below the SWEPAM-E aperture is also made of this ITO coated material. Finally, the original (non-conductive) thermal material on the spacecraft radiator near SWEPAM-E was removed and replaced by a conductive ITO coated silver/Teflon film.

4.2. SWEPAM-E ELECTRONICS

The electron instrument electronics perform the same functions and have a significant amount of commonality with those of the ion instrument. In particular, we determined that it was most cost effective to use a single design for the new electronics hardware for the two instruments. Since the ion instrument's processing requirements were more demanding, they drove the processing and memory

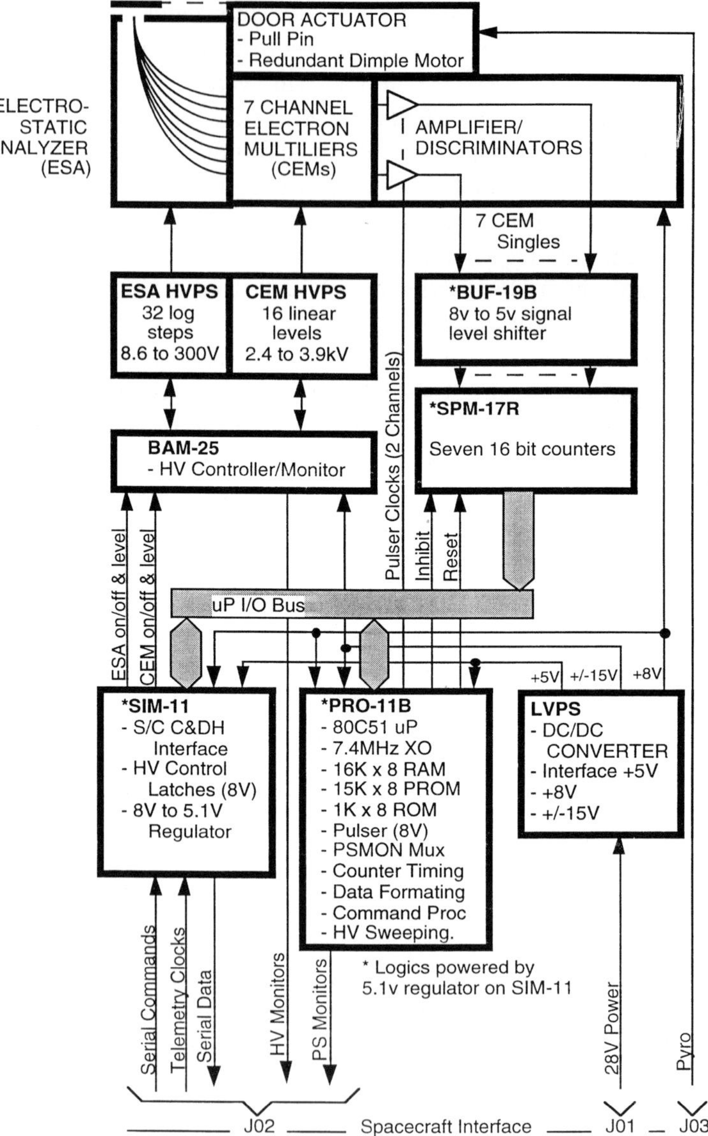

Figure 18. Block diagram of the SWEPAM-E electronics.

resources built into SWEPAM-E as well as SWEPAM-I. SWEPAM-E dissipates 2.7 W when it is fully functional at expected count rates. Its thermal interface is the same as the ion instrument's interface: no heaters, one temperature sensor, and hard mounted to the spacecraft. Figure 18 provides a block diagram of the SWEPAM-E electronics.

This electronics box houses the high and low voltage power supplies, motherboard, wiring, and main electronics. The main electronics cavity contains five boards or modules: the processor (PRO-11), counters (SPM-17), signal buffer/level-shifter module (BUF-19), spacecraft interface (SIM-11), and high voltage controller (BAM-25) module. Of these, only the SPM-17 and BAM-25 are unique to the electron instrument. The rest of the boards are identical to those used in the ion instrument and described above. SWEPAM-E has the very same four connector interfaces as the ion instrument right down to the pin out and signal ground arrangement with one exception. Since the pyro connector is mounted to the sensor housing rather than the electronics box, its shell is grounded to signal rather than chassis ground. Special precautions were taken to assure that this arrangement did not cause a loss of isolation between the grounds.

While the SWEPAM-E heritage low voltage power supplies are of a different design than those of SWEPAM-I (the SWOOPS ion sensor placed additional requirements on the supply for its aperture stepper motor), they are functionally equivalent, including the same 28 V isolation, free running chopper frequency and secondary powers and grounds. The two encapsulated high voltage supplies are a +2.4 kV to +3.9 kV CEM bias supply and a +8.6 V to +300 V electrostatic analyzer field supply. Each supply interfaces identically with the BAM-25 module and sensor head as it did in SWEPAM-I. The CEM bias supply is identical to that of the ion instrument except for output polarity. The ESA field supply assembly is very much like that of the ion instrument except for its voltage range and number of levels. The SWEPAM-E ESA supply output is 32 logarithmic levels from +8.6 V to +300 V. The first eight levels are not used for the ACE mission. Conversely, in order to extend the top-end range of electron science observations, SWEPAM-E uses the top two levels which were not used by the SWOOPS.

The BAM-25 is the complement of the BAM-16 in the ion instrument and functionally differs only in the number of ESA step levels it controls and its reference levels. In order to get the full range out of this ESA supply, the 1 V reference had to be set slightly low ($\sim$0.9 V) so that the supply would not sag at its highest level. This also decreased the output voltage for all other levels proportionately. Consequently, the SWEPAM-E ESA levels are offset from those on SWOOPS. The SPM-17 counter module is the equivalent to the SPM-14 module in the ion instrument except that it has only seven counters instead of 16. Since the electron instrument did not need as much RAM memory for data buffering, the second RAM chip was omitted during assembly.

4.3. SWEPAM-E CALIBRATION

Calibration of the electron analyzer was performed at the Los Alamos plasma instrument calibration facility in a similar manner to that of the ion instrument. Because our facility produces ions rather than electrons, we chose to also calibrate SWEPAM-E using a 1.05 keV proton beam. This was made possible simply by using a negative high voltage on the inner analyzer plate and using ions with energies well above the several hundred volt post-acceleration bias on the CEM

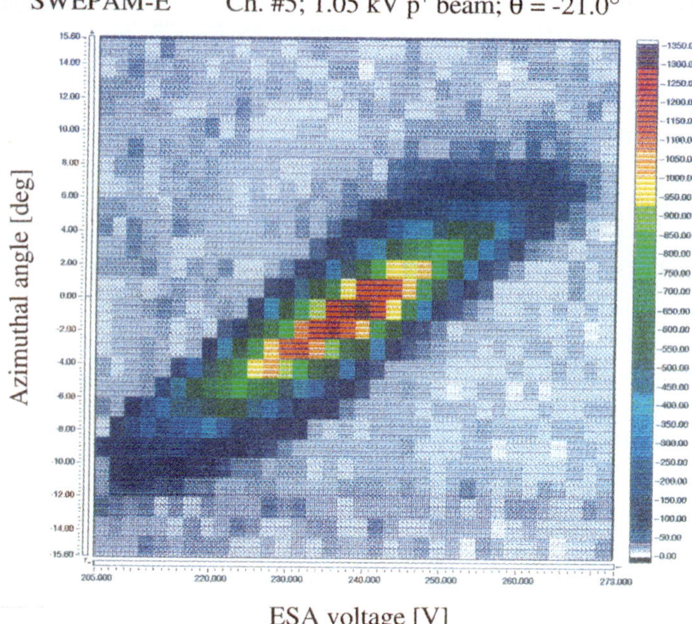

Figure 19. Similar to Figure 13, but for channel 5 of SWEPAM-E, calibrated with a 1.05 kV beam and an incident polar angle of −21°.

funnels. This procedure also has some advantages over using electrons as the ions are far less affected by the Earth's magnetic field than are electrons.

The calibration procedure was essentially the same as for the ion instrument. A 2-dimensional (energy/azimuthal) angle cut through the response function for CEM 5, at its polar angle centroid (−21°), is shown in Figure 19. Figure 20 displays the central 1-dimensional polar, azimuthal, and energy cuts for this channel in the same format as Figure 14. The distribution of counts for Channel 5 is shown as a function of analyzer plate voltage for 1.05 keV protons in a look direction given by $\phi = 0°$ and $\theta = −21°$ in (a); the central azimuthal and polar response are shown in (b) and (c). Again, the responses in (a) and (b) are approximated by simple gaussians. For electrons, the response in (c) is approximated by a skewed gaussian. Finally, Figure 21 shows the polar cut at 0° azimuth through all seven CEMs in SWEPAM-E. The ratio of electron energy to ESA voltage, or analyzer constant, for SWEPAM-E is ∼4.3. The channel-specific geometric factors were derived by the same method as was done for SWEPAM-I and are given in Table V.

Since the SWEPAM instruments, and particularly SWEPAM-E, bear a considerable resemblance to the SWOOPS instruments onboard *Ulysses*, comparisons of calibration results obtained from SWOOPS prior to launch have been made with the SWEPAM calibration; overall the comparison is excellent. As an example, in Figure 22 we compare the geometric factors calculated for the ACE plasma

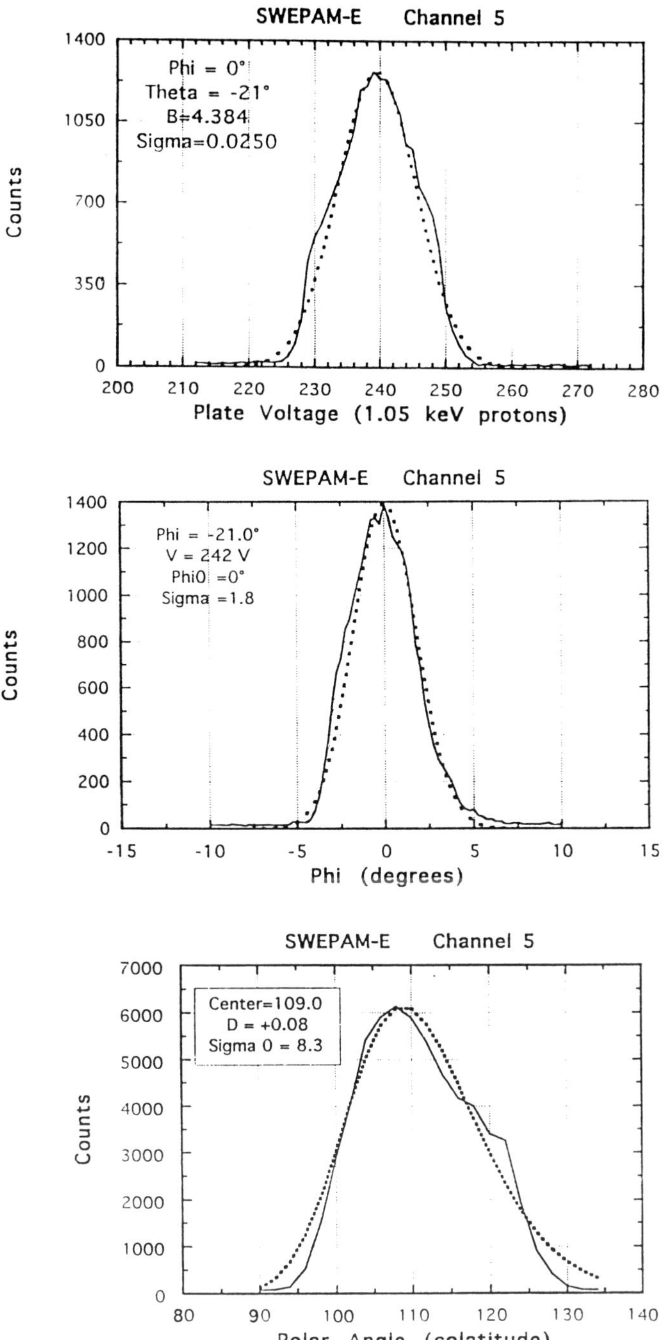

Figure 20. Central energy (a), azimuthal angle (b), and polar angular (c) cuts through the channel 5 response function of SWEPAM-E along with theoretical fits, similar to Figure 14.

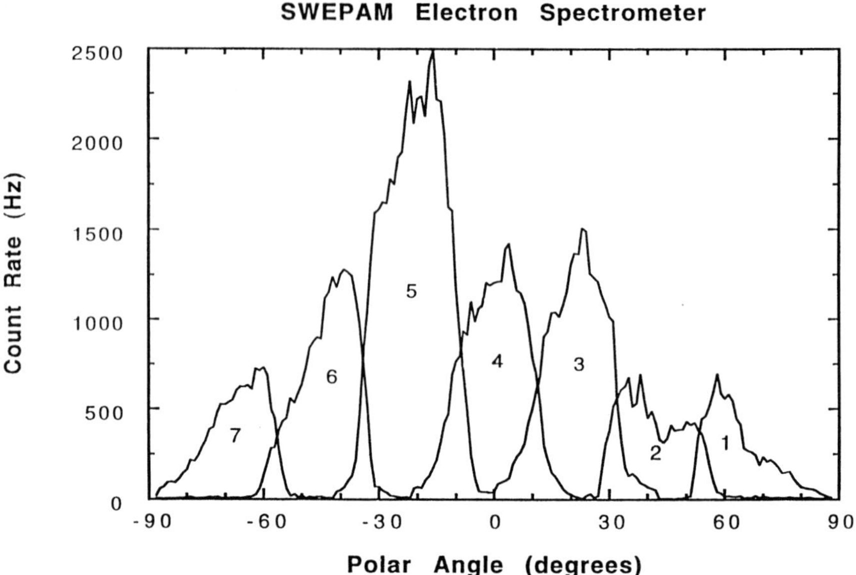

Figure 21. Polar angular cut through all seven channels of SWEPAM-E at 0° azimuth and the peak of the E/q transmission curve, similar to Figure 15.

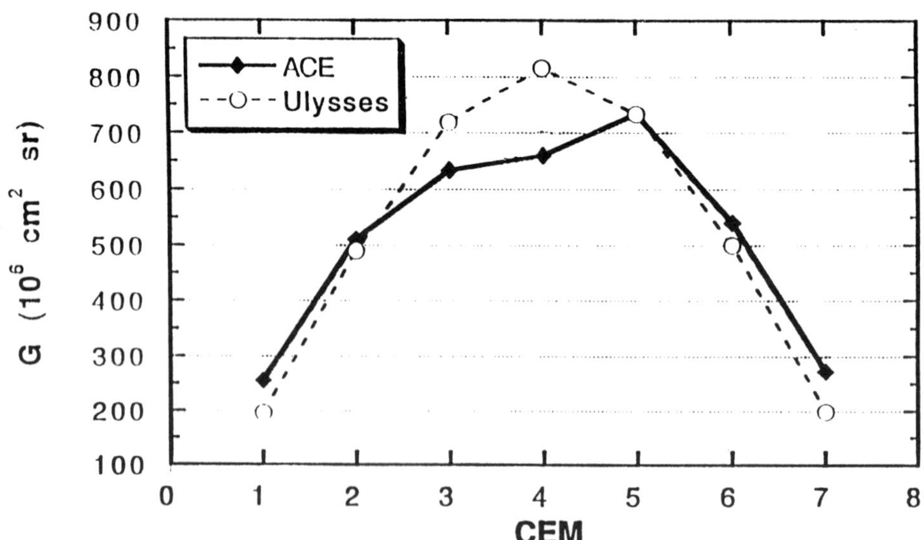

Figure 22. Comparison of SWOOPS and SWEPAM-E geometric factors. The differences in channels 3 and 4 are probably traceable to slight differences in the actual hardware and CEMs.

TABLE V

Geometric factors for SWEPAM-E
(10^{-6} cm^2 sr eV eV^{-1})

CEM	G	CEM	G
1	255.6	5	733.5
2	511.0	6	540.3
3	633.4	7	272.9
4	659.5		

instruments with the geometric factors calculated for *Ulysses*. With the exception of CEMs 3 and 4, the efficiencies are very close. The excellent overall quantitative agreement between these curves, measured in different decades in different calibration facilities, provides support for the accuracy of both sets of measurements. Since the ACE instruments are essentially made up of *Ulysses* spare parts, the flattening of the geometric factor at CEMs 3 and 4 is probably due to differences in the sensor hardware and/or CEMs. Nevertheless, such differences are minor and should not impact the science.

5. Instrument Operations and Data Handling

Owing to the large differences between the *Ulysses* and ACE orbits, as well as the large leaps forward in computing capability over the years between the two instruments' development phases, the SWEPAM data modes and operations have been completely reworked. As such, this section describes a completely new development for the ACE program. Several of the features of the new operations allow SWEPAM-I and SWEPAM-E to provide substantially improved data products over their *Ulysses* counterparts.

Throughout this section we will use the term 'cycle' to describe the top-level instrument process. Cycles are comprised of sequences of 'modes' which are further comprised of sequences of 'segments' which occur over intervals that are roughly equivalent to the spacecraft spin period. The nominal ACE spin rate is 5 rpm or 12 s per spin. Operation of SWEPAM is synchronized with the telemetry rate. This choice yields a nominal data accumulation period (or segment) of 12.8 s which is also just slightly longer than the longest possible spacecraft spin period forecast for the mission (12.2 s). The difference between the actual spin period (which will not be known until after launch, and may change slightly over the course of the mission) and 12.8 s causes the data sampling window to precess around from one spin to the next. As will be described below, this is not a concern because the

SWEPAM microprocessor selects out the most interesting samples in the full data matrix for telemetering.

Table VI summarizes all science data sets for both SWEPAM-I and SWEPAM-E; details are given in the following subsections. The first five products are derived from the ion instrument and the last three are derived from the electron instrument. The data mode from which the data set were derived is shown in column 2. Column 3 provides a brief description of the data set. The fourth column describes the data set matrix in terms of energy (E), polar angle or CEM (θ), and azimuth or spin angle (ϕ). The notation $\theta - \phi$ is used to indicate that a mask has been applied and only a subset of $\theta - \phi$ angles are telemetered. The last column provides the time resolution and cadence for each data product. Thus all science data sets have a basic time resolution of 64 s. The cadences indicated are based on expected data mode cycles, but can be easily changed during the mission.

5.1. COMMANDING AND HOUSEKEEPING

Once the SWEPAM instruments have been powered by the spacecraft, they are operated by two distinct classes of ground commands: hardware commands (hard commands) and software commands (soft commands). Both classes are transmitted to the instruments from flight operations ground computers through the spacecraft as serial data commands. The hard commands are dispatched by a hardware decoder to control critical hardware redundancies and interlocks. The soft commands are dispatched by one of two command processors in the flight software and can either update the EEPROM flight codes and RAM data base or select and control the data acquisition functions. All SWEPAM operational commands are defined in the flight operations data base and are issued from command procedure files which verify the commands and instrument responses. The spacecraft has provisions for stored and autonomous instrument commands. There are currently no plans to use stored commands to operate the SWEPAM instruments. The only autonomously implemented commands for SWEPAM are power shedding if either a spacecraft or instrument power fault is detected.

The SWEPAM instruments each have three housekeeping 'channels'. The analog telemetry consists of eight low voltage monitors multiplexed on one of the three analog signals and 11 high voltage control monitors multiplexed on two other analog signals. These signals are converted by the spacecraft and provided at a rate of one sample per major frame (i.e., 16 s). Eight major frames are required to collect a full set of these monitors. In addition, housekeeping data are provided in the first 40 bits of each major frame to enable operation and monitoring of the instrument by the flight operations team. Twenty-four bits of this same header are output in the RTSW data link. The header is the key to data set formats, identifying the data set type and frame count. Operating parameters required to interpret the science data are embedded in the science data sets, in data set headers, and spectral segment headers as appropriate.

TABLE VI

SWEPAM science data products

Acronym	Data mode	Species	Description	Data matrix	Time res./ cadence (s)
DSWI	SWI	Ion	Plasma ion velocity distribution	$40\,(E) \times 96\,(\theta\text{-}\phi)$	64/64
DSTI	SWI	Ion	Suprathermal ion velocity distribution	$20\,(E) \times 4\,(\theta) \times 6\,(\phi)$	64/64
DSSTI	SSTI	Ion	Plasma ion velocity distribution	$40\,(E) \times 96\,(\theta\text{-}\phi)$	64/1984
DSTI2	SSTI	Ion	Suprathermal ion velocity distribution	$20\,(E) \times 4\,(\theta) \times 6\,(\phi)$	64/1984
DRTSW	SWI	Ion	Real time solar wind	$40\,(E) \times 33\,(\theta\text{-}\phi)$	64/64
DNSWE	NSWE	Electron	Plasma electron velocity distribution	$20\,(E) \times 5\,(\theta) \times 30\,(\phi)$	
				$2\,(\theta) \times 15\,(\phi)^*$	64/128
DSTEA	STEA	Electron	Suprathermal electron distribution	$10\,(E) \times 5\,(\theta) \times 60\,(\phi)$	
				$2\,(\theta) \times 30\,(\phi)^*$	64/128
DPHE	PHE	Electron	Photoelectron distribution	$20\,(E) \times 5\,(\theta) \times 30\,(\phi)$	
				$2\,(\theta) \times 15\,(\phi)^*$	64/1984

* First set of angles refers to CEMs 2 through 6. Second set for CEMs 1 and 7.

5.2. ION MODES AND OPERATIONS

SWEPAM-I has four commandable data modes and two commandable data cycles. The data modes are: engineering (ENGI), calibration (CALI), solar wind ion (SWI), and search suprathermal ion (SSTI). Data modes can be run individually or as part of a data cycle. The data cycles are; normal and calibration. The normal cycle is a sequence of one SSTI (search) mode followed, nominally, by 30 SWI (track) modes. The number of SWI modes, and therefore the duration of this cycle, is command adjustable. Once selected, the normal cycle runs continuously until a command is sent to select a new data mode or cycle. The calibration cycle consists of a sequence of two CALI modes to ensure that a change in the solar wind does not corrupt the data set. This cycle will run once after being commanded and then the operation returns to the data cycle (or data mode if no cycle is running) that was in effect at the time the calibration cycle command was received. Except in the event of a 'cycle abort' command, a cycle cannot be interrupted. Therefore, a newly commanded data mode or data cycle will not take effect until the current cycle has completed.

With the exception of the ENGI data mode, which becomes a contingency mode after instrument turn-on and initialization in space, all other modes collect data for four major frames or 64 s. The data collected during this period comprises one full spectrum subdivided into five 12.8 s spectral segments. Since the nominal spin period is 12 s and the worst case predicted spin period is 12.2 s (a 2% error), each spectral segment contains data from slightly more than one complete spin. For all data modes, count accumulation begins at the start of a major frame, and readout of a spectrum begins one major frame later.

Telemetry rate restrictions necessitate the use of several techniques to reduce the number of bits actually transmitted. One of these techniques is a compression algorithm that is used to compress the 16-bit counts and count sums to eight bits. The algorithm, which makes use of a compression table, introduces errors that are typically 1–2% and in the worst case, no larger than 3%.

5.2.1. *Solar Wind Ions (SWI)*

The SWI data mode is often called the 'Track Mode' since it uses the level with the maximum counting rate (L_{mx}) in the previous SWI or SSTI data set to adjust the E/q range of the current collection. Azimuthal angle samples are taken from each of the 16 CEMs (polar angles, θ) at 40 (8 per each of 5 spectral segments) consecutive even or odd E/q levels starting with the level $L_{mx} - 25$. In a repeating sequence of SWI modes, it is this subtraction of an odd number from the previous L_{mx} that results in the alternating even-numbered/odd-numbered starting energy level. Since the E/q spacing between adjacent levels is 2.5%, every other level corresponds to an E/q spacing of 5%. Thus it is possible to interleave adjacent data sets and provide the highest possible resolution of 2.5% (at the expense of a factor of two in temporal resolution). There are 61 ϕ-angle sectors in a spectral

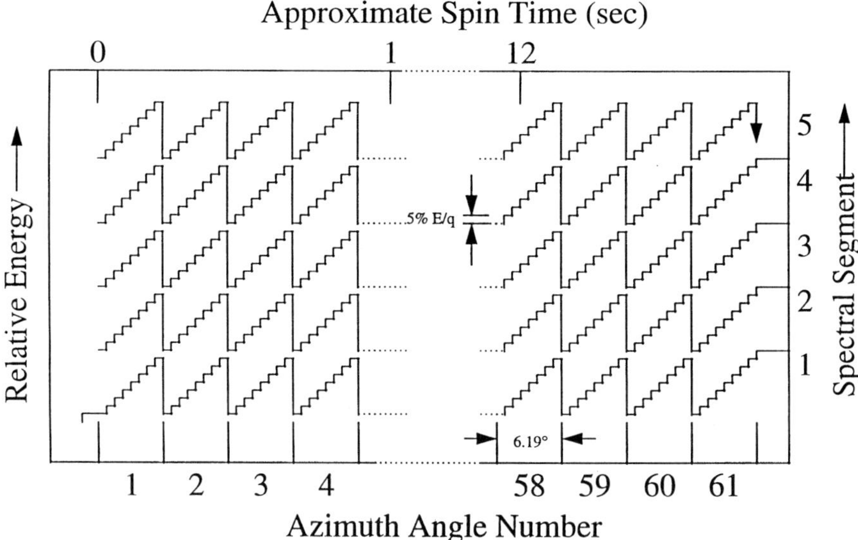

Figure 23. Analyzer plate voltage sequence for SWI (track) mode. Azimuth angle (or spin) is plotted along the *x*-axis and energy is plotted along the *y*-axis. Forty levels are stepped through (5 spectral segments × 8 steps during each ϕ-angle). This subset of 200 possible levels is controlled by continually tracking the proton peak. The stepping sequence for the SSTI (search) mode is qualitatively the same.

segment and at the nominal 5 rpm spin rate the ϕ-angle sectors at a given E/q level are spaced by 6.19° and are 0.59° wide. Since there are 58 ϕ-angles per spin at this nominal spin rate, the SWI mode intentionally over-samples. The sequence of energy level (voltage) stepping is illustrated schematically in Figure 23. Each horizontal profile corresponds to a spectral segment, which is roughly equal to one spin.

The SWI mode produces three data sets. The first (DSWI) consists of counts from CEMs 1 through 12. The second (DSTI) consists of counts from the unmasked CEMs 13 through 16. The third (DRTSW) is a subset of the DSWI data set used for real-time solar wind (see Section 5.4).

For DSWI, the telemetry rate for ACE precludes the transmission of all 732 pixels (12 θ-angles × 61 ϕ-angles) for each of 40 E/q levels, thus a set of $\theta - \phi$ angle masks have been defined that select those pixel locations which well characterize the solar wind peak and adequately cover the suprathermal wings of the distribution function. There are 12 masks, one for each θ-angle (CEM). The mask used for a given spectral segment is determined by the CEM with the peak count rate during that segment. Briefly, the masks were chosen by calculating the angular offset of each pixel look direction and discarding those pixels with an offset greater than 25° from the peak. Of the remaining pixels, 96 were selected such that the pixels were more finely spaced for smaller angular offsets. Thus the output consists of counts at each of 40 energy levels for the 96 $\theta - \phi$ angle positions which best define the solar wind beam. Although this represents only 13% of the total number of pixels, tests with simulated data sets (based on measurements by the ion plasma instrument onboard *Ulysses*) indicate that even under extreme solar wind conditions, where the ion beam is extremely broad (i.e., slow and hot), almost none of the non-zero data pixels will end up being discarded by this scheme. As noted above, the SWI mode collects more than one spin's worth of data. The instrument timing function allows for adequate ESA high voltage settling time at the start of each spectral segment and then collects 61 phi-angle sectors. The mask associated with the theta angle with the maximum counts is applied to the first 58 sectors (a command adjustable number to allow for variations in the spin rate) to preclude seeing the solar wind twice in one data set.

Samples from the outer four θ-angles, representing the suprathermal population from CEMs 13 through 16, are presented in the DSTI data set. Again, telemetry restrictions preclude the transmission of the entire array and two averaging procedures are performed onboard. First, the eight E/q levels measured in each spectral segment are summed in pairs to form four E/q levels per spectral segment. Second, the first 60 of the 61 ϕ-angles per pair of E/q levels are further summed to form 6 ϕ-angles per spectral segment. Thus the final DSTI data set contains 29 E/q levels, 4 θ-angles, and 6 ϕ-angles.

5.2.2. *Search/Suprathermal Ions (SSTI)*

During the SSTI mode, a fixed voltage stepping scheme is used with a step interval of 5 at low energies (first two spectral segments) and 4 at higher energies (last three segments). This mode allows construction of a complete suprathermal ion array and enables a robust determination of the solar wind energy peak, L_{mx}, in the event that the peak-tracking is lost during a SWI mode. The resulting energy range covered by these steps ranges from $\sim$500 eV q^{-1} to 35.2 keV q^{-1}. At low energies (segments 1 and 2), the E/q spacing is 12.5% and at higher energies (segments 3–5) the E/q spacing is 10%.

Two data sets are produced from the samples collected in the SSTI mode; DSSTI and DSTI2. These are exactly the same in format as DSWI and DSTI, respectively, although the energy range covered is broader and the energy resolution is less. Thus the DSSTI contains a complete but coarse energy spectrum from 500 eV q^{-1} to 35.2 keV q^{-1}, while DSSTI provides a spectra at about one quarter of the energy resolution of the DSWI mode, but over a much broader range of energies.

5.2.3. *Ion Calibration (CALI)*

The CALI mode allows the determination of the gain saturation level of the CEMs. The E/q level is fixed at the peak of the distribution (L_{mx} level) from the previous search or track mode. The CALI mode data collection consist of 16 CEMs $\times$ 5 CEM bias levels $\times$ 108 ϕ-angles $\times$ 2 threshold levels. The discriminator threshold is alternated between A and B in successive counting intervals of the same ϕ-angle, with the transition occurring as the counters are read. The operating level of the CEM bias supply (CEMLev) is set by command. During the CALI mode the actual CEM bias level is varied about this operating level for each of the five spectral segments in this sequence; CEMLev−2, CEMLev−1, CEMLev, CEMLev+1 and CEMLev+1.

The data set produced by the CALI mode consists of counts from all 16 CEMs. Again bandwidth restrictions require some averaging of the data. The telemetered data set is created by forming separate sums for the A levels and B levels from four consecutive ϕ-angle sectors. Thus, this data set consists of 16 CEMs × 5 CEM HV levels × 2 threshold levels × 27 ϕ-angle sums.

5.3. ELECTRON MODES AND OPERATIONS

SWEPAM-E has five commandable data modes and two commandable data cycles. The data modes are: engineering (ENGE), calibration (CALE), normal solar wind electron (NSWE), suprathermal electron angle scan (STEA), and photoelectrons (PHE). Data modes can be run individually or as part of a data cycle. The data cycles are: normal and calibration. The normal cycle is a sequence of 15 alternating NSWE/STEA modes (totaling 30 modes) concluding with a single PHE. The number of NSWE/STEA mode pairs, and therefore the duration of this cycle, is command adjustable. Like the ion, the normal cycle runs continuously once selected. The calibration cycle consists of a sequence of two CALE modes and works in exactly the same way as the ion calibration cycle.

With the exception of the ENGE data mode, all modes collect data for four major frames or 64 s. The data collected during this period comprises one full spectrum subdivided into five 12.8 s spectral segments. As with the ion modes, this data collection interval ensures that each spectral segment covers more than a full spin period and ground analysis software is used to orient the data. A 16 bit to 8 bit count compression algorithm is also implemented as described above. Because SWEPAM-I and -E are run independently, the data modes are not synchronized between the two instruments.

5.3.1. *Solar Wind Electrons (NSWE)*
The analyzer plate voltage for the electron sensor has 32 discrete voltage levels (0–31). The first eight levels (0 to 7) correspond to the photoelectron mode on *Ulysses* and are not used on ACE. The NSWE data mode utilizes the highest twenty of these levels (12 through 31), resulting in electron energy steps centered at 3.2 eV to 1377 eV. During each 12.8 s spectral segment four E/q levels are sampled, at each of 30 ϕ-angles, and for each of the seven CEMs. At the nominal 5 rpm spin rate, angle sectors at a given E/q level are spaced by 12.8° and are 3.0° wide.

The resulting DNSWE data set consists of 20 energy levels at each of 30 ϕ-angles for the innermost five CEMs (2–6) and 15 ϕ-angles for the two outermost (largest polar angle) CEMs (1 and 7). The 15 ϕ-angles of the outer two CEMs are the result of summing counts from adjacent ϕ-angle sectors after the data collection.

5.3.2. *Suprathermal Electron Angle Scan (STEA)*

The STEA data mode utilizes the highest ten voltage levels (22 through 31) corresponding to electron energies centered at 84 through 1377 eV. During each 12.8 s spectral segment two E/q levels are sampled, at each of 60 ϕ-angles, and for each of the seven CEMs. At the nominal 5 rpm spin rate, angle sectors at a given E/q level are spaced by 6.4° and are 3.0° wide.

The resulting DSTEA data set consists of ten energy levels at each of 60 ϕ-angles for the innermost five CEMs (2–6) and 30 ϕ-angles for the outermost two CEMs (1 and 7). The 30 ϕ-angles of the outer two CEMs are the result of summing counts from adjacent ϕ-angle sectors after the data collection. Comparison of the NSWE and STEA data sets shows how energy resolution and angular resolution are traded off. In the NSWE data set, 20 energy levels are measured, but only 30 ϕ-angles. In the STEA data set, ten energy levels are measured, but the resolution in ϕ is increased by a factor of two. This higher angular resolution is sometimes necessary for resolving narrow electron beams.

5.3.3. *Photoelectrons (PHE)*

This data mode is identical to the NSWE, with the exception that the analyzer plate voltages 8 through 27 are used, corresponding to electron energy steps centered at 0.86 to 429 eV. Thus the PHE mode includes four low energy steps not used in the NSWE mode. The DPHE data set is identical in format to the DNSWE data set.

5.3.4. *Electron Calibration (CALE)*

The CALE data mode performs the same function as the CALI mode performs for SWEPAM-I. In this case the E/q level is fixed on nominally 10 eV electrons. The CALE mode data collection consists of 7 CEMs × 5 CEM bias levels × 51 ϕ-angles × 2 threshold levels. The discriminator threshold is alternated between A and B in successive counting intervals (ϕ-angles). As with SWEPAM-I, the operating level of the CEM bias supply (CEMLev) is set by command but varied around this set point by the CALE data collection routine.

The DCALE data consists of counts from all seven CEMs but unlike SWEPAM-I, no averaging is required. Thus, the telemetered data set consists of 7 CEMs × 5 CEM bias levels × 2 threshold levels × 51 ϕ-angle counts.

5.4. REAL TIME SOLAR WIND

The Real Time Solar Wind (RTSW) data set is intended to provide a rough measure of the solar wind plasma conditions once every 64 s, in real time, to the international space physics community. To complement this, magnetic field vectors from the MAG instrument, as well as low-energy particle data from the EPAM instrument and high-energy particle data from the SIS instrument are also being provided in real time. By mutual agreement of the ACE project and lead investigators for these four instruments, collection and real-time distribution of these data is being performed by the Space Environment Center (SEC/NOAA) in Boulder. It is anticipated that both plots of current solar wind conditions as well as data sets covering some recent interval of time will be provided via the World Wide Web (see Zwickl et al., this issue). Together, these observations should be very useful for real-time monitoring and prediction purposes and should greatly enhance the nation's ability to predict and monitor the space weather environment near Earth.

The SWEPAM Real Time Solar Wind data set consists of a subset of the DSWI data set. These data are only provided when SWEPAM-I is operating in the SWI mode, and thus will typically provide measurements for 30 out of every 31 data sets. Of the 96 pixels telemetered as the DSWI data set in the ACE Science Format, 33 are used for the ACE RTSW format. All pixels chosen for this purpose have offset angles less then $18°$. Since this represents 34% of the total number of telemetered pixels (or 4.4% of the total number of measured pixels) any derived parameters (e.g. density, temperature, speed) must be viewed as incomplete values. While still extremely valuable for monitoring and prediction, these values will not be up to full, scientific analysis and publication purposes. Thus, we advocate the use of a prefix such as 'pseudo-' or 'real-time-' to delineate these parameters from the full density, temperature, and speed parameters. Researchers should contact the SWEPAM science team for complete and properly verified data sets for any scientific purposes.

Acknowledgements

This paper is dedicated to John Paul Glore who developed analog electronics at Los Alamos National Laboratory starting in 1946 and ending with the development of the *Ulysses* solar wind plasma experiments. After retiring, JP continued to help out on SWEPAM, which still contains his amplifier/discriminator circuits, until his death in 1996. The design, development, construction, and testing of space instrumentation can only be carried out with the combined technical expertise and dedication of a substantial team effort; we gratefully acknowledge such critical contributions to SWEPAM from Nancy Baca, Juan Baldonado, Richard Bramlett, Leroy Cope, Randy Edwards, Danny Everett, Paul Glore, Irma Gonzales, Jim Lake, Gerrard Martinez, Ignacio Medina, and Nick Olivas, all at Los Alamos National Laboratory and T. A. Ashlock, J. C. Chavez, J. W. Daniels, T. J. Ellis, K. M. Olsberg, D. O. Smallwood, J. H. Temple, and F. J. Wymer at Sandia National Laboratory. This work was carried out under the auspices of the United States Department of Energy with financial support from the NASA, Advanced Composition Explorer program.

References

Axford, W. I.: 1985, 'The Solar Wind', *Solar Phys.* **100**, 575.

Bame, S. J., Asbridge, J. R., Feldman, W. C., Fenimore, E. E., and Gosling, J. T.: 1979, 'Solar Wind Heavy Ions from Flare-Heated Coronal Plasma', *Solar Phys.* **62**, 179.

Bame, S. J., McComas, D. J., Barraclough, B. L., Phillips, J. L., Sofaly, K. J., Chavez, J. C., Goldstein, B. E., and Sakurai, R. K.: 1992, 'The *Ulysses* Solar Wind Plasma Experiment', *Astron. Astrophys. Suppl. Ser.* **92**, 237.

Barraclough, B. L., Feldman, W. C., Gosling, J. T., McComas, D. J., Phillips, J. L., and Goldstein, B. E.: 1996, in D. Winterhalter, J. T. Gosling, S. R. Habbal, W. S. Kurth, and M. Neugebauer (eds.), 'He Abundance Variations in the Solar Wind: Observations from *Ulysses*', *Solar Wind Eight*, AIP Proc. 382, New York, p. 277.

Borrini, G., Gosling, J. T., Bame, S. J., Feldman, W. C., and Wilcox, J. M.: 1981, 'Solar Wind Helium and Hydrogen Structure Near the Heliospheric Current Sheet: a Signal of Coronal Streamers at 1 AU', *J. Geophys. Res.* **86**, 4565.

Borrini, G., Gosling, J. T., Bame, S. J., and Feldman, W. C.: 1982, 'Helium Abundance Enhancements in the Solar Wind', *J. Geophys. Res.* **87**, 7370.

Dere, K. P.: 1994, 'Explosive Events, Magnetic Reconnection, and Coronal Heating', *Adv. Space Res.* **14**, 13.

Feldman, W. C. and Marsch, E.: 1997, in J. R. Jokipii, J. R. Sonett, C. P. Giampapa, and M. S. Mathews (eds.), 'Kinetic Phenomena in the Solar Wind', *Cosmic Winds and the Heliosphere*, University Arizona Press, Tucson, in press.

Feldman, W. C., Asbridge, J. R., Bame, S. J., and Gosling, J. T.: 1977, in O. R. White (ed.), 'Plasma and Magnetic Fields from the Sun', *The Solar Output and its Variation*, Colorado Associated University Press, Boulder, pp. 351.

Feldman, W. C., Asbridge, J. R., Bame, S. J., Fenimore, E. E., and Gosling, J. T.: 1981, 'The Solar Origins of Solar Wind Interstream Flows Near Equatorial Coronal Streamers', *J. Geophys. Res.* **86**, 5408.

Feldman, W. C., Phillips, J. L., Gosling, J. T., and Isenberg, P. A.: 1996, in D. Winterhalter, J. T. Gosling, S. R. Habbal, W. S. Kurth, and M. Neugebauer (eds.), 'Electron Impact Ionization Rates for Interstellar H and He Atoms Near Interplanetary Shocks: *Ulysses* Observations', *Solar Wind Eight*, AIP Proc. 382, New York, p. 622.

Feldman, W. C., Phillips, J. L., Barraclough, B. L., and Hammond, C. M.: 1996, in K. C. Tsinganos (ed.), '*Ulysses* Observations of the Solar Wind Out of the Ecliptic Plane', *Solar and Astrophysical Magnetohydrodynamic Flows*, Kluwer Academic Publishers, Dordrecht, Holland, pp. 265.

Gosling, J. T.: 1996, in D. Winterhalter, J. T. Gosling, S. R. Habbal, W. S. Kurth, and M. Neugebauer (eds.), 'Magnetic Topologies of Coronal Mass Ejection Events: Effects of 3-Dimensional Reconnection', *Solar Wind Eight*, AIP Proc. 382, New York, pp. 438.

Gosling, J. T. and McComas, D. J.: 1987, 'Field Line Draping About Fast Coronal Mass Ejecta: a Source of Strong out-of-the Ecliptic Interplanetary Magnetic Fields', *Geophys. Res. Letters* **14**, 355.

Gosling, J. T., Pizzo, V., and Bame, S. J.: 1973, 'Anomalously Low Proton Temperatures in the Solar Wind Following Interplanetary Shock Waves: Evidence for Magnetic Bottles', *J. Geophys. Res.* **78**, 2001.

Gosling, J. T., Hildner, E., Asbridge, J. R., Bame, S. J., and Feldman, W. C.: 1977, 'Noncompressive Density Enhancements in the Solar Wind', *J. Geophys. Res.* **82**, 5005.

Gosling, J. T., Asbridge, J. R., Bame, S. J., and Feldman, W. C.: 1978, 'Effects of a Long Entrance Aperture upon the Azimuthal Response of Spherical Section Electrostatic Analyzers', *Rev. Sci. Inst.* **49**, 1260.

Gosling, J. T., Baker, D. N., Bame, S. J., Feldman, W. C., and Zwickl, R. D.: 1987, 'Bidirectional Solar Wind Electron Heat Flux Events', *J. Geophys. Res.* **92**, 8519.

Gosling, J. T.: 1990, in C. T. Russell, E. R. Priest, and L.C. Lee (eds.), 'Coronal Mass Ejections and Magnetic Flux Ropes in Interplanetary Space', *Physics of Magnetic Flux Ropes*, Geophys. Monogr. 58, American Geophys. Union, pp. 343.

Gosling, J.T., Bame, S. J., Feldman, W. C., McComas, D. J., Phillips, J. L., and Goldstein, B. E.: 1993, 'Counterstreaming Suprathermal Electron Events Upstream of Corotating Shocks in the Solar Wind Beyond 2 AU: *Ulysses*', *Geophys. Res. Letters* **20**, 2335.

Gosling, J. T., McComas, D. J., Phillips, J. L., Pizzo, V., Goldstein, B. E., Forsyth, R. J., and Lepping, R. P.: 1995, 'A CME-Driven Solar Wind Disturbance Observed at Both Low and High Heliographic Latitudes', *Geophys. Res. Letters* **22**, 1753.

Hammond, C. M., Feldman, W. C., Phillips, J. L., Goldstein, B. E., and Balogh, A.: 1995, 'Solar Wind Double Ion Beams and the Heliospheric Current Sheet', *J. Geophys. Res.* **100**, 7881.

Hirshberg, J., Bame, S. J., and Robbins, D. E.: 1972, 'Solar Flares and Solar Wind Helium Enrichments: July 1965–July 1967', *Solar Phys.* **23**, 467.

Hundhausen, A. J.: 1977, in J. B. Zirker (ed.), 'An Interplanetary View of Coronal Holes', *Coronal Holes and High Speed Wind Streams*, Colorado Associated University Press, Boulder, p. 225.

Hundhausen, A. J.: 1988, in V. Pizzo, T. E. Holzer, and D.G. Sime (eds.), 'The Origin and Propagation of Coronal Mass Ejections', *Proceedings of the Sixth International Solar Wind Conference*, TN 306+Proc., NCAR, Boulder, pp. 181.

Isenberg, P. A. and Feldman, W. C.: 1995, 'Electron-Impact Ionization of Interstellar Hydrogen and Helium at Interplanetary Shocks', *Geophys. Res. Letters* **22**, 873.

Kennel, C. F., Scarf, F. L., Coroniti, F. V., Russell, C. T., Wenzel, K. P., Sanderson, T. R., Van Ness, P., Feldman, W. C., Anderson, R. R., Scudder, J. D., and Scholer, M.: 1984, 'Plasma and Energetic Particle Structure Upstream of a Quasi-Parallel Interplanetary Shock', *J. Geophys. Res.* **89**, 5419.

Klein, L. W. and Burlage, L. F.: 1982, 'Magnetic Clouds at 1 AU', *J. Geophys. Res.* **87**, 613.

Low, B. C.: 1996, in K. C. Tsinganos (ed.), 'Magnetohydrodynamic Processes in the Solar Corona: Flares, Coronal Mass Ejections and Magnetic Helicity', *Solar and Astrophysical Magnetohydrodynamic Flows*, Kluwer Academic Publishers, Dordrecht, Holland, pp. 133.

Marsden, R. G., Sanderson, T. R., Tranquille, C., Wenzel, K.-P., and Smith, E. J.: 1987, 'ISEE 3 Observations of Low-Energy Proton Bidirectional Events and Their Relation to Isolated Magnetic Structures', *J. Geophys. Res.* **92**, 11009.

McComas, D. J. and Bame, S. J.: 1984, 'Channel Multiplier Compatible Materials and Lifetime Tests', *Rev. Sci. Inst.* **55**, 463.

McComas, D. J., Gosling, J. T., Phillips, J. L., Bame, S. J., Luhmann, J. G., and Smith, E. J.: 1989, 'Electron Heat Flux Dropouts in the Solar Wind: Evidence for Interplanetary Magnetic Field Reconnection?', *J. Geophys. Res.* **94**, 6907.

McComas, D. J., Gosling, J. T., Hammond, C. M., Moldwin, M. B., and Phillips, J. L.: 1994, 'Magnetic Reconnection Ahead of a Coronal Mass Ejection', *Geophys. Res. Letters* **21**, 1751.

McComas, D. J., Balogh, A., Bame, S. J., Barraclough, B. L., Feldman, W. C., Forsyth, R., Funsten, H. O., Goldstein, B. E., Gosling, J. T., Neugebauer, M., Riley, P., and Skoug, R.: 1998, '*Ulysses'* Return to the Slow Solar Wind', *Geophys. Res. Letters* **25**, 1.

Montgomery, M. D., Asbridge, J. R., Bame, S. J., and Feldman, W. C.: 1974, 'Solar Wind Electron Temperature Depressions Following Some Interplanetary Shock Waves: Evidence for Magnetic Merging', *J. Geophys. Res.* **79**, 3103.

Parker, E. N.: 1958, 'Dynamics of the Interplanetary Gas and Magnetic Fields', *Astrophys. J.* **128**, 664.

Petschek, H. E.: 1964, 'Magnetic Field Annihilation, AAS-NASA Symposium on the Physics of Solar Flares', *NASA Spec. Publ.*, SP-50, pp. 425.

Phillips, J. L., Gosling, J. T., McComas, D. J., Bame, S. J., and Feldman, W. C.: 1992, in S. Fischer and M. Vandes (eds.), 'Magnetic Topology of Coronal Mass Ejections Based on ISEE-3 Observations of Bidirectional Electron Fluxes at 1 AU', in *Proc. First SOLTIP Symp.*, Vol. 2, Astron. Inst. Czech Academy of Science Press, Prague, pp. 165.

Phillips, J. L., Bame, S. J., Barnes, A., Barraclough, B. L., Feldman, W. C., Goldstein, B. E., Gosling, J. T., Hoogeveen, G. W., McComas, D. J., Neugebauer, M., and Suess, S. T.: 1995, *Geophys. Res. Letters* **22**, 3301.

Sheeley, N. R., Bohlin, J. D., Brueckner, G. E., Purcell, J. D., Scherrer, V. E., and Tousey, R.: 1975, 'The Reconnection of Magnetic Field Lines in the Solar Corona', *Astrophys. J.* **196**, 129.

Tsuneta, S.: 1996, in K. C. Tsinganos (ed.), 'The Dynamic Solar Corona in X-rays with *Yohkoh*', *Solar and Astrophysical Magnetohydrodynamic Flows*, Kluwer Academic Publishers, Dordrecht, Holland, pp. 85.

Von Steiger, R. and Geiss, J.: 1989, 'Supply of Fractionated Gases to the Corona', *Astron. Astrophys.* **225**, 222.

Zwickl, R., Doggett, K., Sahm, S., Barrett, W., Grubb, R., Detman, T., Raben, V., Smith, C., Riley, P., Gold, R., Mewaldt, R., and Maruyama, T.: 1998, 'The NOAA Real-Time Solar-Wind (RTSW) System Using ACE Data', *Space Sci. Rev.* **86**, 633.

THE ACE MAGNETIC FIELDS EXPERIMENT

C. W. SMITH, J. L'HEUREUX and N. F. NESS
Bartol Research Institute, University of Delaware, Newark, DE 19716, U.S.A.

M. H. ACUÑA, L. F. BURLAGA and J. SCHEIFELE
Code 695, NASA/Goddard Space Flight Center, U.S.A.

Abstract. The magnetic field experiment on ACE provides continuous measurements of the local magnetic field in the interplanetary medium. These measurements are essential in the interpretation of simultaneous ACE observations of energetic and thermal particles distributions. The experiment consists of a pair of twin, boom-mounted, triaxial fluxgate sensors which are located 165 inches (= 4.19 m) from the center of the spacecraft on opposing solar panels. The electronics and digital processing unit (DPU) is mounted on the top deck of the spacecraft. The two triaxial sensors provide a balanced, fully redundant vector instrument and permit some enhanced assessment of the spacecraft's magnetic field. The instrument provides data for Browse and high-level products with between 3 and 6 vector s^{-1} resolution for continuous coverage of the interplanetary magnetic field. Two high-resolution snapshot buffers each hold 297 s of 24 vector s^{-1} data while on-board Fast Fourier Transforms extend the continuous data to 12 Hz resolution. Real-time observations with 1-s resolution are provided continuously to the Space Environmental Center (SEC) of the National Oceanographic and Atmospheric Association (NOAA) for near-instantaneous, world-wide dissemination in service to space weather studies. As has been our team's tradition, high instrument reliability is obtained by the use of fully redundant systems and extremely conservative designs. We plan studies of the interplanetary medium in support of the fundamental goals of the ACE mission and cooperative studies with other ACE investigators using the combined ACE dataset as well as other ISTP spacecraft involved in the general program of Sun–Earth Connections.

1. Introduction

The ACE Magnetic Field Experiment (MAG) will establish the time-varying, large-scale structure of the interplanetary magnetic field (IMF) near the L_1 point as derived from continuous measurement of the local field at the spacecraft. Measurements of the IMF will permit interpretation of particle distribution functions; determination of the source location for solar wind thermal and solar energetic particles through extrapolation of the local IMF; inference of the path taken by galactic cosmic rays in traversing the heliosphere; and analysis of the local source for *in situ* acceleration mechanisms. The MAG experiment will also measure the characteristics of the IMF fluctuations over a wide range of frequencies from the multi-year variability of the solar source to frequencies an order of magnitude greater than the proton gyrofrequency so that particle-scattering and transport dynamics can be better understood. The MAG experiment will establish the temporal variability of the large-scale structure of the IMF; MAG will correlate the IMF fluctuations

Space Science Reviews **86**: 613–632, 1998.
© 1998 *Kluwer Academic Publishers. Printed in the Netherlands.*

with the large-scale variability; and MAG will serve future studies of the IMF turbulence that will in turn refine our understanding of cosmic-ray propagation in the heliosphere.

This instrument will provide continuous measurements of the local IMF with 3 to 6 vector s^{-1} resolution for use in (1) automated, nonvalidated production of Browse data to provide quick-look values in support of preliminary scientific analyses and (2) timely production of best-quality data products for refined scientific analysis. In addition, snapshot memory buffer data and onboard Fast Fourier Transform (FFT) data based on 24 vectors s^{-1} will be provided as final data products. Both Browse and final data products will be available through the ACE Science Center (Garrard et al., 1997). The measurements will be precise (0.025%), accurate (± 0.1 nT), ultra-sensitive (0.008 nT/step quantization and 0.0005 nT/Hz in the most sensitive range), and have low noise with < 0.006 nT r.m.s. for $0 - 12$ Hz. One vector s^{-1} values will be provided for real-time solar wind monitoring at a separate SEC/NOAA facility (Zwickl et al., 1998).

The instrument configuration consists of a pair of boom-mounted, twin, triaxial fluxgate magnetometers with a shared DPU mounted on the top deck of the spacecraft. Each of the two sensors is mounted at the end of separate booms which are themselves mounted on the $+$Y and $-$Y solar panels (Chiu et al., 1998). Both triaxial sensors are located 165 inches $= 4.19$ m from the center of the spacecraft with the sensor axes aligned with the axes of the spacecraft. The two sensor assemblies are identical and either can serve as the principal or sole sensor for the instrument. Deployment of the booms occurs shortly after instrument activation, approximately 1 hour after launch while the spacecraft is within the Earth's magnetosphere. The MAG instrument remains on throughout the mission, during the transit to L_1, and while the observatory is on station.

MAG is the reconditioned flight spare of the WIND Magnetic Field Investigation (MFI) (Lepping et al., 1995). Modifications were made to interface the instrument to the ACE data bus and to accomodate the reduced telemetry rate allocation provided for the instrument. The digital processing unit uses a 12-bit A/D converter that is microprocessor controlled to accurately resolve small amplitude fluctuations of the field. It also incorporates a dedicated FFT processor, developed around high-performance digital signal processor (DSP) integrated circuits, which produces a 32-channel logarithmic spectrum for each axis, synthesized from a 'raw' 256-point linear spectrum. All components of the power spectral matrices corresponding to the 32 estimates are transmitted to the ground in 80 s providing both power and phase information. Each of the two snapshot memory buffers holds 5 min of high-resolution, 24 vector s^{-1} data and requires 5440 s (90.7 min) to download into spacecraft memory. Twin snapshot memory buffers permit one to be searching for a trigger event and recording high-resolution measurements while the other is downloading a captured event to spacecraft memory. As with other instruments of this heritage, high reliability is obtained by the use of fully redundant systems and extremely conservative designs. The intrinsic zero drift of the sensors is expected to be <0.1 nT over periods of up to 6 months. Electrical 'flippers' designed to simulate a $180°$ mechanical rotation of the sensors will be used to monitor the zero-level drift associated with any possible aging of electronic components. The use of advanced statistical techniques for estimating absolute zero levels, including spacecraft fields, is also planned. The instrument features a very wide range of measurement capability, from ± 4 nT up to $\pm 65\,536$ nT per axis in eight discrete ranges. The upper range permits complete testing in the Earth's magnetic field. Automatic switching between the ranges is a capability of the instrument as is range setting by command.

In the following section we describe some of the science we plan to pursue with the MAG instrument in cooperation with the ACE science team. Following that, we describe the MAG instrument in detail and conclude by describing the various data products that the MAG instrument team will deliver to the ACE Science Center for dissemination and use by the ACE team and the space physics community.

2. Scientific Objectives of the MAG Investigation

The ACE mission is first an investigation of the composition and abundances of energetic particles in near-Earth orbit. Following this, there is a great opportunity to use the ACE observations to study the propagation and evolution of thermal and energetic particle populations so that their characteristics at their respective sources and the dynamics of their origins and transport are better understood. The particles observed by ACE will have origins that range from the sun, through interplanetary space and planetary magnetospheres, to the interstellar medium and other stars of our Galaxy. The particle populations are accelerated by a variety of source dynamics including coronal heating, wave resonances, magnetic reconnection, first- and second-order Fermi acceleration, drift acceleration and, in the case of pickup ions, photoionization and charge exchange. The particle populations evolve significantly between the source location and the measurement and are modified in the interplanetary medium by many of the same general processes: wave resonances, magnetic reconnection (in the case of magnetospheric leakage events), first- and second-order Fermi acceleration, shock processing, and particle drift. It will be the task of ACE to measure charged particle populations at 1 AU and attempt to gain a better understanding of the source location and dynamics at these remote sites.

ACE is designed to study thermal and energetic particle populations including galactic cosmic rays, the anomalous component of cosmic radiation, interstellar pickup ions, solar cosmic rays, energetic storm particles, and particles originating at or behind the Earth's and Jupiter's bow shocks. In virtually every instance, with the possible exception of pickup ions, charged particles observed at L_1 originate at remote locations such as the sun or interstellar space. (Pickup ions are formed from interstellar neutrals that are locally ionized within the interplanetary medium.) In no instance are these particles observed with their original source distribution.

All of the above distributions are processed in some manner by interaction with the solar wind as they stream from their source location (the interstellar medium, the sun, the Earth's bow shock, the Jovian magnetosphere, or the early formation region of interplanetary transients) to the point of observation. The magnetic field provides the coupling between the thermal and energetic charged particle distributions and the large-scale gradients of the solar wind. For this reason, detailed knowledge of the IMF and its fluctuations is necessary in order to understand how the interplanetary medium processes the particle distributions.

Energetic charged particles interact with the solar wind through the influence of the IMF. The magnetic field of interplanetary space is supported by the collective behavior of the charged particles that form the solar wind and interplanetary plasma. In a collisionless plasma, such as the solar wind, individual particle collisions are rare, but 'collisions' with the ambient magnetic field are continuous. When energetic charged particles encounter a large-scale (nearly global) disturbance which, either through a redirection of the IMF or through enhanced scattering by elevated turbulence levels, redirects the velocity of the energetic particles, these large-scale disturbances alter the particle distributions over a scale that is larger than the size of the disturbance. Likewise, particles interact with the ambient, undisturbed interplanetary medium in a continuous, diffusive manner due to the presence of low-level fluctuations, or turbulence. In this way, energetic particles such as cosmic rays interact with thermal particle populations via the ambient magnetic field in a self-consistent and deterministic manner.

The spatial and temporal scales that characterize the interplanetary magnetic field span the widest possible range and every scale affects the scattering, modulation and evolution of charged particle distributions. At the very largest scales, the *Parker* (1963) spiral and reversals of the spiral field associated with solar magnetic dipole reversals define the paths taken by galactic cosmic rays entering the heliosphere. Long-term solar variability (solar magnetic dipole reversals, multi-year trends in the solar activity cycle, etc.) as well as short-term (stream structures, the eruption of solar fields with different magnetic polarity, etc.) and individual transients (interplanetary shocks, coronal mass ejections (CMEs), etc.) alter these pathways and modulate the streaming of cosmic rays to varying degrees. Over the anticipated lifetime of the ACE spacecraft, the MAG instrument will record:

(1) The Parker spiral field with continuous variations due to solar wind speed.

(2) At least one solar magnetic dipole reversal in approximately the year 2003.

(3) The emergence of sunspots and regions of opposing magnetic polarity leading to a complex structure of multiple interplanetary current sheets.

(4) The passage of numerous and assorted transient structures including CMEs and interplanetary shocks.

(5) Many stream interface regions.

(6) Many heliospheric current sheet crossings.

(7) Many recurrent interaction regions which may lead to shock pairs at greater heliocentric distance.

(8) Periods of nearly radial IMF when the spacecraft is magnetically connected to the Earth's bow shock.

All of these observable features and processes define the magnetic field of the interplanetary medium, modify the particle distributions observed at L_1, and alter the evolution of the charged particle populations that are ultimately observed by the ACE spacecraft. By combining observations at L_1 with theoretical models for the evolution of the solar wind, we can extend the measurements beyond the observation point, sunward, and into the outer heliosphere.

The solar magnetic dipole and associated Parker spiral define the magnetic polarity and long-term global structure of interplanetary space, and thereby control the diffusion and drift of galactic and anomalous cosmic rays toward the inner heliosphere (Jokipii et al., 1977). Reversal of the solar magnetic dipole reverses the direction of particle drift, which leads to a 22-year cycle in the galactic cosmic-ray intensity at 1 AU. Recent observations by the *Ulysses* spacecraft of cosmic-ray intensities over the solar poles that are lower than predicted by drift theory (Simpson et al., 1995a, b) have brought into question the role of particle drifts in galactic cosmic ray propagation. New ACE observations, in coordination with a continued *Ulysses* mission, may reveal new insights into this problem so long as detailed magnetic field measurements are available at both locations.

A frequent feature of the inner heliospheric solar wind is the regular and periodic observation of recurrent interaction regions which form at the leading edge of corotating streams. Between 1 and 15 AU forward and reverse shock pairs form and corotating interaction regions (CIRs) merge to form corotating merged interaction regions (CMIRs) (Burlaga et al., 1983, 1993). Interactions among CMIRs, transient ejecta, and shocks occasionally form the global merged interaction regions (GMIRs) in the outer heliosphere that have a significant influence on the solar cycle-dependent transport of galactic cosmic rays. By making extended measurements of the IMF at 1 AU, it is possible to model the evolution of the interaction regions and compare these results with the time-dependent measurements of cosmic rays at 1 AU which originate from outside the heliosphere.

The large-scale magnetic field, as defined by whatever spatial scales are sampled by a particle of given energy over several gyroperiods, defines the average magnetic field followed by a particle as it moves through the heliosphere. Fluctuations of the IMF relative to that average are generally regarded to be sources for scattering the particle to other pitch angles or other field lines. At near-Earth orbit, the IMF is nominally oriented at 45° to the radial (Sun–Earth) direction, but extended instances of nearly radial IMF and periods of nearly azimuthal fields are not uncommon. On average, the large-scale IMF at L_1 is nearly confined to the solar equatorial plane, but individual measurements almost always reveal a significant departure from this prediction that lasts for several hours. Averages over longer timescales are more nearly zero, but persistent, small, non-zero multi-year averages are seen and remain unexplained (Smith and Bieber, 1993; Smith and

Phillips, 1997). Any attempt to trace interplanetary observations of solar cosmic rays to the source on the sun must take into account the geometry of the large-scale field between 1 AU and the sun. That geometry can only be inferred, approximately, after observing the local field and solar wind for days and attempting to reconstruct the IMF of the inner heliosphere over that time.

Additional transient phenomenon, such as CMEs and interplanetary shocks, can both alter the large-scale field and serve as a source of acceleration or scattering centers for particles streaming along the local IMF. CMEs represent additional and identifiable sources of solar wind particles with unique characteristics unlike the 'open' field lines neighboring it. Examination of the geometry of CME field lines helps to determine the source dynamics at the point where the CME originates and may shed some light on the associated process of particle energization that appears to take place in association with plasmoid ejection. The magnetic fluctuations within the CME can process the energetic and thermal particles contained within it while the closed field geometry of CMEs and magnetic clouds (Burlaga, 1995) acts as a barrier to external particles that would penetrate its surface and dilute the native particle population internal to the disturbance. In this way, CME thermal particles remain a distinct sample of the solar plasma with different parameters than the normal solar wind. Examination of this separate population, together with an improved understanding of the expansion and evolution of the magnetically contained plasma, should shed new insights into near-sun processes such as magnetic reconnection and some solar wind acceleration processes.

CME propagation through the interplanetary medium frequently produces shocks in the upstream region. Interplanetary shocks due to CMEs and other sources accelerate both thermal and suprathermal particles from the solar wind to form energetic storm particles upstream and downstream of the disturbance (Forman and Drury, 1983; Lee, 1984; Sanderson et al., 1985) and as such process the *in situ* interplanetary particles in a manner that is both charge- and species-dependent. While significant strides have been made in understanding the acceleration processes, the initial injection mechanism for thermal particles remains a source of many questions. The scale of CME and interplanetary shocks is approximately 1 AU at near-Earth orbit, but the spatial scale of the boundary of these disturbances where significant interaction with the ambient particle population occurs can be as small as a few gyroradii of a thermal ion. By themselves, transient phenomena span both the largest and the smallest spatial scales in the solar wind.

While the ambient IMF fluctuations scatter charged particle populations, the exact nature of the IMF fluctuations remains a source of many questions and controversies. There is abundant evidence that the solar wind forms a turbulent magnetofluid that evolves in a nonlinear fashion from the source to its termination at $\sim$ 100 AU (Coleman, 1968; Matthaeus and Goldstein, 1982; Roberts et al., 1987; Tu and Marsch, 1995; Zank et al., 1996). A component of that turbulence is outward-propagating Alfvén waves, which are particularly evident within the trailing regions behind high-speed streams (Belcher and Davis, 1971). The theory of charged particle scattering by Alfvén waves is now well-developed (Hasselmann and Wibberenz, 1968). However, there is growing evidence that $\sim 80\%$ of the magnetic fluctuation energy is carried by two-dimensional (2-D) turbulence that is self-organized and oriented at right angles relative to the large-scale magnetic field (Matthaeus et al., 1990; Zank and Matthaeus, 1992; Bieber et al., 1996) and there is some evidence that interaction between 2-D turbulence and thermal particles is significant (Ambrosiano et al., 1988). Preliminary analyses show that such 2-D fluctuations show very little interaction with energetic particles in the quasilinear limit (Bieber et al., 1994), but quite compelling evidence exists that even a small percentage of 2-D turbulence can lead to a significant degree of field line diffusion (Matthaeus et al., 1995; Gray et al., 1996) which, in turn, leads to energetic charged particle diffusion. Consideration of such 2-D fluctuations have brought quasilinear theory into agreement with observed cosmic-ray mean-free paths while magnetodynamic revisions of the familiar magnetostatic quasilinear theory coupled to consideration of the smallest scale fluctuations that form the dissipation range of IMF turbulence has proven successful in accounting for the observed charge-dependent scattering rates of protons and electrons (Bieber et al., 1994).

The MAG investigation will continue to study these and other interplanetary processes that have a direct effect on the evolution of thermal and energetic charged particle distributions. It is neither practical nor possible to list all such investigations, but this much is clear: Past efforts to advance the theory of cosmic-ray propagation did not always anticipate the discoveries made by researchers in the field of solar wind research working on the theory and observations of the interplanetary magnetic field. Future studies of the heliospheric plasma will undoubtedly yield new insights into the nature of solar wind turbulence and the multi-spacecraft studies combining ACE with ISTP missions will be at the heart of many of these efforts. If past efforts are any indication, it is reasonable to expect that these new discoveries will have significant effects on our understanding of charged particle propagation in the heliosphere and will alter our view of cosmic-ray physics. For this reason, the MAG investigation will continue to study the solar wind and IMF in the hope of uncovering new results of direct applicability to the fundamental mission of ACE.

3. MAG Instrument Description

The ACE MAG instrument is the reconditioned flight spare of the WIND/MFI experiment (Lepping et al., 1995). The only changes made to the unit were to accommodate the ACE data bus (the MFI I/O board was replaced) and to change the sampling rate of the instrument so that it better met the telemetry allocation for ACE (down from 44 vector s^{-1} for WIND to 24 vector s^{-1} for ACE). The WIND/MFI and ACE/MAG instruments are based on the magnetometers previously developed for the Voyager, ISPM, GIOTTO, Mars Observer, and Mars Global Surveyor missions which represent state-of-the-art instruments with unparalleled performance. Table I summarizes the principal characteristics of this instrument. The basic configuration consists of twin, wide-range (± 0.001 to $\pm 65\,536$ nT) tri-axial fluxgate magnetometers mounted on two deployable titanium booms, a 12-bit resolution A/D converter system, and a microprocessor controlled data processing and control unit (DPU). Both magnetometer sensors are deployed 165 inches ($= 4.19$ m) from the center of the spacecraft along the $\pm Y$ axes of the spacecraft.

There was no magnetics requirement for the ACE spacecraft, but an informal process of screening and design review was maintained by the Johns Hopkins Applied Physics Laboratory (APL) and individual experimenters (Chiu et al., 1998). Screening of AC and static magnetic field characteristics of spacecraft components (both instruments and subsystems) was performed by members of the MAG team in cooperation with instrument and spacecraft system engineers throughout the spacecraft assembly process. Compensation of one subsystem unit was necessary in order to correct a large static field. Spacecraft-level testing of AC magnetic field characteristics was performed by MAG team personnel at APL prior to shipment of the observatory. No significant signals were detected. A simple mapping of the spacecraft's static field was performed at the Goddard Space Flight Center (GSFC) shortly after delivery of the observatory for environmental testing using portable magnetometers while the observatory was suspended from a crane. The static magnetic field of the ACE observatory prior to launch is estimated to be <0.35 nT for each component at the deployed position of the MAG sensors. This value for the spacecraft field is too small to force a range change in the instrument and can easily be removed from the measurements by means of familiar data analysis methods that are designed to detect slowly changing instrumental offsets.

A block diagram of the MAG instrumentation is shown in Figure 1. The twin magnetometer system is supported by the fully redundant DPU which interfaces with the spacecraft data and power systems. The use of full redundancy in the MAG instrument is an important feature that emphasizes the critical nature of the magnetic field measurements for integrating the ACE observations into a consistent and cohesive interpretation of particle dynamics in the solar wind and for the achievement of the overall ACE science goals. Moreover, the fact that MAG is a fully redundant instrument, combined with the low mass, low power and low cost of the instrument, emphasizes the advanced state of development for this experiment. Electronic redundancy significantly reduces the probability of failures and takes advantage of the inherent redundancy provided by the twin magnetometer sensor configuration. This redundancy includes assigning by ground command the role of primary or secondary sensor to either sensor triad A or B. Each magnetometer system incorporates self-resetting electronic 'fuses' which isolate the common subsystems in case of catastrophic problems.

Each sensor assembly consists of an orthogonal triaxial arrangement of ring-core fluxgate sensors plus additional elements required for autonomous thermal

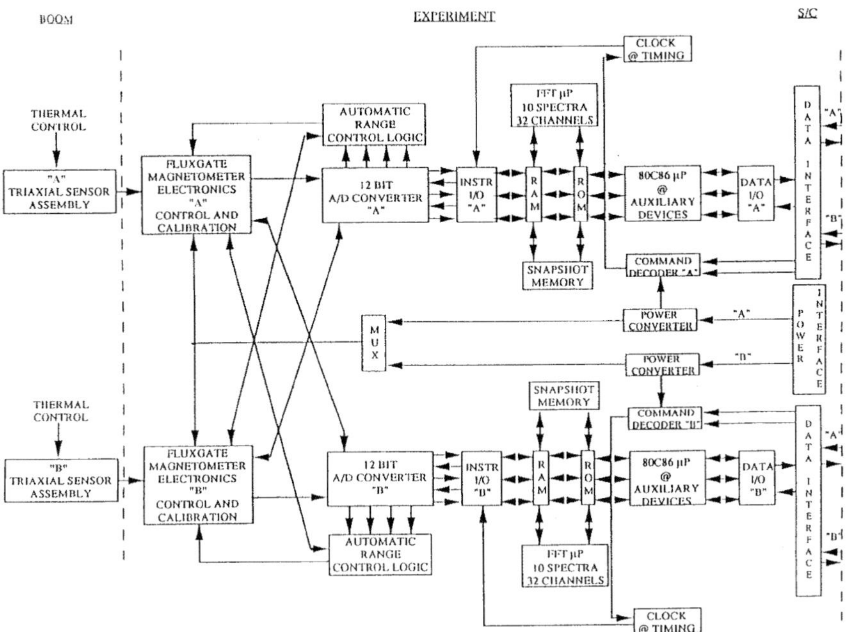

Figure 1. Block diagram of the ACE/MAG magnetic field experiment.

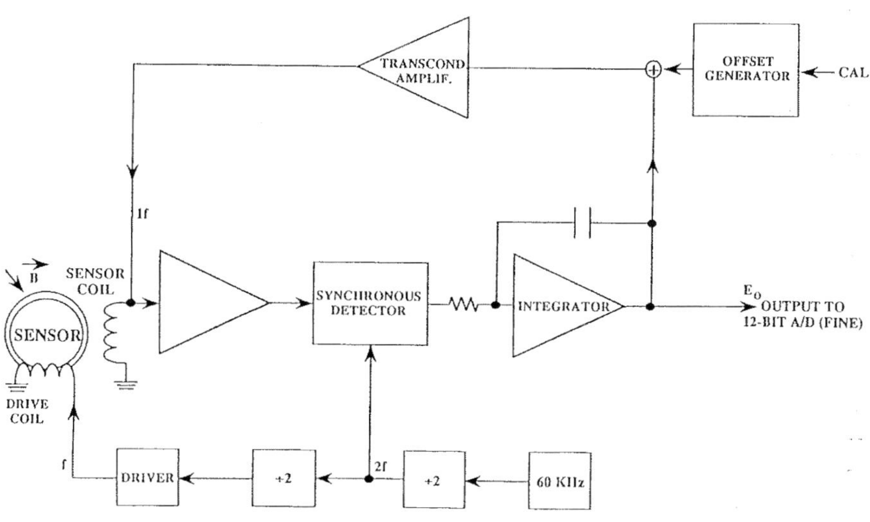

Figure 2. Schematic of standard fluxgate operation.

TABLE I

Summary of instrument characteristics

Instrument type	Twin, triaxial fluxgate magnetometers
	(boom mounted)
Dynamic ranges (8)	± 4 nT (Range 0); ± 16 nT (Range 1);
	± 64 nT (Range 2); ± 256 nT (Range 3);
	± 1024 nT (Range 4); ± 4096 nT (Range 5);
	$\pm 16\,384$ nT (Range 6); $\pm 65\,536$ nT (Range 7)
Digital Resolution (12-bit)	± 0.001 nT (Range 0); ± 0.004 nT (Range 1);
	± 0.016 nT (Range 2); ± 0.0625 nT (Range 3);
	± 0.25 nT (Range 4); ± 1.0 nT (Range 5);
	± 4.0 nT (Range 6); ± 16.0 nT (Range 7)
Bandwidth	12 Hz
Sensor noise level	<0.006 nT RMS, 0–10 Hz
Sampling rate	24 vector samples s^{-1} in snapshot memory and
	3, 4 or 6 vector samples/s continuous data stream
Signal processing	FFT Processor, 32 logarithmically spaced channels,
	0 to 12 Hz. Full spectral matrices generated every
	80 s for four time series $(B_x, B_y, B_z, |B|)$
FFT windows/filters	Full despin of spin plane components, 10% cosine
	taper, Hanning window, first difference filter
FFT dynamic range	72 dB, μ-law log-compressed, 13-bit normalized to
	7-bit with sign
Sensitivity threshold	$\sim 0.5 \times 10^{-3}$ nT Hz^{-1} in Range 0
Snapshot memory capacity	256 Kbits
Trigger modes (3)	Overall magnitude ratio, directional max.-min.
	peak-to-peak change, Spectral increase across
	frequency band (RMS)
Telemetry modes	Three, selectable by command
Mass	Sensors (2): 450 g total
	Electronics (redundant): 2100 g total
Power consumption	2.4 W, electronics – regulated 28 V $\pm$ 2%
	1.0 W, heaters – unregulated 28 V

control. The fluxgate sensors are the latest in a series developed for weak magnetic field measurements by Acuña (1974) which have been used extensively in missions like Voyager, AMPTE, GIOTTO, Mars Observer, Mars Global Surveyor, etc. due to their superior performance and low power consumption. The detailed principles of the operation of fluxgate magnetometers are well known and will not be repeated here. It is sufficient to refer the reader to Figure 2 for a simplified

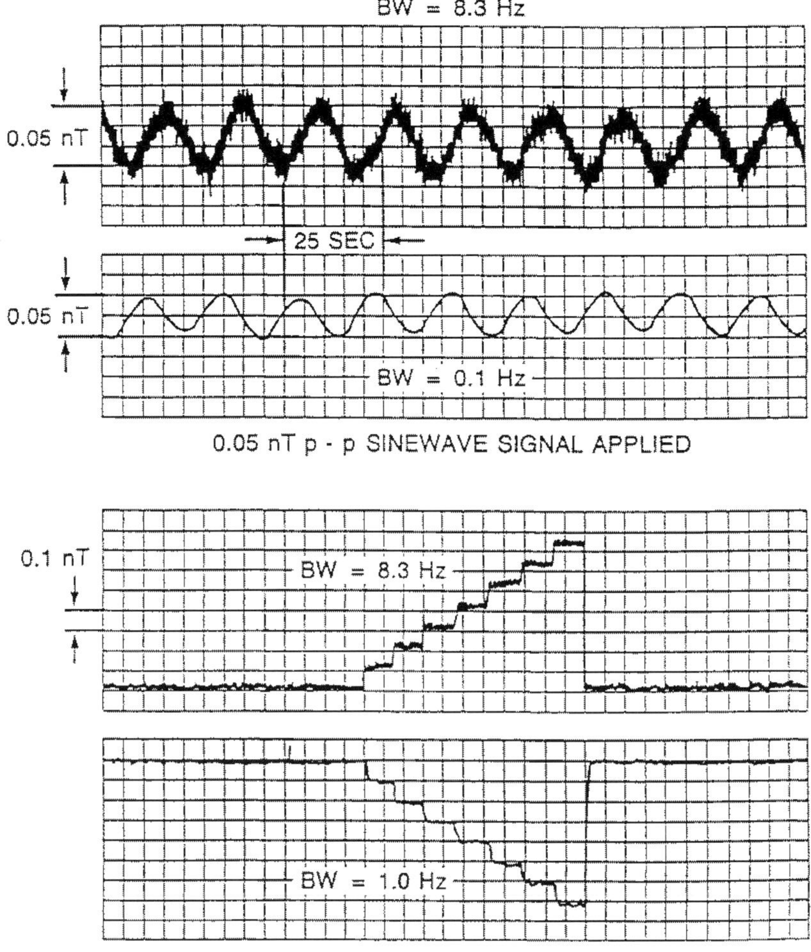

Figure 3. ACE/MAG fluxgate noise performance graphs.

description of their operation. (For additional information the reader is referred to Ness (1970), Acuña (1974), and Acuña and Ness (1976a, b).) As shown in the figure the fluxgate sensors are driven to saturation by a 15 KHz signal derived from the DPU master clock. The sensor drive signals originate from an efficient high-energy storage system which is capable of driving the ring core sensors to peak excitations that are more than 100 times the coercive saturation force of the cores. This type of excitation eliminates from consideration many 'perming' problems which have been attributed to fluxgate sensors in the past.

In the absence of an external magnetic field, the fluxgate sensors are 'balanced' and no signal appears at the output terminals. When an external field is applied, the sensor balance is disturbed and a signal containing only even harmonics of

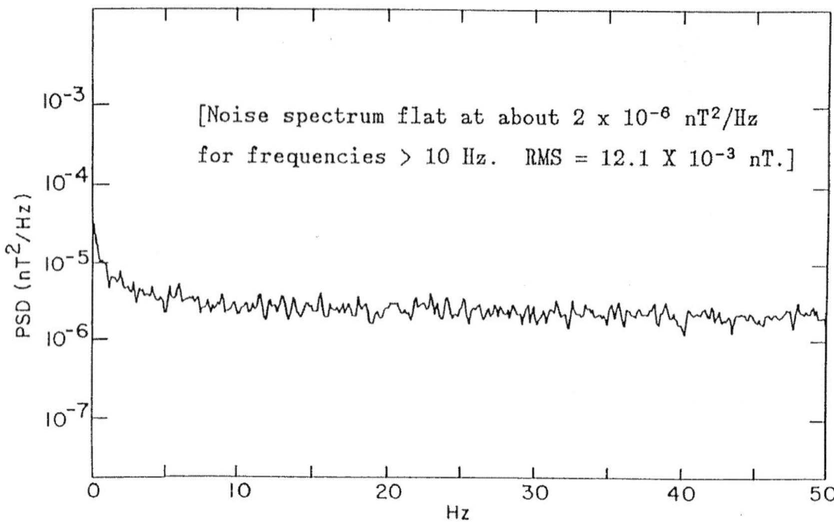

Figure 4. Noise power spectrum for MAG prototype sensor.

the drive frequency appears at the output of the sensors. After amplification and filtering, this signal is applied to a synchronous detector and high-gain integrating amplifier which is used to generate a current proportional to the magnitude of the applied field which is fed back to the sensors to null the effective magnetic field as seen by it. The output of a single sensor axis is then a voltage proportional to the magnitude and direction of the ambient magnetic field with respect to the sensor axis orientation. A triaxial magnetometer is thus created when three single axis sensors are arranged in an orthogonal configuration and three sets of signal processing electronics are used to produce three output voltages proportional to the orthogonal components of the ambient magnetic field.

The noise performance of the MAG fluxgate sensors is shown in Figures 3 and 4. Total r.m.s. noise level over the 0–10 Hz band does not exceed 0.006 nT. This noise level is several orders of magnitude below the lowest recorded levels of IMF fluctuations in this frequency range at 1 AU and is more than adequate to properly detect and identify all magnetic field phenomena of interest to ACE. Preflight tests of the spacecraft and early flight data revealed no measurable instrument- or spacecraft-associated AC signals.

The six analog signals generated by the two sensor units are digitized by the 12-bit successive approximation A/D converter. The 12-bit resolution allows the recovery of a very large dynamic range of signals spanning 72 dB. To further increase the measurement dynamic range and to accommodate simplified integration and test requirements during spacecraft testing, the dynamic range of the magnetometers can be changed automatically if the magnitude of the measured signals exceeds or drops below established digital thresholds illustrated in Figure 5. In this

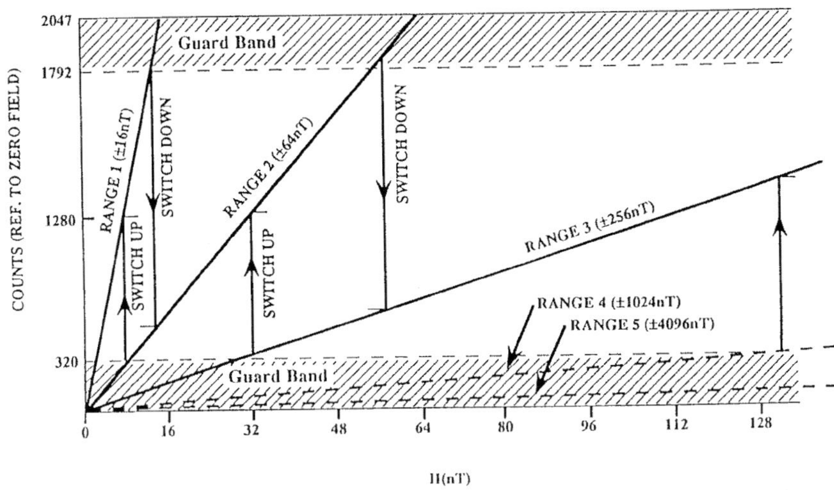

Figure 5. Range switching scheme.

fashion, the MAG instrumentation can cover eight orders of magnitude in magnetic field measurement capability, from 0.001 to 65 536 nT per axis.

The operation of the automatic ranging system is controlled by the microprocessor and allowed only at clearly defined times in the telemetry (once each half major frame) to avoid ambiguities in the interpretation of the data. When the digitized output of any magnetometer axis exceeds $\frac{7}{8}$th of full scale, the microprocessor generates a command to step up (increase) the magnetometer to the next, less sensitive, range. Conversely, when the output of all axes drops below $\frac{1}{8}$th of full scale, the DPU commands the appropriate sensor unit to step down (decrease) to the next most sensitive range. A 'guard band' of $\frac{1}{8}$th scale is provided to avoid the loss of measurements due to saturation until the range is updated. The factor of 4 in each range makes frequent flip-flops or oscillations between adjacent ranges unlikely. The decision to increase or decrease dynamic range is made using the basic internal sampling rate of 24 vectors s^{-1}, although the periods when a range change is permitted is limited. (The Primary sensor magnetic field data is contributed to the snapshot and FFT buffers at this same rate of 24 vectors s^{-1}.) However, the MAG telemetry allocation is not sufficient to allow the transmission of all the data to the ground, so on-board data averaging, compression, and decimation must be used to reduce the 'raw' data rate to an acceptable value. These operations are described in more detail later in this paper. The DPU also controls calibration sequences which provide the necessary currents to determine the scale factor of each of the Primary and Secondary magnetometer axes for various dynamic ranges as well as the determination of zero offsets associated with electronics by implementing a 180° phase reversal of the signals processed (electronic 'flipping') (Lepping et al., 1995).

4. Digital Processing Unit

The DPU, including the analog-to-digital converter, is based on the concept of a 'smart system' which performs all required operations: data manipulation and formatting, averaging, compaction and decimation, etc. The basic microprocessor used is a radiation hardened version of the popular 80C86, provided by the ISTP Project Office for development of the MFI spare, and a block diagram of the DPU architecture is shown in Figure 1. All core operations performed by the system are carried out under the control of interrupt-driven software synchronized to the telemetry system clock, subframe and frame rates. The system design is based on a default executive and processing program which is stored and executes in Read-Only-Memory (ROM). All subsequent operations are carried out from ROM and no commands or memory loads are required to obtain valid data from the instrument after initial turn-on. All default parameter values for the system are stored in tables in ROM which, once mapped into RAM during initialization, can later be modified by ground commands to update calibrations, alignments, sampling rates, zero levels, etc. Program changes are possible in-flight through memory uploads.

The execution of the executive and auxiliary programs is monitored by hardware and software watchdog timers. The external hardware watchdog timer is normally reset by proper execution of the executive program; in the absence of a reset pulse, the watchdog timer will reset the DPU and restart the default ROM program, reloading all default parameters from ROM.

In addition to the core DPU functions described above, to better enable the study of rapid changes in the ambient magnetic field, the MAG instrument includes two additional functional elements designed specifically for this purpose: (a) a pair of 256 Kbit snapshot memory buffers and (b) an onboard FFT implemented around a TI320C10 dedicated digital signal processor and associated memory. This enables the study of the physics of the fine-scale structure of shock waves; directional discontinuities; boundary structures including those of CMEs, magnetic clouds, current sheets, and CIRs; and other transient phenomena associated with the acceleration and modulation of energetic charged particle populations, as well as the dissipation range of IMF fluctuations and other various wave modes and non-coherent fluctuations occurring regularly in the solar wind. The snapshot memory can be programmed to trigger upon the occurrence of one or more of the three classes of conditions:

(1) A magnetic field magnitude jump, especially important for some kinds of shock ramp measurements.

(2) A directional change (peak to peak), useful for measurements of transition regions of directional discontinuities (tangential and rotational).

(3) Changes in the characteristics of field fluctuations over time, useful for some kinds of wave studies on the kinetic scale.

Two independently operating buffers, each with 256 Kbits of memory, provide the snapshot memory function of this instrument. When the selected trigger condition is satisfied, the contents of the snapshot memory are 'frozen' such that the triggering event is centrally located within the buffer and an equal interval of data both prior to and following the event is captured for later analysis. Under normal conditions, Primary sensor data sampled at the highest possible rate (24 vectors s^{-1}) is circulated through the 256 Kbit snapshot memory with cyclical overwriting once the memory is full. Thus a maximum of 7140 vector measurements can be stored in memory. This corresponds to approximately 297.5 s of data, or 4 min 57.5 s. The use of memory pointers in the DPU software allows the recovery of data acquired 148.75 s prior to the occurrence of the trigger (i.e., one half of the buffer). Thus, it is possible to study precursor events in high time resolution and the trigger event is located midway through the snapshot buffer record. If no trigger is observed prior to the scheduled beginning of a snapshot buffer download, the last 7140 vector measurements are sent to ground as a possible sample of high-resolution quiet-time data. The timing of the measurement is established on the ground through the use of a frame counter and comparison with the averaged data.

The FFT processor complements the snapshot memory by providing full spectral estimation capabilities in the frequency range of 0-12 Hz for Primary sensor magnetometer data. The basic FFT engine produces raw spectral estimates of the three components of the field in 256 spectral bands using 512 samples of the ambient magnetic field data (i.e., 21.3 s). In addition, the processor computes the magnitude of the magnetic field vectors in the time series being analyzed, its Fast Fourier Transform and the cross-spectral estimates associated with the three orthogonal components. In order to reduce the effects of the large signals associated with the spacecraft spin, the TI320C10 processor can be used to de-spin the spin plane components of the data prior to the computation of the FFT for these axes. Other functions included in the FFT processor are pre-whitening of input data, windowing (cosine taper and Hanning), and data compression. The latter is required to reduce the volume of raw data produced by the spectral analysis (256 spectral estimates for each element of a 3×3 spectral matrix plus a 10th spectrum, also with 256 elements, which relates to the magnitude of the IMF) to a manageable size that can be accommodated by the allocated telemetry rate for the MAG instrument. In the frequency domain, the 256 spectral estimates are compressed into 32 logarithmically spaced frequency bands of constant fractional bandwidth (or 'Q' filters). In the amplitude domain, the 12-bit data are logarithmically compressed to 7-bits plus sign using two alternate schemes: (a) a variable most-significant-bit (MSB) truncation approach and (b) an algorithm based on the μ-Law commonly used in communication systems. The net result is a set of 32 full spectral matrices for the components and 32 spectral estimates for the field magnitude, transmitted to ground using 8-bit words but representing the original dynamic range of 12-bits. Further details of the FFT processor can be found in Panetta and Acuña (1991), Panetta (1992), and Smith (1996).

TABLE II

MAG telemetry modes

Data stream	Mode 0 (BPS)	Mode 1 (BPS)	Mode 2 (BPS)
Primary[1]	108 (3 vectors s^{-1})	144 (4 vectors s^{-1})	216 (6 vectors s^{-1})
Secondary[1]	108 (3 vectors s^{-1})	72 (2 vectors s^{-1})	0 (0 vectors s^{-1})
FFT[2]	32 (4 comp s^{-1})	32 (4 comp s^{-1})	32 (4 comp s^{-1})
Snapshot[3]	48 (4 comp s^{-1})	48 (4 comp s^{-1})	48 (4 comp s^{-1})
HK and status	8	8	8
Total	304	304	304

[1] The role of primary and secondary sensor is assigned to either sensors A or B by command.

[2] FFT vector components are 8 bits. Thirty-two bits represent 4 vector components.

[3] Snapshot vector components are 12 bits. Forty-eight bits represent 4 vector components.

Finally, the DPU supports the distribution of 1 vector s^{-1} Primary sensor data to the NOAA Real-Time Solar Wind (RTSW) processor for immediate transmission to ground throughout the lifetime of the mission (Zwickl et al., 1998). All Status and Housekeeping bytes from each major frame of data are also sent to ground with the RTSW data to aid in data processing and interpretation. The MAG team is committed to support the processing of this data.

The three telemetry modes of the MAG instrument are illustrated in Table II. All three modes transmit a total of 304 bits s^{-1} of MAG data, status and housekeeping variables. All three modes transmit 216 bits of combined Primary and Secondary data totaling 6 vectors s^{-1}), but allocate the 216 bits differently between the two sensors according to the mode: 3 vectors s^{-1} for both Primary and Secondary sensor measurements in mode 0, 4 vectors s^{-1} for Primary sensor measurements and 2 vectors s^{-1} for Secondary sensor measurements in mode 1, and 6 vectors s^{-1} for Primary sensor measurements and 0 vectors s^{-1} for Secondary sensor measurements in mode 2. All three modes also transmit 32 bits s^{-1} of FFT data (4 compressed, 8-bit vectors s^{-1}) and require the same number of major frames of data to download the FFT buffer. All three modes transmit 48 bits s^{-1} of snapshot data (4 non-compressed, 12-bit vectors s^{-1}) and require the same number of major frames of data to download the snapshot buffer. All three modes transmit 8 bits s^{-1} of either status or housekeeping data and 16 s of status and housekeeping data (1 major frame of ACE telemetry) is required to fully describe the state of the MAG instrument. The only difference between the 3 telemetry modes is the allocation for Primary and Secondary sensor data.

5. Power Converter and Thermal Control

The MAG instrumentation derives power from the 28 V regulated spacecraft bus through two redundant power converters. Only one subsystem is powered at any time. The converters are high-efficiency units which operate at 50 KHz and are synchronized to the master crystal clock of the DPU to minimize interference with other experiments onboard the spacecraft. Selection of the active converter is simply accomplished by powering the desired unit.

To maintain the fluxgate sensors within their optimum operating temperature range, it is necessary to provide heater power to the boom-mounted and blanketed triaxial sensor assemblies during periods of sun occultation. (Occultation periods are not anticipated, except perhaps for a single passage through Earth's shadow on the first orbit after launch. Nevertheless, autonomous control of the MAG sensors' temperature is a feature of this magnetometer design.) Since it is extremely difficult to reduce the stray magnetic field associated with the operation of D.C. powered foil heaters to acceptable levels for MAG, a magnetic amplifier operating at 50 KHz is used to obtain automatic, proportional control of AC power supplied to the heating elements. The nominal power required to maintain the sensors at the desired temperatures in shadow is estimated to lie in the range of 0.3 to 0.5 W. No heater power should be required when the spacecraft is on station at L_1.

6. MAG Ground Data Processing

A major consideration in the design of the experiment software and operating modes is the general requirement by the ACE Project for multi-instrument, multi-user investigations. This requirement is reflected in the existence of the ACE Science Center (ASC) (Garrard et al., 1998), the generation of preliminary Browse data by the ASC in collaboration with the instrument teams, and the deposition of final, high-quality data in the ASC for dissemination to the community. In fulfilling this responsibility, Browse data will consist of 1 m, 5 m, 1 hr, and daily averages of the IMF. The MAG experiment will provide averages of the IMF over 16 s, 4 min, and 1 hr intervals for Level-2 data products, as well as graphical representations of the data.

Time ordering of the raw telemetry (real time data transmission is interleaved with recorder playback during ground contact) is performed by the Flight Operations Team (FOT) at the GSFC along with verification of Reed-Solomon coding. The resulting time-ordered and validated Level-0 data is sent to the ASC, which reformats the data, reconstitutes MAG FFT and snapshot buffer dumps which span multiple major frames, and then forwards the resulting Level-1 data along with spacecraft position and attitude files to the Bartol Research Institute (BRI) for analysis. Data transmission to BRI will be by Internet and CD-ROM.

This same Level-1 data is processed at the ASC within one week of receipt to produce the Browse datasets (Garrard et al., 1998). The ASC uses routines provided by the MAG team for this purpose, but Browse data is not validated by the MAG team prior to release. Sensor offset files at the ASC will be periodically updated by the MAG team in order to maintain the highest standard of browse data possible for an automated and unverified analysis. (Likewise, the MAG team will update offset files at NOAA for use in RTSW processing (Zwickl et al., 1998).

The Level-1 MAG data, along with useful spacecraft subsystem and attitude data, is then processed at BRI to produce the final, validated and released MAG data. Level-1 data contains MAG measurements in their raw, uncalibrated, and uncorrected form. In processing the MAG data, it is necessary to evaluate the offsets for each axis of each sensor in each range for which there is data. The data is despun and rotated to physically useful coordinate systems (heliocentric (R, T, N) and Earth-centered GSE coordinates, for instance). This constitutes Level-2 data. Visualization software written in Interactive Data Language (IDL) is then used to create plots useful in scanning the data for relevant periods of interest to the investigator. These plots include both time series and color spectrograms of the data at various time resolutions. The resulting Level-2 and above data and data products, including graphs and documentation of instrument performance, are then stored on CD-ROM and transferred to the ASC for incorporation into the ASC data structure (Garrard et al., 1998). The estimated turn around of final MAG data products is 12 weeks.

A further, Level-3 data product is planned in cooperation with the Solar Wind Electron Proton Alpha Monitor (SWEPAM) team (McComas et al., 1998) wherein magnetic field and thermal particle data will be combined in a manner that is especially useful to studies of the solar wind conditions and disturbances that alter energetic particle distributions. A delivery time for this product has not yet been set, but the production of this dataset must follow the production of the Level-2 data. Snapshot memory and onboard FFT buffer data will also be included as a Level-3 data product. The data processing philosophy used for the MAG data is similar to that used for the IMP, Voyager, Mars Observer and Mars Global Surveyor data processing and will not be reviewed in detail here.

7. Summary

The ACE Magnetic Fields Experiment is a state-of-the-art, fully redundant, triaxial fluxgate magnetometer with two matching sensors flown on opposing sides of the spacecraft. Booms mounted off the $\pm Y$ solar panels place the sensors at 165 inches (= 4.19 m) from the center of the spacecraft. The MAG instrument incorporates onboard processing of snapshot and FFT buffers to enhance the continuous 3 to 6 vectors s^{-1} measurements of the IMF by providing enhanced resolution of interplanetary transients and extension of the measured IMF fluctuation spectra to 12 Hz.

The MAG instrument will support a wide range of species-, charge- and isotope-dependent measurements that will be made by the ACE spacecraft, providing essential measurements of the *in situ* magnetic field that will permit refined analyses of the advanced particle data. MAG will also continue the long-standing tradition of studying the solar wind and interplanetary magnetic field, gaining new insights into the nature of interplanetary magnetic turbulence and serving new studies of the physics of cosmic-ray propagation and acceleration.

Acknowledgements

The MAG team would like to thank P. Panetta for his efforts in reprogramming the operating system of the WIND/MFI spare for flight on ACE. We would also like to acknowledge the consideration and efforts of the managers, scientists, engineers and technicians who participated in the unofficial and voluntary, no-cost magnetics containment program. Preflight and post-launch tests indicate that magnetic contamination was adequately limited through the diligent attention and cooperation of the many ACE personnel without the imposition of a rigorous and expensive magnetics program.

References

Acuña, M. H.: 1974, *IEEE Trans. Magnetics* **MAG-10**, 519.

Acuña, M. H. and Ness, N. F.: 1976a, in T. Gehrels (ed.), *Jupiter*, University of Arizona Press, Tucson, p. 830.

Acuña, M. H. and Ness, N. F.: 1976b, *J. Geophys. Res.* **81**, 2917.

Ambrosiano, J., Matthaeus, W. H., Goldstein, M. L., and Plante, D.: 1988, *J. Geophys. Res.* **93**, 14 383.

Belcher, J. W. and Davis, L., Jr.: 1971, *J. Geophys. Res.* **76**, 3534.

Bieber, J. W., Matthaeus, W. H., Smith, C. W., Wanner, W., Kallenrode, M.-B., and Wibberenz, G.: 1994, *Astrophys. J.* **420**, 294.

Bieber, J. W., Wanner, W., and Matthaeus, W. H.: 1996, *Solar Wind 8*, Am. Inst. of Phys., p. 355.

Burlaga, L. F.: 1995, *Interplanetary Magnetohydrodynamics*, Oxford University Press, New York.

Burlaga, J. F., Schwenn, R., and Rosenbauer, H.: 1983, *Geophys. Res. Letters* **10**, 413.

Burlaga, L. F., McDonald, F. B., Goldstein, M. L., and Lazarus, A. J.: 1993, *J. Geophys. Res.* **98**, 1.

Chiu, M. C. et al.: 1998, *Space Sci. Rev.* **86**, 257.

Coleman, P. J., Jr.: 1968, *Astrophys. J.* **153**, 371.

Forman, M. A. and Drury, L.O'C.: 1983, *18th Conf. Pap. Int. Cosmic Ray Conf.* **2**, 267.

Garrard, T. L., Davis, A. J., Hammond, J. S., and Sears, S. R.: 1998, *Space Sci. Rev.* **86**, 649.

Gray, P. C., Pontius, D. H., Jr., and Matthaeus, W. H.: 1996, *Geophys. Res. Letters* **23**, 965.

Hasselmann, K. and Wibberenz, G.: 1968, *Z. Geophys.* **34**, 353.

Jokipii, J. R., E. H. Levy, and W. B. Hubbard: 1977, *Astrophys. J.* **213**, 861.

Lee, M. A.: 1984, *Adv. Space Res.* **4**, 295.

Lepping, R. P. et al.: 1995, *Space Sci. Rev.* **71**, 207.

Matthaeus, W. H. and Goldstein, M. L.: 1982, *J. Geophys. Res.* **87**, 6011.

Matthaeus, W. H., Goldstein, M. K., and Roberts, D. A.: 1990, *J. Geophys. Res.* **95**, 20 673.

Matthaeus, W. H., Gray, P. C., Pontius, D. H., Jr., and Bieber, J. W.: 1995, *Phys. Rev. Letters* **75**, 2136.

McComas, D. J., Bame, S. J., Barker, P., Feldman, W. C., Phillips, J. L., Riley, P., and Griffee, J. W.: 1998, *Space Sci. Rev.* **86**, 563.

Ness, N. F.: 1970, *Space Sci. Rev.* **11**, 459.

Panetta, P. V. and Acuña, M. H.: 1991, *WIND MFI FFTP Processor Requirements Document*, Rev. 10.

Panetta, P. V.: 1992, *WIND MFI Flight Software Description Document*, Rev. 2.

Parker, E. N.: 1963, *Interplanetary Dynamical Processes*, Wiley-Interscience, New York.

Roberts, D. A., Goldstein, M. L., Klein, L. W., and Matthaeus, W. H.: 1987, *J. Geophys. Res.* **92**, 12 023.

Sanderson, T. R., Reinhard, R., van Nes, P., Wenzel, K. P., Smith, E. J., and Tsurutani, B. T.: 1985, *J. Geophys. Res.* **90**, 3973.

Simpson, J. A. et al.: 1995a, *Science* **268**, 1019.

Simpson, J. A., Connell, J. J., Lopate, C., McKibben, R. B., and Zhang, M.: 1995b, *Geophys. Res. Letters* **22**, 3337.

Smith, C. W.: 1996, *ACE MAG Data Formats and Algorithms*, Rev. 3 (BRI document BRI-ACE-005).

Smith, C. W. and Bieber, J. W.: 1993, *J. Geophys. Res.* **98**, 9401.

Smith, C. W. and Phillips, J. L.: 1997, *J. Geophys. Res.* **102**, 249.

Tu, C.-Y. and Marsch, E.: 1995, *MHD Structures, Waves and Turbulence in the Solar Wind*, Kluwer Academic Publishers, Dordrecht, Holland. (Reprinted from 1995: *Space Sci. Rev.* **73**, 1–2.)

Zank, G. P. and Matthaeus, W. H.: 1992, *J. Geophys. Res.* **97**, 17 189.

Zank, G. P., Matthaeus, W. H., and Smith, C. W.: 1996, *J. Geophys. Res.* **101**, 17 093.

Zwickl, R. D. et al.: 1998, *Space Sci. Rev.* **86**, 633.

THE NOAA REAL-TIME SOLAR-WIND (RTSW) SYSTEM USING ACE DATA

R. D. ZWICKL, K. A. DOGGETT, S. SAHM, W. P. BARRETT, R. N. GRUBB,
T. R. DETMAN and V. J. RABEN
Space Environment Center, NOAA, Boulder, Colorado 80303, U.S.A.

C. W. SMITH
Bartol Research Institute, University of Delaware, Newark, Delaware 19716, U.S.A.

P. RILEY
Los Alamos National Laboratory, Los Alamos, New Mexico 87545, U.S.A.

R. E. GOLD
Johns Hopkins University, Applied Physics Laboratory, Laurel, Maryland 20707, U.S.A.

R. A. MEWALDT
California Institute of Technology, Pasadena, California 91125, U.S.A.

T. MARUYAMA
Communications Research Laboratory, Tokyo 184, Japan

Abstract. The Advanced Composition Explorer (ACE) RTSW system is continuously monitoring the solar wind and produces warnings of impending major geomagnetic activity, up to one hour in advance. Warnings and alerts issued by NOAA allow those with systems sensitive to such activity to take preventative action. The RTSW system gathers solar wind and energetic particle data at high time resolution from four ACE instruments (MAG, SWEPAM, EPAM, and SIS), packs the data into a low-rate bit stream, and broadcasts the data continuously. NASA sends real-time data to NOAA each day when downloading science data. With a combination of dedicated ground stations (CRL in Japan and RAL in Great Britain), and time on existing ground tracking networks (NASA's DSN and the USAF's AFSCN), the RTSW system can receive data 24 hours per day throughout the year. The raw data are immediately sent from the ground station to the Space Environment Center in Boulder, Colorado, processed, and then delivered to its Space Weather Operations center where they are used in daily operations; the data are also delivered to the CRL Regional Warning Center at Hiraiso, Japan, to the USAF 55th Space Weather Squadron, and placed on the World Wide Web. The data are downloaded, processed and dispersed within 5 min from the time they leave ACE. The RTSW system also uses the low-energy energetic particles to warn of approaching interplanetary shocks, and to help monitor the flux of high-energy particles that can produce radiation damage in satellite systems.

1. Introduction

The late 1950s and early 1960s saw the dawn of the space exploration era, with each new mission producing vast amounts of information about our local space environment. By the late 1960s and early 1970s, scientists began to look at how

Space Science Reviews **86**: 633–648, 1998.
© 1998 *Kluwer Academic Publishers. Printed in the Netherlands.*

the pieces of the space puzzle fit together. The study by Arnoldy (1971), for example, examined the relationship between the southward component of the interplanetary magnetic field, Bz, and geomagnetic activity. The Arnoldy study integrated the southward Bz component of the field, in the Geocentric Solar Magnetospheric (GSM) coordinate system, and compared it with geomagnetic activity; a very strong positive correlation was found. Within a few years many other studies were completed and numerous suggestions made for various functional forms relating observed solar wind parameters and measures of geophysical disturbance (Crooker, 1975). Today, it is taken for granted that whenever the interplanetary field turns southward (compared to the Earth's magnetic field) geomagnetic activity will increase, and the longer it stays southward the higher the probability of increased activity. There is also general agreement that the magnitude of the field and the speed of the solar wind play a role.

By the early 1970s the scientific community accepted that the solar wind (the plasma and magnetic field) continually perturbed the Earth's magnetosphere. To make further progress in understanding the space environment near Earth, it would be necessary to fly a multi-probe mission, where one of the satellites stayed outside of the magnetosphere to monitor the solar wind conditions while others measured the environment within the magnetosphere. In 1977 the first two International Sun–Earth Explorer satellites (ISEE-1 and ISEE-2) were launched. In August 1978 the third (ISEE-3) and final satellite was launched to take up a new orbit around a point in space called L1, the place where the force of gravity of the Sun balances both the force of gravity from the Earth and the force induced from the orbital motion about the Sun. In addition to having a unique orbit, ISEE-3 also broadcast over S-Band all of the science data in real-time. Although there was no onboard storage, ISEE-3 enjoyed rather full tracking coverage with about 80 to 90% coverage during the early years.

The unique attributes of the ISEE-3 satellite led to the suggestion of establishing a real-time warning system (Tsurutani and Baker, 1979). In early 1979, scientists from NOAA, NASA and four of the science teams on ISEE-3 (Principal Investigators: E. Smith, JPL; S. Bame, LANL; F. Scarf, TRW; and K. Anderson, UC Berkeley) started working on the first Real-Time Solar Wind (RTSW) warning system. On March 21, 1980 the data started flowing to the NOAA Space Environment Laboratory. The magnetic field and solar wind plasma data had a time resolution of about 1 min and would appear ...'on continuous paper strip charts' and on a 'monitoring screen' (Joselyn et al., 1981). In October 1982 the RTSW experiment came to an end when ISEE-3 was moved from L1 into an Earth orbit, before sending it to the comet Giacobini–Zinner. Before the end of the ISEE-3 L_1 mission, however, they had shown the system worked: Through simple algorithms major geomagnetic activity was predicted 30 to 40 min before it was observed at Earth.

With the success of ISEE-3 and the need by the USAF to have similar information for their space weather operations, the Air Force Geophysics Laboratory commissioned a conceptual design study of a real-time solar wind mission for deployment on an operational satellite at L_1. The study, carried out by the Johns Hopkins University/Applied Physics Laboratory in 1985, established the concept and described the necessary components to build an entire system from satellite to ground stations for an RTSW system (Report: Conceptual Design of a Synoptic Interplanetary Monitor Platform at L1, SIMPL, AFGL-TR-85–0310).

Shortly after the study was completed, the Air Force Geophysics Laboratory started working with NASA to modify the WIND mission to have the ability to send real-time data to the Space Weather Operations Centers at USAF and NOAA. An option was built into WIND, called Solar Wind Interplanetary Monitor (SWIM) that involved sending data for two hours each day from three instruments, measuring the solar wind plasma (A. Lazarus, MIT), magnetic field (R. Lepping, GSFC) and low-energy energetic particles (R. Lin, UC Berkeley). While SWIM has limited data delivery and WIND never went into the L_1 orbit, the concept was once again established. The RTSW data was included in the daily download of the science data, stripped off and sent to GSFC for processing, and then to NOAA and the USAF 55th Space Weather Squadron.

In 1989, shortly after the selection of ACE, NOAA began working with the ACE Science Team to have a RTSW capability included in the design. In 1993 a study was commissioned by NOAA to develop the concept of processing the data on board, and then sending the RTSW data by an independent x-band transmitter on ACE. In 1994 a simpler design, using only existing resources on ACE, was accepted. This design and the resulting ground-station transfer of data, ingest, and processing are detailed in the sections below.

Starting in the 1960s and on through the 1980s, science made great progress in understanding the space environment of Earth. In parallel with this, and almost unique among disciplines in space science, it became possible to apply some of this extensive science knowledge to practical problems. Many government and commercial systems were impacted by extreme geomagnetic activity caused by large and sudden changes in the space environment. This can be summarized best by examining the growth in Space Weather Customers for the Space Environment Center, shown in Figure 1. The time line at the bottom shows the sunspot cycle starting in 1940 and continuing through the end of the century. As our use of electronic technology has grown, the number of Users of Space Weather Services has grown. Some of the more notable users include those involved in the communication industry, navigation systems, satellite systems, geologic exploration, and electric power companies.

2. ACE RTSW Satellite Components

The RTSW data stream is transmitted continuously at 434 bits per second using a low-rate engineering format, except when science data are downloaded. It is

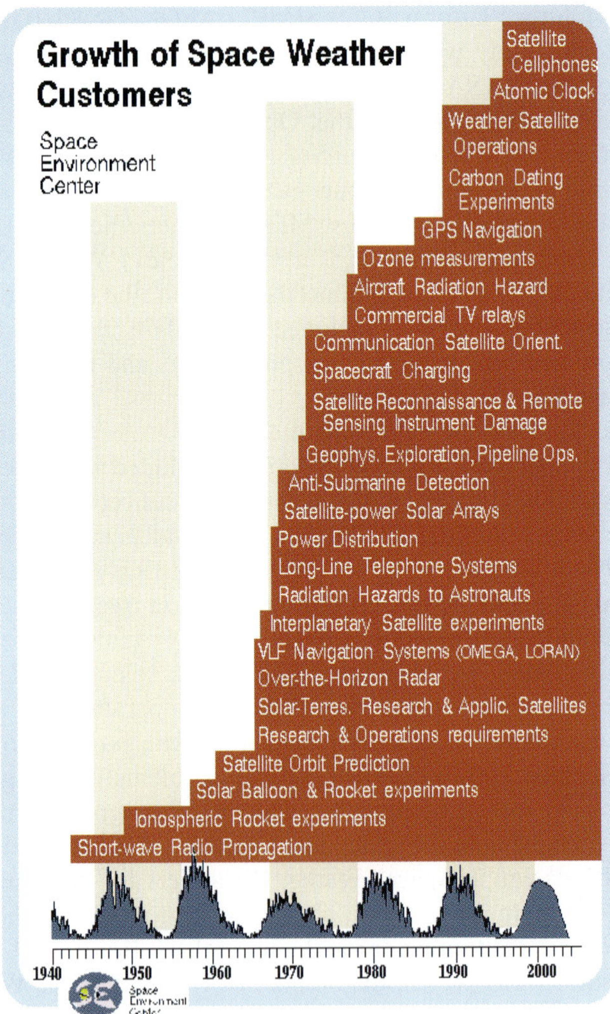

Figure 1. A partial list of customers served by the NOAA Space Environment Center; the left hand edge of each name represents the time the customer started receiving support from SEC. The SEC customer list grew as the reliance on technology increased. The monthly sunspot number since 1940 is plotted along the bottom of the figure.

composed of data from four instruments and selected housekeeping data. NASA sends real-time data at 6944 bits per second to SEC when downloading the science data each day; these data are reduced to mirror the 434 bit stream, and are processed by the same codes used to process the RTSW. Each major part is described below, as it pertains to the RTSW system.

2.1. MAG

The ACE magnetic fields experiment (MAG) is a triaxial, fluxgate magnetometer, with twin sensors and a redundant data processing unit (Smith et al., 1998). The MAG instrument provides one vector measurement each second to the RTSW data stream, along with the 8-bit status or housekeeping variable.

On the ground, the MAG data values are combined to produce a 1-min average of the interplanetary magnetic field vector. This ground analysis includes converting telemetry values to physical units using the calibration files, subtracting the computed offsets and spacecraft fields and allowing for the mounting alignments. It also includes despinning the measurement and rotating the field into heliocentric RTN (Radial Tangential Normal) and magnetospheric GSM coordinates. A statistical filter is used prior to averaging the measurements to detect and remove bad points resulting from various possible sources.

Rotation of the data to heliocentric or magnetospheric coordinates requires spacecraft pointing information that is derived from predict values, in keeping with the requirement for prompt analysis of real-time data. No corrections are made for nutation following orbit maneuvers. It is important to note the limitations on the quality of the RTSW-MAG data imposed by the need to process the data in real-time; there is a short period of time following each maneuver when the MAG data can have large errors. Each major frame (16 s) of RTSW-MAG data are analyzed independently, and without requiring availability of the preceding or following major frame. Each resulting 1-min average constitutes an approximation to the interplanetary magnetic field, derived from an automated real-time analysis. The RTSW-MAG data is a low-resolution approximation to the browse data available to the public through the ACE Science Center (Garrard et al., 1998). The RTSW-MAG data should only be used for space weather prediction and related applications.

2.2. SWEPAM ION DATA

The Ion Instrument of the Solar Wind Electron Proton Alpha Monitor (SWEPAM-I) provides data for the RTSW system only when it is configured in the Solar Wind Ion (SWI) mode, typically 97% of the time. For a complete description of the SWEPAM instrument see McComas et al. (1998). The SWI mode makes a complete measurement of the solar wind spectrum in 64 s, corresponding to 4 major frames. During this cycle the SWI mode fills a 5 by 8 by 61 by 12 element array of 16 bit scalar values. The dimensions of this array are 5 spectral segments (similar to spins, but not synchronized with the spin clock), 8 energy levels per

TABLE I

RTSW parameters

Instrument	Values calculated	Range	Units	Base time resolution	Operational time resolution
MAG	Bx, By, Bz	-200 to 200^1	nT, in GSM	1 s	1 min ave
SWEPAM	V	200 to 2000^1	km s^{-1}	64 s	1 min
	n	0 to 200	cm^{-3}	snapshot	snapshot[2]
	T	10^4 to 10^7	K		
EPAM	Electrons	38–53 keV	(s sr cm^2 MeV)$^{-1}$	32 s ave	5 min ave
		175–315 keV			
	Ions	47–65 keV			
	(Protons)	112–187 keV			
		310–580 keV			
		1.06–1.91 MeV			
		480–970 keV			
	Anisotropy	0 to 2	Dimensionless		
SIS	Protons	>10 MeV	(s sr cm^2 MeV)$^{-1}$	32 s ave	5 min ave
		>30 MeV			
Location[3]	X	0 to 300	R_e (1 = 6378 km) in GSE	1 hr	1 hr
	Y, Z	-150 to 150			

[1]Range set by RTSW coding; Instrument range larger.
[2]Data placed into nearest 1-min UT value, no averaging or interpolation.
[3]Predict value used, accurate to 0.1 R_e in GSE.

spectral segment, 61 sectors per spectral segment and 12 of the 16 data channels corresponding to the 16 channel electron multipliers. The output bandwidth precludes transmission of this entire array so a number of compression processes are used. A mask is applied to the array, based on the peak of the solar wind protons. The RTSW data set is a subset of this array, consisting of 5 Real-Time solar wind spectral segment count arrays, 8 energies, and 33 sectors, or roughly one third of the total number of transmitted pixels. Under normal conditions, however, this reduced set typically contains 70 to 90% of all the transmitted counts.

The raw data are processed on the ground to provide the solar wind speed, density and temperature every 64 s (see Table I). Since the RTSW SWEPAM-I data are an approximation to the full SWI data set, the parameters derived from them are intended primarily for space weather prediction and related operational applications. For scientific studies, data can be obtained from the SWEPAM team at Los Alamos National Laboratory.

2.3. EPAM

The Electron Proton and Alpha Monitor (EPAM) includes all of its data in the RTSW data stream. For a complete description of the EPAM instrument see Gold, et al. (1998). The data are not a subset of the original, or degraded in resolution. The only limitations imposed on this data set are those due to running in real-time, such as having the most precise attitude vector, or small adjustments in the time header. Only a very small subset of their energy channels were chosen for the RTSW system (see Table I). Briefly, the RTSW system includes two low-energy electron channels and four low-energy ion (proton) channels, spread over a wide energy range, plus one additional proton channel used to calculate an anisotropy magnitude index. They can be reprogrammed in the future if needed.

The EPAM data are collected every spin (about 12 s) but are read out only every 32 s. At SEC, the data are averaged into 5-min time intervals with the time indicating the starting time of the average interval.

2.4. SIS

The Solar Isotope Spectrometer (SIS) provides two count rates with energy ranges of 10–29 MeV and 30–80 MeV. For a complete description of the SIS instrument see Stone et al. (1998). The rates are read out every 32 s and passed directly into the RTSW data stream. At SEC the count rate data are transformed to two integral channels; the count rate from each individual channel is converted to integral fluxes and added together to produce the >10 MeV data. The >30 MeV channel is the 30–80 MeV data, converted to units of integral flux. The >10 MeV channel should be a reasonable approximation of the standard >10 MeV integral flux channel from GOES, while the higher energy channel will slightly underestimate the actual integral flux. No interpolations or extrapolations were performed to calculate either data set. After calculating the integral fluxes, the data are summed into 5-min averages.

2.5. RTSW DATA MODE

The ACE satellite was designed with three general types of downlink format: an engineering rate of 434 bps used for the RTSW; the real-time science and engineering rate of 6944 bps; and, the real-time solid-state recorder playback rate of 76 384 bps. The RTSW format is one of five engineering (434 bit) formats on ACE. During the life of the ACE satellite the RTSW format will run continuously, except during downloading of the science data once each day for about three and one-half hours, and during maneuvers or other non-routine operations of ACE. The RTSW operation is a secondary payload on ACE, and is run as long as it does not interfere with the science mission.

The RTSW mode is composed of data from the four instruments listed above, and some limited housekeeping data. As science data from each instrument passes through the data-handling electronics to be stored in the onboard solid-state recorders, the RTSW data are stripped off and sent to the RF subsystem for downlinking. The data are formed into 16 s records before sending to the transmitter for broadcasting.

3. Ground Stations

Because ACE is in a halo orbit at the L1 libration point, it appears to continuously circle the Sun as seen by ground-based observers. ACE has an orbital period of approximately 177 days with a maximum extension above and below the Sun (north/south) of 6 deg and 10 deg east/west. To acquire continuous real-time data then requires tracking stations located around the world to assure visibility of the ACE spacecraft; to a first approximation, one station is always in daylight. This condition could ideally be satisfied with three dedicated tracking stations, equally spaced in longitude and at low latitude, to provide overlapping coverage during all seasons and for all ACE spacecraft positions within its orbit. Reception of the ACE RTSW telemetry transmission requires, minimally, a 10 m diameter S band antenna. The current operational RTSW ground system shown in Figure 2 is based on the use of preexisting resources (with the exception of the CRL station) for both tracking stations and the necessary communications to SEC.

The first part of the ground-system network is the NASA Deep Space Network (DSN). The scientific mission of ACE requires daily support from the DSN for a period of about three and one-half hours per day for downloading scientific data and orbit determination and maintenance. This support has usually been provided during the normal working hours of the GSFC ACE Flight Operations Team, using the DSN stations at either Madrid (Spain) or Goldstone (California). The DSN support in UT time is shown at the bottom of Figure 3. The thin line indicates DSN could track ACE anytime during the day, while the thicker box on top of the thin line indicates a typical tracking pass. This box can slide anywhere along this line. During these passes the real-time science data are fed from the DSN through the dedicated NASCOM network to GSFC, and then through the NASA Internet (NI) to SEC (see Figure 2).

The second part of the RTSW ground system is the dedicated station at the Communication Research Laboratory (CRL) in Tokyo, Japan. Early in the project CRL agreed to set up a dedicated ground station for ACE at their facility. CRL is the parent organization in Japan for the Space Weather Operations Center located in Hiraiso, Japan, similar to NOAA SEC in Boulder. A Scientific Atlanta 11-m antenna was available to CRL from the Japanese Space Agency, and was moved to Tokyo expressly for ACE support. This antenna is dedicated to tracking ACE during daylight hours every day. The range of tracking coverage of CRL in UT hours can be seen in Figure 3. Note the large difference between summer and winter coverage. This is due to the seasonal change for a northern latitude station. Data are fed back to SEC in Boulder using a socket connection over the general Internet.

Ground Stations

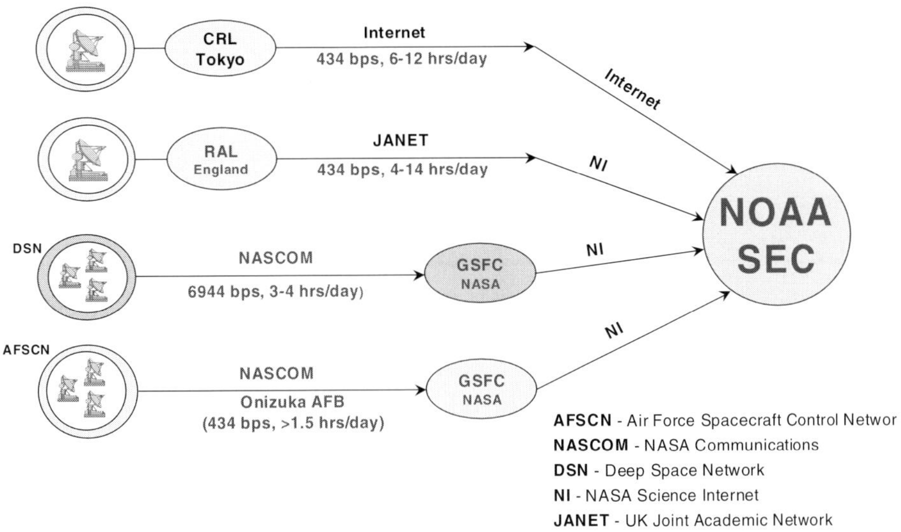

Figure 2. The ground-system configuration for the RTSW project. The stations at the Communication Research Laboratory (CRL) in Japan and the Rutherford Appleton Laboratory (RAL) in the United Kingdom are single dish antennas dedicated to tracking ACE full time, every day. The NASA Deep Space Network (DSN) and the United States Air Force Satellite Control Network (AFSCN) are multi-antenna networks that track numerous satellites. DSN will track and send real-time data during the science download period each day (about $3\frac{1}{2}$ hours). AFSCN will track ACE during times not covered by the other ground systems. The raw data received at each antenna is captured and immediately sent over NI or the Internet to SEC for processing.

The third part of the ground-system network is the United States Air Force Satellite Control Network (AFSCN). In early planning it was assumed that periods not covered by either the DSN or dedicated CRL coverage would be filled in by stations of the AFSCN. This is a worldwide network of stations, with dedicated communications that converge on Onizuka Air Force Station in California. These stations are shown just below the time line in Figure 3. The thin lines indicate when coverage is possible from each of the sites and the three letter initials indicate the station; the center initial represents local noon. The larger box indicates a typical coverage pass from one of the stations. NASA communications are utilized to bring ACE data from Onizuka to SEC via the NI during AFSCN passes (see Figure 2).

The fourth, and currently final, part of the ground-system network is at the Rutherford Appleton Laboratory (RAL) in the United Kingdom. In order to provide a more robust tracking network, NOAA and the USAF worked together to enable RAL to operate their 12-meter station in a mode dedicated to tracking ACE throughout the day, every day. The UT time coverage of RAL for summer and winter conditions is shown in Figure 3. As with CRL, there is a large difference

Tracking Station Coverage
Noon is midpoint
(Estimated Coverage)

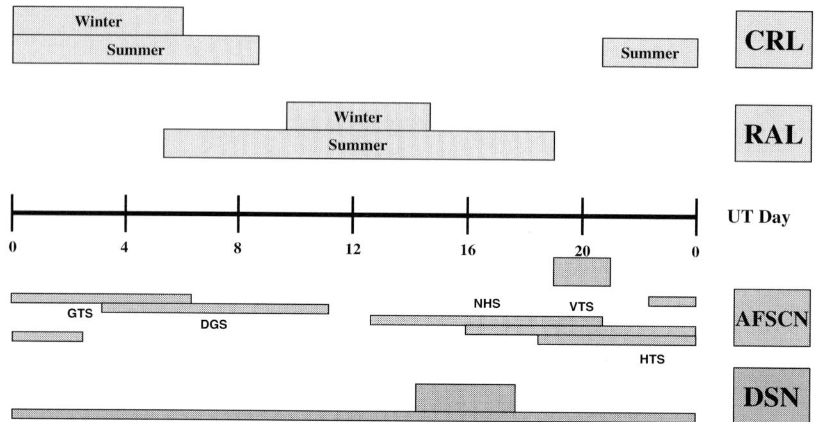

Figure 3. Tracking coverage for the four parts of the RTSW ground station system. The two horizontal bars for CRL and RAL give the winter-time minimum coverage and summer-time maximum expected coverage for the two northern latitude stations. The thin lines below the UT time line indicate possible tracking coverage, while the thicker horizontal bar gives a representative example of a tracking pass. NASA, for example, can track ACE any time, but usually will track ACE late in the UT day. The three letter initials indicate the names of the AFSCN tracking stations and the middle letter corresponds to its location at noon.

in coverage from summer to winter. Data are fed from RAL over the UK Joint Academic Network (JANET) to NI and then to the Space Environment Center, as shown in Figure 2.

The combination of the DSN coverage and the dedicated stations at CRL and RAL make it practical for the AFSCN to attempt to provide complete fill-in for 24-hour coverage. The amount of fill required varies from less than an hour at Northern Hemisphere summer solstice, to 6 hours or more at winter solstice.

4. SEC Preprocessor System

The main component of the RTSW system is the preprocessor. As shown in Figure 4, raw data flows in from the ground systems, and processed data flows out to the Data Management System (DMS) and to CRL. Details of these processes are divided into three sections: Ingest (bringing the raw data into the machine), base processing (high time-resolution processing of the raw data) and averaging of the base data.

Ground System Data Flow

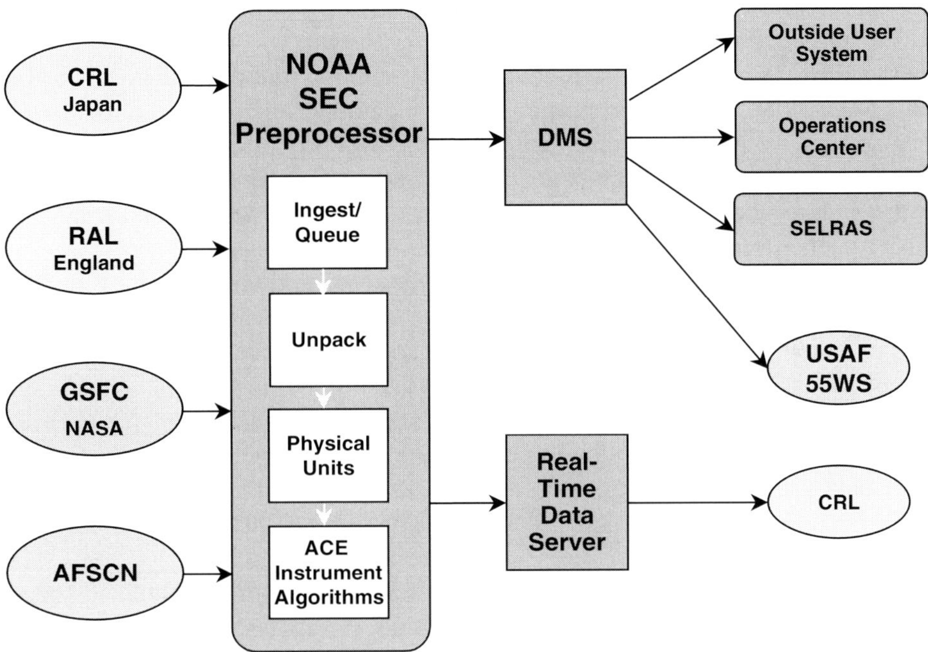

Figure 4. A diagram showing the general flow of data through the Space Environment Center pre-processor and data distribution system. The outside user system is located on the World Wide Web for easy access to the RTSW data.

4.1. INGEST

The RTSW preprocessor receives its input in real-time over the Internet from the DSN, AFSCN, CRL, and RAL ground stations(see Figures 2 and 4) using the socket software developed at Berkeley for their version of the UNIX operating system. Data from the DSN, taken during the science download, arrives in 1 s blocks of 6944 bits, while all the other ground stations receive and send RTSW data in blocks of 16 s at a rate of 434 bits per second. The 6944 bit stream contains data from all instruments on ACE, while the 434 bit stream contains only the RTSW data. The SEC RTSW preprocessor is a server that listens passively for connections from external clients, such as the DSN. Because coverage from the tracking sites may overlap at times, the RTSW preprocessor is capable of handling multiple simultaneous connections. When one of these clients connects with the SEC server, the RTSW software reads in the data, assembles it into frames, and checks these frames for validity. One of the tests, for example, is to verify that the spacecraft identifier is in its proper place in the assembled frame. If the assembled frame meets our criteria for validity, then it is passed on to the base level processing code.

4.2. BASE PROCESSING

There are five separate programs continuously running that process the ACE Real-Time Solar Wind data. These programs run in parallel; one program processes the data for each instrument independently, and the fifth program saves information on which ground station sent the data. The programs are designed to be independent so if a failure occurs in the processing of data from one instrument, there is no impact on the processing of the data from the other instruments.

Each program processes data from both the 6944 and 434 bit-per-second data streams. The programs parse the bit streams and save the pertinent telemetry bytes (in the appropriate order) that are needed to convert the satellite data into science data in their natural units (see Table I). Both general and instrument specific error checking is done for all processing, and a status word is derived for each data set. When sufficient information to perform the calculations has been acquired, the processing programs calculate the science data and write these values to the SEC Data Management System (DMS) database. The data are also written to an area of memory on the preprocessor for use in the averaging programs.

The base data for MAG is calculated when values have been obtained for 16 s of data. There can be no missing values within an eight second time period for the calculations to be computed. The values calculated are Bx, By, Bz, B-total, latitude, and longitude in both the GSE and GSM coordinate systems. They are put into the DMS database in 1 s and 16 s time increments. The 1-s data are used by the 1-min and 1-hour averaging routines.

The base data for SWEPAM-I is calculated when a complete matrix of data spanning 64 s has been collected. If any of the information needed within a 64-s time frame is not available, the data are not processed. In this case the status word is changed to indicate the type of error and a fill value is generated (-99999). The parameters calculated are density, temperature, and solar wind speed. These 64-s values are put into the DMS and are used within the preprocessor to compute the 1-min and 1-hour averages.

The base data for SIS is calculated for each of two count rates from a single 32-s value. The algorithm converts the differential flux channel data to integral flux data for protons >10 MeV and protons >30 MeV. The 32-s averaged values are written to the DMS and are used within the preprocessor to calculate the 5 min and hourly averaged data.

The base data for EPAM is calculated by taking a small subset of the available energy channels (see Table I) for each 32 s data sample; these include two electron and four ion (proton) channels. The differential flux is calculated by summing the data first by spin for each sector, then summing over the sectors to get a simple average over time. An anisotropy magnitude index, based on the cosine approximation and ranging from 0 to 2, is calculated from a fifth proton channel along with its differential flux. The 32-s averages, along with the number of points used in computing each value and a status word, are written to the DMS and used in calculating the 5-min and 1-hour averages.

A fifth program provides ground-station tracking information. This program determines and then stores the ground-station location information in a database with 16-s time resolution. Also stored is a status word indicating if a complete physical frame of data has been received. Both values are written to the DMS.

4.3. AVERAGING AND DERIVED PARAMETERS

Operational use of the ACE RTSW base data requires a set of standard time headers. At SEC the time header for averaged data represents the beginning of the average interval. The plasma and magnetic field data must be time synchronized in order to compute solar-wind-to-magnetosphere coupling functions that involve products of plasma and field parameters. For both the plasma (SWEPAM) and magnetic field (MAG) data the operational time resolution is 1 min. The choice for these two data sets was determined by the need for rapid response to changing conditions and the highest time resolution of the SWEPAM instrument (64 s). For the two energetic particle instruments, EPAM and SIS, a time resolution of 5 min best meets operational requirements. This time interval is long enough to reduce the fluctuations of the data during quiet times, and short enough to reveal the important features of the evolving time profile during large energetic particle events. The characteristics of each RTSW parameter are shown in Table I. Four nearly identical algorithms, running in parallel, process the base data into the operational 1- and 5-min averages.

An output time interval is defined, and all data falling into that interval are range-checked and spike-filtered. Data failing these tests are flagged as "bad" and excluded from the computed average. For 3 of the instruments (MAG, EPAM, and SIS) the output time intervals exactly match the output time cadence. For SWEPAM each 64-s data point is fit into the nearest 1-min time interval. The result is that periodically a single 64-s input record is represented in two consecutive 1-min outputs.

5. SEC Distribution of Processed Data

ACE RTSW data are routinely distributed to several destinations for operational purposes, including the SEC Space Weather Operations (SWO), the SEC Outside User System (OUS) for general operational user community access, the USAF 55th Space Weather Squadron (55 SWXS), and the Communications Research Laboratory (CRL) in Tokyo, Japan.

5.1. SEC SPACE WEATHER OPERATIONS

ACE RTSW is a significant new operational data stream for SWO. Real-time solar wind data from all of the RTSW instruments (MAG, SWEPAM, EPAM, and SIS) are continuously displayed in SWO, and tabulated lists of the data are readily available to SWO forecasters. In addition to real-time use, all of the RTSW data are retained in the SEC internal data store, SELRAS, for long-term operational analysis and future development projects.

SWO forecasters incorporate solar wind data into their products in a variety of ways. These data are particularly useful as subjective input to the daily synopsis of the space environment. Current solar wind conditions are routinely used to either strengthen forecaster confidence in previous forecasts, or to assist in forecast modifications. Knowing if current solar wind characteristics are normal or abnormal is crucial information to the forecast process. SWO also provides qualitative summaries of this solar wind information in several of its text products, including the daily *Report of Solar and Geophysical Activity*, the *Solar Coronal Disturbance Report*, and the weekly *Preliminary Report and Forecast of Solar Geophysical Data*.

Continuous tracking of ACE RTSW will allow SWO to use these data reliably in its forecasts and alerts. The first quantitative application of this type will be to drive a short-term warning for geomagnetic disturbances. Solar wind and magnetic field information from ACE will be used to continuously generate a predicted geomagnetic activity index, comparable to the geomagnetic *Kp index*, that is subsequently used as guidance for issuing a warning of an expected disturbance. Future applications will involve a similar process by which ACE real-time data are used as input to algorithms that generate new space weather products or continuously supply guidance for alerts and warnings.

5.2. SEC OUTSIDE USER SYSTEM

SEC provides access to ACE RTSW data to the general operational user community through its Outside User System (OUS). These data are provided for operational use and are not for research publications or citation. The OUS provides several avenues for accessing the data, including the World Wide Web (WWW) and anonymous ftp/gopher servers. These methods and access notes are listed in Table II.

The ACE RTSW WWW pages are designed for real-time operational purposes. They include real-time displays of the physical parameters measured by the RTSW instruments (see Table I) at operational time resolutions of 1 or 5 min. There are also links to data lists and to related ACE WWW sites (ACE experiment homepages, etc.).

Users that wish access to the real-time ACE data itself are best supported by the SEC anonymous ftp and gopher servers. These methods provide access to format-

TABLE II

SEC OUS access to ACE RTSW data

Access method	Comments
World Wide Web (WWW)	URL: http://www.sec.noaa.gov/ace/ACErtsw_home.html
	Gopher access through WWW: gopher://gopher.sec.noaa.gov
Anonymous ftp	Address: ftp.sec.noaa.gov
	Login: ftp (or anonymous)
	Password: your e-mail address
Gopher	Address: gopher.sec.noaa.gov

ted ASCII files that can be easily downloaded for use in operational algorithms or displays.

5.3. COMMUNICATIONS RESEARCH LABORATORY AND USAF 55TH SPACE WEATHER SQUADRON

CRL receives both base data and the operational RTSW data from SEC. The operational data are used for the daily space weather prediction at Hiraiso Solar Terrestrial Research Center (CRL/HSTRC), which is one of the Regional Warning Centers (RWCs) of the international network of space environment services. CRL incorporates ACE RTSW data into their operations in a fashion similar to SEC SWO, using the data to drive operational algorithms and as guidance to their space weather forecasters for producing products and services.

The USAF 55th Space Weather Squadron receives the RTSW operational data from SEC. They also have an operational center similar in nature to SEC SWO, and incorporate the ACE RTSW data to drive operational algorithms and as guidance to their space weather forecasters for producing products and services.

6. Future Plans for ACE RTSW System

The ACE RTSW System is up and running full time and is expected to continue operating throughout the life of the ACE mission, now believed to be at least 5 years. As experience is gained with the current system, new products will be developed and moved into operations. Through the National Space Weather Program many members of the science community will be working on products, including numerical models, that we hope to someday add to the operational system.

Acknowledgements

The authors thank R. Conde for designing the low-mode RTSW data stream, P. Mulligan for help in obtaining the USAF tracking, P. Vaughan at RAL for making the arrangements to utilize their Ground Station in the program, K. Pendergast for work on the database system, M. Forman at NASA, A. Thomas and D. Josephson at NOAA for help in supporting this project, J. Joselyn and E. Hildner for helpful input over the course of this project, and L. Puga for help in developing the figures and comments on the text.

References

Arnoldy, R. L.: 1971, 'Signature in the Interplanetary Medium for Substorms', *J. Geophys. Res.* **76**, 5189.

Crooker, N. U.: 1975, 'Solar Wind-Magnetosphere Coupling', *Rev. Geophys. Space Phys.* **13**, 955.

Garrard, T. L. et al.: 1998, 'The ACE Science Center', *Space Sci. Rev.* **86**, 649.

Gold et al.: 1998, 'Electron, Proton, and Alpha Monitor on the Advanced Composition Explorer Spacecraft', *Space Sci. Rev.* **86**, 541.

Joselyn, J. A., Hirman, J., and Heckman, G. R.: 1981, 'ISEE 3 in Real-time: An Update', *Eos* **62**, 617.

McComas et al.: 1998, 'Solar Wind Electron Proton Alpha Monitor (SWEPAM) for the Advanced Composition Explorer', *Space Sci. Rev.* **86**, 563.

Smith, C. W., Ness, N. S., Acuña, M., Burlaga, L. F., L'Heureux, J., and Scheifele, J.: 1998, 'The ACE Magnetic Fields Experiment', *Space Sci. Rev.* **86**, 613.

Stone, E. C. et al.: 1998, 'The Solar Isotope Spectrometer for the Advanced Composition Explorer', *Space Sci. Rev.* **86**, 357.

Tsurutani, B. T. and Baker, D. N.: 1979, 'Substorm Warnings: An ISEE-3 Real-Time Data System', *Eos* **60**, 702.

THE ACE SCIENCE CENTER

T. L. GARRARD[†], A. J. DAVIS, J. S. HAMMOND and S. R. SEARS
California Institute of Technology, Pasadena, CA 91125, U.S.A.

Abstract. The Advanced Composition Explorer (ACE) mission is supported by the ACE Science Center for the purposes of processing and distributing ACE data, and facilitating collaborative work on the data by instrument investigators and by the space physics community at large. The Science Center will strive to ensure that the data are properly archived and easily available. In particular, it is intended that use of a centralized science facility will guarantee appropriate use of data formatting standards, thus easing access to the data, will improve communications within and to the ACE science working team, and will reduce redundant effort in data processing. Secondary functions performed by the Science Center include acting as an interface between the scientists and the mission operations team.

1. Introduction

The Advanced Composition Explorer, ACE, will perform comprehensive studies of the elemental, isotopic, and ionic charge-state composition of energetic nuclei in interplanetary space, at energies ranging from ~ 1 keV nucl^{-1} (solar wind) to ~ 0.5 GeV nucl^{-1} (cosmic radiation), including ions accelerated in the Sun, in interplanetary space, at the edge of the heliosphere, and in the Galaxy. These measurements are being made from orbit about the L1 Lagrangian point, ~ 0.01 AU sunward of the Earth. The spacecraft was launched successfully on August 25, 1997. ACE includes six high-resolution spectrometers and three monitoring instruments that characterize the environment in which a given composition measurement is made. Many of the instruments take advantage of the spacecraft's spin to scan for particle arrival direction distributions. The mission, the spacecraft, and each of the nine instruments are described in detail in a series of companion papers (Stone et al., 1998a–c; Chiu et al., 1998; Gold et al., 1998; McComas et al., 1998; Smith et al., 1998; Gloeckler et al., 1998; Mason et al., 1998; Möbius et al., 1998).

The following sections describe the flow of the data from the spacecraft to the end users, the processing and the contents of the data, the standard interchange formats used to store and transmit the data, and other data processing tools. The emphasis is on the role of the ACE Science Center (referred to hereafter as 'the Science Center') in coordinating the data flow and formats.

[†]Deceased.

 *Space Science Reviews* **86**: 649–663, 1998.

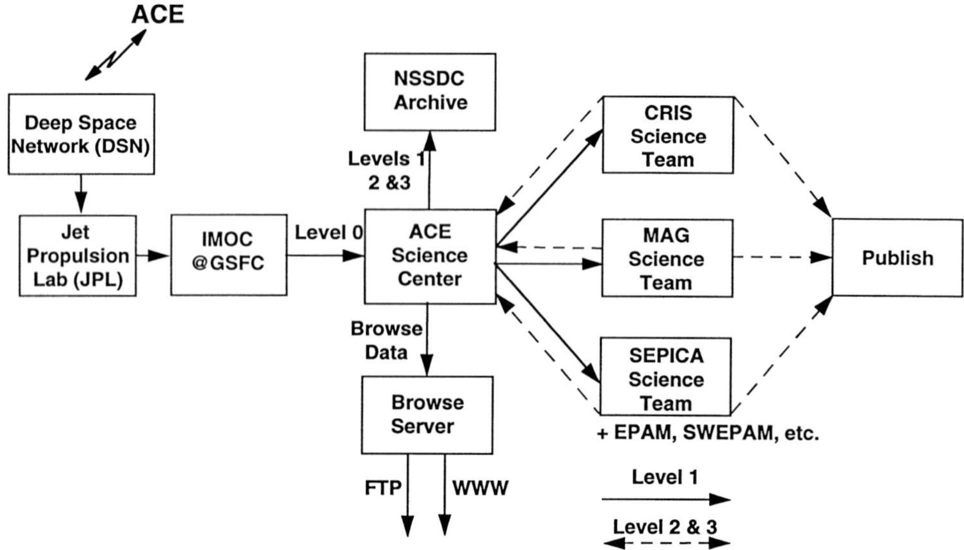

Figure 1. Flow of ACE data from the spacecraft to the scientific community.

2. The Data Flow and Processing

The data flow from the instruments to the scientific community involves spacecraft hardware, a number of NASA institutional facilities, and ACE facilities including the Science Center. It is illustrated schematically in Figure 1 and described below.

2.1. DATA TELEMETRY AND LEVEL-ZERO PROCESSING

The ACE spacecraft Command and Data Handling (C&DH) system gathers data from the instruments and formats the data into minor and major frames. One minor frame (996 bytes) is read into the C&DH system each second and there are 16 minor frames per major frame. Section 3.1 describes the data read out from each of the nine instruments. The C&DH system also gathers data from various analog sensors and digital telltales, from the sun sensors and star sensor, and from the command system, etc. Most of the time the spacecraft is not in touch with the ground facilities and these data are stored in an onboard Solid State Data Recorder (SSDR). Typically one contact per day is initiated by ground facilities and lasts roughly two to four hours. The SSDR is large enough to allow contacts to be spaced by more than 50 hours when necessary. The SSDR contents are read out to the ground at a rate exceeding 10 minor frames per second while current data are being simultaneously telemetered to the ground and stored in the SSDR for the next contact. The telemetry is formatted into two virtual channels (CCSDS '89) (real-time and playback) and received by the Caltech Jet Propulsion Laboratory Deep Space Network (DSN). The telemetry is then forwarded via the Internet to the ACE Integrated Mission Operations Center (IMOC) at the Goddard Space

Flight Center (GSFC). There the data are reviewed in near real time for purposes of monitoring spacecraft and instrument status. The data then undergo level-zero processing (per NASA's standard terminology) as soon as all the data contained within the current 24-hour time frame have been received. In level-zero processing, duplicate data are removed from the data stream, data are time ordered, and data quality and accounting summaries are appended. The data are formatted into a 24-hour Science Routine Data Set File, and forwarded via the Internet to the Science Center, accompanied by a Standard Formatted Data Unit (CCSDS '92) header file.

2.2. LEVEL ONE AND BROWSE PROCESSING

At the Science Center, the data undergo level one processing, usually within a few days of receipt. In level-one processing, the data are separated out by instrument and each instrument data set is formatted (using the NCSA HDF standard, see Sections 4.1 and 4.2) in a fashion which is both consistent with the other instruments and customized to meet the special requirements of that data set and team. At this point in the processing, i.e., in level one, the data are supplemented with ancillary data including position, attitude, and spin phase of the spacecraft; command history and comments; calibration of the spacecraft clock; and documentation of the data items. Excepting the documentation, these ancillary data are all received by the Science Center from the IMOC. The level-one data are archived at the Science Center, which is a Cosmic and Heliospheric discipline node of NASA's Space Physics Data System, and a copy is transmitted to the National Space Science Data Center (the NSSDC) for long-term archiving. Each instrument team receives a copy of all the level-one data, including, of course, that from their own instrument.

In addition to formatting, level-one processing includes those data-processing steps which are judged to be of sufficient simplicity that they can be understood, defined, and coded before launch, and do not require iterated processing with increasing experience. Examples of such steps include decompression of compressed rate scaler data and proper time labeling of data which are buffered for a number of minor frames within the instrument before readout. A counter example (a process which clearly does not belong in level one) is application of calibration data to convert digital pulse heights from detector signals to engineering units. Experience indicates that calibrations are often adjusted repeatedly to improve resolution based on extended iterative study of the instrument response.

In parallel with the level-one processing, the level-zero data is processed to yield browse parameters. Browse parameters are a subset of ACE measurements which allow monitoring of the solar wind and large-scale particle and magnetic-field behavior. They also allow the selection of time intervals of particular interest for more intensive study. Since it is considered important to distribute first-order ACE results as soon as possible, the browse parameters are delivered to the public domain immediately, at the expense of full verification. A description of the browse parameters and the forms in which they are made available to the public is provided in Section 3.2.

2.3. LEVEL TWO AND HIGHER LEVEL DATA PROCESSING

Data processing beyond level one is the responsibility of the individual instrument teams. Level-two processing includes such operations as application of calibration data and detector response maps, organization of data into appropriate energy and time bins, and application of ancillary data (for example, conversion of magnetic field vectors to useful coordinate systems using the spacecraft attitude data). The Science Center attempts to facilitate these efforts within its resources, especially when high-level processing involves multiple instrument teams. For example, much of the anisotropy/flow data for the particle instruments, in particular for the Electron, Proton, and Alpha-Particle Monitor (EPAM), will be computed in terms of the direction of the magnetic field. Thus the EPAM team will need high-level results from the MAG team to do high-level EPAM analysis. The Science Center can facilitate data sharing and communications with its substantial data storage capabilities and its data-formatting experience. Another example is the high level processing for the Cosmic Ray Isotope Spectrometer, CRIS. Four institutions are involved in this processing, each contributing expertise and experience in a different subassembly of this very complex instrument. Communications and iteration of the data processing are being facilitated by the Science Center for this team.

Each instrument team is required to deliver level-two data back to the Science Center, which will then make the data available to the other instrument teams, the space science community (as required by NASA), and the NSSDC for long-term archiving. Delivery of level-two data back to the Science Center is expected to begin about three months after the spacecraft enters orbit about the L1 Lagrangian point. Thereafter, roughly a two-month lag time is expected between receipt of level-one data by the instrument teams and delivery of level-two data back to the Science Center. However, these delivery schedules may require revision if instrument checkout and debugging take longer than expected. In addition, the level-two dataset is expected to be evolutionary, in the sense that an instrument team may enhance their level-two data with additional products in the future, as the sophistication of their analysis increases.

Data processing beyond level two consists of publication or presentation-quality items, such as data plots and graphics, and the contents of talks and journal articles. These items will also be archived at the Science Center and the NSSDC.

2.4. REAL-TIME SOLAR WIND DATA

A parallel data-flow scheme is mentioned here for convenient reference, although the Science Center plays a very minor role in this parallel flow. In addition to the normal telemetry, a small, selected subset of the data is being telemetered in real time from the spacecraft to ground stations operated for the Space Environment

Center of NOAA, the National Oceanic and Atmospheric Administration (Zwickl et al., 1998). These data are being made available by NOAA in near real time for purposes of monitoring interplanetary space weather and predicting geomagnetic activity. (It takes ~ 1 hour for the solar wind and embedded magnetic field observed at ACE to propagate to the Earth, while the raw telemetry reaches NOAA in seconds.) These space-weather data products may, like the browse parameters, also be considered useful by many people in the space science community. They are also available through the Science Center, but not in near real time.

3. The Contents of the Data

The 'raw' data, as telemetered from the spacecraft, are the ultimate description of the instruments and all higher level data products. They are described here, and, in more detail, in the instrument papers. The browse data are described here because they are expected to be the most popular product of the Science Center for the larger space science community. Level-one data are not described; they contain little beyond the raw data and are not likely to be of use outside the ACE team.

3.1. THE RAW DATA

As mentioned above, each of the nine instruments is described in detail in a companion paper. Presented here is a uniform view of the data so that they may be compared. This overview is primarily given in terms of types of data and time resolution. For an overview of of the elemental, isotope and energy ranges covered by ACE, see Stone et al. (1998a).

Among the particle detectors there is a great deal of commonality in the raw data, although the analysis of the data from the solar wind instruments (SWEPAM, SWICS, SWIMS) frequently differs substantially from the analysis of the other particle instruments. The magnetometer (MAG) data are, of course, rather different from the data from the eight particle-detecting instruments. In order to maximize and take advantage of the commonality of science and data processing, the instrument data can be organized in terms of the following four data types:

Housekeeping and status data. These data include the digitized readouts of analog parameters such as temperature, voltage, and current and the digital indicators of parameters such as command state, subsystem power on or off, etc. Some of these parameters are monitored by the instruments and included in their data output to the spacecraft; others are monitored by the spacecraft and added to the telemetry by the C&DH system. Since they describe the instrument or spacecraft rather than the physical phenomena measured by the instruments, they are generally of interest only to the instrument team and are not detailed here.

Rate data.   These data specify a count of the number of times a particular logical condition in the instrument electronics was satisfied during a particular time interval, usually the interval since that counter was last read out. The use of the word 'rate' implies that the counter readout will eventually be normalized to the time interval. ACE rates can be subdivided into three major categories – singles, coincidence, and matrix rates – as detailed below.

Any of these three rate types can be sectored or multiplexed. Sectored rates are counted according to the phase of the spin of the spacecraft, i.e., the pointing direction of the telescope. Multiplexing is used to share valuable telemetry resources for several rates at the cost of less time resolution or less than 100% coverage. Multiplexing is very common for singles rates, but is also used for some coincidence and matrix rates on ACE.

Singles rates.   These rates typically specify a count of particle-detection events as seen in a single individual detector, as opposed to a rate of some logical coincidence of several detectors within a telescope or instrument. These rates are generally intended primarily for monitoring the health of a detector and are frequently multiplexed (sub-commutated) to avoid using too much telemetry. They usually reflect the particle environment (when the detector is healthy) and are of some general interest.

Coincidence rates.   These rates typically specify a count of particle-detection events as identified by some combination of detectors and are less subject to background due to detector noise. They also generally respond to a better defined range of particle charges and energies.

Matrix rates.   These rates are counts of events identified by both a combination of detectors triggered and the signal sizes (pulse heights) in those detectors. The use of pulse-height information allows these rates to be even more specifically identified with particular particle species and energies.

Pulse-height events.   These are telemetry items containing pulse-height information describing one particular ion as observed in one or more (frequently three or more) detectors. All ACE instruments observe more events than can be telemetered; thus the instruments employ priority systems to select the most interesting events for telemetry and it is therefore necessary to use rate information to calculate the flux of ions from the pulse-height event data.

Other

The MAG instrument's measurements of magnetic field can be thought of as similar to pulse-height events for level-zero processing, but processing at higher levels is very different for the two types of data. In addition, MAG occasionally mea-

TABLE I

ACE data summary

Instrument	Matrix rates	Coincidence rates	Singles rates	Event types
CRIS		78	32	64
EPAM	$12s_8$	$15s_8+19s_4$	$2s_4$	1
MAG				6 vectors s^{-1}
SEPICA	$36s_8+49$		$3s_8+3$	14
SIS		118	24	96
SWEPAM			23	
SWICS	$27s_8$	2	4	3
SWIMS	$3s_8$	1	7	3
ULEIS	$76s_8$	$3s_8$	$13s_8$	5

The table entries for rates specify the number of rates telemetered; for events, the number of kinds of events as determined by onboard priority buffers. The s_N after some rate numbers indicate that particular rate is sectored into N sectors. For example, $12s_8$ is a rate consisting of 12 individual items with 8 sectors each (a total of 96 values).

sures and telemeters power spectra (Fourier transforms) of the magnetic field as a function of time for very short time intervals.

In Table I we report numbers of rate readouts and numbers of types of events for the various instruments. The table contents are explained briefly below.

Using the terminology described above, CRIS has 64 coincidence rates which are tied to the 64 CRIS event priority buffers, and 14 coincidence rates which are not, for a total of 78 coincidence rates. CRIS also has 32 singles rates. Similarly, SIS has 96 coincidence rates which are tied to the 96 SIS event priority buffers, and 20 coincidence rates which are not, for a total of 116 coincidence rates. SIS also has 2 programmable coincidence rates and 24 singles rates. The CRIS and SIS event priority buffers are defined in Stone et al. (1998b, c). CRIS and SIS rates are not sectored, i.e., no spacecraft spin-phase information is recorded.

EPAM has sectoring information for all rates. Matrix rates select particular ions and energies and are subdivided into 8 sectors per spacecraft spin period. Some coincidence rates are sectored by 4, others by 8. The coincidence rates include separate rates of ions, and electrons at various energies from multiple telescopes directed at various angles from the spacecraft spin axis. The singles rates are multiplexed. EPAM pulse-height events are prioritized using 8 of the matrix rates. They are sectored by 8, with 2 events being reported per sector.

SEPICA reports 16 coarse and 20 fine mesh matrix rates, each with 8 sectors, and 49 unsectored fine mesh matrix rates. The coarse rates normalize the event selection in the priority system. The fine mesh matrix rates furnish more detail about the ion species and energy. SEPICA pulse-height events are sectored by 4, with up to 33 events being reported per sector, prioritized by 14 of the coarse matrix rates.

All SWICS and SWIMS matrix rates are sectored into 8 bins. Each of the two instruments reports 3 basic matrix rates, which normalize the PHA event selection. SWICS also reports 24 fine mesh matrix rates.

SWEPAM has 16 ion rates and 7 electron rates which are (technically) singles rates. These rates are read out frequently as SWEPAM scans the voltage (which corresponds to particle energy per charge) and the azimuthal space (due to spacecraft spin). This parametric information is analyzed on the ground to yield a science result which looks like sectored matrix rates (and then analyzed further to yield solar wind velocity, density, etc). SWEPAM telemeters no events.

ULEIS has 76 matrix rates of ions of various species and energies, each of which are sectored by 8. Six pulse-height events are reported per sector, for a total of 48 events per spacecraft spin period. Five onboard event priority buffers determine the events selected for telemetry.

MAG magnetic field vectors are crudely analogous to particle detector PHA events. The instrument reports a continuous data stream of 6 vectors per second. There are no rate equivalents.

3.2. THE BROWSE PARAMETERS

Browse parameters are a subset of measurements by the ACE instruments which are created at the Science Center during level-one processing. They are delivered to the public domain as soon as possible. Their purpose is to allow monitoring of the solar wind and large-scale particle and magnetic field behavior, and selection of interesting time periods for more intensive study. Interesting time periods might include solar energetic particle events or the passage of an interplanetary shock. An additional use of the browse parameters is to investigate relationships between the data from the various ACE instruments, and between ACE data and data from other sources.

The browse parameters include unsectored fluxes of ions at many different energies and electrons at a few energies. They also include the interplanetary magnetic field and solar wind parameters such as proton speed and temperature. They therefore furnish a very abbreviated description of what is being observed by the ACE instruments, without the relatively high cost of storing and analyzing all the level-one data. Eventually they may be supplemented with event data from the particle detectors, but experience with the flight data is a prerequisite for delivering useful products of that type.

Because the browse parameters are intended to be delivered to the public domain within a few days of receipt of the raw data from the spacecraft, they are not subjected to any prior scrutiny by the science teams. Their production is automatic, and the data are not routinely checked for accuracy before release. Therefore the browse parameters are not suitable for serious scientific work and should not be cited without first consulting the appropriate ACE instrument team. However, the algorithms used to create the browse parameters are subject to revision, and their reliability is expected to improve with time. The browse parameters will probably be the most popular Science Center product for the larger community outside the instrument teams, particularly during the early stages of the mission, so early delivery is considered more important than full verification.

The best time resolution for the browse parameters is generally limited by data collection cycles in the instruments. CRIS and SIS have separate 256-s cycles and SWICS has a 12-min cycle. EPAM, ULEIS and SEPICA have separate 128-s cycles, each cycle containing data for 10 consecutive spacecraft spins. SWEPAM has a 64-s cycle and MAG browse data is reported with 16-s time resolution. The SWIMS instrument does not contribute to the browse parameters.

In addition to the cycle/averaging periods noted above, all the browse parameters are averaged to common one-hour and one-day periods, and the data from EPAM, MAG, SEPICA, SWEPAM and ULEIS are also averaged to a common 5-min period. These common periods are in time phase with UTC clock, i.e., at integral 5-min, hour, and day values.

The charged particle fluxes in the browse data include H, He, C, O, Mg+Si, Ne–Fe, and iron-group fluxes in various energy bands. The current list is shown in Table II. This list may be augmented in the future, and the energy bands may be revised by the instrument teams as the data analysis proceeds.

The solar wind parameters include the proton speed, proton density, radial component of the proton temperature tensor, and the He^{++}/proton ratio, all from SWEPAM, and the following parameters from SWICS: He speed, He and oxygen thermal speed, coronal temperature, and the He/O and Fe/O density ratios. The interplanetary magnetic field vector and magnitude from MAG are reported in both RTN and GSE coordinate systems. It should be noted that the attitude, position and velocity of the ACE spacecraft are also made available to the public by the Science Center, in various coordinate systems.

A selection of the browse parameter data is shown in Figure 2, for a period of high solar activity in November 1997.

4. Science Techniques

4.1. DATA INTERCHANGE STANDARDS

The use of data interchange standards is an important tool in making data freely available to the ACE team or to the space science community. Some standards are imposed by NASA regulations; in other cases a choice from a plethora of possible standards had to be made by the team. Different standards are optimal for different levels of processing of the data, but we have striven to compromise between using a minimal number of standards and supporting a heterogeneous community.

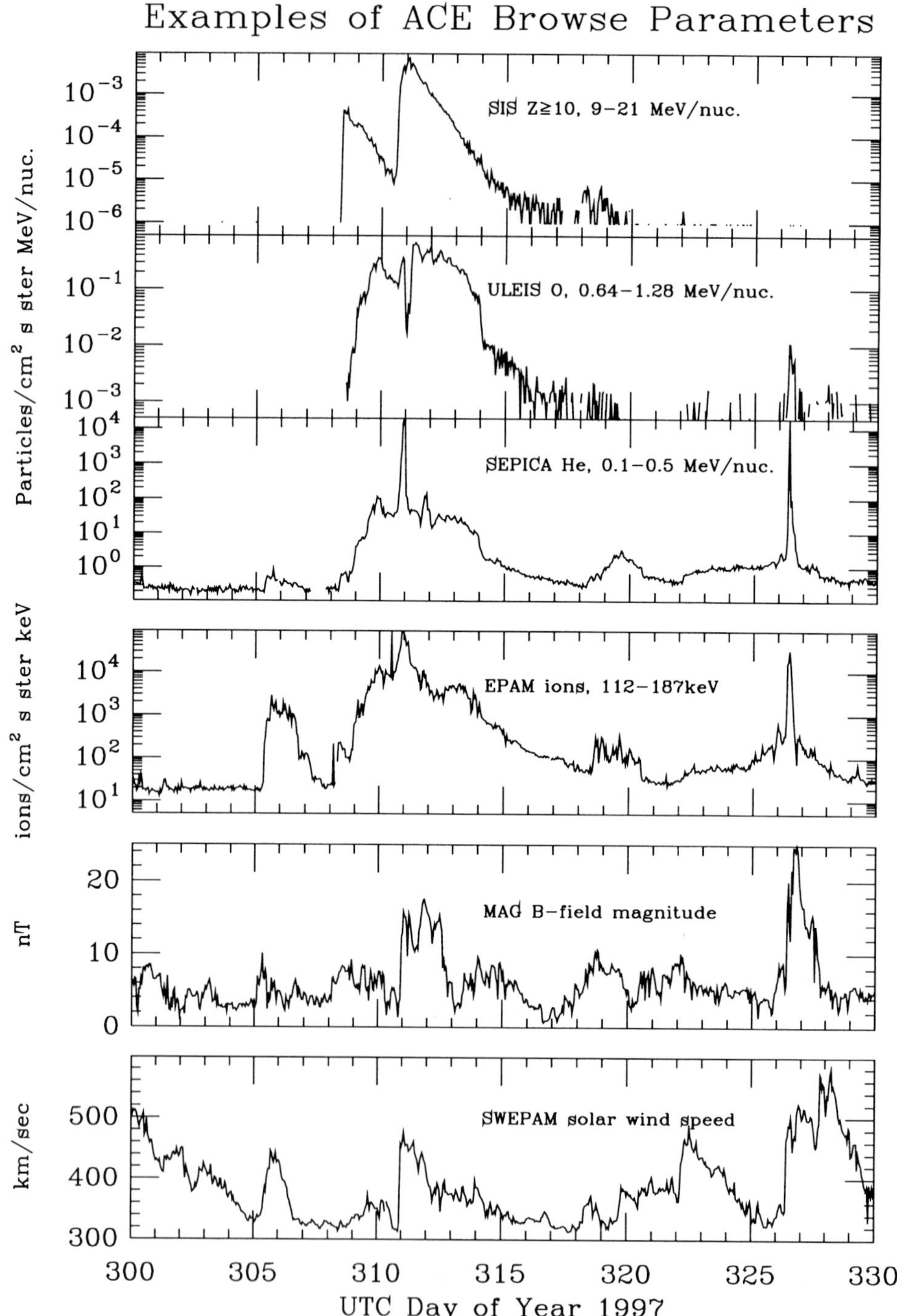

Figure 2. A selection of browse parameters from the ACE instruments for the period around the November 4 and 6 solar particle events. Each instrument contributes at least several additional browse parameters. For instance, the components of the magnetic field vector are also available from MAG. The CRIS instrument is not designed to function during periods of high solar activity, so CRIS data are not shown. SWICS browse parameters were not yet available.

TABLE II

Browse parameter charged particle fluxes and energy bands

Instrument	Electrons (MeV)	Ions (MeV)
EPAM	0.04–0.05	0.05–0.07
	0.18-0.32	0.11–0.19
		0.31–0.58
		1.06–1.91

Instrument	H	He	C	O (MeV nucl^{-1})	CNO	Ne–Fe	Fe group
CRIS						100–400	
						>300	
SIS					7-10	9–21	
					10–15		
SEPICA	0.1–0.6	0.1–0.5	1.0–15	0.8–17		0.5–11*	0.3–4.9
	0.6–5.4	0.5–8.4					
ULEIS		0.64–1.28^{3}					
	0.64–1.28	0.64–0.91^{4}		0.64–1.28			0.64–0.91
	0.16–0.32	0.08–0.11^{4}		0.09–0.16			0.08–0.16

ULEIS reports three helium browse parameters; one for ^{3}He, and two for ^{4}He. The superscripts on the energy ranges indicate the isotope. The SEPICA entry in the Ne–Fe column is really a Mg+Si flux. All the energy ranges quoted are subject to revision by the instrument teams.

As noted in Section 2 above, the data flow from the spacecraft to the ground is based on standards specified by the Consultative Committee for Space Data Systems (CCSDS) of which NASA is a member. Each spacecraft minor frame is encapsulated in a CCSDS packet (CCSDS '89) and transmitted (eventually) via a CCSDS virtual channel (CCSDS '89). The DSN checks and removes the error protection coding attached to each packet and forwards the data to the IMOC using the TCP/IP standard made familiar by the Internet. Uplink transmission of commands to the spacecraft from the ground also follows CCSDS standards. References to CCSDS standards can be obtained from the Consultative Committee for Space Data Systems Secretariat, Communications and Data Systems Division, Code OS, NASA Headquarters, Washington, D. C. 20546. All of these standards are largely invisible to the ACE team including the Science Center.

At the IMOC the level-zero processing encapsulates the spacecraft packets in Standard Formatted Data Units (SFDUs) per another CCSDS standard (CCSDS '92). After the data are received by the ACE team, all further data sets (e.g., Level 1, 2, 3, and browse) are stored and transmitted per the HDF (Hierarchical Data

Format) standard of the NCSA (the National Center for Supercomputing Applications). Some data sets are translated into other standards for the convenience of particular user communities that have settled on other standards. In particular, the browse parameter files are expected to be of use to a large community, so it is important to make them available easily. They are being translated into the Common Data format (CDF)(NSSDC '92) (with the assistance of the National Space Science Data Center) for the convenience of the ISTP (International Solar Terrestrial Physics) program, which uses CDF for their equivalent Key Parameter files. They are also being made available in ASCII via the Internet (see Section 4.2), since we expect that to be the easiest mode to access for the largest possible community.

The ACE team imposes additional standards of self-documentation on top of the facilities furnished with HDF. These rules are inspired by the Caltech Tennis standard (Garrard, 1993) and by experience with earlier missions, such as Voyager and HEAO. These rules demand self-documentation of each data item within a data set and a record of the pedigree of the data (i.e., what program created a data file, what other data files were input to that program, etc.). In Tennis, these rules were both enforced and facilitated by the tennis library of input/output functions (i/o). In the HDF i/o library, the enforcement function is missing, but adequate tools are present to facilitate these rules. The major advantage of HDF over Tennis, in the judgement of the ACE team, is substantial support for a wide variety of operating systems and computer types. It also has a much larger tool library.

4.2. TOOLS AND DATA DELIVERY

The HDF standard is supported by the National Center for Supercomputing Applications at University of Illinois, Champaign-Urbana. It comes with an i/o library which is supported for a large variety of operating systems and computers and a library of tools for browsing, displaying and indexing of HDF data sets. The standard is sufficiently popular that tools are being created and made available by users as well as the NCSA. For instance, the Science Center has created and contributed tools for mapping C language structure declarations into descriptions of HDF data sets. It is also noteworthy that Research Systems, Inc. and Fortner Software LLC, have both incorporated an HDF interface into their popular data analysis and visualization tools, IDL and Noesys (reg. trademarks).

The Science Center has adopted Unix as its preferred operating system, with most of the machines being 64-bit Sun workstations running the Solaris dialect of Unix. Many of the ACE instrument teams are using similar hardware and operating systems, but not all. The intention is to use standards which are not operating system dependent, while, at the same time, doing whatever is reasonable to reduce variety among the team to simplify system administration. The Science Center does have several Hewlett Packard Unix workstations and a number of Apple MacIntosh and Windows compatible personal computers available for guest investigators and for communications with investigators who prefer those systems. As noted above, the HDF i/o library is supported for a wide variety of operating systems, so data communications should not be OS dependent.

Of course, the Internet and the World Wide Web are the tools of choice at this time for interfacing users to the data and the documentation. At this time the Science Center Web address is

http://www.srl.caltech.edu/ACE/ASC/index.html.

The Web site provides documentation for the data, catalogs of the data files, plots of browse parameters to facilitate the selection of data files, and ancillary data such as spacecraft position and attitude. As explained below, the ACE data files themselves (except for the browse parameter data) are not available to anonymous users, but identified users will be granted access promptly. Immediate access to the level-one data is via the Internet (ftp), and level-one CD-ROMs are distributed to the instrument teams roughly every three weeks.

Although the primary responsibility for mission operations, including commanding and health monitoring, rests with the Flight Operations Team at GSFC, the Science Center has a secondary goal of providing flexibility in monitoring instrument health. Each instrument team has the option of monitoring the health of their instrument from their home institution, in real time, using the same ground-support equipment they used for integration and testing. To achieve this, the Science Center receives a copy of the spacecraft telemetry from GSFC via the Internet during each DSN contact and makes it available to each instrument team, also via the Internet. This is possible because the Science Center is running a subset of the MOC (Mission Operations Center) system software used both at the IMOC and at the spacecraft Integration and Test facility at APL. The use of common software for these three purposes has saved a great deal of money and is described by Stone et al. (1998a) and Snow et al. (1996).

Catalog tools for the ACE data are not yet well defined – it is preferable to wait until some examples of level-two data are available before defining the requirements for catalog tools. At this time, these tools are expected to resemble the 'incremental data set' tools of the Planetary Data System (King et al., 1993), which have the very useful feature that they link data files to the relevant documentation and calibration files. Another possibility is the development of additional tools by NCSA, specifically aimed at HDF files.

4.3. POLICY ISSUES

The ACE team has chosen to emphasize collaborative science and sharing of data within the team and with the larger space physics community. The browse data in particular are being made available very promptly, even at the risk of inadequate verification. Some evolution of browse parameter definitions/computation will almost certainly occur over the life of the mission as a result of user feedback and extended verification activities. In many cases, this evolution will be handled by adding new parameters; in some cases, improved parameters may be substituted. Users are advised to maintain close contact with instrument teams when analyzing ACE data, especially if attempting to do careful science based on browse parameters.

In order to facilitate communications with users, the Science Center will make a substantial effort to keep track of all users and will discourage anonymous data transfers.

The Science Center has a limited allocation of office space and computer facilities available to Guest Investigators, either formally designated and funded by NASA or selected by informal negotiations with the instrument teams.

The Science Center has actively coordinated with the Space Physics Data System organization (Garrard et al., 1995) and will continue to do so, either with SPDS or the potential Space Science Data System which might succeed it.

5. Conclusions

The ACE team, working through the ACE Science Center, are planning to investigate the composition and dynamics of the interplanetary medium and all the various energetic particle populations permeating the interplanetary medium in a coordinated and collaborative fashion. The team is making every effort to allow a larger community to participate in these studies. The tools used in that effort include data-interchange standards, standard visualization tools, Web interfaces for data access, a variety of storage and/or communications media, and open channels of communication with scientists.

Acknowledgements

The authors would particularly like to thank the members of the ACE Flight Operations and Flight Dynamics Teams at NASA/GSFC for their invaluable help and cooperation in defining the interface between the IMOC and the Science Center before launch and for their sustained efforts to deliver the highest quality data possible in the months since launch. We also thank the members of the ACE instrument teams for their efforts and help during the development and testing of the ACE level-one and browse processing software.

References

CCSDS : 1989, 'Advanced Orbiting Systems, Networks and Data Links', *CCSDS Blue Book* **701.0-B-1**

CCSDS : 1992, 'Standard Formatted Data Units – Structure and Construction Rules', *CCSDS Blue Book*, **620.0-B-2**

Chiu, M. C. et al.: 1998, 'ACE Spacecraft', *Space Sci. Rev.* **86**, 257.

Garrard, T. L.: 1993, 'Tennis: A Standard For Data Formatting, Input, and Output', *Proc. Am. Inst. Phys.* **283**, 648.

Garrard, T. L. for the C and H Discipline Coordination Team: 1995, 'The Space Physics Data System — Cosmic and Heliospheric Nodes', *Proc. 24th Int. Cosmic Ray Conf.* **3**, 627.

Gloeckler, G. et al.: 1998, 'Investigation of the Composition of Solar and Interstellar Matter Using Solar Wind and Pickup Ion Measurements with SWICS and SWIMS on the ACE Spacecraft', *Space Sci. Rev.* **86**, 497.

Gold, R. et al.: 1998, 'Electron, Proton and Alpha Monitor on the Advanced Composition Explorer Spacecraft', *Space Sci. Rev.* **86**, 541.

King, T. A., Joy, Steven P., and Walker, Raymond J.: 1993, 'The Design and Implementation of Scalable Data Systems and Incremental Data Sets', *Proc. Am. Inst. Phys.* **283**, 461.

Mason, G. M. et al.: 1998, 'The Ultra Low Energy Isotope Spectrometer (ULEIS) for the ACE Spacecraft', *Space Sci. Rev.* **86**, 409.

McComas, D. J. et al.: 1998, 'Solar Wind Electron Proton Alpha Monitor (SWEPAM) for the Advanced Composition Explorer', *Space Sci. Rev.* **86**, 563.

Möbius, E. et al.: 1998, 'The Solar Energetic Particle Ionic Charge Analyzer (SEPICA) and the Data Processing Unit (S3DPU) for SWICS, SWIMS and SEPICA', *Space Sci. Rev.* **86**, 449.

NSSDC: 1992, 'NSSDC CDF User's Guide', NSSDC/WDC-A-R&S 92-04, Version 2.6.

Smith, C. W. et al. : 1998, 'The ACE Magnetic Fields Experiment', *Space Sci. Rev.* **86**, 613.

Snow, F., Garrard, T. L., Steck, J. A., and Maury, J. L.: 1996, 'Enhancing a Control Center System for the Multiple Uses of Spacecraft Integration and Testing, Mission Operations, and Science Center Operations', *Proc. Fourth Int. Symp. Space Mission Operations Ground Data Systems*, in press.

Stone, E. C. et al.: 1998a, 'The Advanced Composition Explorer', *Space Sci. Rev.* **86**, 1.

Stone, E. C. et al.: 1998b, 'The Cosmic Ray Isotope Spectrometer for the Advanced Composition Explorer', *Space Sci. Rev.* **86**, 285.

Stone E. C. et al.: 1998, 'The Solar Isotope Spectrometer for the Advanced Composition Explorer', *Space Sci. Rev.* **86**, 357.

Zwickl, R. D. et al.: 1998, 'NOAA Real-Time Solar-Wind (RTSW) System Using ACE Data', *Space Sci. Rev.* **86**, 633.